BIOPOLYMER NANOCOMPOSITES

WILEY SERIES ON POLYMER ENGINEERING AND TECHNOLOGY

Richard F. Grossman and Domasius Nwabunma, Series Editors

Polyolefin Blends / Edited by Domasius Nwabunma and Thein Kyu

Polyolefin Composites / Edited by Domasius Nwabunma and Thein Kyu

Handbook of Vinyl Formulating, Second Edition / Edited by Richard F. Grossman

Total Quality Process Control for Injection Molding, Second Edition / M. Joseph Gordon, Jr.

Microcellular Injection Molding / Jingyi Xu

Poly(lactic acid): Synthesis, Structures, Properties, Processing, and Applications / Edited by Rafael Auras, Loong-Tak Lim, Susan E. M. Selke, and Hideto Tsuji

Hyperbranched Polymers: Synthesis, Properties, and Applications / Edited by Deyue Yan, Chao Gao, and Holger Frey

Advanced Thermoforming: Methods, Machines and Materials, Applications and Automation / Sven Engelmann

Biopolymer Nanocomposites: Processing, Properties, and Applications / Edited by Alain Dufresne, Sabu Thomas, and Laly A. Pothan

BIOPOLYMER NANOCOMPOSITES

PROCESSING, PROPERTIES, AND APPLICATIONS

Edited by

Alain Dufresne
Grenoble Institute of Technology (Grenoble INP)
The International School of Paper
Print Media, and Biomaterials (Pagora)
Saint Martin d'Hères Cedex, France

Sabu Thomas
Centre for Nanoscience and Nanotechnology
School of Chemical Sciences
Mahatma Gandhi University
Kottayam, Kerala, India

Laly A. Pothan
Department of Chemistry
Bishop Moore College
Mavelikara, Kerala, India

WILEY

Published by John Wiley & Sons, Inc., Hoboken, New Jersey
Published simultaneously in Canada

For general information on our other products and services or for technical support, please contact our Customer Care Department within the United States at (800) 762-2974, outside the United States at (317) 572-3993 or fax (317) 572-4002.

Wiley also publishes its books in a variety of electronic formats. Some content that appears in print may not be available in electronic formats. For more information about Wiley products, visit our web site at www.wiley.com.

Library of Congress Cataloging-In-Publication Data:

Biopolymer nanocomposites : processing, properties, and applications / edited by Alain Dufresne, Sabu Thomas, Laly A. Pothan.

pages cm

Includes index.

ISBN 978-1-118-21835-8 (hardback)

1. Biopolymers. 2. Nanocomposites (Materials) I. Dufresne, Alain, 1962– editor of compilation. II. Thomas, Sabu, editor of compilation. III. Pothan, Laly A, editor of compilation.

TP248.65.P62B5457 2013

572–dc23

2013002843

Printed in the United States of America

10 9 8 7 6 5 4 3 2 1

CONTENTS

FOREWORD

It is important to minimize the environmental impact of materials production by decreasing the environmental footprint at every stage of their life cycle. Therefore, composites where the matrix and reinforcing phase are based on renewable resources have been the subject of extensive research. These efforts have generated environmental friendly applications for many uses such as for automotive, packaging, and household products to name some.

Cellulose is the most abundant biomass on the earth and its use in the preparation of biobased nanomaterials has gained a growing interest during the last ten years. This interest can be illustrated by how the number of scientific publications on the cellulose nanomaterial research has grown very rapidly and reached more the 600 scientific publications during 2011. The research topics have been extraction of cellulose nanofibers and nanocrystals from different raw material sources, their chemical modification, characterization of their properties, their use as additive or reinforcement in different polymers, composite preparation, as well as their ability to self-assemble.

Nanocelluloses, both fibers and crystals, have been shown to have promising and interesting properties, and the abundance of cellulosic waste residues has encouraged their utilization as a main raw material source. Cellulose nanofibers have high mechanical properties, which combined with their enormous surface area, low density, biocompatibility, biodegradability, and renewability make them interesting starting materials for many different uses, especially when combined with biobased polymers. Since bionanocomposites are a relatively new research area, it is necessary to further develop processing methods to make these nanomaterials available on a large scale, so that new applications based on them can be developed.

Information about this emerging research field could also prove to be a catalyst and motivator not only for industries but also to a large number of students and young scientists. A matrix of tools that could aid such work could be developed through research enterprise. The book *Biopolymer Nanocomposites: Processing, Properties, and Applications* by Alain Dufresne, Sabu Thomas, and Laly A. Pothan, as the authors themselves have pointed out elsewhere, "is an attempt to introduce various biopolymers and bionanocomposites to a student of materials science. Going beyond mere introduction, the book delves deep into the characteristics of various biopolymers and bionanocomposites and discusses the nuances of their preparation with a view to helping researchers find out newer and novel applications." Students, researchers, and industrialists in the field of biocomposites will be greatly benefitted by this book since its chapters are authored by an impressive array of prominent current researchers in this field. Sincere attempts like this at promoting the use of green materials for sustainable growth of humanity should be lauded indeed.

Kristiina Oksman

Luleå University of Technology

CONTRIBUTORS

Debbie P. Anderson, Bioproducts and Bioprocesses National Science Program, Agriculture and Agri-Food Canada, Saskatoon, SK, Canada

Hélène Angellier-Coussy, UnitéMixte de RechercheIngénierie des Agropolymères et Technologies Emergentes, INRA/ENSA.M/UMII/CIRAD, Université Montpellier II, Montpellier Cedex, France

Mirta I. Aranguren, INTEMA-CONICET, Facultad de Ingeniería-Universidad Nacional de Mar del Plata, Mar del Plata, Argentina

Joe Aspler, FPInnovations, Pointe Claire, QC, Canada

Stephanie Beck, FPInnovations, Pointe Claire, QC, Canada

Richard Berry, FPInnovations, Pointe Claire, QC, Canada

Jean Bouchard, FPInnovations, Pointe Claire, QC, Canada

John A. Carver, School of Chemistry and Physics, The University of Adelaide, Adelaide, SA, Australia; Research School of Chemistry, Australian National University, ACT, Australia

Georgeta Cazacu, "Petru Poni" Institute of Macromolecular Chemistry, Physical Chemistry of Polymers Department, Ghica Voda Alley, Iasi, Romania

Pascale Chalier, Unité Mixte de Recherche Ingénierie des Agropolymères et Technologies Emergentes, INRA/ENSA.M/UMII/CIRAD, Université Montpellier II, Montpellier Cedex, France

Peter R. Chang, Bioproducts and Bioprocesses National Science Program, Agriculture and Agri-Food Canada, Saskatoon, SK, Canada; Department of Chemical and Biological Engineering, University of Saskatchewan, Saskatoon, SK, Canada

B. Deepa, Department of Chemistry, Bishop Moore College, Mavelikara, Kerala, India; Department of Chemistry, C.M.S. College, Kottayam, Kerala, India

François Drolet, FPInnovations, Pointe Claire, QC, Canada

Alain Dufresne, Grenoble Institute of Technology (Grenoble INP), The International School of Paper, Print Media, and Biomaterials (Pagora), Saint Martin d'Hères Cedex, France

Heath Ecroyd, School of Biological Sciences, University of Wollongong, NSW, Australia

Megan Garvey, Institute of Molecular Biotechnology, RWTH Aachen University, Aachen, Germany

Emmanuelle Gastaldi, Unité Mixte de Recherche Ingénierie des Agropolymères et Technologies Emergentes, INRA/ENSA.M/UMII/CIRAD, Université Montpellier II, Montpellier Cedex, France

Juliet A. Gerrard, Biomolecular Interaction Centre, University of Canterbury, Christchurch, New Zealand; School of Biological Sciences, University of Canterbury, Christchurch, New Zealand; MacDiarmid Institute, University of Canterbury, Christchurch, New Zealand

Nathalie Gontard, Unité Mixte de Recherche Ingénierie des Agropolymères et Technologies Emergentes, INRA/ENSA.M/UMII/CIRAD, Université Montpellier II, Montpellier Cedex, France

Valérie Guillard, Unité Mixte de Recherche Ingénierie des Agropolymères et Technologies Emergentes, INRA/ENSA.M/UMII/CIRAD, Université Montpellier II, Montpellier Cedex, France

Carole Guillaume, Unité Mixte de Recherche Ingénierie des Agropolymères et Technologies Emergentes, INRA/ENSA.M/UMII/CIRAD, Université Montpellier II, Montpellier Cedex, France

Youssef Habibi, Center of Innovation and Research in Materials and Polymers, University of Mons, Belgium

Wadood Hamad, FPInnovations, Pointe Claire, QC, Canada

Jin Huang, College of Chemical Engineering, Wuhan University of Technology, Wuhan, China; and State Key Laboratory of Pulp and Paper Engineering, South China University of Technology, Guangzhou, China

Qing Huang, College of Chemical Engineering, Wuhan University of Technology, Wuhan, China

Shinsuke Ifuku, Department of Chemistry and Biotechnology, Graduate School of Engineering, Tottori University, Koyama-cho Minami, Tottori, Japan

Marco Iotti, Research Scientist, Paper and Fibre Research Institute, Trondheim, Norway

Jun-Ichi Kadokawa, Graduate School of Science and Engineering, Kagoshima University, Korimoto, Kagoshima, Japan

Rekha Rose Koshy, Department of Chemistry, Bishop Moore College, Mavelikara, Kerala, India

Marie-Pierre Laborie, Institute of Forest Utilization and Work Sciences, Albert-Ludwigs University of Freiburg, Freiburg, Germany, and Freiburg

Materials Research Centre—FMF, Albert-Ludwigs University of Freiburg, Freiburg, Germany

Déborah Le Corre, University of Canterbury, New Zealand

Liyuan Lu, Engineering Research Center of Nano-Geomaterials of Ministry of Education, China University of Geosciences, Wuhan, China

Norma E. Marcovich, INTEMA-CONICET, Facultad de Ingeniería-Universidad Nacional de Mar del Plata, Mar del Plata, Argentina

Siji K. Mary, Bishop Moore College, Mavelikara, Kerala, India

Vikas Mittal, Chemical Engineering Department, The Petroleum Institute, Abu Dhabi, United Arab Emirates

Mirna A. Mosiewicki, INTEMA-CONICET, Facultad de Ingeniería-Universidad Nacional de Mar del Plata, Mar del Plata, Argentina

Tiffany Omokanwaye, Catholic University of America, BONE/CRAB Lab, Department of Biomedical Engineering, Washington, DC

Anayancy Osorio-Madrazo, Institute of Forest Utilization and Work Sciences, Albert-Ludwigs University of Freiburg, Freiburg, Germany, and Freiburg Materials Research Centre—FMF, Albert-Ludwigs University of Freiburg, Freiburg, Germany

Stéphane Peyron, Unité Mixte de Recherche Ingénierie des Agropolymères et Technologies Emergentes, INRA/ENSA.M/UMII/CIRAD, Université Montpellier II, Montpellier Cedex, France

Saumya S. Pillai, Department of Chemistry, Bishop Moore College, Mavelikara, Kerala, India

David Plackett, Department of Chemical and Biochemical Engineering, Technical University of Denmark, Kgs. Lyngby, Denmark

Laly A. Pothan, Department of Chemistry, Bishop Moore College, Mavelikara, Kerala, India

Shiva P. Rao, New Zealand Institute of Plant and Food Research, Christchurch, New Zealand; Biomolecular Interaction Centre, University of Canterbury, Christchurch, New Zealand

Hiroyuki Saimoto, Department of Chemistry and Biotechnology, Graduate School of Engineering, Tottori University, Koyama-cho Minami, Tottori, Japan

Robert A. Shanks, School of Applied Sciences, RMIT University, Melbourne, Vic., Australia

Zameer Shervani, Department of Chemistry and Biotechnology, Graduate School of Engineering, Tottori University, Koyama-cho Minami, Tottori, Japan

Deepa Thomas, Department of Chemistry, Bishop Moore College, Mavelikara, Kerala India

Merin Sara Thomas, Centre for Nanoscience and Nanotechnology, M.G. University, Kottayam, Kerala, India

Sabu Thomas, Centre for Nanoscience and Nanotechnology, School of Chemical Sciences, M.G. University, Kottayam, Kerala, India

Eliane Trovatti, CICECO and Department of Chemistry, University of Aveiro, Aveiro, Portugal

Cornelia Vasile, "Petru Poni" Institute of Macromolecular Chemistry, Physical Chemistry of Polymers Department, Ghica Voda Alley, Iasi, Romania

Madhusudan Vasudevamurthy, New Zealand Institute of Plant and Food Research, Christchurch, New Zealand; Biomolecular Interaction Centre, University of Canterbury, Christchurch, New Zealand

Otto C. Wilson, Jr., Catholic University of America, BONE/CRAB Lab, Department of Biomedical Engineering, Washington, DC

Qingyu Xu, Hubei Research Institute of Chemistry, Wuhan, China, and Haiso Technology Co., Ltd, Wuhan, China

Jiahui Yu, Institute of Biofunctional Materials and Devices, East China Normal University, Shanghai, China

Ming Zeng, Engineering Research Center of Nano-Geomaterials of Ministry of Education,China University of Geosciences, Wuhan, China, and State Key Laboratory of Polymer Materials Engineering, Sichuan University, Chengdu, China

Xuejun Zou, FPInnovations, Pointe Claire, QC, Canada

CHAPTER 1

Bionanocomposites: State of the Art, Challenges, and Opportunities

ALAIN DUFRESNE, SABU THOMAS, and LALY A. POTHAN

1.1 INTRODUCTION

Researchers are currently developing and modifying biobased materials that have various applications in different fields. Ecological concerns are the main reasons behind this renewed interest in natural and compostable materials. Tailoring new products with the perspective of sustainable development is a philosophy that is applied to more and more materials now. The importance gained by natural polymers recently should be viewed from this perspective. Compared with their synthetic counterparts, natural polymers are renewable, biocompatible, and biodegradable. Production of nanocomposites from natural polymers, such as starch, chitin, and cellulose, and specific research in this field aimed at increasing the properties of the products and developing newer techniques are the order of the day. Polysaccharide polymers that are abundant in nature are increasingly being used for the preparation of nanocomposites.

Biopolymers are polymers that are biodegradable. They are designed to degrade through the action of living organisms. They are the best alternatives to traditional nonbiodegradable polymers whose recycling is unpractical or not economical. The input materials for the production of such biodegradable polymers may be either renewable (based on agricultural plant or animal products) or synthetic. Biopolymers from renewable resources are more important than others for obvious reasons [1]. Biopolymers are said to be from renewable sources because they are made from materials that can be grown each year, indefinitely. Plant-based biopolymers usually come from agricultural nonfood crops. Therefore, the use of biopolymers would create a

Biopolymer Nanocomposites: Processing, Properties, and Applications, First Edition. Edited by Alain Dufresne, Sabu Thomas, and Laly A. Pothan.

sustainable industry. In contrast, the feedstock of synthetic polymers derived from petrochemicals will eventually run out. Biopolymers have also been reported to be close to carbon-neutral. When a biodegradable material (neat polymer, blended product, or composite) is obtained completely from renewable resources, we may call it a green polymeric material.

Nature provides an impressive array of polymers that are generally biodegradable and that have the potential to replace many current polymers, as biodegradation is part of the natural biogeochemical cycle. Natural polymers, such as proteins, starch, and cellulose, are examples of such polymers. Polymer nanocomposites represent a new alternative to conventional polymers. Polymer nanocomposites are materials in which nanoscopic inorganic or organic particles, typically 10–1000 Å in at least one dimension, are dispersed in an organic polymer matrix in order to improve the properties of the polymer dramatically. Owing to the nanometer length scale, which minimizes scattering of light, nanocomposites are usually transparent and exhibit properties that are markedly improved over those of pure polymers or their traditional composites. They have increased modulus and strength, outstanding barrier properties, improved solvency, heat resistance, and generally lower flammability, and they do not have detrimental effects on ductility.

1.2 NANOCRYSTALLINE CELLULOSE

The hierarchical structure and semicrystalline nature of polysaccharides (cellulose, starch, and chitin) allow nanoparticles to be extracted from naturally occurring polymers. Native cellulose and chitin fibers are composed of smaller and mechanically stronger long thin filaments, called microfibrils, consisting of alternating crystalline and noncrystalline domains. Multiple mechanical shearing actions can be used to release these microfibrils individually.

The extraction of crystalline cellulosic regions, in the form of nanowhiskers, can be accomplished by a simple process based on acid hydrolysis. Samir et al. have described cellulose whiskers as nanofibers that have been grown under controlled conditions that lead to the formation of high purity single crystals [2]. Many different terms have been used in the literature to designate these rod-like nanoparticles. They are mainly referred to as whiskers or cellulose nanocrystals. A recent review from Habibi et al. gives a clear overview of such cellulosic nanomaterials [3].

Nanocrystalline cellulose (NCC) derived from acid hydrolysis of native cellulose possesses different morphologies depending on the origin and hydrolysis conditions. NCCs are rigid rod-like crystals with a diameter in the range of 10–20 nm and lengths of a few hundred nanometers (Figure 1.1). Acid treatment (acid hydrolysis) is the main process used to produce NCC, which are smaller building blocks released from the original cellulose fibers. Native cellulose consists of amorphous and crystalline regions. The amorphous regions have lower density than the crystalline regions. Therefore, when cellulose

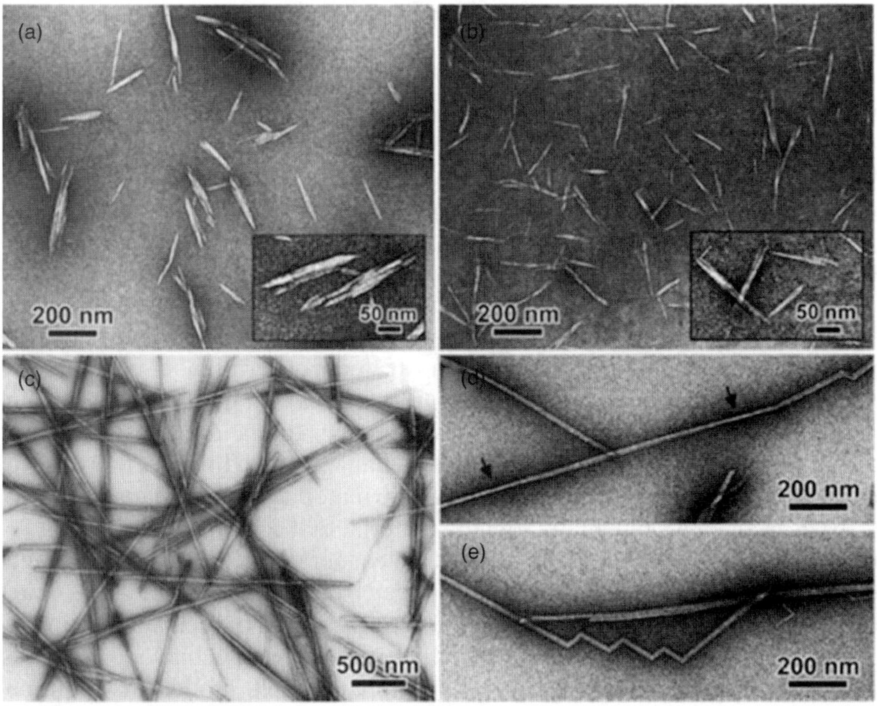

Figure 1.1 NCCs are rigid rod-like crystals with diameter in the range of 10–20 nm and lengths of a few hundred nanometers Reproduced with the permission from Reference [5].

fibers are subjected to harsh acid treatment, the amorphous regions break up, releasing the individual crystallites. The properties of NCC depend on various factors, such as cellulose sources, reaction time and temperature, and types of acid used for hydrolysis.

Polysaccharide nanoparticles are obtained as aqueous suspensions, and most investigations have focused on hydrosoluble (or at least hydrodispersible) or latex-form polymers. However, these nanocrystals can also be dispersed in nonaqueous media using surfactants or chemical grafting. The hydroxyl groups present on the surface of the nanocrystals make extensive chemical modification possible. Even though this improves the adhesion of nanocrystals with nonpolar polymer matrices, it has been reported that this strategy has a negative impact on the mechanical performance of the composites. This unusual behavior is ascribed to the reinforcing phenomenon of polysaccharide nanocrystals resulting from the formation of a percolating network due to hydrogen bonding forces.

As a result of its distinctive properties, NCC has become an important class of renewable nanomaterials, which has many useful applications, the most important of which is the reinforcement of polymeric matrices in nanocomposite materials. Favier et al. were the first to report the use of NCC as reinforcing fillers in poly(styrene *co*-butyl acrylate) (poly(S-*co*-BuA))-based nanocomposites [4]. Since then, numerous nanocomposite materials have been developed by incorporating NCC into a wide range of polymeric matrices. Owing to their abundance, high strength and stiffness, low weight, and biodegradability, nanoscale polysaccharide materials can be used widely for the preparation of bionanocomposites. In fact, a broad range of applications of these nanoparticles exists. Many studies show its potential, though most focus on their mechanical properties and their liquid crystal self-ordering properties. The homogeneous dispersion of cellulosic nanoparticles in a polymer matrix is challenging. In addition, there are many safety concerns about nanomaterials, as their size allows them to penetrate into cells of humans and to remain in the system. However, finding newer applications for nanocellulose will have a very positive impact on organic waste management. To date, there is no consensus about categorizing nanocellulosic materials as new materials.

NCC is an environmentally friendly material that could serve as a valuable renewable resource for rejuvenating the beleaguered forest industry. New and emerging industrial extraction processes need to be optimized to achieve more efficient operations, and this will require active research participation from the academic and industrial sectors. The application of nanotechnology in developing NCC from the forest industry to more valuable products is required because the availability of materials based on NCC is still limited. Increasing attention is devoted to producing NCC in larger quantities and to exploring various modification processes that enhance the properties of NCC, making it attractive for use in a wide range of industrial sectors [5]. As the second most abundant biopolymer after cellulose, chitin is mainly synthesized via a biosynthetic process by an enormous number of living organisms such as shrimp, crab, tortoise, and insects and can also be synthesized by a nonbiosynthetic pathway through chitinase-catalyzed polymerization of a chitobiose oxazoline derivative [6, 7].

Chitosan, as the most important derivative of chitin, can be prepared by deacetylation of chitin. Chitin and chitosan have many excellent properties including biocompatibility, biodegradability, nontoxicity, and absorption, and thus they can be widely used in a variety of areas such as biomedical applications, agriculture, water treatment, and cosmetics. Chitin has been known to form microfibrillar arrangements in living organisms. These fibrils with diameters from 2.5 to 25 nm, depending on their biological origins, are usually embedded in a protein matrix [8]. Therefore, they intrinsically have the potential to be converted to crystalline nanoparticles and nanofibers and to find application in nanocomposite fields. The structure of chitin is very analogous to cellulose. Chitin and cellulose are both supporting materials for living bodies and are found in living plants or animals with sizes increasing from simple molecules and highly crystalline fibrils on the nanometer level to composites on the micrometer level upward [9]. Therefore, they intrinsically have the potential to be converted to crystalline nanoparticles and nanofibers and to find application in nanocomposite fields. Chitin has been known to form microfibrillar arrangements in living organisms [10, 11].

Chitin whiskers (CHW) can be prepared from chitins isolated from chitin-containing living organisms by a method similar to the preparation of cellulose whisker through hydrolysis in a strong acid aqueous medium. On the basis of preparation of cellulose crystallite suspension, Marchessault et al. [11] for the first time reported a route for preparing suspension of chitin crystallite particles in 1959. In this method, purified chitin was first treated within 2.5 N hydrochloric acid (HCl) solutions under reflux for 1 hour; the excess acid was decanted; and then distilled water was added to obtain the suspension. Acid-hydrolyzed chitin was found to be spontaneously dispersed into rod-like particles that could be concentrated to a liquid crystalline phase and self-assemble to a cholesteric liquid crystalline phase above a certain concentration [12].

CHWs are attracting attention from both the academic field and industry since it is a renewable and biodegradable nanoparticle. CHWs have numerous advantages over conventional inorganic particles such as low density, nontoxicity, biodegradability, biocompatibility, easy surface modification, and functionalization. Figure 1.2 shows the transmission electron microscopy (TEM) and atomic force microscopy (AFM) images of CHWs obtained by the 2,2,6,6-tetramethylpiperidine-1-oxyl radical (TEMPO) mediated oxidation method. CHWs, with or without modification, are hoped to have extensive application in many areas such as reinforcing nanocomposites, the food and cosmetics industries, drug delivery, and tissue engineering. However, recent studies have focused mainly on the preparation and nanocomposite application of CHWs, and less attention has been paid to other application areas. It is hoped that in the future, more attention will be focused on developing novel applications of CHWs. Even for the CHW-reinforced nanocomposites there will still be much valuable work to be done, for example, developing new simple and effective processing methods so as to commercialize high performance polymer/CHWs composites, producing polymer nanocomposites filled with individual CHWs that would create higher reinforcing efficiency than the conventional CHW due to the high aspect ratio of individual CHWs. Thus, there are abundant opportunities combined with challenges in CHW-related scientific and industrial fields [13].

Starch is the second most studied organic material for producing nanocrystals. Starch nanocrystals are the nanoscale biofillers derived from native starch granules and are a suitable candidate for the preparation of semicrystalline polymers for preparing renewable and potentially biodegradable nanoparticles. As a natural biopolymer, starch is abundant, renewable, inexpensive, biodegradable, environmentally friendly, and easy to chemically modify, making it one of the most attractive and promising bioresource materials. Several techniques for preparing starch nanoparticles (SNP) have been developed over the years and render different kinds of SNP, which are described in this book. Acid hydrolysis and precipitation methods are the two main methods employed for the preparation of SNPs. Starch nanocrystals obtained by acid

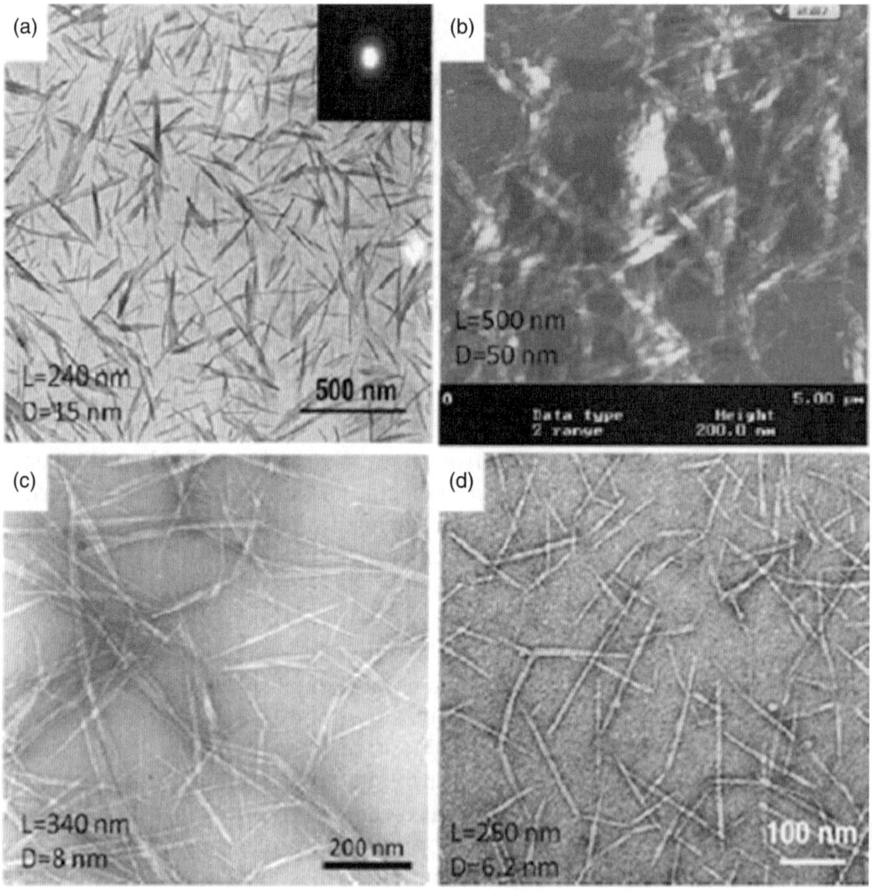

Figure 1.2 TEM (a) and AFM (b) images of a dilute suspension of chitin whiskers and TEM images of individual chitin whiskers obtained by the TEMPO method (c) and surface cationization (d). Reproduced with permissions from Reference [9].

hydrolysis of starch have been used as fillers in natural and synthetic polymeric matrices and appear to be an interesting reinforcing agent. Figure 1.3 shows the TEM image of starch nanocrystals [14]. Nanoreinforced starch-based nanocomposites generally exhibit enhanced mechanical and thermal properties when nanofillers are well dispersed, while the nature of the matrix and/or nanofiller contributes to its biological properties.

Nanocellulose produced by the bacterium *Gluconacetobacter xylinus* (bacterial cellulose, BC), is an another emerging biomaterial with great potential as a biological implant, wound and burn dressing material, and scaffold for tissue regeneration. This BC is quite different from plant celluloses and is defined by high purity (free of hemicelluloses, lignin, and alien functionalities

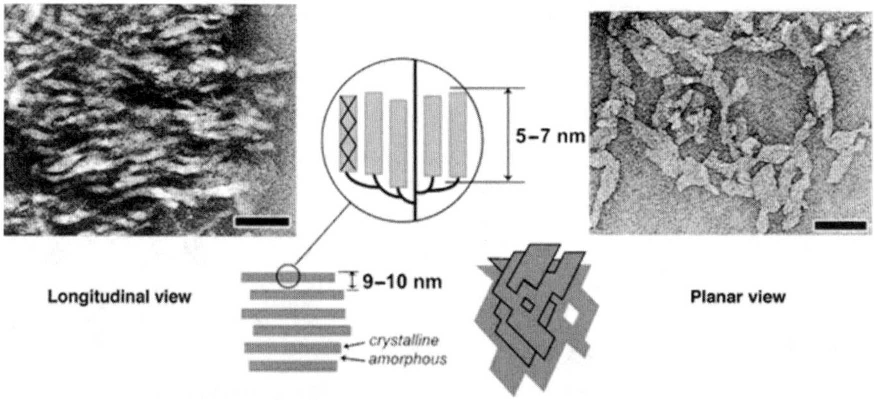

Figure 1.3 TEM observations of starch nanocrystals: longitudinal view and planar view. Reproduced with permission from Reference [14]. Copyright 2003 American Chemical Society.

such as carbonyl or carboxyl groups) and a high degree of polymerization (up to 8000) [15]. BC has remarkable mechanical properties despite the fact that it contains up to 99% water. The water-holding ability is the most probable reason why BC implants do not elicit any foreign body reaction. Fibrosis, capsule formation, or giant cells were not detected around the implants, and connective tissue was well integrated with the BNC structures. Moreover, the nanostructure and morphological similarities with collagen make BC attractive for cell immobilization, cell migration, and the production of extracellular matrices [16, 17]. Figure 1.4 shows BNC fleeces formed by different *Gluconacetobacter* strains and their network structure.

The advanced natural fiber-reinforced polymer composite contributes to enhancing the development of bionanocomposites with regard to performance and sustainability. In the future, these biocomposites will see increased use in optical, biological, and engineering applications. But there are still a number of problems that have to be solved before biocomposites become fully competitive with synthetic fiber composites. These include extreme sensitivity to moisture and temperature, expensive recycling processes, high variability in properties, nonlinear mechanical behavior, poor long-term performance, and low impact strength. As of now, the methods for extracting nanocrystals of these various biomaterials are expensive, and more economical methods will have to be sorted out in future.

The poor interfacial adhesion between natural fibers and polymeric matrix is the key issue that dictates the overall performance of the composites. Interaction of two or more different materials with each other depends on the nature and strengths of the intermolecular forces of the components involved. The mechanical performance of composites is dependent on the degree of

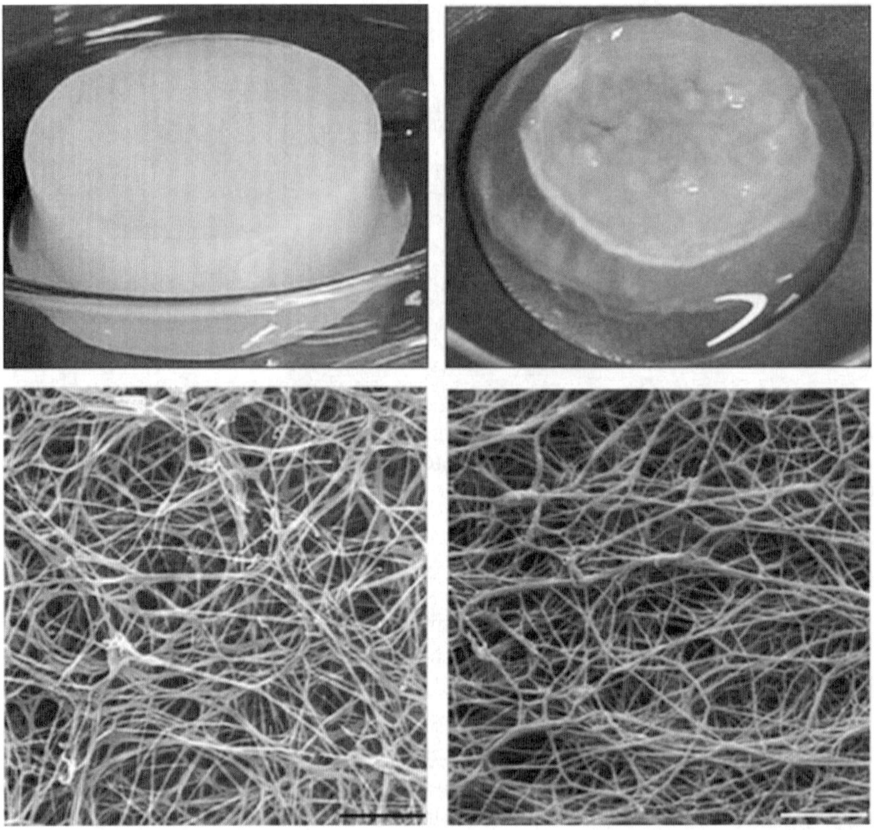

Figure 1.4 Fleeces of bacterial nanocellulose produced by two different *Gluconacetobacter* strains and their network structure. Reproduced with permissions from Reference [17].

dispersion of the fibers in the matrix polymer and the nature and intensity of fiber–polymer adhesion interactions. Therefore, the selection of appropriate matrices and filler with good interfacial interaction is of great importance. The irreversible aggregation of the nanofiller (hornification) in the matrix, which prevents its redispersion in the matrix, is another hurdle to be overcome. This irreversible aggregation results in a material with ivory-like properties that can neither be used in rheological applications nor be dispersed for composite applications. Therefore, it is necessary to continue research in this area to obtain a better understanding of the adhesion interactions including mechanical interlocking, interpenetrating networks, and covalent linkages on a fundamental level to improve interfacial properties with thermoplastics, thermosets, and biopolymers.

This book is an attempt to introduce various biopolymers and bionanocomposites to students of material sciences. Going beyond a mere introduction, the book delves deep into the characteristics of various biopolymers and bionanocomposites and discusses the nuances of their preparation with a view to helping researchers discover newer and novel applications. Chapter 2, for instance, describes the preparation of chitin nanofibers and their composites and discusses the basics, such as isolation of chitin nanofibersfrom different sources. Chapter 3 discusses chemical modification of chitosan and its biomedical application. While biometric lessons for processing chitin-based composites are provided in Chapter 4, Chapter 5 deals with morphological and thermal investigations of chitin-based nanocomposites. Mechanical properties of chitin-based nanocomposites are discussed in Chapter 6, and preparation and applications of chitin nanofibers/nanowhiskers is the topic of Chapter 7. Thus, Chapters 2 to 7 are allotted to chitin and related topics.

Various aspects of starch-based composites, such as preparation of SNPs (Chapter 8), chemical modification of SNPs (Chapter 9), processing techniques of starch-based bionanocomposites (Chapter 10), morphological and thermal investigations of starch-based nanocomposites (Chapter 11), mechanical properties of starch-based nanocomposites (Chapter 12), and applications of SNPs and starch-based bionanocomposites (Chapter 13), are the subject matter of Chapters 8 to 13.

Preparation of nanofibrillated cellulose and cellulose whiskers are dealt with in Chapter 14. Chapter 15 is exclusively set apart for BC. It examines the details of production of microorganisms, production of BC, production of BC from food and agro-forestry residues, and the structure of BC. Chemical modification of nanocelluloses is discussed in Chapter 16, and processing techniques of cellulose-based nanocomposites are dealt with in Chapter 17. Chapter 18 is on morphological and thermal investigations of cellulosic bionanocomposites, and Chapter 19 discusses mechanical properties of cellulose-based bionanocomposites. A review of nanocellulosic products and their applications is provided in Chapter 20. In Chapter 21 spectroscopic characterization of renewable nanoparticles and their composites are dealt with. Chapter 22 deals with barrier properties of renewable nanomaterials. Chapter 23 is set apart for biocomposites and nanocomposites containing lignin. While Chapter 24 deals with preparation, processing, and applications of protein nanofibers, Chapter 25 deals with protein-based nanocomposites for food packaging. Thus, this book is a sincere attempt at promoting the use of green materials for sustainable growth of humanity.

REFERENCES

[1] Kaplan, D.L., ed. (1998) *Biopolymers from Renewable Resources; Macromolecular Systems—Materials Approach*. Berlin, Heidelberg: Springer-Verlag.

[2] Samir, M.A.S.A., Alloin, F., and Dufresne, A. (2005) Review of recent research into cellulosic whiskers, their properties and their application in nanocomposite field. *Biomacromolecules*, *6*, 612–626.

[3] Habibi, Y., Lucia, L.A., and Rojas, O.J. (2010) Cellulose Nanocrystals: Chemistry, self assembly, and applications. *Chemical Reviews*, *110*(6), 3479–3500.

[4] Favier, V., Chanzy, H., and Cavaille, J.Y. (1995) Polymer nanocomposites reinforced by cellulose whiskers. *Macromolecules*, *28*, 6365–6367.

[5] Peng, B.L., Dhar, N., Liu, H.L., and Tam, K.C. (2011) Chemistry and applications of nanocrystalline cellulose and its derivatives: A nanotechnology perspective. *Canadian Journal of Chemical Engineering*, *89*(5), 1191–1206.

[6] Kobayashi, S., Kiyosada, T., and Shoda, S.I. (1996) Synthesis of artificial chitin: Irreversible catalytic behavior of a glycosyl hydrolase through a transition state analogue substrate. *Journal of American Chemical Society*, *118*, 13113–13114.

[7] Kadokawa, J. (2011) Precision polysaccharide synthesis catalyzed by enzymes. *Chemical Reviews*, *111*, 4308–4345.

[8] Revol, J.F., Marchessault, R.H. (1993) *In vitro* chiral nematic ordering of chitin crystallites. *International Journal of Biological Macromolecules*, *15*, 329–335.

[9] Fan, Y., Saito, T., and Isogai, A. (2008) Preparation of chitin nanofibers from squid pen β-chitin by simple mechanical treatment under acid conditions. *Biomacromolecules*, *9*, 1919–1923.

[10] Carlstrom, D. (1957) The crystal structure of α-chitin (poly-*n*-acetyl-d-glucosamine). *The Journal of Biophysical and Biochemical Cytology*, *3*, 669–683.

[11] Marchessault, R.H., Morehead, F.F., and Walter, N.M. (1959) Liquid crystal systems from fibrillar polysaccharides. *Nature*, *184*, 632–633.

[12] Li, J., Revol, J.F., Naranjo, E., and Marchessault, R.H. (1996) Effect of electrostatic interaction on phase separation behaviour of chitin crystallite suspensions. *International Journal of Biological Macromolecules*, *18*, 177–187.

[13] Zeng, J.B., He, Y.S., Li, S.L., and Wang, Y.Z. (2012) Chitin whiskers: An overview. *Biomacromolecules*, *13*, 1–11.

[14] Putaux, J.L., Molina-Boisseau, S., Momaur, T., and Dufresne, A. (2003) Platelet nanocrystals resulting from the disruption of waxy maize starch granules by acid hydrolysis. *Biomacromolecules*, *4*(5), 1198–1202.

[15] Kramer, F., Klemm, D., Schumann, D., Heßler, N., Wesarg, F., Fried, W., and Stadermann, D. (2006) Nanocellulose polymer composites as innovative pool for (Bio) material development. *Macromolecular Symposia*, *244*, 136–148.

[16] Gatenholm, P., Klemm, D. (2010) Bacterial nanocellulose as a renewable material for biomedical applications. *MRS Bulletin*, *35*, 208–213.

[17] Klemm, D., Kramer, F., Moritz, S., Lindstrom, T., Ankerfors, M., Gray, D., and Dorris, A. (2011) Nanocelluloses: A new family of nature-based materials. *Angew. Chem. Int. Ed.*, *50*, 5438–5466.

CHAPTER 2

Preparation of Chitin Nanofibers and Their Composites

SHINSUKE IFUKU, ZAMEER SHERVANI, and HIROYUKI SAIMOTO

2.1 INTRODUCTION

Biodegradable chitin nanofibers (CNFs) have attracted the attention of researchers worldwide in recent years, as they constitute an important part of the rapidly growing field of nanotechnology, which deals with nanometer-sized (1–100 nm) composites. Bionanofibers have superiority over their synthetic counterparts because of their biocompatible nature. They have applications in the medical, cosmetics, pharmaceutical, and chemical industries. If doped with inorganic metals, the hybrid organic–inorganic composites can have vast applications in electronics, electrical, and optical fields, and, most important, in much needed renewable energy production. Nanofibers (NFs) have a large surface-to-mass ratio, making them promising candidates for advanced material devices. A number of methods are employed to spin NFs. Electrospinning is one of the methods that use electrical charge to draw nanoscale fibers from polymer liquid solutions. As synthetic NFs are environmentally toxic, natural NFs are preferred products over synthetic fibers. Natural NFs are known to exist in nature in various forms: collagen fibrils, silk fibroin, double helical deoxyribonucleic acid, and so on. Apart from natural chitin NFs, there are cellulose microfibrills, which are more abundant natural NFs.

Abe et al. [1] achieved efficient extraction of wood cellulose NFs, which existed in a cell wall, of a uniform width of 15 nm, using a simple mechanical treatment. Wood powder of size <60 mesh from the Radiata pine tree was used. First, organic solvent extraction was conducted to remove wax and other small organic components. The larger complex lignin moiety was separated by

Biopolymer Nanocomposites: Processing, Properties, and Applications, First Edition. Edited by Alain Dufresne, Sabu Thomas, and Laly A. Pothan.

acidified sodium chlorite solution. A number of extraction cycles were used until the product turned white. Hemicellulose was leached by the treatment of 6 wt% potassium hydroxide. Cellulose thus obtained as α-cellulose constituted 85% of whole cellulose. Finally, 1 wt% slurry of cellulose was passed through a grinder at a speed of 1500 rpm. Field emission scanning electron microscopy (FE-SEM) measurement confirmed the formation of 15 nm width cellulose NFs.

All cellulose composites of consisting fibers and matrix were studied [2, 3] for structural, mechanical, and thermal properties. The elastic modulus, in the direction parallel to the polymer molecule chain axis, was found to be 138 GPa for the crystalline regions. This value is comparable to that of high performance synthetic fibers. The tensile strength of cellulose is 17.8 GPa, which is seven times higher than that of steel. The linear thermal expansion coefficient is about 10^{-7}/K. Due to their high elastic modulus and tensile strength and low thermal expansion, fibers are promising candidates as reinforcement agent for a number of composites.

Chitin is the second most abundant biopolymer after cellulose that occurs in nature [4]. Its annual production worldwide is 10^{10} to 10^{11} tons, mostly produced from external skeletons of shellfish, crabs, shrimp, insects, mushrooms, and algae. The fibrous material of cellular walls of mushrooms and algae and external skeletons in shellfish and insects are composed of chitin. Chitin content is in the range of 8–33%, which is disposed of as industrial waste in shellfish canning industries. Chitin and chitin compounds find application in various fields, including cosmetics and chemical industries, engaging researchers from around the world. Chitin and their hybrid inorganic composites are also expected to have applications in electrical, electronics, and optical devices. Natural chitin is highly crystalline (mostly α-chitin), though the distribution among α- and -β-chitin depends on the source. Among the applications discovered so far, chitin serves as an effective reinforcement for the preparation of composites; there are reports in the literature on chitin whisker-reinforced nanocomposites [4]. The chitin structure comprises the repeating units along the *N*-acetylglucosamine structure. It has two hydroxyl and an acetamide groups per unit [5], which make the molecule reactive for a number of applications. A dominant feature of arthropod exoskeletons is that they are well organized, arranged in different structural levels. Considering molecular levels, there are long chain polysaccharide chitin fibrils with dimensions of 3 nm in width and 300 nm in length. The fibrils are wrapped in proteins and aggregated into bundles of fibers of about 60 nm in diameter. Step-by-step breakup of these assemblies has been shown by Chen et al. [6], who have described the structural and mechanical properties of crab exoskeleton in detail. Chitin whiskers were prepared from crab shells by using chemical treatment followed by mechanical treatment [4]. The proteins were removed with 5% KOH; $NaClO_2$ and a small amount of sodium acetate buffer were used as bleaching agents. The residual proteins were again removed by using 5% KOH.

The purified chitin sample was hydrolyzed with 3 N HCl followed by centrifuging the hydrolyzed suspension. The centrifugated suspension was dialyzed overnight in distilled water until the pH of the preparation reached 4. The dispersion was sonicated for 5 minutes, and prepared whiskers were finally stored at 6°C. There are other methods as well to extract CNFs from natural materials, such as ultrasonic methods [7] and electrospinning [8]. However, NFs obtained by these methods are different from the high quality CNFs in terms of width, aspect ratio, crystallinity, chemical structure, and narrow size distribution. Researchers have extracted NFs by 2,2,6,6-tetramethylpiperidine-1-oxy radical (TEMPO)-mediated oxidation of natural cellulose, followed by ultrasonic treatment. This method was then tested for isolating α-chitin from crab shells, but the average nanocrystal length was considerably low [9], and instead of chitin, the derivatives of chitin were also obtained. Fan et al. [10] obtained CNFs of 3–4 nm size from less chitin content biomass squid pen by ultrasonication treatment under acidic conditions; however, the crystallinity of the NFs from the source was relatively low. In addition to the method [1] of isolating CNFs from a cellulose source, one can isolate CNFs from a number of sources. In this chapter we describe isolation and characterization of natural CNFs from different sources and the composite preparations from these NFs. Most of the work described in this chapter has been conducted by our group, but we also describe and compare research carried out in other laboratories.

2.2 ISOLATION OF CHITIN NANOFIBERS FROM DIFFERENT SOURCES

2.2.1 Processing of Chitin Nanofibers from Crab Shells

Extraction of CNFs from crab shells was conducted following a series of chemical treatments followed by mechanical grinding. Low cost commercial dried crab shell flakes of the species *Paralithodes camtschaticus*, commonly known as "red king crab," were used as raw material. These crabs are in abundance and used as fertilizers for crops. Crab shells were milled to powder and purified according to the established procedure. Chemical treatment was as follows. Powdered crab shell flakes were treated with 2 N HCl for 2 days at room temperature to remove the mineral salts, then the suspension was filtered and washed thoroughly with distilled water. Afterward, the mixture was refluxed in 2 N NaOH for 2 days to remove the various proteins. The NaOH-treated solution was mixed with ethanol to extract pigments and lipids. The purified chitin was then removed by filtration and rinsed with water. The processed chitin cake was kept wet, which helped fibrillation by mechanical treatment. It was established that all types of proteins and minerals such as $CaCO_3$ can be removed by NaOH and HCl treatment stages. The chitin content in crab shells was 12.1 wt% after purification. Various authors [4, 11, 12] established these individual steps separately, which helped us to extract chitin from complicated and tightly bonded fibril bundles. The different steps adopted for separation are based on reports in the literature. The purified wet chitin from dry crab shells obtained via the above process was dispersed in distilled water at 1 wt% content, and hereafter called slurry of chitin. The slurry was passed through a grinder (MKCA6-3, Masuko Sangyo Co., Kawaguchi, Japan) under a neutral pH condition. The grinding stones were cleaned and adjusted carefully. In reality, the presence of the chitin suspension between the grinder stones does not allow direct contact between the stones; thus, they remain safe during operation. After passing through the grinder the uneven slurry suspension of chitin changed to a highly viscous homogeneous suspension of CNFs, which remained wet and stable indefinitely if stored in a tightly closed container.

2.2.1.1 Structure of Chitin Matrix after Chemical Treatment before Grinding Sheets of chemically isolated chitin were prepared without mechanical grinding. FE-SEM of the sheet was recorded after coating with a 2-nm layer of platinum by an ion sputter coating apparatus. Figure 2.1 shows the SEM image of pre-ground chitin. The image is of the crab shell surface after removal of the calcium carbonate mineral phase. From the SEM image of endocuticle, it is apparent that the cuticle takes up about 90% of the volume of the crab exoskeleton. One can observe that the chitin was made up of regularly structured chitin fiber networks strongly stacked into flat ribbon structured networks composed of bundles of CNFs.

2.2.1.2 Nanofibrillation of Chitin by Grinding and Effect of pH Chemically treated slurry (1 wt%) described in Section 2.2.1 was passed for one cycle through a grinder under two pH of slurry 7 and 3. At neutral pH, the widths of the fibers were more widely distributed in the range of 10–100 nm (Fig. 2.2a). The bundles of embedded chitin–protein fibers were fibrillated successfully by

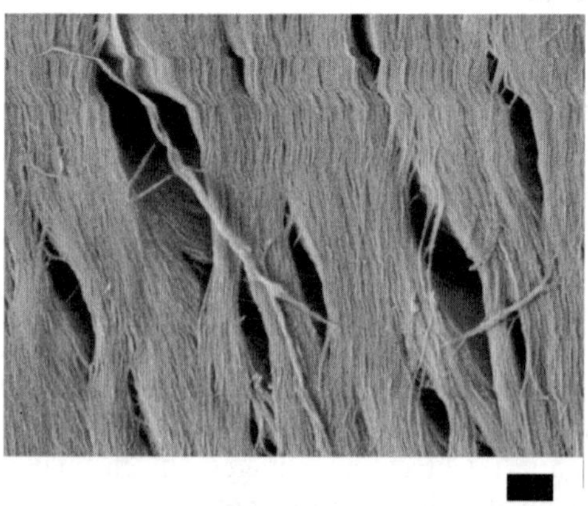

Figure 2.1 FE-SEM image of crab shell surface after the removal of matrix.

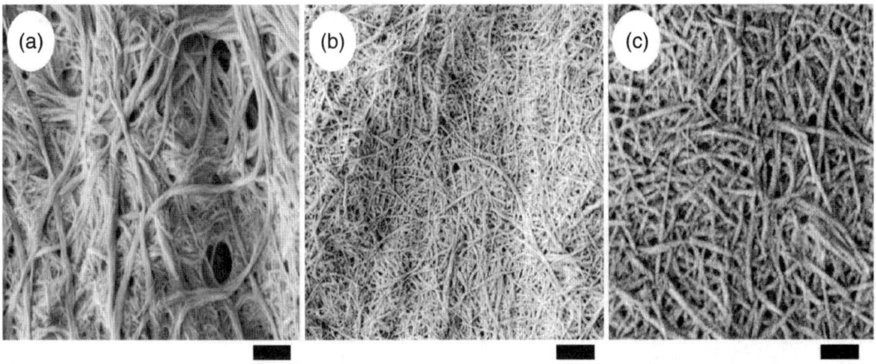

Figure 2.2 SEM pictures of chitin nanofibers (CNFs) obtained from crab shell after one cycle of grinding at two pH values of slurry: (a) pH = 7; (b) and (c) pH = 3. The scale is (a) and (b) 400 nm, (c) 200 nm.

grinding of wet chitin. It was easy to remove protein from water-soaked chitin to isolate chitin fibrils. Fan et al. [9] reported a preparation method of CNFs from wet squid pen β-chitin at pH 3–4. At low pH in acidic conditions, cationization of C2 amino groups in β-chitin occurred, which maintained a more dispersed and stable phase due to electrostatic repulsions. Similar electrostatic phenomena on cationization of amino groups was applied in purified α-chitin produced from crab shells by our research group [5] in acidic conditions (pH = 3) to produce very fine fibrils in the narrow range of 10–20 nm (Fig. 2.2b,c). The low pH was adjusted by the addition of acetic acid followed by grinding process. It is noteworthy that the chitin slurry of 1 wt% became a highly viscous gel phase after one cycle of grinding treatment due to the large surface area of NFs. In FE-SEM images, the unbroken high aspect ratio of homogeneously dispersed chitin nanocomposite was noticed in widely scanned areas. The physical appearance of the composite was a highly viscous gel phase, which is an indication [1] that fibrillation was successfully achieved and facilitated in acidic medium. Thus, CNFs (10–20 nm) were successfully isolated from crab shells in our laboratory [5], similar to the process applied to isolate cellulose NFs from wood cell wall [1].

2.2.2 Chitin Nanofibers from Prawn Shells

We successfully isolated α-chitin NFs from crab shells with a uniform width of 10–20 nm and a very high aspect ratio, as described in Section 2.2.1. From crab shells thin fibers were prepared in acidic conditions of pH 3, which cationized amino groups of chitin before grinding the slurry. However, acidic pools and extra acetic acid (AcOH) in the CNF preparation is a matter of great concern when it comes to application of NF composites in pharmaceutical, cosmetic, biomedical, electric, electronics, and optical devices, as well as other applications. The above products are very sensitive to acid content, which as a contaminant can be toxic. Moreover, removal of acid from NFs is difficult, making these products expensive. Therefore, preparation of CNFs in normal conditions of neutral pH is preferable when used for the above products.

The following describes the extraction of CNFs from prawn shells conducted under neutral conditions without addition of any acid. Fresh shells of the species *Penaeus monodon*, commonly known as "black tiger prawn," was used to prepare CNFs. This prawn is cultivated worldwide and its shell is typically thrown away as waste. Using NaOH and HCl aqueous solutions, proteins and minerals were removed, leaving the chitin and pigments in the shell [11]. The pigment from the sample was then removed with ethanol extraction. The yield of dry chitin from the wet prawn shells was approximately 16.7%. The degree of deacetylation (DDA) of the samples determined by elemental analysis was 7%. The SEM micrograph of the black tiger prawn shell surface after removal of the matrix components (without grinding treatment) is shown in Figure 2.3. The exocuticle, which is the main part of the prawn shell, is shown in the SEM picture. The prawn shell is still intact after removal of the matrix by chemical treatment; it is very important that uniform CNFs with an elaborate interwoven design are clearly visible.

The chemically treated 1 wt% chitin suspension was crushed by a domestic blender followed by passing through a grinder for fibrillation without addition of acid. The chitin slurry thus obtained was viscous after a single grinding treatment, similar to the CNFs from the crab shells. The resultant fibrous slurry was examined by FE-SEM (Fig. 2.4). In the crab shell preparation, the width of the fibers was widely distributed in the range of 10–100 nm by grinder treat-

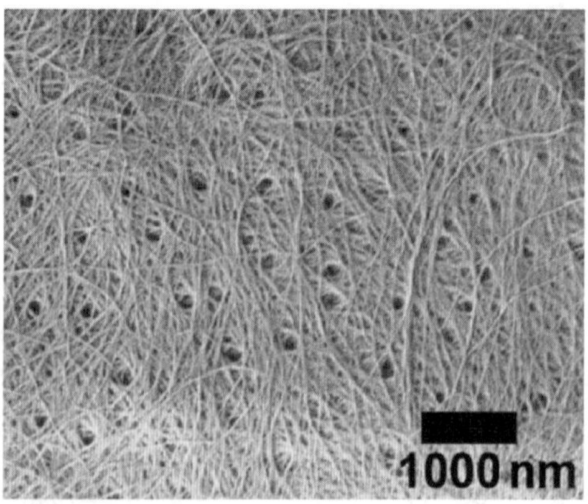

Figure 2.3 FE-SEM image of the surface of the black tiger prawn after removal of matrix components.

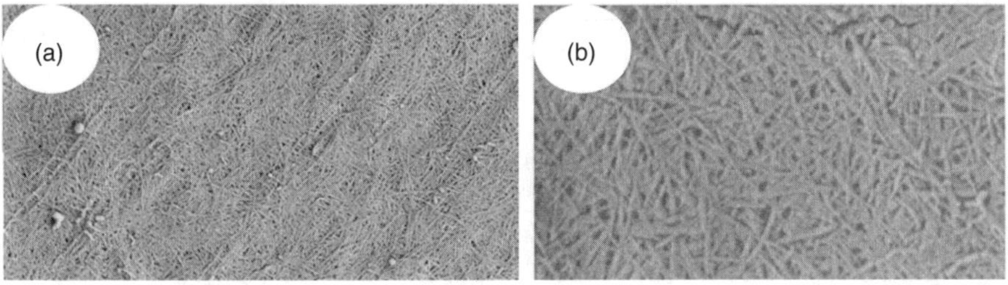

Figure 2.4 FE-SEM recording of CNFs from black tiger prawn shell after one pass through the grinder.

ment under neutral conditions. In the prawn shell preparation, we obtained a uniform shape of CNFs using the same extraction treatment. The CNFs were uniform (Fig. 2.4a), as observed by scanning over a wide area. The width of the NFs was approximately 10–20 nm (Fig. 2.4b), similar to that obtained in crab shells in acidic condition [5]. Hence, using prawns as the source, thin, homogeneous, uniformly distributed, well separated, and large aspect ratio CNFs were successfully prepared in neutral medium with much superiority over acidic crab shell preparations.

A possible explanation for this observation is as follows. The outermost skeleton (exoskeleton) of prawn or crab shells is made up of two parts, the exocuticle and the endocuticle. The exocuticle has a very fine interwoven plywood-type structure; endocuticle is rather more coarse and has thick fibers in the form shown in Figure 2.1. About 90% of the crab shell is made up of these thicker endocuticular fibers [6]. Thus, a low pH of 3 is used to obtain nanofibrils in the crab shell. On the other hand, the exoskeleton of *Natantia* prawn, including black tiger prawn, is made up of mostly semitransparent soft shell of fine (Fig. 2.3) exocuticle [13–15]; thus, their fibrillation occurs at neutral pH and is easier than crab shell. The preparation for CNFs from prawn shells in neutral pH can also be applied to other species of prawn. Figure 2.5 shows SEM images of the CNFs prepared from *Marsupenaeus japonicus* (Japanese tiger prawn) and *Pandaluseous* (Alaskan pink shrimp). These prawn also exist in abundance as a food source. These CNFs were prepared by the same method described for prawn in general in neutral pH. Figure 2.5 shows the uniform width of NFs in the range of 10–20 nm, as from black tiger prawn described previously. These results suggest that CNFs can be obtained from other prawn species having a very fine exocuticle structure by nanofibrillation under neutral pH conditions. Since many materials are sensitive to acid chemicals, this study will expand the application of CNFs.

2.2.3 Facile Preparation of Chitin Nanofibers from Dry Chitin

The isolation method of α-CNFs from crab and shrimp shells conducted by our group [5] is described in Section 2.2.1. CNFs with homogeneous thickness

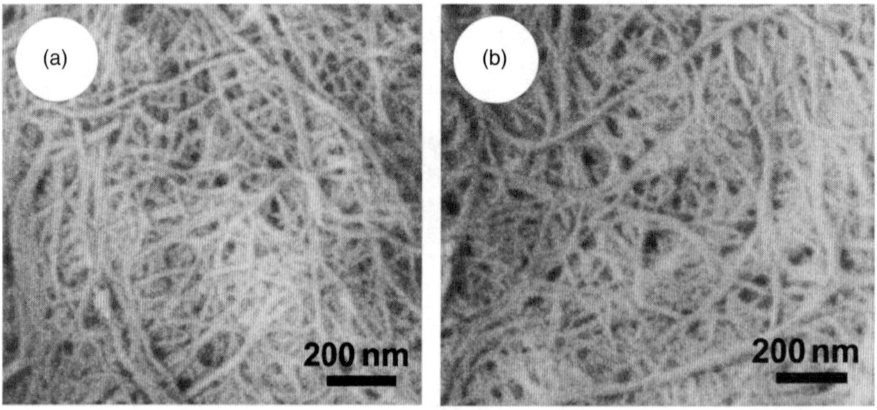

Figure 2.5 FE-SEM images of chitin nanofibers (a) from Japanese tiger prawn shell and (b) from Alaskan pink shrimp shell.

of 10–20 nm and a high aspect ratio were produced. Until now we have discussed the chemical and mechanical treatment of wet CNFs where hydrogen bonding between NFs was relatively weak compared with dried chitin or cellulose fibers; in the latter there is strong hydrogen bonding, which makes fibrillation difficult. Therefore, chitin and cellulose must be kept wet after removal of the matrix to facilitate the process of nanofibrillation [1, 5, 16–18]. However, this requirement is a disadvantage in commercial production of NFs. From the viewpoint of industrial production, the preparation of NFs from dry chitin or cellulose dry powder has an advantage since it is easier to store, preserve, and transport than wet materials. Thus, for industrial production of NFs, priority should be given to developing NFs from powdered chitin. As described above in Section 2.2.1, treatment under acidic conditions is necessary [10] to fibrillate the chitin strongly embedded in the matrix, as shown in Figure 2.1. Amino groups of chitin are cationized by the addition of acid, which facilitates the fibrillation of chitin into NFs due to electrostatic repulsion. Similarly, if the electrostatic repulsion among the cationized amino groups can break the strong hydrogen bonds in chitin bundles of dry chitin, fibrillation of dry chitin can be achieved. The method has been tested by Ifuku et al. [19] in our laboratory by successfully fibrillating strongly embedded CFs in crab shells. The method was then applied to fibrillate commercial dry chitin powder (from Nacalai Tesque, Kyoto, Japan). In Figure 2.6a, we can see that commercially available dry chitin powder from crab shells is also composed of NFs. Figure 2.6b,c shows FE-SEM micrographs of chitin fibers after one pass through the grinder with and without acetic acid. We can see that the chitin powder was not fibrillated (Fig. 2.6b) at all because of the strong interfibrillar hydrogen bonding. While the powder was completely fibrillated (Fig. 2.6c) into uniform NFs (10–20 nm width) by grinder treatment at pH 3, the degree of

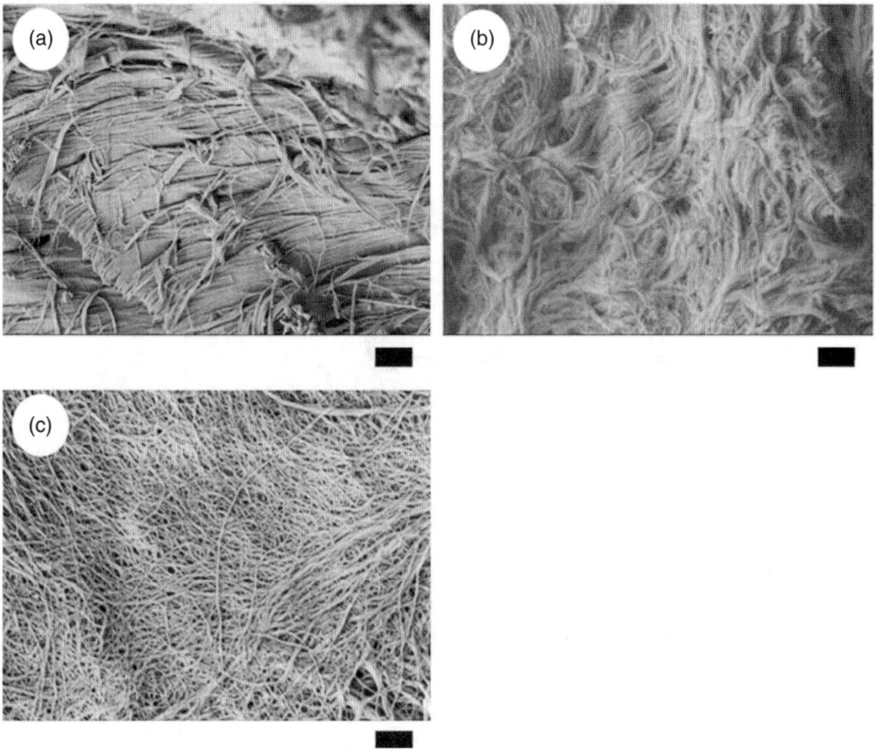

Figure 2.6 FE-SEM images of (a) commercially available dry chitin powder, (b) chitin fibers after one pass through the grinder without acetic acid, and (c) passing through grinder with acetic acid. The length of the scale bar in (a) is 1000 nm and (b) and (c) 300 nm, respectively.

substitution of amino groups was just 3.9% and the slurry obtained was a gel phase. Such fine NFs of chitin were obtained because of electrostatic repulsion resulting from the cationic charge on the chitin fiber surface that overcame the interfiber hydrogen bonding. The preparation of CNFs from commercially pre-purified dry chitin powder is advantageous for laboratory study and commercial production of CNFs, as CNFs can be made available in a few hours by following the established method of grinding in acidic conditions rather than by purifying crab shells for removal of proteins, minerals, lipids, and pigments, which may take about a week as described earlier.

2.3 CHARACTERIZATION OF CHITIN NANOFIBERS OBTAINED FROM CRAB, PRAWN, AND DRY CHITIN POWDER

The degree of *N*-acetylation of CNFs obtained from the crab shell was 95% as worked out from C and N elemental analysis, rendering a DDA of only 5%

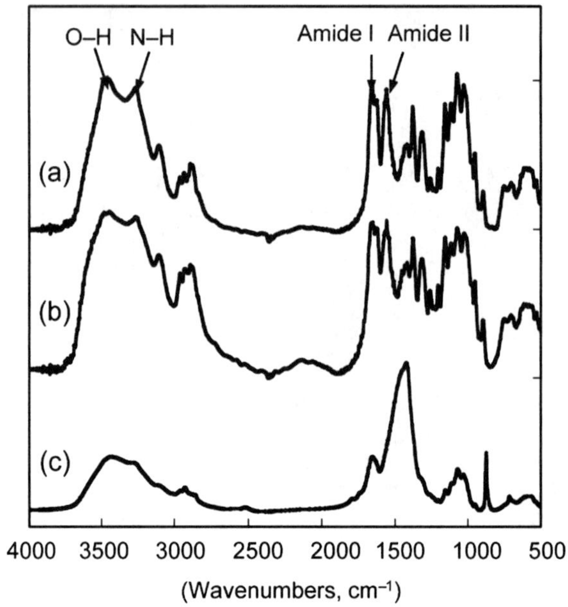

Figure 2.7 FT-IR spectra of (a) commercially available chitin powder, (b) prepared CNFs, and (c) crab shell flakes.

even after several chemical and mechanical grinding treatments. Fourier transform infrared (FT-IR) spectra of three samples—commercially available pure chitin, newly prepared CNFs from crab shell, and unpurified dried crab shell flakes—were recorded, as shown in Figure 2.7. The spectra of unpurified dried crab shell flakes were different from the other two chitins due to the matrix component presence in the shells. The spectral features of purified CNFs was in good agreement with the spectrum of commercial pure α-chitin [5, 19]. It may be assumed that the current chemical processing has removed the matrix, proteins, and minerals by the series of chemical processing steps. The protein band at 1420/cm completely disappeared in prepared CNFs. The OH stretching band at 3482/cm, NH stretching band at 3270/cm, amide I bands at 1661 and 1622/cm, and amide II band at 1559/cm of the CNFs are characteristics of pure α-chitin [4]. Figure 2.8 shows the X-ray diffraction (XRD) spectrum of commercially available pure α-chitin, newly prepared CNFs from crab shell, and dried crab shell flakes from red king crabs. The band at 29.6°, characteristic of calcium carbonate, disappeared completely from the spectrum of CNFs,confirming that the chemical treatment of crab shells has completely washed away minerals from processed CNFs. The four diffraction bands of 9.5°, 19.5°, 20.9°, and 23.4° correspond to planes 020, 110, 120, and 130, respectively. The bands are characteristic of α-chitin crystal and correspond to the pattern of commercial α-chitin [20]. Thus, X-ray investigation proved that even

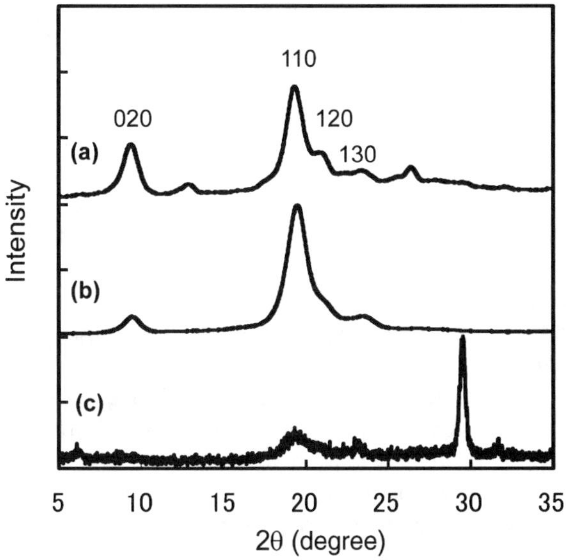

Figure 2.8 XRD profiles of (a) commercially available chitin powder, (b) chitin nanofibers, and (c) crab shell flakes.

after difficult chemical and mechanical treatment of crab shells, the resultant purified CNFs maintained the α-chitin crystalline structure. The application of FT-IR and XRD techniques of analysis of newly prepared CNFs from a number of sources has justified the success of this method.

2.4 PREPARATION OF CHITIN NANOFIBERS FROM EDIBLE MUSHROOMS

CNFs were isolated [21] from the cell walls of mushrooms by a number of chemical treatments to remove glucans, minerals, and proteins associated with mushrooms followed by grinding treatment in acidic conditions. NF widths ranged from 20 to 28 nm, depending on the type of mushroom used. The goal of extraction of CNFs from edible mushrooms was to produce a novel functional food ingredient. The detailed extraction method and final SEM images of extracted NFs and methods employed to characterize them are described below. The mushroom species *Pleuotuseryngii* (king trumpet mushroom), *Agaricus bisporus* (common mushroom), *Lentinula edodes* (shiitake), *Grifola frondosa* (maitake), and *Hypsizygus marmoreus* (buna-shimeji), commonly used as human food, were used in this study. The purification was carried out by a series of chemical treatments to remove associated compounds (proteins, pigments, glucans, and minerals) according to the procedure describe in the literature [21, 22]. In brief, sodium hydroxide was used to dissolve, hydrolyze, and remove proteins and alkali-soluble glucans. Hydrochloric acid was used to remove minerals. At this stage partial neutral saccharides and acid-soluble protein compounds were also removed. The extraction step with sodium chlorite and acetic acid removed pigments from the sample. At the final stage, the sample was treated with sodium hydroxide again to eliminate and remove the residual glucans, including trace amounts of proteins. After chemical treatment, if the extracted mass is allowed to dry, it causes strong hydrogen bonding between CNFs when all matrix substances are washed away, making it difficult to fibrillate chitin to NFs [1]. Thus, the sample was kept wet after removal of the matrix for preparation of CNFs. The purified sample with 1 wt% content of chitin was passed through a grinder for nanofibrillation in acetic acid medium at pH 3. After grinder treatment, the chitin slurry thus obtained formed a gel after a single grinder treatment, suggesting nano-fibrillation was accomplished, because of its high dispersion property in water and high surface-to-volume ratio of nanofiber. Figure 2.9 shows SEM images of CNFs from five mushrooms after removal of matrix components and one pass though the grinder. The isolated chitins are well fibrillated and uniform. The width of the fibers was in the range of 20–28 nm depending on the species of mushroom. The appearance of the fibers was similar to that of CNFs prepared from crab and prawn shells. Since damaged fibers are not observed after chemical and mechanical treatments, aspect ratios of the NFs are good. The width of fibers varied slightly according to the type of mushrooms used. It is expected that preparation of CNFs from crab and prawn shells and vegetable source mushrooms will be applicable to other cell wall or skeleton-containing vegetable or animal sources. The yield of CNF contents in mushrooms was not as high as in crab or prawn shells; it was in the range of 1.3–3.5 wt% depending on the species of mushrooms. The details of elemental analysis and species-wise composition of elements and NFs have been described by Ifuku et al. [21]. FT-IR and XRD spectrometry were employed to characterize the CNFs from mushrooms. FT-IR spectra of commercially available chitin derived from crab shell and CNFs from five types of mushroom were compared for analysis. The major bands of the spectra of CNFs are in agreement with commercial chitin. The characteristic bands of chitin molecule, O–H stretching band at 3450/m, N–H stretching band at 3270/cm, amide I band at 1660 and 1620/cm, and amide II band at 1560/cm, were noticed from CNFs, indicating that the α-chitin was isolated from mushrooms successfully. Similarly, XRD of commercially available chitin and the CNFs prepared from five types of mushrooms were compared. The four diffraction bands of CNFs were observed at 9.4°, 19.3°, 20.6°, and 22.5°, corresponding to 020, 110, 120, and 130 planes, respectively. These are typical crystal patterns of α-chitin. Thus, CNFs extracted from mushrooms maintained α-chitin crystalline structures after removal of matrix substances, and the grinding treatment.

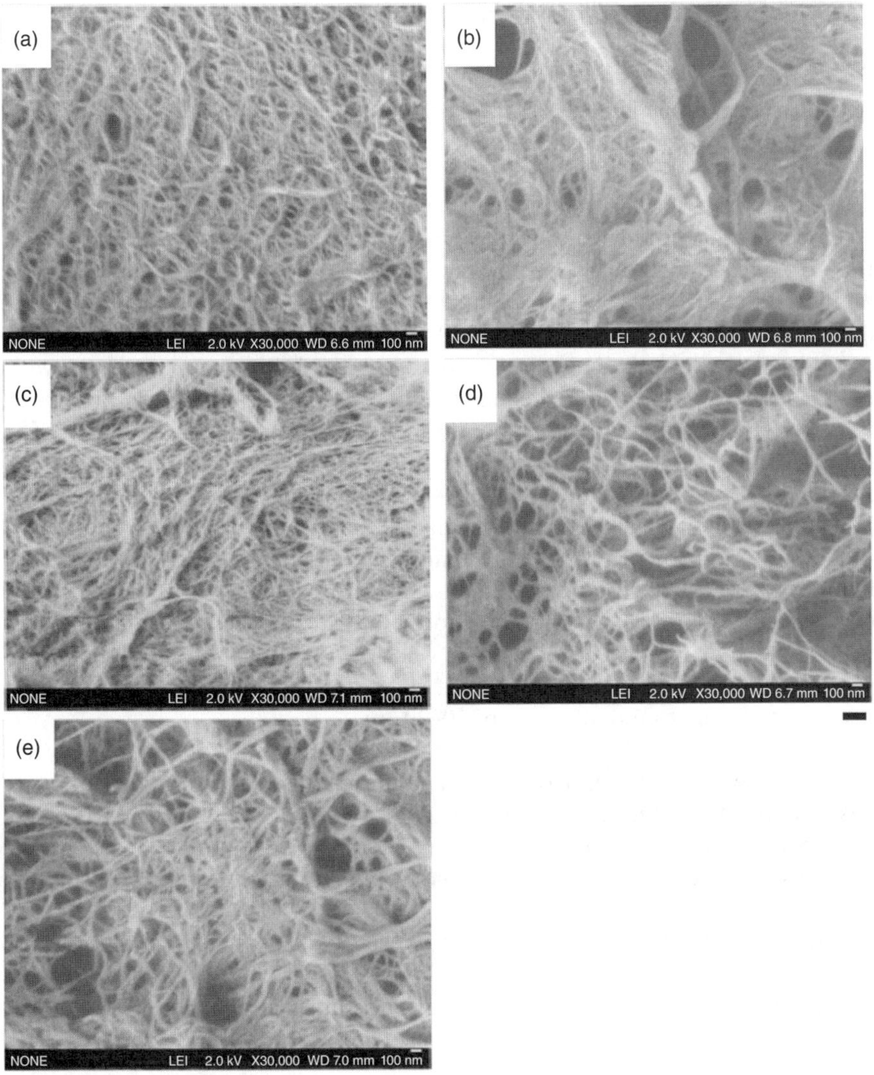

Figure 2.9 FE-SEM images of CNFs prepared from (a) *Pleuotus eryngii*, (b) *Agaricus bisporus*, (c) *Lentinula edodes*, (d) *Grifola frondosa*, and (e) *Hypsizygus marmoreus*. The scale bars are 200 nm.

2.5 PREPARATION OF CHITIN NANOFIBER NANOCOMPOSITES

CNFs are composed of an antiparallel extended crystalline structure; thus, they have excellent mechanical properties: high Young's modulus and fracture strength and low thermal expansion [23,24]. Their minutely small size and good physical properties make them strong candidates for reinforcement agents or reinforced materials for making high performance nanocomposites.

CNFs composed of two different types of acrylic resins were prepared and their properties, including transparency, Young's modulus, mechanical strength, and thermal expansion, were characterized. The aim was to use CNFs as advanced nanocomposite materials. One such material, a CNF–acrylic resin composite transparent sheet, was prepared by the following method: 0.1 wt% of fibrillated CNFs was dispersed in water. The suspension was vacuum filtered using a polytetrafluoroethylene membrane filter to produce a CNF sheet. The CNF sheet was dried by hot pressing, and cut into 3 × 4 cm dimensions. The sheet was 45 μm thick and weighed 40 mg. The sheet was impregnated by acrylic resins poly(propylene glycol) diacrylate (A-600) and tricyclodecanedimethanoldimethacrylate (DCP). The molecular structures of resins are shown in Figure 2.10. The resin-impregnated sheets were polymerized using ultraviolet (UV) curing equipment. Finally, the impregnated CNF composite film obtained was 60 μm thick, and the fiber content was 40 wt%. Both prepared nanocomposite films were optically transparent (Fig. 2.11) despite the high (40 wt%) fiber content due to the fiber size of 10–20 nm.

Figure 2.12 shows the regular light transmittance spectra of nanocomposites reinforced with CNFs and other neat components. Transmittance of DCP nanocomposites is higher than that of A-600. The higher transparency of the DCP composite is due to the refractive index of the DCP (1.5), which is higher and closer to neat CNFs (1.56) compared with that of the (1.46) resin. The transmittance of CNFs at 800/cm increased from opaque (0%) to 80–85% on reinforcing with A-600 and DCP resins, respectively. The mechanical and thermal properties of resin-reinforced chitin NFs were more suitable for application purposes. Young's modulus (GPa) of resins increased from 0.02 to 2.03

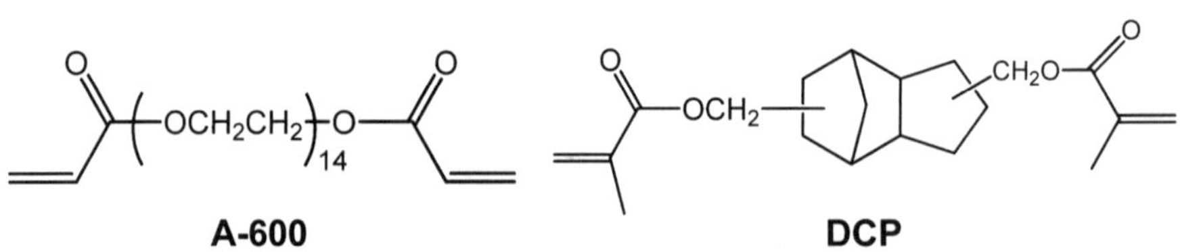

Figure 2.10 Chemical structures of resins A-600 and DCP.

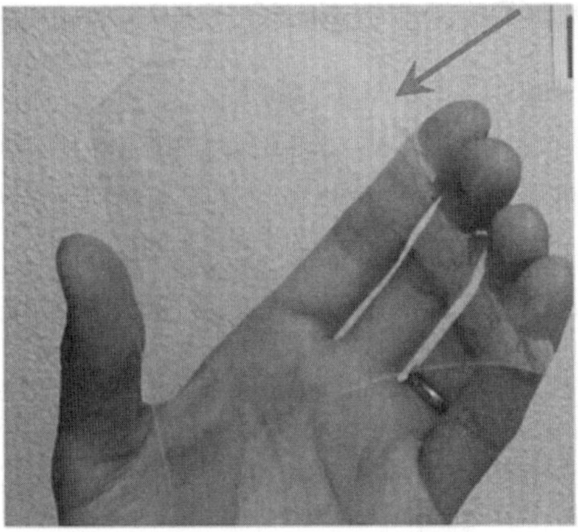

Figure 2.11 Transparent thin (60 μm) DCP film reinforced with CNFs. See color insert.

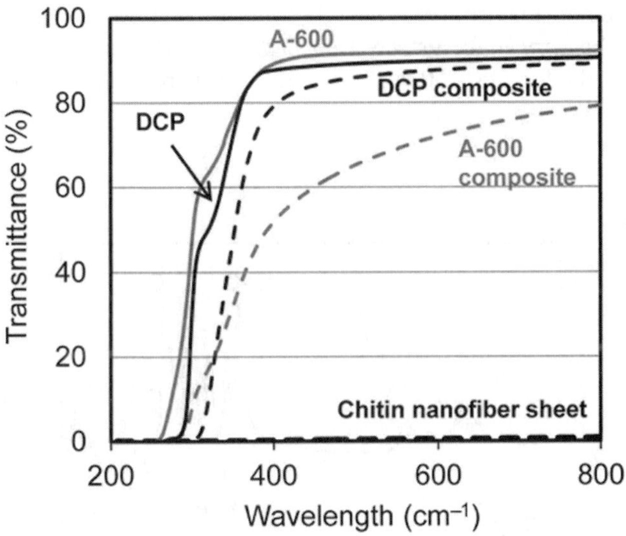

Figure 2.12 Light transmittance spectra of nanocomposites and other neat components.

for A-600 and from 2.3 to 5.34 for DCP by reinforcing the resins with CNFs, respectively. The fracture stress (MPa) increased from 4 to 41 for A-600 and from 11 to 56 for DCP. Fracture strain (%) decreased from 16.5 to 9.0 for A-600 but increased from 0.6 to 1.2 for DCP resins. Coefficient of thermal expansion (CTE; ppm/K) decreased in composites compared with neat resins: from 184 to 19 in the case of A-600 and from 100 to 24 for DCP.

2.6 ACETYLATION OF CHITIN NANOFIBERS

2.6.1 Study of Degree of Substitution

Green nanomaterial CNFs prepared by our group were developed to have wider scope and application; this was possible by chemical modification of the CNF surface. Introducing hydrophobic functional groups into polar moieties of fibers is expected to improve the fiber dispersion as well as the adhesion properties with hydrophobic matrices. Acetylation is considered to be a simple chemical modification. It is a popular and inexpensive approach to change the surface property [25]. Until Ifuku et al. modified the CNF surface with an acetyl group, there was no report in the literature of chemical modification of CNFs. Until then the effect of the reaction behavior of highly crystalline CNFs and the relationship between the acetylation degree of substitution (DS) and the various properties of the NFs remained unclear. Ifuku et al. modified CNFs by acetylation and prepared their nanocomposites, and then characterized them. The method of acetylation has been described in Reference [26]. The acetyl DS ranged from 0.99 to 2.96 in a reaction time of 50 minutes. This is the highest degree of substitution of acetyl groups in CNFs. The high reaction rate was due to the very high surface area of fibers. FT-IR spectra (Fig. 2.13) of acetylated CNFs for DS of 0.99, 1.81, and 2.96 were recorded. As the DS of

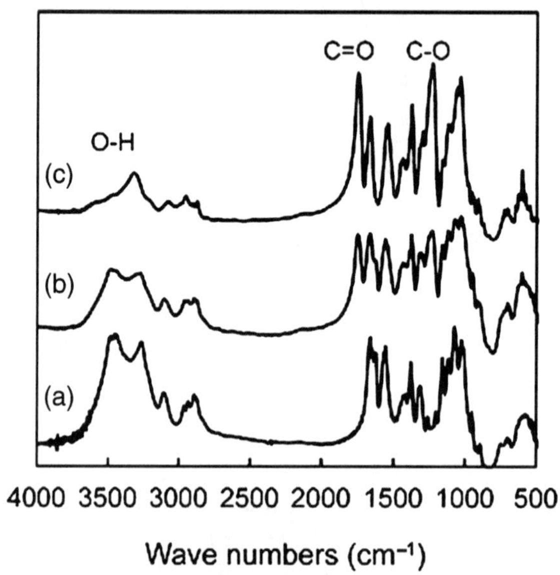

Figure 2.13 FT-IR spectra of acetylated CNFs of (a) DS 0.99, (b) DS 1.81, and (c) DS 2.96.

acetyl groups increased, two major bands at 1231 and 1748/cm increased, corresponding to the C–O and C=O stretching vibration modes of the acetyl group. Simultaneously, the O–H stretching band at 3972/cm decreased with increasing DS and almost disappeared at a DS value of 2.96, indicating that a complete substitution of acetyl groups in the CNFs had occurred.

XRD profiles of a series of acetylated CNFs are shown in Figure 2.14. In original CNFs (DS 0.99), the four diffraction peaks of CNFs observed at 9.5, 19.4, 20.9, and 23.4° are charateristic of 020, 110, 120, and 130 planes, respectively. The spectra show typical antiparallel crystal pattern of α-chitin. The α-chitin diffraction pattern completely disappeared at DS 2.96, and the sample showed a well-defined uniform pattern of di-*O*-acetylated chitin (chitin diacetate) at $2\theta = 7.4$ and 17.7°. While the XRD pattern of α-chitin still remained at DS 1.81, here about 50% of OH groups were substituted.

2.6.2 SEM Images of Substituted Chitin Nanofibers

FE-SEM images of acetylated CNFs of three DS NFs are shown in Figure 2.15. The images remained unchanged even in DS 2.96 preparation, which indicates that chitin diacetate is insoluble in the reaction mixture acetic anhydride. In all of the images of CNF sheets, individual isolated NFs existed; however, as

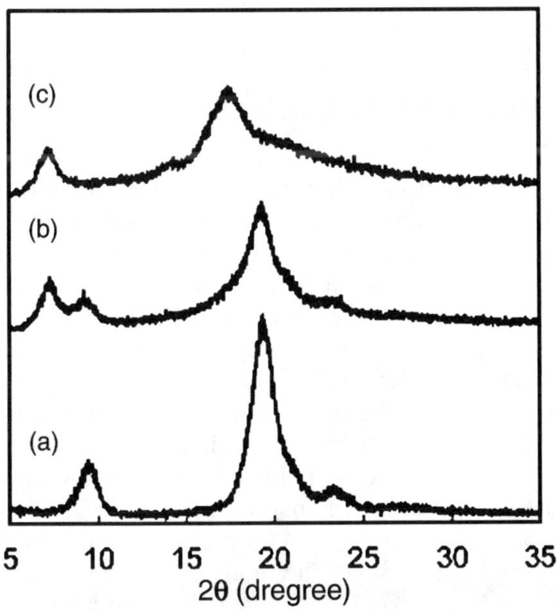

Figure 2.14 XRD spectra of acetylated CNFs at (a) DS 0.99, (b) DS 1.81, and (c) DS 2.96.

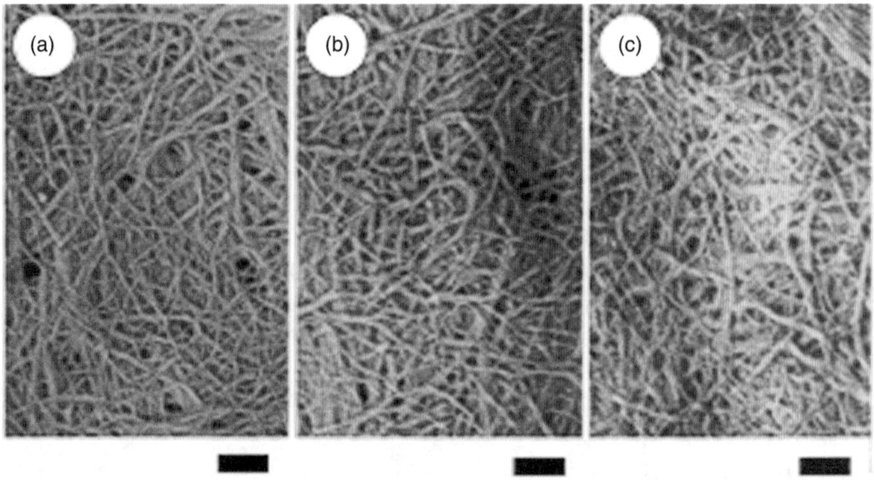

Figure 2.15 FE-SEM pictures of acetylated CNFs samples of three DS values: (a) DS 0.99, (b) DS 1.81, and (c) DS 2.96. Scale 200 nm.

the DS increased, the thickness of the NFs also increased. The average thickness of NFs with DS 0.99, 1.81, and 2.96 were 21.6, 28.9, and 32.1 nm, respectively. The thickness of substituted NFs increased as a result of the attachment of bulky acetyl groups to CNFs.

2.6.3 Acetylated Chitin Nanofiber Composites

Acetylated CNFs were reinforced with acrylic resin DCP. The method of reinforcement has been cited elsewhere [26]. The final reinforced sheet has a thickness of 150 μm and chitin content of 25 wt%. Acetylated nanocomposites have high transparencies in all acetyl DS from 0.99 to 2.96. At the transmittance measuring wavelength of 700 nm, the DS 0.99 sheet had a transparency of 77%. As the acetyl DS increased, the transmittance decreased. At 2.96 DS, the transmittance was 73%. This decrease can likely be attributed to the change in the refractive index of the NFs. A reinforced CNF composite can have promising applications as it is an optically functional composite. It is hygroscopic, unlike neat CNFs, which are hydroscopic. Absorption of moisture in CNFs can deform the composite in humid conditions. Acetylation seems to reduce the moisture content of the nanocomposite. The moisture content of the original sample with DS 0.99 was 4.0 wt%, which is higher than the 0.33% of hygroscopic acrylic resin filler. The moisture absorption of the nanocomposite with DS 1.30 decreased to 2.2% due to the introduction of hydrophobic acetyl groups to the hydroxyl moiety of chitin molecules. Acetylation of the NFs improved the compatibility with the acrylic resin, and improvement of miscibility at the resin–chitin interface thus decreased water adsorption onto reinforced NFs. With further acetylation, the moisture content did not change much; the change was only in the range of 2.0–2.5%. This suggests that most hydroxyl groups on the CNF surface were acetylated in the first minute of the acetylation reaction. The CTE of reinforced resin DCP by CNFs was also studied. DCP resin has a CTE of 64/ppmK; on reinforcement by chitin, the CTE of the CNF/DCP composite decreased to 23/ppmK. The CNFs with a low thermal expansion of 92/ppm K reduced the CTE of the DCP by reinforcement with a fiber content of 25 wt%. On the other hand, the CTE of CNF samples and their composites increased proportionally and gradually with increases in the acetyl DS. This increase due to the acetylation reduces the degree of crystallinity of CNFs, which increases the CTE of CNFs and the nanocomposites.

2.7 CONCLUSION

Homogeneous CNFs of 10–20 nm width were prepared from dried crab shell by combination of chemical treatment and mechanical grinding. The grinding treatment was in wet state; removal of matrix from wood and crab shell to isolate CNFs were similar. CNFs were also prepared from dried chitin. Grinding in acidic conditions is important for fibrillation of dried chitin. In acidic conditions, calibrated addition of acetic acid cationized amino groups of chitin and broke the strong intermolecular hydrogen bonding due to development of charge followed by electrostatic repulsion. The method of preparation of NFs from dry chitin powder has advantage for commercial production, and raw dry chitin is easy to store, transport, and supply. Using dry chitin powder, we have successfully prepared stable and homogeneous CNFs with a high surface-to-volume ratio. CNFs were prepared from a variety of prawn shells without addition of any acid. Prawn shell has a finer structure than crab shell; thus, CNFs were isolated more easily. Two different types of transparent acrylic resin sheets reinforced with CNFs were also prepared. Due to nanosized chitin fibers, the prepared composites were highly transparent and flexible and had low thermal expansion, high Young's modulus, and high tensile strength. The study of preparation of CNF-reinforced resins has expanded the application of CNFs; chitin raw materials until now were thrown away as industrial waste. Moreover, chitin is a green, environmentally benign biocomposite. Chemical modification of CNFs was also done for making the hydrophilic chitin surface hydrophobic to disperse it in more hydrophobic solvents. Stepwise acetylation of CNFs was carried out successfully, adjusting the DS value at different reaction times. Thinner (60–150 μm) transparent films were obtained by reinforcement of CNFs. Reinforcement of CNFs by resins or reinforced resins by CNFs improved physical properties such as Young's modulus, fracture stress and strain, CTE and flexibility, transparency, and hygroscopicity. Favorably modified properties show that neat CNFs or blended fibers have vast applications in the future as advanced biologically benign nanocomposites.

REFERENCES

[1] Abe, K., Iwamoto, S., and Yano, H. (2007) Obtaining cellulose nanofibers with a uniform width of 15 nm from wood. *Biomacromolecules*, *8*, 3276–3278.

[2] Nishino, T., Takano, K., and Nakamae, K. (1995) Elastic-modulus of the crystalline regions of cellulose polymorphs. *J. Polym. Sci.[B]*, *33*, 1647–1651.

[3] Nishino, T., Matsuda, I., and Hirano, K. (2004) All cellulose composite. *Macromolecules*, *37*, 7683–7687.

[4] Gopalan, N.K., Dufresne, A. (2003) Crab shell chitin whisker reinforced natural rubber nanocomposites. 1. Processing and swelling behavior. *Biomacromolecules*, *4*, 657–665.

[5] Ifuku, S., Nogi, M., Abe, K., Yoshioka, M., Morimoto, M., Saimoto, H., and Yano, H. (2009) Preparation of chitin nanofibers with a uniform width as α-chitin from crabshells. *Biomacromolecules*, *10*, 1584–1588.

[6] Chen, P.-Y., Lin, A.Y.-M., McKittrick, J., and Meyers, M.A. (2008) Structure and-mechanical properties of crab exoskeletons. *Acta Biomater.*, *4*, 587–596.

[7] Zhao, H., Feng, X., and Gao, H. (2007) Ultrasonic technique for extracting nanofibers from nature materials. *Appl. Phys. Lett.*, *90*, 073112.

[8] Min, B.M., Lee, S.W., Lim, J.N., You, Y., Lee, T.S., Kang, P.H., and Park, W.H. (2004) Chitin and chitosan nanofibers: Electrospinning of chitin and deacetylation of chitinnanofibers. *Polymer*, *45*, 7137–7142.

[9] Fan, Y., Saito, T., and Isogai, A. (2008) Chitin nanocrystals prepared by TEMPO mediated oxidation of α-chitin. *Biomacromolecules*, *9*, 192–198.

[10] Fan, Y., Saito, T., and Isogai, A. (2008) Preparation of chitin nanofibers from squid pen β-chitin by simple mechanical treatment under acid conditions. *Biomacromolecules*, *9*, 1919–1923.

[11] Shimahara, K., Takiguchi, Y. (1998) Preparation of crustacean chitin. In: *Methods in enzymology*, Vol. 161, eds. W.A. Wood, S.T. Kellogg. Academia Press, pp. 417–423.

[12] BeMiller, J.N., Whistler, R.L. (1962) Alkaline degradation of amino sugars. *J. Org. Chem.*, *27*, 1161–1164.

[13] Yano, I. (1972) A historical study on the exocuticle with respect to its calcification and associated epidermal cells in a shore crab. *Bull. Jpn. Soc. Sci. Fish.*, *38*, 733–739.

[14] Yano, I. (1975) An electron microscope study on the calcification of the exoskeleton in a shore crab. *Bull. Jpn. Soc. Sci. Fish.*, *41*, 1079–1082.

[15] Yano, I. (1977) Structures of exoskeletons of prawn and crab shells. *Kagaku to Seibutsu*, *15*, 328–336.

[16] Fan, Y., Saito, T., and Isogai, A. (2009) TEMPO-mediated oxidation of β-chitin to prepare individual nanofibrils. *Carbohydr. Poly.*, *77*, 832–838.

[17] Iwamoto, S., Abe, K., and Yano, H. (2008) The effect of hemicelluloses on wood pulpnanofibrillation and nanofiber network characteristics. *Biomacromoles*, *9*, 1022–1026.

[18] Saito, T., Nishiyama, Y., Putaux, J.-L., Vignon, M., and Isogai, A. (2006) Homogeneoussuspensions of individualized microfibrils from TEMPO-catalyzed oxidation ofnative cellulose. *Biomacromolecules*, *7*, 1687–1691.

[19] Ifuku, S., Nogi, M., Yoshioka, M., Morimoto, M., Yano, H., and Saimoto, H. (2010) Fibrillation of dried chitin into 10–20 nm nanofibers by a simple grinding method under acidic conditions. *Carbohydr. Polym.*, *81*, 134–139.

[20] Minke, R., Blackwell, J. (1978) The structure of α-chitin. *J. Mol. Biol.*, *120*, 167–181.

[21] Ifuku, S., Nomura, R., Morimoto, M., and Saimoto, H. (2011) Fabrication of chitin nanofibers with a width of 20 nm from mushrooms. *Materials*, *4*, 1417–1425.

[22] Ivshina, T.N., Artamonova, S.D., Ivshin, V.P., and Sharnina, F.F. (2009) Isolation of the chitin-glucan complex from the fruiting bodies of mycothallus. *Appl. Biochem. Microbiol.*, *45*, 313–318.

[23] Vincent, J., Wegst, U. (2004) Design and mechanical properties of insect cuticle. *Arthropod Struct. Dev.*, *33*, 187–199.

[24] Wada, M., Saito, Y. (2001) Lateral thermal expansion of chitin crystals. *J. Polym. Sci. [B]*, *39*, 168–174.

[25] Ifuku, S., Nogi, M., Abe, K., Handa, K., Nakatsubo, F., and Yano, H. (2007) Surface modification of bacterial cellulose nanofibers for property enhancement of optically transparent composites: Dependence on acetyl-group DS. *Biomacromolecules*, *8*, 1973–1978.

[26] Ifuku, S., Morooka, S., Morimoto, M., and Saimoto, H. (2010) Acetylation of chitin nanofibers and their transparentnanocompositefilms. *Biomacromolecules*, *11*, 1326–1330.

CHAPTER 3

Chemical Modification of Chitosan and Its Biomedical Application

DEEPA THOMAS and SABU THOMAS

3.1 INTRODUCTION

Chitin is the second most abundant natural polymer in the world after cellulose. It is the only pseudo-natural cationic biopolymer that has good properties, including biodegradability, biocompatibility, nontoxicity, and antibacterial activity, which makes it an interesting novel functional material [1]. Chitosan is composed primarily of glucosamine and *N*-acetylglucosamine residues with a 1,4-β-linkage. Chitosan offers remarkable biological properties, which have paved the way for its application in the pharmaceutical and biomedical fields, such as biomedicine, membranes, drug delivery systems, and hydrogels, and as a scaffold for tissue engineering. The presence of -NH_2 groups in chitosan is the reason it exhibits much greater potential than chitin for use in different applications. Nevertheless, the properties of chitosan, such as its poor solubility in water or in organic solvents, can limit its utilization for a specific application. The solubilization of chitosan occurs by protonation of the -NH_2 functional group on the C-2 position of the D-glucosamine repeating unit, whereby the polysaccharide is converted to a polyelectrolyte in acidic media. An elegant way to improve or to impart new properties to chitosan is the chemical modification of the chain, generally by grafting of functional groups, without modification of the initial skeleton in order to preserve the original properties. The functionalization is carried out on the primary amine group, generally by quaternization, or on the hydroxyl group.

Biopolymer Nanocomposites: Processing, Properties, and Applications, First Edition. Edited by Alain Dufresne, Sabu Thomas, and Laly A. Pothan.

3.2 STRUCTURE OF CHITOSAN

Chitin and chitosan are heteropolymers. Chitosan is the *N*-deacetylated derivative of chitin and can be obtained by deacetylation of chitin, which is produced from shells of crustaceans and insects and from other sources. The degree of *N*-acetylation can be defined as the ratio of 2-acetamido-2-deoxy-D-glucopyranose to 2-amino-2-deoxy-D-glucopyranose structural units [2]. This ratio has a striking effect on the solubility and solution properties of chitin and chitosan. In chitin, the degree of acetylation is typically 0.90, while in chitosan the degree of acetylation is less than 0.35. To determine this ratio, attempts have been made with many analytical tools, which include infrared (IR) spectroscopy, pyrolysis gas chromatography, and ultraviolet (UV) spectroscopy. Chitosan is a polycationic polymer that has one amino group and two hydroxyl groups in the repeating glucosidic residue. The carbohydrate backbone is very similar to that of cellulose, which consists of β-1, 4-linked D-glucosamine with a variable degree of *N*-acetylation, except that the acetyl amino group replaces the hydroxyl group on the C2 position. Thus, chitosan is a copolymer consisting of *N*-acetyl-2-amino-2-deoxy-D-glucopyranose and 2-amino-2-deoxy-D-glucopyranose, where the two types of repeating units are linked by (1→4)-β-glycosidic bonds. After refinement, chitosan has a rigid crystalline structure through inter- and intramolecular hydrogen bonding. The repeating units of chitosan, cellulose, and chitin are shown in Figure 3.1 and Figure 3.2 [3]. The active primary amino groups on the molecule being reactive provide sites for a variety of side group attachment employing mild reaction conditions (Fig. 3.3). The attached side groups on chitosan provide versatile materials with specific functionality, alter biological properties, or modify physical properties [4]. Because of chitosan's biological and chemical properties, it has the ability to bind to particular materials including cholesterols, fats, proteins, metal ions, and even tumor cells. This allows chitosan to be used as a chelating agent in various applications. The facile derivatization makes chitosan an ideal candidate for biofabrication. In addition, the characteristic features of chitosan, such as being cationic, haemostatic, and insoluble at high pH, can be reversed by sulfating the amine, which makes the molecule anionic and water-soluble, with the introduction of anticoagulant properties.

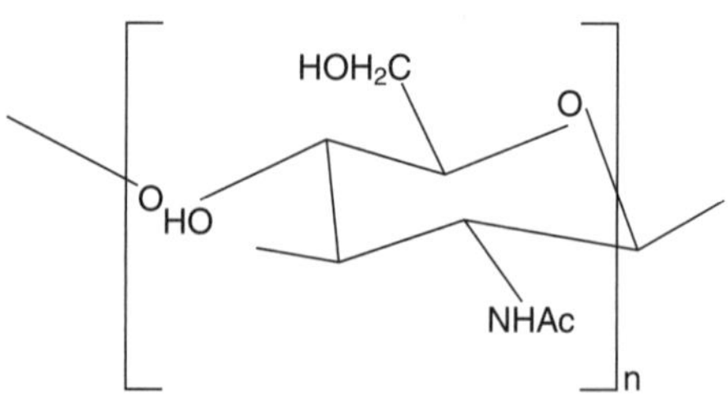

Figure 3.1 Structure of chitosan.

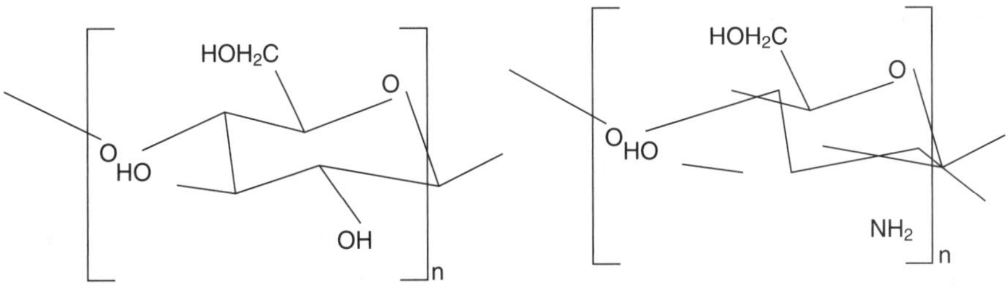

Figure 3.2 Repeating unit of cellulose and chitin.

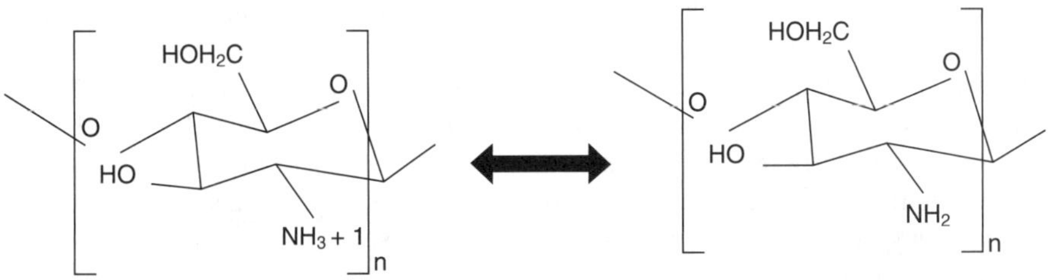

Figure 3.3 Schematic illustration of chitosan's versatility. At low pH (less than about 6), chitosan's amine groups are protonated, conferring polycationic behavior to chitosan. At a higher pH (above about 6.5), chitosan's amines are deprotonated and reactive [4].

3.3 CHEMICAL MODIFICATIONS OF CHITOSAN

In spite of the potential applications of chitin and chitosan, it is necessary to establish efficient and appropriate modifications to explore fully the high potential of these biomacromolecules. Chemical modifications of chitin are generally difficult owing to the lack of solubility, and the reactions under heterogeneous conditions are accompanied by various problems such as the poor extent of reactions, difficulty in selective substitution, structural ambiguity of the products, and partial degradation due to severe reaction conditions. Therefore, with regard to developing advanced functions, much attention had been paid to modification of chitosan rather than chitin. The advantage of chitosan over other polysaccharides (such as cellulose, starch, and galactomannans) is that its chemical structure allows easier modifications at the C2 position. Specific groups can be introduced to produce novel polymers for selected applications. The practical use of chitosan has been largely limited to its unmodified forms, and thus modified chitosans are required for use in a wide range of biomedical applications including interaction and intracellular delivery of genetic materials. In order to improve or impart new properties to chitosan, chemical modification of the chitosan chains, generally either by grafting of small molecules or polymer chains onto the chitosan backbone or by quaternization of the amino groups, has been investigated [5]. Chitosan chains possess three attractive reactive sites for chemical modification: two hydroxyl groups (primary or secondary) and one primary amine. The site of modification is dictated by the desired application of the final chitosan derivative [6]. In many reactions, the nonbonding pair of electrons on the primary amino groups of chitosan units performs the role of acceptor protons. Thus, the primary amino group becomes cationically charged. These nonbonding pairs of electrons on the primary amino group also make chitosan a potent nucleophile, reacting readily with most aldehydes to form imines [7]. Although the majority of reactions involving chitosan involve the primary amino group, it is possible to modify selectively the hydroxyl groups. This can be accomplished by protecting the amine group through the formation of the polysaccharide formate or acetate and subsequent reaction of the salt with an electrophile. The hydroxyl group pendant at C6 is more reactive than that at C3 and therefore is derivatized preferentially [8]. Chitin and chitosan have been modified via a variety of chemical modifications such as nitration, phosphorylation, sulphation, xanthation, acylation, hydroxyalkylation, Schiff's base formation and alkylation, and graft copolymerization [9].

3.3.1 Chitosan-Grafted Copolymers

Graft copolymerization is expected to be one of the most promising approaches to develop novel chitin/chitosan-based materials with improved properties. In recent years, a number of initiator systems such as ammonium per sulfate (APS), potassium per sulfate (PPS), ceric ammonium nitrate (CAN), thiocarbonationpotassium bromate (TCPB), potassium diperiodatocuprate (III) (PDC), 2,2-azobisisobutyronitrile (AIBN), and ferrous ammonium sulfate (FAS) have been developed to initiate grafting copolymerization [10, 11]. Graft copolymerization can also be initiated by γ-irradiation and enzymes. The grafting parameters such as grafting percentage and grafting efficiency are greatly influenced by several parameters such as the type and concentration of initiator, monomer concentration, reaction temperature, and time. The properties of the resulting graft copolymers are widely controlled by the characteristics of the side chains, including molecular structure, length, and number. Many researchers have studied the effects of these variables on the grafting parameters and the properties of the resultant grafted chitosan [10, 11].

3.3.1.1 Grafting Initiated by Free Radicals Graft copolymerization of vinyl monomers onto chitosan using free radical initiation has attracted the interest of many scientists in the last two decades. For example, Sun et al. prepared carboxymethyl chitosan-grafted methacrylic acid (MAA) by using APS as an initiator in aqueous solution [12]. The effects of APS, MAA, reaction temperature, and time on graft copolymerization were analyzed by determining

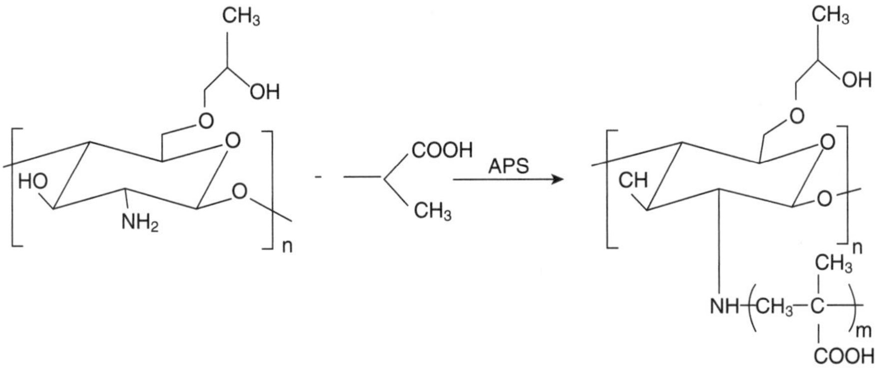

Figure 3.4 Graft copolymerization of MAA on hydroxypropyl chitosan [13].

the grafting percentage and grafting efficiency. After grafting, the chitosan derivatives had much improved water solubility. Similarly, Xie et al. prepared hydroxypropyl chitosan-grafted MAA by using APS as an initiator (Fig. 3.4), obtaining a derivative that also had good solubility in water [13].

3.3.1.2 Grafting Using Radiation Recently, there has been great interest in graft natural polymers using the radiation method. Grafting of polystyrene onto chitin and chitosan using ^{60}Co γ-irradiation at room temperature was investigated [14]. Researchers analyzed the effect of adsorbed dose, solvent, and oxygen and found that the grafting yield increases with an increase in the adsorbed dose. Singh and Roy investigated the radiation grafting of chitosan with *N*, *N*_-dimethylaminoethylmethacrylate (DMAEMA) [15] and found that parameters such as solvent composition, monomer concentration, radiation dose rate, and total dose/time affect the rate of grafting and homopolymerization. They also found that a desired level of grafting of DMAEMA onto chitosan films was achieved by appropriate selection of these grafting conditions. Yu et al. reported the graft copolymerizationof butyl acrylate onto chitosan by using γ-irradiation [16]. They found that the grafting percentage increases with increase in the monomer concentration and total dose. Singh et al. grafted poly(acrylonitrile) onto chitosan using the microwave irradiation technique under homogeneous conditions. They studied the effect of monomer and chitosan concentration, microwave power, and exposure time on the graft-copolymerization and found that grafting increases with an increase in the monomer concentration [17].

3.3.1.3 Enzymatic Grafting There are several potential advantages of using enzymes in polymer synthesis and modification [18]. With respect to health and safety, enzymes offer the potential of eliminating the hazards associated with reactive reagents. A potential environmental benefit from using enzymes is that their selectivity may be exploited to eliminate the need for waste full protection and deprotection steps. Finally, enzymes specificity may offer the potential for precisely modifying macromolecular structure to better control polymer function [19]. For instance, enzymatic modification can yield chitosan derivatives with unique pH-sensitive water solubility and adhesive properties.

3.3.1.4 Cationic Graft Polymerization In the late 1990s, Yoshikawa et al. showed that grafting reactions onto chitosan can also be performed by using living cationic polymerization [20]. The viscosity of the resulting polymer was found to increase with the increasing percentage of grafting. This grafted polymer was also found to be soluble in water.

3.3.2 Cyclodextrin-Linked Chitosan

The cyclic oligosaccharides, namely α-, β-, γ- cyclodextrins (CD), are important because of their ability to encapsulate hydrophobic molecules in their toroidal hydrophobic cavity, whose selectivity depends on the number of glucose units (respectively 6, 7, 8 D-glucose units). CD- linked chitosan is interesting from the viewpoint of pharmaceutics, including drug delivery, cosmetics, and analytical chemistry. Although functionalization at the 6-position of OH in CD is relatively easy, the secondary 2 or 3 position is shown to be the more important site of CD in binding studies. Sakairi prepared α-CD-linked chitosan (Fig. 3.5) using 2-O-formylmethyl-β-CD by reductive N-alkylation. Tosylated β-CD is also useful for linking chitosan at the 2-position of CD, as reported by Chen et al. [24]. CD-linked chitosan could also be prepared via the intermediate of its monochlorotriazinyl derivative [25]. This compound was used for decontamination of water containing textile dyes. An insoluble cross-linked chitosan bearing β-CD was prepared using N-succinyl chitosan and aminated-β-CD via amide bond formation [26]. The β-CD linked chitosan using 1,6-hexamethylene diisocyanate as a spacer was prepared by Sreenivasan [27]. This material showed interaction with cholesterol and was useful as an adsorbent matrix.

3.3.3 Crown Ether Bound Chitosan

Crown ethers have molecular structures that cause them to exhibit complexing selectivity for metal ions. Crown ether bound chitosan not only had good adsorption capacities for noble metal ions Pd^{2+}, Au^{3+}, and Ag^{+}, but also had high selectivity for adsorption of Pd^{2+}, in the presence of Cu^{2+} and Hg^{2+}. Cross-linked crown ether bound chitosan was also reported [28]. These cross-linked derivatives have space net structures with embedded crown ethers, and each mesh has a certain spacevolume. When the original chitosan was reacted with4,4′-dibromobenzo-18-crown-6-crown ether, the cross-linked product between 6-OH and NH_2 was obtained (Fig. 3.6). However, this product included heterogeneous cross-linked structures between 6-OH and 6-OH, or NH_2 and

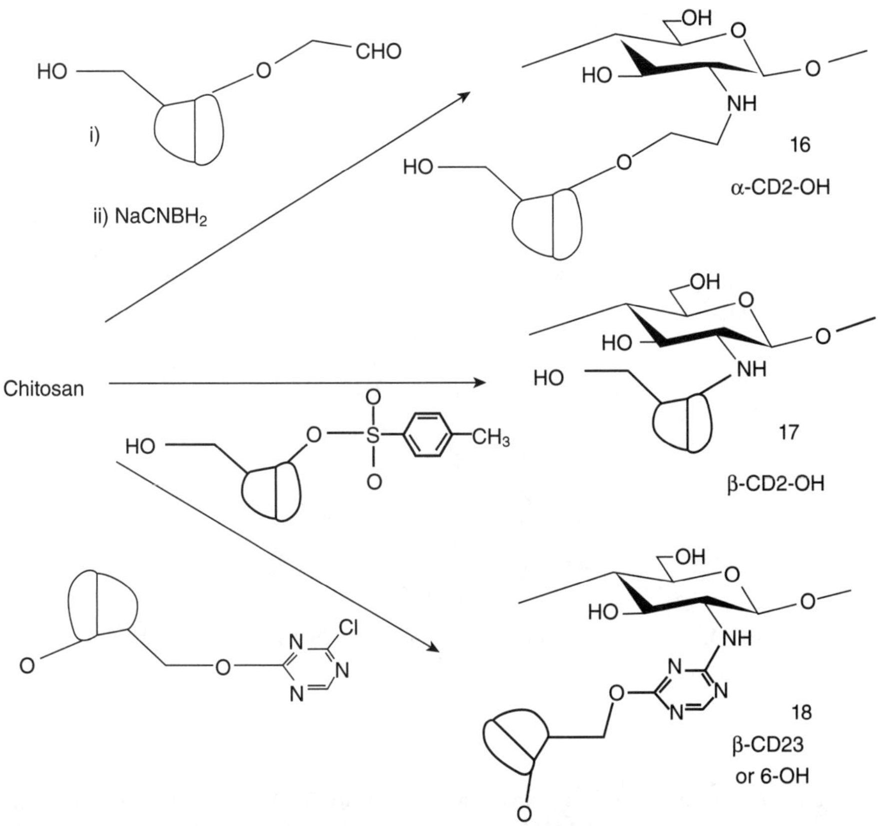

Figure 3.5 Cyclodextrin-linked chitosan [21–23].

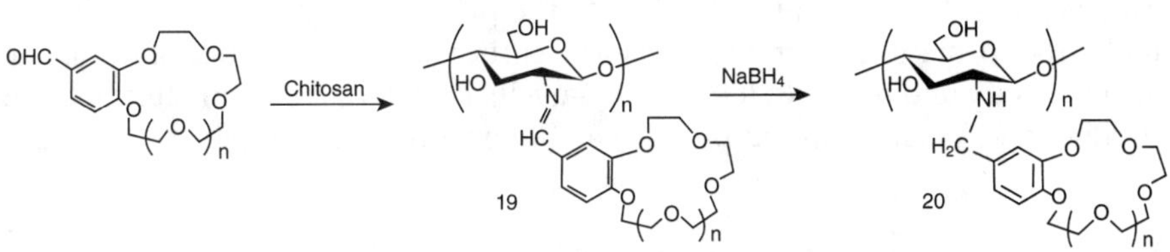

Figure 3.6 Crown ether bound chitosans [22, 28].

NH_2, while benzylidene protected chitosan (CTB) produced homogeneous cross-linked structures between 6-OH and 6-OH (Fig. 3.7).

3.3.4 Thiol-Containing Chitosan

Thiol-containing chitosan, also called thiolated chitosan, is obtained through the reaction between chitosan and thiolactic acid. In this reaction,

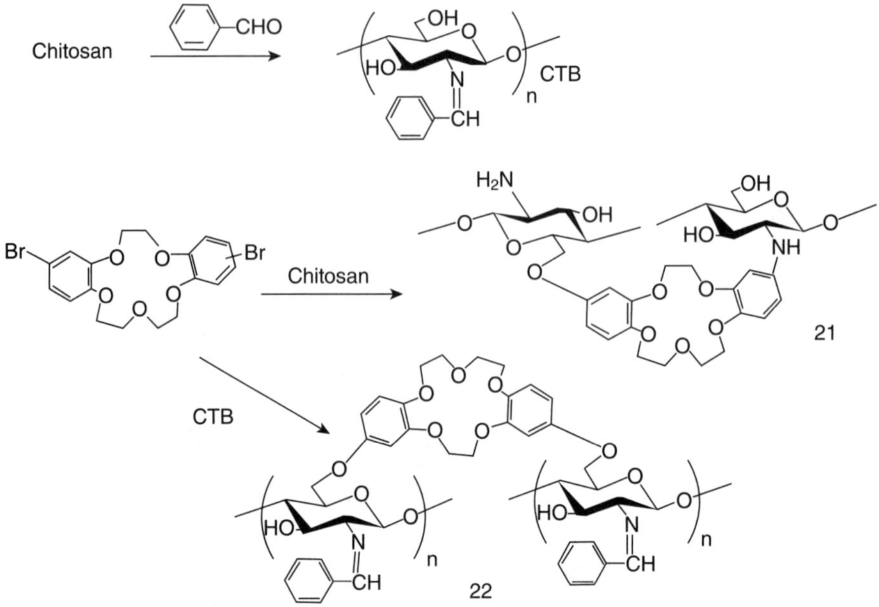

Figure 3.7 Cross-linked type of crown ether bound chitosan [22, 28].

1-ethyl-3-(3-dimethylaminopropyl)carbodiimide (EDC) can be used to graft these two materials [29]. EDC is a water-soluble carbodiimide that is typically employed in the 4.0–6.0 pH range. It is a zero-length cross-linking agent that has been widely used to couple carboxylic acid groups to primary amines. Thiolactic acid is covalently attached to the primary amino group of chitosan under the formation of amide bonds. The carboxylic acid moieties of thiolactic acid are activated by EDC, forming a O-acylurea derivative as an intermediate product that reacts with the primary amino groups of chitosan.

3.3.5 Carbohydrate Branched Chitosans

Carbohydrates can be grafted on the chitosan backbone at the C2 position by reductive alkylation. For that purpose, disaccharides (cellobiose, lactose, etc.) having a reducing end group are introduced, in the presence of a reductant, on chitosan in the open chain form [30]. Synthesis of galactosylated chitosan is shown in Figure 3.8. Carbohydrates can also be introduced without ring opening on the C6 position [1]. These derivatives are important as they are recognized by the corresponding specific lectins and thus could be used for drug targeting.

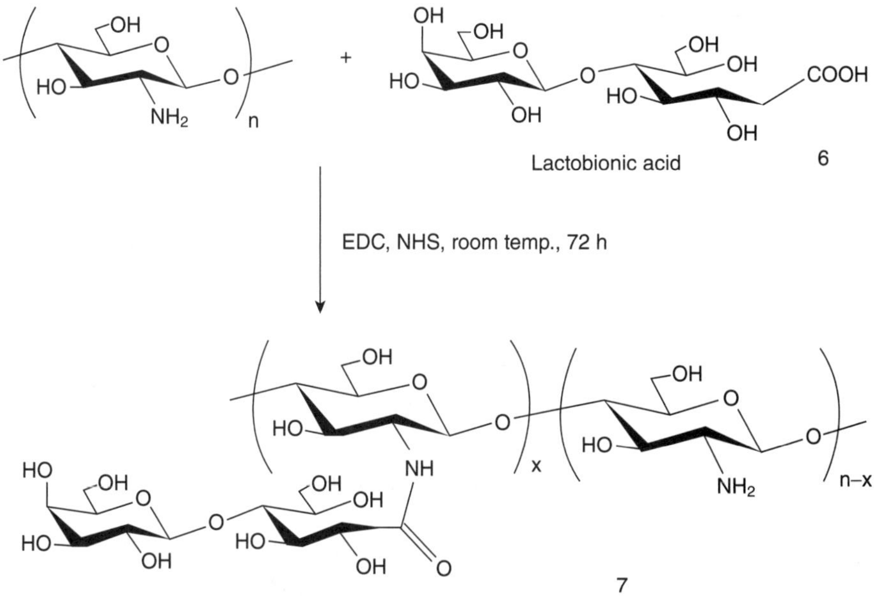

Figure 3.8 Synthesis of galactosylated chitosan [22, 31].

3.3.6 Carboxymethylated Chitosans

Carboxymethylated chitosans (CMCs) are some of the most extensively explored derivatives of chitosan [32]. The carboxylation, which can be carried out by exploiting amino and/or hydroxyl groups (N-, O-, or N, O-CMC), extends the water solubility of chitosan to basic pH values and provides the polymer with amphotheric characteristics.

3.3.7 Alkylated Chitosans

Alkylated chitosans are very important as amphiphilic polymers based on polysaccharides. Using the reductive amination, a series of amphiphilic derivatives was produced with different chain lengths. This technique was also used to introduce n-lauryl chains. Alkylated chitosans with good solubility in acidic conditions ($pH < 6$) have a number of interesting properties such as low surface activity and high viscosity due to hydrophobic interchain interactions. Alkyl chitosans are compatible with neutral andcationic surfactants; cationic surfactant adsorbed on the alkyl chain grafted on chitosan promotes its solubilization [33]. Grafting of hydrophobic alkyl chains onto chitosan is the most straightforward method to impart amphiphilic properties to chitosan. For this purpose, alkyl aldehydes or alkyl ketones were selectively grafted onto the

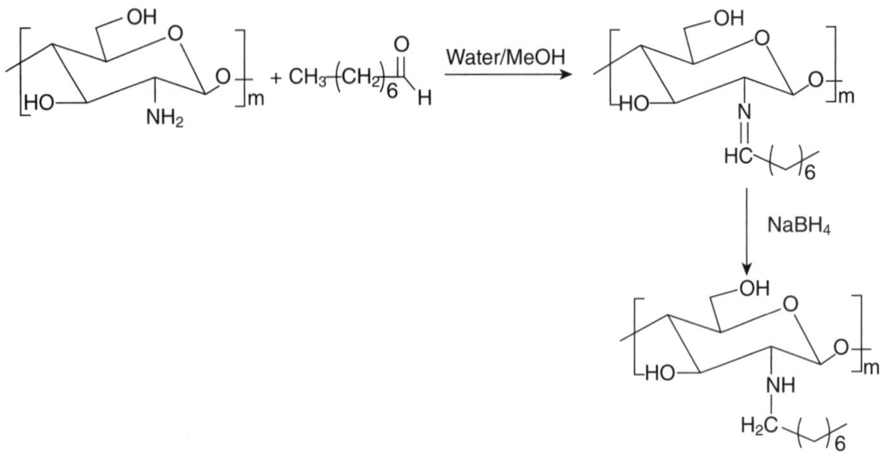

Figure 3.9 Grafting of alkyl chain by reductive amination [34].

primary amino groups of chitosan with formation of the corresponding Schiff base [34]. A reduction step mediated by sodium/potassium borohydride ($NaBH_4$/KBH_4) or sodium cyanoborohydride ($NaBH_3CN$) converted the imine group into the more stable amine with formation of the corresponding alkylchitosan derivatives (Fig. 3.9).

3.3.8 Quaternized Chitosan Derivatives

Chemical modification increased the solubility of chitosan in water [35], keeping chitosan soluble over a wide pH range. In addition, the cationic character can be controlled and kept pH independent, which is desirable for improving the stability of ionic complexes. The reaction of chitosan with methyliodide under basic conditions is the most straightforward route for quatenizing chitosan [36]. With excess methyl iodide and sodium hydroxide, the amino group of chitosan was directly trimethylated, and the cationized chitosan was soluble in water in a wide pH range. Ammonium groups were introduced as side chains by the reaction with quaternized epoxides such as glycidyltrimethylammonium chloride. The product is an analogue of cationized cellulose, which is a common polycation component of shampoos. On mixing solutions of chitosan and polyanions, polyelectrolyte complexes precipitate. The resulting chitosan-based polyelectrolyte complexes are attracting increasing interest as biomedical materials; they may be useful for the repair of damaged blood vessels, gene delivery, and anticoagulation and procoagulation. Among all the quaternized chitosans described in the literature, N,N,N-trimethyl chitosan chloride (TMC) is the most widely used in gene therapy applications. The quaternization maintained and improved the muco-adhesive properties of chitosan, depending on the quaternization degree, which makes this chitosan derivative an ideal candidate for gene delivery [37]. To enhance the delivery properties of TMC, Verheul et al. developed a synthetic route for the preparation of thiol-bearing TMC that has enhanced delivery properties [38, 39]. Indeed, the presence of thiol increased the muco-adhesion of chitosan derivatives by formation of a disulfide bond with mucin proteins of the cell membrane. Toh et al. grafted succinic acid onto chitosan, improving the water solubility of chitosan [40].

3.3.9 Chitosan Hydrogels

Hydrogels are three-dimensional (3D) networks that swell in water and aqueous solutions. Consequently, they are soft, pliable, wet materials with a wide range of potential biomedical applications. They play a crucial role in current strategies to remedy malfunctions in and injuries to living systems. The high water content of hydrogels renders them compatible with most living tissue and their viscoelastic nature minimizes damage to the surrounding tissue when implanted in the host. In addition, their mechanical properties parallel those of soft tissue, making them particularly appealing with the host tissues, assisting and improving the healing process, and mimicking functional and morphological characteristics of organ tissue. By its hydrophilic nature, chitosan can be applied as a starting material for the elaboration of biodegradable and biocompatible hydrogels. Chitosan readily forms hydrogels that have been used in a number of genes and drug delivery applications. Chitosan hydrogels can be divided into two classes: physical and chemical. Chemical hydrogels are formed by irreversible covalent links, whereas physical hydrogels are formed by various reversible links. For various reasons, physically cross-linked hydrogels have attracted increasing attention as bioactive compounds. Physically cross-linked chitosan hydrogels have been synthesized by grafting lactic acid (LA) and glycolic acid (GA) [41]. Chitosan derivatives were then used for the synthesis of improved chitosan-based hydrogels. Thus, chitosan-g-polyethyleneglycol (PEG) graft copolymer, synthesized by nucleophilic attack of the primary amine of chitosan onto chloride-terminated PEG, was able to form a stable physical hydrogel [42]. Chemical chitosan-based hydrogel was successfully prepared by Michael addition of a thiol-terminated six-armed star-shaped PEG onto acrylate-bearing chitosan [43]. The residual amino groups of chitosan are ionized in acidic buffers, which contributes to the electrostatic repulsion between adjacent ionized residual NH_2 groups of chitosan, leading to chain expansion, and consequently increases the water uptake of the gel. The hydrophobic side chains aggregate and physical cross-linking is formed. The crystallinity of original chitosan decreases by grafting lactic acid (LA) and glycolic acid (GA). PEG–chitosan hydrogels may have potential for application in the biomedical field. Thermo- and pH-sensitive semi-interpenetrating networks (IPN) polyampholyte hydrogels have been prepared by using carboxymethyl chitosan and poly (N-isopropylacrylamide) with N,N′-methylenebis (acryl amide) as a cross-linking agent [44]. The results indicate that the semi-IPN hydrogels are pH and temperature responsive and have good reversibility. By increasing the carboxymethyl chitosan content in the hydrogel, the release rate is increased. These semi-IPN hydrogels can be used in pH–temperature oral drug delivery systems. A full-IPN hydrogel with temperature response has been prepared from chitosan and N-isopropylacrylamide in the presence of a suitable cross-linker and an initiator [45]. Chitosan–cellulose hydrogel beads have been synthesized using ethylene glycol diglycidyl ether as a cross-linker for Cu adsorption from aqueous solutions [46]. The physical properties, including flexibility, of cross-linked chitosan hydrogels can be improved by blending chitosan with pre-gelatinized starch. The presence of starch in the system increased the water absorption of the hydrogel when compared to the system without starch.

3.4 BIOMEDICAL APPLICATIONS OF CHITOSAN DERIVATIVES

3.4.1 Tissue Engineering

Tissue engineering is a highly interdisciplinary field that combines the principles and methods of life sciences and engineering to utilize structural and functional relationships in normal and pathological tissue to develop biological substitutes for restoring, maintaining, or improving biofunction. Chitosan and its derivatives are suitable for tissue engineering applications because of their porous structure, gel-forming properties, ease of chemical modification, biodegradability, biocompatibility, antibacterial activity, and high affinity to *in vivo* macromolecules. It is one of the most important biomaterials in tissue engineering and shows very good physicochemical and biological properties. Various types of chitosan derivatives have been used in skin, bone cartilage, liver, nerve, and blood vessel. Chitosan has been extensively used in bone tissue engineering since it was shown to promote cell growth and mineral-rich matrix deposition by osteoblasts cells in culture. The biocompatibility of chitosan minimizes additional local inflammation, and it can be molded into porous structures to allow osteoconduction [4]. Chitosan/glycerophosphate has also been complemented with hydroxyethyl cellulose [47] for cartilage reconstruction or for improving the myocardial performance in infarcted heart. For bone regeneration, the chitosan/polyvinyl alcohol (PVA) blend was supplemented with hydroxyapatite (HA), which significantly enhanced the gel strength. Hu et al. prepared a chitosan–HA multilayer nanocomposite with high strength and bending modulus, rendering the material suitable for possible application for internal fixation of long bone fractures [48]. Park et al. [49] designed an injectable cell delivery chitosan–Pluronic hydrogel for articular cartilage regeneration and bone regeneration [50]. Chitosan's functional groups allow it to interact with many materials, thus it can be used in conjunction with materials such as hydroxyapatite or other calcium-based minerals to form composites that have multiple applications in the orthopedic and periodontal industries. These calcium–chitosan composites can be used as a coating in conjunction with joint prostheses. As the chitosan is degraded, new bone can be deposited adjacent to the prosthesis to stabilize the implant within the bone. An additional use for chitosan in orthopedics includes a direct replacement of bone or hard tissue. It is also a natural bioadhesive used to improve bone cement, which is used to secure implants as well as to fill bone cavities [51]. Chitosan is also used as an adjuvant with bone cements to increase their injectability while keeping the chemicophysical properties suitable for surgical use with respect to setting time and mechanical properties [4]. Chitosan is also used to modify the surface properties of prosthetic materials for the attachment of osteoblasts [52]. Li et al. [53] reported the possibility of making films of mPEG-g-chitosan by preparing a composite film with suitable hollow and high capacity of water adsorption, which could have potential application in wound healing and tissue engineering. Chitosan was chosen as a scaffolding material in articular cartilage engineering due to its structural similarity to various glycosaminoglycans (GAGs) found in articular cartilage. An alginate–chitosan hybrid based on polymer fibers showed increased tensile strength, implying a possible use in developing a 3D load-bearing scaffold for cartilage regeneration. The chitosan surface modified with fructose induced the formation of cellular aggregates and enhanced liver-specific metabolic activities and cell density to a satisfactory level [4].

3.4.2 Wound Healing

In wound healing, an ideal dressing should protect the wound from bacterial infection and promote healing. It must be capable of absorbing the exuded liquid from the wounded area and should permit water evaporation at a certain rate and allow no microbial transport. Because of its hydrogel-forming properties chitosan has been considered to be advantageous as a wound dressing because it produces less scarring. Its hemostatic properties also make chitosan a good candidate for wound dressing. Typically, there are four forms in which chitosan provides antimicrobial effects to wound dressing materials: fiber, membrane, sponge, and hydrogel. Chitosan derivatives remarkably accelerate the healing of various types of wounds [54]. The improvement in the healing process is caused by the hydrolytic activity of lysozyme and *N*-acetyl-β-D-glucosaminidase, which makes available *N*-acetylglucosamine, a common amino sugar in human and animal bodies. The released hydrolysis products are incorporated into glycoproteins or enter in different metabolic pathways. It causes macrophage and fibroplast activation, enhances the hyaluronic acid synthesis, and its deposition in regenerated connective tissue, as well as influences on collagen deposition into extracellular matrix during the rebuilding of the valid tissues. The effectiveness of these actions is increased, when low molecular weight chitosan, isolated from fungi, is applied. Hydrogels are applied to the burns and wound dressings in the form of films, bandages, cotton-like materials, and nonwoven napkins. These dressings have good hygroscopicity, show a high bacteriostatic effect, and are completely biodegradable in the human body. A significant advantage also consists in the fact that repeated dressings are usually not needed [55]. CMC prevent against pathogen infections of the wound tissues of animal and human bodies [4]. Chitosan-containing polyurethane (PU)/poly(N-isopropylacrylamide), that is, poly(NIPAAm), thermosensitive membranes have very low cytotoxicity, support the growth of 3T3 fibroblasts, and are antibacterial. Therefore, these materials may be beneficial for wound dressing [3]. A chitosan hydrogel scaffold impregnated with β-FGF-loaded microspheres were developed by Park et al.; it accelerates wound closure in the treatment of chronic ulcers [56]. Chitosan was complemented with PVA for wound healing [54]. The chitosan/PVA wound dressing was more swellable, flexible, and elastic because of its cross-linking interaction with PVA. The hydrogel significantly improved the wound healing effect compared with a gauze control and the conventional product.

3.4.3 Drug Delivery

Chitosan has been effectively used in drug delivery as a hydrogel system, drug conjugate, biodegradable release system, and PEC for many components. Chitosan-based systems are used for the delivery of proteins/peptides, growth factors, anti-inflammatory drugs, antibiotics, as well as, in gene therapy and bioimaging applications [4]. N-succinyl chitosan/alginate hydrogelbeads are suitable polymeric carrier for controlled drug delivery in the intestinal tract. Chitosan solution (C) β-glycerophosphate (GP) hydrogel is thermosensitive and has been used for sustained release of paclitaxel at tumor resection sites in order to prevent local tumor recurrence. The *in vitro* release profiles have shown controlled delivery over 1 month. (2-Hydroxypropyl-3-butoxy) propyl into succinyl–chitosan is an amphiphilic derivative of chitosan and may be used for controlled release of hydrophobic drugs. n-Lauryl-carboxymethyl chitosan is another amphiphilic polymer. It forms micelles that solubilize taxol, making it more effective therapeutically, and it is found to be safe in terms of membrane toxicity. This type of derivative is generally useful as a carrier for hydrophobic cancer drugs [57]. N-carboxymethyl chitosan impregnated with flurbiprofen and timolol maleate using the supercritical solvent impregnation (SSI) is used as ophthalmic drug release systems. These N-chitosan-derivative-based ophthalmic drug delivery systems can be easily and efficiently used for patients. Recently, water-soluble N-(c-bromopropanoyl amino acid)–chitosan derivatives have been synthesized. Compared with the parent unmodified chitosan, the four peptide–chitosans have higher thermosensitivity, porosity, and water-holding capacity, and these effects increase with the hydrophilicity of the peptide ligands. These new peptide– chitosans possess physicochemical properties that make them useful for drug delivery systems. Carboxymethyl chitosan (CM-Ch) nanoparticles have been prepared by gelification with calcium ions. These particles have been used as colloidal carriers for the delivery of doxorubicin (DOX), a commonly used cationic anticancer drug [3]. N-phthaloylchitosan (PLC) is a typical aryl-modified chitosan developed to improve the solubility of poorly soluble drugs such as camptothecin, retinoicacid, and prednisone acetate [32]. A chitosan-based hydrogel with 131I-norcholesterol (131I-NC) was tested in a breast cancer xenograft mouse model by Azab et al. This hydrogel, cross-linked with glutaraldehyde, reduced the progression of the tumor and prevented 69% of tumor recurrence and metastatic spreading. Most important, there was little or no systemic distribution of the radioisotope after hydrogel implantation [58]. Insulin-loaded chitosan nanoparticles have a good loading capacity (65–80%) and a fast release of insulin [59]. Chitosan hydrogels coupled with bone morphogenetic protein (BMP)-7 have shown the ability to enhance lesion repair [60].

In addition to the above applications, chitosan has a wide range of application in the medical field such as ophthalmology, gene therapy, and bioimaging. Chitosan has replaced the synthetic polymer in ophthalmological application. Chitosan possesses all the characteristics required for an ideal contact lens: optical clarity, mechanical stability, sufficient optical correction, gas permeability partially toward oxygen, wettability, and immunological compatibility. The antimicrobial and wound healing properties of chitosan, along with excellent film-forming capability, make it suitable for development of ocular bandage lens. Chitosan membranes have been proposed as an artificial kidney membrane because of their suitable permeability and high tensile strength. Chitosan and its derivatives have been used for gene transfection for N-alkylated chitosan. Quaternized chitosan can be used for the same purpose [1]. A photo-cross-linkable chitosan to which both azide and lactosemoieties were introduced (Az-CH-LA) was prepared as a biological adhesive for soft tissues and its effectiveness was compared with that of fibrin glue. Because of their biocompatible properties such as good blood compatibility and cell growth efficiency, grafted chitosan materials have potential use in cardiovascular applications [61].

3.5 CONCLUSION

Chitosan is a nature-based polymer, obtained by alkaline deacetylation of chitin, which presents excellent biological properties such as biodegradability and immunological, antibacterial, and wound healing activity. Recently, there has been a growing interest in the chemical modification of chitosan in order to improve its solubility and widen its applications. The main chemical modifications of chitosan that have been proposed in the literature are reviewed in this chapter. Moreover, these chemical modifications lead to a wide range of derivatives with a broad range of applications. Recent and relevant examples of the distinct applications, with particular emphasis on tissue engineering, drug delivery, and wound healing, are presented. Chitosan's versatility is a great challenge to the scientific community and to industry. The efficient utilization of marine biomass resource not only augurs in converting wastes from marine food-processing industries into useful products, but it has also become an environmental priority. It is anticipated that more specific applications of chitosan will be realized in the near future.

REFERENCES

[1] Rinaudo, M. (2006) Chitin and chitosan: Properties and applications. *Progress in Polymer Science*, *31*, 603.

[2] Zeng, J.B., He, Y.S., Li, S.L., and Wang, Y.Z. (2012) Chitin whiskers: An overview. *Biomacromolecules*, *13*, 1–11.

[3] Hengameh, H., Mehdi, B. (2009) Applications of biopolymers I: Chitosan. *Monatsh Chemistry*, *140*, 1403–1420.

[4] Dash, M., Chiellini, F., Ottenbrite, R.M., and Chiellini, E. (2011) Chitosan—A versatile semi-synthetic polymer in biomedical applications. *Progress in Polymer Science*, *36*, 981–1014.

[5] Li, J., Ming, K., Xiao, J.C., Jing, J.L., Wei, F.L., and Xi, G.C. (2011) A facile method for preparing biodegradable chitosan derivatives with low grafting degree of poly lactic acid. *International Journal of Biological Macromolecules*, *49*, 1016–1102.

[6] Mark, E.F., Anthony, A., Mark, E., and Van, D. (2007) Smart biomaterials design for tissue engineering andregenerative medicine. *Biomaterials*, *28*, 5068–5073.

[7] Kurita, K., Ishiguro, M., and Kitajima, T. (1988) Studies on chitin: Introduction of long alkylidene groups and their influence on the properties. *International Journal of Biological Macromolecules*, *10*, 124–125.

[8] Tharanathan, R.N., Farooqahmed, S.K. (2003) Chitin: the undisputed biomolecule of great potential. *Critical Reviews in Food Science and Nutrition*, *43*, 61–87.

[9] Jalal ZohuriaanMehr, M. (2005) Advances in chitin and chitosan modification through graft copolymerization: A comprehensive review. *Iranian Polymer Journal*, *14*, 235–265.

[10] Manoj, P., Nayak, P.L. (2012) Graft copolymerization of methyl acrylate on chitosan: Initiated by ceric ammonium nitrate as initiator-characterisation and antimicrobial activity. *Advances in Applied Science Research*, *3*, 1646–1654.

[11] Kim, S.Y., Cho, S.M., Lee, Y.M., and Kim, S.J. (2000) Thermo and pH responsive of graft copolymer and blend based on chitosan and N-isopropylacrylamide. *Journal of Applied Polymer Science*, *78*, 1381–1391.

[12] Sun, T., Xu, P., Liu, Q., Xue, J., and Xie, W. (2003) Graft polymerization of methacrylic acid onto carboxymethyl chitosan. *European Polymers Journal*, *39*, 189–192.

[13] Wang, Y., Xi, G.C., Hui, Y.L., Hui, Y.Z., and Zhang, J. (2007) Effects of molecular structural parameters of carboxymethyl chitosan on the growth of fibroblasts *in vitro*. *Journal of Applied Polymer Science*, *50*, 3136–3142.

[14] Pengfei, L., Maolin, Z., and Jilan, W. (2001) Study of radiation induced grafting of styrene onto chitin and chitosan. *Radiation Physics and Chemistry*, *61*, 149–153.

[15] Singh, D.K., Roy, A.R. (1997) Graft copolymerization of 2-hydroxyethylmethacrylate onto chitosan films and their blood compatibility. *Journal of Applied Polymer Science*, *66*, 869–877.

[16] Yu, L.L., He, Y., Bin, L., and Yue, F. (2003) Study of radiation induced graft copolymerization of butyl acrylate onto chitosan in acetic acid aqueous solution. *Journal of Applied Polymer Science*, *90*, 2855–2860.

[17] Singh, V., Tripathi, D.N., Tiwari, A., and Sanghi, R. (2005) Studies on K2S2O8/ascorbic acid initiated synthesis of Ipomoeadasysperma seed gum-g-poly (acrylonitrile): A potential industrial gum. *Journal of Applied Polymer Science*, *95*, 820–825.

[18] Kaplan, D.L., Dordick, J.S., Gross, R.A., and Swift, G. (1998) Enzymes in polymer science: An introduction. *Proceedings of ACS Symposium Series*, *684*, 2–16.

[19] Kumar, G., Smith, P.J., and Payne, G.F. (1999) Enzymatic modification of the synthetic polymer polyhydroxystyrene. *Biotechnology and Bioengineering*, *63*, 154–165.

[20] Yoshikawa, S., Takayama, T., and Tsubokawa, N. (1998) Graft copolymerization of hydroxyethyl methacrylate onto chitosan. *Journal of Applied Polymer Science*, *68*, 1883–1889.

[21] Hamodrakas, S.J., Jones, C.W., and Kafatos, F.C. (1982) Secondary structure predictions for silkmoth chorion proteins. *Biochimica et Biophysica Acta*, *700*, 42–51.

[22] Sashiwa, H., Aiba, S. (2004) Chemically modified chitin and chitosan as biomaterials. *Progress in Polymer Science*, *29*, 887–908.

[23] Nishi, N., Ebina, A., Nishimura, S., Tsutsumi, O., Hasegawa, A., and Tokura, S. (1986) Macrophage attraction with multiporous beads prepared from practically deacetylated chitin. *International Journal of Biological Macromolecules*, *20*, 8311–8317.

[24] Chen, S., Wang, Y. (2001) Study of α-cyclodextrin grafting with chitosan and slow release of its inclusion complex with radioactive iodine. *Journal of Applied Polymer Science*, *82*, 2414–2421.

[25] Martlet, B., Devassin, M., Crini, G., Weltrowski, M., Bourdonneau, M., and Morcellet, M. (2001) Preparation and sorption properties of α-cyclodextrin linked chitosan derivatives. *Journal of Polymer Science*, *39*, 169–176.

[26] Aoki, N., Nishikawa, M., and Hattori, K. (2003) Synthesis of chitosan derivatives bearing cyclodextrin and adsorption of p-nonylphenyl and bisphenol. *Carbohydrate Polymer*, *52*, 219–223.

[27] Sreenivasan, K. (1998) Synthesis and preliminary studies on a β-cyclodextrin-coupled chitosan as a novel adsorbent matrix. *Journal of Applied Polymer Science*, *69*, 1051–1055.

[28] Wan, L., Wang, Y., and Qian, S. (2002) Study on the adsorption property of novel crown ether cross linked chitosan for metal ions. *Journal of Applied Polymer Science*, *84*, 29–34.

[29] Jayakumar, R., Reis, R.L., and Mano, J.F. (2007) Synthesis and characterization of pH sensitive thiol containing chitosan beads for controlled drug delivery applications. *Drug Delivery*, *14*, 9–17.

[30] Hall, L.D., Yalpani, M.D. (1980) Formation of branched-chain, soluble polysaccharides from chitosan. *Journal of the Chemical Society, Chemical Communications*, *38*, 1153–1154.

[31] Rudall, K.M. (1969) Chitin and its association with other molecules. *Journal of Polymer Science*, *28*, 83–102.

[32] Ye, W., Xi Guang, C., Ying Hui, L., Hui Yun, Z., and Jing, Z. (2007) Effects of molecular structural parameters of carboxymethyl chitosan on the growth of fibroblasts *in vitro*. *Journal of Applied Polymer Science*, *106*, 3136–3142.

[33] Desbrieres, J., Rinaudo, M., Babak, V., and Vikhoreva, V. (1997) Surface activity of water soluble amphiphilic chitin derivatives. *Polymer Bulletin*, *39*, 209–215.

[34] Raphael, R., Heloıse, R., Annedes, R., Nicolas, D., Christine, J., and Veronique, P. (2011) Chitosan and chitosan derivatives in drug delivery and tissue engineering. *Advances in Polymer Science*, *244*, 19–44.

[35] Verheul, R.J., Amidi, M. (2008) Synthesis, characterization and *in vitro* biological-properties of O-methyl free N, N, N-trimethylated chitosan. *Biomaterials*, *29*, 3642–3649.

[36] Sieval, A.B., Thanou, M. (1998) Preparation and NMR characterization of highly substituted N-trimethyl chitosan chloride. *Carbohydrate Polymer*, *36*, 157–165.

[37] Mourya, V.K., Inamdar, N.N. (2008) Chitosan-modifications and applications: Opportunities. *Reactive and Functional Polymes*, *68*, 1013–1051.

[38] Verheul, R.J., vander Wal, S. (2010) Tailorable thiolated trimethyl chitosans for covalently stabilized nanoparticles. *Biomacromolecules*, *11*, 1965–1971.

[39] Varkouhi, A.K., Verheul, R.J. (2010) Gene silencing activity of siRNA polyplexes based on thiolated N, N, N-trimethylated chitosan. *Bioconjugate Chemistry*, *21*, 2339–2346.

[40] Toh, E.K.W., Chen, H.Y. (2011) Succinated chitosan as a gene carrier for improved chitosan solubility and gene transfection. *Nanomedicine*, *7*, 174–183.

[41] Qu, X., Wirsen, A., and Albertsson, A.C. (2000) Synthesis, characterization and swelling of a chemically cross-linked PVA hydrogel. *Polymer*, *41*, 4589.

[42] Bhattarai, N., Ramay, H.R. (2005) PEG-grafted chitosan as an injectable thermosensitive hydrogel for sustained protein release. *International Journal of Nanomedicine*, *103*, 609–624.

[43] Kim, M.S., Choi, Y.J. (2007) Synthesis and characterization of *in situ* chitosan-based hydrogel via grafting of carboxyethyl acrylate. *Journal of Biomedical Materials Research. Part A*, *83*, 674–682.

[44] Guo, B.L., Gao, Q.Y. (2007) Preparation and properties of a pH/temperature-responsive carboxymethyl chitosan/poly(N-isopropylacrylamide)semi-IPN hydrogel for oral delivery of drugs. *Carbohydrate*, *342*, 2416.

[45] Wang, M., Fang, Y., and Hu, D. (2001) Preparation and properties of chitosan-poly (*N*-isopropylacrylamide) full-IPN hydrogels. *Reactive and Functional Polymers*, *48*, 215.

[46] Li, N., Bai, R. (2005) Copper adsorption on chitosan–cellulose hydrogel beads: Behaviors and mechanisms. *Separation and Purification Technology*, *42*, 237.

[47] Yan, J.H., Yang, L. (2010) Biocompatibility evaluation of chitosan-based injectable hydrogels for the culturing mice mesenchymal stem cells *in vitro*. *Journal of Biomaterials Applications*, *24*, 625–637.

[48] Hu, Q., Li, B., Wang, M., and Shen, J. (2004) Preparation and characterization of biodegradable chitosan/hydroxyapatite nanocomposite rods via *in situ* hybridiza-

tion: A potential material as internal fixation of bone fracture. *Biomaterials*, *25*, 779–785.

[49] Park, K.M., Lee, S.Y. (2009) Thermosensitive chitosan-pluronic hydrogel as an injectable cell delivery carrier for cartilage regeneration. *Acta Biomaterialia*, *5*, 1956–1965.

[50] Tang, Y., Du, Y. (2009) A thermosensitive chitosan/poly(vinyl alcohol) hydrogel containing hydroxyapatite for protein delivery. *Journal of Biomedical Materials Research. Part A*, *91*, 953–963.

[51] Duck, W.L., Hosun, L. (2009) Advances in chitosan material and its hybrid derivatives: A review. *The Open Biomaterials Journal*, *1*, 10–20.

[52] Lee, B.H., Lee, Y.M., Sohn, Y.S., and Song, S.C. (2002) A thermosensitivepoly(org anophosphazene) gel. *Macromolecules*, *35*, 3876–9229.

[53] Li, X., Kong, X., Shi, S., Gu, Y., Yang, L., Guo, G., Luo, F., Zhao, X., Wei, Y.Q., and Qian, Z.Y. (2009) Biodegradable MPEG-g-chitosan and methoxypoly(ethylene glycol)-b-poly([epsilon]-caprolactone) composite films: Part 1. Preparation and characterization. *Carbohydrate Polymers*, *79*, 429–436.

[54] Sung, J.H., Hwang, M.R. (2010) Gel characterisation and *in vivo* evaluation of minocycline loaded wound dressing with enhanced wound healing using polyvinyl alcohol and chitosan. *International Journal of Pharmaceutics*, *392*, 232–240.

[55] Jozef, S., Nadia, A. (2003) Production, properties, and some new applications of chitin and its derivatives. *Critical Reviews in Food Science and Nutrition*, *43*, 145–171.

[56] Park, C.J., Clark, S.G., Lichtensteiger, C.A., Jamison, R.D., and Johnson, A.J. (2009) Accelerated wound closure of pressure ulcers in aged mice by chitosan scaffolds with and without βFGF. *Acta Biomaterialia*, *5*, 1926–1936.

[57] Liu, W., Sun, S.J., Zhang, X., and DeYao, K. (2003) Self-aggregation behaviour of alkylated chitosan and its effect on the release of a hydrophobic drug. *Journal of Biomaterials Science, Polymer Edition*, *14*, 851–859.

[58] Azab, A.K., Kleinstern, J., Doviner, V., Orkin, B., Srebnik, M., Nissan, A., and Rubinstein, A. (2007) Prevention of tumor recurrence and distant metastasis formation in a breast cancer mouse model by biodegradable implant of 131I norcholesterol. *Journal of Controlled Release*, *123*, 116–122.

[59] Grenha, A., Seijo, B., and Remunan, L.C. (2005) Microencapsulated chitosan nanoparticles for lung protein delivery. *European Journal of Pharmaceutical Sciences*, *25*, 427–437.

[60] Mattioli, B.M., Gigante, A., Muzzarelli, R.A., Politano, R., De Benedittis, A., Specchia, N., Buffa, A., Biagini, G., and Greco, F. (1999) N,N-dicarboxymethyl chitosan as delivery agent for bone morphogenetic protein in the repair of articular cartilage. *Medical & Biological Engineering & Computing*, *3*, 7130–7134.

[61] Alves, N.M., Mano, J.F. (2008) Chitosan derivatives obtained by chemical modifications for biomedical and environmental applications. *International Journal of Biological Macromolecules*, *43*, 401–414.

CHAPTER 4

Biomimetic Lessons for Processing Chitin-Based Composites

OTTO C. WILSON, JR. and TIFFANY OMOKANWAYE

4.1 INTRODUCTION

Nanocomposites are the most exciting class of materials in nature because atomic and molecular interactions at interfaces are utilized to achieve extraordinary properties [1, 2]. A unique level of promise is offered by naturally derived nanocomposites for the synthesis of advanced materials. This arises because there are a number of novel strategies that are employed in the synthesis of natural nanocomposites that could greatly enhance our ability to fabricate advanced nanocomposites with unique microstructures that exhibit exceptional properties. Bone tissue and arthropod integument are two of the most widely known and exciting natural nanocomposites from which we can learn many lessons. Bone is a nanocomposite that consists of a collagen fiber and extracellular matrix network with hydroxyapatite as the reinforcing phase, while arthropod integument is a chitin–protein nanocomposite matrix with calcium carbonate as the reinforcing phase. These two natural hard tissue nanocomposites have been the subject of a number of fascinating studies that have been directed at elucidating the intricate structures and mechanisms that account for their unique structural and biological properties [3–13].

Advances in nanocomposite processing technologies rely on discerning the fundamental relationships among nanoscale dimensions, nanoscale microarchitecture, and physical properties. There are a great number of challenges in processing optimized nanocomposites. Bioinspired insights can provide great strategies for addressing and improving nanocomposite processing concerns such as particle size and phase distribution, interfacial bonding, and other

Biopolymer Nanocomposites: Processing, Properties, and Applications, First Edition. Edited by Alain Dufresne, Sabu Thomas, and Laly A. Pothan.

relevant parameters that can be used to tailor properties. Gaining more processing capabilities to tailor microstructures at the nanoscale level will enable the synthesis of a wider range of multifunctional nanocomposites with enhanced properties [14–20]. Current advances in nanotechnology have made it possible to understand, create, characterize, and use material structures, devices, and systems with new properties derived from their nanostructure [21, 22]. For example, recent achievements in electrospinning have enabled the production of ultrafine, solid, and continuous polymer fibers with diameters as small as a few nanometers, and good structural integrity [23]. Compared with natural polymers, synthetic polymer nanofibers pose more challenges with regard to processibility, biocompatibility, and biodegradability. Hence, there is increased activity in producing nanocomposites from natural polymers, namely chitin and cellulose. These abundant, natural, polysaccharide polymers serve as promising candidates for the preparation of nanocomposites by introducing additional functionalities such as biodegradability, biocompatibility, and renewability [24].

The aim of this chapter is to provide an overview of chitin nanocomposites. The basics of chitin physicochemical characteristics will be presented first as an introduction to the unique characteristics of chitin that make it desirable for nanocomposite processing. Next, the focus will shift to natural chitin-based bionanocomposites such as crab shell integument, nacre, and jumbo squid beaks to gain biomimetic insights that can be useful in developing advanced processing strategies for chitin nanocomposite synthesis. Topics covered will include chitin biosynthesis, chitin–protein nanocomposites, biomineralization, liquid crystal (LC) order parameters, and multilevel hierarchical structure development. The final section will focus on practical applications of bioinspiration that are currently being used for synthesizing chitin nanocomposites. Various types of chitin nanocomposites will be considered, including chitin matrix nanocomposites, chitin dispersed as a reinforcing phase (whiskers, fibers, particulates), chitin chitosan copolymer-based nanocomposites, and chitosan-based composites. Chitosan nanocomposites are included due to the interrelationship between chitin and its deacetylated relative chitosan [25]. Chitin and chitosan can be interchanged by varying the degree of acetylation, and this makes some aspects of processing easier because of the greater ease in solubilizing chitosan. Various types of chitin nanocomposites will be covered, including chitin/organic phase, chitin/ceramic (bomineralized Ca phases), and chitin/metal (Ag, Au, etc.) based nanocomposites. A number of techniques have been reported that use chitosan as a reducing agent in synthesizing precious metal nanoparticles or utilize chitosan as a matrix to nucleate silver nanoparticles within chitosan films [26–29].

4.2 PHYSICOCHEMICAL PROPERTIES OF CHITIN

Chitin possesses a number of physical and chemical properties that make it a very attractive component for nanocomposite processing. Chitin is highly crys-

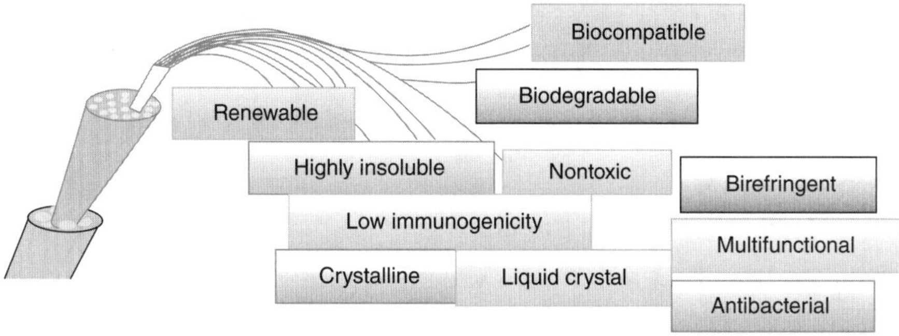

Figure 4.1 Chitin fiber schematic diagram depicting the hierarchical structure and properties of chitin fibers (modified from Jones and Bartlett, Nutrition resources: chemistry review: carbohydrates; http://www.nutrition.jbpub.com/resources/chemistry review9.cfm).

talline, biodegradable, and biocompatible; it plays a role in mechanical support, is highly insoluble, birefringent, and possesses antimicrobial characteristics as diagrammed in Figure 4.1 [30, 31]. Chitin is very abundant in nature and its simple yet robust chemistry lends itself to a variety of modifications that enhance the processing and the ultimate properties that can be exhibited by nanocomposites. One of chitin's most appealing characteristics relates to its ability to form multilevel hierarchical structural units that are reminiscent of molecular-level building blocks. This feature adds various levels of scalability that can be integrated into the chitin nanocomposite design. Chitin can serve as a discrete reinforcing phase, or as a "continuous" matrix phase for dispersing secondary phase(s). Chitin may also act as a true matrix (in partnership with associated proteins) by directing the nucleation and growth of secondary phases that are polymeric, ceramic, or metallic in nature [5, 9, 11, 13, 26–29].

Chitin was first isolated from mushrooms in 1811 by the French scientist Henri Braconnot [32]. The same compound was found in the cuticles of insects in 1823 by Odier. The name chitin derives from the ancient Greek word "chiton," which means a coat of mail or body armor [33, 34]. Chitin is the second most abundant natural polymer on earth while cellulose is the most abundant. Chitin is found in arthropod exoskeleton (crustacean and insect) and in the cell wall of fungi [30]. Chitin also plays an important role in the hierarchical control of biomineralization processes used by mollusks and crustaceans [4, 9, 11, 13].

Chitin is a linear polymer consisting of β-1,4-linked *N*-acetylglucosamine and has high crystallinity and poor reactivity because of highly ordered hydrogen bonding between chains. This hydrogen bonding makes chitin a rigid material exhibiting poor solubility, swelling behavior, and processing characteristics. The basic units are joined together by glycosidic bonds. The glycosidic link joins the number 1 carbon of an *N*-acetylglucosamine to the number four carbon of the adjacent *N*-acetylglucosamine. One of the structural features distinguishing polysaccharides from proteins is the great variation possible in the configuration of the glycosidic bond between adjacent sugar residues. β-1-4-linked residues tend to form extended ribbon-like structures and a range of unique molecular shapes may be generated. A unique facet of polysaccharide structure lies in the uniform and repeating shapes and surfaces, which are synthesized via cellular processes [35].

4.2.1 Chitin Hierarchical Structure

Chitin has been known to form a variety of microfibrillar arrangements in living organisms, and the chitin characteristics vary based on the origin and method of biosynthesis [36]. These fibrils are usually embedded in a protein matrix and have diameters that are typically reported to range from 2.5 to 2.8 nm. However, measurements as low as 2 nm have been reported for arthropod cuticles [37]. Crustacean cuticles possess chitin microfibrils with some of the largest dimensions in the range of 25 nm [38]. Even larger dimensions of 50 nm in width have been reported for chitin crystallites obtained from vestimentiferan tubes [39]. Arthropod exoskeletons are complex, multifunctional nanophase composites that are hierarchically structured. The linear chitin chains align antiparallel and form α-chitin crystals at the molecular level. Several α-chitin crystals with associated proteins form nanofibrils about 2–5 nm in diameter and 300 nm in length. A bundle of chitin–protein nanofibrils then form chitin–protein fibers of about 50–100 nm in diameter. These chitin–protein fibers align together, forming planar layers that stack up helicoidally. Chitin fibers align in a parallel fashion within individual layers. Each successive underlying layer has a 3–5 degree shift in the angle of the parallel in plane chitin–protein fibers. This repetitive pattern is the origin of the helicoid stacking sequence, which is called the twisted plywood or Bouligand structure [40]. The basic hierarchical sequence for chitin fabrication proceeds along the pathway from a molecular unit to a chain of molecules to microfibril to fibril aggregates to fibrous bundle to twisted plywood structure.

4.2.2 Chitin Crystallinity

Three polymorphic forms of chitin (α-, β-, and γ-chitins) have been distinguished based on differences in packing and the polarities of adjacent chains in successive sheets, which influence crystal structure as shown in Figure 4.2. α-Chitin is the most abundant form found in nature and is arranged in an antiparallel configuration. β-Chitin is organized in a parallel configuration. The antiparallel configuration of α-chitin gives a highly ordered crystalline structure with strong hydrogen bonding between chitin chains, which leads to the rigid and insoluble properties [9]. α-Chitin types vary in terms of the length of their *N*-acetylglucosamine chains, degree of crystallinity, and presence and identity of integrated proteins. As a consequence, the consistency of the chitinous layers in organisms like arthropods differs considerably [30]. A chitinous exoskeleton is diagnostic of the crustaceans. It is secreted by a single layer of

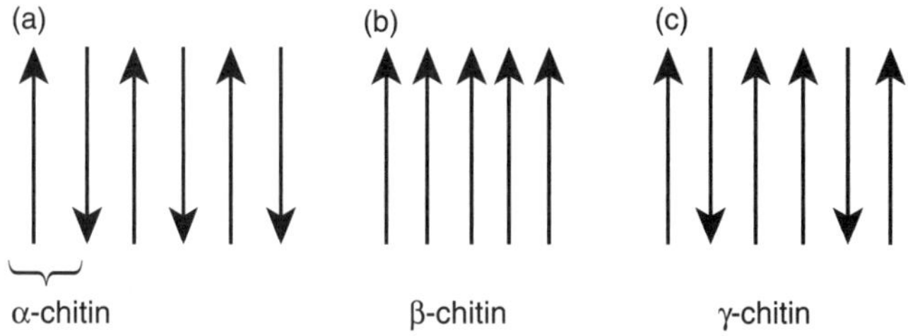

Figure 4.2 Diagram showing the direction of the polymer chains in the three crystallographic forms of chitin: (a) α-chitin, (b) β-chitin, (c) γ-chitin (modified from Wainwright et al., *Mechanical Design in Organisms*, Princeton, NJ: Princeton University Press, 1982).

cells within the epidermis and gains its stiffness and structural complexity by being folded and curved in many complex ways [41, 42].

4.2.3 Liquid Crystal Behavior of Chitin

Chitin is a polymeric LC consisting of a linear chain of N-acetyl-glucosamine repeating units. Chemical entities (-H, -OH, and -CH_2OH) are attached to the central part of the monomer and these functional groups participate in reactions. The linear chain adopts a helical structure and gains structural rigidity and the capacity to produce LC phases. The functional groups on the polysaccharide backbone play vital roles in the distinctive qualities that are exhibited by chitin-based nanocomposites and give rise to its wide range of applications [43].

The inherent symmetry and order exhibited by biological materials such as chitin, collagen, cellulose, and other fibrous composites suggests the strong presence of liquid crystalline ordering phenomena [5, 40, 42, 44]. After studying fibrous structures such as the cuticles of crabs, Bouligand [40] found that several fibrous biological materials exhibited similar characteristic features such as (1) fibrils arranged in highly organized geometric patterns (twisted plywood model), (2) solids with elongated molecules showing strong anisotropic behaviors, (3) fluid phases between well-defined temperatures, and (4) the ability to form LC phases in solutions. These factors led Bouligand to postulate that the secretions of fibrous structures pass through a brief transient liquid phase. This provides an intriguing mechanism for the morphogenesis of certain fibrous nanocomposite materials based on unique LC to solid phase transformations. Neville also suggested that biological fibrous composites may develop via self-assembly using a liquid crystalline intermediate phase [44, 45]. This idea was further corroborated by the work of Giraud-Guille [46–48]. While there are some similarities, notable differences exist between biological fibrous composites and LCs. The fibrous component consists of small parallel bundles of molecular entities, while LCs consist of individual molecular chains. Another difference is that biological composites that have reached structural maturity exist in a solid state, while LCs exist in a state with properties intermediate between those of a solid and liquid.

4.2.4 Chitin and Proteins

The intimate association between chitin and proteins has been the subject of a number of comprehensive studies that focus on ascertaining the nature of the molecular interactions [37, 49–57]. Much of the knowledge of the properties and structure of chitin has been derived from studies on "purified" chitin. Removal of the associated protein, for example, with alkali, increases the degree of orientation of the chitin chains, as shown by X-ray diffraction patterns, and the attraction between them is such that they aggregate to give a continuous sheet of chitin. Consequently, the results of studies on "purified" chitins may have to be modified when considering chitin as it actually occurs in cuticles in close association with proteins and other components [49].

Chitin always occurs associated with proteins [50] and no sample of purified chitin has been found to be free of amino acids [49]. A few models have been developed that attempt to describe the chitin–protein complex. Blackwell and Wieh [51] have developed a model for the chitin–protein complex based on observations of the ovipositors of the ichneumon flies *Megarhyssa lunator* and *Megarhyssa atrata*. This widely adopted model enlists a sheath of protein subunits that are arranged in a helical pattern with six protein subunits per turn around the chitin core. Not all cuticular proteins are directly bound to chitin. Some proteins help form the protein matrix in which the chitin–protein complex is embedded. The chitin–protein complexes are ordered structures [49, 51]. More elaborate modeling and experimental work is needed to uncover the most accurate description of the true nature of the chitin–protein complex interactions within cuticular systems of arthropods, mollusks, and other species. It is also evident from the overall data that each species has its own characteristic protein binding matrix. Apparently there is no simple relation between chitin content and the amount of chemically bound protein in the chitin–protein complex [58].

4.3 BIOMIMETIC LESSONS FROM NATURAL CHITIN NANOCOMPOSITES

There are many lessons to learn from the various ways in which living organisms synthesize chitin and use it to form elaborate chitin-based nanostructures [59]. In this section, a simplified description of how chitin is synthesized is presented along with an introduction to a few of the many chitin-based nanocomposites in nature including nacre, crab shell, and jumbo squid beak. The biosynthesis and degradation of chitin is a stepwise procedure governed by enzymatic activity. The established pathways include the synthesis of precursor and catalytic units, biotransformations (including phosphorylation, amination, and acetylation reactions), polymerization reactions, hydrolysis, and fibrilogenesis. Chitin has components formed in the intracellular and in the extracellular space during the synthesis process. Its synthesis pathway involves changes in the cytoplasmic matrix, the rough endoplasmic reticulum, processing in the Golgi apparatus, and transport through vesicles culminating in exocytosis into an assembly zone. In the assembly zone, chitin molecules form fibrous, hierarchical structures. These extracellular fibrous molecules are synthesized in a precursor form (*N*-acetylglucosamine-6-phosphate and uridine diphosphate (UDP)-N-acetylglucosamine). Furthermore, the resulting chitin tissue formation process follows a common developmental course in which the major organic constituent chitin is synthesized, extruded from the cell, and then self-assembled in the extracellular space before mineralization begins. For this reason, cuticles are good examples of "organic matrix-mediated" mineralization processes in comparison to mineralized collagen composites [40, 42, 48, 60].

4.3.1 Chitin Synthesis in Mollusk and Crustacean Hard Tissue

Nacre is a chitin-based bionanocomposite that is made up of mineral (aragonite) and a polysaccharide–protein complex organic phase composed of chitin and glycoprotein-based components (glycoproteins, silk fibroin, and other proteins) that are intimately mixed. Cartwright and Checa [61] highlighted the dynamics of nacre self-assembly in terms of a genetically controlled process. This process proceeds down a pathway with steps including cellular synthesis of organic building blocks (chitin protein nanocomposite) and multilevel self-assembly and biomineralization processes. The key steps in nacre biosynthesis involve the cellular template-directed synthesis of the chitin building blocks, the release of the units to the extrapallial space, and the self-assembly of the building blocks into a successively more complex series of structures that contribute to the hierarchical characteristics of chitin-based structures [62, 63].

Chitin synthesis is directly governed by an enzyme called chitin synthase. A cascade of cytoplasmic biotransformations of sugar precursors takes place to provide a source of substrate molecules for the functional catalytic units. Chitin synthesis in the epithelial cell begins with the glycosylation of a protein acceptor to form a primer molecule for chitin synthase or chitin–UDP–acetylglucosaminyltransferase [64]. Catalytic units synthesized at the endoplasmic reticulum compartment are packaged in clusters that involve the Golgi apparatus. The assembled multiunits are translocated and integrated into the plasma membrane at the apical compartments of epidermal cells. Trehalose, glucose, or the polysaccharide glycogen, which circulates in the hemolymph, could be the starting points for the chain of bioconversions that include major steps such as phosphorylation, amination, and acetylation. This metabolic pathway culminates in the formation of UDP–GlcNAc, the ultimate substrate of chitin synthase. Transformation of fructose-6-phosphate to glucosamine-6-phosphate by glutamine-fructose-6-phosphate aminotransferase can be considered as the first turning point on the pathway to chitin synthesis. Next, there is molecular organization of the chitin synthase units in the cell membrane, and a coordinated process that involves the translocation of chitin polymers across this barrier into the extracellular space where chemical modification of the noncrystallized chitin occurs [65]. The synthesis of chitin occurs either in the intracellular region or at the interphase with the extracellular medium.

The modified chitin associates with other molecules like protein. The last step in chitin biotransformation involves the formation of microfibrils at the cell surface, following the coagulation of adjacent developing chitin chains via extensive hydrogen bonding. Chitin polymerization and crystallization are coupled as consecutive processes. The resulting chitin–protein structure matures through formation of secondary covalent bonds among its components, creating chitin fibers [30].

The biosynthesis of the crustacean cuticle or exoskeleton involves the synthesis and assembly of component lipids, carbohydrates, and proteins by the epithelial cell layer, which lies just underneath the cuticle. Studies carried out by Horst [53] support the conclusion that crustacean chitin is secreted as a glycoprotein complex and that concurrent protein synthesis is required for chitin deposition to continue. LC mobility may account for changes in orientation of the chitin microfibrils and, thus, for the helicoid appearance. The cyclic phenomenon of molting in crustaceans involves a concerted and coordinated flow of hormonally controlled biochemical events governed by enzymatic activity. One such major event is the degradation of the endocuticular chitin, followed by the subsequent recycling of the amino sugar monomers into newly formed polymer units [65]. In biologically processed chitin, glycoprotein surface modification results in amphiphilic character, which influences LC ordering that occurs in natural chitin nanocomposites [62].

4.3.2 Jumbo Squid Beak

The jumbo squid (*Dosidicus gigas*) beak is one of the hardest known organic composites with hardness values around 1 GPa. The hardness and modulus decrease as you move away from the beak tip toward the soft tissue adjacent to the beak in a functionally graded fashion. The beak is a unique material because its design integrates local macromolecular compositional variation with large mechanical property gradients to withstand the stresses that are generated at the tip during predation. These stresses are gracefully translated to the underlying compliant tissue of the muscular buccal mass, which controls the movement of the two opposing parts of the squid beak. The structure of this chitin–protein nanocomposite consists of lamellar layers of chitin fibers, which are oriented perpendicular to the external beak surface and in the direction of the beak tip. There are a number of mechanisms that have been attributed to the extraordinary mechanical properties of jumbo squid beak including control of hydration, cross-link chemistry and density via catechol chemistry, and functional grading [66–69]. This microstructure arrangement makes the squid beak exceptionally hard and tough, with the fracture toughness attributed to laminated structure with weak interfaces. This feature is similar to the laminated microstructure in nacre, except that mineral bridges are not present. The fracture toughness values for dry and wet jumbo squid beak are 3.2 +/− 1.5 and 3.5 +/− 1.1 MPa $m^{0.5}$, respectively. Materials usually exhibit a tradeoff between hardness and toughness due to contrasting microstructural features. However, the hardness of the jumbo squid beak is well above the hardest engineering polymers while its fracture toughness is comparable to the toughest engineering polymers. Antler is the only known organic biological material that exhibits higher toughness values [70].

4.4 BIOINSPIRED LESSONS FOR PROCESSING CHITIN NANOCOMPOSITES

The examples of naturally derived chitin nanocomposites illustrate in an introductory fashion the great magnitude of design ideas that are available for adaptation to processing chitin nanocomposites. The use of nanoscale (<100 nm) dimensions offers the opportunity to optimize and create novel material properties [71]. With the common thread of nanoscale dimensionality, nature offers an excellent guide and resource for the design and construction of synthetic nanocomposites with markedly better properties. Biomimetics uses the lessons learned from biology to form the basis for novel materials. It involves investigation of both structure and physical functions of biological composites of interest with the goal of designing and synthesizing new and improved materials [72]. Unquestionably, natural materials such as chitin have exquisite properties that cannot be found in synthetic materials. Biological systems produce these extraordinary materials around room temperature in aqueous environments. The benefits of adapting biomimetic and bioinspired processing concepts for designing and manufacturing chitin nanocomposites encompass microstructural and economic issues. These issues include imparting the ability to achieve controlled hierarchical ordering and phase distribution at lower temperatures and reduced processing times. Gaining a fundamental understanding of the underlying principles that control these features in natural chitin nanocomposites will result in advances in nanocomposite processing capabilities. Table 4.1 provides a list of biomimetic design features that are exhibited in biologically derived chitin nanocomposites and may be useful in materials design and synthesis.

4.4.1 Chitin Nanocomposite Processing

The old saying, "One person's trash is another person's treasure," is very appropriate in regard to crab-, shrimp-, and lobster shell-based waste products. An accumulation of large quantities of discards from processing of crustaceans has become a major concern in the seafood processing industry. Out of the

TABLE 4.1 Key Bioinspired/Biomimetic Lessons to Learn from Chitin Nanocomposites

1. Multilevel, hierarchical architectural design and control
2. Liquid crystal ordering phenomena coupled with liquid to solid phase transformation
3. Phase transformations (liquid crystal to solid state) and self-assembly phenomena utilized to develop dynamic microstructure features
4. Control over organic phase crystalline polymorphs
5. Genetically controlled algorithm for chitin synthesis
6. Cell-directed synthesis of chitin structural building blocks
7. Surface modification via protein and functional group presentation used to enhance interactions
8. Tailored interfacial chemistry for optimized interfacial bonding, liquid crystal, and colloid interactions
9. Optimized chitin–protein distribution and interactions
10. Controlled distribution of inorganic phases
11. Variation in water content to control hydration/dehydration state
12. Cross-link density
13. Chitin bulk chemistry (degree of acetylation)

different species of crustaceans, shrimp and crab shell wastes are the most easily accessible and have been widely used for the isolation of chitin [34]. In industrial processing, chitin is extracted from crustaceans by acid treatment to remove calcium carbonate mineral phase. Alkaline extraction is used to degrade and solubilize proteins. A depigmentation step is often added to remove leftover pigments to obtain a colorless product. Owing to differences in the ultrastructure of the initial materials, these treatments can be adapted to each chitin source. The resulting chitin needs to be assessed in terms of purity since residual protein and pigment can cause problems for further utilization. There are a number of techniques that can be utilized to characterize the chemical and physical properties of chitin such as electron microscopy, X-ray diffraction (wide angle and small angle), spectroscopic techniques, mechanical properties, and a host of other techniques as described in the literature [37, 57, 73–79]. There are several different forms of chitin that can be produced at the nanoscale level. These forms include nanocrystals, whiskers, and fibers.

4.4.1.1 Chitin Crystals and Whiskers Whiskers or crystalline micro- or nanofibrils have received a great deal of interest in the nanocomposite research community due to their high aspect ratio and highly crystalline nature [80]. These whiskers were used as reinforcing fillers in both synthetic and natural polymeric matrices. Aqueous suspensions of chitin crystals can be prepared by acid hydrolysis of raw chitin. Chitin nanofibril-based nanocomposites have been prepared for wound healing applications. Muzzarelli et al. [81] prepared chitin nanofibril–chitosan glycolate composites for wound healing. Watthanaphanit et al. [82] prepared alginate–chitin whisker nanocomposites for biomedical applications. Chitin whiskers were used to achieve two main objectives: (1) to improve the mechanical properties of the nanocomposites and (2) to enhance wound healing. The enzymatic hydrolysis of chitin whiskers by lysosome releases chitin oligomers, which can play a role in wound healing. The whiskers were prepared by acid hydrolysis of shrimp shell chitin and to synthesize a chitin–alginate nanocomposite via the wet spinning process. Incorporation of a low amount of the whiskers in the nanocomposite fibers improved both the mechanical and the thermal properties of the fibers, possibly as a result of specific interactions, such as hydrogen bonding and electrostatic interactions, between the alginate molecules and the homogeneously dispersed chitin whiskers.

Chitin whiskers have been successfully prepared from crab shells, squid pen, and tubes of *Riftia pachyptila* tube worms by Junkasem et al. [76]. Regardless of the chitin sources, the commonly used hydrolytic condition to arrive at chitin whiskers is 3 N HCl at the boil for 90 minutes under vigorous stirring. Junkasem et al. fabricated α-chitin whisker-reinforced poly(vinyl alcohol) (PVA) nanocomposite nanofibres by electrospinning. The as-prepared α-chitin whiskers consisted of slender rods having a broad distribution in both length and width. The average length and width of these whiskers were about 549 and 31 nm, respectively, with an average aspect ratio of approximately 18. The α-chitin whiskers were prepared from α-chitin flakes from shrimp shells by acid hydrolysis. The incorporation of chitin whiskers within the as-spun nanocomposite fiber mats increased Young's modulus by about 4–8 times over that of the neat as-spun PVA fiber mat.

Wongpanit et al. [83] improved the dimensional stability of silk fibroin sponge by incorporating chitin whiskers as nanofillers using a freeze-drying technique. Their chitin whiskers exhibited an average length and width of 427 and 43 nm, respectively. The dispersion of chitin whiskers embedded in the silk fibroin matrix was found to be homogeneous. The presence of chitin whiskers embedded into silk fibroin sponge not only improved its dimensional stability but also enhanced its compression strength. Mouse fibroblast (L929) cells were seeded onto the surface of these sponges to investigate the biocompatibility. The positive test results indicated that silk fibroin sponges both with and without chitin whiskers were cytocompatible. Moreover, when compared to the neat silk fibroin sponge, the incorporation of chitin whiskers into the silk fibroin matrix was found to promote cell spreading.

Nanocomposites were synthesized from a colloidal suspension of high aspect ratio α-chitin whiskers interspersed in a poly(caprolactone) (PCL) matrix. The chitin whiskers, prepared by acid hydrolysis of *Riftia* tubes, consisted of slender parallel-piped rods with an aspect ratio close to 120. A procedure was optimized to prepare a latex of PCL. After mixing and stirring the two aqueous suspensions, solid films were obtained by either freeze-drying and hot-pressing or casting and evaporating the preparations [84].

It is often difficult to achieve a homogeneous dispersion of chitin whickers within a polymeric matrix. One way of processing polysaccharide-based nanocomposites is to transform them into a co-continuous material through long chain surface chemical modification. This involves the surface chemical modification of the nanoparticles based on the use of grafting agents bearing a reactive end group and a long hydrocarbon tail. Due to the hierarchical structure and semicrystalline nature of chitin and related polysaccharides, nanoparticles can be extracted from these naturally occurring polymers. Native cellulose and chitin fibers are made up of smaller and mechanically stronger long thin filaments, called microfibers, consisting of alternating crystalline and noncrystalline domains. Multiple mechanical shearing actions can be used to release individual microfibrils. Longitudinal cutting of these microfibrils can be performed by submitting the biomass to a strong acid hydrolysis treatment, which promotes dissolution of amorphous domains. The ensuing nanoparticles occur as rod-like nanocrystals or whiskers. Polysaccharide nanocrystals possess a reactive surface covered with hydroxyl groups, providing the possibility of extensive chemical modification [24].

Nair and Dufresne [85–87] chemically modified the surface of chitin whiskers to investigate the effect of surface modification on the properties of chitin whisker reinforced natural rubber (NR) matrix nanocomposites. The various chemical treatments were shown to improve the adhesion between the filler and the matrix. However, the mechanical performances of the composites strongly decreased after the chemical modification. This loss of performance, more pronounced for the isocyanate treatments, could be due to the partial or total destruction of the three-dimensional network of chitin whiskers assumed to be present in the unmodified composites.

Yánez et al. [88] studied the preparation and physical properties of composites, employing chitin as organic fillers to produce new biomaterials with controlled biodegradability and chemical functionalities. Composites of PCL–chitin were prepared by melt mixing. Aqueous and organic suspensions of chitin whiskers were prepared from shrimps shell waste. Blends of PCL–chitin whiskers were prepared by solvent casting. Nanocomposites represent the preferred way to produce a substantial improvement in the mechanical properties of a polymer matrix as compared to conventional composites because of the use of very low filler contents with great impact on mechanical properties.

4.4.1.2 Fibers Nanofibers provide a connection between the nanoscale world and the macroscale world because the diameters can be in the nanometer range and the lengths can be measured in meters [23]. Chitin fiber-based filaments have a relatively long history and can be processed by wet spinning chitin in nondegrading solvent systems [58]. The chitin fibers are nonallergenic, deodorizing, antibacterial, and moisture controlling [89]. Recent studies have demonstrated that nanofibers can be produced from a wide range of polymers. The unique characteristics and the functionalities of the polymers themselves impart nanofibers with many desirable properties for advanced applications. However, the processability, biocompatibility, and biodegradability of the obtained composites are limited compared to those of natural polymers. Chitin nanofibers offer outstanding potential as reinforcement in the fabrication of transparent flexible composites. These polymer nanofibers, which are derived from completely sustainable and renewable natural, chitin materials, are expected to create new opportunities for the application of transparent substrates in future electronic devices such as flexible displays and solar cells [22]. Other new products include barrier fabrics, microaerodynamic accelerators based on permeable nanofiber mats, nanofiber-based filters, biosensors, and protective clothing [23].

The incorporation of chitin nanofibers contributes to the significant improvement of the thermal expansion and mechanical properties of the neat acrylic resin. The properties of high light transmittance and low thermal expansion make chitin nanocomposites promising candidates for the substrate in a continuous roll-to-roll process in the manufacturing of various optoelectronic devices such as flat panel displays. Ifuku et al. acetylated chitin nanofibers to develop transparent acrylic resin nanocomposite films [90]. Chitin nanofiber-reinforced plastics have high transparency. Acetylation reduced the moisture absorption of the chitin nanocomposite due to the introduction of hydrophobic acetyl groups. This application of hydrophobic chitin nanofibers with a uniform structure and a very high surface-to-volume ratio has promise for novel green nanomaterials. Shams et al. prepared chitin nanofibers from crab shells and developed a chitin plastic composite that exhibited a lower coefficient of thermal expansion and higher transparency comparable to that of cellulose plastic composites [22]. The thermal characteristics of chitin–poly(e-caprolactone) copolymer nanocomposites have also been measured [91].

4.4.1.3 Chitin Derivatives There are a number of synthetic procedures that are described in the literature to chemically modify chitin [92]. A few examples include di-butyryl chitin [93, 94], carboxymethyl chitin, hydroxyalkyl chitin, fluorinated chitin, mercaptochitin, tosyl chitin, iodochitin, triphenylsilyl chitin, 6-oxychitin [95, 96], and 6-deoxychitin. While there are a few liquids that allow chitin to be derivatized such as imidazolinium-based ionic liquids [97], the most widely utilized and significant derivative of chitin is chitosan [89, 92]. The deacetylation of chitin under alkaline conditions yields chitosan. In most cases, the reaction of deacetylating chitin cannot reach completion even under harsh treatment in an alkaline solution. Depending on the method used, the degree of deacetylation usually ranges from 50% to 95% [98–103]. The degree of *N*-acetylation is defined as the average number of *N*-acetyl-D-glucosamine units per 100 monomers expressed as a percentage. When this average number is less than 50%, chitin becomes soluble in aqueous acidic solutions ($pH < 6.0$) and is then called chitosan. Thus, chitosan is a collective name for a group of fully and partially deacetylated chitins.

Chitosan has been the most researched version of the chitin biopolymer because of its ready solubility in dilute acids. This solubility renders chitosan more accessible for utilization and chemical reactions [104]. Some of the salts of chitosan that are water soluble include formate, acetate, lactate, malate, citrate, tartarate, glyoxylate, pyruvate, glycolate, malonate, and ascorbate [92]. Chitosan can also be derivatized to exhibit amphiphilic characteristics [105, 106]. Chitosan has also been used in conjunction with cellulose whiskers and nanowhiskers to make nanocomposites [107, 108].

One of the more interesting implications of the interchangeable relationship between chitin and chitosan is that it is possible to make chitin-based nanocomposites with a high degree of dispersion in which chitin plays the role of both matrix and reinforcing phase. Nanocomposites that exhibit molecular-level dispersion of the reinforcing phase in the matrix, tailored interfacial bonding, and aqueous processing routes are typically most desirable. Molecular-level dispersion can most easily be accomplished when the matrix phase can be processed in a low viscosity liquid form for dispersing the reinforcing phase. When nonaqueous solvents are needed to dissolve the chitin-based matrix, the surface of the reinforcing phase may need to be modified to enhance the degree of dispersion. The surface modification treatment depends on the solvent and characteristics of the reinforcing phase. Esterification is one of the more versatile surface modification methods for imparting colloid stability in various nonaqueous solvents [109–111]. Muzzarelli et al. [112–114] have reported on novel ways to make chitin nanofibrils [112] and nanochitins and nanochitosans [113, 114] which have optimized surface areas and porosities. These chitin forms can be used to add an additional level of process control in synthesizing chitin-based nanocomposites with enhanced levels of mixing between the phases.

4.4.2 Chitin Nanocomposites in Biomedical Engineering

If chitin is put in the body, it is immediately hydrolyzed and metabolized by the action of hydrolases that exist in body fluids [115]. Chitin has low toxicity and is inert in the gastrointestinal tract of mammals. It is biodegradable owing to the presence of chitinases, which are widely distributed in nature and found in bacteria, fungi and plants, and in the digestive systems of many animals. Chitin oligomers have been claimed as anticancer drugs and as active in controlling the photosynthesis of maize and soybeans [89]. Chitin has been used in cosmetics, water purification, as separation material, and as food additive. Attempts have been made to use chitin for various biomedical applications such as wound dressings and as scaffolds in tissue engineering due to their wound healing, antibacterial and anti-inflammatory properties [115, 116]. Other biomedical applications include dental restoration, drug delivery systems based on nanotubes, and structural elements in artificial organs, multifunctional membranes, biomedical structures, protective shields in specialty fabrics, filter media for submicron particles in the separation industry, composite reinforcement, and structures for nanoelectronic machines [23].

4.4.2.1 Antimicrobial The antimicrobial properties of chitin expand its degree of utility. The exact mechanism of the antimicrobial action of chitin and its derivatives is still unknown, but different mechanisms have been proposed. Interaction between positively charged chitin or derivative molecules and negatively charged microbial cell membranes may cause leakage of protein and other intracellular constituents [23]. The fungicidal and bactericidal action of chitin appears to be mediated by electrostatic forces between protonated -NH_2 groups and negative cell surface residues. The number of protonated -NH_2 groups present increases with increased degree of deacetylation and this factor influences the antimicrobial activity of chitosan [32]. In addition to being antimicrobial, chitin resists chemical attack in biological fluids. The chemical inertness of chitin in the presence of biological fluids was highlighted in a study that showed chitin sutures resisted chemical attack in bile, urine, and pancreatic juice [117].

4.4.2.2 The Influence of Chitin-Based Nanocomposites in Tissue Engineering Tissue engineering holds the promise of developing replacement tissues and organs on demand to improve the quality of life of people who suffer from conditions related to organ failure. Tissue engineering is a rapidly developing interdisciplinary field that utilizes aspects of cell and tissue biology and physiology, materials science and engineering, and physics to develop scaffolds that exhibit the requisite porosity, surface and bulk chemistry, surface texture, and architecture that can support the attachment and development of cells into fully functioning tissues and organs [118–124]. There are a host of polymers that have been used for the development of tissue engineering scaffolds, and it is surprising to note that chitin-based scaffolds have a relatively long history. In fact, a case can be made that chitin nanocomposites are the oldest tissue engineering scaffolds based on the Mayan practice of using nacre to replace missing teeth [125]. Westbroek and Marin reported that an archaeological dig in the 1930s uncovered a cache of Mayan skulls that had nacre implants in the jawbone that could not be easily removed. Further inspection including X-ray analysis revealed that the nacre had integrated into the jawbone. Studies by Lopez et al. in the 1990s demonstrated that nacre possessed the ability to induce osteoblasts to generate new bone in osteoblast cell cultures *in vitro* [126, 127] and in human patients with periodontal defects *in vivo* [125, 128]. Later research on the mechanism for the osteoinductive properties of nacre revealed that nacre contained water-soluble matrix proteins that stimulate osteoblast activity and even promote the differentiation of nonbone cells to the osteogenic line [129]. At least one of the nacre matrix proteins, Perlustrin, has been found to share a number of similarities with human insulin-like growth factors [125].

There has been a number of chitin-based tissue engineering scaffolds that show much promise for addressing some of the questions involved in developing fully functioning tissues. Chitin composite scaffolds containing nanosilica were prepared using chitin hydrogel and their bioactivity, swelling ability, and cytotoxicity were analyzed *in vitro*. These scaffolds were found to be bioactive in simulated body fluid (SBF) and biocompatible when tested with MG 63 cell line. These results suggest that chitin/nanosilica composite scaffolds can be useful for bone tissue engineering applications [116].

β-Chitin scaffolds were prepared by mixing β-chitin in saturated calcium chloride alcoholic solutions and using dialysis and lyophilization to produce the scaffolds. It seems that the β-chitin scaffolds are useful in the tissue engineering field. Chitin possesses not only the excellent wound healing, antibacterial, and anti-inflammatory abilities, but also the bioactive functions of enzyme immobilization on the amino groups. These properties were beneficial to the biomedical applications of chitin-containing tissue engineering scaffolds [115].

Biological familiarity is a major attraction for the use of natural polymers in processing nanocomposites. It is important to utilize polymers with collagen-like characteristics. Chitin is a unique polysaccharide that shares a remarkable number of structural and chemical similarities with collagen [5]. Chitosan and chitin have been used in a number of biomedical applications including skin grafting templates, hemostatic agents, drug and DNA delivery vehicles, and as a wound healing material. There have been great advances in our understanding of the underlying processes related to the storage, transfer, and propagation of genetic information via DNA and RNA. Developments in the field of biotechnology and genetic engineering have been phenomenal, and chitin-based nanocomposites have played a key role in the development of vectors for transporting DNA across the cell membrane and into the nucleus for delivery of instructions for protein synthesis [130–133]. The cationic character of chitin has been used in developing hemostatic dressings to stop blood flow from wounds, and a number of studies have focused on how chitosan functions to enhance wound healing [104, 134–137].

Chitosan has demonstrated promise as a scaffold component due to its ability to support specific aspects of cell function such as extracellular matrix protein production in human osteoblast and chondrocyte cell cultures [136]. Chemical functional groups can be grafted onto the surface of chitosan to influence bioactivity. Muzzarelli et al. [138] determined that methylpyrrolidone chitosan exhibited osteoconductive properties as evidenced by new bone formation in a rat tibia model. Fructose-modified chitosan scaffolds and galactosylated chitosan have been used to study hepatocytes cell adhesion in culture [139, 140]. Chitosan has also been mixed with other polymers such as PAA [141] and lecithin [142] to form nanocomposites. Chitosan and chitosan–hydroxyapatite composites have been developed to fabricate bone filling pastes, which can be used in the treatment of periodontal bony defects [89, 143, 144].

Chitosan nanocomposites have been used in a number of novel ways in tissue engineering. The use of chitin/chitosan-based nanocomposites as scaffolds is well known and developed [122, 145–157]. This idea follows from direct analogy to chitin's role in natural tissues. A novel approach for synthesizing chitin-based nanocomposites involves the synthesis of complex coacervate mixtures in which cationic chitosan gels can electrostatically interact with anionic gels. Interaction between chitosan gel and alginate gel is a popular example of this complex coacervation process and this system is being used to encapsulate living and viable cells [158–160]. This opens up many exciting possibilities in tissue engineering because now it is possible not only to address issues related to scaffold structure, but the chitosan chemistry can also be enhanced to integrate pharmaceutical agents, growth factors, viable cells, and genes for directing cell function.

Additional research on chitosan-based complex coacervates was performed by Espinoza et al. [161], who prepared gum Arabic–chitosan-based complex coacervates. In these polyelectrolyte complexes, chitosan serves as a cationic phase to coacervate with anionic polymeric phases such as gum Arabic. This complex coacervate composition has been developed into a bioactive nanocomposite scaffold by cross-linking the dried complex coacervate with a base [162]. The addition of hydroxyapatite to the gum Arabic–chitosan system improves the bioactivity of films and three dimensional porous scaffolds as assessed by promising *in vitro* [162] and *in vivo* (R. Silva et al., unpublished results) results using a rat subcutaneous implantation model.

There are a number of additional efforts to utilize chitin-based nanocomposites in both natural and synthetic forms. Wilson et al. conducted *in vitro* and *in vivo* studies on whole and demineralized crab shell integument from *Callinectes sapides* [163]. They found that the subcutaneous implantation of both whole crab shell and demineralized crab shell in the abdominal region of 28-day-old Sprague Dawley rats influenced the formation of mineralized collagen bundles in the abdominal implant region. The mineralized collagen bundles were observed after 1 week for the demineralized crab shell, while mineralized collagen bundles were seen after 4 weeks for whole crab shell.

4.4.3 Inorganic Chitin-Based Nanocomposites

Biomimetic inspiration from natural chitin composites has been carried over into studies of biomineralization phenomena using chitosan and chitin [164, 165] to form nanocomposites with silica, clays, silver, and calcium carbonate-based nacre-like structures. Chitin and chitosan exhibit unique macromolecular conformations and liquid crystalline behaviors [5, 7, 73, 116, 166], which may be used to influence biomineralization in conjunction with soluble polymer molecules [167]. Chitosan has been used as a template to synthesize meso- or macroporous silica-based nanocomposites [166] while self-assembly processes have been used to make chitin–silica nanocomposites [168]. Chitosan-based clay nanocomposites have been synthesized using montmorillonite clay [169] and organoclay-based bionanocomposites for removal of heavy metals in waste water treatment [170]. Methods for the synthesis of chitin–bentonite-based polyurethane bionanocomposites have been reported for potential applications in medical sutures [171, 172]. Silver possesses unique antimicrobial characteristics and tissue healing potential that when coupled with chitin and or chitosan may yield extremely useful nanocomposites for tissue healing and antimicrobial effects. Chitosan–silver nanocomposites have been prepared where the chitosan plays a dual role. Chitosan serves as a green, environmentally benign reducing agent, and also serves as the matrix [26–29].

The development of layered nanomaterials with structures reminiscent of nacre has provided concrete demonstrations of the versatility of using biomimetic design and synthesis principles [173, 174]. A number of researchers have utilized novel freeze processing techniques, which allow the attainment of porous layered structures that can be backfilled with polymers to attain nacre-like materials [175–177]. The mechanical properties of these materials have been phenomenal with toughness values approaching 35 MPa $m^{0.5}$. These results indicate the great promise of adopting bioinspired approaches that allow the development of structures reminiscent of biologically derived chitin-based nanocomposites.

4.5 CONCLUSIONS

The wide variety of chitin-based nanocomposites in nature provides a rich palette for designing novel nanocomposites that utilize the unique characteristics of chitin. The bioactivity and adaptable chemistry of chitin makes it very suitable for a many biomedical applications including drug, gene, and viable cell delivery systems, tissue engineering scaffolds, bone pastes, wound healing bandages, and antimicrobial coatings. Its effectiveness in these areas can be enhanced by integrating inorganic phases such as calcium carbonate to emulate natural mineralized chitin-based tissues and hydroxyapatite to mimic bone tissue. Other inorganic phases including silica, silver nanoparticles, and even magnetic nanoparticles have been integrated into chitin-based nanocomposites to develop targeted material's properties. As we learn more from structural, chemical, and biological insights from natural chitin-based nanocomposites such as nacre, crab shell, and jumbo squid beak, we will gain more biomimetic-based design concepts to advance our capabilities in fully utilizing chitin in the next generation of functionally advanced bioinspired chitin-based nanocomposites.

ACKNOWLEDGMENTS

The authors would like to acknowledge the financial support of the National Science Foundation (NSF CAREER Award—DMR Biomaterials [DMR-0645675]).

REFERENCES

[1] Varadan, V., Pillai, A.S., Mukherji, D., Dwivedi, M., and Chen, L. (2010) *Nanoscience and Nanotechnology in Engineering*. Hackensack, NJ: World Scientific Publishing Co. Pte. Ltd.

[2] Twarowski, T.E. (2007) *Introduction to Nanocomposite Materials: Properties, Processing, Characterization*. Lancaster: DEStech Publication, Inc.

[3] Roer, R.D., Dillaman, R. (1984) The structure and calcification of the crustacean cuticle. *Am. Zool.*, *24*, 893–909.

[4] Bonucci, E. (1992) *Calcification in Biological Systems*. Boca Raton: CRC Press.

[5] Giraud-Guille, M.M. (1994) Liquid crystalline order of biopolymers in cuticles and bones. *Microsc. Res. Tech.*, *27*(5), 420–428.

[6] Giraud-Guille, M.M. (1996) Twisted liquid crystalline supramolecular arrangements in morphogenesis. *Int. Rev. Cytol.*, *166*, 59–101.

[7] Giraud Guille, M.M., Mosser, G., Helary, C., and Eglin, D. (2005) Bone matrix like assemblies of collagen: From liquid crystals to gels and biomimetic materials. *Micron*, *36*, 602–608.

[8] Meyers, M.A., Lin, A.Y., Seki, Y., Chen, P.-Y.K., and Bodde, S. (2006) Structural biological composites: An overview. *J. Oper. Manage.*, *58*(7), 35–41.

[9] Meyers, M.A., Chen, P.-Y., Lin, A.Y.-M., and Seki, Y. (2008) Biological materials: Structure and mechanical properties. *Prog. Mater. Sci.*, *53*, 1–206.

[10] Chen, P.-Y., Lin, A.Y.M., Lin, Y.-S., Seki, Y., Stokes, A., Peyras, J., et al. (2008) Structure and mechanical properties of selected biological materials. *J. Mech. Behav. Biomed. Mater.*, *1*, 208–226.

[11] Chen, P.-Y., Lin, A.Y.-M., McKittrick, J., and Meyers, M.A. (2008) Structure and mechanical properties of crab exoskeletons. *Acta Biomater.*, *4*, 587–596.

[12] Hing, K.A. (2004) Bone repair in the twenty-first century: Biology, chemistry or engineering? *Philos. Transact. A. Math Phys Eng. Sci.*, *362*, 2821–2850.

[13] Raabe, D., Sachs, C., and Romano, P. (2005) The crustacean exoskeleton as an example of a structurally and mechanically graded biological nanocomposite material. *Acta Mater.*, *53*, 4281–4292.

[14] Whitesides, G.M., Grzybowski, B. (2002) Self-assembly at all scales. *Science*, *295*, 2418–2421.

[15] Gao, H., Ji, B., Jager, I.L., Arzt, E., and Fratzl, P. (2003) Materials become insensitive to flaws at nanoscale: Lessons from nature. *Proc. Natl. Acad. Sci. U. S. A.*, *100*(10), 5597–6000.

[16] Currey, J.D. (2005) Hierarchies in biomineral structures. *Science*, *309*, 253–254.

[17] Fratzl, P. (2007) Biomimetic materials research: What can we really learn from nature's structural materials? *J. R. Soc. Interface*, *4*, 637–642.

[18] Vo-Dinh, T. (2007) *Nanotechnology in Biology and Medicine: Methods, Devices, and Applications*. Boca Raton, FL: CRC Press.

[19] Wiwanitkit, V. (2008) *Advanced Nanomedicine and Nanobiotechnology*. New York: Nova Science Publishers, Inc.

[20] Reisner, D.E. (2009) *Bionanotechnology: Global Prospects*. Boca Raton, FL: CRC Press.

[21] Stroscio, M., Dutta, M. (2004) *Biological Nanostructures and Applications of Nanostructures in Biology: Electrical, Mechanical, & Optical Properties*. New York: Kluwer Academic/Plenum Publishers.

[22] Shams, M.I., Ifuku, S., Nogi, M., Oku, T., and Yano, H. (2011) Fabrication of optically transparent chitin nanocomposites. *Appl. Phys. A Mater. Sci. Process.*, *102*, 325–331.

[23] Venugopal, J., Ramakrishna, S. (2005) Applications of polymer nanofibers in biomedicine and biotechnology. *Appl. Biochem. Biotechnol.*, *125*, 147–157.

[24] Dufresne, A. (2010) Processing of polymer nanocomposites reinforced with polysaccharide nanocrystals. *Molecules*, *15*, 4111–4128.

[25] Phongying, S., Aiba, S., and Chirachanchai, S. (2007) Direct chitosan nanoscaffold formation via chitin whiskers. *Polymer*, *48*, 393–400.

[26] Huang, H., Yuan, Q., and Yang, X. (2004) Preparation and characterization of metal–chitosan nanocomposites. *Colloid Surf. B*, *39*, 31–37.

[27] Huang, H., Yang, X. (2004) Synthesis of polysaccharide-stabilized gold and silver nanoparticles: A green method. *Carbohydr Res.*, *339*, 2627–2631.

[28] Sun, C., Qu, R., Chen, H., Ji, C., Wang, C., Sun, Y., and Wang, B. (2008) Degradation behavior of chitosan chains in the "green" synthesis gold nanoparticles. *Carbohydr. Res.*, *343*, 2595–2599.

[29] Wei, D., Qian, W. (2008) Facile synthesis of Ag and Au nanoparticles utilizing as a mediator agent. *Colloid Surf. B*, *62*, 136–142.

[30] Jolles, P., Muzzarelli, R. (1999) *Chitin and Chitinases*. Boston: Birkhauser Verlag.

[31] Tsai, G.-J., Su, W.-H., Chen, H.-C., and Pan, C.-L. (2002) Antimicrobial activity of shrimp chitin and chitosan from different treatments and applications of fish preservation. *Fish. Sci.*, *68*, 170–177.

[32] Muzzarelli, R.A.A., Boudrant, J., Meyer, D., Manno, N., DeMarchis, M., and Paoletti, M.G. (2012) A tribute to Henri Braconnot, precursor of the carbohydrate polymers science, on the chitin bicentennial. *Carbohydr. Poly.*, *87*, 995–1012.

[33] Stephen, A.M. (1995) *Food Polysaccharides and Their Applications*. New York: Marcel Dekker.

[34] Shahidi, F., Arachchi, J.K., and Jeon, Y.-J. (1999) Food applications of chitin and chitosan. *Trends Food Sci. Technol.*, *10*, 37–51.

[35] Smith, C.A., Wood, E.J. (1991) *Biological Molecules: Molecular and Cell Biochemistry*. New York: Chapman & Hall.

[36] Brine, C.J., Austin, P.R. (1981) Chitin variability with species and method of preparation. *Comp. Biochem. Physiol. B.*, *69B*, 283–286.

[37] Giraud Guille, M.M., Chanzy, H., and Vuomt, R. (1990) Chitin crystals in arthropod cuticles revealed by diffraction contrast transmission electron microscopy. *J. Struct. Biol.*, *103*, 232–240.

[38] Kumar, M.N. (2000) A review of chitin and chitosan applications. *React. Funct. Polym.*, *46*, 1–27.

[39] Gaill, F., Chanzy, H., and Vuong, R. (1989) Proceedings of the 1989 French Society for Electron Microscopy conference. *Biol Cell.*, *66*, 8a.

[40] Bouligand, Y. (1972) Twisted fibrous arrangements in biological materials and cholesteric mesophases. *Tissue Cell*, *4*, 189–217.

[41] Vincent, J.F. (2002) Arthropod cuticle: A natural composite shell system. *Compos. A*, *33*, 1311–1315.

[42] Bouligand, Y. (2008) Liquid crystals and biological morphogenesis: Ancient and new questions. *C. R. Chimie*, *11*, 281–296.

[43] Kurita, K. (2006) Chitin and chitosan: Functional biopolymers from marine crustaceans. *Mar. Biotechnol.*, *8*, 203–226.

[44] Neville, A.C., Berg, C.W. (1971) Cuticle ultrastructure of a Jurassic crustacean (Eryma stricklandi). *Paleontology*, *14*, part 2, 201–205.

[45] Neville, A.C. (1993) *Biology of Fibrous Composites: Development Beyond the Cell Membrane*. Cambridge, UK: Cambridge University Press.

[46] Giraud-Guille, M.-M. (1992) Liquid crystallinity in condensed type I collagen solutions: A clue to the packing of collagen in extracellular matrices. *J. Mol. Biol.*, *224*, 861–873.

[47] Giraud-Guille, M.-M., Besseau, L. (1998) Banded patterns in liquid crystalline phases of type I collagen: Relationship with crimp morphology in connective tissue architecture. *Connect. Tissue Res.*, *37*(3), 183–193.

[48] Belamie, E., Mosser, G., Gobeaux, F., and Giraud-Guille, M.M. (2006) Possible transient liquid crystal phase during the laying out of connective tissues: α-chitin and collagen as models. *J. Phys. Condens. Matter*, *18*, S115–S129.

[49] Wright, J.E., Arthur, R. (1987) *Chitin and Benzoylphenyl Ureas*. Boston: W. Junk.

[50] Trim, A.R. (1941) Studies in the chemistry of the insect cuticle I. Some general observations on certain arthropod cuticles with special reference to the characterization of the proteins. *Biochem. J.*, *35*, 1088–1098.

[51] Blackwell, J., Weih, M.A. (1980) Structure of chitin-protein complexes: Ovipositor of the ichneumon fly megarhyssa. *J Mol. Biol.*, *137*, 49–60.

[52] Hunt, S., Nixon, M. (1981) A comparative study of protein composition in the chitin protein complexes of the beak, pen, sucker disc, radula, and oesophogeal cuticle of cephalopods. *Comp. Biochem. Physiol.*, *68B*, 535–546.

[53] Horst, M.N. (1989) Association between chitin synthesis and protein synthesis in the shrimp penaeus vannamei. *J. Crustacean Biol.*, *9*(2), 257–265.

[54] Coblentz, F.E., Shafer, T.H., and Roer, R.D. (1998) Cuticular proteins from the blue crab alter *in vitro* calcium carbonate mineralization. *Comp. Biochem. Physiol. B.*, *121B*, 349–360.

[55] Suetake, T., Tsuda, S., Kawabata, S., Miura, K., Iwanaga, S., Hikichi, K., et al. (2000) Chitin-binding proteins in invertebrates and plants comprise a common chitin-binding structural motif. *J. Biol. Chem.*, *275*(24), 17929–17932.

[56] Iconomidou, V.A., Willis, J.H., and Hamodrakas, S.J. (2005) Unique features of the structural model of "hard" cuticle proteins: Implications for chitin–protein interactions and cross-linking in cuticle. *Insect Biochem. Mol. Biol.*, *35*, 553–560.

[57] Diaz-Rojas, E.I., Arguelles-Monal, W., Higuera-Ciapara, I., Hernandez, J., Lizardi-Mendoza, J., and Goycoolea, F.M. (2006) Determination of chitin and protein contents during the isolation of chitin from shrimp waste. *Macromol. Biosci.*, *6*, 340–347.

[58] Austin, P.R., Brine, C.J., Castle, J.E., and Zikakis, J.P. (1981) Chitin: New facets of research. *Science*, *212*, 749–753.

[59] Muzzarelli, R.A.A. (2012) Chitin nanostructures in living organisms. In: *Chitin: Formation and Diagenesis*, Vol. 34, ed. N.S. Gupta. Dordrecht: Springer, pp. 1–34.

[60] Cui, F., Li, Y., and Ge, J. (2007) Self-assembly of mineralized collagen composites. *Mater. Sci. Eng.*, *57*, 1–27.

[61] Cartwright, J.H.E., Checa, A.G. (2007) The dynamics of nacre self-assembly. *J. R. Soc. Interface*, *4*, 491–504.

[62] Meyers, M.A., Lim, C.T., Li, A., Hairul Nizam, B.R., Tan, E.P.S., Seki, Y., and McKittrick, J. (2009) The role of organic inter-tile layer in abalone nacre. *Mater. Sci. Eng. C*, *29*(2009), 2398–2410.

[63] Meyers, M.A., Yu, A., Lin, M., Chen, P.Y., and Muyco, J. (2008) Mechanical strength of abalone nacre: Role of the soft organic layer. *J. Mech. Behav. Biomed. Mater.*, *1*, 76–85.

[64] Palli, S.R., Retnakaran, A. (1999) Molecular and biochemical aspects of chitin synthesis inhibition. In: *Chitin and Chitinases*, eds. P. Jolles, R. Muzzarelli. Boston: Birkhauser Verlag.

[65] Cohen, E. (1993) Chitin synthesis and degradation as targets for pesticide action. *Arch. Insect Biochem. Physiol.*, *22*, 245–261.

[66] Miserez, A., Li, Y., Waite, J.H., and Zok, F. (2007) Jumbo squid beaks: Inspiration for design of robust organic composites. *Acta Biomater.*, *3*, 139–149.

[67] Miserez, A., Schneberk, T., Sun, C.J., Zok, F.W., and Waite, J.H. (2008) The transition from stiff to compliant materials in squid beaks. *Science*, *319*, 1816–1819.

[68] Miserez, A., Rubin, D.J., and Waite, J.H. (2010) Cross-linking chemistry of squid beak. *J. Biol. Chem.*, *285*, 38115–38124.

[69] Moses, D.N., Pontin, M.G., Waite, J.H., and Zok, F.W. (2008) Effects of hydration on mechanical properties of a highly sclerotized tissue. *Biophys. J.*, *94*, 3266–3272.

[70] Broomell, C.C., Khan, R.K., Moses, D.N., Miserez, A., Pontin, M.G., Stucky, G.D., Zok, F.W., and Waite, J.H. (2007) Review: Mineral minimization in natures alternative teeth. *J. R. Soc. Interface*, *4*, 19–31.

[71] Koch, C.C. (2002) *Nanostructured Materials: Processing, Properties, and Applications*. Norwich: Noyes Publications.

[72] Sarikaya, M. (1994) An introduction to biomimetics: A structural viewpoint. *Microsc. Res. Tech.*, *27*, 360–375.

[73] Ratto, J., Hatakeyama, T., and Blumstein, R. (1995) Diffrential scanning calorimetry investigation of phase transitions in water/chitosan systems. *Polymer*, *36*(15), 2915–2919.

[74] Cardenas, G., Cabrera, G., Taboada, E., and Miranda, S.P. (2004) Chitin characterization by SEM, FTIR, XRD, and 13C cross polarization mass angle spinning NMR. *J. Appl. Polym. Sci.*, *93*(4), 1876–1885.

[75] Oduor-Odote, P.M., Struszczyk, M.H., and Peter, M.G. (2005) Characterization of chitosan from blowfly larvae and some crustacean species from Kenyan marine waters prepared under different conditions. *West. Indian Ocean J. Mar. Sci*, *4*, 99–107.

[76] Junkasem, J., Rujiravanit, R., and Supaphol, P. (2006) Fabrication of alpha-chitin whisker reinforced poly(vinyl alcohol) nanocomposites nanofibers by electrospinning. *Nanotechnology*, *17*, 4519.

[77] Wanchoo, R.K., Thakur, A., and Sweta, A. (2008) Viscometric and rheological behaviour of chitosan-hydrophilic polymer blends. *Chem. Biochem. Eng. Q.*, *22*(1), 15–24.

[78] Yen, M.-T., Yang, J.-H.Y., and Mau, J.-L. (2009) Physicochemical characterization of chitin and chitosan from crab shells. *Carbohydr. Poly.*, *75*, 15–21.

[79] Kumirska, J., Czerwicka, M., Kaczyński, Z., Bychowska, A., Brzozowski, K., Thöming, J., and Stepnowski, P. (2010) Application of spectroscopic methods for structural analysis of chitin and chitosan. *Marine Drugs*, *8*, 1567–1636.

[80] Zeng, J.B., He, Y.S., Li, S.L., and Wang, Y.Z. (2012) Chitin whiskers: An overview. *Biomacromolecules*, *13*(1), 1–11.

[81] Muzzarelli, R.A.A., Morganti, P., Morganti, G., Palombo, P., Palombo, M., Biagini, G., Belmonte, M.M., Giantomassi, F., Orlandi, F., and Muzzarelli, C. (2007) Chitin nanofibrils/chitosan glycolate composites as wound medicaments. *Carbohydr. Poly.*, *70*(3), 274–284.

[82] Watthanaphanit, A., Supaphol, P., Tamura, H., Tokura, S., and Rujiravanit, R. (2008) Fabrication, structure, and properties of chitin whisker-reinforced alginate nanocomposite fibers. *J. Appl. Polym. Sci.*, *110*, 890–899.

[83] Wongpanit, P., Sanchavanakit, N., Pavasant, P., Bunaprasert, T., Tabata, Y., and Rujiravanit, R. (2007) Preparation and characterization of chitin whisker-reinforced silk fibroin nanocomposite sponges. *Eur. Polym. J.*, *43*, 4123–4135.

[84] Morin, A., Dufresne, A. (2002) Nanocomposites of chitin whiskers from riftia tubes and poly(caprolactone). *Macromolecules*, *35*, 2190–2199.

[85] Nair, K.G., Dufresne, A. (2003) Crab shell chitin whisker reinforced natural rubber nanocomposites. 1. Processing and swelling behavior. *Biomacromolecules*, *4*(3), 657–665.

[86] Nair, K.G., Dufresne, A. (2003) Crab shell chitin whisker reinforced natural rubber nanocomposites. 2. Mechanical behavior. *Biomacromolecules*, *4*(3), 666–674.

[87] Nair, K.G., Dufresne, A. (2003) Crab shell chitin whiskers reinforced natural rubber nanocomposites. 3. Effect of chemical modification of chitin whiskers. *Biomacromolecules*, *4*, 1835–1842.

[88] Yánez, J., Castillo, V., Matos, M., Rosales, C., Arnal, M.L., and Müller, A.J. (2005) Polycaprolactone composites and nanocomposites containing chitin or chitosan. *E-polymer*. Available at http://www.e-polymers.org/journal/PAT2005ePolymers/page/Posters/Section%20B/Rosales_Carmen.pro.240652613.proceeding budapest3.pdf.

[89] Rinaudo, M. (2006) Chitin and chitosan: Properties and applications. *Prog. Polym. Sci.*, *31*, 603–632.

[90] Ifuku, S., Morooka, S., Morimoto, M., and Saimoto, H. (2010) Acetylation of chitin nanofibers and their transparent nanocomposite films. *Biomacromolecules*, *11*, 1326–1330.

[91] Jayakumar, R., Tamura, H. (2008) Synthesis, characterization and thermal properties of chitin-g-poly(e-caprolactone) copolymers by using chitin gel. *Int. J. Biol. Macromol.*, *43*, 32–36.

[92] Mourya, V.K., Inamdar, N.N. (2008) Chitosan-modifications and applications: Opportunities galore. *React. Funct. Polym.*, *68*, 1013–1051.

[93] Muzzarelli, C., Francescangeli, O., Tosi, G., and Muzzarelli, R.A.A. (2004) Susceptibility of dibutyryl chitin and regenerated chitin fibres to deacylation and depolymerization by lipases. *Carbohydr. Poly.*, *56*, 137–145.

[94] Muzzarelli, R.A.A., Guerrieri, M., Goteri, G., Muzzarelli, C., Armeni, T., Ghiselli, R., and Cornelissen, M. (2005) Biocompatibility of dibutyryl chitin, and use in wound dressings together with 5-methylpyrrolidinone chitosan. *Biomaterials*, *26*(29), 5844–5854.

[95] Muzzarelli, R.A.A., Biagini, G., DeBenedittis, A., Mengucci, P., Majni, G., and Tosi, G. (2001) Chitosan-oxychitin coatings for prosthetic materials. *Carbohydr. Poly.*, *45*, 35–41.

[96] Genta, I., Perugini, P., Modena, T., Pavanetto, F., Castelli, F., Muzzarelli, R.A.A., Muzzarelli, C., and Conti, B. (2003) Myconazole-loaded 6-oxychitin-chitosan microcapsules. *Carbohydr. Poly.*, *52*, 11–18.

[97] Muzzarelli, R.A.A. (2011) Biomedical exploitation of chitin and chitosan via mechano-chemical disassembly, electrospinning, dissolution in imidazolium ionic liquids, and supercritical drying. *Marine Drugs*, *9*(9), 1510–1533.

[98] Muzzarelli, R.A. (1973) *Natural Chelating Polymers*. London: Pergamon Press.

[99] Muzzarelli, R.A. (1977) *Chitin*. London: Pergamon Press.

[100] Goosen, M.F. (1997) *Application of Chitin and Chitosan*. Boca Raton: CRC Press.

[101] Venugopal, V. (2011) *Marine Polysaccharides: Food Applications*. Boca Raton: CRC Press.

[102] Casas, J., Simpson, S.J. (2010) *Advances in Insect Physiology: Insect Integument and Colour*. London: Academic Press.

[103] Dai, T., Tanaka, M., Huang, Y.Y., and Hamblin, M.R. (2011) Chitosan preparations for wounds and burns: Antimicrobial and wound-healing effects. *Epert Rev. Anti. Infect. Ther.*, *9*(7), 857–879.

[104] Khor, E., Lim, L.Y. (2003) Implantable applications of chitin and chitosan. *Biomaterials*, *24*, 2339–2349.

[105] Aranaz, I., Harris, R., and Heras, A. (2010) Chitosan amphiphilic derivatives. Chemistry and applications. *Curr. Org. Chem.*, *14*, 308–333.

[106] Desbrieres, J., Babak, V.G. (2008) Interfacial properties of amphiphilic systems on the basis of natural polymers-chitin deratives. *Russ. J. Gen. Chem.*, *78*(11), 2230–2238.

[107] de Mesquita, J.P., Donnici, C.L., and Pereira, F.V. (2010) Biobased nanocomposites from layer-by-layer assembly of cellulose nanowhiskers with chitosan. *Biomacromolecules*, *11*(2), 473–480.

[108] Li, Q., Zhou, J.P., and Zhang, L.N. (2009) Structure and properties of the nanocomposite films of chitosan reinforced with cellulose whiskers. *J. Polym. Sci. [B]*, *47*(11), 1069–1077.

[109] Ballard, C.C., Broge, E.C., Iler, R.K., St. John, D.S., and McWhorter, J.R. (1961) Esterification of the surface of amorphous silica. *J. Phys. Chem.*, *65*, 20–25.

[110] Van Helden, A.K., Jansen, J.W., and Vrij, A. (1981) Preparation and characterization of monodisperse silica dispersions in non-aqueous solvents. *J Colloid Interface Sci.*, *81*, 354–368.

[111] Borum Nicholas, L., Wilson Jr., O.C. (2003) Surface modification of hydroxyapatite. Part 1. Dodecyl alcohol. *Biomaterials*, *24*, 3671–3679.

[112] Muzzarelli, R.A.A., Muzzarelli, C. (2005) Chitin nanofibrils. In: *Chitin and Chitosan, Opportunities and Challenges*, ed. P.M. Dutta. Contai, India: SSM International Publ.

[113] Muzzarelli, R.A.A. (2011) New techniques for optimization of surface area and porosity in nanochitins and nanochitosans. In: *Advances in Polymer Science: Chitosan for Biomaterials*, Vol. 2, eds. R. Jayakumar, A. Prabaharan, and R.A.A. Muzzarelli. Berlin: Springer-Verlag, pp. 167–186.

[114] Muzzarelli, R.A.A. (2011) Nanochitins and nanochitosans, paving the way to eco-friendly and energy-saving exploitation of marine resources. In: *Comprehensive Polymer Science: Vol 10. Polymers for Sustainable Environment and Energy*, eds. J. McGrath, R. Hoefer. Heidelberg: Springer.

[115] Maeda, Y., Jayakumar, R., Nagahama, H., Furuike, T., and Tamura, H. (2008) Synthesis, characterization and bioactivity studies of novel Beta-chitin scaffolds for tissue-engineering applications. *Int. J. Biol. Macromol.*, *42*, 463–467.

[116] Madhumathi, K., Sudheesh Kumar, P.T., Kavya, K.C., Furuike, T., Tamura, H., Nair, S.V., et al. (2009) Novel chitin/nanosilica composite scaffolds for bone tissue engineering applications. *Int. J. Biol. Macromol.*, *45*, 289–292.

[117] Dutta, P.K., Ravikumar, M., and Joydeep, D. (2002) Chitin and chitosan for versatile applications. *J. Mcromol. Sci. C-Polym. Rev.*, *C42*(3), 307–354.

[118] Jones, J.R., Lee, P.D., and Hench, L.L. (2006) Hierarchical porous materials for tissue engineering. *Philos. Trans. R. Soc. Lond. A.*, *364*, 263–281.

[119] Van Blitterswijk, C. (2007) *Tissue Engineering*. Amsterdam: Elsevier.

[120] Atala, A., Lanza, R., Thomson, J.A., and Nerem, R.M. (2008) *Principles of Regenerative Medicine*. Boston: Academic Press.

[121] Enderle, J., Blanchard, S., and Bronzino, J. (2005) *Introduction to Biomedical Engineering*, 2nd ed. Amsterdam: Elsevier Academic Press.

[122] Leong, K.F., Cheah, C.M., and Chua, C.K. (2003) Solid freeform fabrication of three-dimensional scaffolds for engineering replacement tissues and organs. *Biomaterials*, *24*(13), 2363–2378.

[123] Ratner, B.D. (2001) Replacing and renewing: Synthetic materials, biomimetics, and tissue engineering in implant dentistry. *J. Dent. Educ.*, *65*(12), 1340–1347.

[124] Ratner, B.D., Hoffman, A.S., Schoen, F.J., and Lemons, J.E. (1996) *Biomaterials Science: An Introduction to Materials in Medicine*. San Diego: Academic Press.

[125] WestBroek, P., Marin, F. (1998) A marriage of bone and nacre. *Nature*, *392*, 861–862.

[126] Lopez, E., Vidal, B., Berland, S., Camprasse, S., Camprasse, G., and Silve, C. (1992) Demonstration of the capacity of nacre to induce bone formation by human osteoblasts maintained in vitro. *Tissue Cell*, *24*(5), 667–679.

[127] Silve, C., Lopez, E., Vidal, B., Smith, D.C., Camprasse, S., Camprasse, G., and Couly, G. (1992) Nacre initiates biomineralization by human osteoblasts maintained in vitro. *Calcif. Tissue Int.*, *51*(5), 363–369.

[128] Atlan, G., Balmain, N., Berland, S., Vidal, B., and Lopez, E. (1997) Reconstruction of human maxillary defects with nacre powder: Histological evidence for bone regeneration. *C. R. Acad. Sci. Paris/Life Sci.*, *320*, 253–258.

[129] Mouries, L.P., Almeida, M.J., Milet, C., Berland, S., and Lopez, E. (2002) Bioactivity of nacre water soluble matrix from the bivalve mollusk *Pinctada maxima* in three mammalian cell types: Fibroblasts, bone marrow stromal cells and osteoblasts. *Comp. Biochem. Physiol. B Biochem Mol. Biol.*, *132*(1), 217–229.

[130] Erbacher, P., Zou, S., Bettinger, T., Steffan, A.M., and Remy, J.S. (1998) Chitosan-based vector/DNA complexes for gene delivery: Biophysical characteristics and transfection ability. *Pharm. Res.*, *15*(9), 1332–1339.

[131] Mao, H.Q., Roy, K., Troung-Le, V.L., Janes, K.A., Lin, K.Y., Wang, Y., August, J.T., and Leong, K.W. (2001) Chitosan-DNA nanoparticles as gene carriers: Synthesis, characterization and transfection efficiency. *J. Control. Release*, *70*(3), 399–421.

[132] de Campos, A.M., Sanchez, A., and Alonso, M.J. (2001) Chitosan nanoparticles: A new vehicle for the improvement of the delivery of drugs to the ocular surface. Application to cyclosporin A. *Int. J. Pharm.*, *224*(1–2), 159–168.

[133] Kim, T.H., Jiang, H.L., Jere, D., Park, I.K., Cho, M.H., Nah, J.W., Choi, Y.J., Akaike, T., and Cho, C.S. (2007) Chemical modification of chitosan as a gene carrier in vitro and in vivo. *Prog. Polym. Sci.*, *32*, 726–753.

[134] Muzzarelli, R.A., Mattioli-Belmonte, M., Pugnaloni, A., and Biagini, G. (1999) Biochemistry, histology and clinical uses of chitins and chitosans in wound healing. *EXS*, *87*, 251–264.

[135] Kojima, K., Okamoto, Y., Kojima, K., Miyatake, K., Fujise, H., Shigemasa, Y., et al. (2004) Effects of chitin and chitosan on collagen synthesis in wound healing. *J. Vet. Med. Sci*, *66*, 1595–1598.

[136] Lahiji, A., Sohrabi, A., Hungerford, D., and Frondoza, C. (2000) Chitosan supports the expression of extracellular matrix protein in human osteoblasts and chondrocytes. *J. Biomed. Mater. Res.*, *5*, 586–595.

[137] Tanodekaewa, S., Prasitsilpa, M., Swasdison, S., Thavornyutikarna, B., Pothsreea, T., and Pateepasen, R. (2004) Preparation of acrylic grafted chitin for wound dressing application. *Biomaterials*, *25*, 1453–1460.

[138] Muzzarelli, R.A., Zucchini, C., Ilari, P., Pugnaloni, A., Mattioli Belmonte, M., Biagini, G., and Castaldini, C. (1993) Osteoconductive properties of methylpyrrolidinone chitosan in an animal model. *Biomaterials*, *14*(12), 925–929.

[139] Li, J., Pan, J., Zhang, L., and Yu, Y. (2003) Culture of hepatocytes on fructose modified chitosan scaffolds. *Biomaterials*, *24*(13), 2317–2322.

[140] Park, I.K., Yang, Y., Jeong, H.J., Bom, H.S., Harada, I., Akaike, T., Kim, S.I., and Cho, C.S. (2003) Galactosylated chitosan as a synthetic extracellular matrix for hepatocytes attachment. *Biomaterials*, *24*(13), 2331–2337.

[141] Hu, Y., Jiang, X., Ding, Y., Ge, H., Yuan, Y., and Yang, C. (2002) Synthesis and characterization of chitosan-poly(acrylic acid) nanoparticles. *Biomaterials*, *23*(15), 3193–3201.

[142] Magdassi, S., Bach, U., and Mumcuoglu, K.Y. (1997) Formation of positively charged microcapsules based on chitosan-lecithin interactions. *J. Microencapsul.*, *14*(2), 189–195.

[143] Ito, M. (1991) *In vitro* properties of a chitosan-bonded hydroxyapatite bone-filling paste. *Biomaterials*, *12*(1), 41–45.

[144] Ito, M., Yamagishi, T., Yagasaki, H., and Kafrawy, A.H. (1996) In vitro properties of a chitosan-bonded bone filling paste: Studies on solubility of calcium phosphate compounds. *J. Biomed. Mater. Res.*, *32*(1), 95–98.

[145] Ito, M., Hidaka, Y., Nakajima, M., Yagasaki, H., and Kafrawy, A.H. (1999) Effect of hydroxyapatite content on physical properties and connective tissue reaction to a chitosan-hydroxyapatite composite membrane. *J. Biomed. Mater. Res.*, *45*(3), 204–208.

[146] Yamaguchi, I., Tokuchi, K., Fukuzaki, H., Koyama, Y., Takakuda, K., Monma, H., and Tanaka, J. (2001) Preparation and microstructure analysis of chitosan/hydroxyapatite nanocomposites. *J. Biomed. Mater. Res.*, *55*(1), 20–27.

[147] Zhao, F., Yin, Y., Lu, W.W., Leong, J.C., Zhang, W., Zhang, J., Zhang, M., and Yao, K. (2002) Preparation and histological evaluation of biomimetic three-dimensional hydroxyapatite/chitosan-gelatin network composite scaffolds. *Biomaterials*, *23*(15), 3227–3323.

[148] Tachaboonyakiat, W., Serizawa, T., and Akashi, M. (2002) Inorganic-organic polymer hybrid scaffold for tissue engineering II: Partial enzymatic degradation of hydroxyapatite-chitosan hybrid. *J. Biomater. Sci. Polym. Ed.*, *13*(9), 1021–1032.

[149] Zhang, Y., Ni, M., Zhang, M., and Ratner, B. (2003) Calcium phosphate-chitosan composite scaffolds for bone tissue engineering. *Tissue Eng.*, *9*(2), 337–345.

[150] Tuzlakoglu, K., Alves, C.M., Mano, J.F., and Reis, R.L. (2004) Production and characterization of chitosan fibers and 3-D fiber mesh scaffolds for tissue engineering applications. *Macromol. Biosci.*, *4*, 811–819.

[151] Yamaguchi, I., Itoh, S., Suzuki, M., Sakane, M., Osaka, A., and Tanaka, J. (2003) The chitosan prepared from crab tendon I: The characterization and the mechanical properties. *Biomaterials*, *24*, 2031–2036.

[152] Yamaguchi, I., Itoh, S., Suzuki, M., Sakane, M., Osaka, A., and Tanaka, J. (2003) The chitosan prepared from crab tendons II: The chitosan/apatite composites and their application to nerve regeneration. *Biomaterials*, *24*, 3285–3292.

[153] Li, Z., Ramay, H., Hauch, K., Xiao, D., and Zhang, M. (2005) Chitosan-aliginate hybrid scaffolds for bone tissue engineering. *Biomaterials*, *25*, 1059–1067.

[154] Ge, Z., Baguenard, S., Lim, L.Y., Wee, A., and Khor, E. (2004) Hydroxyapatite-chitosan materials as potential tissue engineered bone substitutes. *Biomaterials*, *25*(6), 1049–1058.

[155] Hu, Q., Li, B., Wang, M., and Shen, J. (2004) Preparation and characterization of biodegradable chitosan/hydroxyapatite nanocomposite rods via *in situ* hybridization: A potential material as internal fixation of bone fracture. *Biomaterial*, *25*(5), 779–785.

[156] Khanna, R., Katti, K.S., and Katti, D.R. (2011) Bone nodules on chitosan–polygalacturonic acid–hydroxyapatite nanocomposite films mimic hierarchy of natural bone. *Acta Biomater.*, *7*, 1173–1183.

[157] Ma, P.X., Elisseeff, J. (2006) *Scaffolding in Tissue Engineering*. Boca Raton: Taylor & Francis.

[158] Wan, A.C.A., Yim, E.K.F., Liao, I.C., Visage, C.L., and Leong, K.W. (2004) Encapsulation of biologics in self-assembled fibers as biostructural units for tissue engineering. *J. Biomed. Mater. Res. A*, *71A*(4), 586–595.

[159] Baruch, L., Machluf, M. (2006) Alginate–chitosan complex coacervation for cell encapsulation: Effect on mechanical properties and on long-term viability. *Biopolymers*, *82*(6), 570–579.

[160] Wu, M.Y., Chen, N., Liu, L.K., Yuan, H., Li, Q., and Chen, S.H. (2009) Chitosan/alginate multilayer scaffold encapsulating bone marrow stromal cells *in situ* on titanium. *J. Bioact. Compat. Polym.*, *24*, 301–315.

[161] Espinosa, H., Baez, J., Cruz, F., and Vernon, E. (2007) Gum Arabic–chitosan complex coacervation. *Biomacromolecules*, *8*, 1313–1318.

[162] Silva, R.A., Mehl, P., and Wilson Jr., O.C. (2010). Gum Arabic modified chitosan composite biopolymer scaffolds for bone tissue engineering. *IFMBE Proceedings 32, 26th Southern Biomedical Engineering Conference (SBEC 2010)* (Edited by Herold, K., Bentley, W.E., and Vossoughi, J.).

[163] Wilson Jr., O.C., Gugssa, A., Mehl, P., and Anderson, W. (2012) An initial assessment of the biocompatibility of crab shell for bone tissue engineering. *Mater. Sci. Eng. C*, *32*, 78–82.

[164] Muzzarelli, C., Muzzarelli, R.A. (2002) Natural and artificial chitosan-inorganic composites. *J Inorg. Biochem.*, *92*(2), 89–94.

[165] Falini, G., Fermani, S. (2004) Chitin mineralization. *Tiss Eng.*, *10*(1/2), 1–6.

[166] Pedroni, V., Schulz, P.C., Gschaider de Ferreira, M.E., and Morini, M.A. (2000) A chitosan-templated monolithic siliceous mesoporous-macroporous material. *Colloid Polym. Sci.*, *278*, 964–971.

[167] Wilson Jr., O.C. (2002) Soluble polymers in bioceramic systems. In: *Polymers in Particulate Systems: Properties and Applications*, Surfactant Science Series, Vol. 104, eds. V.A. Hackley, P. Somasundaran, and J.A. Lewis. New York: Marcel Dekker, Inc, pp. 325–360.

[168] Alonso, B., Belamie, E. (2010) Chitin–silica nanocomposites by self-assembly. *Angew. Chem. Int. Ed.*, *49*, 8201–8204.

[169] Paluszkiewicz, C., Stodolak, E., Hasik, M., and Blazewicz, M. (2011) FT-IR study of montmorillonite-chitosan nanocomposite materials Spectrochim. *Acta A. Mol. Biomol. Spec.*, *79*(4 SP ISS), 784–788.

[170] Pandey, S., Mishra, S.B. (2011) Organic-inorganic hybrid of chitosan/organoclay bionanocomposites for hexavalent chromium uptake. *J. Colloid Interface Sci.*, *361*(2), 509–520.

[171] Zuber, M., Zia, K.M., Mahboob, S., Hassan, M., and Bhatti, I.A. (2010) Synthesis of chitin-bentonite clay based polyurethane bio-nanocomposites. *Int. J. Biol. Macromol.*, *47*(2), 196–200.

[172] Zia, K.M., Zuber, M., Barikani, M., Hussain, R., Jamil, T., and Anjum, S. (2011) Cytotoxicity and mechanical behavior of chitin-bentonite clay based polyurethane bionanocomposites. *Int. J. Biol. Macromol.*, *49*(5), 1131–1136.

[173] Sellinger, A., Weiss, P.M., Nguyen, A., Lu, Y., Assink, R.A., Gong, W., and Brinker, C.J. (1998) Continuous self assembly of organic-inorganic nanocomposite coatings that mimic nacre. *Nature*, *394*, 256–260.

[174] Tang, Z., Kotov, N.A., Magonov, S., and Ozturk, B. (2003) Naostructured artificial nacre. *Nat Mater*, *2*(6), 413–418.

[175] Luz, G.M., João, F., and Mano, J.F. (2009) Biomimetic design of materials and biomaterials inspired by the structure of nacre. *Philos. Trans. R. Soc. Lond. A.*, *367*(1893), 1587–1605.

[176] Deville, S., Saiz, E., Nalla, R.K., and Tomsia, A.P. (2006) Freezing as a path to build complex composites. *Science*, *311*(5760), 515–518.

[177] Wegst, U.G.K., Schecter, M., Donius, A.E., and Hunger, P.M. (2010) Biomaterials by freeze casting. *Philos. Trans. R. Soc. Lond. A.*, *368*(1917), 2099–2121.

CHAPTER 5

Morphological and Thermal Investigations of Chitin-Based Nanocomposites

MING ZENG, LIYUAN LU, and QINGYU XU

5.1 MORPHOLOGICAL INVESTIGATIONS OF CHITIN-BASED NANOCOMPOSITES

The techniques of light and electron microscopy have been well described in the literature. Therefore, optical microscopy, scanning electron microscopy, transmission electron microscopy (TEM), and atomic force microscopy are briefly reviewed in the following subsections since they are the most commonly employed techniques. We concentrate on their applications in the field of chitin-based nanocomposites.

5.1.1 Optical Microscopy

Optical microscopy (OM) is used to examine the texture of chitin-based nanocomposites. Two common techniques are used: (1) polarized-light microscopy, in which advantage is taken of the ability of crystalline material to rotate the plane of polarized-light, and (2) phase-contrast microscopy, which allows observations of structural features involving differences in refractive index rather than absorption of light as in the former case.

OM is a powerful technique for monitoring the formation and microstructure of nanocomposites. The dispersion of nanosized fillers within chitin or chitosan has been characterized by OM [1]. Usually, OM determines the nature of the nanocomposites and provides additional information that is helpful in the interpretation of the X-ray diffraction results [2].

Biopolymer Nanocomposites: Processing, Properties, and Applications, First Edition. Edited by Alain Dufresne, Sabu Thomas, and Laly A. Pothan.

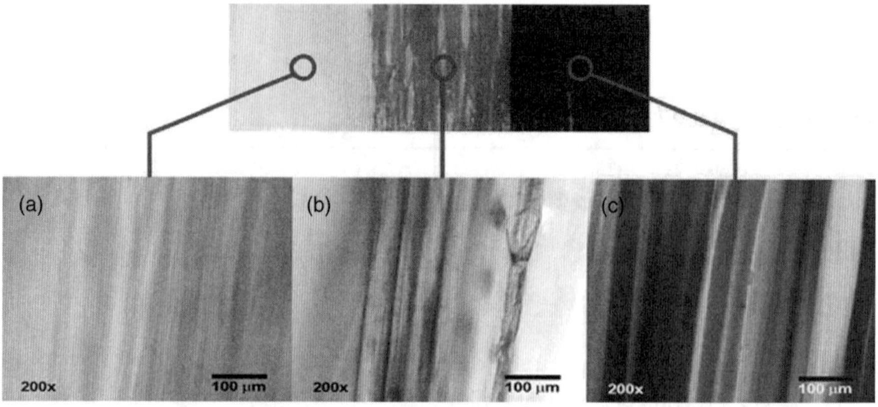

Figure 5.1 Representative photographic (top panel) and corresponding optical microscopic (bottom panel; magnification 200×) images of (a) neat alginate yarns, (b) alginate nanocomposite yarns containing 0.6% w/w of chitosan whiskers, and (c) neat chitosan yarns, after they had been stained with 0.01% w/v Amido Black 10B (Wako Pure Chemical Industries, Japan) aqueous solution for 12 hours [4]. See color insert.

OM could also provide complementary evidence of grafting reactions. OM observation reveals that chitosan-grafted carbon nanotubes (CNTs) were uniformly dispersed in the chitosan matrix with a size of the dark area less than 100 μm, whereas the ungrafted CNTs aggregated into a dark area size much larger than 100 μm [3].

To further examine the dispersion and/or distribution of chitosan whiskers within the individual nanocomposite fibers, OM could be applied with the aid of a staining method. The dye molecules would selectively adsorb onto the whisker surfaces, owing to the positively charged nature of the whiskers [4]. Figure 5.1 illustrates photographic and corresponding OM images of neat alginate yarns, neat chitosan yarns, and nanocomposite yarns containing 0.6% w/w of the whiskers, after they had been immersed in an aqueous solution of the dye for 12 hours. Closer examination of the OM image of the nanocomposite fibers revealed that not all parts of the fibers were stained positive for the dye, indicating that parts of the incorporated chitosan whiskers existed as aggregated entities.

5.1.2 Scanning Electron Microscopy and Transmittance Electron Microscopy

Resolution of smaller objects can be achieved by scanning electron microscopy (SEM) and TEM. Electron microscopy has been a powerful tool in the study of the morphology of crystalline polymers and the dispersion level of filler and fibers. Furthermore, the adverse applications of the chitin-based nanocomposites have been examined by SEM and TEM.

Nanosized chitin particles, fibers, whiskers, and chemically modified whiskers have been reported as reinforcing elements in matrices such as styrene and butyl acrylate copolymers, poly(caprolactone) (PCL), chitosan, starch, and natural rubber (NR) [5–7]. SEM and TEM are used to observe the morphology relative to the compatibility between the components.

The SEM images (Fig. 5.2a) of NR reinforced with 10 wt% unmodified chitin whiskers show that the chitin whiskers appear as white dots and are distributed uniformly throughout the NR matrix [8]. This uniform distribution of chitin whiskers was assumed to be due to the formation of a rigid chitin–chitin network.

After whiskers are chemically modified with phenyl isocyanate (PI) and alkenyl succinic anhydride (ASA), nanocomposite films were obtained using a toluene NR solution in which the whiskers were dispersed [9]. In Figure 5.2b,c, the nonuniform dispersion of chitin whiskers is clear, and aggregated chitin whiskers can be observed for chemically modified whisker-based nanocomposites. Because the PI and ASA are reactive, the electrostatic interactions between chitin whiskers were made negligible, thereby causing the complete destruction of colloidal stability of the whiskers in toluene, which ultimately leads to the aggregation of chitin whiskers in the NR matrix.

The nanosized filler of glass, metal and metal oxides, clay and modified clay, nanotubes, and modified nanotubes has been reported to reinforce the chitin or chitosan matrix [10, 11]. SEM and TEM are powerful methods for monitoring the formation and micro- and nanostructure of nanocomposites. The dispersion level of nanosized fillers within chitin or chitosan has been characterized by SEM and TEM [12].

Nanocomposite films comprising nanometer-sized metal oxide, semiconductor, and magnetic particles dispersed in polymer matrices have received increased attention due to their broad range of exceptional properties, namely for process imaging, as biosensors, in disease diagnosis, and so on [12–25]. Chitin and chitosan are good dispersants for a variety of different nanoparticles, including those of platinum, gold, and silver, as well as iron oxide [26–32].

The SEM and TEM results showed that the hexagonal and spherical monodispersed α-Fe_2O_3 nanoparticles were encapsulated into the spherical dumbbell-shaped chitosan (CH)-α-Fe_2O_3 nanocomposite film with a mean diameter of ~87 and ~110 nm, respectively (Fig. 5.3a,a′) [33]. The TEM image of CH-α-Fe_2O_3 nanocomposite film is shown in Figure 5.3b. The structure of CH-coated magnetite α-Fe_2O_3 nanoparticles was looser, leading to a larger size; the average diameter of such a structure was 110 nm. This revealed that the coating process that results in the agglomeration and the change in size of the particles is due to the cross-linking among the different nanoparticles with CH.

Clays such as montmorillonite (MMT), rectorite, silica, magnetite, attapulgite, and bentonite have been used to fabricate chitin or chitosan-based

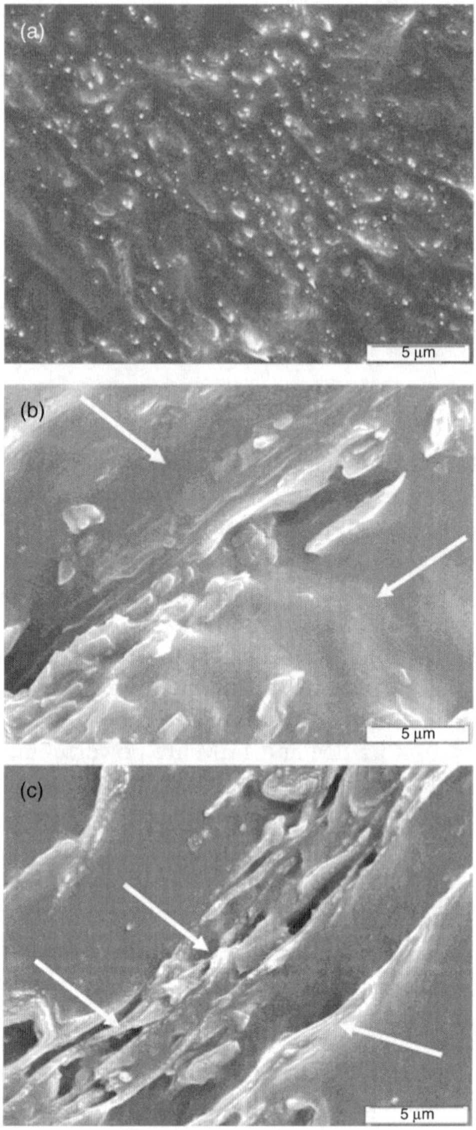

Figure 5.2 Scanning electron micrographs of the cryofractured surfaces of 10 wt% (a) unmodified, (b) ASA-modified, and (c) PI-modified chitin whisker-filled NR films [9].

nanocomposites. The incorporation of these mineral powders not only can reduce production cost, but also can improve the mechanical and thermal properties of the corresponding composites [34, 35]. The microscopic morphology of the inner intercalated or exfoliated structure of clay or modified clays has been examined using SEM and TEM [36, 37].

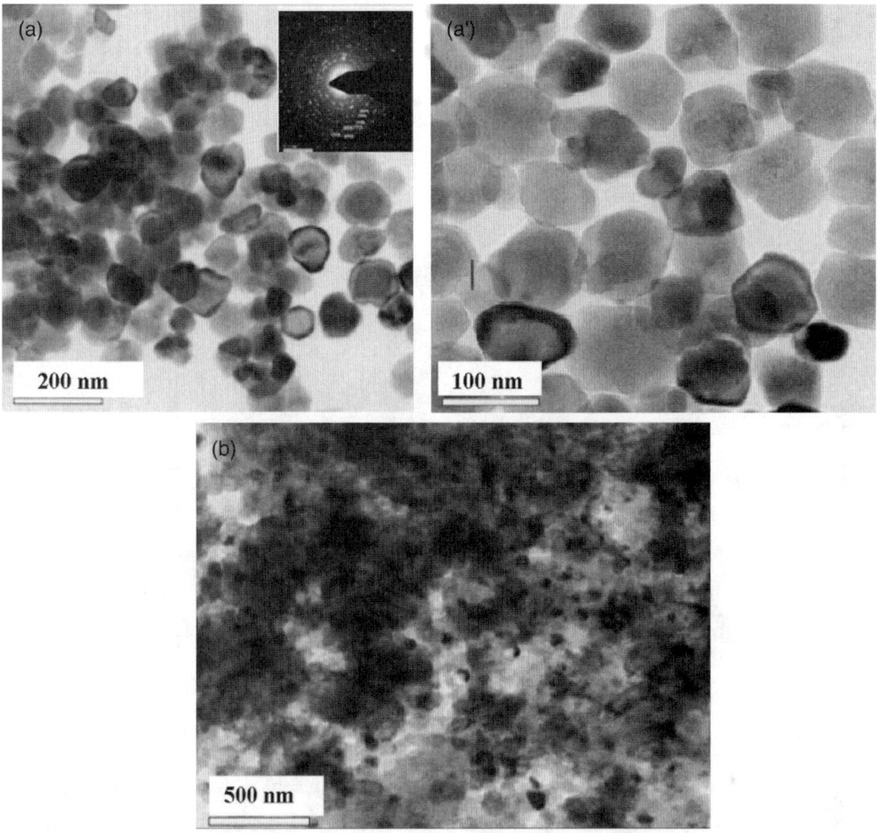

Figure 5.3 TEM micrograph and (selected area electron diffraction in the inset) pattern of α-Fe_2O_3 nanoparticles (a and a′) and CH-α-Fe_2O_3 nanocomposite film (b) [33].

Recently, layered silicates, including MMT and rectorite (REC), were used to improve the properties of quaternized chitosan. Quaternized carboxymethyl chitosan (QCMC), consisting of β-(1,4)-2-acetamido-2-deoxy-D-glucopyranosyl and β-(1,4)-2-amino-2-deoxy-D-glucopyranosyl units, is a novel, promising amphoteric polymer. It is obtained by introducing a carboxymethyl group and quaternary ammonium salt on chitosan. The nanocomposites showed better antibacterial activity, drug-controlled abilities, gene transfection efficacy, and thermal stability, among others, in comparison with chitosan or quaternized chitosan.

REC was dispersed in quaternized chitosan–N-(2-hydroxyl) propyl-3-trimethyl ammonium chitosan chloride (HTCC) and QCMC, and coded as HR and QR, respectively [38]. TEM results (Fig. 5.4) revealed that REC was well dispersed in the polymer matrix and obtained the largest interlayer distance

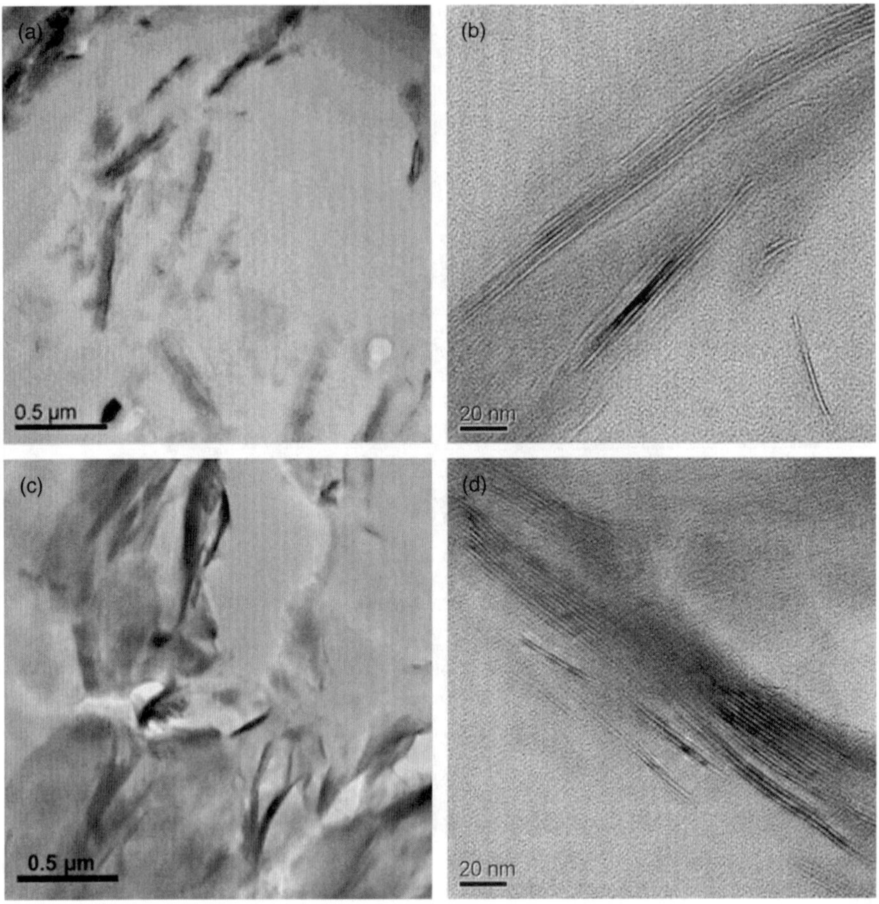

Figure 5.4 TEM micrographs of HR3 nanocomposite (a,b) and QR3 nanocomposite (c, d) [38].

when the mass ratio of QCMC to REC was 2:1. TEM images exhibited the typical morphology of the layered materials.

Among the nanostructured filler materials, CNTs are attractive due to their excellent properties. It is well known that CNTs possess extraordinarily high aspect ratios, thermal and electro-activities, unique atomic structures, and remarkably good mechanical properties, and have potential in reinforced polymer composites [39]. SEM and TEM are used to study the morphology and homogeneity of the CNTs or multiwalled carbon nanotubes (MWCNTs) dispersed in the chitin or chitosan matrix [40–45].

The SEM images in Figure 5.5 show the pristine MWCNTs, pure chitosan (CS), and MWCNT-CS nanocomposite [46]. The pure MWCNTs were

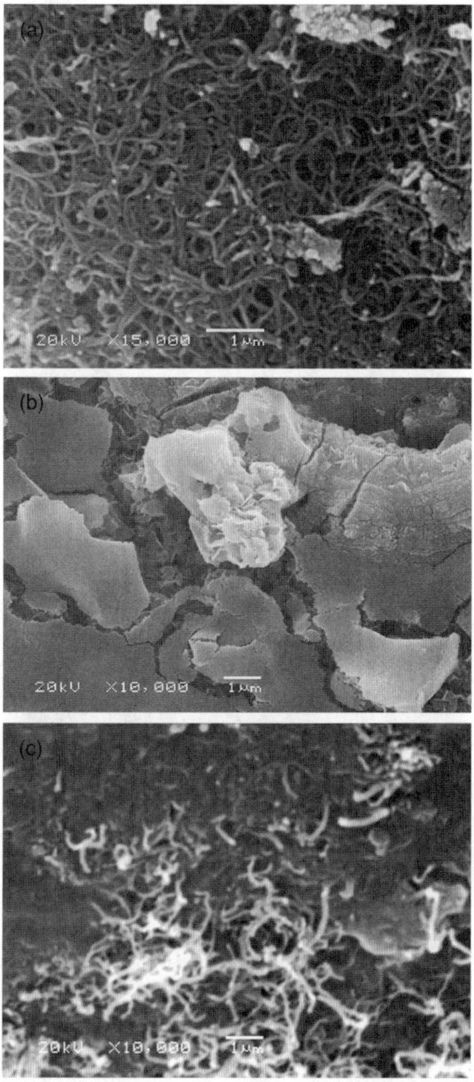

Figure 5.5 SEM images for pure MWCNTs (a), pure chitosan (b), and the MWCNTs–CS nanocomposite (c) [46].

cylindrical, with an average diameter of 200 nm, and were tangled together, while pure chitosan samples had no common structure. Figure 5.5c shows the MWCNT–CS nanocomposite where the MWCNTs were embedded and well dispersed within the chitosan matrix.

In fact, the agglomeration of CNTs limits their application because this behavior could lead to many nanoscale and/or microscale defects, which would diminish the mechanical performance and physical properties of the composites. There are some useful and effective ways of preparing well-dispersed CNTs in the chitin matrix: (1) chemical modification to attach a functional group on the surface of the CNT, (2) using ultrasonication for dispersing CNTs, and (3) using a surfactant for a well-dispersed CNT solution [46].

Recently, the chitosan ternary nanocomposites with two-dimensional (2D) clay platelets and one-dimensional (1D) CNTs have been successfully prepared by a simple solution-intercalation/mixing method in acid media [47]. To see directly the dispersion of clay and CNTs in chitosan matrix, the TEM image of chitosan nanocomposite with 2 wt% clay and 1 wt% CNTs is shown in Figure 5.6. Figure 5.6a shows an exfoliated structure of clay platelets, which

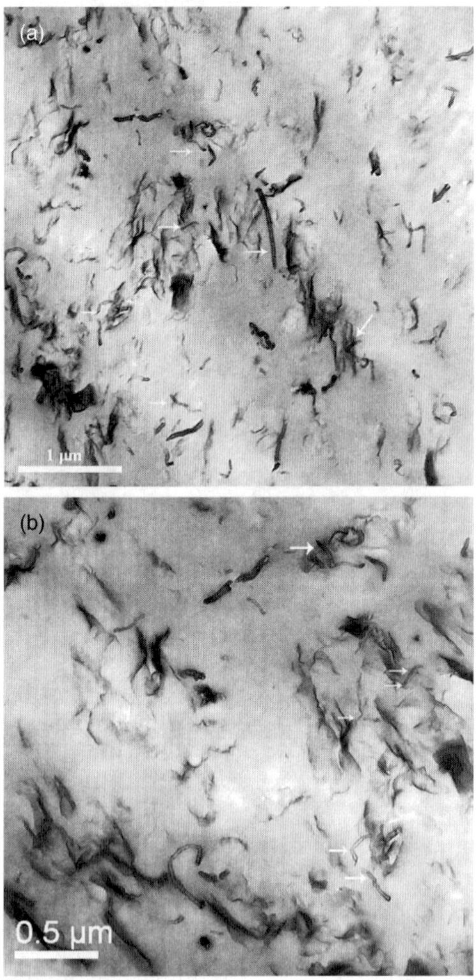

Figure 5.6 TEM images of chitosan/2% clay-1% CNT composite: (a) at low magnification and (b) at high magnification [47].

are randomly distributed in the chitosan matrix, and meanwhile CNTs are observed to be well dispersed in chitosan. In Figure 5.6b, some CNTs, indicated by white arrows, are located around the clay platelets and form a kind of clay–CNT network, which could result in strong interaction between chitosan chains and the fillers.

Chitosan–hydroxyapatite (HA) nanocomposites show good biocompatibility and favorable bonding ability with surrounding host tissues. The approaches currently used to obtain CS-HA composite materials are based on mechanical mixing, co-precipitation, *in situ* hybridization, and an alternate soaking process [48–54]. With all of these methods, there is a common disadvantage that inorganic particles cannot be distributed within the organic matrices at the nanolevel.

Much research has focused on the effect of the interface between the HA and CS matrix on the resultant morphology and structure. Biomorphic mineralization synthesis and biological template synthesis are possible approaches for obtaining the fine biological nanostructure [45–57]. A mild hydrothermal reaction achieved the biomorphic mineralization synthesis of chitosan–HA nanocomposite from crab shell [55]. In this reaction, calcium carbonate crystals in the crab shell were converted into HA crystals; at the same time, the crab shell's chitin is partially deacetylated into chitosan. The fine nanocomposite structure of the crab shell was almost exactly reproduced (shown in Fig. 5.7).

In addition, chitin or chitosan-based nanocomposites have been widely used as matrices in drug release systems and as vehicles in the treatment of waste waters [58–67]. The chitin-based drug nanoparticle carrier was subjected to examination by SEM before and after drug release [64]. A comparison of the micrographs suggests detachment of the drug doxorubicin hydrochloride (DOX), clay, and chitosan during drug release, with a consequent increase in the size of pores. The pore size increased from 1.1 ± 0.1 μm before drug release (Fig. 5.8) to 2.3 ± 0.2 μm after drug release (Fig. 5.9).

Other chitin-based nanocomposites include the blend fibers and membranes [4, 65–71]. The SEM images were taken to analyze any morphological changes in the fibrous and membrane structures.

Submicron fibers of the composite of poly(vinyl alcohol) (PVA), chitosan oligosaccharide [COS, (1→4)2-amino-2-deoxy-β-D-glucose], and MMT clay were prepared using electrospinning with aqueous solutions [70]. TEM observation reveals the formation of the composite nanofibers and the distribution of the MMT nanoparticles in the nanofiber matrix. It can be clearly observed that each silicate platelet forms a dark line in the nanofiber compared with the pure PVA nanofiber (Fig. 5.10a). The TEM images in Figure 5.10b–d indicate the nanosized MMT in the nanofibers electrospun from the solution containing 5 wt% MMT. The size of the dark line is about 1–3 nm in width and 100–200 nm in length, indicating the good dispersion and exfoliation of MMT layers in the nanofibers.

Moreover, typical morphology observed by field emission scanning electron microscopy (FE-SEM) for the electrospun chitosan-PVA nanofibrous

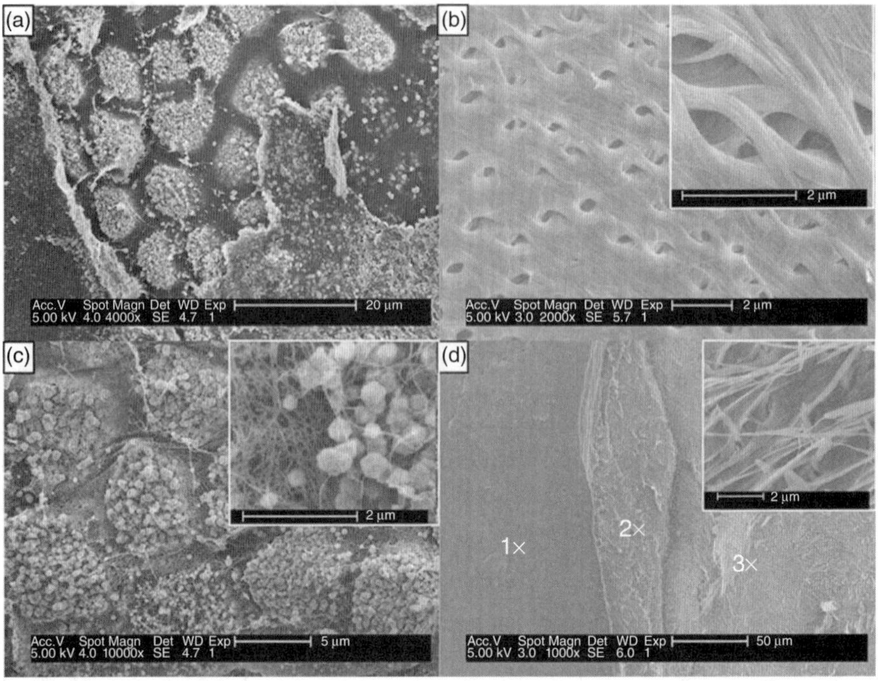

Figure 5.7 SEM images of CS-HA: the outer surfaces' structures (a); the inner surfaces' structures (b); the circular ball-like aggregations (c); and the breaking faces of inner layers (d) [55].

membrane is an almost homogenous network with fiber diameter distribution of between 150 and 300 nm [72]. To remove PVA from the blended nanofibrous membrane, the original nanofibrous membrane was treated with 0.5 M NaOH aqueous solution for 4 hours. It was found that the fiber diameter was decreased from 150–300 nm to 80–150 nm. This treatment consequently resulted in over 50 wt% weight loss, which was nearly identical to the PVA content in the original chitosan-PVA nanofibrous membrane.

At present, graphene has attracted increasing attention worldwide, owing to its structure, single atomic monolayer of sp^2-bonded carbon atoms, and unique properties. Graphene oxide (GO) is a precursor of graphene-based nanocomposites with similar one-atom thickness but with a large number of hydrophilic oxygenated functional groups including hydroxyl (-OH), epoxy (-C–O–C-), carbonyl (-C=O), and carboxyl (-COOH) groups. These groups make GO sheets hydrophilic and dramatically improve their miscibility with polymer matrix [73, 74]. Biopolymer nanocomposites were prepared from CS as the matrix and GO as reinforcing nanofillers by a simple self-assembly. A uniform distribution and fine dispersion for GO in the CS matrix are shown in the SEM images in Figure 5.11 [74].

Figure 5.8 Scanning electron micrographs of DOX-loaded chitosan–clay particles before drug release. (a) Low magnification; (b) high magnification [64].

5.1.3 Atomic Force Microscopy

Atomic force microscopy (AFM) has been applied to determine the structure of chitin-based nanocomposites and to provide three-dimensional morphology. Because of its high spatial and vertical resolution, AFM is considered to be a powerful tool for imaging the topography of the surfaces.

The nanosized filler dispersion within chitin or chitosan has been characterized by AFM in the study of the nanostructure of composites [13, 38, 75, 76]. AFM is also used to characterize the compatibility of chitin- or chitosan whisker-incorporated blended films or blended fibers [5, 74].

The morphology of fractured and polished surfaces of chitin-based nanocomposites can be studied by AFM in "Tapping Mode." In tapping mode, the AFM tip oscillates at a frequency close to its resonance and the tip is allowed to make contact with the sample for only a short duration in each oscillation cycle. During oscillation of the tip over the sample surface, the sample–tip

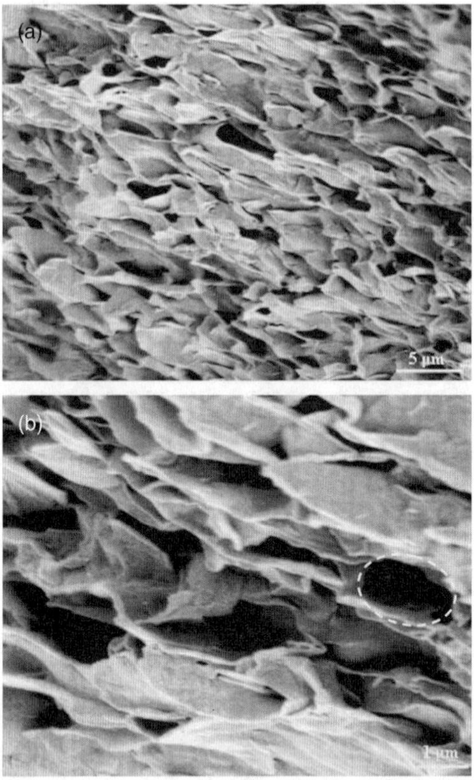

Figure 5.9 Scanning electron micrographs of DOX-loaded chitosan–clay particles after drug release at pH 5.3 and 37°C. (a) Low magnification; (b) high magnification [64].

interaction may alter the amplitude, resonance frequency, and phase angle of the oscillating cantilever. Detection of phase angle changes of the cantilever probe during scanning provides an image, called a phase image. Phase angle change is associated with energy dissipation during sample–tip interaction. Several parameters, such as topography of the sample, sample–tip interactions, deformation of sample–tip contact area, and experimental conditions, can cause energy dissipation. The phase image is very useful for compositional mapping of surfaces and interfaces of polymeric materials and generally provides better contrast than the topographic images.

AFM images in tapping mode illustrate fractured and polished surfaces of nanocomposites of chitosan–HA in a 50:50 ratio (ChiHAP50), polygalacturonic acid–HA in 50:50 ratio (PgAHAP50), and chitosan–HA–polygalacturonic acid (ChiPgAHAP50) [77]. In Figure 5.12, it is clear that the building blocks of ChiPgAHAP50 are chitosan-rich and polygalacturonic-rich domains. These chitosan-rich and polygalacturonic-rich domains appear to have differ-

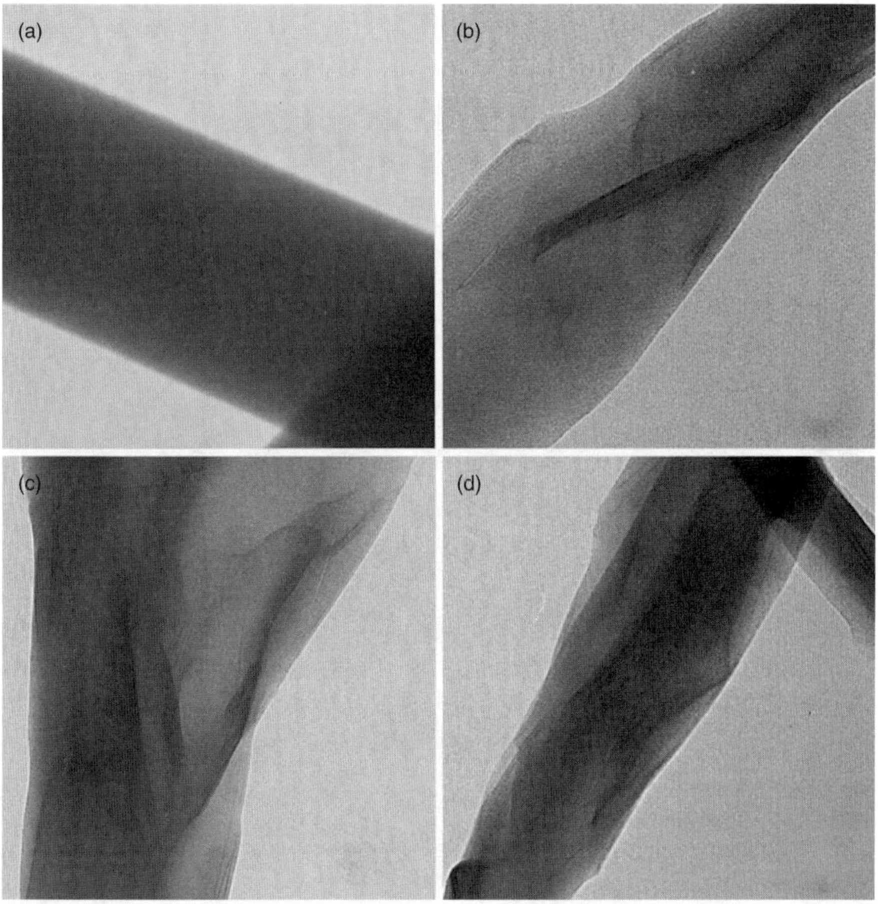

Figure 5.10 TEM images of PVA-COS-MMT nanofiber mats that were electrospun with various PVA/COS mass ratios: (a) pure PVA, (b) PVA/COS = 6/4, (c) PVA/COS=8/2, (d) PVA/COS = 10/0 (MMT = 5 wt%, voltage = 15 kV, tip to collector distance = 15 cm) [70].

ent morphologies in ChiHAP50 and PgAHAP50, respectively. Chitosan-rich domains and polygalacturonic-rich domains consist of smaller globular shaped particles. The AFM phase images of polished samples suggest that these globular shaped particles have HA particles embedded in the biopolymer matrix, implying the multilevel organization in HA–biopolymer composites.

In addition to the morphological analysis methods mentioned above, confocal laser scanning microscope (CLSM) [21] and analytical electron microscopy (AEM) [78] have been applied to examine the structure and applications of chitin-based nanocomposites.

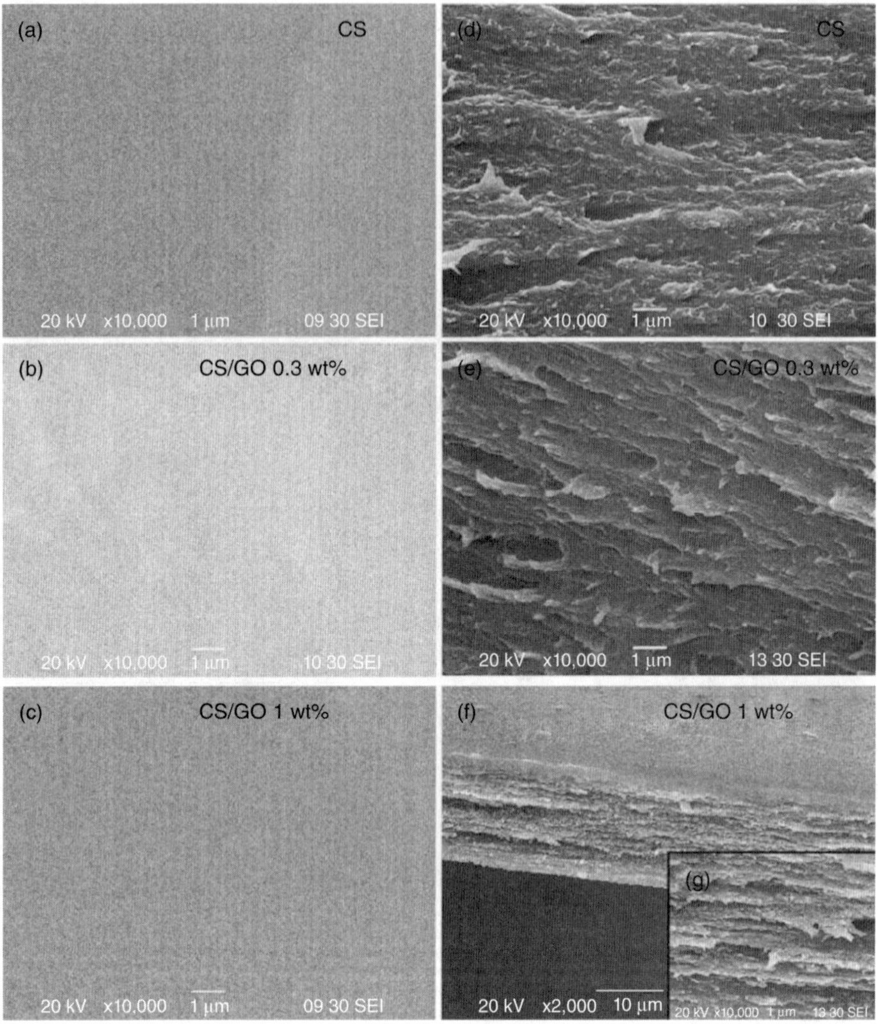

Figure 5.11 SEM film-surface images (a–c) and fracture-surface images (d–g) of CS and CS-GO nanocomposites. Image (g) is the magnification of (f) [74].

5.2 THERMAL INVESTIGATIONS OF CHITIN-BASED NANOCOMPOSITES

The main thermal techniques applied for chitin-based nanocomposite measurements include differential scanning calorimetry (DSC), thermogravimetric analysis (TGA), thermomechanical analysis (TMA), and dynamic mechanical analysis (DMA). The application possibilities of these methods in the field of chitin-based nanocomposites are so diverse that only a brief review can be given here.

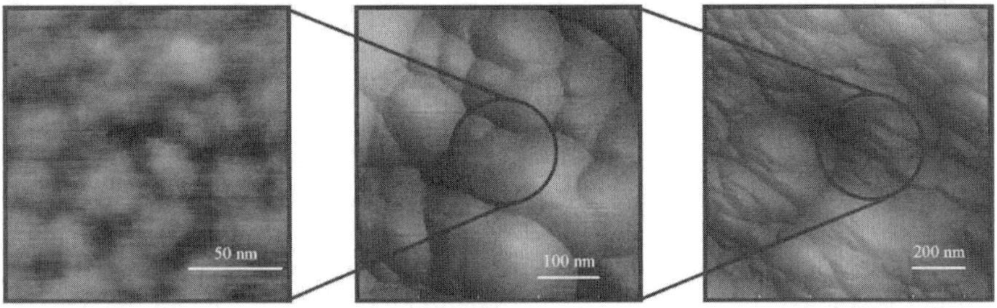

Figure 5.12 Atomic force microscope images in tapping mode illustrating multilevel structure in ChiHAP50, PgAHAP50, and ChiPgAHAP50 nanocomposites. Images are illustrative of the multilevel structure and are not to scale [77]. See color insert.

5.2.1 Differential Scanning Calorimetry

DSC can be used to investigate thermal events such as physical transitions (the glass transition, crystallization, melting, and vaporization of volatile compounds) and chemical reactions. Thermal properties such as heat capacity, glass transition temperature (Tg), melting temperature (Tm), and thermal stability can be determined. The information obtained characterizes the chitin-based nanocomposites with regard to their thermal behavior, composition, compatibility, and process method.

DSC was used to measure Tg values and to investigate the compatibility of natural chitosan nanofibers and polyacrylamide (PAM) in dry gels [7]. The Tg of PAM in the composite gel was higher than that of pure PAM due to the interaction between PAM and chitosan, which limited the chain movement of PAM in the composite.

DSC was also used to study the influence of the fabricating method on the thermal properties of chitin whisker-based nanocomposites. Nanocomposites were prepared from PCL as the matrix and chitin microcrystals, or whiskers, as the reinforcing phase [79]. Nanocomposite films were obtained either by casting and evaporating or by freeze-drying and hot-pressing the mixture. For all the chitin whisker–PCL composites, both Tg and Tm seem to be almost independent of the whisker concentration. The crystallinity of semicrystalline PCL polymers is strongly dependent on the thermal history of the sample. A significant reinforcing effect was observed at T > Tg of the matrix. This effect can be described in this temperature range by a "hybrid" percolation model, equivalent to a mean-field model, considering only the nonpercolating fraction of microcrystals.

Normally, DSC results indicated that heat treatment causes a reduction of intramolecular hydrogen bonding of chitosan and also may induce thermal depolymerization of chitosan, resulting in a decrease in viscosity of chitosan dispersions after heat treatment. Surprisingly, incorporation of nanosized magnesium aluminum silicate (MAS) caused an increase in the decomposition temperature of chitosan in composite films prepared using heated dispersions [80]. This suggests that the interaction of chitosan with MAS by flocculation in the composite dispersion and the intercalation of chitosan into the MAS layers could possibly protect the chitosan molecules from thermal depolymerization when treated by heating.

The influence of addition of chitin or chitosan whiskers on the crystallinity of polymer matrix was investigated by DSC. The DSC profiles of all the composites exhibited endothermic peaks due to the melting point of PVA being shifted to a lower temperature accompanied by broadening when the content of chitin in the composites increased [81]. The crystallinities of PVA were decreased in the composites with increasing chitin content. Therefore, chitin and PVA might be partially miscible at the interfacial area between the two polymers in the composites by the formation of hydrogen bonding. On the contrary, an increase in nanoclay content definitely improved the crystalline pattern of the synthesized chitin-based polyurethane bionanocomposite samples [82].

The thermal stability and thermal decomposition behavior of chitin-based nanocomposites were also studied by DSC. Figure 5.13 shows the DSC thermograms of the methanol-treated chitin whisker–silk fibroin sponges at different C:S ratios, and the freeze-dried chitin whiskers within the temperature range of 50–450°C. The decomposition temperature of neat silk fibroin appeared to be at the broader endothermic peak of 287°C and at the sharper endothermic peak of 304°C. The presence of the whiskers at the C:S ratio of 1:8 (Fig. 5.13b) induced the lowering of the sharp peak of the decomposition temperature from 304°C to 298°C with the change in thermal transition from a rough to a smooth curve, but the wide endothermic peak of the decomposition temperature (288°C) was still similar. The further increase in the amount of chitin whiskers (Fig. 5.13c,d) resulted in the absence of the sharp endothermic peak of the decomposition temperature. Analogous to the C:S ratio of 1:8, the broad endothermic peak of the decomposition temperature (287°C) was still unchanged. By careful consideration of the thermograms present in Figure 5.13c,d, there was a new exothermal curve, which could not be observed in the thermograms of both neat silk fibroin and chitin whiskers, due to the shift of the baselines in the range of 385–415°C. It should be noted that the decomposition of the chitin whiskers caused the curve shifting (exothermic) of the baseline in the temperature range of 260–410°C (Fig. 5.13e).

Compared with the neat silk fibroin and chitin whiskers, the noticeable improvement in the thermal decomposition behavior of the nanocomposites might imply that interfacial interactions (mainly hydrogen bonding) form between polar groups present in the chemical structures of silk fibroin and chitin whiskers in the nanocomposite sponges [6]. These interactions restricted the mobility of silk fibroin chains and also reduced inter- and intramolecular interactions between silk fibroin chains. In addition, a short single crystal fiber, that is, chitin whisker, is a highly crystalline material so its volume is still unchanged after methanol treatment. Hence, the incorporation of dimension-

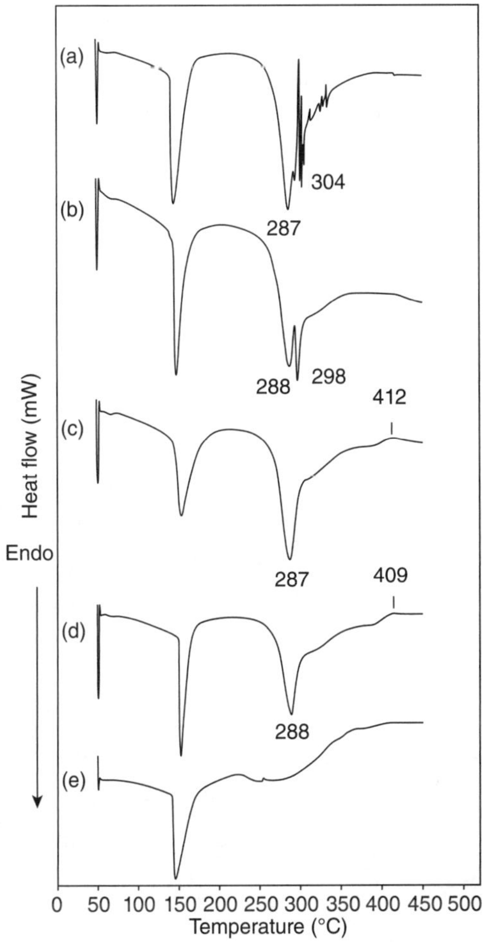

Figure 5.13 DSC thermograms of the methanol-treated silk fibroin sponge (a), the methanol-treated chitin whisker–silk fibroin sponges having different C/S ratios, 1/8 (b), 2/8 (c), 4/8 (d), and the freeze-dried chitin whisker (e) [6].

stable fillers into the matrix resulted in a reduction in the percentage shrinkage of the sponges, which also decreased with an increasing amount of stable filler.

5.2.2 Dynamic Thermal Mechanical Analysis

In dynamic thermal mechanical analysis (DMA), a mechanical modulus is determined as a function of temperature, frequency, and amplitude. The important properties of storage moduli, loss factor, Tg, and, especially, the influence of fillers and reinforcing effect on the modulus and Tg can be investigated by DMA. Since the mechanical modulus depends on the molecular conditions, information can be gained about the cross-linking and compatibility of polymer blends and additives. The Tg is often measured by DSC, but the DMA technique is more sensitive and yields more precise data.

The influence of chitin whisker effect on the modulus and Tg of nanocomposites could be studied by DMA [83]. Figure 5.14 shows the plot of log(E'/Pa) (storage tensile modulus, Fig. 5.14a) and tan δ (loss angle tangent, Fig. 5.14b) at 1 Hz as a function of temperature for both unfilled and filled unvul-

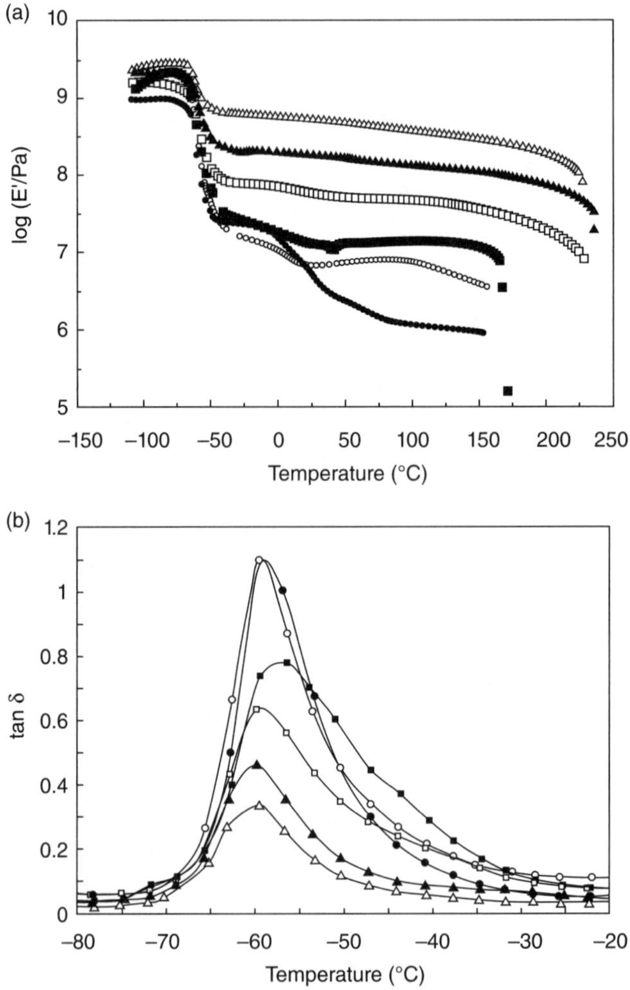

Figure 5.14 (a) Logarithm of the storage tensile modulus E′ and (b) loss angle tangent tan α versus temperature at 1 Hz for chitin whisker–unvulcanized NR composites obtained by evaporation: NRev (●), NCH2ev (○), NCH5ev (■), NCH10ev (□), NCH15ev (▲), and NCH20ev (△) [83].

canized NR films prepared by evaporation. Increasing the amount of chitin whiskers from 0 wt% to 20 wt% (coded as NRev, NCH2ev, NCH5ev, NCH10ev, NCH15ev, and NCH20ev) successively increases the values of E′, and the highest modulus is observed for the composite containing 20 wt% whiskers. This sort of enhancement in modulus even below the Tg is good evidence for the strong reinforcing tendency of chitin whiskers in the NR matrix. A sharp modulus drop appears for all samples around −60°C, that is, in the glass–rubber transition zone. This relaxation process involves cooperative motions of long chain sequences. Above Tg the modulus becomes roughly constant over a wide temperature region, called the rubbery plateau region (rubbery modulus), except in the case of unfilled NR and low filler content composites. For chitin whisker-filled NR, a significant increase in rubbery modulus is observed with increasing chitin content. For instance, the relaxed modulus at Tg + 150°C of a film containing only 2 wt% (around 1.34 vol%) of chitin is seven times higher than that of the unfilled matrix. For the 20 wt% chitin whisker-filled composites, the rubbery modulus is more than 350 times higher than that of the unfilled matrix. In addition to this high reinforcing effect, for filler contents higher than 5 wt%, a significant improvement in the thermal stability of the composite is also noticed up to 220–230°C. Normally, chitin starts to degrade at this temperature. The cause of both the high reinforcing effect and the improvement of the thermal stability of these materials can be attributed to the formation of a rigid network of cellulose whiskers within the polymer matrix, formed as a result of hydrogen bonding.

However, the chemically modified whiskers opposed the reverse influence on the NR-based nanocomposites [9]. The increase in the filler/matrix interactions results in a decrease in the chitin–chitin interactions responsible for the high mechanical characteristics of the unmodified whisker-filled NR film. Above Tg, the modulus of all chemically modified whisker-based composites was found to be much lower than that of the unmodified one and more similar to that of the unfilled matrix.

DMA also reveals the intensity of the tan δ peak relative to the crystalline type of the matrix. The intensity of the tan δ peak decreased as the crystallinity of the chitin-based polyurethane bionanocomposite samples increased [82]. Therefore, it can be concluded that an increase in nanoclay content definitely improves the crystalline pattern of the synthesized chitin-based polyurethane bionanocomposite samples.

DMA is a technique that is also helpful for estimating the increase in stiffness of the composites caused by the addition of nanosized filler. In the region corresponding to the maximum loss factor (tan δ), the decrease in storage modulus was usually rapid. The loss factor is sensitive to molecular motion and its peak represents the Tg. Nanocomposites based on the novel chitin nanoparticles (CNP) and glycerol plasticized potato starch (GPS) were detected by DMA [84]. When CNP was introduced, the storage modulus of GPS–CNP composites increased. In GPS–CNP composites, both the upper transition and the lower transition shifted to higher temperatures, which indicated that both the starch-rich and starch-poor phases interacted well with CNP. Serving as junctions, CNP improved the intermolecular interaction of GPS in the starch-rich phase, bringing adjacent chains of starch close, restraining chain mobility, and reducing the free volume, thereby raising the Tg of composites as the CNP content increased.

5.2.3 Thermogravimetric Analysis

In TGA, the change in mass of a sample is measured as a function of temperature or time under an inert atmosphere (nitrogen) or an oxidative atmosphere (air). TGA is used to quantitatively determine the main constitutes of the chitin-based nanocomposites. The moisture and other volatile compounds, polymer content through pyrolysis or combustion, fillers or reinforcing fibers, and ash as decomposition and combustion residues were measured respective to the intermolecular interactions of composites.

TGA measurements are often displayed as the first derivative of the TGA curve, the so-called DTG curve. Steps due to loss of mass in the TGA curve then appear as peaks in the DTG curves. The DTG curve corresponds to the rate of change of sample mass.

The influence of nanosized fillers on the thermal behavior of chitin or chitosan has been studied by TGA. The thermal stability of chitosan is usually improved by incorporation of the nanoparticles. The improved thermal stability is a common advantage of inorganic fillers [15, 60, 61, 85].

However, slight or reverse changes also occurred when certain inorganic fillers were added into the chitin or chitosan matrix [86]. The intercalation by polymers is usually preferred to that by cationic surfactants due to a decrease in thermal stability of the organoclays that have a surfactant [37].

The combination of 2D clay platelets and 1D CNTs has a significantly synergistic effect on the thermal properties of chitosan [47]. The large improvement in thermal stability of chitosan may arise from three causes: (1) good heat barrier properties of CNTs and clay for the polymer matrix during formation of chars; (2) formation of a carbonaceous layer on the polymer surface at lower temperatures, which could hinder the flux of degradation product and heat flowing into underlying materials; and (3) a greatly jammed and conjugated 3D clay–CNT network, or strong interaction between chitosan and nanofillers, which could also restrict the polymer motion during heating.

Recently, the effects of ultrasonification method and chitin whisker content on the thermal stability were measured by TGA [87]. Two series of nanocomposite films were prepared from waterborne poly(ester-urethane) and chitin whisker with and without ultrasound treatment, coded as CW/WPU and CHW/WPU, respectively. Thermal degradation patterns of the films are shown in Figure 5.15 and Figure 5.16. Data related to the temperature corresponding to the weight losses of the initial weights are summarized in Table 5.1. Chitin whiskers exhibited two degradation peaks at about 260°C and 350°C. The CW/WPU and CHW/WPU films exhibited much higher thermal stability than that

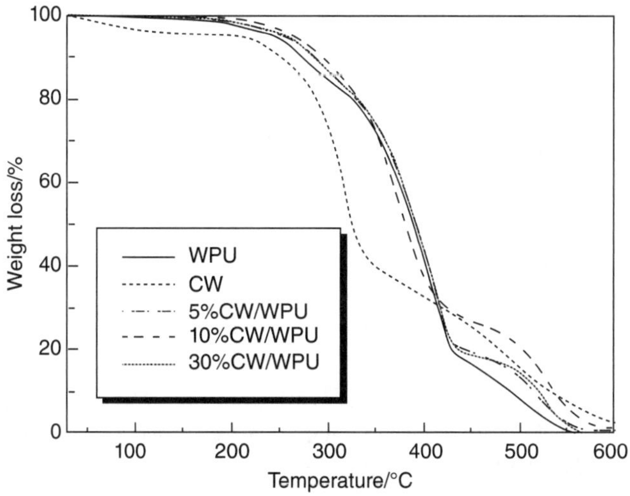

Figure 5.15 Thermogravimetric analyses of WPU, CW, 5%CW/WPU, 10%CW/WPU, and 30%CW/WPU [87].

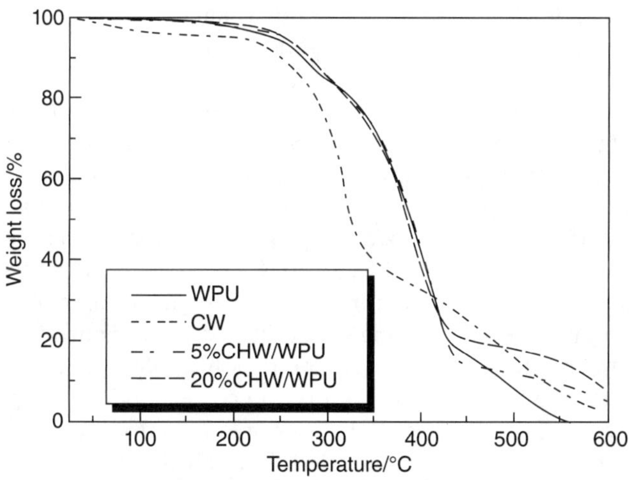

Figure 5.16 Thermogravimetric analyses of WPU, CW, 5%CHW/WPU, and 20%CHW/WPU [87].

of the pure WPU film and chitin whisker. The hydrogen-bonded chitin whisker network induced a thermal stabilization of the composite up to 250°C, which is higher than the temperature of 200°C at which chitin whisker starts to decompose, indicating the reinforcing effect of nanometer-sized chitin whiskers in the WPU matrix. Moreover, the film 5%CW/WPU appeared as having

TABLE 5.1 Thermogravimetric Analyses Data of the Samples: Temperatures (°C) According to the Weight Loss (%) [87]

Weight Loss/%	WPU	CW	5% CW/ WPU	5% CHW/ WPU	10% CW/ WPU	20% CHW/ WPU	30% CW/ WPU
5	245	208	256	252	257	256	264
10	273	249	284	280	286	277	293
20	325	286	330	321	329	322	333
30	355	304	359	351	358	353	353
50	388	324	392	390	391	385	380
80	430	479	442	429	434	429	503

a higher thermal stability along the wide temperature range, compared with the film 5%CHW/WPU without the ultrasound treatment. The above results suggest that more effective reinforcement and stronger interactions between WPU and CW occurred with the ultrasound treatment.

5.2.4 Thermomechanical Analysis

TMA measures the dimensional changes of a sample as it is heated. The linear coefficient of thermal expansion (CTE) was calculated from the measurement curve.

TMA is used to determine important properties and temperature effects by measuring the dimensions of a sample as a function of temperature. It is noted that chitin nanofiber-reinforced acrylic resin films exhibited considerably low thermal expansion [88]. By adding 40 wt% of chitin nanofibers, the CTE of the acrylic resin dropped to 15.6 ppm/K, approximately 1/14 of its original value, a value comparable to the CTE of cellulose nanocomposites. The fact that the CTE of the chitin crystal in the axial direction is probably close to zero and the axial Young's modulus of chitin is around 80 GPa, in combination with the inplane random nanofiber orientation distribution, apparently leads to the very low CTE of the corresponding composites.

To further interpret the thermal measurement results of chitin-based nanocomposites, more information is needed. In addition to the thermal analysis methods already mentioned, this can be obtained by combining two or more suitable techniques. Techniques often used include simultaneous differential thermal analysis (DTA), simultaneous DSC, and various methods for online gas analysis (mass spectrometry and infrared spectroscopy).

REFERENCES

[1] Huang, H., Yuan, Q., and Yang, X. (2004) Preparation and characterization of metal–chitosan nanocomposites. *Colloids and Surfaces. B, Biointerfaces*, *39*, 31–37.

[2] Zuber, M., Mahmood Zia, K., Mahboob, S., Hassan, M., and Ahmad Bhatti, I. (2010) Synthesis of chitin–bentonite clay based polyurethane bio-nanocomposites. *International Journal of Biological Macromolecules*, *47*, 196–200.

[3] Shieh, Y., Yang, Y.-F. (2006) Significant improvements in mechanical property and water stability of chitosan by carbon nanotubes. *European Polymer Journal*, *42*, 3162–3170.

[4] Watthanaphanit, A., Supaphol, P., Tamura, H., Tokura, S., and Rujiravanit, R. (2010) Wet-spun alginate/chitosan whiskers nanocomposite fibers: Preparation, characterization and release characteristic of the whiskers. *Carbohydrate Polymers*, *79*, 738–746.

[5] Lu, Y., Weng, L., and Zhang, L. (2004) Morphology and properties of soy protein isolate thermoplastics reinforced with chitin whiskers. *Biomacromolecules*, *5*, 1046–1051.

[6] Wongpanit, P., Sanchavanakit, N., Pavasant, P., Bunaprasert, T., Tabata, Y., and Rujiravanit, R. (2007) Preparation and characterization of chitin whisker-reinforced silk fibroin nanocomposite sponges. *European Polymer Journal*, *43*, 4123–4135.

[7] Zhou, C., Wu, Q. (2011) A novel polyacrylamide nanocomposite hydrogel reinforced with natural chitosan nanofibers. *Colloids and Surfaces. B, Biointerfaces*, *84*, 155–162.

[8] Gopalan Nair, K., Dufresne, A. (2003) Crab shell chitin whisker reinforced natural rubber nanocomposites. 1. Processing and swelling behavior. *Biomacromolecules*, *4*, 657–665.

[9] Gopalan Nair, K., Dufresne, A., Gandini, A., and Naceur Belgacem, M. (2003) Crab shell chitin whiskers reinforced natural rubber nanocomposites. 3. Effect of chemical modification of chitin whiskers. *Biomacromolecules*, *4*, 1835–1842.

[10] Sowmya, S., Sudheesh Kumar, P.T., Chennazhi, K.P., Nair, S.V., Tamura, H., and Jayakumar, R. (2011) Biocompatible β-chitin Hydrogel/nanobioactive glass ceramic nanocomposite scaffolds for periodontal bone regeneration. *Trends in Biomaterials and Artificial Organs*, *25*, 1–11.

[11] Madhumathi, K., Sudheesh Kumar, P.T., Binulal, N.S., Nair, S.V., Tamura, H., and Jayakumar, R. (2009) Development of novel chitin/nano bioactive glass ceramic nanocomposite scaffolds for tissue engineering applications. *Carbohydrate Polymers*, *78*, 926–931.

[12] Chen, J., Loo, L.S., and Wang, K. (2011) Enhanced mechanical properties of novel chitosan nanocomposite fibers. *Carbohydrate Polymers*, *86*, 1151–1156.

[13] Rana, V.K., Pandey, A.K., Singh, R.P., Kumar, B., Mishra, S., and Ha, C.-S. (2010) Enhancement of thermal stability and phase relaxation behavior of chitosan dissolved in aqueous *l*-lactic acid: Using "silver nanoparticles" as nano filler. *Macromolecular Research*, *18*, 713–720.

[14] Madhumathi, K., Sudheesh Kumar, P.T., Abhilash, S., Sreeja, V., Tamura, H., Manzoor, K., Nair, S.V., and Jayakumar, R. (2010) Development of novel chitin/nanosilver composite scaffolds for wound dressing applications. *Journal of Materials Science: Materials in Medicine*, *21*, 807–813.

[15] Krishna Rao, K.S.V., Ramasubba Reddy, P., Lee, Y.-I., and Kim, C. (2010) Synthesis and characterization of chitosan-PEG-Ag nanocomposites for antimicrobial application. *Carbohydrate Polymers*, *7*, 28–38.

[16] Tripathi, S., Mehrotra, G.K., and Dutta, P.K. (2011) Chitosan–silver oxide nanocomposite film: Preparation and antimicrobial activity. *Bulletin of Materials Science*, *34*, 29–35.

[17] Chen, J.-Y., Zhou, P.-J., Li, J.-L., and Wang, Y. (2008) Studies on the photocatalytic performance of cuprous oxide/chitosan nanocomposites activated by visible light. *Carbohydrate Polymers*, *72*, 128–132.

[18] Malhotra, B.D., Kaushik, A. (2009) Metal oxide–chitosan based nanocomposite for cholesterol biosensor. *Thin Solid Films*, *518*, 614–620.

[19] Yang, D., Li, J., Jiang, Z., Lu, L., and Chen, X. (2009) Chitosan/TiO_2 nanocomposite pervaporation membranes for ethanol dehydration. *Chemical Engineering Science*, *64*, 3130–3137.

[20] Jayakumar, R., Ramachandran, R., Sudheesh Kumar, P.T., Divyarani, V.V., Srinivasan, S., Chennazhi, K.P., Tamurab, H., and Nair, S.V. (2011) Fabrication of chitin–chitosan/nano ZrO_2 composite scaffolds for tissue engineering applications. *International Journal of Biological Macromolecules*, *49*, 274–280.

[21] Sanpui, P., Murugadoss, A., Durga Prasad, P.V., Ghosh, S.S., and Chattopadhyay, A. (2008) The antibacterial properties of a novel chitosan–Ag-nanoparticle composite. *International Journal of Food Microbiology*, *124*, 142–146.

[22] Namdeo, M., Bajpai, S.K. (2008) Chitosan–magnetite nanocomposites (CMNs) as magnetic carrier particles for removal of Fe(III) from aqueous solutions. *Colloids and Surfaces. A, Physicochemical and Engineering Aspects*, *320*, 161–168.

[23] Jayakumar, R., Roshini, R., Divya Rani, V.V., Chennazhi, K.P., Tamura, H., and Nair, S.V. (2011) Fabrication of chitin-chitosan/nano TiO_2-composite scaffolds for tissue engineering applications. *International Journal of Biological Macromolecules*, *48*, 336–344.

[24] Sudheesh Kumar, P.T., Abhilash, S., Manzoor, K., Nair, S.V., Tamura, H., and Jayakumar, R. (2010) Preparation and characterization of novel *β*-Chitin/nano silver composite scaffolds for wound dressing applications. *Carbohydrate Polymers*, *80*, 761–767.

[25] Madhumathi, K., Sudhessh Kumar, P.T., Abhilash, S., Nair, S.V., Tamura, H., Manzoor, K., Nair, S.V., and Jayakumar, R. (2010) Development of novel chitin/nanosilver composite scaffolds for wound dressing applications. *Journal of Materials Science: Materials in Medicine*, *21*, 807–813.

[26] Chen, Y., Wang, J. (2011) Preparation and characterization of magnetic chitosan nanoparticles and its application for Cu(II) removal. *Chemical Engineering Journal*, *168*, 286–292.

[27] Huang, H., Yuan, Q., and Yang, X. (2005) Morphology study of gold–chitosan nanocomposites. *Journal of Colloid and Interface Science*, *282*, 26–31.

[28] Xu, Q., Mao, C., Liu, N.-N., Zhu, J.-J., and Sheng, J. (2006) Direct electrochemistry of horseradish peroxidase based on biocompatible carboxymethyl chitosan–gold nanoparticle nanocomposite. *Biosensors and Bioelectronics*, *22*, 768–773.

[29] Li, B., Jia, D., Zhou, Y., Hu, Q., and Cai, W. (2006) *In situ* hybridization to chitosan/magnetite nanocomposite induced by the magnetic field. *Journal of Magnetism and Magnetic Materials*, *306*, 223–227.

[30] Cao, S., Mishra, R., Pilla, S., Tripathi, S., Pandey, M.K., Shah, G., Mishra, A.K., Prabaharan, M., Mishra, S.B., Jin, X., Pandey, R.R., Wu, W., Pandey, A.C., and

Tiwari, A. (2010) Novel chitosan/gold-MPA nanocomposite for sequence-specific oligonucleotide detection. *Carbohydrate Polymers*, *82*, 189–194.

[31] Esumi, K., Takei, N., and Yoshimura, T. (2003) Antioxidant-potentiality of gold/chitosan nanocomposites. *Colloids and Surfaces. B, Biointerfaces*, *32*, 117–123.

[32] Yang, K.-H., Liu, Y.-C., Yu, C.-C., and Chen, B.-C. (2011) Fabrication of chitosan/silver nanocomposites based on electrochemical methods for removing formaldehyde in air. *Materials Chemistry and Physics*, *126*, 993–997.

[33] Singh, J., Srivastava, M., Dutta, J., and Dutta, P.K. (2011) Preparation and properties of hybrid monodispersed magnetic α-Fe_2O_3 based chitosan nanocomposite film for industrial and biomedical applications. *International Journal of Biological Macromolecules*, *48*, 170–176.

[34] Witoon, T., Chareonpanich, M., and Limtrakul, J. (2009) Effect of acidity on the formation of silica–chitosan hybrid materials and thermal conductive property. *Journal of Sol-Gel Science and Technology*, *51*, 146–152.

[35] Tan, W., Zhang, Y., Szeto, Y., and Liao, L. (2008) A novel method to prepare chitosan/montmorillonite nanocomposites in the presence of hydroxy-aluminum oligomeric cations. *Composites Science and Technology*, *68*, 2917–2921.

[36] Wang, S.F., Shen, L., Tong, Y.J., Chen, L., Phang, I.Y., Lim, P.Q., and Liu, T.X. (2005) Biopolymer chitosan/montmorillonite nanocomposites: Preparation and characterization. *Polymer Degradation and Stability*, *90*, 123–131.

[37] Zhang, K., Xu, J., Wang, K.Y., Cheng, L., Wang, J., and Liu, B. (2009) Preparation and characterization of chitosan nanocomposites with vermiculite of different modification. *Polymer Degradation and Stability*, *94*, 2121–2127.

[38] Wang, X., Liu, B., Ren, J., Liu, C., Wang, X., Wu, J., and Sun, R. (2010) Preparation and characterization of new quaternized carboxymethyl chitosan/rectorite nanocomposite. *Composites Science and Technology*, *70*, 1161–1167.

[39] Sun, F., Cha, H.-R., Bae, K., Hong, S., Kim, J.-M., Kim, S.H., Lee, J., and Lee, D. (2011) Mechanical properties of multilayered chitosan/CNT nanocomposite films. *Materials Science and Engineering A*, *528*, 6636–6641.

[40] Qiu, J.-D., Xie, H.-Y., and Liang, R.-P. (2008) Preparation of porous chitosan = carbon nanotubes film modified electrode for biosensor application. *Microchim Acta*, *162*, 57–64.

[41] Tiwari, A. (2009) A novel nanocomposite matrix based on silylated chitosan and multiwall carban nanotubes for the immobilization of urease. *Journal Inorganic and Organometallic Polymers*, *19*, 361–366.

[42] Liu, Y.-L., Chen, W.-H., and Chang, Y.-H. (2009) Preparation and properties of chitosan/carbon nanotube nanocomposites using poly(styrene sulfonic acid)-modified CNTs. *Carbohydrate Polymers*, *76*, 232–238.

[43] Yang, H., Yuan, R., Chai, Y., and Zhuo, Y. (2011) Electrochemically deposited nanocomposite of chitosan and carbon nanotubes for detection of human chorionic gonadotrophin. *Colloids and Surfaces. B, Biointerfaces*, *82*, 463–469.

[44] Tiwari, A., Dhakate, S.R. (2009) Chitosan–SiO_2–multiwall carbon nanotubes nanocomposite: A novel matrix for the immobilization of creatinc amidinohydrolase. *International Journal of Biological Macromolecules*, *44*, 408–412.

[45] Salam, M.A., Makki, M.S.I., and Abdelaal, M.Y.A. (2011) Preparation and characterization of multi-walled carbon nanotubes/chitosan nanocomposite and its

application for the removal of heavy metals from aqueous solution. *Journal of Alloys and Compounds*, *509*, 2582–2587.

[46] Wu, T., Pan, Y., Bao, H., and Li, L. (2011) Preparation and properties of chitosan nanocomposite films reinforced by poly(3,4-ethylenedioxythiophene)-poly(styrenesulfonate) treated carbon nanotubes. *Materials Chemistry and Physics*, *129*, 932–938.

[47] Tang, C., Chen, N., Zhang, Q., Wang, K., Fu, Q., and Zhang, X. (2009) Preparation and properties of chitosan nanocomposites with nanofillers of different dimensions. *Polymer Degradation and Stability*, *94*, 124–131.

[48] Shen, X., Tong, H., Jiang, T., Zhu, Z., Wan, P., and Hu, J. (2007) Homogeneous chitosan/carbonate apatite/citric acid nanocomposites prepared through a novel *in situ* precipitation method. *Composites Science and Technology*, *67*, 2238–2245.

[49] Cai, X., Tong, H., Shen, X., Chen, W., Yan, J., and Hu, J. (2009) Preparation and characterization of homogeneous chitosan–polylactic acid/hydroxyapatite nanocomposite for bone tissue engineering and evaluation of its mechanical properties. *Acta Biomaterialia*, *5*, 2693–2703.

[50] Hu, Q., Li, B., Wang, M., and Shen, J. (2004) Preparation and characterization of biodegradable chitosan/hydroxyapatite nanocomposite rods via *in situ* hybridization: A potential material as internal fixation of bone fracture. *Biomaterials*, *25*, 779–785.

[51] Liji Sobhana, S.S., Sundaraseelan, J., Sekar, S., Sastry, T.P., and Mandal, A.B. (2009) Gelatin–chitosan composite capped gold nanoparticles: A matrix for the growth of hydroxyapatite. *J Nanopart Res*, *11*, 333–340.

[52] Pang, X., Zhitomirsky, I. (2008) Electrodeposition of hydroxyapatite–silver–chitosan nanocomposite coatings. *Surface & Coatings Technology*, *202*, 3815–3821.

[53] Sudheesh Kumar, P.T., Sowmya, S., Vinoth, K.L., Tamura, H., Nair, S.V., and Jayakumar, R. (2011) Synthesis, characterization and cytocompatibility studies of α-chitin hydrogel/nano hydroxyapatite composite scaffolds. *International Journal of Biological Macromolecules*, *49*, 20–31.

[54] Sudheesh Kumar, P.T., Sowmya, S., Vinoth, K.L., Tamura, H., Nair, S.V., and Jayakumar, R. (2011) β-Chitin hydrogel/nano hydroxyapatite composite scaffolds for tissue engineering applications. *Carbohydrate Polymers*, *85*, 584–591.

[55] Ge, H., Zhao, B., Lai, Y., Hu, X., Zhang, D., and Hu, K. (2010) From crabshell to chitosan-hydroxyapatite composite material via a biomorphic mineralization synthesis method. *Journal of Materials Science: Materials in Medicine*, *21*, 1781–1787.

[56] Jiang, L., Li, Y., and Xiong, C. (2009) Preparation and biological properties of a novel composite scaffold of nano-hydroxyapatite/chitosan/carboxymethyl cellulose for bone tissue engineering. *Journal of Biomedical Science*, *16*, 65–75.

[57] Zhang, Y., Venugopal, J.R., El-Turki, A., Ramakrishna, S., Su, B., and Lim, C.T. (2008) Electrospun biomimetic nanocomposite nanofibers of hydroxyapatite/chitosan for bone tissue engineering. *Biomaterials*, *29*, 4314–4322.

[58] Wan Ngah, W.S., Md Ariff, N.F., and Megat Hanafiah, M.A.K. (2010) Preparation, characterization, and environmental application of crosslinked chitosan-coated bentonite for tartrazine adsorption from aqueous solutions. *Water, Air, and Soil Pollution*, *206*, 225–236.

[59] Wang, X., Du, Y., Luo, J., Lin, B., and Kennedy, J.F. (2007) Chitosan/organic rectorite nanocomposite films: Structure, characteristic and drug delivery behaviour. *Carbohydrate Polymers*, *69*, 41–49.

[60] Wang, X., Du, Y., Yang, J., Wang, X., Shi, X., and Hu, Y. (2006) Preparation, characterization and antimicrobial activity of chitosan/layered silicate nanocomposites. *Polymer*, *47*, 6738–6744.

[61] El-Sherif, H., El-Masry, M. (2011) Superabsorbent nanocomposite hydrogels based on intercalation of chitosan into activated bentonite. *Polymer Bulletin*, *66*, 721–734.

[62] Khunawattanakul, W., Puttipipatkhachorn, S., Rades, T., and Pongjanyakul, T. (2011) Novel chitosan–magnesium aluminum silicate nanocomposite film coatings for modified-release tablets. *International Journal of Pharmaceutics*, *407*, 132–141.

[63] Liu, T.-Y., Chen, S.-Y., Li, J.-H., and Liu, D.-M. (2006) Study on drug release behaviour of CDHA/chitosan nanocomposites—Effect of CDHA nanoparticles. *Journal of Controlled Release*, *112*, 88–95.

[64] Yuan, Q., Shah, J., Hein, S., and Misra, R.D.K. (2010) Controlled and extended drug release behavior of chitosan-based nanoparticle carrier. *Acta Biomaterialia*, *6*, 1140–1148.

[65] Tang, Y., Zhao, Y., Li, Y., and Du, Y. (2010) A thermosensitive chitosan/poly(vinyl alcohol) hydrogel containing nanoparticles for drug delivery. *Polymer Bulletin*, *64*, 791–804.

[66] Zhang, F.-Y., Zhou, Y.-M., Sun, Y., Chen, J., Ye, X., and Huang, J. (2010) Preparation and characterization of Chitosan/Konjac glucomannan/CdS nanocomposite film with low infrared emissivity. *Materials Research Bulletin*, *45*, 859–862.

[67] Niamsa, N., Srisuwan, Y., Baimark, Y., Phinyocheep, P., and Kittipoom, S. (2009) Preparation of nanocomposite chitosan/silk fibroin blend films containing nanopore structures. *Carbohydrate Polymers*, *78*, 60–65.

[68] Nayak, P.L., Sahoo, D. (2011) Chitosan-alginate composites blended with cloisite 30B as a novel drug delivery system for anticancer drug paclitaxel. *International Journal of Plastics Technology*, *15*, 68–81.

[69] Paipitak, K., Pornpra, T., Mongkontalang, P., Techitdheera, W., and Pecharapa, W. (2011) Characterization of PVA-chitosan nanofibers prepared by electrospinning. *Procedia Engineering*, *8*, 101–105.

[70] Park, J.H., Lee, H.W., Chae, D.K., Oh, W., Yun, J.D., Deng, Y., and Yeum, J.H. (2009) Electrospinning and characterization of poly(vinyl alcohol)/chitosan oligosaccharide/clay nanocomposite nanofibers in aqueous solutions. *Colloid and Polymer Science*, *287*, 943–950.

[71] Shalumon, K.T., Binulal, N.S., Selvamurugan, N., Nair, S.V., Menon, D., Furuike, T., Tamura, H., and Jayakumar, R. (2009) Electrospinning of carboxymethyl chitin/poly(vinyl alcohol) nanofibrous scaffolds for tissue engineering applications. *Carbohydrate Polymers*, *77*, 863–869.

[72] Huang, X.-J., Ge, D., and Xu, Z.-K. (2007) Preparation and characterization of stable chitosan nanofibrous membrane for lipase immobilization. *European Polymer Journal*, *43*, 3710–3718.

[73] Han, D., Yan, L., Chen, W., and Li, W. (2011) Preparation of chitosan/graphene oxide composite film with enhanced mechanical strength in the wet state. *Carbohydrate Polymers*, *83*, 653–658.

[74] Yang, X., Tu, Y., Li, L., Shang, S., and Tao, X. (2010) Well-dispersed chitosan/graphene oxide nanocomposites. *Applied Materials and Interfaces*, *6*, 1707–1713.

[75] Wang, S., Tan, Y., Zhao, D., and Liu, G. (2008) Amperometric tyrosinase biosensor based on Fe_3O_4 nanoparticles–chitosan nanocomposite. *Biosensors and Bioelectronics*, *23*, 1781–1787.

[76] Hsu, S., Chang, Y.-B., Tsai, C.-L., Fu, K.-Y., Wang, S.-H., and Tseng, H.-J. (2011) Characterization and biocompatibility of chitosan nanocomposites. *Colloids and Surfaces. B, Biointerfaces*, *85*, 198–206.

[77] Verma, D., Katti, K.S., Katti, D.R., and Mohanty, B. (2008) Mechanical response and multilevel structure of biomimetic hydroxyapatite/polygalacturonic/chitosan nanocomposites. *Materials Science and Engineering C*, *28*, 399–405.

[78] Twu, Y.-K., Chen, Y.-W., and Shih, C.-M. (2008) Preparation of silver nanoparticles using chitosan suspensions. *Powder Technology*, *185*, 251–257.

[79] Morin, A., Dufresne, A. (2002) Nanocomposites of chitin whiskers from *Riftia* tubes and poly(caprolactone). *Macromolecules*, *35*, 2190–2199.

[80] Khunawattanakul, W., Puttipipatkhachorn, S., Rades, T., and Pongjanyakul, T. (2010) Chitosan–magnesium aluminum silicate nanocomposite films: Physicochemical characterization and drug permeability. *International Journal of Pharmaceutics*, *393*, 219–229.

[81] Kadokawa, J., Takegawa, A., Mine, S., and Prasad, K. (2011) Preparation of chitin nanowhiskers using an ionic liquid and their composite materials with poly(vinyl alcohol). *Carbohydrate Polymers*, *84*, 1408–1412.

[82] Mahmood Zia, K., Zuber, M., Barikani, M., Jabbar, A., and Kaleem Khosa, M. (2010) XRD pattern of chitin based polyurethane bio-nanocomposites. *Carbohydrate Polymers*, *80*, 539–543.

[83] Gopalan Nair, K., Dufresne, A. (2003) Crab shell chitin whisker reinforced natural rubber nanocomposites. 2. Mechanical behavior. *Biomacromolecules*, *4*, 666–674.

[84] Chang, P.R., Jian, R., Yu, J., and Ma, X. (2010) Starch-based composites reinforced with novel chitin nanoparticles. *Carbohydrate Polymers*, *80*, 420–425.

[85] Chang, P.R., Jian, R., Yu, J., and Ma, X. (2010) Fabrication and characterisation of chitosan nanoparticles/plasticised-starch composites. *Food Chemistry*, *120*, 736–740.

[86] Sriupayo, J., Supaphol, P., Blackwell, J., and Rujiravanit, R. (2005) Preparation and characterization of α-chitin whisker-reinforced chitosan nanocomposite films with or without heat treatment. *Carbohydrate Polymers*, *62*, 130–136.

[87] Zeng, M., Gao, H., Wu, Y., Fan, L., and Li, A. (2010) Preparation and characterization of nanocomposite films from chitin whisker and waterborne poly(ester-urethane) with or without ultra-sonification treatment. *Journal of Macromolecular Science Part A*, *47*, 867–876.

[88] Shams, M.I., Ifuku, S., Nogi, M., Oku, T., and Yano, H. (2011) Fabrication of optically transparent chitin nanocomposites. *Applied Physics A*, *102*, 325–331.

CHAPTER 6

Mechanical Properties of Chitin-Based Nanocomposites

MERIN SARA THOMAS, LALY A. POTHAN, and SABU THOMAS

6.1 INTRODUCTION

Nature is gifted with several biomaterials that could be obtained from animals and plants. Chitin is one of such abundant biomacromolecules occurring mainly in the exoskeletons of shellfish and insects and are extracted at a rate of 10^{10} to 10^{11} tons per year [1]. Although native chitin is a semicrystalline biopolymer with microfibrillar morphology and excellent material properties, most of the biomass is thrown away as industrial waste (shrimp and crab shells) without effective utilization. Thus, it is important to make efficient use of this biomass resource as a green and environmentally friendly material [2].

Chitin occurs in nature as ordered crystalline microfibrils, forming structural components in animals, algae, and fungi in which chitin acts as a supportive and protective component (Fig. 6.1). In animals, chitin occurs essentially in crustacean, mollusks, and insects, where it is the main constituent of their exoskeleton, associated with organic substances, mainly proteins, and impregnated with inorganic substances, such as calcium salts.

Chitin is a copolymer of *N*-acetylglucosamine and *N*-glucosamine units randomly or block distributed throughout the biopolymer chain, depending on the processing method used to derive the biopolymer (Fig. 6.2). When the number of *N*-acetylglucosamine units is higher than 50%, the biopolymer is termed chitin. Conversely, when the number of *N*-glucosamine units is higher, the term chitosan is used. Chitosan (CS) has been the better researched version of the biopolymer because of its ready solubility in dilute acids [3].

Biopolymer Nanocomposites: Processing, Properties, and Applications, First Edition. Edited by Alain Dufresne, Sabu Thomas, and Laly A. Pothan.

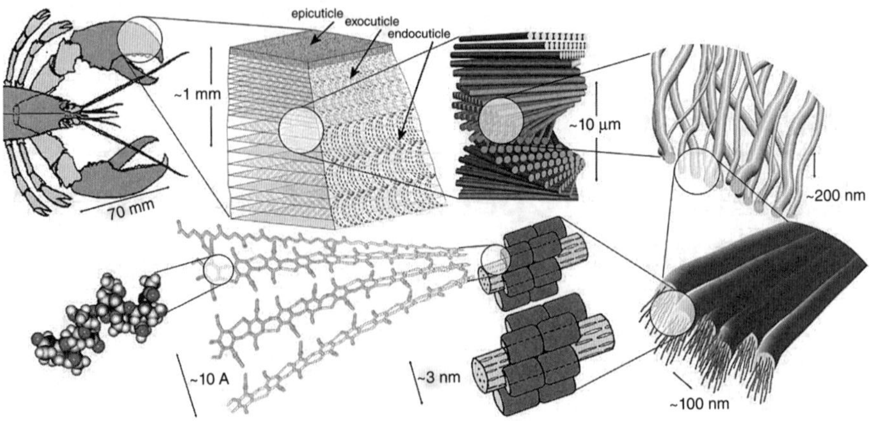

Figure 6.1 Source and hierarchial crustaceans cuticle showing the ordered structural chitin. Reproduced and reprinted with permission from Reference [3].

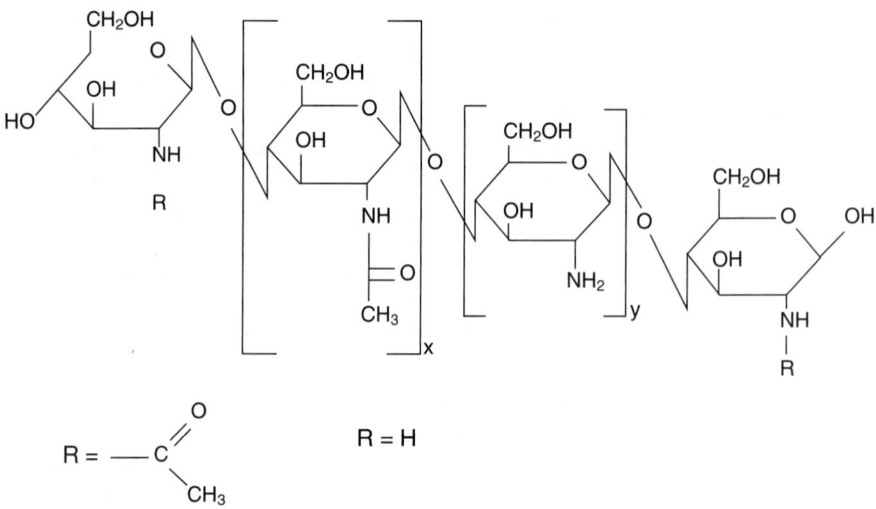

Figure 6.2 Schematic representation of the chitin and chitosan depicting the copolymer character of biopolymers.

Chitin and CS possess unique properties such as biocompatibility, biodegradability, antibacterial properties, and nontoxicity. Thus, they have been extensively investigated for several decades for filtration, food packaging film, artificial skin, bone substitutes, water engineering, and so on [4]. However, regardless of the numerous advantages and exceptional properties of CS, its mechanical properties are not good enough to satisfy a wide range of applications. The formation of organic–inorganic hybrids through incorporation of fillers is an effective approach for improving the physical and mechanical properties of CS. Hence, for reinforcing CS, several reinforcing agents such as hydroxyapatite [5–8], clay [9–11], carbon nanotubes [12–16], and graphene oxide [17, 18] have been used.

6.2 MECHANICAL PROPERTIES OF CHITIN/CHITOSAN NANOCOMPOSITES

6.2.1 Chitosan–Hydroxyapatite Nanocomposites

Hydroxyapatite [$Ca_{10}(PO_4)_6(OH)_2$] (HAp) is one of the most stable forms of calcium phosphate. It occurs in bones as a major component (60–65%), possesses various useful properties such as biocompatibility, bioactivity, and osteoconductivity, and is nontoxic, noninflammatory, and nonimmunogenic. The mechanical properties of HAp can be improved by forming a composite of polymer layer on its surface. In recent years, considerable attention has been given to CS–HAp composite materials and their applications in the field of bone tissue engineering due to their minimal foreign body reactions, antibacterial nature, biocompatibility, biodegradability, and the ability to be molded into various geometries and forms such as porous structures, suitable for cell ingrowth, and osteoconduction. The composite of CS including HAp is very popular because of their biodegradability and biocompatibility in nature [19].The mechanical properties of the CS-HAp composites play a significant role in bone tissue engineering [20]. The intermolecular hydrogen bond and chelate interaction between the CS and HAp contribute to good mechanical properties. There is a possible interaction between the NH_2 group and primary and secondary –OH group of CS with Ca2+ (metal coordination interaction) of HAp (Fig. 6.3). This interaction might be responsible for the

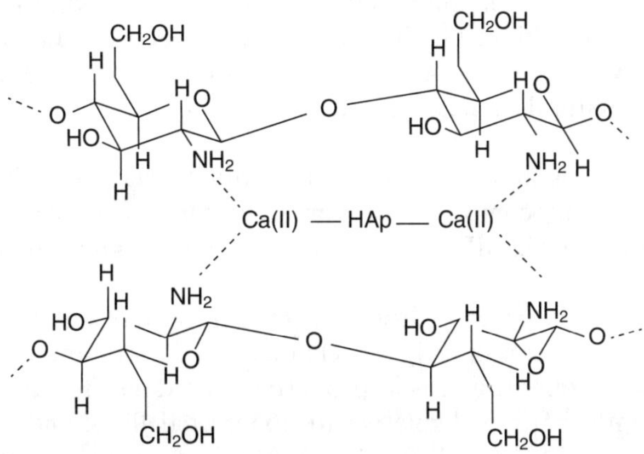

Figure 6.3 Chemical interaction between chitosan –HAp.

higher mechanical strength of the composite scaffolds as compared to CS and HAp alone. Compressive strength has been a widely used parameter to measure the mechanical strength of porous scaffolds [19, 20].

Li et al. [5] compared the compressive strengths of different composite ratios of CS/HAp and found the maximum compressive strength to be 119.86 MPa (30:70 CS/HAp ratio). Increasing the HAp ratio leads to an increase in the compressive strength. The incorporation of CS into the HAp matrix via blending methods would result in the decrease of mechanical properties of the composite material due to the weaker interfacial bonding between CS and the HAp matrix, and as a result the compressive strength decreases to 47.8 MPa [21, 22]. The molecular weight of CS also contributes significantly to the mechanical properties of the composite material. In general, high molecular weight CS scaffolds have a higher compression modulus than medium molecular weight CS. The compression moduli of high molecular weight CS and CS/nanoHAp (nHAp) (1%) nanocomposite three-dimensional scaffold were found to be 6.0 ± 0.3 kPa and 9.2 ± 0.2 kPa, respectively [22]. The bending strength and modulus of the composites were 86 MPa and 3.4 GPa, respectively. Due to these good mechanical properties, CS/nHAp composite can be used as a potential scaffold material for bone regeneration.

The mechanical properties of CS composite scaffold also depend on the temperature and they increase with an increase in the temperature. A possible justification to this may be that with an increase in the temperature, the interfacial bonding between CS and HAp increases [23]. The water content of the scaffold may also have a major role to play in mechanical strength. In an interesting study, it was found that the nHAp/CS/carboxymethyl cellulose (40/30/30% w/v) has the highest mechanical property of 40 MPa in dry and 12 MPa in wet state [24].

Cai et al. [6] synthesized homogeneous nanocomposites composed of HAp and CS in the presence of polylactic acid. The mechanical properties of the composites were evaluated by measuring their compressive strength and elastic modulus. Significant improvements in the elastic modulus and compressive strength were achieved by incorporating polylactic acid and CS biopolymer together with HAp. The CS–PLA/HAp composites may even match the properties of natural bones and therefore may be suitable for surgical applications.

Young's modulus and ultimate tensile strength (UTS) of formaldehyde-treated CS/HAp nanocomposite film containing 66 wt% (47 vol%) HAp reach up to 17.3 GPa and 222 MPa, respectively, which are significantly higher than those of pure CS [7].

Pradal et al. [8] prepared hydrophobic modification of CS/HAp composites (by grafting n-alkyl acids on CS), and characterized for their tensile properties under wet conditions. The tensile properties of the grafted composite films were significantly improved relative to nongrafted films, and a modulus of 393 ± 68 MPa and a UTS of 18.7 ± 1.5 MPa were reached for composite films grafted for 1 day. This study provides a simple method to improve the wet tensile properties of CS/HAp composite films, making them more suitable for biomedical applications. One way to improve the mechanical properties of CS/HAp biocomposites under wet conditions is to reduce the swelling of these nanocomposites, perhaps through hydrophobic modification of CS.

In the case of CS/HAp composites, the improvement in mechanical properties is due to metal coordination interaction. The addition of HAp can reduce the water absorption, which postponed the retention of mechanical properties of CS/HAp composite under moisture conditions.

6.2.2 Chitosan–Clay Nanocomposites

In recent years, it has been observed that the addition of just a small quantity of nanosized layered silicates greatly improved the properties of virgin polymers without affecting their processability. Layered silicate clays have been found to be effective in reinforcing fillers for polymers due to their lamellar structure and high surface area of about 750 m^2/g. Figure 6.4 shows the different

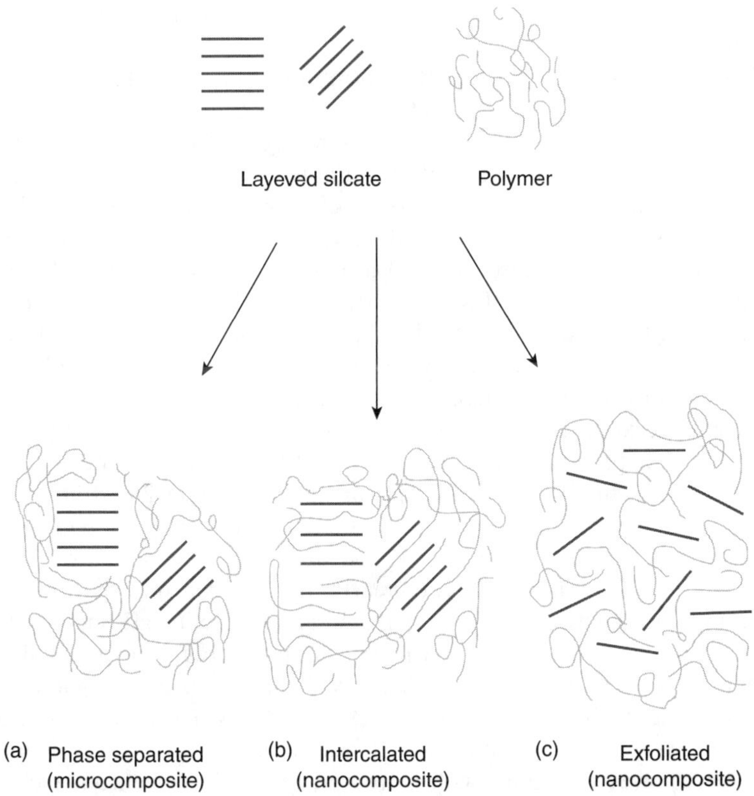

Figure 6.4 Illustration of different types of composite that can arise from the interaction between layered silicates and polymers. (a) Phase-separated composite (microcomposite). (b) Intercalated composite (nanocomposite). (c) Exfoliated composite (nanocomposite). Reproduced and reprinted with permission from Reference [25].

types of composites that can arise from the interaction between layered silicates and polymers.

Chitosan can interact with nanoclays via hydrogen bonding and/or electrostatic attraction interactions due to the presence of both amino and hydroxyl groups [26, 27]. As a result, one can blend CS with nanoclays in aqueous solutions for well-dispersed CS/clay nanocomposites. Several efforts have been made to prepare CS/nanoclay composites with improved mechanical and thermal properties [28–31]. Increased interfacial compatibility via hydrogen bonding and/or electrostatic attraction between CS and clays is responsible for the enhanced performance of the nanocomposites [32, 33].

Polymer–clay nanocomposites have become a subject of intensive research because of their ability for nanoscale dispersion, which brings significant improvement in mechanical and physical properties compared with microscale polymer composites. The complete dispersion of clay nanolayers in a polymer optimizes the number of available reinforcing elements that carry an applied load and deflect the evolving cracks. The coupling between the large surface area of the clay and the polymer matrix facilitates the stress transfer to the reinforcing phase, allowing for the improvement of the tensile stress and toughness.

Abdollahi et al. [11] incorporated montmorillonite (MMT) nanoclay and rosemary essential oil (REO) into CS film. The tensile strength (TS) of chitosan film increased significantly by incorporating MMT into chitosan up to 3 wt%, but adding more MMT could not improve it. This improvement in the TS of nanocomposite in a low amount of MMT may be attributed to the uniform dispersion of MMT in the CS matrix and a strong interaction between CS and MMT. The decrease in TS at a high concentration of MMT may be caused by the aggregation of MMT particles with high surface energies [34].

Zia et al. [10] studied the mechanical behavior of chitin–bentonite clay-based polyurethane bionanocomposites. The UTS of chitin-based polyurethane bionanocomposites increased with increasing Delite HPS® (Laviosa Chemica Mineraria, Italy) bentonite clay content. The prepared bionanocomposites show a plateau behavior of up to 4% bentonite nanoclay content (PUBNC4), yielding a constant UTS of approximately 35 MPa. However, there is tremendous decrease in the tensile strength using 8% nanoclay contents (PUBNC5). This behavior may be due to the incompatibility of the higher contents of nanoclay. It can clearly be observed that there is a linear reduction of elongation at break with increasing Delite HPSbentonite clay contents. This could be attributed to the larger aggregations of Delite HPS bentonite clay contents, as well as the formation of more aggregated centers with increased Delite HPS bentonite clay contents. These increased aggregated centers would probably act as additional stress centers leading to increased chances of failure with increased Delite HPS bentonite clay contents. The higher values in elongation at break could be attributed to the surface characteristics of the lower Delite HPS bentonite clay particles, which cause a better dispersion of these particles in the matrix, ultimately strengthen-

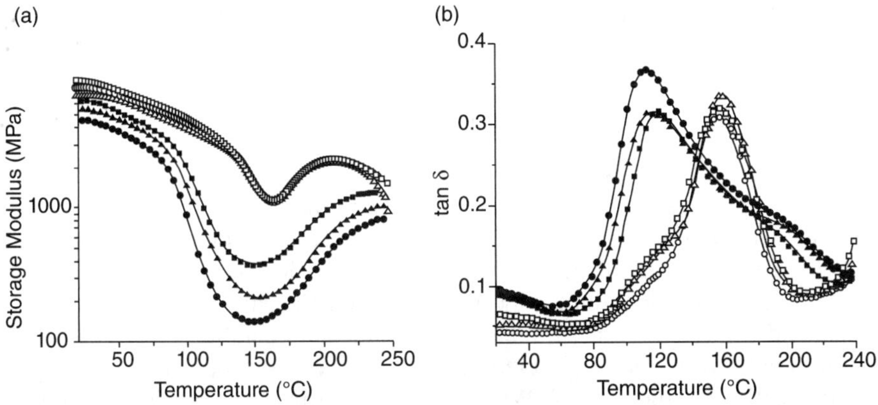

Figure 6.5 DMA curves including storage modulus (a) and tan δ (b) as a function of temperature for chitosan and its nanocomposites. (○) CS, (△) CS/3MMT and (□) CS/10MMT, (●) CS/GLY, (▲) CS/3MMT/GLY and (■) CS/10MMT/GLY. Reproduced and reprinted with permission from Reference [9].

ing the interfacial interactions between the dispersed phase and the matrix phase.

Lavorgna et al. [9] studied the combined effect of both clay and glycerol plasticizer on the properties of CS films (Fig. 6.5). Dynamic mechanical analysis (DMA), tensile, and nanoindentation tests show that, as far as the effect of clay loading is concerned, the mechanical properties of nanocomposite films without glycerol are not significantly affected by the presence of sodium–montmorillonite (Na–MMT) stacks. On the contrary, for nanocomposites containing glycerol, the mechanical properties are improved as clay loading increases. This is due to the combined effect of clays and plasticizer. Glycerol modifies the hydrogen-bonding network within the material and allows for better interaction between the filler and the matrix, thus facilitating the stress transfer to the reinforcement phase and improving its mechanical properties.

The effects of sodium montmorillonite (Na–MMT) and organoclaymontmorillonite (ODA-MMT) on mechanical and thermal properties and surface morphology of poly(caprolactone) (PCL)/CS composites were studied by Abdolmohammadi et al. [35]. The tensile modulus increased by adding ODA-MMT and Na–MMT into the composite. Elongation at break decreases for both micro- and nanocomposites. The TS increases by loading up to 3wt% of ODA-MMT, and then decreases with the increase of the clay content, the effect of good dispersion of organoclay into the matrix and intercalation of organoclay particles by PCL/CS at a low filler content. However, increasing Na–MMT into the composite gradually decreases the TS.

Chitosan/halloysite nanotubes (HNT) bionanocomposites were prepared [36] by Liu et al. The tensile properties of CS were enhanced significantly by

HNTs. The storage modulus and glass transition temperature of CS/HNTs films were also increased significantly with HNTs.

6.2.3 Chitosan–Carbon Nanotube Nanocomposites

Carbon nanotubes (CNTs) are long cylinders of covalently bonded carbon atoms that possess extraordinary electronic and mechanical properties. CNTs exhibit excellent mechanical properties with Young's modulus as high as 1.2 TPa and TS of 50–200 GPa. The combination of these exceptional mechanical properties, along with the nanometer size, low density, high aspect ratio, and high surface area, makes CNTs an ideal candidate for reinforcement in composite materials. Generally, there are two types of CNTs: single-walled carbon nanotubes (SWCNTs) and multiwalled carbon nanotubes (MWCNTs). However, the difficulty in separating SWCNTs from bundles into individual nanotubes is a big challenge to the use of SWCNT in real applications. CNT-reinforced nanocomposites can be considered as a kind of particulate composites with filler dimensions on the nanometer scale and high aspect ratios. Unlike the microscopic particulate composites, the mechanical properties of nanocomposites depend strongly on the dispersion state of nanofillers, apart from the properties of fillers and matrix material. In addition to dispersion, there are other major requirements for effective reinforcement of CNTs in composites [11]: they include a high aspect ratio and alignment and interfacial interactions between CNTs and the polymer matrix. The aspect ratio must be sufficiently large to maximize the load transfer between the CNTs and matrix material and thus to achieve enhanced mechanical properties.

Raw CNTs are mainly hydrophobic and poorly miscible with water. Acid-treated CNTs contain many defects and hydrophilic groups, such as –OH and –COOH, which help improve the solubility of CNTs in water. Chitosan possesses three kinds of functional groups, namely. amino, primary, and secondary hydroxyl groups in a glucosamine unit, and the functionalized CNTs contain carboxylic and hydroxyl groups. There is great possibility that strong hydrogen bonds may form between CS and the CNTs (Fig. 6.6). Compatibility and strong interaction between MWCNTs fillers and the matrix greatly enhances the dispersion as well as the interfacial adhesion, thus significantly increasing the mechanical properties of the matrix [12]. Covalent functionalization of CS and CNT improves the interaction between CS with CNTs, reduces the damage of CNTs, and increases their mechanical properties.

Incorporation of super strong light weight CNT structures into the CS matrix offers a novel approach to the design of high performance composite materials with superior mechanical properties. However, the atomically smooth graphene surface of nanotubes can provide only limited load transfer from the matrix to nanotubes across the nanotube–polymer interface because of weak van der Waals interfacial bonding through covalent grafting of CNTs to the CS to form the CNT–CS nanocomposites [13].

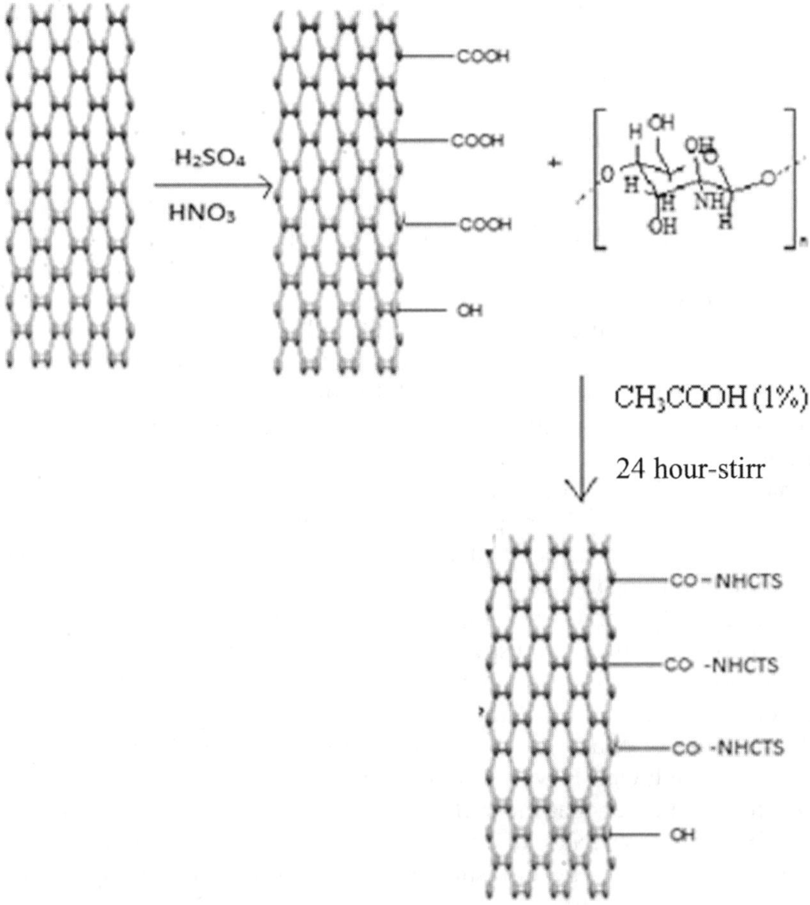

Figure 6.6 Chemical interactions between chitosan and CNT.

Wang et al. [12] reported homogeneous dispersion of CNTs in matrix. When 0.8% of CNTs was introduced in the CS matrix, the mechanical properties, including the tensile modulus and strength of the nanocomposite, were greatly improved by about 93–99%. A small addition of CNTs significantly improves the tensile properties of the CS matrix, and the mechanical properties increase with the increase of CNT loading (0.4 wt% of CNTs filler), whereas the tensile modulus and strength of the nanocomposite increases dramatically by about 78–94%. Due to the aggregation of CNTs within the CS matrix at higher concentrations of CNTs, the tensile modulus only increases slightly, while the TS remains stable.

In another study, the tenacity of neat CS fiber was recorded as 96 MPa, whereas that of CS reinforced with purified but nonfunctionalized single-walled carbon nanotubes (p-SWCNTs) increased from 132 MPa for 0.01 wt% SWCNT addition to 180 MPa for 0.4 wt% SWCNT addition. Even at a very low SWCNT concentration (0.01 wt%), a significant increase in tenacity was observed, but at higher concentrations the increase in tenacity was not proportional. On the other hand, on incorporation of functionalized carbon nanotubes (f-CNTs), the tenacity increased from 136 MPa on addition of 0.01 wt% SWCNT to 226 MPa on addition of 0.4 wt% SWCNT. This continuous improvement in tenacity (even at higher SWCNT concentrations) may be ascribed to better dispersion as well as interaction of functionalized nanotubes in CS [14]. This might be due to the strong chemical interaction of the COOH group of f-CNTs with the amine group of CS matrix [15].

Fan et al. [16] used unzipped multiwalled carbon nanotube oxides (UMCNOs) as mechanical reinforcement materials for CS. With the increase of the available interface areas of MWNTs after unzipping, the UMCNOs are more efficient for enhancing the mechanical property of the CS matrix than the pristine MWCNTs. With the addition of only 0.2 wt% UMCNOs into the CS matrix, the TS and Young's modulus rise from 69.3 to 142.7 MPa, 2.6 to 6.9 Gpa, respectively. The relatively small amount of UMCNOs results in a significant improvement of mechanical properties in the UMCNOs/CS nanocomposite film (Fig. 6.7).

He et al. [37] fabricated composites composed of CS and MWCNTs by milling and ultrasonication dispersion methods. A comparative study of these two composite films were done. From dynamic mechanical analysis (DMA), the storage modulus (E′) of the composites was always higher than that of pure CS and, especially within the temperature range of 35°C–100°C, the composites display an E′ value, which was almost twice higher than that for pure CS. This implies that the stiffness of CS was improved significantly with the addition of MWCNTs. In comparison with the film prepared by milling, the film prepared by the ultrasonic method had a higher storage modulus in the lower temperature range, but the value become lower once the temperature was beyond ~100°C as the E′ dropped precipitously. It should be noted that the upturn in the E′ for all samples when the temperature was beyond about 180°C is due to the onset of degradation (Fig. 6.8).

Ong et al. reported green nanocomposite membranes consisting of poly(3-hydroxybutyrate) (PHB)–functionalized MWCNTs into a CS matrix. The presence of PHB–MWCNTs greatly enhances the TS of the CS matrix by 41.8% and increases Young's modulus by 24.1%. However, a decrease in the flexibility of these membranes was observed. These superior mechanical properties of the PHB–MWCNT/CS nanocomposite membrane can be attributed to the good dispersion of the PHB-MWCNT in the CS matrix and their strong compatibility with the CS matrix. Additionally, the bulk alignment of the PHB–MWCNTs plays a role in enhancing the mechanical properties of the nanocomposite membrane. The addition of 0.1 wt% bulk-aligned PHB–MWCNTs to the CS matrix significantly enhances the TS of the PHB–MWCNT/CS nanocomposite membrane [38].

Wu et al. [39] prepared CS nanocomposite films reinforced by poly(3,4-ethylenedioxythiophene) (PEDOT)–poly(styrenesulfonate) (PSS)-treated

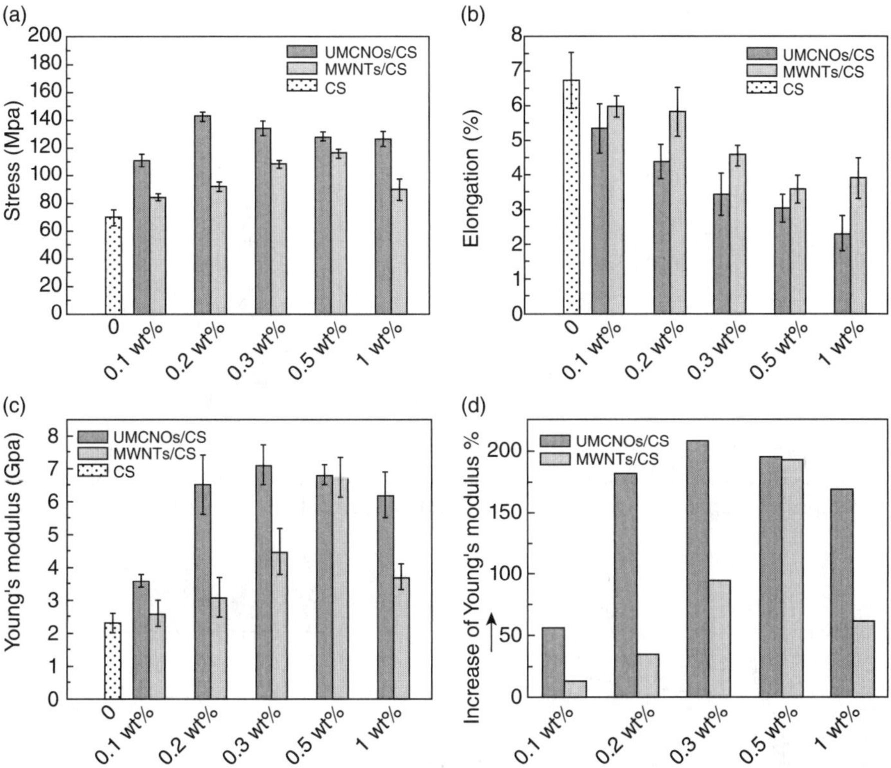

Figure 6.7 Averaged results for the ultimate tensile strength for the baseline CS and nanocomposites with various nanofiller weight fractions; (b) Percent elongation at failure for the baseline CS and nanocomposites with various nanofiller weight fractions; (c) Absolute value in Young's modulus of the CS and nanocomposites with various nanofiller weight fractions; (d) Percentage changes of the Young's modulus of the CS and nanocomposites with various nanofiller weight fractions. Reproduced and reprinted with permission from Reference [16].

carbon nanotubes. The addition of MWCNTs significantly improved the mechanical properties of CS. When compared with the neat CS, the mechanical properties, such as TS and modulus, of the nanocomposites were greatly improved by about 61% and 34%, respectively, with incorporation of only 0.5 wt% of MWCNTs. The reinforcement caused by PEDOT–PSS/MWCNTs was higher than that by pristine MWCNTs at lower contents of MWCNTs. The hardness of the nanocomposite with 0.5 wt% MWCNTs increased by about 31%, as compared with those of neat CS (Fig. 6.9).

Shieh et al. [14] reported the improvement in the mechanical properties of CS by incorporating CS-grafted carbon nanotubes (CNTs-g-CS). Significant improvement in the storage modulus. The storage modulus was significantly

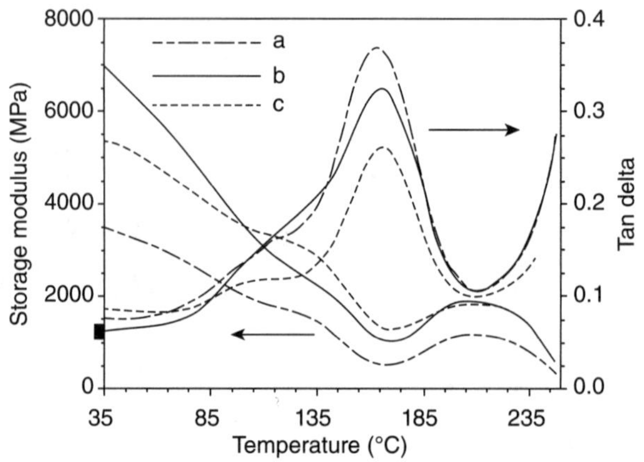

Figure 6.8 Storage modulus (E′) and tan δ curves of (a) pure chitosan, (b) chitosan/1% MWCNTs composite prepared by ultrasonication, and (c) chitosan/1%MWCNT scomposite prepared by milling. Reproduced and reprinted with permission from [37].

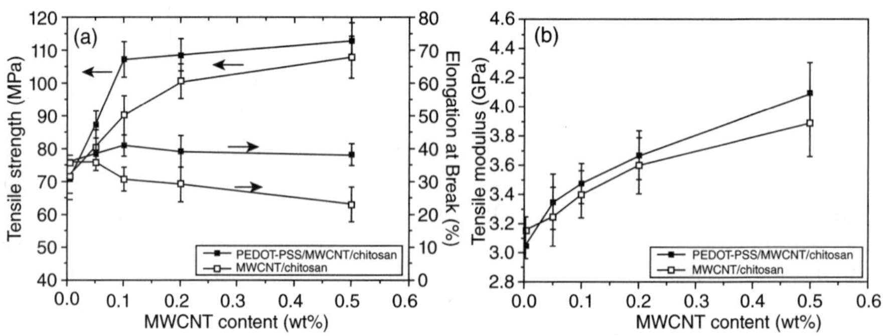

Figure 6.9 Tensile strength and elongation at break and (b) tensile modulus for chitosan nanocomposites as a function of MWCNT content. Reproduced and reprinted with permission from Reference [39].

up by 134% from 6.4 GPa for the pure CS to 15 GPa for the CS nanocomposite containing 40 wt% CNTs-g-CS (Fig. 6.10).

6.2.4 Chitosan–Graphene Oxide Nanocomposites

Graphene is an attractive new type of carbon material with a honeycomb and one-atom-thick structure. Interest in graphene is largely attributed to its excellent mechanical and physicochemical properties [40]. Recent studies have indicated that graphene oxide (GO), the oxidized form of graphene, proves to

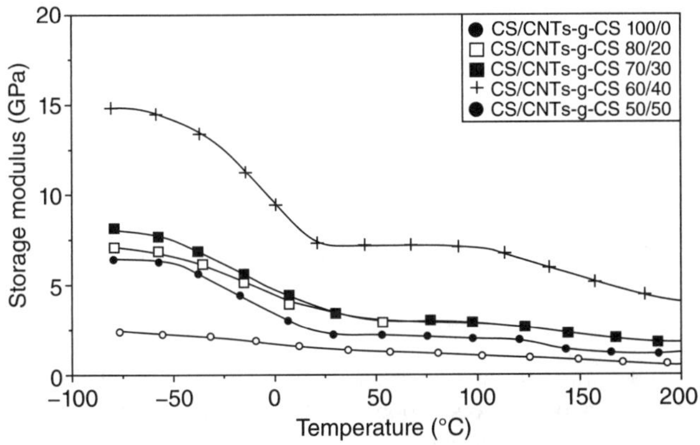

Figure 6.10 Storage moduli of the CS/CNTs-g-CS nanocomposites as a function of temperature Reproduced and reprinted with permission from Reference [14].

be a promising material for adsorbing dyes and supporting catalysts due to its extraordinary mechanical strength and relatively large specific area [41–43].

One of the major advantages with GO is that it is hydrophilic with a very high negative charge density arising from the oxygen-containing functional groups. In solution phase, GO exists as a single layer. GO can act as weak acid cation exchange resin because of the ionizable carboxyl groups, which allow ion exchange with metal cations or positively charged organic molecules [42]. The intrinsic van der Waals interactions between layers of graphene easily results in agglomeration, which leads to insolubility. The occurrence of such agglomerations inevitably lowers the reinforcement effectiveness of nanofillers because the maximal load transfer and lowest filler loading can only be attained when the nanofillers are dispersed on the molecular scale in the polymer matrix. If the nanofillers in the polymer matrix could be dispersed on a molecular scale and interacted with the matrix by chemical bonding or strong intermolecular forces, significant improvements in the mechanical properties of the material or unexpected new properties might be achieved. These chemical functionalizations have been found to be a feasible and effective way for improving the dispersion of graphene materials and interfacial bonding between the graphene and the matrix [43, 44] (Fig. 6.11).

Yang et al. studied the mechanical properties of well-dispersed CS–GO nanocomposites. Compared with the pure CS, the TS, and Young's modulus of the graphene-based materials are significantly improved by about 122% and 64%, respectively, with incorporation of 1 wt% GO. At the same time, the elongation at the break point increases remarkably [17]. In another study, the mechanical properties of composite films were tested by nanoindentation method. With the addition of a small amount of graphene in CS (0.1–0.3 wt%), the elastic modulus of CS increased over ~200%.

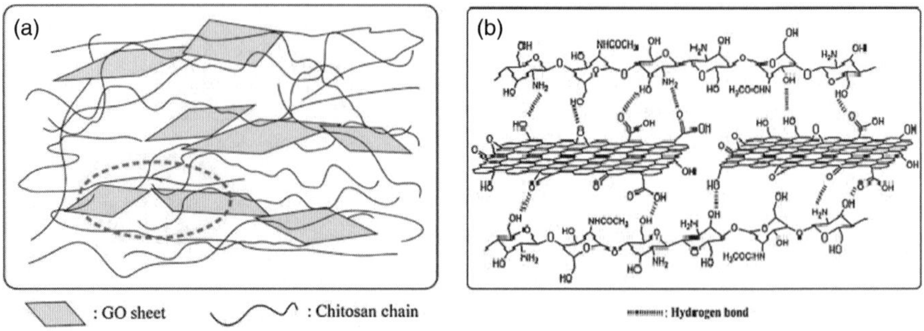

Figure 6.11 (a) Scheme for possible microstructure present in GO/chitosan nanocomposites; (b) formation of hydrogen bonds between GO sheets and chitosan chains. Reproduced and reprinted with permission from Reference [45].

Pan et al. [45] presented green fabrication of CS films reinforced with parallel aligned graphene oxide. The nanocomposite films containing 1 wt% GO sheets are strong and ductile. Young's modulus, TS, and elongation at break are found to increase by 51% and 93% and 41%, respectively, which are much higher than those of neat CS films. Homogeneous dispersion of GO sheets in the CS matrix and strong interfacial adhesion between them significantly enhanced the mechanical properties of nanocomposites.

6.2.5 Chitin/Chitosan as Nanofiller in PolymerNanocomposites

Junkasem et al. [46] fabricated α-chitin whisker (ChW)-reinforced poly(vinyl alcohol) (PVA) nanocompositenanofibers by electrospinning. The TS of the as-spun PVA/chitin whisker nanocomposite fiber mats increased from that of the neat as-spun fiber mat (i.e., 4.3 ± 0.7 MPa) with initial addition of the chitin whiskers to reach a maximum value (i.e., 5.7 ± 0.6 MPa) at the chitin whisker to PVA ratio of about 5.1% and decreased with further increasing the whisker content. The presence of ChWs within the as-spun nanocompositefiber mats increased Young's modulus by about 4–8 times over that of the neat as-spun fiber mat [46].

ChW-reinforced nanocomposites films were prepared by Sriupayo et al. [47]. The TS of α-ChW-reinforced CS films increased from that of the pure CS film with initial increase in the whisker content to reach a maximum value at the whisker content of 2.96 wt% and decreased gradually with further increase in the whisker content, while the percentage of elongation at break decreased from that of the pure CS with initial increase in the whisker content, and leveled off when the whisker content was greater than or equal to 2.96 wt% (Fig. 6.12). The increase in the TS of the nanocomposite films with increasing α-ChW content could be due to the interaction between CS molecules and

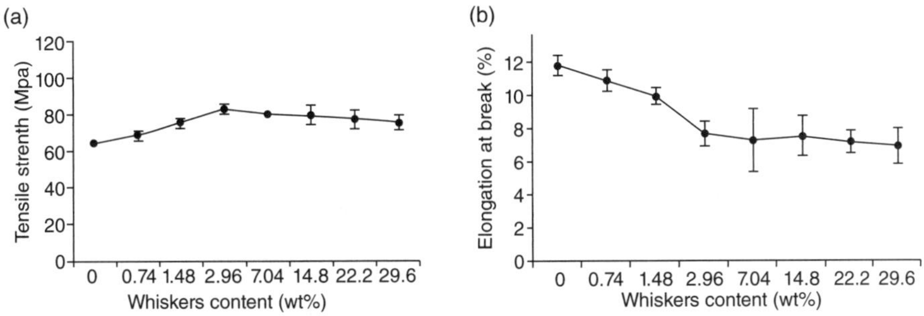

Figure 6.12 (a) Tensile strength and (b) percentage of elongation at break of pure chitosan and α-chitin whisker-reinforced chitosan films as a function of whisker content. Reproduced and reprinted with permission from Reference [47].

α-ChWs via hydrogen bonding. Such interaction, however, caused the nanocomposite films to be more rigid as the whisker content increased and, as a result, the percentage of elongation at break decreased [47].

A similar behavior was observed for soy protein isolate films reinforced with α-ChWs from crab shells, in which the TS of the films initially increased with increasing whisker content, and leveled off when the whisker content was about 20 wt% or greater, while the percentage of elongation at break decreased steadily with increasing whisker content [48].

Uddin et al. [49] presented the outstanding reinforcing effect of highly oriented ChWs in PVA. Compared with neat PVA, significant improvements in TS, initial modulus, and toughness were observed for the composite fibers (Fig. 6.13). The highest TS (1.88 GPa) and toughness (68 J/g) were obtained for the PVA–ChW 5% composites. It implies that the 5% ChWs loading is the optimum loading in the context of TS and toughness. Young's modulus increased almost proportionally with the ChW content. PVA–ChW 30% fiber showed a Young's modulus of 50 GPa, much higher than that of neat PVA, 28 GPa.

The dynamic mechanical properties of these composites have been strongly depend on the filler orientation Figure 6.14a,b show, respectively the storage modulus (E′), and the E′ values at glassy (–100°C) and rubbery (100°C) regions of neat PVA and composite fibers. The significant enhancement of E′ in both glassy and rubbery regions were observed after incorporating ChWs. For PVA–ChW 30% fiber, the E′ at –100°C(116 GPa) was 38% higher than that of neat PVA (84 GPa). For the same composite, the E′ at 100°C (27 GPa) was 125% higher than that of neat PVA fiber (12 GPa).

CS nanoparticles/plasticized-starch composites were fabricated by Chang et al. [50]. With increasing CS content, the TS of the composites increased, but the elongation at break decreased (Fig. 6.14). When the CS content varied

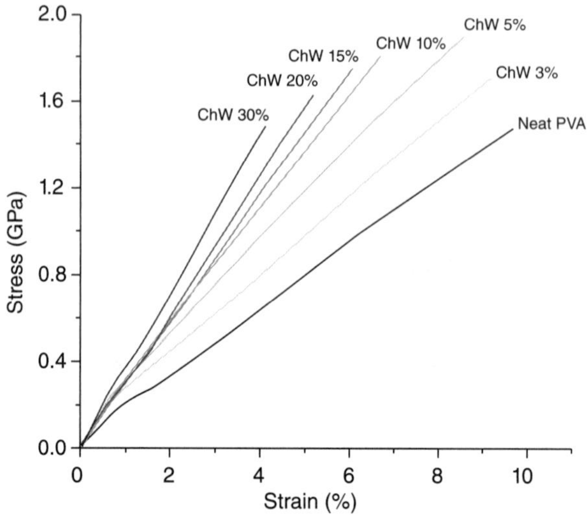

Figure 6.13 Representative stress–strain curves of neat PVA and PVA–ChW fibers. Reproduced and reprinted with permission from Reference [49].

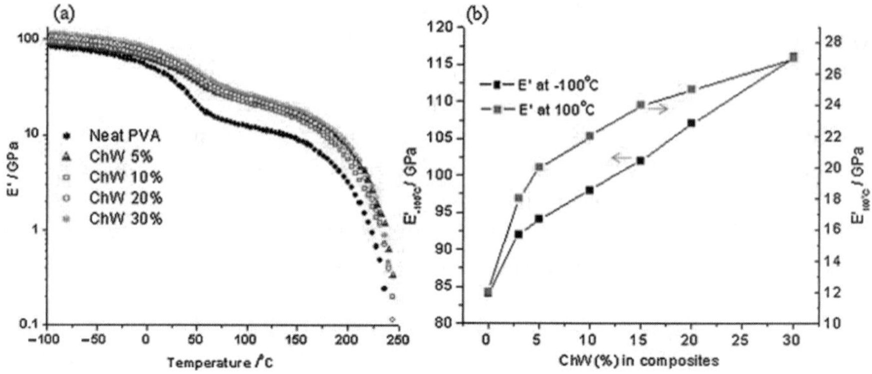

Figure 6.14 (a) Storage modulus (E′), and (b) E′ at −100°C (left) and 100°C (right) of the neat PVA and PVA–ChW fibers. Reproduced and reprinted with permission from Reference [49].

from 0 to 6 wt%, the TS increased from 2.84 to 10.80 MPa, while the elongation at break decreased from 59.3% to 22.7%. This may be ascribed to the interfacial interaction between chitosan nanoparticles (CN) and the glycerol plasticized starch (GPS) matrix because of the similar polysaccharide structures of CS and starch. CN could serve as a junction, which improved the TS, but decreased the flexibility of starch molecules. When more CS (8 wt%) was added to the matrix, the TS of the composites decreased to 8.9 MPa, which may be ascribed to the agglomeration of CS.

The storage modulus of GPS-CS composites increased with the introduction of CS particles. This improvement was likely due to the interaction between the CS filler and the GPS matrix. In GPS-CS composites, the glass transition shifted to a higher temperature, indicating that the GPS matrix could form composites with CS. Functioning like a junction, CS improved the intermolecular interaction of GPS in the GPS matrix, bringing adjacent chains of starch closer, and reducing the free volume, thereby raising the Tg of composites as the CS content increased.

6.3 CONCLUSION

There has been an increased interest in developing new strategies of obtaining composite materials with desired properties for various applications. Chitin and CS can be used as a natural polymer matrix as well as a nanofiller for the fabrication of bionanocomposites because of their natural origin, biocompatibility, biodegradability, and bioactivity. These features make it a suitable candidate for the fabrication of new generation of biomaterials. Regardless of these numerous advantages and exceptional properties of CS, its mechanical properties are not good enough to satisfy a wide range of applications. In order to improve the mechanical properties of composites, reinforcing processes might be employed. The mechanical properties of the composite depend not only on the nature of filler but also on the interaction between matrix and filler, temperature, and filler content.

REFERENCES

[1] Nair, K.G., Dufresne, A. (2003) Crab shell chitin whisker reinforced natural rubber nanocomposites. 1. processing and swelling behavior. *Biomacromolecules*, *4*, 657–665.

[2] Ifuku, S., Nogi, M., Abe, K., Yoshioka, M., Morimoto, M., Saimoto, H., and Yano, H. (2009) Preparation of chitin nanofibers with a uniform width as α-chitin from crab shells. *Biomacromolecules*, *10*, 1584–1588.

[3] Raabe, D., Sachs, C., and Romano, P. (2005) The crustacean exoskeleton as an example of a structurally and mechanically graded biological nanocomposite material. *Acta Mater.*, *53*, 4281–4292.

[4] Kumar, M.N.V.R. (2000) A review of chitin and chitosan applications. *React. Func. Polym.*, *46*, 1–27.

[5] Li, Z., Yubao, L., Aiping, Y., Xuelin, P., Xuejiang, W., and Xiang, Z. (2005) Preparation and *in vitro* investigation of chitosan/nano-hydroxyapatite composite used as bone substitute materials. *J. Mater. Sci. Mater. Med.*, *16*, 213–219.

[6] Cai, X., Tong, H., Shen, X., Chen, W., Yan, J., and Hu, J. (2009) Preparation and characterization of homogeneous chitosan–polylactic acid/hydroxyapatite

nanocomposite for bone tissue engineering and evaluation of its mechanical properties. *Acta Biomater.*, *5*, 2693–2703.

[7] Kithva, P., Grndahl, L., Martin, D., and Trau, M. (2010) Biomimetic synthesis and tensile properties of nanostructured high volume fraction hydroxyapatite and chitosan biocomposite films. *J. Mater. Chem.*, *20*, 381–389.

[8] Pradal, C., Kithva, P., Martin, D., Traub, M., and Grøndahl, L. (2011) Improvement of the wet tensile properties of nanostructured hydroxyapatite and chitosan biocomposite films through hydrophobic modification. *J. Mater. Chem.*, *21*, 2330–2337.

[9] Lavorgna, M., Piscitelli, F., Mangiacapra, P., and Buonocore, G.G. (2010) Study of the combined effect of both clay and glycerol plasticizer on the properties of chitosan films. *Carbohydr. Polym.*, *82*, 291–298.

[10] Zia, K.M., Zuber, M., Barikani, M., Hussain, R., Jamil, T., and Anjum, S. (2011) Cytotoxicity and mechanical behavior of chitin–bentonite clay based polyurethane bio-nanocomposites. *Int. J. Biol. Macromol.*, *49*, 1131–1136.

[11] Abdollahi, M., Rezaei, M., and Farzi, G. (2012) A novel active bionanocomposite film incorporating rosemary essential oil and nanoclay into chitosan. *J. Food Eng.*, *111*, 343–350.

[12] Wang, S.F., Shen, L., Zhang, W.D., and Tong, Y.J. (2005) Preparation and mechanical properties of chitosan/carbon nanotubes composites. *Biomacromolecules*, *6*, 3067–3072.

[13] Carson, L., Kelly-Brown, C., Stewart, M., Oki, A., Regisford, G., and Luo, Z. (2009) Bakhmutov VI. Synthesis and characterization of chitosan–carbon nanotube composites. *Mater. Lett.*, *63*, 617–620.

[14] Shieh, Y.T., Yang, Y.F. (2006) Significant improvements in mechanical property and water stability of chitosan by carbon nanotubes. *Eur. Polym. J.*, *42*, 3162–3170.

[15] Ozarkar, S., Jassal, M., and Agrawal, A. (2008) pH and electrical actuation of single walled carbon nanotube/chitosan composite fibers. *Smart Mater. Struct.*, *17*, 055016.

[16] Fan, J., Shi, Z., Ge, Y., Wang, Y., Wang, J., and Yin, J. (2012) Mechanical reinforcement of chitosan using unzipped multiwalled carbon nanotube oxides. *Polymer*, *53*, 657–664.

[17] Yang, X., Tu, Y., Li, L., Shang, S., and Tao, X. (2010) Well-dispersed chi. *Appl. Mater. Interfaces*, *2*, 1707–1713.

[18] Du, X., Yu, Z.Z., Dasari, A., Ma, J., Mo, M., Meng, Y., and Mai, Y.W. (2008) New method to prepare graphite nanocomposites. *Chem. Mater.*, *20*, 2066–2068.

[19] Venkatesan, J., Kim, S.-K. (2010) Chitosan composites for bone tissue engineering—An overview. *Mar. Drugs*, *8*, 2252–2266.

[20] Kikuchi, M., Ikoma, T., Itoh, S., Matsumoto, H., Koyama, Y., Takakuda, K., Shinomiya, K., and Tanaka, J. (2004) Biomimetic synthesis of bone like nanocomposites using the self-organization mechanism of hydroxyapatite and collagen. *Compos. Sci. Technol.*, *64*, 819–825.

[21] Li, B., Hu, Q., Qian, X., Fang, Z., and Shen, J. (2002) Bioabsorbable chitosan/hydroxyapatite composite rod for internal fixation of bone fracture prepared by *in situ* precipitation. *Acta Polym. Sin.*, 828–833.

[22] Thein-Han, W., Misra, R. (2009) Biomimetic chitosan–nanohydroxyapatite composite scaffolds for bone tissue engineering. *Acta Biomater.*, *5*, 1182–1197.

[23] Yamaguchi, I., Tokuchi, K., Fukuzaki, H., Koyama, Y., Takakuda, K., Monma, H., and Tanaka, J. (2001) Preparation and microstructure analysis of chitosan/hydroxyapatite nanocomposites. *J. Biomed. Mater. Res.*, *55*, 20–27.

[24] Liuyun, J., Yubao, L., and Chengdong, X. (2009) A novel composite membrane of chitosan-carboxymethyl cellulose polyelectrolyte complex membrane filled with nano-hydroxyapatite I. Preparation and properties. *J. Mater. Sci. Mater. Med.*, *20*, 1645–1652.

[25] Alexandre, M., Dubois, P. (2000) Polymer-layered silicate nanocomposites: Preparation, properties and uses of a new class of materials. *Mater. Sci. Eng.*, *28*, 1–63.

[26] Wang, X., Tang, Y., Li, Y., Zhu, Z., and Du, Y. (2010) The rheological behaviour and drug-delivery property of chitosan/rectoritenanocomposites. *J. Biomater. Sci. Polym. Ed.*, *21*, 171–184.

[27] Darder, M., Colilla, M., and Ruiz-Hitzky, E. (2005) Chitosan–clay nanocomposites: Application as electrochemical sensors. *Appl. Clay Sci.*, *28*, 199–208.

[28] Katti, K.S., Katti, D.R., and Dash, R. (2008) Synthesis and characterization of a novel chitosan/montmorillonite/hydroxyapatite nanocomposite for bone tissue engineering. *Biomed. Mater.*, *3*, 034122.

[29] Chang, M.Y., Juang, R.S. (2004) Adsorption of tannic acid, humic acid, and dyes from water using the composite of chitosan and activated clay. *J. Colloid Interface Sci.*, *278*, 18–25.

[30] Wang, X., Du, Y., and Luo, J. (2008) Biopolymer/montmorillonitenanocomposite: Preparation, drug-controlled release property and cytotoxicity. *Nanotechnology*, *19*, 065707.

[31] Wang, X., Du, Y., Luo, J., Yang, J., Wang, W., and Kennedy, J.F. (2009) A novel biopolymer/rectoritenanocomposite with antimicrobial activity. *Carbohydr. Polym.*, *77*, 449–456.

[32] Wang, S.F., Shen, L., Tong, Y.J., Chen, L., Phang, I.Y., Lim, P.Q., and Liu, T.X. (2005) Biopolymer chitosan/montmorillonitenanocomposites: Preparation and characterization. *Polym. Degrad. Stab.*, *90*, 123–131.

[33] Monvisade, P., Siriphannon, P. (2009) Chitosan intercalated montmorillonite: Preparation, characterization and cationic dye adsorption. *Appl. Clay Sci.*, *42*, 427–431.

[34] Xu, Y., Ren, X., and Hanna, M. (2006) Chitosan/clay nanocomposite film preparation and characterization. *J. Appl. Polym. Sci.*, *99*, 1684–1691.

[35] Abdolmohammadi, S., Yunus, W.M.Z.W., Rahman, M.Z.A.B., and Ibrahim, N.A. (2011) Effect of organoclay on mechanical and thermal properties of polycaprolactone/chitosan/montmorillonitenanocomposites. *J. Reinf. Plast. Compos.*, *30*, 1045–1054.

[36] Liu, M., Zhang, Y., Wu, C., Xiong, S., and Zhou, C. (2012) Chitosan/halloysite nanotubes bionanocomposites: Structure, mechanical properties and biocompatibility. *Int. J. Biol. Macromol.*, *51*, 566–575.

[37] He, L., Yao, L., Yang, D., Cheng, Q., Sun, J., Song, R., and Hao, Y. (2011) Preparation and characterization of chitosan-blended multiwalled carbon nanotubes. *J. Macromol. Sci. B Phys.*, *50*, 2454–2463.

[38] Ong, Y.T., Ahmad, A.L., Sharif Zein, S.H., Sudesh, K., and Tan, S.H. (2011) Poly(3-dydroxybutyrate)-functionalised multi-walled carbon nanotubes/chitosan green

nanocomposite membranes and their application in pervaporation. *Sep. Purif. Technol.*, *76*, 419–427.

[39] Wu, T., Pan, Y., Bao, H., and Li, L. (2011) Preparation and properties of chitosan nanocomposite films reinforced by poly(3,4-ethylenedioxythiophene)-poly(styrenesulfonate) treated carbon nanotubes. *Mater. Chem. Phys.*, *129*, 932–938.

[40] Rao, C.N.R., Sood, A.K., Subrahmanyam, K.S., and Govindaraj, A.A. (2009) Graphene: The new two dimensional nanomaterial. *Angew. Chem. Int. Ed. Engl.*, *48*, 7752.

[41] Ramesha, G.K., Vijaya Kumara, A., Muralidhara, H.B., and Sampath, S. (2011) Graphene and graphene oxide as effective adsorbents toward anionic and cationic dyes. *J. Colloid Interface Sci.*, *361*, 270–277.

[42] Fan, L., Luo, C., Sun, M., Li, X., Lu, F., and Qiu, H. (2012) Preparation of novel magnetic chitosan/graphene oxide composite as effective adsorbents toward methylene blue. *Bioresour. Technol.*, *114*, 703–706.

[43] Verdejo, R., Barroso-Bujans, F., Rodriguez-Perez, M.A., Sajab, J.A., and Lopez-Manchado, M.A. (2008) Functionalized graphene sheet filled silicone foam nanocomposites. *J. Mater. Chem.*, *18*, 2221–2226.

[44] Yang, Y., Wang, J., Zhang, J., Liu, J., Yang, X., and Zhao, H. (2009) Exfoliated graphite oxide decorated by PDMAEMA chains and polymer particles. *Langmuir*, *25*, 11808–11814.

[45] Pan, Y., Wu, T., Bao, H., and Li, L. (2011) Green fabrication of chitosan films reinforced with parallel aligned graphene oxide. *Carbohydr. Polym.*, *83*, 1908–1915.

[46] Junkasem, J., Rujiravanit, R., and Supaphol, P. (2006) Fabrication of α-chitin whisker-reinforced poly(vinyl alcohol) nanocompositenanofibres by electrospinning. *Nanotechnology*, *17*, 4519–4528.

[47] Sriupayo, J., Supaphol, P., Blackwell, J., and Rujiravanit, R. (2005) Preparation and characterization of α-chitin whisker-reinforced chitosan nanocomposite films with or without heat treatment. *Carbohydr. Polym.*, *62*, 130–136.

[48] Lu, Y., Weng, L., and Zhang, L. (2004) Morphology and properties of soy protein isolate thermoplastics reinforced with chitin whiskers. *Biomacromolecules*, *5*, 1046–1051.

[49] Uddin, A.J., Fujie, M., Sembo, S., and Gotoh, Y. (2012) Outstanding reinforcing effect of highly oriented chitin whiskers in PVA nanocomposites. *Carbohydr. Polym.*, *87*, 799–805.

[50] Chang, P.R., Jian, R., Yu, J., and Ma, X. (2010) Fabrication and characterisation of chitosan nanoparticles/plasticised-starch composites. *Food Chem.*, *120*, 736–740.

CHAPTER 7

Preparation and Applications of Chitin Nanofibers/Nanowhiskers

JUN-ICHI KADOKAWA

7.1 INTRODUCTION

Natural polysaccharides provide two important roles in nature: as an energy source and as a structural material [1]. Cellulose and chitin are two representative structural polysaccharides, which are considered to be the most important biomass resources. Chitin is an aminopolysaccharide consisting of a chain of β-(1→4)-linked 2-acetamido-2-deoxyl-D-glucopyranose (*N*-acetyl-D-glucosamine) residues (Fig. 7.1) [2–4]. Despite its huge production in nature and easy accessibility, chitin remains an unutilized biomass resource primarily because of its intractable bulk structure and insolubility in water and common organic solvents; thus, only limited attention has been paid to chitin, principally to its biological properties. Because chitin is the second most abundant biological macromolecule after cellulose, there is major interest in its conversion into various useful materials after proper dissolution in suitable solvents. Chitin occurs mainly in the exoskeletons of crustacean shells, such as crab and shrimp shells. Native chitin in crustacean shells is arranged as microfibrils embedded in a protein matrix (Fig. 7.2) [5, 6]. The microfibril consists of nanofibers 2–5 nm in diameter. Moreover, three types of crystalline forms, that is, α-, β-, and γ-chitins, are known (Fig. 7.1). The most abundant form is α-chitin (e.g., from crab and shrimp shells), where the polymeric chains are aligned in an antiparallel fashion. This arrangement is favorable for the formation of strong intermolecular hydrogen bonding, leading to the most stable form of the three types of crystalline structures. In β-chitin (e.g., from squid pen), the polymeric

Biopolymer Nanocomposites: Processing, Properties, and Applications, First Edition. Edited by Alain Dufresne, Sabu Thomas, and Laly A. Pothan.

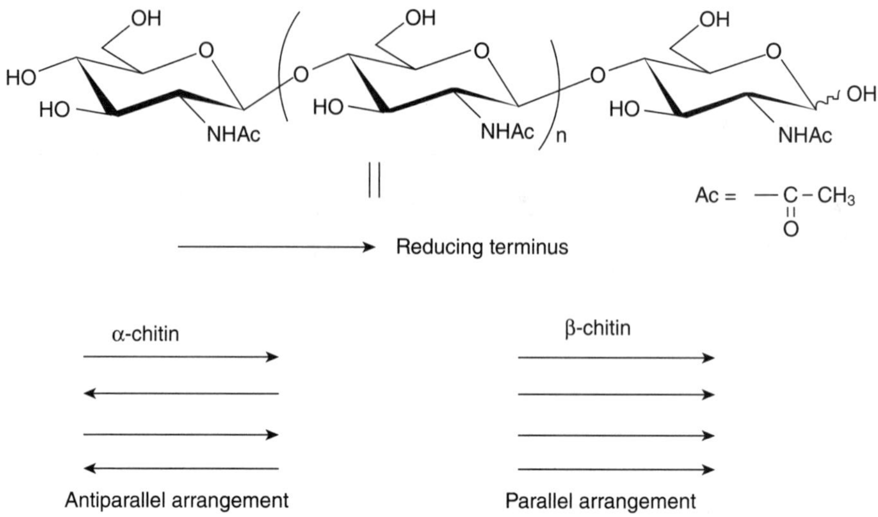

Figure 7.1 Chemical and crystalline structures of chitin.

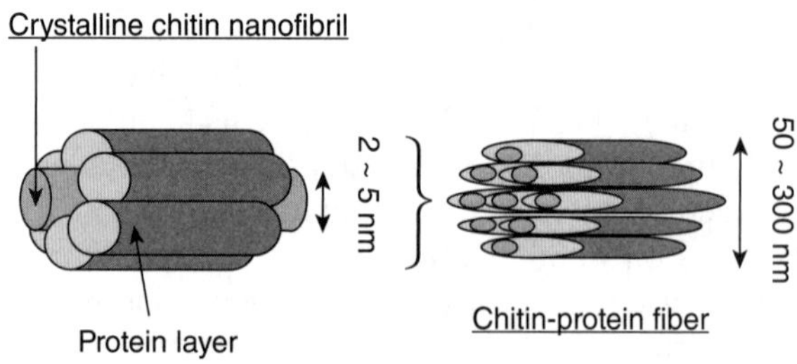

Figure 7.2 Presence of native chitin in crustacean shells. See color insert.

chains are packed in a parallel arrangement, resulting in weaker intermolecular forces. Accordingly, β-chitin is considered to be less stable than α-chitin.

Preparation of nanoscaled polymeric assemblies such as nanofibers and nanowhiskers is one of the most useful methods to practically utilize polymeric materials as observed in the case of cellulose [7, 8]. Furthermore, cellulose nanowhiskers were used as reinforcing fillers for natural polymeric matrices [9]. Efficient methods have also been developed for the preparation of chitin nanofibers and nanowhiskers (nanocrystals). There are two major processes to produce the chitin nanofibers or nanowhiskers from the chitin source: one is disentanglement of chitin microfibrils and the other is regeneration from the chitin solution in appropriate solvents. As examples of the first technique, acid hydrolysis and 2,2,6,6-tetramethylpiperidine-1-oxyl radical (TEMPO)-

mediated oxidation were previously reported. On the other hand, selected conditions for the regeneration of the chitin solutions simply gave chitin nanofibers/nanowhiskers. Electrospinning technique was also employed in the formation of chitin nanofibers from chitin solutions. This chapter deals with the above four methods as efficient processes for the preparation of chitin nanofibers/nanowhiskers. The applications of the resulting nanoscaled chitin assemblies to the practical composite materials are also discussed. Besides these methods, a simple grinding technique was also developed for the preparation of chitin nanofibers from chitin microfibrils [10, 11]. The series on this study is covered in Chapter 2 of this book.

7.2 CHITIN NANOWHISKERS BY ACID HYDROLYSIS

Formation of some nanoscaled chitin assemblies (crystallites) by acidic treatment was reported by Revol and Marchessault in 1993 (Fig. 7.3) [12]. Microfiller fragments of crab and shrimp chitin were prepared by hydrolysis in 3 mol/L HCl aqueous solution at its boiling point. After removal of the acid by centrifugal washing and dialysis, ultrasonication converted the residual product into a colloidal suspension stabilized by $-NH^{3+}$ charges. The positive charge arose from protonation of the few amino groups that were already present in a chitin sample, as well as those that might have been formed by additional deacetylation during the acid hydrolysis. In water, the protonated amino groups and their counterions formed an electrical double layer around the crystallites, which prevented flocculation to yield a stable colloidal suspension. The transmission electron micrograph (TEM) image of a dispersion of chitin crystallites after 90 minutes of acid hydrolysis showed the rod-like

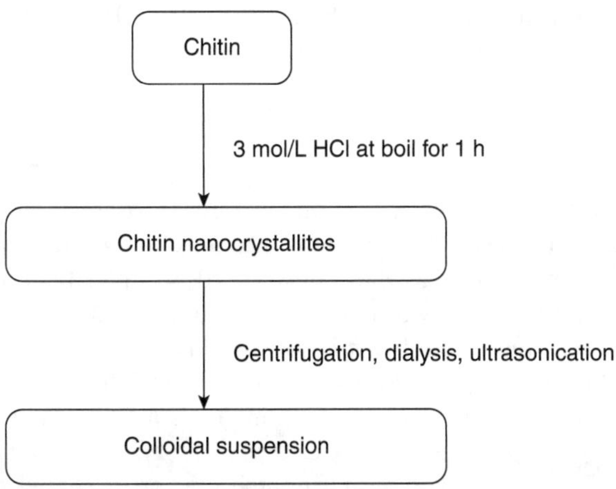

Figure 7.3 Typical procedure for preparation of chitin nanocrystallites by acid hydrolysis.

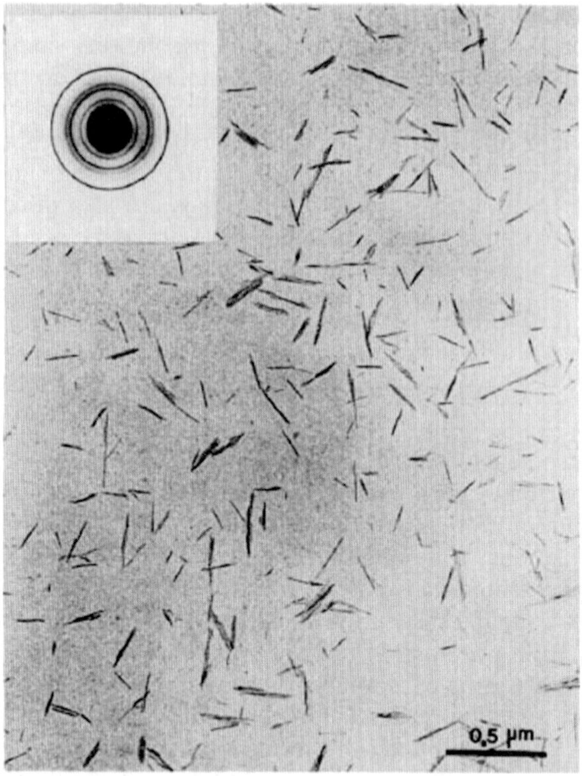

Figure 7.4 Transmission electron micrograph of a dispersion of chitin crystallites after 90 minutes of acid hydrolysis. The micrograph was obtained by diffraction contrast in the bright field mode and under low dose conditions. Underfocusing was about 3 μm. (Inset) Electron diffraction pattern from such a preparation. It corresponds to the α-chitin crystal structure (reprinted with permission from Reference [12]. Copyright 1993 Elsevier).

particles ranging from 50 to more than 300 nm in length (Fig. 7.4). The electron diffraction pattern from the dispersion showed an α-chitin crystalline structure. Thus, the acid hydrolysis treatment has not changed the original crystalline structure. Furthermore, when dewatered to a critical concentration, spontaneous formation of a two-phase equilibrium system occurred. The upper phase was isotropic and the lower phase was anisotropic. The latter displayed a chiral nematic order and dried to a solid film, which mimicked the helical organization characteristic of the chitin microfibrils in the cuticle of arthropods.

Figure 7.5a illustrates a rod with a noncircular cross section. The effective electric double layer thickness, which was of the same order as the crystallite lateral size, was expected to retain the original geometry. Thus, any helical

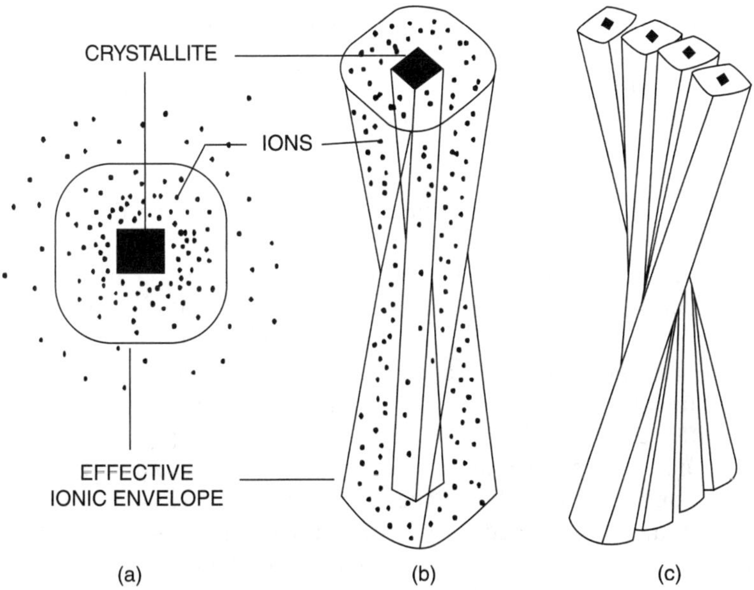

Figure 7.5 (a) Cross-sectional view of a crystallite and its electrical double layer. The effective envelope represents the distance from the crystallite surface at which there is impenetrability between adjacent rods in colloidal suspension. Adjacent rods will approach each other beyond this distance when the suspension is concentrated. (b) A twist in the rigid rod generates a twist in the effective ionic envelope since the original noncylindrical geometry of the rod is still present in the envelope. (c) Chiral nematic interaction between four crystallites in close contact. The scale is smaller than in (a) and (b) (reprinted with permission from Reference [12]. Copyright 1993 Elsevier).

geometry of the rod was still present in the effective electrical double layer, as shown in Figure 7.5b. As soon as a nematic prealignment of the rods was reached, when the space between them was small enough, there was contact and interaction between the electrical double layers of adjacent asymmetric rods. If the twist along the crystallites always had the same handedness, the chiral interaction generated a helicoidal organization of opposite handedness. Figure 7.5c shows such a chiral ordering between four rods.

The effect of electrostatic interaction on phase separation behavior of chitin crystallite suspensions was investigated [13]. Optical and electron microscopies were used to observe the dynamics of the phase separation in aqueous chitin suspensions prepared by HCl hydrolysis of crab chitin. Freeze-fracture transmission electron microscopy revealed that chitin crystallites were partially aggregated in the suspension and had an average length of 200 nm and an average width of 8 nm. They showed a positive surface charge of ~0.5 e/nm^2 when they were fully protonated. The zeta potential measurements, which were made in relatively dilute conditions, showed the screening effect due to the double layer compression when the acid and salt concentrations were increased. The phase diagrams, however, which were obtained with concentrated liquid crystalline suspensions, did not show significant change with added NaCl. This was attributed to the contribution of the chitin crystallites themselves to the ionic strength.

The effect of degree of deacetylation (DD) of chitin on the properties of chitin crystallites was also studied [14]. Chitin samples with different DD were first prepared by alkaline hydrolysis through control reaction time. It was found that the treatment using a constant strength of alkali reduced the crystallinity of the sample, but the crystalline structure of α-chitin was retained. The weight loss of the partially saponified and neutralized samples after HCl hydrolysis increased linearly with the DD. The crystallinity of the samples was found to increase after acid hydrolysis. The increase of surface charge density resulting from the increased surface amino groups was found to affect the boundary concentration for the formation of the anisotropic phase. However, the observed change of boundary concentration with surface charge was opposite to that predicted by theories. This was attributed to the contribution of the charged crystallites to the ionic strength. The suspensions of crystallites prepared from the chitin samples treated with alkali for 6–7 hours did not phase-separate due to high viscosity. Furthermore, chiral nematic structures were not observed for the suspensions of crystallites beyond 4 hours' treatments.

Rheologically, suspensions of chitin crystallites were found to behave as other molecular liquid crystalline polymers (LCPs) [15]. The average hydrodynamic diameter of the crystallites in the suspension at pH 4 was determined to be approximately 80 nm using dynamic light scattering. Conductimetric and pH titration results showed that the pKa of the crystallites was 6.3, which was the same as that reported for chitosan. In combination with phase diagrams, flow properties of isotropic, biphasic, and anisotropic chitin suspensions were investigated. For isotropic suspensions, consequently, a classical shear thinning behavior of rod-like particles was observed. In the case of biphasic suspensions, on the other hand, a two-regime curve was observed where tactoids first oriented, deformed, and then broke up under a shearing force. Furthermore, similar to other LCPs, a three-regime flow curve was found for the anisotropic suspensions. The viscosity–concentration curve showed a maximum at the phase separation concentration, and the maximum was less distinct with a decrease of the ionic strength. Moreover, secondary electroviscous effects due to strong particle–particle interactions influenced the viscosity of the suspensions at higher concentrations.

The shrimp shell chitin has been subjected to extensive treatments of multiple acid hydrolysis processes and subsequent mechanical disruption to yield nanocrystals [16]. The produced chitin nanocrystals were characterized in terms of crystalline properties and the specific surface area of the chitin nanoparticles. The X-ray diffraction (XRD) data indicated an increase in chitin crystallinity after hydrolysis, as less-ordered chitin domains were digested. Dye absorption with Congo red was used to measure the specific surface area of the particles, indicating values near 350 m^2/g. This value was supported with calculations derived from X-ray crystallite size measurements.

Chitin nanowhisker colloidal suspensions, which were produced based on the aforementioned chitin crystallites, have been used as the reinforcing phase in nanocomposites with some polymeric materials. For example, chitin whisker-reinforced thermoplastic nanocomposites were prepared [17]. A latex (average diameter of the particle around 150 nm) obtained by the copolymerization of styrene (34% [w/w]) and butyl acrylate (64% [w/w]) and containing 1% acrylic acid and 1% acrylamide was used as the thermoplastic component. The colloidal chitin fragments dispersion was mixed with the latex with various amounts in order to obtain nanocomposite films with a good level of dispersion and with a weight fraction of chitin ranging from 0 wt% to 20 wt%. After stirring, the air in the suspension was removed by vacuum prior to casting in a Teflon mold. Homogeneous and bubble-free 1-mm-thick films were obtained by storing the casting at 35°C, allowing water evaporation and polymer particle coalescence. It was observed from the differential scanning calorimetry (DSC) measurements that the glass transition temperature (T_g) of the matrix was not modified upon chitin whiskers addition. When reinforced by a low weight fraction (lower than 10 wt%) of chitin whiskers, the polymer films did not show any significant improvement in mechanical properties over the whole temperature range. For higher percentages of chitin whiskers, the composite glassy shear modules increased to up to ~2 GPa for the 20 wt% filled material.

It was shown in the above study that chitin could be used as an environmentally friendly particulate filler, and it was useful for the processing of stiff small-size wares. Therefore, other chitin whisker-reinforced polymer nanocomposites have been prepared. For example, nanocomposite materials from a colloidal suspension of high aspect ratio β-chitin whiskers as the reinforcing phase and poly(ε-caprolactone) as the matrix were obtained [18]. A latex of poly(ε-caprolactone) using a copolymer of poly(ethylene oxide) and poly(propylene oxide) as the surfactant was used. After mixing and stirring the two aqueous suspensions, solid films were obtained by either freeze-drying and hot-pressing or casting and evaporating the preparations. A significant reinforcing effect was observed at $T > T_g$ of the matrix. This effect was described in this temperature range by a "hybrid" percolation model, equivalent to a mean-field mode, considering only the nonpercolating fraction of microcrystals. The relatively weak interactions between chitin whiskers resulted in a partially formed microcrystals network within the rubbery matrix. At a higher temperature ($T >$ melting temperature (T_m)) and above 5 wt% whiskers, the network was restored, resulting in the stabilization of the mechanical properties over a large temperature range.

α-Chitin whisker-reinforced poly(vinyl alcohol) (PVA) nanocomposite films were prepared by solution casting technique [19]. Incorporation of chitin whiskers helped to improve the thermal stability of the resulting nanocomposite films, but it did not have any effect on the crystallinity of the PVA matrix. The tensile strength of the nanocomposite films initially increased with increasing whisker content and leveled off when the whisker content was greater than or equal to 2.96 wt%. On the contrary, the percentage of elongation at break only decreased monotonically with increasing whisker content. Both the presence of chitin whiskers and heat treatment decreased the affinity to water of the nanocomposite films, as reflected by the observed decrease in both the percentage of weight loss and the percentage degree of swelling of the films with either increase in the whisker content or the heat treatment duration. The reduction in the affinity to water caused the PVA/chitin whisker nanocomposite films to be more stable when used in aqueous environments.

Other nanocomposite materials were obtained from a colloidal suspension of chitin whiskers as the reinforcing phase and latex of both unvulcanized and prevulcanized natural rubber as the matrix [20]. The composites were obtained according to the processes similar to those using poly(ε-caprolactone) as the matrix. The processing and swelling behaviors of composite films were evaluated. Consequently, the whiskers formed a rigid network assumed to be governed by a percolation mechanism in the evaporated samples only. Comparatively, better resistance of evaporated samples than hot-pressed ones against swelling in an organic solvent medium was good evidence for the existence of a rigid chitin network. The values of diffusion coefficient, bound rubber content, and relative weight loss also supported the presence of a three-dimensional chitin whisker network within the evaporated samples. The mechanical behavior of the resulting nanocomposite films was also analyzed [21]. Dynamic mechanical analysis revealed the presence of a small percentage of crystallinity in unvulcanized natural rubber prepared by the evaporation method, whereas no such evidence of crystallinity was detected either in vulcanized rubber prepared by evaporation or in unvulcanzied natural rubber prepared by hot-pressing. Successive tensile test experiments helped to understand the existence of a three-dimensional chitin network formed as a result of hydrogen bonding within the evaporated samples. A clear decrease in modulus values of the evaporated samples up to first three or four cycles of the cyclic test indicated the progressive damaging of the chitin network, which was initially present in the samples. All the results revealed that the main factor that governed the mechanical behavior of the chitin whisker-reinforced natural rubber nanocomposites was the processing technique.

Chemical modification of chitin whiskers was performed to investigate the effect of the incorporation of the modified whiskers into a natural rubber matrix on the properties of the ensuing nanocomposite [22]. The surface of chitin whiskers, prepared by acid hydrolysis of chitin from crab shells, was chemically modified using different coupling agents, namely phenyl isocyanate, alkenyl succinic anhydride, and 3-isopropenyl-α,α′-diemthylbenzyl isocyanate (Fig. 7.6). Fourier transform infrared (FT-IR) spectroscopy, TEM, and contact angle measurements were performed to prove the occurrence of the surface

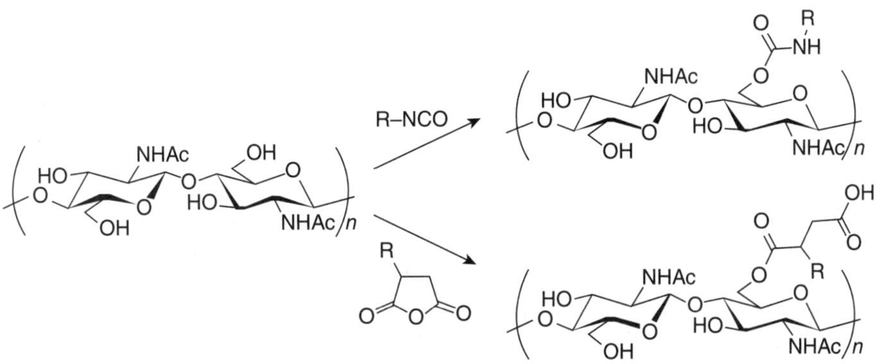

Figure 7.6 Chemical modification of chitin whiskers by isocyanate or succinic anhydride.

modification without any major morphological changes associated with the various treatments applied. Stable suspensions of the chemically modified chitin whiskers were obtained in toluene. Nanocomposites were prepared using a toluene natural rubber solution in which the whiskers were dispersed. After removal of the solvents, the resulting solid films were characterized by scanning electron microscopy (SEM), swelling experiments, and mechanical analysis, in both the linear and nonlinear ranges. All the results led to the conclusion that the various chemical treatments improved the adhesion between the filler and the matrix. However, the mechanical performances of the composites strongly decreased after the chemical modification. This loss of performance, more pronounced for the isocyanate treatments, was due to the partial or total destruction of the three-dimensional network of chitin whiskers assumed to be present in the unmodified composites.

Environmentally friendly thermoplastic nanocomposites were successfully developed using a colloidal suspension of chitin as a filler to reinforce soy protein isolate (SPI) plastics [23]. A suspension of chitin whiskers having an average length of about 500 ± 50 nm and a diameter of around 50 ± 10 nm was used as a reinforcing agent for glycerol plasticized SPI to obtain nanocomposites by hot-pressing. Nanocomposites with lower whisker content exhibited a relatively uniform dispersion in the SPI matrix than those with higher chitin content. Compared with a glycerol plasticized SPI sheet, the chitin-filled SPI composites increased in Young's modulus and tensile strength from 26 to 158 MPa and 3.3 to 8.4 MPa with increasing chitin content from 0 wt% to 20 wt%. As the chitin whiskers increased in the SPI matrix, the composites showed greater water resistance. The improvement in all of the properties of these novel SPI/chitin whisker composites was ascribed to three-dimensional networks of intermolecular hydrogen bonding interactions between filler and filler and between filler and matrix.

Figure 7.7 Preparation of chitosan scaffold from chitin flake via chitin whiskers.

Chitosan in a nanoscaffold structure was successfully prepared via the one-pot deacetylation of chitin whisker (Fig. 7.7) [24]. The structural analysis by the ^{1}H nuclear magnetic resonance (NMR) measurement confirmed that the alkaline treatment changed chitin whisker to chitosan nanoscaffold with as high as 98% DD. The SEM and TEM images confirmed that the short fiber of chitin whiskers developed to form a network in nanoscale of chitosan or chitosan nanoscaffold. The increase in viscosity of chitosan nanoscaffold as compared to that of the chitin whiskers suggested that the scaffold network performed as a polymer bulk in the solution. The increase in surface area, pore volume, and pore size as studied by the Brunauer–Emmett–Teller gas adsorption informed the development of porous structure in chitosan. Furthermore, the good dispersion of nanoscaffold in the solution was easily controlled by the solvent polarity and salt concentration.

7.3 CHITIN NANOFIBERS/NANOCRYSTALS BY TEMPO-MEDIATED OXIDATION

Highly crystalline cellulose nanofibers dispersed in water have been efficiently provided by TEMPO-mediated oxidation of native celluloses and the following mild mechanical treatment [25–27]. For example, the cellulose nanofibers obtained from wood cellulose had a high density of carboxylate groups at C6 positions of glucose residues on surface and were typically 3–4 nm in width and a few micrometers in length. The carboxylate content in the TEMPO-oxidized native celluloses was a key factor in the preparation of individualized cellulose nanofibers.

Therefore, TEMPO-mediated oxidation of α-chitin in water at pH 10 under specific conditions and subsequent ultrasonic treatment were performed to prepare chitin nanowhiskers (nanocrystals) dispersed in water [28]. The outline of the reaction is shown in Figure 7.8. Since general α-chitins originating from crab and shrimp shells had high crystallinity like native celluloses, it might be possible to convert into chitin nanocrystals by formation of the C6 carboxylate groups selectively on the surface of the chitin crystallites by TEMPO-mediated oxidation. The formation of C6 carboxylate groups in chitin was controlled depending on the amount of NaClO added in the TEMPO-mediated oxidation of α-chitin. Particularly, when 5.0 mmol of NaClO per gram of chitin was used, the weight percentage of the water-insoluble fraction in the TEMPO-oxidized chitin remained at as high as 90%, and the carboxylate content reached 0.48 mmol/g. Besides, no *N*-deacetylation occurred on the TEMPO-oxidized

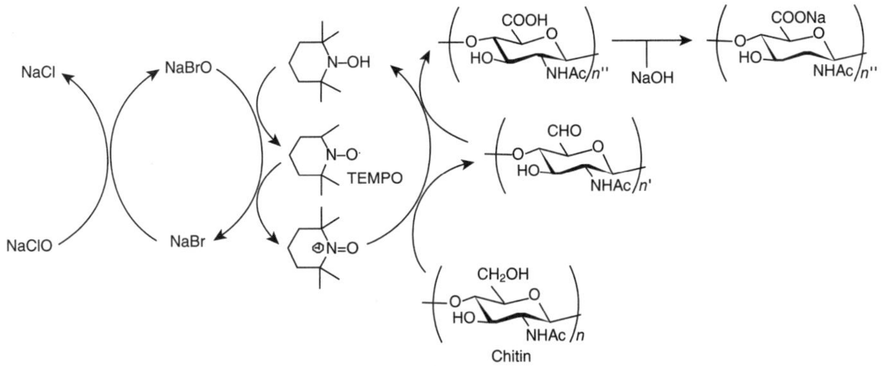

Figure 7.8 TEMPO-mediated oxidation of chitin.

chitins, irrespective of the amount of NaClO added in the oxidation. Moreover, the crystal structure and high crystallinity of the original α-chitin were maintained during TEMPO-mediated oxidation, showing that the C6 carboxylate groups were present only on the chitin crystallite surfaces. When the TEMPO-oxidized chitin was subjected to ultrasonic treatment in water, mostly individualized chitin nanocrystals were obtained, and the dispersion was translucent. The average nanocrystal length and width were 340 and 8 nm, respectively.

TEMPO-mediated oxidation was applied to β-chitins, originating from tubeworm and squid pen, to prepare chitin nanofibers [29]. In the case of squid pen β-chitin, although the water-insoluble fractions were obtained from the TEMPO-oxidized products by controlling the NaClO addition level, the oxidized products were not converted to individual nanofibers by disintegration in water. On the other hand, when tubeworm β-chitin was oxidized by the TEMPO-mediated system with 2.5–10 mmol NaClO per gram of chitin, the water-insoluble fractions with carboxylate contents of 0.18–0.25 mmol/g were obtained in the yields of more than 70%. The analytical data revealed that crystallinities, crystal sizes, and DD values of the produced nanofibers were similar to those of the original chitin. Furthermore, homogeneous, highly viscous, and translucent gels were obtained by disintegration of the water-insoluble TEMPO-oxidized β-chitin in water. The gels consisted of nanofibers 20–50 nm in width and several micrometers in length. When the TEM images of the dispersion of the water-insoluble fraction of TEMPO-oxidized tubeworm β-chitin prepared with 10 mmol NaClO per gram of chitin were measured after drying (Fig. 7.9), long TEMPO-oxidized tubeworm β-chitin nanofibrils with widths ranging from 20 to 50 nm and lengths of more than several micrometers were clearly observed. Thus, the tubeworm β-chitin fibrils were mostly converted to highly crystalline and individual nanofibrils dispersed

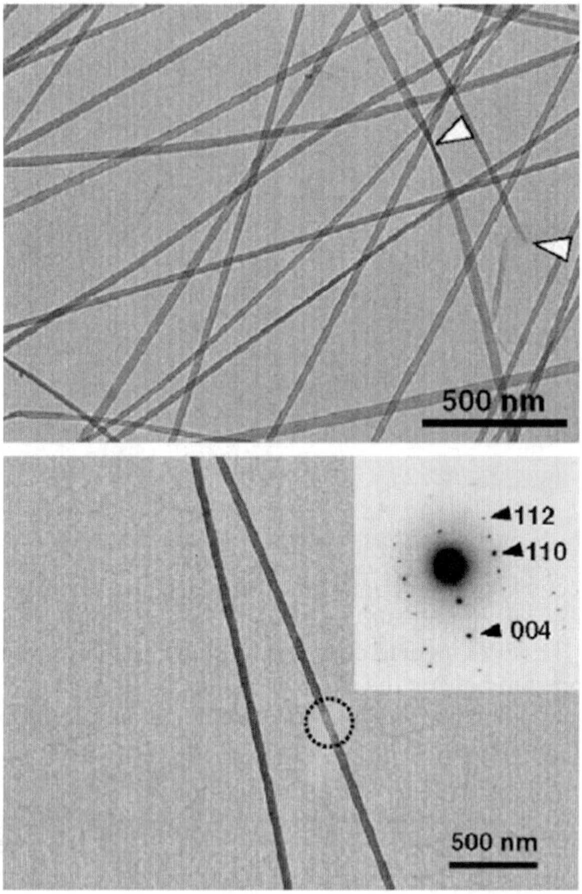

Figure 7.9 Transmission electron microphotographs and electron diffraction diagram of nanofibrils of TEMPO-oxidized tubeworm β-chitin prepared with 10 mmol NaClO per gram of chitin. White arrows show twisted and cut parts of the fibrils (reprinted with permission from Reference [29]. Copyright 2009 Elsevier).

in water by the TEMPO-mediated oxidation and the following mechanical disintegration. The fibril had a twisted part, indicating that each fibril had a flat and ribbon-like morphology 20–50 nm in width and ca. 15 nm in thickness. Furthermore, an edge of the nanofibrils probably formed by scission during disintegration in water. Because each individual nanofibril had a clear electron diffraction spots at any locations (an inset figure in Fig. 7.9), one TEMPO-oxidized tubeworm β-chitin nanofibril did not consist of multiple elementary fibrils but formed a perfect crystallite.

7.4 CHITIN NANOFIBERS/NANOWHISKERS BY REGENERATION FROM CHITIN SOLUTIONS

Self-assembled chitin nanofibers were prepared by regeneration from the solutions of chitin (Fig. 7.10) [30]. Hexafluoro-2-propanol (HFIP) and LiCl/*N,N*-dimethylacetamide (DMAc), well-known solvents for chitin, were used [4]. The self-assembly process was initiated through either solvent evaporation for the HFIP system or addition of water for the LiCl/DMAc system. Self-assembly of chitin in HFIP was remarkably simple; that is, slowly drying solutions of appropriate concentrations led to long (10–100 μm) nanofibers with a small diameter (2.8 ± 0.7 nm). For the low volatility of LiCl/DMAc, on the other hand, simple drying was not a practical approach. However, fibers were easily precipitated out upon addition of ample amounts of water.

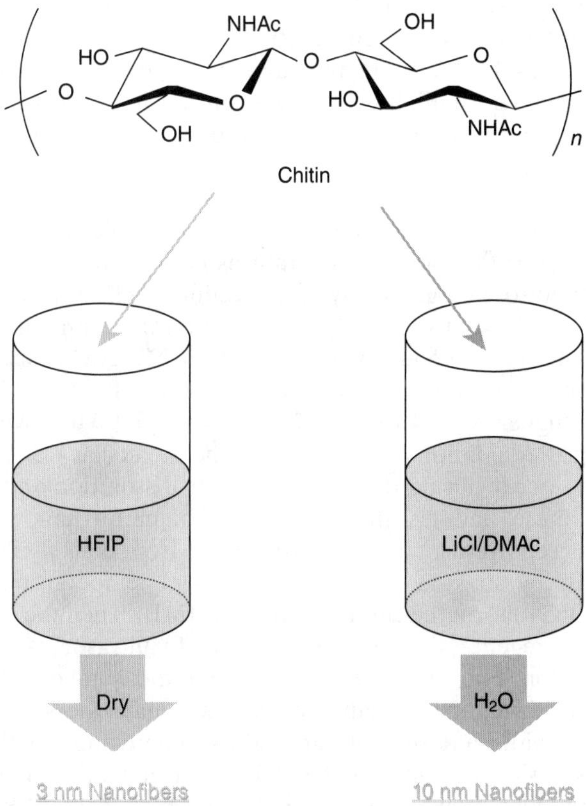

Figure 7.10 Preparation of chitin nanofibers by regeneration from the solutions of chitin.

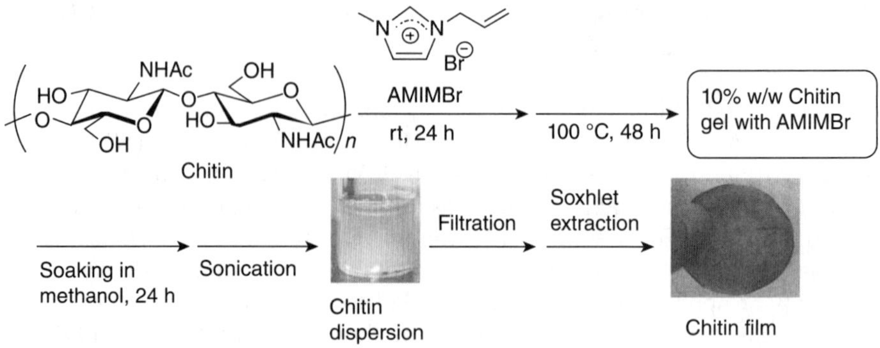

Figure 7.11 Procedures for preparation of chitin dispersion and film.

Nanofibers prepared in this fashion had generally a larger diameter (10.2 ± 2.9 nm) than those prepared from HFIP. Both types of nanofibers had similar lengths. Furthermore, the nanofibers showed an α-chitin structure and retain the high crystallinity. Deacetylation was not observed and the mild conditions employed to produce the nanofibers were unlikely to cause depolymerization.

An ionic liquid was also used as the solvent for the preparation of chitin nanowhiskers by the regeneration method [31]. Ionic liquids, low-melting point salts that form liquids at temperatures below a boiling point of water, have been found to be a good solvent for cellulose [32–34]. However, only a few examples have been reported on dissolution of chitin with ionic liquids [35, 36]. 1-Allyl-3-methylimidazolium bromide (AMIMBr) is one of such ionic liquids that dissolves chitin (5% w/w), and further, it forms gel-like materials with chitin at higher concentrations (7–12% w/w) [37]. Moreover, it has been confirmed that degradation and decrease in the molecular weight of chitin do not frequently occur during the procedures for dissolution and gelation. On the basis of these findings, the preparation of chitin nanowhiskers using AMIMBr was reported. First, chitin (10% w/w) from crab shells was swollen with AMIMBr by soaking the mixture at room temperature, followed by heating at 100°C to form the chitin gel with AMIMBr. Then the gel was soaked in methanol for regeneration of chitin, followed by ultrasonication to produce a chitin dispersion (Fig. 7.11). The SEM image of the spin-coated sample from the dispersion showed the formation of the chitin nanowhiskers ranging from 20 to 60 nm in width and several hundred nanometers in length (Fig. 7.12a). When the dispersion was filtered, the residue formed a film. The SEM image of the resulting film showed the morphology of highly condensed nanowhiskers (Fig. 7.12b). Such condensed structure of the nanowhiskers probably contributed to formability of the film. The XRD pattern of the nanowhiskers showed the diffraction pattern corresponding to an α-chitin crystalline struc-

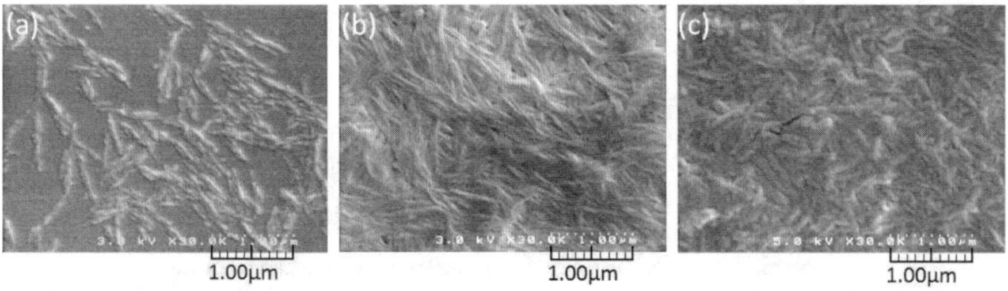

Figure 7.12 SEM images of (a) chitin dispersion, (b) chitin nanowhisker film, and (c) composite film of chitin nanowhiskers with PVA.

ture. These data indicated that the crystalline structure of α-chitin was reconstructed by the above regeneration process during the formation of nanowhiskers.

As one of the possible applications of the present chitin nanowhiskers obtained using AMIMBr, the whiskers were compatibilized with PVA to give chitin nanowhisker/PVA composite materials. First, 10% w/w chitin gel with AMIMBr was prepared according to the aforementioned procedure and solutions of the desired amounts of PVA in hot water were added to the gel. Then, the same regeneration and filtration procedures as aforementioned resulted in the chitin nanowhiskers/PVA composite films. The SEM image of the composite showed that the nanowhisker-like morphology was maintained and PVA components probably filled in spaces between the nanowhiskers (Fig. 7.12c). This result indicated relative immiscibility of chitin and PVA in the composite. Indeed, the DSC profiles of all the composites exhibited endothermic peaks due to a melting point of PVA, suggesting that PVA in the composites had some crystallinity. However, the melting points of PVA were shifted to lower temperatures, accompanied by broadening when the contents of chitin in the composites increased. From the DSC results, therefore, it was supposed that chitin and PVA might be partially miscible at the interfacial area between the two polymers in the composites by the formation of hydrogen bonding between them or by the presence of a little amount of AMIMBr. Furthermore, the tensile testing of the composite films indicated the enhancement of the mechanical properties with increasing the ratios of PVA to chitin.

7.5 CHITIN NANOFIBERS BY ELECTROSPINNING

An electrospinning method was used to fabricate chitin nanofibous matrix [38, 39]. Chitin was first depolymerized by gamma irradiation to improve its solubility [40]. The electrospinning of chitin was then performed with HFIP as a spinning solvent. The SEM images of nanofibers electrospun from the chitin solutions with different concentrations or viscosities ranging from 1% to 6%

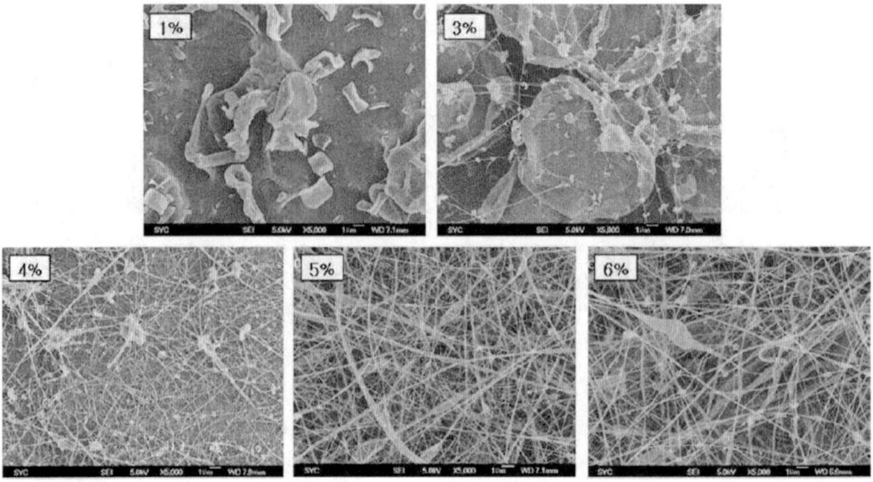

Figure 7.13 SEM images of chitin nanofibers electrospun at different concentrations (reprinted with permission from Reference [40]. Copyright 2004 Elsevier).

by weight are shown in Figure 7.13. At the concentration of up to 3% by weight, large irregular beads or beaded fibers were generated by electrospinning. The continuous nanofibers were obtained at the concentration of more than 4% by weight. Therefore, it was considered that extensive chain entanglements were necessary to produce the continuous fibers of chitin by electrospinning. However, at 6 wt% concentration, the continuous and uniform electrospinning was inhibited because chitin solution had a very high viscosity (5310 cP) at that concentration. The resultant fibrous structure containing irregular, but small, beads was observed again. Although as-spun chitin nanofibers had the broad fiber diameter distribution, most of the fiber diameters were less than 100 nm. From the image analysis, they had an average diameter of 110 nm. For deacetylation, the as-spun chitin nanofiber matrix was chemically treated with a 40% aqueous NaOH solution at 60°C or 100°C. By the deacetylation for 120 minutes at 100°C or for 1 day at 60°C, the chitin matrix was transformed into a chitosan matrix with DD ~85% without dimensional change.

Chitin nanofibrous matrix produced by electrospinnin, was introduced for potential use in wound dressings or as scaffolding for tissue engineering [41]. The chitin nanofibers were approximately two orders of magnitude smaller than chitin microfibers, and thus the former degraded faster than the latter. Morphological changes during *in vitro* and *in vivo* degradation were investigated with SEM. In the cytocompatibility assessment, the chitin nanofibers were found to promote cell attachment and spreading of normal human keratinocytes and fibroblasts compared with the chitin microfibers. This was probably a result of the high surface area available for cell attachment due to chitin nanofibers' three-dimensional features and high surface area-to-volume ratios, which were favorable parameters for cell attachment, growth, and proliferation. Additionally, chitin nanofibers coated with type I collagen were found to promote cellular response in all the cells tested.

To mimic the natural extracellular matrix (ECM), chitin/silk fibroin blend nanofibrous scaffolds were obtained by electrospinning with the average fiber diameter being in the range of 340–920 nm, whereas electrospinning of pure silk fibroin produced an average fiber diameter of 1260 nm [42]. Electrospinning of chitin/silk fibroin blend solutions produced blend nanofibers with phase-separated structure because chitin and silk fibroin were immiscible in the as-spun nanofibrous structure. From water vapor after-treatment, it was found that silk fibroin in the blend nanofibers could be dimensionally stabilized by the solvent-induced crystallization. The hydrophilicity of the blend nanofibrous matrices was improved by blending with more hydrophilic chitin. The cell attachment and spreading of normal human epidermal keratinocytes and fibroblasts onto the blend nanofibrous matrices were also evaluated. However, no such a tendency in cellular attachment and spreading depending on the blend composition or hydrophilicity, except the one that contained 75% chitin and 25% silk fibroin in normal human epidermal fibroblasts, was observed.

Electrospinning of poly(glycolic acid) (PGA)/chitin solutions in HFIP was investigated to fabricate biodegradable and biomimetic nanostructured scaffold for tissue engineering [43]. The morphology of the electrospun PGA/chitin blend nanofibers was investigated with a field emission scanning electron microscope. The blend fibers had an average diameter of around 140 nm, and their diameters had a distribution in the range of 50–350 nm. The blend nanofibers had phase-separated structure because PGA and chitin were immiscible in the as-spun nanofibrous structure. From *in vitro* degradation results, it was found that PGA in the blend nanofibers was degraded faster than pure PGA nanofibers. This accelerating effect might be attributed to the hydrophilicity of chitin. The cell attachment and spreading onto the blend nanofibrous scaffolds showed relatively promising results for the normal human fibroblasts tested. The PGA/chitin blend matrix, particularly the one that contained 25% PGA and 75% chitin with bovin serum albumin coating, could be a good candidate for tissue engineering scaffolds because it had both biomimetic three-dimensional structure, resembling the collagen–glycosaminoglycans composite structure in the ECM, and an excellent cell attachment and spreading for normal human fibroblasts.

Chitin whisker-reinforced nanocomposite nanofibers were also fabricated by electrospinning a mixture between an aqueous solution of PVA and a suspension of α-chitin whiskers, which were prepared by the acid hydrolysis method from chitin flakes from shells of *Penaeus merguiensis* shrimps [44]. Electrospinning of the PVA/chitin whisker mixtures was straightforward and the average diameter of the as-spun nanocomposite nanofibers ranged between

185 and 218 nm, regardless of the whisker content. Incorporation of the chitin whiskers within the as-spun nanocomposite nanofibers was confirmed by the IR spectroscopic and thermogravimetric analysis (TGA) measurements. The tensile strength of the as-spun PVA/chitin whisker nanocomposite mats increased from that of the neat as-spun fiber mat (i.e., 4.3 ± 0.7 MPa) with initial addition of the chitin whiskers to reach a maximum value (i.e., 5.7 ± 0.6 MPa) at the chitin whisker to PVA ratio of about 5.1%, and decreased with further increasing the whisker content. Furthermore, the presence of chitin whiskers within the as-spun nanocomposite fiber mats increased Young's modulus by about 4–8 times over that of the neat as-spun fiber mat.

7.6 OTHER METHODS FOR CHITIN NANOFIBERS

A simple and versatile approach for extracting bionanofibers from various natural materials using the ultrasonic technique has been attempted [45]. By this technique, chitin nanofibers that had uniform diameters were obtained The authors claimed that this methodology might be valuable to providing a convenient, versatile, and environmentally benign fabrication method for producing bionanofibers at an industrial scale.

The procedure for preparing individualized chitin nanofibers of 3–4 nm in cross-sectional width and at least a few micrometers in length from squid pen β-chitin was successfully established by simple mechanical treatment in water at pH 3–4 without any chemical modification [46]. Transparent and highly viscous dispersions of squid pen β-chitin nanofibers in water were obtained by this method. No *N*-deacetylation occurred on the chitin molecules during the procedure. Moreover, the original crystal structure of β-chitin was maintained, although crystallinity index decreased from 0.51 to 0.37 as a result of the nanofiber conversion. Cationization of the C2 amino groups present on the crystallite surfaces of the squid pen β-chitin under acidic conditions was necessary to prepare the nanofibers.

Mostly individualized α-chitin nanowhiskers were also obtained in yields of 85–90% by partial deacetylation with 33% NaOH at 90°C for 2–4 hours and subsequent disintegration in water at pH 3–4 [47]. Transparent and highly viscous liquids were obtained by the disintegration procedure of the partially deacetylated chitins. The degrees of *N*-acetylation of the products were 0.74–0.70, which corresponded to C2-NH_2 contents of 1.34–1.56 mmol/g. XRD analysis showed that the crystallinity and crystal size of the original α-chitin were mostly maintained after the NaOH treatment. This result indicated that partial deacetylation took place selectively on the α-chitin crystallite surfaces. The TEM measurement revealed that the resulting liquids consisted mostly of individual nanowhiskers with average width and length of 6.2 ± 1.1 nm and 250 ± 140 nm, respectively. Some α-chitin nanofibers more than 500 nm in length were also detected in the TEM images. Because conversion to nanowhiskers was achieved in water at pH 3–4, protonation of the C2-NH_2 groups in the partially deacetylated chitins to provide cationic charges in high density on the crystalline fibril surfaces, associated with partial mechanical scission of the fibrils during the disintegration, was considered to be the key driving force for the individualization of α-chitin fibrils.

REFERENCES

[1] Schuerch, C. (1986) Polysaccharides. In: *Encyclopedia of Polymer Science and Engineering*, Vol. 13, 2nd ed, eds. H.F. Mark, N. Bilkales, and C.G. Overberger. New York: John Wiley & Sons, pp. 87–162.

[2] Rinaudo, M. (2006) Chitin and chitosan: Properties and applications. *Progress in Polymer Science*, *31*, 603–632.

[3] Kurita, K. (2006) Chitin and chitosan: Functional biopolymers from marine crustaceans. *Marine Biotechnology (New York, N.Y.)*, *8*, 203–226.

[4] Pillai, C.K.S., Paul, W., and Sharma, C.P. (2009) Chitin and chitosan polymers: Chemistry, solubility and fiber formation. *Progress in Polymer Science*, *34*, 641–678.

[5] Raabe, D., Romano, P., Sachs, C., Fabritius, H., Al-Sawalmih, A., Yi, S.B., Servos, G., and Hartwig, H.G. (2006) Microstructure and crystallographic texture of the chitin-protein network in the biological composite material of the exoskeleton of the lobster *Homarus americanus*. *Materials Science and Engineering A-Structural Materials Properties Microstructure and Processing*, *421*, 143–153.

[6] Chen, P.Y., Lin, A.Y.M., McKittrick, J., and Meyers, M.A. (2008) Structure and mechanical properties of crab exoskeletons. *Acta Biomaterialia*, *4*, 587–596.

[7] Abe, K., Iwamoto, S., and Yano, H. (2007) Obtaining cellulose nanofibers with a uniform width of 15 nm from wood. *Biomacromolecules*, *8*, 3276–2378.

[8] Fukuzumi, H., Saito, T., Iwata, T., Kumamoto, Y., and Isogai, A. (2009) Transparent and high gas barrier films of cellulose nanofibers prepared by tempo-mediated oxidation. *Biomacromolecules*, *10*, 162–165.

[9] Anglès, M.N., Dufresne, A. (2000) Plasticized starch/tunicin whiskers nanocomposites. 1. Structural analysis. *Macromolecules*, *33*, 8344–8353.

[10] Ifuku, S., Nogi, M., Abe, K., Yoshioka, M., Morimoto, M., Saimoto, H., and Yano, H. (2009) Preparation of chitin nanofibers with a uniform width as alpha-chitin from crab shells. *Biomacromolecules*, *10*, 1584–1588.

[11] Ifuku, S., Nogi, M., Yoshioka, M., Morimoto, M., Yano, H., and Saimoto, H. (2010) Fibrillation of dried chitin into 10–20 nm nanofibers by a simple grinding method under acidic conditions. *Carbohydrate Polymers*, *81*, 134–139.

[12] Revol, J.F., Marchessault, R.H. (1993) In-vitro chiral nematic ordering of chitin crystallites. *International Journal of Biological Macromolecules*, *15*, 329–335.

[13] Li, J., Revol, J.F., Naranjo, E., and Marchessault, R.H. (1996) Effect of electrostatic interaction on phase separation of chitin crystallite suspensions. *International Journal of Biological Macromolecules*, *18*, 177–187.

[14] Li, J., Revol, J.F., and Marchessault, R.H. (1997) Effect of degree of deacetylation of chitin on the properties of chitin crystallites. *Journal of Applied Polymer Science*, *65*, 373–380.

[15] Li, J., Revol, J.F., and Marchessault, R.H. (1996) Rheological properties of aqueous suspensions of chitin crystallites. *Journal of Colloid and Interface Science*, *183*, 365–373.

[16] Goodrich, J.D., Winter, W.T. (2007) α-Chitin nanocrystals prepared from shrimp shells and their specific surface area measurement. *Biomacromolecules*, *8*, 252–257.

[17] Paillet, M., Dufresne, A. (2001) Chitin whisker reinforced thermoplastic nanocomposites. *Macromolecules*, *34*, 6527–6530.

[18] Morin, A., Dufresne, A. (2002) Nanocomposites of chitin whiskers from *Riftia* tubes and poly(caprolactone). *Macromolecules*, *35*, 2190–2199.

[19] Sriupayo, J., Supaphol, P., Blackwell, J., and Rujiravanit R. (2005) Preparation and characterization of α-chitin whisker-reinforced poly(vinyl alcohol) nanocomposite films with or without heat treatment. *Polymer*, *46*, 5637–5644.

[20] Gopalan Nair, K., Dufresne, A. (2003) Crab shell chitin whisker reinforced natural rubber nanocomposites. 1. Processing and swelling behavior. *Biomacromolecules*, *4*, 657–665.

[21] Gopalan Nair, K., Dufresne, A. (2003) Crab shell chitin whisker reinforced natural rubber nanocomposites. 2. Mechanical behavior. *Biomacromolecules*, *4*, 666–674.

[22] Gopalan Nair, K., Dufresne, A. (2003) Crab shell chitin whisker reinforced natural rubber nanocomposites. 3. Effect of chemical modification of chitin whiskers. *Biomacromolecules*, *4*, 1835–1842.

[23] Lu, Y., Weng, L., and Zhang, L. (2004) Morphology and properties of soy protein isolate thermoplastics reinforced with chitin whiskers. *Biomacromolecules*, *5*, 1046–1051.

[24] Phongying, S., Aiba, S. (2007) Chirachanchai S. Direct chitosan nanoscaffold formation via chitin whiskers. *Polymer*, *48*, 393–400.

[25] Saito, T., Isogai, A. (2004) TEMPO-mediated oxidation of native cellulose. The effect of oxidation conditions on chemical and crystal structures of the water insoluble fractions. *Biomacromolecules*, *5*, 1983–1989.

[26] Saito, T., Kimura, S., Nishiyama, Y., and Isogai, A. (2007) Cellulose nanofibers prepared by TEMPO-mediated oxidation of native cellulose. *Biomacromolecules*, *8*, 2485–2491.

[27] Saito, T., Nishiyama, Y., Putaux, J.L., Vignon, M., and Isogai, A. (2006) Homogeneous suspensions of individualized microfibrils from TEMPO-catalyzed oxidation of native cellulose. *Biomacromolecules*, *7*, 1687–1691.

[28] Fan, Y., Saito, T., and Isogai, A. (2008) Chitin nanocrystals prepared by TEMPO-mediated oxidation of α-chitin. *Biomacromolecules*, *9*, 192–198.

[29] Fan, Y., Saito, T., and Isogai, A. (2009) TEMPO-mediated oxidation of β-chitin to prepare individual nanofibers. *Carbohydrate Polymers*, *77*, 832–838.

[30] Zhong, C., Cooper, A., Kapetanovic, A., Fang, Z., Zhang, M., and Rolandi, M. (2010) A facile bottom-up route to self-assembled biogenic nanofibers. *Soft Matter*, *6*, 5298–5301.

[31] Kadokawa, J., Takegawa, A., Mine, S., and Prasad, K. (2011) Preparation of chitin nanowhiskers using an ionic liquid and their composite materials with poly(viny alcohol). *Carbohydrate Polymers*, *84*, 1408–1412.

[32] Feng, L., Chen, Z.I. (2008) Research progress on dissolution and functional modification of cellulose in ionic liquids. *Journal of Molecular Liquids*, *142*, 1–3.

[33] Pinkert, A., Marsh, K.N., Pang, S., and Staiger, M.P. (2009) Ionic liquids and their interaction with cellulose. *Chemical Reviews*, *109*, 6712–6828.

[34] Zakrzewska, M.E., Bogel-Lukasik, E., and Bogel-Lukasik, R. (2010) Solubility of carbohydrates in ionic liquids. *Energy and Fuels*, *24*, 737–745.

[35] Wu, Y., Sasaki, T., Irie, S., and Sakurai, K. (2008) A novel biomass-ionic liquid platform for the utilization of native chitin. *Polymer*, *49*, 2321–2327.

[36] Qin, Y., Lu, X., Sun, N., and Rogers, R.D. (2010) Dissolution or extraction of crustacean shells using ionic liquids to obtain high molecular weight purified chitin and direct production of chitin films and fibers. *Green Chemistry*, *12*, 968–971.

[37] Prasad, K., Murakami, M., Kaneko, Y., Takada, A., Nakamura, Y., and Kadokawa, J. (2009) Weak gel of chitin with an ionic liquid, 1-allyl-3-methylimidazolium bromide. *International Journal of Biological Macromolecules*, *45*, 221–225.

[38] Schiffman, J.D., Stulga, L.A., and Schauer, C.L. (2009) Chitin and chitosan: Transformations due to the electrospinning process. *Polymer Engineering and Science*, *49*, 1918–1928.

[39] Jayakumar, R., Prabaharan, M., Nair, S.V., and Tamura, H. (2010) Novel chitin and chitosan nanofibers in biomedical applications. *Biotechnology Advances*, *28*, 142–150.

[40] Min, B.M., Lee, S.W., Lim, J.N., You, Y., Lee, T.S., Kang, P.H., and Park, W.H. (2004) Chitin and chitosan nanofibers: Electrospinning of chitin and deacetylation of chitin nanofibers. *Polymer*, *45*, 7137–7142.

[41] Noha, H.K., Lee, S.W., Kim, J.M., Oh, J.E., Kim, K.H., Chung, C.P., Choi, S.C., Park, W.H., and Min, B.M. (2006) Electrospinning of chitin nanofibers: Degradation behavior and cellular response to normal human keratinocytes and fibroblasts. *Biomaterials*, *27*, 3934–3944.

[42] Park, K.E., Jung, S.Y., Lee, S.J., Min, B.M., and Park, W.H. (2006) Biomimetic nanofibrous scaffolds: Preparation and characterization of chitin/silk fibroin blend nanofibers. *International Journal of Biological Macromolecules*, *38*, 165–173.

[43] Park, K.E., Kang, H.K., Lee, S.J., Min, B.M., and Park, W.H. (2006) Biomimetic nanofibrous scaffolds: Preparation and characterization of PGA/chitin blend nanofibers. *Biomacromolecules*, *7*, 635–643.

[44] Junkasem, J., Rujiravanit, R., and Supaphol, P. (2006) Fabrication of α-chitin whisker-reinforced poly(vinyl alcohol) nanocomposite nanofibers by electrospinning. *Nanotechnology*, *7*, 4519–4528.

[45] Zhao, H.P., Feng, X.Q., and Gao, H. (2007) Ultrasonic for extracting nanofibers from nature materials. *Applied Physics Letters*, *90*, 73112–73113.

[46] Fan, Y., Saito, T., and Isogai, A. (2008) Preparation of chitin nanofibers from squid pen β-chitin by simple mechanical treatment under acid conditions. *Biomacromolecules*, *9*, 1919–1923.

[47] Fan, Y., Saito, T., and Isogai, A. (2010) Individual chitin nano-whiskers prepared from partially deacetylated α-chitin by fibril surface cationization. *Carbohydrate Polymers*, *79*, 1046–1051.

CHAPTER 8

Preparation of Starch Nanoparticles

DÉBORAH LE CORRE and ALAIN DUFRESNE

8.1 INTRODUCTION

Engineered nanoscaled materials are used globally in numerous industrial applications and the list of proposed commercial applications continues to grow [1]. Indeed, these nanomaterials exhibit outstandingly enhanced properties even when used at very low content. Moreover, fossil depletion and growing environmental concerns have led to the development of bio-based materials.

Thanks to their semicrystallinity, polysaccharides offer the opportunity to prepare bio-based nanoparticles. Indeed, since pioneer work on polymer nanocomposites reinforced with nanoclay, at Toyota in the early 1990s, work on nanocrystalline cellulose (NCC) has exponentially increased as reviewed recently [2]. It was soon followed by investigations on chitin [3–5] and starch nanocrystals (SNCs) as described in Section 8.4.1.

Starch is an abundant, renewable, biodegradable polymer produced in nature. It is found in many plants roots, stalks, crop seeds, and staple crops such as rice, corn, wheat, tapioca, and potato [6], as a source of storage energy. Worldwide the main sources of starch are maize (82%), wheat (8%), potatoes (5%), and cassava (5%) from which tapioca starch is derived [7]. Starch is either used as extracted from the plant and is called "native starch," or it undergoes one or more chemical modifications to reach specific properties and is called "modified starch." The preparation of SNC is based on native starches.

8.2 STARCH

After its extraction from plants, starch occurs as a white flour-like powder, consisting of microscopic granules with diameters ranging from 2 to 100 μm,

Biopolymer Nanocomposites: Processing, Properties, and Applications, First Edition. Edited by Alain Dufresne, Sabu Thomas, and Laly A. Pothan.

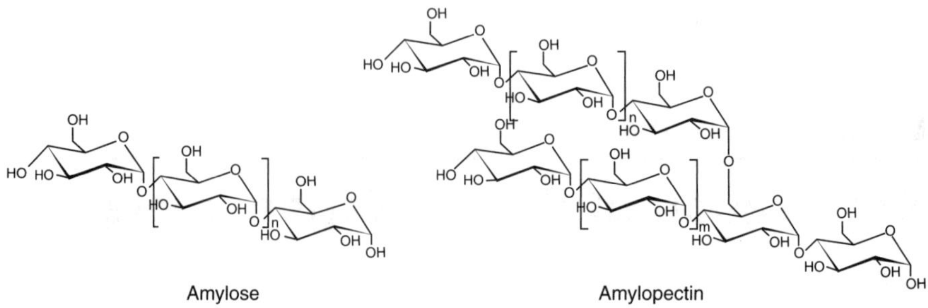

Figure 8.1 Chemical structure of amylose and amylopectin.

depending on the botanic origin. It is insoluble in cold water. The basic formula of this polymer is $(C_6H_{10}O_5)_n$ and the glucose monomer is called α-D-glycopyranose (or α-D-glycose) when in cycle.

8.2.1 Content

Starch's composition was first determined by studying the residue of its total acid hydrolysis. It consists mainly of two families of glucosidic macromolecules, presented in Figure 8.1: (1) a mostly linear one called amylose, consisting of glucose units linked by (1-4) α-D-glycoside bonds, and slightly branched by (1-6) α-linkages; and (2) a highly branched one called amylopectin, consisting of relatively short branches of α-D-(1-4)-glycopyranose that are interlinked by α-D-(1-6)-glycosidic linkages approximately every 22 glucose units [8].

In most common types of starch the weight percentages of amylopectin ranges between 72% and 82%, and the amylose content ranges from 18% to 28%. However, some mutant types of starch have a very high amylose content (up to 70% and more for amylomaize) and some have a very low amylose content (0–2% for waxy maize).

Minor components of starch granules are of three types: (1) cell-wall fragments, (2) surface components, and (3) internal components. The main constituents of surface components are proteins, enzymes, amino acid, and nucleic acid, whereas internal components are composed mainly of lipids. Thus, minor components of starch are often described as proteins, lipids, and minerals [9, 10]. The proportion of these components depends on the botanical origin of starch (species and varieties). Starch contains several different minerals in small amounts, but the most important mineral is phosphorus [9].

8.2.2 Structure

Starch structure has been under investigation for years. Because of its complexity, a universally accepted model has been lacking [9]. However, these last

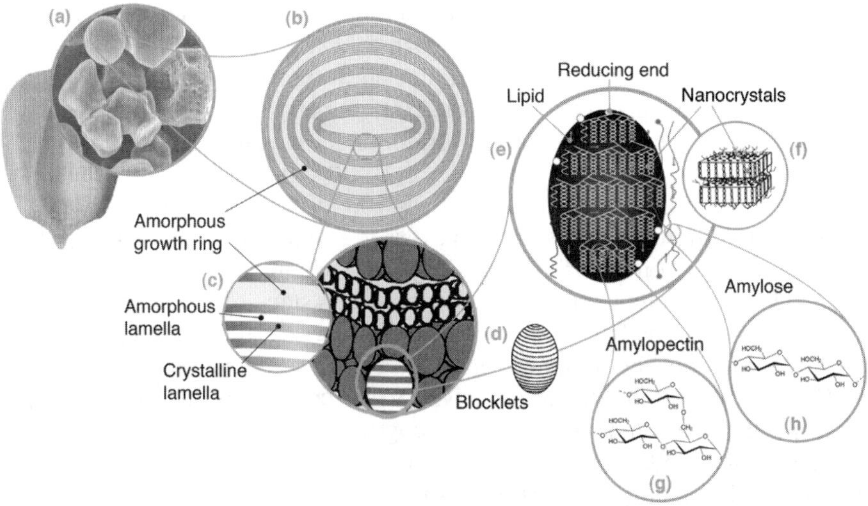

Figure 8.2 Starch multiscale structure. (a) Starch granules from normal maize (30 μm); (b) amorphous and semicrystalline growth rings (120–500 nm); (c) amorphous and crystalline lamellae (9 nm): magnified details of the semicrystalline growth ring; (d) blocklets (20–50 nm): constituting unit of the growth rings; (e) amylopectin double helices forming the crystalline lamellae of the blocklets; (f) nanocrystals: other representation of the crystalline lamellae called SNC when separated by acid hydrolysis; (g) amylopectin's molecular structure; (h) amylose's molecular structure (0.1–1 nm). Reproduced with permissions from Gallant et al. [11], Copyright 1997 Elsevier; Donald et al. [12], Copyright Science and Technology Facilities Council 1997; and Tang et al. [13], Copyright 2006 Elsevier.

decades a predominant model seems to have emerged. It consists in a multiscale structure as presented in Figure 8.2. The granule (a, 2–100 μm) consists in alternating amorphous and semicrystalline growth rings (b, 120–400 nm [14]), rendering an onion-like structure when observed under a scanning electron microscope. These growth rings consist respectively of small and larger blocklets (d, 10–100 nm) made of alternating crystalline and amorphous lamellae (c, 9 nm [11]) containing (g) amylopectin, and (h) amylose chains (0.1–1 nm).

On the surface of starch granules, pores can be observed as seen in Figure 8.2a. They are believed to be channels going through the growth rings to the center of the granule (called hilum). Observed under a microscope and polarized light, starch shows birefringence. The refracted "Maltese cross" corresponding to the crystalline region is characteristic of a radial orientation of the macromolecules [9]. Indeed, starch granules consist of concentric alternating amorphous and semicrystalline growth rings growing by apposition from the hilum. The shape and particle size of granules, as well as the number and thickness of the growth rings, is strongly dependent on botanic origin.

Although the blocklet concept is not commonly mentioned in the literature, it was heavily supported and brought back to discussion by Gallant et al. [11] They suggested that both semicrystalline and amorphous growth rings are subdivided into respectively large (ca. 100 nm) and small (25 nm) spherical blocklets. Depending on the botanic origin of starch, authors report different blocklet sizes. Long axial diameters are 10–30 nm for maize starches [15], 20–50 nm for potato starches [16], and 25–100 nm for wheat starches [17].

Given that these blocklets have an average size of 100 nm in diameter, they are proposed to contain 280 amylopectin side chain clusters [18]. Schematically, the semicrystalline growth rings consist of a stack of repeated crystalline and amorphous lamellae (Fig. 8.2c). The thickness of the combined layers is 9 nm regardless of the botanic origin [11].

The crystalline lamellae are believed to be created by the interwining of amylopectin side chains with a linear length above 10 glucose units to form double helices [19]. These double helices are packed together to form the crystallites. The amorphous regions correspond to branching points of amylopectin. Gallant et al. [11] calculated that assuming that an amylopectin side chain cluster is 10 nm, a small blocklet (20–50 nm) is composed of about 2 to 5 side chain clusters. Tang et al. [13] illustrated this model as shown in Figure 8.2d, making amylopectin the backbone of the blocklet structure.

Amylose molecules are thought to occur in the granule as individual molecules, randomly interspersed among amylopectin molecules and in close proximity with one another, in both the crystalline and the amorphous regions [19]. Depending on the botanic origin of starch, amylose is preferably found in the amorphous region [20] (e.g., wheat starch), interspersed among amylopectin clusters in both the amorphous and the crystalline regions (e.g., normal maize starch), in bundles between amylopectin clusters [20], or co-crystallized with amylopectin (e.g., potato starch) [20].

8.2.3 Crystalline Type

Native starches usually contain between 15% and 45% of crystalline material. Depending on their X-ray diffraction (XRD) pattern, starches are categorized in three crystalline types called A, B, and C. Imberty et al. [21, 22] proposed a model for the double helices packing configuration to explain the difference between A and B types starches. A-type structures are closely packed with water molecules between each double helical structure, whereas B-types are more open and water molecules are located in the central cavity formed by six double helices as shown in Figure 8.3.

It was later envisaged that branching patterns of the different types of starch may also differ [23]. It was recently confirmed [24] that the branching density inside each cluster and the distance between two α(1-6) linkages are determining factors for the development of crystallinity in starch granules. Clusters with numerous short chains and short linkage distance produce densely packed structures, which crystallize into the A allomorphic type.

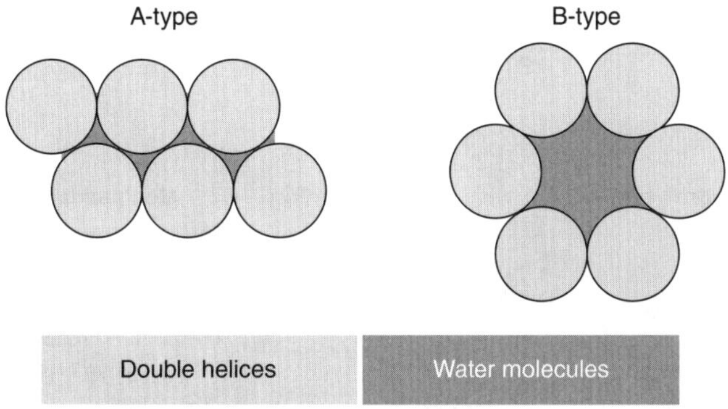

Figure 8.3 Double helices packing configuration according to crystalline type.

Longer chains and distances lead to a B-type. C-type starch pattern has been considered to be a mixture of both A- and B-types since its XRD pattern can be resolved as a combination of the previous two. It has been suggested by Bogracheva et al. [25] that C-type starch granules contain both types of polymorph: the B-type at the center of the granule and the A-type at the surrounding.

8.3 STARCH NANOPARTICLES

8.3.1 Preparation

Starch nanoparticles (SNP) can schematically be divided into two kinds: (1) those exploiting the semicrystalline property of starch, such as SNC; and (2) those exploiting the gelatinizing/melting property of starch referred to as regenerated or colloidal SNP. They can further be categorized by their production process as presented in Figure 8.4. SNC (discussed later) are mainly prepared by hydrolysis, while SNP are prepared by "regeneration" (chemical treatment or solubilization/precipitation) or mechanical treatment.

In recent studies, new chemical processes have been tried out to produce regenerated SNP. Ma et al. [32] and Tan et al. [31] prepared SNP by (1) precipitation of amorphous starch and cross-linking, while (2) Kim and Lim [30] combined complex formation and enzymatic hydrolysis to yield V-type nanocrystals (i.e., complexed with lipids). Only one attempt to prepare mechanically treated SNP by microfluidization, most likely to mimic treatments performed on cellulose to obtain microfibrillated cellulose (MFC), was reported by Liu et al. [35].

Most recently, studies have been focusing on combining mechanical and chemical treatments to yield SNP. Bloembergen et al. [33, 36] and Song et al. [34] prepared SNP by reactive extrusion (REx) (the cross-linking chemical

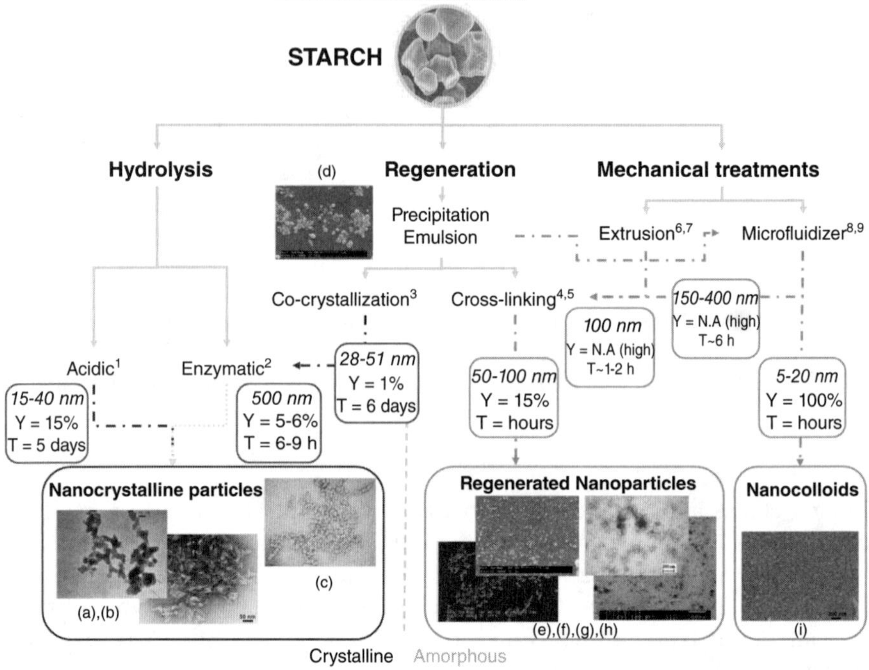

Figure 8.4 Different ways of producing crystalline and amorphous starch nanoparticles: hydrolysis leads to nanocrystals, whereas regeneration and mechanical treatment lead to both amorphous and crystalline particles [26]. **(1)** References to SNC production by *acid hydrolysis*: (a) TEM micrograph of SNC from Garcia et al. [27] and (b) TEM micrograph of SNC from Putaux et al. [28]. **(2)** Intent to produce SNC by *enzymatic hydrolysis* as reported by Kim et al. [29]. It is believed that the process leads to blocklets rather than nanoparticles. **(3)** Kim and Lim [30] reported the preparation of nanocrystals by *complex formation* between amylose and n-butanol. Enzymatic hydrolysis is used to selectively keep crystalline particles. (c) TEM micrograph of SNC after co-crytallization and enzymatic hydrolysis. **(4)** Tan et al. [31] and **(5)** Ma et al. [32] report the production of starch nanoparticles by *precipitation* of gelatinized starch in nonsolvent followed by cross-linking reaction: (e) starch nanoparticles from Tan et al. [31], (d) precipitated starch nanoparticles before cross-linking from Ma et al. [32], and (f) citric acid cross-linked starch nanoparticles from Ma et al. [32]. **(6)** Bloembergen et al. [33] and **(7)** Song et al. [34] report the production of starch nanoparticles by *reactive extrusion* (REx) of plasticized starch with a cross-linker. (g) SEM picture of extruded starch particles at 85°C with 2% glyoxal at 200 rpm (water and glycerol as plasticizers). **(8)** Starch nanoparticles prepared by *mini-emulsion cross-linking* as reported by Shi et al. and (h) the corresponding reflects TEM micrograph. **(9)** By analogy with microfibrillated cellulose (MFC), Liu et al. [35] intended to produce starch nanoparticles by *fluidization/homogenization*. It leads to crystalline microparticles turning into amorphous nanoparticles with an increasing number of run. (i) TEM micrograph of starch nanocolloid from Liu et al. [35]. Photos reproduced with permission from References [27, 28, 31], Copyright 2003, 2008, 2009 American Chemical Society; Reference [27], Copyright 2009 Wiley-VCH Verlag GmbH & Co. KGaA; Reference [30], Copyright 2008 Elsevier; Reference [35], Copyright 2009 Elsevier; Reference [34], Copyright 2011 Elsevier.

reaction takes place in the extruder) and Shi et al. [37] by a "mini-emulsion cross-linking" technique where the water-in-oil (*w/o*) mini-emulsion is obtained through high pressure homogenization and subsequently cross-linked.

In more details, Ma et al. [32] prepared SNP by precipitating a starch solution with ethanol as the precipitant. Native starch was mixed with water and completely gelatinized (90°C, 1 hour). Then ethanol was added dropwise to the solution at room temperature and stirred for 50 minutes. The suspension was then centrifuged with ethanol to remove water and the settled material was dried at 50°C to remove ethanol. The resulting nanoparticles were used to prepare a bionanocomposite.

Kim and Lim [30] studied alternative ways to obtain SNP. They proposed a method of preparing nanoscale starch particles by complex formation with other components. Experiments were conducted with n-butanol. Starch was dissolved in an aqueous dimethyl sulphoxide (DMSO) solution heated for 1 hour and stirred for 24 hours. The aliquot was passed through a polytetrafluoroethylene (PTFE) membrane filter with 10 μm pore size into the bottom compartment filled with n-butanol (6 days at 70°C). Maximum starch precipitation after 72 hours was 6.78% of the initial amount. Complex formation involved mainly amylose rather than amylopectin. The precipitate (starch–butanol complex) was collected by centrifugation and washed three times in n-butanol. However, the complex contained a large portion of amorphous matrix, so that its selective removal by enzymatic hydrolysis was needed. Since most of the starch was hydrolyzed (85–90%), the resulting yield of the nanoparticles was extremely low. SNP display spherical or oval shapes with diameters in the range of 10–20 nm.

SNP prepared by REx as reported by Bloembergen et al. [38] are patented and produced by Ecosynthetix as commercial product Eco-sphere™ (discussed later). However, their SNP formation mechanism during REx was not reported. Thus, Song et al. [34], based on Giezen et al. [39] 's anterior patent (2000), investigated the influence of extrusion conditions (such as temperature, screw speed, torque, the content proportion of starch and water) and cross-linker addition on particle size. Roughly, native starch granules are premixed with plasticizers (typically water and/or glycerol) and fed into an extruder. A cross-linker (typically glyoxal) is added in a downstream hopper of the extruder. The extrusion is conducted under a given temperature to allow for starch melting and cross-linking (initiation and curing) reaction. It was found that the addition of glyoxal during extrusion led to a higher torque and reduced particle size (from 300 nm to 160 nm). In this condition, an increase in reaction temperature and/or screw speed (i.e., strain) also strongly contributed to reducing particle size and even size distribution (for temperature). As expected, resulting nanoparticles were almost amorphous due to extrusion, and suspension viscosity was much lower than for gelatinized starch.

The most innovative approach is that of Shi et al. [37], referred to as "miniemulsion cross-linking." It combines techniques of *w/o* emulsion cross-linking (starch and cross-linker are the aqueous phase; and cyclohexane and surfactant are the oil phase) with a subsequent mini-emulsion. Mini-emulsion can typically be obtained chemically by creating chemical potential through the use of surfactants, or mechanically by emulsifiers, sonicators, or, as in this study, high pressure homogenization (1 to 5 passes). The cross-linking reaction, obtained at low temperatures using sodium trimetaphosphate (STMP), takes place during the stable phase of mini-emulsion. The sphere-like nanoparticles are precipitated using acetic acid and centrifugation. The influence of process parameters (*w/o* ratio, cross-linker concentration, homogenization pressure, and number of cycles) on particle size is also investigated. It was found that there was an optimal surfactant concentration in the oil phase, which was at about 6% (w/v), and in this concentration the particle size was the lowest (a minimum value). Also, it was found that pressure and number of cycles had the least influence on particle size, whereas an increase in water/oil and starch/STMP ratios remarkably increased particle size. Prepared SNP could be swelled in water (like those prepared by Rex) and were stable for several days at body temperature, allowing applications as drug carrier.

Liu et al. [35] applied the method of producing MFC to the production of starch colloids. A 5% slurry of high amylose corn starch was run through a Microfluidizer for several passes (up to 30). The particles size of samples obtained from more than 10 passes was below 100 nm, and the starch slurry changed from sol into gel. The gel-like suspension remained stable for more than a month when diluted to a concentration of 0.5 wt%. The thermal stability was not affected, and since no chemical or thermal degradation occurred during the treatment, the reported yield was almost 100%. However, the resultant starch colloids were obtained from breaking down both amorphous and crystalline domains, rendering an amorphous diffraction pattern after 10 passes.

8.3.2 Patented Starch Nanoparticles

To the best of our knowledge, there are no commercial SNC available on the market. In fact, only two patents—in Chinese and by the same authors—dealing with SNC have been found. They both deal with the preparation of polyurethanes from starch nanocrystal-grafted polyesters [40, 41]. All other patents and commercial products deal with regenerated SNP (presented Fig. 8.4).

Most popular commercial products are the Mater-Bi™ products (a wide family of grades) from Novamont and Eco-Sphere from Ecosynthetix.

An application of Novamont SNP as fillers was developed in collaboration with Goodyear and the product is called BioTRED. It is a biotire in which Mater-Bi is used to replace part of the lampblack and silica, which are usually contained in the tire mixture [42]. Instead of adding nanoscaled materials to the tire, the biopolymer itself was nanosized and used as a bionanofiller.

Eco-Sphere is a starch-based biolatex substitute for oil-based latexes providing adhesive applications such as paper coating and tissue complexing, or even as a replacement of polyvinyl acetate [42]. Starch granules are extruded and cross-linked simultaneously by REx. The process converts starch into a thermoplastic melt that is transformed into an agglomerate of dry cross-linked nanoparticles. The product is commercialized dried and has to be re-dispersed to form the latex [43]. WO Patent "Process for producing biopolymer nanoparticles" [39], written in English, describes the REx process used by EcoSynthetix.

Four other patents dealing with SNP have been identified. Two are Korean and from the same authors, Lim and Kim [29, 44]. The first one claims starch/alcohol complex nanoparticles as in the article of Kim and Lim [30] and the second, enzymatically hydrolyzed nanoparticles as reported in the article of Kim and Lim [29], and both are represented in Figure 8.3. The last two patents are in Chinese and relate to the production of SNP for water treatment and medical application, respectively. The first one describes cross-linked SNP grafted with lignin quaternary ammonium salt to be used as antichlor [45]. The second claims polylysine-SNP with cationic starch as a core and polylysine as a shell [46].

8.4 STARCH NANOCRYSTALS

First of all, it is important to clarify the terms commonly used. Starch crystallite, starch nanocrystal, microcrystalline starch, and sometimes hydrolyzed starch all refer to the crystalline part of starch obtained by hydrolysis but at different extents (from the most to the least). It has to be distinguished from SNP which can be amorphous.

8.4.1 General Description

First interest in SNC has been studied by analogy with cellulose nanocrystals (whiskers) to be used as reinforcing fillers in a matrix. In 1996, Dufresne et al. [47] reported the preparation of what was called "microcrystalline starch" at the time and which they described as agglomerated particles of a few tens of nanometers in diameters. The procedure consisted in hydrolyzing starch (5 wt%) in a 2.2N HCl suspension for 15 days. It was shown that classical models for polymers containing spherical particles could not explain the reinforcing effect of microcrystals. Further studies on the morphology of these microcrystals were conducted by Dufresne and Cavaillé [48] in light of aggregate formation and percolation concept. In 2003, Putaux et al. [28] revealed the morphology of SNC. Shapes and lateral dimensions were derived from observation of individual platelets in planar view: a marked 60–65°acute angles for parallelepiped blocks with a length of 20–40 nm and a width of 15–30 nm. However, more recent publications report bigger SNC (40–70 nm for potato SNC [49]; 30–80 nm [50] and 60–150 nm [51, 52] for pea SNC; and 50 nm [53] and 70–100 nm [54] for waxy SNC), with round edges [55] and found as grape-like aggregates of 1–5 μm. The heterogeneity in particle size could be explained not only by the differences in starch types but also by the difficulty of obtaining well-defined pictures of nonaggregated nanocrystals. Contrary to cellulose nanocrystals, SNC are not almost 100% crystalline but are rather 45–50% crystalline with variations depending on the botanic origin as recently presented by Le Corre et al. [56].

8.4.2 Acid Hydrolysis Kinetics

Acid hydrolysis has been used for a long time to modify starch and its properties. Nageli [57] reported the obtaining of a low molecular weight acid resistant fraction (known as Nageli amylodextrin) after the hydrolysis of potato starch, at room temperature during 30 days in a 15% H_2SO_4 suspension. Lintner [58] also gave his name to a hydrolysis process consisting in a 7.5% (w/v) HCl suspension of potato starch at 30–40°C to produce a high molecular weight starch suspension called "lintnerized starch." Most recent publications use either one of these two acidic conditions. In the industry, starch slurry is treated with dilute HCl or H_2SO_4 at 25–55°C for various periods of time, to produce "acid-thinned" starch used as sizing agents, in gum candies production, in paper and paper board production. Nowadays, most existing acid modifications aim at producing "soluble" starch or glucose syrup for biotechnologies with production of polylactic acid (PLA) or biofuels, for example. Only few studies have reported the use of acid hydrolysis to produce microcrystalline starch and SNC.

Despite differences in hydrolysis mechanisms, both enzymatic and acidic hydrolysis studies showed evidence of concentric "soft" layers which are more readily hydrolyzed. Although the origin and structure of the soft layer led to several debates in the 1960s, it was later accepted that it is always the less organized structure that degrades first [59] (corresponding to amylose), leaving more resistant amylopectin-based crystallites.

Hydrolysis has been described at different scales. At an atomic scale, it was found that α-1,6 bonds (found in amylopectin) are more resistant to hydrolysis than α-1,4 bonds [60]. The activation energy for hydrolyzing an α-1,6 (determined for an isomaltose) is 33 kcal/mol, whereas that of α-1,4 is 30 kcal/mol [61].

When observed at the microscopic scale, with an optical or scanning electron microscope, acid hydrolysis is decomposed in three steps [62]: (1) erosion of the surface, (2) collapse of the granule's wall due to radial diffusion of the acid, and (3) fragmentation of the growth rings. Indeed between the second and third step, hollow shells suggest that the outer layer is more acid resistant. Angellier [62] suggested that the acid diffuses from the surface of the granule to the hilum (center of the granule) through pores. This is in agreement with French's conclusions [14] assessing that the hilum is less acid resistant than the outer shell, and that the hydrolysis starts from the center of the granule.

At the macroscopic scale, techniques for following the extent of hydrolysis are numerous. The most common one consists in following the total solubilized

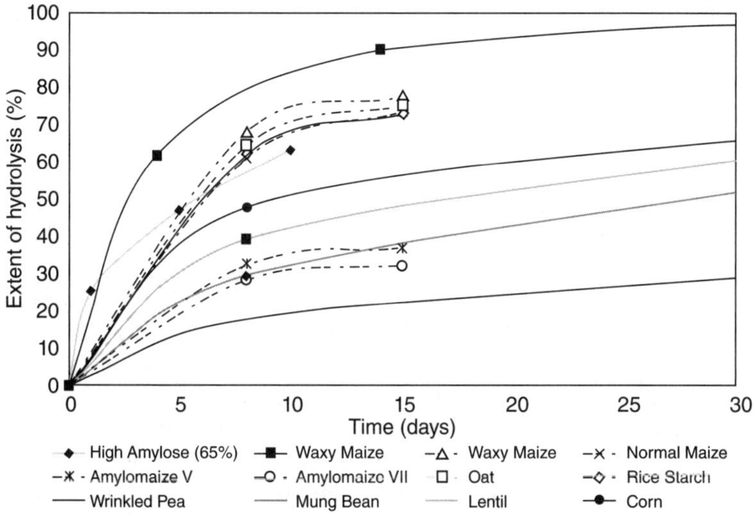

Figure 8.5 Hydrolysis kinetic of different starches. High amylose (65%) from Li et al. [63] (dotted line ····); waxy maize (■) from Angellier et al. [67] (plain line); waxy maize, normal maize, amylomaize V, amylomaize VII, oat, and rice starch from Jayakody and Hoover [64] (dash line----); wrinkled pea, mung bean, lentil, and corn from Robin et al. [68] (dot filled line —•—).

carbohydrates content [63]. It is sometimes expressed as a percentage of initial starch [64] or analyzed as molecular weight fractions [65]. To produce acid-thinned starch, Wang et al. [66] evaluated the extent of hydrolysis with viscosity of the resulting slurry, measured by Brabender Viscograph-E. For the preparation of SNC, Angellier et al. [7] followed the dry weight of nonsolubilized particles as a percentage of initial dry starch.

Figure 8.5 shows the hydrolysis kinetics for different starches from different research groups.

For all starches a two-stage hydrolysis profile can be evidenced: (1) an initial fast hydrolysis step, presumably due to the hydrolysis of the amorphous regions of starch granules, and (2) a second slower step, presumably due to the hydrolysis of the crystalline regions [64, 69]. Some authors distinguish three steps of hydrolysis: a rapid one, a slow one, and a very slow one [62], presumably corresponding to the hydrolysis of amorphous layers, semicrystalline layers, and crystalline ones, respectively. There are two common hypotheses to account for the slower hydrolysis rate of the crystalline domain. The first one is that the dense packing of starch in the crystalline regions does not readily allow the penetration of H_3O^+ [63]. The second one is that the hydrolysis of the glucosidic bonds requires a change in conformation from chair to half-chair [14].

The obvious factors influencing hydrolysis kinetic are: time, acid type, acid concentration, and temperature. The influence of acid and starch type has been studied extensively by Singh and Ali [65, 70–72]. They showed, through different characterization techniques (alkali fluidity number, number average molecular weight, intrinsic viscosity), that for the same equal normality, HCl and HNO_3 gave the highest thinning effect followed respectively by H_2SO_4 and H_3PO_4. Angellier et al. [7] also obtained a lower yield of hydrolysis with H_2SO_4 than with HCl for the production of nanocrystals but showed that final suspensions were more stable with H_2SO_4 due to the presence of sulfate groups at the surface. The difference in the rate and yield of hydrolysis among starch types was attributed to the difference in granule sizes and number of pores on the granule surface [71]. However, Biliaderis and Grant [69] thought that the granular organization had probably more influence. Jayakody and Hoover [64] studied further the susceptibility of different cereal starches to hydrolysis. They concluded that (1) the first stage of hydrolysis (amorphous regions) is influenced by the granule size, pores on the surface, amylose content, and the amount of lipid-complexed amylose chains, and (2) the second step of hydrolysis (crystalline region) is influenced by the amylopectin content, mode of distribution of α(1-6) branches between the amorphous and the crystalline regions, and degree of packing of the double helices within the crystallites (i.e., the parameters also influencing crystallinity). Wang et al. [66] observed as expected that the rate of hydrolysis increased when increasing the concentration of acid while the temperature was kept at 50°C.

Angellier et al. [7] studied the influence of hydrolysis parameters on SNC preparation using a surface response methodology. They concluded that both the acid concentration and temperature should not be too high. Since then hydrolysis conditions have not changed much among researchers.

8.4.3 Preparation Protocol

Only a limited number of articles concerning SNC preparation have been published during the past decade. They are listed in Table 8.1, when nanocrystals were produced for characterization, and in Table 8.2, when nanocrystals were produced for use in composite applications.

Most of them report the use of acid hydrolysis and even though a variety of starch sources has been used, they all refer to the same two processes:

- Dufresne et al.'s process using HCl as previously discussed [47]
- Angellier et al.'s optimized preparation process using H_2SO_4 [7]. A given amount of starch granules (14.69 wt%/acid) was mixed with a given volume of 3.16M H_2O_2 in an Erlenmeyer and stirred constantly at 100 rpm while maintained at 40°C. After 5 days, the suspension was washed by successive centrifugations with distilled water until neutrality using Centrifuge SIGMA 6K15, for 10 minutes at 10,000 rpm. The suspension was then submitted to a mechanical treatment with homogenizer Ultra Turrax

TABLE 8.1 Preparation Conditions of SNC and Their Characterization (Features [28, 47], Optimization [7], Kinetic [63], Molecular Structure [73])

Starch Type	C%	Acid Type	Concentration	Duration (days)	T (°C)	Year	Ref.
Potato	5.0%	HCl	2.2 N	15	35	1996	[47]
Waxy maize	5.0%	HCl	2.2 N	15	36	2003	[28]
Waxy maize	14.1%	H_2SO_4	2.87 M	7	40	2004	[7]
Waxy maize	14.7%	H_2SO_4	3.16M	5	40	2004, 2009 and 2011	[7, 73, 74]
Waxy maize	n.a.	HCl	Mild	42		2005	[75]
Hi-maize/Hylon VII	10.0%	HCl	2.2 N	8	40	2007	[63]
Rhizoma Dioscorea	12.5%	HCl	2.2 N	16/32/40	35	2008	[76]
High amylose corn	Enzymatic hydrolysis					2008	[29]
High amylose corn	Complex formation + enzymatic hydrolysis					2009	[30]
Pea starch	14.7%	H_2SO_4	3.16 M	5	40	2011	[77]

Adapted from Reference [26].
n.a., not available.

TABLE 8.2 Use of SNC in Composites and the Measured Properties

Starch		Acid		Duration	Temperature		Measured Properties									
Type	C%	Type	Concentration	(days)	°C	Matrix	X Ray	FT-IR	DMA	DSC	DMA	Tensile	WU	WVP	Year	Ref.
PS	5.0	HCl	2.2 N	15	30	Poly(S-co-BuA)							*		1998	[48]
WM	5.0	HCl / H_2SO_4			35/40	Poly(S-co-BuA)			*						2005	[67]
WM	14.7	H_2SO_4	3.16 M	5	40	Grafted SNC	*	*							2005	[78]
WM	14.7	H_2SO_4	3.16 M	5	40	Natural rubber							*	*	2005	[79]
WM	14.7	H_2SO_4	3.16 M	5	40	Natural rubber			*						2005	[80]
WM	14.7	H_2SO_4	3.16 M	5	40	Grafted SNC	*	*							2006	[81]
WM	14.7	H_2SO_4	3.16 M	5	40	Natural rubber					*				2006	[82]
WM	14.7	H_2SO_4	3.16 M	5	40	TPS (glycerol)			*			*			2006	[83]
WM	5.0	HCl	2.2 N	15	35	Pullulan (sorbitol)	*		*			*	*	*	2007	[84]
WM	14.7	H_2SO_4	3.16 M	5	40	TPS (sorbitol)			*	*	*	*			2007	[85]
WM	14.7	H_2SO_4	3.16 M	5	40	Grafted SNC	*	*							2007	[86]
NM	14.7	H_2SO_4	3.16 M	5	40	Grafted SNC	*	*							2007	[87]
PS	14.7	H_2SO_4	3.16M	5	40	Waterborne polyurethane (WPU)					*				2008	[49]

Pea	14.7	H_2SO_4	3.16 M	5	40	PCL-grafted SNC PLA matrix	*	*	*	*		*			2008	[88]
Pea	14.7	H_2SO_4	3.16 M	5	40	PCL-grafted SNC WPU matrix	*	*	*	*		*			2008	[89]
PS	5.0	HCl	2.2 N	15	30	PC-grafted SNC	*	*		*					2008	[90]
WM	14.7	H_2SO_4	3.16 M	5	40	Cassava and waxy with glycerol	*	*	*				*	*	2009, 2010	[27, 91]
Pea	14.7	H_2SO_4	3.16 M	5	40	Soy protein isolate (SPI)	*		*	*		*	*		2009	[52]
WM	14.7	H_2SO_4	3.16 M	5	40	Isociyanate cross-linked SNC PPG-based PU	*			TGA					2010	[92]
Pea	14.7	H_2SO_4	3.16 M	5	40	Cyclodextrin-based hydrogel	*				*				2010	[93]
Pea	14.7	H_2SO_4	3.16 M	5	40	Sodium alginate (SA)	*	*					*		2011	[94]
Pea	14.7	H_2SO_4	3.16 M	5	40	PBS/SA	*	*	*	*	*	*			2011	[95]
							*	*		*						[94]

Adapted from Reference [26].

PS, potato; WM, waxy maize; NM, normal maize; PU, polyurethane; TPS, thermoplastic starch; PBS/SA, poly(butylene succinate)/sodium alginate.

> for 2 minutes at 13,000 rpm to disperse aggregates and obtain a "stable" suspension. To avoid bacterial development during storage, a few drops of chloroform were added to the suspension that was kept in a 4°C cell.

Using the above preparation conditions, LeCorre et al. [96] also evidenced two stages when studying the kinetics of SNC preparation. In an attempt to classify SNC using differential centrifugation, they revealed that SNC were already found in the suspension after only 24 hours of hydrolysis, and that at any time during the 5 days, both micro- and nanoparticles could be found in suspension. This led to the very recent development of a continuous extraction technique using ceramic membrane filtration to extract SNC as soon as they "released" [97]. Extracted SNC were more crystalline, more homogeneous in size and smaller. Named "second-generation" SNC, such nanocrystals are very promising for industrial applications because their preparation process strongly limit yield and heterogeneity issues.

To our knowledge, no work on a purely enzymatic hydrolysis method for producing SNC has been developed. However, an attempt has been reported by Kim and Lim [29] rendering 500-nm particles believed to be starch blocklets.

Most recently, LeCorre et al. [98] investigated the use of enzymes as a pretreatment for starch granules before acid hydrolysis to reduce preparation time. Using enzymatic pretreatment, duration was reduced by three times by creating microporous starch having preferential acid diffusion pathways. However, at the current stage of development, yield was also reduced.

All previous observations could not have been done without the development and adaptation of a number of characterization techniques to SNC.

8.4.4 Starch Nanocrystal Characterization

8.4.4.1 Morphology Although first interest in SNC dates back to 1996, first visual characterization was revealed in 2003 by Putaux et al. [28]. They revealed the morphology of "nanocrystals resulting from the disruption of waxy maize starch granules by acid hydrolysis." After 6 weeks of hydrolysis, transmission electron microscopy (TEM) observations showed a longitudinal view of lamellar fragments consisting of stack of elongated elements, with a width of 5–7 nm, and a planar view of individualized platelet after hydrolysis. In this technique, the contrast of the sample is enhanced by negative staining with uranyl acetate. First, a dilute SNC suspension deposited onto a previously treated (by 30 seconds electric discharge to increase surface tension and allow for better deposition) copper carbon-coated microscopy grid. After a minute, the liquid in excess is blotted with filter paper and the staining solution is deposited on top. After another minute, the excess liquid is again blotted with filter paper and the grid is set aside to dry out. As observed materials (SNC) are lighter than the staining agent, SNC are revealed as white elements on a dark field. This technique is still the most commonly used [50, 54, 73, 74, 87, 88, 91, 93, 99–101].

Around the same time, another characterization technique was based on laser granulometry measurements. Angellier et al. [7] reported the use of a Malvern Mastersizer (measuring particles from 0.05 to 900 μm) to determine optimal preparation conditions for SNC. More recently, dynamic light scattering was used for particle size measurements using a Malvern Zetasizer (with maximum size ranging from 0.3 nm to 10 microns). However, results from this granulometry technique can easily be disturbed sedimentation, aggregation, or ill-defined shapes.

More recently, the development and accessibility of new techniques has led to further characterization of SNC. For example, recent studies [98, 102] report the use of atomic force microscopy (AFM), a high resolution scanning probe. This measurement technique requires little preparation and provides quick analysis. A drop of diluted SNC suspension (0.01%) is simply deposited onto a mica substrate (split with adhesive tape) and dried for a few minutes. Then the measurement is performed in both the tapping and the conductive mode (C-AFM).

Another increasingly popular microscopy characterization of SNC is conducted with a field electron gun scanning electron microscope (FEGSEM) [27, 91, 96, 97, 102, 103]. Indeed, this kind of scanning electron microscope (SEM) allows high resolution (0.8 nm) images in both the reflection mode (topography information) and the transmission mode (as in TEM). Samples are simply prepared by depositing 2.5 μL of diluted (0.01–0.001%) and sonicated SNC suspension on a previously treated (as previously described) TEM grid. A thin Au–Pd conductive coating (about 1 nm) is deposited on top of the sample to reduce charge effect. In order to obtain the best possible resolution, the secondary electron imaging mode with the In-lens detector is preferred (reflective mode). However, some samples are better detected in the transmission mode using the scanning transmission electron microscopy (STEM) detector. The average dimensions of SNC can then be determined using digital image analyses.

8.4.4.2 Crystallinity As for other crystals, the crystallinity of SNCs can be verified using XRD. They display the same types of diffraction patterns as native starches. Indeed, SNC correspond to the crystallites of starch granules (as described in Fig. 8.2). Thus, they are made of closely packed amylopectin side chains. As such, depending on the packing configuration of the crystallites, they can display A-type or B-type diffraction patterns or even a combination of the two if the two types of crystallites are present.

Also, X-ray diffractograms of SNC are usually better defined than that of the native starch. The “amorphous background” of the diffractogram is reduced. This is a proof that SNC are more crystalline than their native counterparts and that some amorphous starch has indeed been hydrolyzed [98, 102].

Most recent studies use XRD as a mean to verify that SNC have not lost their crystallinity subsequent to a treatment. In most cases, the treatment is a chemical grafting intended to improve SNC thermal stability, compatibility with matrices and reinforcing effect [78, 81, 86–90, 92, 93, 95], and, recently, hydrophobicity for water treatment purposes [104]. Others used XRD to evidence the presence of SNC in polymeric matrices [84, 91, 100].

8.4.4.3 Thermal Properties Studies on the thermal behavior of SNC are scarce in the literature. To our knowledge, only two studies [62, 81] deal with the thermal behavior of freeze-dried waxy maize SNC. They both involve the use of differential scanning calorimetry (DSC). Angellier [62] reported the existence of two endothermal peaks in excess water with a broad temperature range and the disappearance of the first peak in the dry state. In excess water, the peaks were located at 80°C and 178°C for 83.78% water content, and at 109°C and 158°C for 50.66% water content. In the dry state, the maximum peak was located at 130°C and 139°C. Thielemans et al. [81] presented a DSC thermogram for waxy maize SNC in dry conditions for comparison and evidencing of grafting. The thermogram for freeze-dried SNC also showed two endothermic peaks. The first maximum peak was located around 150°C and the second one, around 200°C. More recently, a study [103] compared the thermal properties of SNC prepared from different starch sources and proposed a mechanism to explain observations.

In all cases [62, 81, 103], with freeze-dried SNC, about 5 to 10 mg of powder is conditioned at 0% relative humidity and placed in a hermetically closed DSC sealed pan. The samples are then tested in the range of –100°C to 350°C at a heating rate of 10°C/min under nitrogen atmosphere. When tested in water [103], deionized water is added to the freeze-dried SNC to reach the targeted water content.

8.4.5 Starch Nanocrystals from Different Starch Sources

Cellulose nanocrystals, discussed in more detail in other chapters, are called whiskers because of their shape. They exhibit very different sizes and properties when they are prepared from different sources. Thus, two recent studies [102, 103] compared five different kinds of starches used to prepare SNC and characterized them to assess potential differences.

In the first study [102], SNC are compared in terms of size, shape, crystallinity, and viscosity. Parameters that had the most influence were amylose content and crystalline type of native starches. They reflected mainly on the morphology of SNC and somewhat on their size. Starch botanic origin showed only a moderate influence on size distribution, thickness, and viscosity, due to differences in amylose distribution and blocklet sizes among starches. The study shows that, depending on the application, some criteria can be found for selecting one type of native starch over another. For example, some SNC show

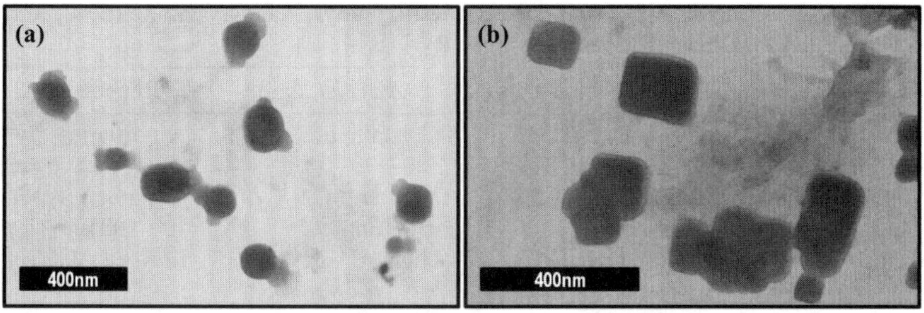

Figure 8.6 Starch nanocrystals from (a) high amylose SNC and (b) wheat SNC.

rather square platelet-like morphology while others are other disk-like, as presented in Figure 8.6.

The second study [103] completes the previous with the analysis of the thermal properties of the same five types of SNC. In both hydrous and anhydrous conditions, SNC exhibited a two-phase melting transition at temperatures corresponding to the melting of perfect crystallites, before depolymerization. The study showed that the amylose content of the native starch has little influence on the final thermal properties of SNC. On the contrary, differences between starches observed in the native state are somewhat compensated when in SNC. However, it seems that, in excess water, B-type crystallites (with longer chain lengths) are more stable and that, in dry conditions, maize SNC showed least homogeneous crystallites. Finally, it was concluded that SNC can be used in wet processes at temperatures lower than 80–100°C, and in dry processes at temperatures below 150–200°C.

Thus, it seems that SNC can be prepared from different starch sources without rendering substantial differences in morphological, rheological, and thermal properties contrary to cellulose nanocrystals. Yet some moderate differences among SNC are observed, giving leads for potential discrimination among sources for a given application.

Information that is often sought for not only compares elements within one type of nanoparticles but also relates to differences between the different types of nanoparticles such as regenerated SNPs (rSNPs) and SNCs.

8.5 COMPARISON OF RSNP AND SNC

As described in previous sections, both types of nanoparticles present interesting and unique properties compared to native starch. Main differences in structural and processing properties of native, regenerated, and nanocrystalline starches are shown in Table 8.3.

As represented in Figure 8.4, the most obvious discriminating property among SNP is crystallinity. As previously described, native starches are

TABLE 8.3 Comparison of Some Properties of Native, Nanocrystalline (SNC) and Regenerated Starches Nanoparticules (rSNP)

Type of Starch	Native	Nanocrystalline	Regenerated
Crystallinity	Semicrystalline	Crystalline	Amorphous
Index %	40	>40	<40 (much lower)
Type	A,B, and C	A, B, and C	None or V-type
Size	20–25 μm	<100 nm	100–500 nm
Density (g/cm^3)	Ref. (1.1) [105]	Higher (1.2) [106]	Lower (1[a])
Thermal stability	Ref.		
T_m	120–150°C [103]	150–250°C [103]	
Degradation	300°C [103] 313–327°C [32]	[a]251°C [103]	285°C–294°C [32]
Viscosity (Pa·s at 100 s^{-1})	0.1 at 3% [33]	0.001–0.1 [102]	0.01 at 9% [34]

[a]H_2SO_4 hydrolysis (as opposed to HCl).

semicrystalline polymers with an average crystallinity index of 40% and that can present different crystalline types (A, B, or C). SNC are prepared by elimination of the amorphous parts of native starch. Thus, SNC present higher crystallinity than their native counterparts and the same crystalline type. On the contrary, rSNP are prepared from gelatinized or melted starch, that is, from amorphous starch. Thus, rSNP present the crystalline properties of, at most, a retrograded (i.e., V-type crystallinity before cross-linking) or extruded starch (crystalline index of 2–4% [107]), or of fully amorphous starch.

Reported particles sizes also differ. rSNP tend to be a few hundreds of nanometers and thus slightly bigger than SNC, which are often less than 100 nm. This difference is mainly due to the preparation processes. SNC prepared using H_2SO_4 present sulfate ($-SO_3^-$) groups at their surface, which favors individual SNC and stabilizes the suspension. rSNP with sizes smaller than 100 nm are difficult to obtain as precipitated, emulsified, or extruded SNP tend to aggregate before cross-linking, thus rendering bigger particles.

Starch density can vary greatly depending on authors, starch sources, and even within a given starch granule. For instance, Perry et al. [106] reported large differences between density of crystalline lamella of waxy maize and potato starch (1.2 and 1.1, respectively) and their amorphous growth rings (0.7 and 0.5, respectively). Therefore, it is suspected but not reported that SNC have higher density than native starch. Density of rSNP is also scarcely reported. Only Bloembergen et al. [38] postulated that their rSNP biolatex has a virtual density (as coated) that approaches 1.0, similar to that of carboxylated styrene butadiene (SB) latex, due to the creation of a nanocellular "foam-like" structure upon drying.

Data concerning viscosity are also scarce. Bloemberg et al. compared the viscosity of rSNP biolatex at 9 wt% and cooked starch at 3 wt%. They were

0.1 Pa· and 0.01 Pa·s at 100 s^{-1}, respectively, achieving a decrease in viscosity of one decade for similar solid contents. Bras et al. [108] compared the suspension viscosity of 5 wt% SNC suspensions prepared from different native starch sources and reported viscosity of 0.001 to 0.1 Pa·s at 100 s^{-1}, thus offering a wider range of viscosity depending on application.

Current applications for SNC and rSNP are mainly focused on nanocomposites or paper coating, showing the importance of the thermal stability of these nanoparticles. rSNP are based on melted starch; thus, their melting temperature is not studied nor reported. However, Ma et al. [32] reported the degradation temperature of citric acid (CA) cross-linked rSNP to be lower (285°C and 294°C for pea and rice starch rSNP, respectively) than that of native starches (313°C and 327°C, respectively). This was ascribed to the ester bonds from the substitution (due to cross-linking) that break at lower temperatures of degradation. The same observations were reported for SNC by LeCorre et al. [103]. The lower thermal stability of SNC (251°C) compared with native starches (about 300°C) was ascribed to the presence of sulfonic acid groups at the surface of SNC [62], which thermally decompose between 200°C and 300°C [109]. In that respect, rSNP and SNC are comparable.

Final deference between rSNP and SNC is linked to their morphology. rSNP are sphere-like nanoparticles and the reported dimension corresponds to the sphere's diameter. SNC are platelet-like nanoparticles with thickness of 5–10 nm and the reported dimension corresponds to the diameter that a disk-like nanoparticles would have. This particularity leads to SNC being preferred for barrier applications, although a comparative assessment of barrier properties brought by SNC and rSNP has never been performed. Some authors have measured the water vapor barrier of a 30% glycerol plasticized peach starch nanocomposite filled with CA modified pea rSNP [32]. They are in the same magnitude as those reported for a 50% glycerol plasticized cassava starch nanocomposite filled with normal maize SNC [27]. Also, it has been shown that SNC can reduce the oxygen permeability of natural rubber nanocomposite. No such measurements were performed on rSNP.

Current available properties and performances do not allow for a proper comparison of rSNP and SNC. Current research is focus on preparation processes and the feasibility of preparing SNP with dimensions below 100 nm. Thus available characterizations focus on structural properties or specific properties linked to targeted application. rSNP are mainly considered for mechanical reinforcement or as latex/adhesive replacement, whereas SNC are developed and optimized for barrier applications.

8.6 CONCLUSIONS AND PROSPECTS

Starch is a suitable semicrystalline polymer for preparing renewable and potentially biodegradable nanoparticles. Thus, several techniques for preparing SNP have been developed over the years and render different types of

SNP. On the one hand, when performing a hydrolysis (of the amorphous parts of starch), nanocrystalline particles (SNC) are prepared. On the other hand, when implementing a precipitation, a chemico-mechanical or a mechanical treatment, regenerated nanoparticles (rSNP), or nanocolloidal SNP, based on amorphous starch are prepared.

SNC combine several advantages. They are renewable, biodegradable [74], nanoscaled, crystalline, and present platelet-like morphology favorable for seeking barrier properties. Also, contrary to cellulose whiskers, little differences are found between SNC from different sources, allowing many raw materials to be considered as sources [102]. However, they are currently only produced in small amounts (250 mL), in five days and in water-based suspension. Very recent study proposes new preparation process that could lead to a "second generation" of SNC with more homogeneous size distribution and thus possibly enhanced reinforcing and barrier properties.

rSNP, on the contrary, can easily be produced at industrial scale, can present a neutral carbon footprint [38] (no study on rSNP biodegradability was found), and are sometimes nanoscaled (depending on authors). However, they are not crystalline, and they do not present a sphere-like morphology, which is the least favorable for barrier properties.

Both types of nanoparticles present increased thermal stability compared to native starch (but lower degradation temperature) and a positive reinforcing effect in nanocomposite applications. For further comparison and possible discrimination, SNC should be tested as a binder replacement and rSNP as a nanocomposite material; and both preparation processes should be compared from an environmental point of view, that is, using comparable methods (life cycle analysis, carbon neutrality, or biodegradability).

REFERENCES

[1] Geraci, C.L. (2011) *2011 TAPPI Intl Conference on Nano for Renewable Materials*, Washington, USA 2011.

[2] Habibi, Y., Lucia, L., and Rojas, O. (2010) Cellulose nanocrystals: Chemistry, self-assembly, and applications. *Chemical Reviews*, *110*, 3479–5000.

[3] Paillet, M., Dufresne, A. (2001) Chitin whisker reinforced thermoplastic nanocomposites. *Macromolecules*, *34*, 6527–6530.

[4] Gopalan Nair, K., Dufresne, A. (2003) Crab shell chitin whisker reinforced natural rubber nanocomposites. 2. Mechanical behavior. *Biomacromolecules*, *4*, 666–674.

[5] Morin, A., Dufresne, A. (2002) Nanocomposites of chitin whiskers from riftia tubes and poly(caprolactone). *Macromolecules*, *35*, 2190–2199.

[6] Whistler, R.L., Paschall, E.F. (1965) *Starch: Chemistry and Technology*. New York: Elsevier.

[7] Angellier, H., Choisnard, L., Molina-Boisseau, S., Ozil, P., and Dufresne, A. (2004) Optimization of the preparation of aqueous suspensions of waxy maize starch

nanocrystals using a response surface methodology. *Biomacromolecules*, *5*, 1545–1551.

[8] Dufresne, A. (2007) Polymer nanocomposites from biological sources. In: *Biopolymers Technology*, ed. A.C. Bertolini. Sao Paulo: Cultura Academica.

[9] Buléon, A., Colonna, P., Planchot, V., and Ball, S. (1998) Starch granules: Structure and biosynthesis. *International Journal of Biological Macromolecules*, *23*, 85–112.

[10] Hizukuri, S. (1996) Starch: Analytical aspects. In: *Carbohydrates in Food*, ed. A.-C. Eliasson. New York: Marcel Dekker, pp. 347–429.

[11] Gallant, D.J., Bouchet, B., and Baldwin, P.M. (1997) Microscopy of starch: Evidence of a new level of granule organization. *Carbohydrate Polymers*, *32*, 177–191.

[12] Donald, A., Jenkins, P., and Waigh, T. (1997) Council for the Central Laboratory of the Research Councils 1997, p. 54.

[13] Tang, H., Mitsunaga, T., and Kawamura, Y. (2006) Molecular arrangement in blocklets and starch granule architecture. *Carbohydrate Polymers*, *63*, 555–560.

[14] French, D. (1984) Starch: Chemistry and technology. In: *Starch: Chemistry and Technology*, eds. R.L. Whistler, J.N. BeMiller, and E.F. Paschall. New York: Academic, pp. 183–247.

[15] Baker, A.A., Miles, M.J., and Helbert, W. (2001) Internal structure of the starch granule revealed by AFM. *Carbohydrate Research*, *330*, 249–256.

[16] Szymonska, J., Krok, F. (2003) Potato starch granule nanostructure studied by high resolution non-contact AFM. *International Journal of Biological Macromolecules*, *33*, 1–7.

[17] Perez, S., Baldwin, P.M., and Gallant, D.J. (2009) Structural features of starch granules I. In: *Starch: Chemistry and Technology*, 3rd ed, eds. J. BeMiller, R. Whistler. New York: Elsevier, p. 176.

[18] Vandeputte, G.E., Delcour, J.A. (2004) From sucrose to starch granule to starch physical behaviour: A focus on rice starch. *Carbohydrate Polymers*, *58*, 245–266.

[19] Oates, C.G. (1997) Towards an understanding of starch granule structure and hydrolysis. *Trends in Food Science and Technology*, *8*, 375–382.

[20] Blanshard, J.M.V. (1987) Starch granule structure and function: A physicochemical approach. In: *Starch: Properties and Potentials*, ed. T. Galliard. London, UK: Society of Chemical Industry, pp. 16–54.

[21] Imberty, A., Chanzy, H., Perez, S., Buleon, A., and Tran, V. (1987) New three-dimensional structure for A-type starch. *Macromolecules*, *20*, 2634–2636.

[22] Imberty, A., Perez, S. (1988) A revisit to the three-dimensional structure of B-type starch. *Biopolymers*, *27*, 1205–1221.

[23] Jane, J.-l., Wong, K.-s., and McPherson, A.E. (1997) Branch-structure difference in starches of A- and B-type X-ray patterns revealed by their Naegeli dextrins. *Carbohydrate Research*, *300*, 219–227.

[24] Gérard, C., Planchot, V., Colonna, P., and Bertoft, E. (2000) Relationship between branching density and crystalline structure of A- and B-type maize mutant starches. *Carbohydrate Research*, *326*, 130–144.

[25] Bogracheva, T.Y., Morris, V.J., Ring, S.G., and Hedley, C.L. (1998) The granular structure of C-type pea starch and its role in gelatinization. *Biopolymers*, *45*, 323–332.

[26] Le Corre, D., Bras, J., and Dufresne, A. (2010) Starch nanoparticles: A review. *Biomacromolecules*, *11*, 1139–1153.

[27] García, N.L., Ribba, L., Dufresne, A., Aranguren, M.I., and Goyanes, S. (2009) Physico-mechanical properties of biodegradable starch nanocomposites. *Macromolecular Materials and Engineering*, *294*, 169–177.

[28] Putaux, J.L., Molina-Boisseau, S., Momaur, T., and Dufresne, A. (2003) Platelet nanocrystals resulting from the disruption of waxy maize starch granules by acid hydrolysis. *Biomacromolecules*, *4*, 1198–1202.

[29] Kim, J.-Y., Park, D.-J., and Lim, S.-T. (2008) Fragmentation of waxy rice starch granules by enzymatic hydrolysis. *Cereal Chemistry*, *85*, 182–187.

[30] Kim, J.-Y., Lim, S.-T. (2009) Preparation of nano-sized starch particles by complex formation with n-butanol. *Carbohydrate Polymers*, *76*(1), 110–116.

[31] Tan, Y., Xu, K., Li, L., Liu, C., et al. (2009) Fabrication of size-controlled starch-based nanospheres by nanoprecipitation. *ACS Applied Materials & Interfaces*, *1*, 956–959.

[32] Ma, X., Jian, R., Chang, P.R., and Yu, J. (2008) Fabrication and characterization of citric acid-modified starch nanoparticles/plasticized-starch composites. *Biomacromolecules*, *9*, 3314–3320.

[33] Bloembergen, S., Lee, D.I. (2010) *TAPPI PaperCon*, TAPPI, Atlanta GA 2010, p. 18.

[34] Song, D., Thio, Y.S., and Deng, Y. (2011) Starch nanoparticle formation via reactive extrusion and related mechanism study. *Carbohydrate Polymers*, *85*(1), 208–214.

[35] Liu, D., Wu, Q., Chen, H., and Chang, P.R. (2009) Transitional properties of starch colloid with particle size reduction from micro- to nanometer. *Journal of Colloid and Interface Science*, *339*, 117–124.

[36] Bloembergen, S., Kappen, F., and Beelen, B. (2002) Environmentally friendly biopolymer adhesives and applications based thereon, p. 18.

[37] Shi, A.-m., Li, D., Wang, L.-j., Li, B.-z., and Adhikari, B.-z. (2011) Preparation of starch-based nanoparticles through high-pressure homogenization and miniemulsion cross-linking: Influence of various process parameters on particle size and stability. *Carbohydrate Polymers*, *83*(4), 1604–1610.

[38] Bloembergen, S., Mc Lennan, I., Lee, D.I., and Van Leeuwen, J. (2008) Paper binder performance with biobased nanoparticles. *TAPPI Journal—paper 360*, *3*(9), 46–48.

[39] Giezen, F.E., Jongboom, R.O.J., Feil, H., Gotlieb, K.F., and Boersma, A. (2000) Process for producing biopolymer nanoparticles, p. 17.

[40] Huang, J., Chen, G., Dufresne, A., Wei, M., and Cui, G. (2007) Method for preparing starch nano-crystal-modified aqueous polyurethane. In: Shuomingshu, F. Z. S. G. (Ed.), China 2007.

[41] Huang, J., Yi, F., Zhang, R., Xia, W., and Wei, M. (2008) Preparation of polyurethanes from polysaccharide nanocrystals-grafted polyesters. China 2008.

[42] Novamont. Mater Bi's application as additives. http://www.novamont.com/default.asp?id=514 (accessed March 2011).

[43] Ecosynthetix. Ecosphere. http://www.ecosynthetix.com/ecosphere.html (accessed June 2009).

[44] Lim, S.T., Kim, J.Y. (2009) Crystalline nanoscale starch-alcohols complex, and method for manufacturing starch nanoparticles with this complex. In: Kongbo,

R. K. K. T. (Ed.), Repub. Korean Kongkae Taeho Kongbo, Republic of Korea 2009.

[45] Ma, B., Zhu, X. (2009) Method for preparing polysaccharide nanoparticles as antichlor used in industrial water purification. In: Shuomingshu, F. Z. S. G. (Ed.), Farming Zhuanli Shenqing Gongkai Shuomingshu, China 2009.

[46] Liu, X., Xiao, S., Liu, B., Liu, J., and Tang, D. (2003) Polylysine/starch nanoparticles, its preparation and application as gene carrier. In: Shuomingshu, F. Z. S. G. (Ed.), Farming Zhuanli Shenqing Gongkai Shuomingshu, China 2003.

[47] Dufresne, A., Cavaille, J.-Y., and Helbert, W. (1996) New nanocomposite materials: Microcrystalline starch reinforced thermoplastic. *Macromolecules*, *29*, 7624–7626.

[48] Dufresne, A., Cavaillé, J.-Y. (1998) Clustering and percolation effects in microcrystalline starch-reinforced thermoplastic. *Journal of Polymer Science Part B: Polymer Physics*, *36*, 2211–2224.

[49] Chen, G., Wei, M., Chen, J., Huang, J., et al. (2008) Simultaneous reinforcing and toughening: New nanocomposites of waterborne polyurethane filled with low loading level of starch nanocrystals. *Polymer*, *49*, 1860–1870.

[50] Chen, Y., Cao, X., Chang, P.R., and Huneault, M.A. (2008) Comparative study on the films of poly(vinyl alcohol)/pea starch nanocrystals and poly(vinyl alcohol)/ native pea starch. *Carbohydrate Polymers*, *73*, 8–17.

[51] Yu, J., Ai, F., Dufresne, A., Gao, S., et al. (2008) Structure and mechanical properties of poly(lactic acid) filled with (starch nanocrystal)-graft-poly(ε-caprolactone). *Macromolecular Materials and Engineering*, *293*, 763–770.

[52] Zheng, H., Ai, F., Chang, P.R., Huang, J., and Dufresne, A. (2009) Structure and properties of starch nanocrystal-reinforced soy protein plastics. *Polymer Composites*, *30*, 474–480.

[53] García, N.L., Famá, L., Dufresne, A., Aranguren, M., and Goyanes, S. (2009) A comparison between the physico-chemical properties of tuber and cereal starches. *Food Research International*, *42*, 976–982.

[54] Namazi, H., Dadkhah, A. (2010) Convenient method for preparation of hydrophobically modified starch nanocrystals with using fatty acids. *Carbohydrate Polymers*, *79*, 731–737.

[55] Wang, Y., Zhang, L. (2008) High-strength waterborne polyurethane reinforced with waxy maize starch nanocrystals. *Journal of Nanoscience and Nanotechnology*, *8*, 5831–5838.

[56] Le Corre, D., Bras, J., and Dufresne, A. (2009) Starch nanoparticles for eco-efficient packaging: influence of botanic origin. In: Jimenez, A. (Ed.), 2nd I*nternational Conference on Biodegradable Polymers and Sustainable Composites (BIOPOL 2009)*, Alicante, Spain 2009.

[57] Nägeli, W. (1874) Beiträge zur näheren Kenntnifs der Stärkegruppe. *Justus Liebigs Annalen der Chemie*, *173*, 218–227.

[58] Lintner, C.J. (1886) Studien über Diastase. *Journal fur Praktische Chemie*, *34*, 378–394.

[59] Gallant, D.J. (1974) Contribution à l'étude de la structure et de l'ultrastructure du grain d'amidon. University of Paris VI, Paris, France 1974.

[60] Wolfrom, M.L., Thompson, A., and Timberlake, C.E. (1963) Comparative hydrolysis rates of the reducing disaccharides of D-glucopyranose. *Cereal Chemistry*, *40*, 82–86.

[61] Bemiller, J.N. (1967) Acid-catalized hydrolysis of glucosides. 5. Hydrolysis of polysaccharides. In: *Advanced Carbohydrate Chemistry*, pp. 85–108.

[62] Angellier, H. (2005) *Material Sciences*, Grenoble, France: University Joseph Fourier.

[63] Li, W., Corke, H., and Beta, T. (2007) Kinetics of hydrolysis and changes in amylose content during preparation of microcrystalline starch from high-amylose maize starches. *Carbohydrate Polymers*, *69*, 398–405.

[64] Jayakody, L., Hoover, R. (2002) The effect of lintnerization on cereal starch granules. *Food Research International*, *35*, 665–680.

[65] Singh, V., Ali, S.Z. (1987) Comparative acid modification of various starches. *Starch/Stärke*, *39*, 402–405.

[66] Wang, Y.-J., Truong, V.-D., and Wang, L. (2003) Structures and rheological properties of corn starch as affected by acid hydrolysis. *Carbohydrate Polymers*, *52*, 327–333.

[67] Angellier, H., Putaux, J.-L., Molina-Boisseau, S., Dupeyre, D., and Dufresne, A. (2005) Starch nanocrystal fillers in an acrylic polymer matrix. *Macromolecular Symposia*, *221*, 95–104.

[68] Robin, J.P., Mercier, C., Charbonniere, R., and Guilbot, A. (1974) Gel filtration and enzymatic studies of insoluble residues from prolonged acid treatment of potato starch. *Cereal Chemistry*, *51*, 389–406.

[69] Biliaderis, C.G., Grant, D.R., and Vose, J.R. (1981) Structural characterization of legume starches I. Studies on amylose, amylopectin, and beta-limit dextrins. *Cereal Chemistry*, *58*, 496–502.

[70] Singh, V., Ali, S.Z., and Divakar, S. (1993) 13C CP/MAS NMR spectroscopy of native and acid modified starches. *Starch/Staerke*, *45*, 59–62.

[71] Singh, V., Ali, S.Z. (2000) Acid degradation of starch. The effect of acid and starch type. *Carbohydrate Polymers*, *41*, 191–195.

[72] Singh, V., Ali, S.Z. (2008) Properties of starches modified by different acids. *International Journal of Food Properties*, *11*, 495–507.

[73] Angellier-Coussy, H., Putaux, J.-L., Molina-Boisseau, S., Dufresne, A., et al. (2009) The molecular structure of waxy maize starch nanocrystals. *Carbohydrate Research*, *344*, 1558–1566.

[74] Kümmerer, K., Menz, J., Schubert, T., and Thielemans, W. (2011) Biodegradability of organic nanoparticles in the aqueous environment. *Chemosphere*, *82*, 1387–1392.

[75] Putaux, J.-L. (2005) Morphology and structure of crystalline polysaccharides: Some recent studies. *Macromolecular Symposia*, *229*, 66–71.

[76] Wang, S., Yu, J., Jin, F., and Yu, J. (2008) The new insight on ultrastructure of C-type starch granules revealed by acid hydrolysis. *International Journal of Biological Macromolecules*, *43*, 216–220.

[77] Le Corre, D., Gallur, M., Bras, J., Ancejo, S., and Dufresne, A. (2011) *2011 TAPPI Intl Conference on Nano for Renewable Materials*, Washington, USA 2011.

[78] Angellier, H., Molina-Boisseau, S., Belgacem, M.N., and Dufresne, A. (2005) Surface chemical modification of waxy maize starch nanocrystals. *Langmuir*, *21*, 2425–2433.

[79] Angellier, H., Molina-Boisseau, S., Lebrun, L., and Dufresne, A. (2005) Processing and structural properties of waxy maize starch nanocrystals reinforced natural rubber. *Macromolecules*, *38*, 3783–3792.

[80] Angellier, H., Molina-Boisseau, S., and Dufresne, A. (2005) Mechanical properties of waxy maize starch nanocrystal reinforced natural rubber. *Macromolecules*, *38*, 9161–9170.

[81] Thielemans, W., Belgacem, M.N., and Dufresne, A. (2006) Starch nanocrystals with large chain surface modifications. *Langmuir*, *22*, 4804–4810.

[82] Angellier, H., Molina-Boisseau, S., and Dufresne, A. (2006) Waxy maize starch nanocrystals as filler in natural rubber. *Macromolecular Symposia*, *233*, 132–136.

[83] Angellier, H., Molina-Boisseau, S., Dole, P., and Dufresne, A. (2006) Thermoplastic starch-waxy maize starch nanocrystals nanocomposites. *Biomacromolecules*, *7*, 531–539.

[84] Kristo, E., Biliaderis, C.G. (2007) Physical properties of starch nanocrystal-reinforced pullulan films. *Carbohydrate Polymers*, *68*, 146–158.

[85] Viguié, J., Molina-Boisseau, S., and Dufresne, A. (2007) Processing and characterization of waxy maize starch films plasticized by sorbitol and reinforced with starch nanocrystals. *Macromolecular Bioscience*, *7*, 1206–1216.

[86] Labet, M., Thielemans, W., and Dufresne, A. (2007) Polymer grafting onto starch nanocrystals. *Biomacromolecules*, *8*, 2916–2927.

[87] Song, S., Wang, C., Pan, Z., and Wang, X. (2008) Preparation and characterization of amphiphilic starch nanocrystals. *Journal of Applied Polymer Science*, *107*, 418–422.

[88] Yu, J., Ai, F., Dufresne, A., Gao, S., et al. (2008) Structure and mechanical properties of poly(lactic acid) filled with (starch nanocrystal)-graft-poly(e-caprolactone). *Macromolecular Materials and Engineering*, *293*, 763–770.

[89] Chang, P.R., Ai, F., Chen, Y., Dufresne, A., and Huang, J. (2009) Effects of starch nanocrystal-graft-polycaprolactone on mechanical properties of waterborne polyurethane-based nanocomposites. *Journal of Applied Polymer Science*, *111*, 619–627.

[90] Namazi, H., Dadkhah, A. (2008) Surface modification of starch nanocrystals through ring-opening polymerization of Îµ-caprolactone and investigation of their microstructures. *Journal of Applied Polymer Science*, *110*, 2405–2412.

[91] García, N.L., Ribba, L., Dufresne, A., Aranguren, M., and Goyanes, S. (2011) Effect of glycerol on the morphology of nanocomposites made from thermoplastic starch and starch nanocrystals. *Carbohydrate Polymers*, *84*, 203–210.

[92] Valodkar, M., Thakore, S. (2011) Isocyanate crosslinked reactive starch nanoparticles for thermo-responsive conducting applications. *Carbohydrate Research*, *345*, 2354–2360.

[93] Zhang, X., Huang, J., Chang, P.R., Li, J., et al. (2010) Structure and properties of polysaccharide nanocrystal-doped supramolecular hydrogels based on Cyclodextrin inclusion. *Polymer*, *51*, 4398–4407.

[94] Lin, N., Huang, J., Chang, P.R., Feng, L., and Yu, J. (2011) Effect of polysaccharide nanocrystals on structure, properties, and drug release kinetics of alginate-based microspheres. *Colloids and Surfaces B: Biointerfaces*, *85*(2), 270–279.

[95] Lin, N., Yu, J., Chang, P.R., Li, J., and Huang, J. (2011) Poly(butylene succinate)-based biocomposites filled with polysaccharide nanocrystals: Structure and properties. *Polymer Composites*, *32*, 472–482.

[96] LeCorre, D., Bras, J., and Dufresne, A. (2011) Evidence of micro- and nanoscaled particles during starch nanocrystals preparation and their isolation. *Biomacromolecules*, *12*, 3039–3046.

[97] LeCorre, D., Bras, J., and Dufresne, A. (2011) Ceramic membrane filtration for isolating starch nanocrystals. *Carbohydrate Polymers*, *86*, 1565–1572.

[98] LeCorre, D., Vahanian, E., Dufresne, A., and Bras, J. (2012) Enzymatic pretreatment for preparing starch nanocrystals. *Biomacromolecules*, *13*, 132–137. Submitted.

[99] Wang, Y., Tian, H., and Zhang, L. (2009) Role of starch nanocrystals and cellulose whiskers in synergistic reinforcement of waterborne polyurethane. *Carbohydrate Polymers*, *80*, 665–671.

[100] Zheng, H., Ai, F., Chang, P.R., Huang, J., and Dufresne, A. (2009) Structure and properties of starch nanocrystal-reinforced soy protein plastics. *Polymer Composites*, *30*, 474–480.

[101] Xu, Y., Ding, W., Liu, J., Li, Y., et al. (2010) Preparation and characterization of organic-soluble acetylated starch nanocrystals. *Carbohydrate Polymers*, *80*, 1078–1084.

[102] LeCorre, D., Bras, J., and Dufresne, A. (2011) Influence of botanic origin and amylose content on the morphology of starch nanocrystals. *Journal of Nanoparticle Research*, *13*(12), 7193–7208.

[103] LeCorre, D., Bras, J., and Dufresne, A. (2012) Influence of native starch's properties on starch nanocrystals thermal properties. *Carbohydrate Polymers*, *87*(1), 658–666.

[104] Alila, S., Aloulou, F., Thielemans, W., and Boufi, S. (2011) Sorption potential of modified nanocrystals for the removal of aromatic organic pollutant from aqueous solution. *Industrial Crops and Products*, *33*, 350–357.

[105] Cheetham, N.W.H., Tao, L. (1998) Variation in crystalline type with amylose content in maize starch granules: An X-ray powder diffraction study. *Carbohydrate Polymers*, *36*, 277–284.

[106] Perry, P.A., Donald, A.M. (2000) SANS study of the distribution of water within starch granules. *International Journal of Biological Macromolecules*, *28*, 31–39.

[107] Farhat, I.A., Mitchell, J.R., Blanshard, J.M.V., and Derbyshire, W. (1996) A pulsed 1H NMR study of the hydration properties of extruded maize-sucrose mixtures. *Carbohydrate Polymers*, *30*, 219–227.

[108] Bras, J., Senechal, T., Le Corre, D., Martin, C., and Dufresne, A. (2011) Cellulose and starch nanocrystals: Influence of the source on rheological, mechanical and barrier properties. American Chemical Society, pp. CELL-25.

[109] Jiang, D.D., Yao, Q., McKinney, M.A., and Wilkie, C.A. (1999) TGA/FTIR studies on the thermal degradation of some polymeric sulfonic and phosphonic acids and their sodium salts. *Polymer Degradation and Stability*, *63*, 423–434.

CHAPTER 9

Chemical Modification of Starch Nanoparticles

JIN HUANG, QING HUANG, PETER R. CHANG, and JIAHUI YU

9.1 INTRODUCTION

Starch is an abundant, inexpensive, naturally renewable, and biodegradable polysaccharide found in the roots, stalks, and seeds of staple crops such as rice, corn, wheat, tapioca, and potato [1–4]. Biosynthesized as semicrystalline granules containing densely packed polysaccharides and a small amount of water [5], starch is a mixture of two main components: amylose, a linear or slightly branched (1→4)-α-D-glucan; and amylopectin, a highly branched macromolecule consisting of (1→4)-α-D-glucan short chains linked through α-(1→6) linkages [5–7]. It has been reported [5, 8] that the starch granule consists of alternating crystalline and amorphous zones that lead to a concentric domain structure. The crystalline regions are arranged as thin lamellar domains, perpendicular to the radial axis, in a kind of "decentralized onion" structure. The amorphous region (starch molecules in a single chain state) and ordered region (starch molecules in a double helix conformation) components [9] are arranged alternately to form a growth ring structure that encircles the point of initiation of the granule, called the hilum [5]. Starch has been widely used in many industrial applications [10, 11] because of its attractive properties such as biodegradability and biocompatibility, as well as its universality, low cost, versatility, and functional attributes.

With the development of biodegradable materials, polysaccharides (especially nanocrystals/nanoparticles) have attracted more and more interest. As a typical biodegradable natural polymer, starch is a good candidate for production of nanoparticles because it has a partly crystalline structure that confers

Biopolymer Nanocomposites: Processing, Properties, and Applications, First Edition. Edited by Alain Dufresne, Sabu Thomas, and Laly A. Pothan.

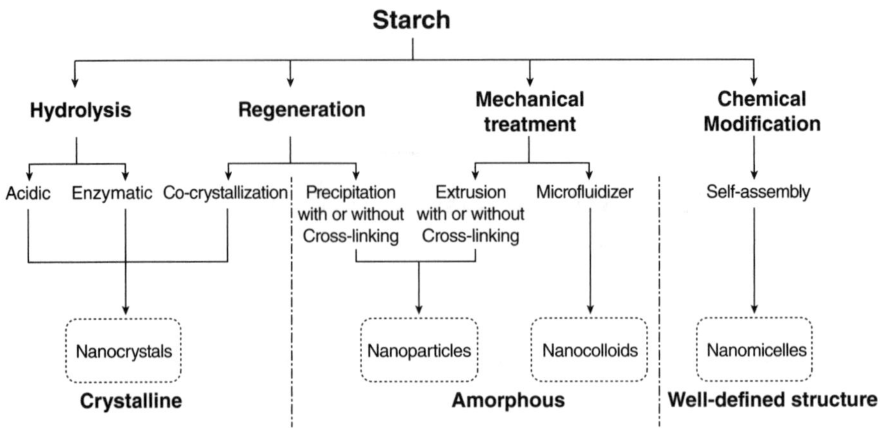

Figure 9.1 Different methods for preparation of starch nanoparticles.

interesting properties. Many interesting and detailed reviews [3, 12–14] and articles [15–17] on the preparation and application of starch nanoparticles (StNP) have been published. StNP have many potential applications in various products such as food, medicine [18], binders [19], implant materials [20], and plastics [21–23]. The most common application is the addition of StNP to a matrix (such as starch or protein) as reinforcement in the preparation of degradable bionanocomposites [24–28].

StNP are formed through the application of alkali, acid and/or enzymes which open and disperse the starch structure, and allow the formation of hydrogen bonds [29]. As shown in Figure 9.1, nanosized starch particles reported in the literature can be subdivided into several categories including: (1) starch nanocrystals extracted from natural resources by hydrolysis (acidic [3, 24, 28, 30, 31] or enzymatic [4, 32–34]) or obtained by co-crystallization during the regeneration process; (2) StNP obtained by regeneration–precipitation [7, 27, 35, 36] or by extrusion [37], where cross-linking can contribute to the stability of the nanoparticle structure [29, 38, 39]; (3) nanocolloids formed by microfluidization [40, 41]; and (4) nanomicelles that are self-assembled from chemically modified starch [42–44]. Acid hydrolysis is the typical method used for generating starch nanocrystals. Practical application of an acidic method is difficult due to its low yield, long treatment period, and use of concentrated acid [37], and hence recent development has turned attention to using enzymes [45] such as α-amylase for the pretreatment to enhance the efficiency of preparation. Meanwhile, both high pressure homogenization and extrusion are simple, environmentally friendly mechanical methods that can also be used to efficiently prepare nanoparticles. It should be noted that nanoparticles prepared using different methods have different properties, crystallinity, and shape. In addition, starch particles may also vary in degree of

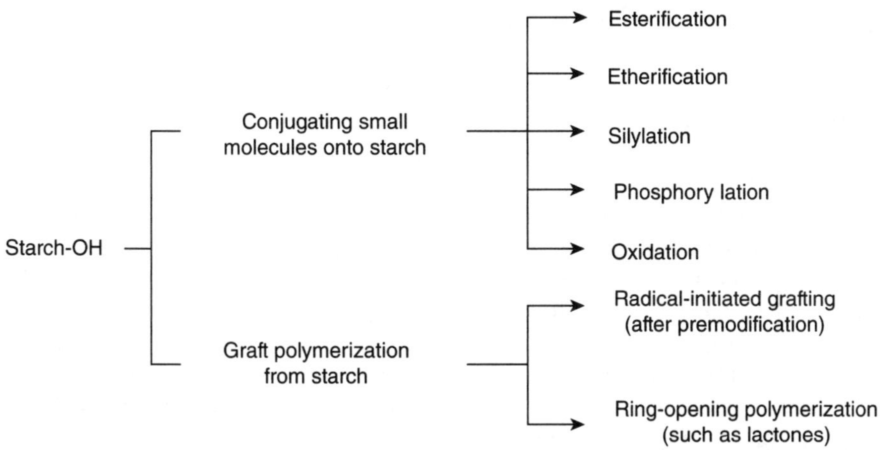

Figure 9.2 General chemical reactions for modification of starch.

substitution (DS) after chemical modification. Nanosized (100–400 nm), smooth, hexagonal, flaked carboxymethyl starches (CMS) with a DS of 1.15 have been synthesized by reacting starch with sodium hydroxide under ultrasonic irradiation [46].

The addition of StNP to degradable bionanocomposites may considerably enhance the mechanical properties of the nanocomposites; however, compatibility between the starch nanofillers and the matrix is a bottleneck in fully utilizing the unique functions of the nanofiller. Over the past decades, surface chemical modification has been largely used on ceramic fillers [47, 48], talc [49, 50], montmorillonite [51, 52], and polysaccharide fibers [53–58] in order to achieve maximum interaction and adhesion between the filler and the matrix [59]. Because StNP possess reactive surfaces covered with hydroxyl groups, chemical modification of the surface is possible in order to obtain specific functions, increase compatibility between the matrix and nanoparticles, and hence expand the application of StNP. Chemical modification of nanosized starch particles also expands the types of polymers that can possibly be used as matrices [5, 12, 59]. In this chapter, general methods for chemical modification of StNP will be outlined and discussed in detail. General chemical reactions for modification of starch are shown in Figure 9.2.

9.2 CHEMICAL MODIFICATION OF STARCH NANOCRYSTALS PERTAINING TO HYDROLYSIS AND REGENERATION–CO-CRYSTALLIZATION

Starch nanocrystals obtained by hydrolysis and regeneration–co-crystallization have surface hydroxyl groups that give them a polar surface; however, polysaccharide nanoparticles with polar surfaces have poor dispersibility and poor compatibility with nonpolar solvents or resins [28]. Also, it is difficult to reduce the surface energy of starch nanocrystals to improve their level of dispersion in a polymer matrix. Chemical modification of starch nanocrystals is an effective method for solving these problems because it alters the surface characteristics of the starch nanocrystals. Functional modification of starch nanocrystals may induce strong interfacial adhesion, which facilitates the transfer of stress and plays a reinforcing role for nanocrystals [5]. There are several strategies for chemical modification of starch nanocrystals including modification by chemical reaction with small molecules and grafting polymer chains onto the surface of starch nanocrystals using a "graft onto" or a "graft from" strategy. It is worth mentioning that the integrity of the starch nanocrystal's crystalline structure must be preserved, or have a low degree of destruction, regardless of the type of modification strategy used.

9.2.1 Conjugation of Small Molecules onto the Surface of Starch Nanocrystals

Many methods have been explored for the modification of starch nanocrystals by chemical reaction with small molecules in order to obtain nanoparticles with differences in morphology, size, solubility, polarity, surface properties, and thermal properties. So far, oxidative modification and ester or ether formation by substitution of hydroxyl groups in starch have been developed industrially [59]. It has been reported that modification by small molecules improved adhesion between the filler and the matrix, offering a good level of dispersion of the starch nanocrystals in the host matrix and a good interface between polymer and filler, both of which play fundamental roles in obtaining good mechanical properties in nanocomposites.

Esterification with organic acid is a typical method for production of thermoplastic and hydrophobic StNP. Esterification of starch nanocrystals occurs via small (–OH) and isocyanate groups (–NCO), and was used in the first attempt at modification of starch nanocrystals. Starch nanocrystals prepared by acid hydrolysis of starch granules have been chemically modified on the surface using isocyanates such as phenyl isocyanate (PI) [59] and 1,4-hexamethylene diisocyanate (HMDI) [38] in organic solvents such as toluene. The platelet-like starch nanocrystals become pseudospherical after modification with HMDI. Compared with the unmodified counterpart, the size of the modified nanocrystals is controllable, depending on the concentration of diisocyanate [38]. The PI-modified starch nanocrystals dispersed well in methylene chloride solution, suggesting a lower polar nature of the surface-modified nanocrystal. The contact angle measurement data showed that surface chemical modification with isocyanate functions weakened the polarity of the original starch nanocrystals, thereby allowing the use of nonpolar polymers as matrices for composite materials [59].

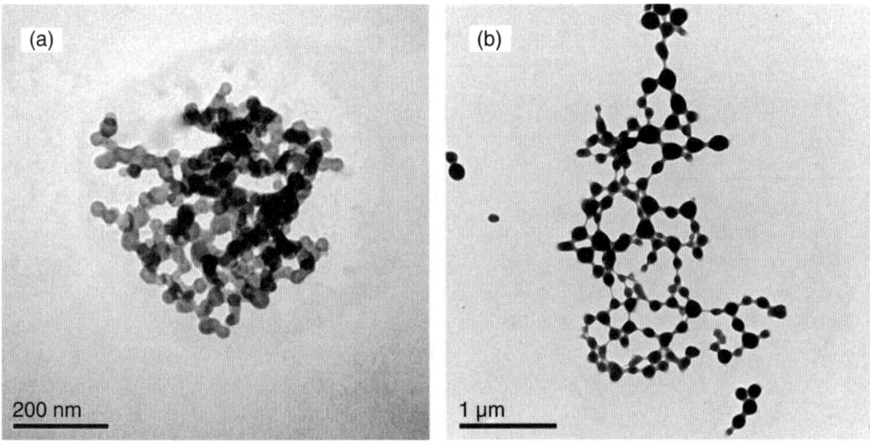

Figure 9.3 Transmission electron micrographs of (a) unmodified and (b) DS = 2.45 acetylated starch nanocrystals. Reprinted with permission from Reference [1].

Acetylated starch nanocrystals, which exhibited much improved solubility in common organic solvents and a significantly increased hydrophobic performance, were prepared with different DS by Xu et al. [1] using acetic anhydride, and by Angellier [59] using alkenyl succinic anhydride (ASA). The crystalline structure of these acetylated starch nanocrystals changed from A-style crystallinity to V-style crystallinity. Figure 9.3 shows the transmission electron micrographs of the unmodified and modified nanocrystals (DS = 2.45). The unmodified nanocrystals (Fig. 9.3a) were found to aggregate due to hydrogen bond interactions via the surface hydroxyl groups, while the modified nanoparticles (Fig. 9.3b) were more individualized and had increased in size [1].

9.2.2 Polymer Grafting onto the Surface of Starch Nanocrystals

Although modification of starch nanocrystals by chemical reaction with small molecules can improve interaction between the nanofiller and the polymer matrix (which may, to some degree, enhance the mechanical performance of the nanocomposites), the partial or total destruction of the nanocomposite's three-dimensional networks would lead to a sharp decrease in the mechanical performance of the nanocomposites. When a polymer was grafted onto the surface of starch nanocrystals, a near-perfect transfer of stress between the matrix and the filler could be achieved. A polymer matrix phase may also form directly at the nanocrystal surface, which may improve the interplay between the modified nanofillers and the matrix, thanks to the entanglement and interaction of covalently linked grafted chains on the surface of starch nanocrystals

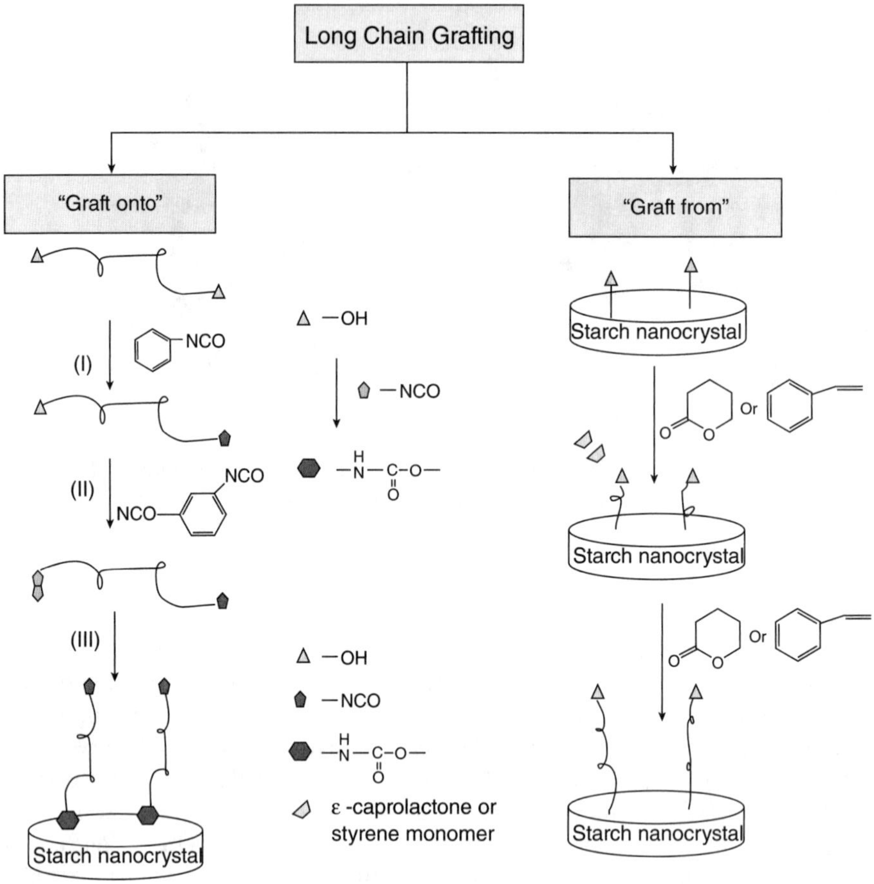

Figure 9.4 Schematic illustration of two strategies used for grafting polymer chains onto a starch nanocrystal. 2,4-toluene diisocyanate (2,4-TDI) is used as an example of a coupling agent for premodification in the "graft onto" strategy; in the "graft from" strategy, ring-opening polymerization (ROP) uses poly(ε-caprolactone) as an example [55–57] and the free radical polymerization (FRP) uses polystyrene (PS) [58].)

with the polymer matrix. A co-continuous structure forms as the grafted chains and matrix continue to twist tightly together. Figure 9.4 shows the schematic illustration of grafting polymer chains onto a starch nanocrystal based on the "graft onto" and "graft from" strategies.

9.2.2.1 "Graft Onto" Strategy for Polymer Conjugation As a reasonable approach to chemical modification based on the strategy of "grafting onto" polymer chains, graft copolymerization has been used extensively in the modification of starch nanocrystals. This method has received a lot of attention for reinforcement of nanocomposites, and also for utilization in applications in the medical field [20,21,26,54,55,60]. Various polymers, such as poly(ethylene glycol) methyl ether (PEGME) [20], poly(propylene glycol) monobutyl ether (PPGBE) [55], poly(tetrahydrofuran) (PTHF) [55], poly(caprolactone) (PCL) [55], and aliphatic chloride [20, 30], have been reported to graft onto the surface of starch nanocrystals using different coupling agents.

2,4-Toluene diisocyanate (2,4-TDI) was used as coupling agent in grafting reactions based on the "graft onto" strategy. The synthetic routes used may vary with the different number of end groups in the grafted polymers, which possess one or two reactive hydroxyl groups, but the routes are similar. When the grafted polymers have only one end-hydroxyl group, such as PEGME or PPGBE, they are grafted onto the surface of the starch nanocrystals with a two-step process that first functionalizes the polymer with an isocyanate from 2,4-TDI, and then reacts the second isocyanate in the 2,4-TDI with the surface hydroxyl groups of the starch nanocrystals to graft the polymer chains. When there are two end-hydroxyl groups on the polymer chains, such as on PTHF and PCL, a capping reaction using PI for one hydroxyl group occurs first, and then the remaining synthesis procedures are the same as for polymers with only one end-hydroxyl group. The morphologies and sizes of modified starch nanocrystals can be regulated and controlled as they are closely related to the relative molecular weights and lengths of the polymer chains. It has been confirmed [20] that grafted polymers efficiently covered the entire surface of the StNP without affecting the starch crystalline structure. Individualization of the modified starch nanocrystals was affected because of reduced hydrogen bonding and polar interactions between the individual particles [5, 20]. Monocrystalline StNP have also been reported to have been successfully grafted with polytetrahydrofuran, polycaprolactone, and poly(ethylene glycol) monobutyl ether chains using 2,4-TDI as the linking agent. These results show that the grafting efficiency of this method was greatly affected by the category and length of the polymer chain. For example, the grafting probability decreased with an increase in polymer chain length [28].

Reaction between the hydroxyl group and the acid chloride group (–COCl) is another method of modification in which starch nanocrystals are grafted with polymer chains, such as stearoyl chloride [20], octanoyl, nonanoyl, and decanoyl chloride [30]. The starch nanocrystals grafted by stearate polymers dispersed as individual nanoparticles more easily than its native counterpart and formed a crystalline brush-like structure from the starch surface outward, which could be attributed to the high level of surface grafting efficiency and to the covalently linked stearates that crystallize at the starch nanocrystal surface. The polarity of the starch–stearate surface disappeared because of efficient coverage of the surface by the stearate aliphatic chain ends [20]. When the octanoyl, nonanoyl, and decanoyl polymers were grafted, the polarity of the starch nanocrystals decreased and dispersion in organic solvent improved [30]. Grafting to alkyl polymer chains also improved the thermal properties because of the protective crystalline layer formed by the oxygen-poor stearate surface, which was attested to by an increase of more than 100°C in the degradation temperature of the stearoyl chloride grafted StNP [20]. Similar conclusions were reached by Namazi and Dadkhah [30].

It is obvious that the "graft onto" strategy can effectively change characteristics of starch nanocrystals such as morphology, size, polarity, surface properties, and thermal properties, while maintaining crystalline integrity. There are still some shortcomings, such as low controllability and grafting efficiency, as well as difficulty in grafting long chain polymers that limit the modification of StNP; as a result, a new method based on the "graft from" strategy has been developed.

9.2.2.2 "Graft From" Strategy for Polymer Growth Besides the previously mentioned weaknesses of the "graft onto" strategy, the isocyanates, such as 2,4-TDI and methylenediphenyldiisocyanate (MDI), used as chain extenders in this method are considered environmentally hazardous materials [61]. To solve these problems, a novel "graft from" strategy has been proposed in which polymerization of monomers and chain propagation of polymers are induced, and ultimately the polymer chains are coupled with the starch nanocrystals [5]. Various polymers have been grafted from the surface of starch nanocrystals based on the "graft from" strategy, together with ring-opening polymerization (ROP) for poly(ε-caprolactone) (PCL) [61–63] and free radical polymerization (FRP) for polystyrene (PS) [64].

The PCL polymer chains prepared through ROP of the caprolactone monomer were grafted onto the surface of starch nanocrystals by acid hydrolysis via bulk polymerization [61]. Results showed that the crystalline structure and morphology of the starch nanocrystals were maintained during chemical modification, and the decomposition temperature range of the modified nanocrystals was broader than that of unmodified starch nanocrystals [61]. Chang et al. [62]. and Yu et al. [63]. also modified starch nanocrystals with PCL chains under different conditions of microwave-assisted ROP. They found that after microwave-assisted chemical modification, most of the modified nanocrystals had a platelet-like structure of less than 100 nm, suggesting that the process of grafting PCL chains facilitates the dispersion of nanocrystals [5].

The FRP method has also been used to modify starch nanocrystals with polymers. An amphiphilic starch nanocrystal copolymer was prepared by grafting styrene chains onto the surface of the starch nanocrystals using acid hydrolysis based on the "graft from" strategy [64]. The modified starch nanocrystals transformed into amphiphilic nanoparticles, exhibiting a size of around 80–100 nm and a spherical morphology. In this case, polymerization of styrene induced by the free radical on the starch backbone occurred as potassium persulfate was used as the radical initiator. A micelle containing a growing free radical chain of polystyrene was then adsorbed onto the starch surface and attached to an oxy radical at the surface. The amphiphilic starch nanocrystals dispersed well in both polar and nonpolar solvents because of the different conformations of the hydrophobic polystyrene side chains and the hydrophilic starch backbone in polar or nonpolar solvents [64].

As reported in the literature, a higher grafting density can be realized with the "graft from" strategy, and the properties of starch nanocrystals can be partially controlled by altering the length and type of the grafted polymer chains. The grafted chains will also improve compatibility between the nanocrystals and matrix, and enhance the mechanical properties of nanocomposites [5].

9.3 CHEMICAL MODIFICATION OF STARCH NANOPARTICLES, NANOMICELLES, AND NANOCOLLOIDS

Apart from starch nanocrystals, many other types of nanosized starch, such as nanoparticles, nanomicelles, and nanocolloids, can be prepared using different methods and techniques. Amorphous StNP can be prepared by regeneration–precipitation of starch molecules or by way of extrusion. Chemical modification of amorphous StNP may be carried out before or after the cross-linking of StNP. Starch nanomicelles can be obtained through assembly of chemically modified starch molecules. Microfluidization can be used to prepare starch nanocolloids.

9.3.1 Nanoparticle Formation from Chemically Modified Starch

Modified starches derived from many methods including redox reaction [65], esterification [66], and even grafting of hydrophobic long chain fatty acids such as oleic acid and stearic acid [18] have been used for the preparation of StNP.

Highly active dialdehyde groups were first introduced into starch using $NaIO_4$ as the oxidizer, and then $POCl_3$ was used as a cross-linking agent to prepare dialdehyde StNP (DASNP) in a water-in-oil (*w/o*) microemulsion [65]. The dialdehyde groups on the StNP made reactions with hydrazine, acid, amine, imine, and so on possible, and as a result, DASNP have great potential for use in the medical field. The obtained DASNP also showed good thermal stability and low biological toxicity. Moreover, the obtained DASNP exhibited superior chemical combinative and physical adsorptive capabilities with drugs such as doxorubicin (DOX) [65].

Allyl and ester groups have also been introduced into starch to make acetylated allylic starch (AAS), which was prepared by first etherizing using allyl chloride and then etherifying using acetic anhydride. The uniform starch-based nanospheres obtained through nanoprecipitation of AAS had a core-shell structure with hydrophobic ester groups in the core and derivable allyl groups in the nanosphere shell [66]. Based on this, a type of thermoresponsive poly(N-isopropylacrylamide) hydrogel, with an extremely high mechanical strength and a remarkably rapid response rate, was prepared using the obtained biodegradable starch-based nanospheres as cross-linkers [66].

In order to broaden the application of nanoparticles in medical fields, starch was grafted with long chain fatty acids, such as oleic acid and stearic acid, to decrease hydrophilicity, and then made into nanoparticles and subsequently cross-linked with sodium tripolyphosphate for stabilization. The grafted starch showed higher swelling power at neutral pH compared with native starch, which is useful in drug delivery applications. Fatty acid-grafted StNP were found to be good vehicles for controlled oral drug delivery [18].

Modified StNP may also exhibit special functions, including the contribution of nutrition [27]. StNP prepared by regeneration–precipitation using ethanol as the precipitant were modified using citric acid (CA), which is recognized as nutritionally harmless as compared to other substances used for starch derivatization. The CA-modified starch nanocrystals (CASN) were then used as a reinforcing agent in a glycerol plasticized pea starch (GPS) matrix. Substitution of CA groups on starch chains formed a highly cross-linked starch, thus limiting the starch chain mobility. At a high temperature, native starch displays mainly thermoplastic behavior because water and glycerol physically break up the starch granules and disrupt intermolecular and intramolecular hydrogen bonds. At the same temperature, CASN did not gelatinize. The reason for the better thermal stability of CASN than that of starch granules and unmodified starch nanocrystals was that the ester bonds in the cross-linked CASN sterically hindered the hydrogen bond interaction. As a result, CASN could not be destroyed by the processing of CASN/GPS composites even at a high temperature. Moreover, the storage modulus, tensile strength, glass transition temperature, and water vapor barrier properties of these CASN/GPS nanocomposites were all improved as compared with pure GPS [27].

9.3.2 Nanomicelle Assembly from Chemically Modified Starch

Nanoparticle self-assembly is a common and efficient way to create nanomicelles with well-defined structure and outstanding properties [33, 42–44]. Most StNP are modified with functional groups or long chain polymers before self-assembly.

A micelle with unique characteristics, including nanosize, core-shell architecture, good thermodynamic stability under physiological conditions, and so on, was self-assembled from amphiphilic starch piperinic ester, which was synthesized by coupling a carboxyl group of piperic acid with a hydroxyl group on the starch backbone [44]. The self-assembled nanoparticles were prepared using dialysis solvent displacement in a dimethyl sulfoxide/water (DMSO/H_2O) system. The hydrophilic shell of the micelles consisted of water-soluble starch chains and the hydrophobic core consisted of piperic ester moieties. This core-shell structure has excellent properties that make the obtained StNP micelles useful for industrial, pharmaceutical, and biomedical applications [44]. Similarly, AAS can self-assemble in water to form core-shell nanospheres. The starch-based nanosphere composite hydrogels (SNH), with starch-based nanospheres as cross-linkers, can sustain a strength of 10.34 MPa, which is 60 times more than that sustained by the normal hydrogel (NH) [43].

It has been reported that hydrophobized polysaccharides cannot self-assemble into vesicles unless there is a hydrophilic spacer molecule between the lipid and the polysaccharide backbone [42]. Besheer et al. hydrophobically modified hydroxyethyl starch (HES) by reacting starch with ethylene oxide in an alkaline medium, in which fatty acids of varying chain lengths and molar substitution ($MS_{fatty\ acid}$) were used. Nanosized polymeric micelles and vesicles were then prepared by adding a tetrahydrofuran (THF) solution of modified HES into water and subsequently evaporating the THF. The results showed that only HES modified with lauric acid and with low $MS_{fatty\ acid}$ formed stable nanodispersions, while the others formed a precipitate. Dynamic light scattering (DLS) and transmission electron microscopy (TEM) showed the formation of micelles and vesicles with *z*-average sizes of between 20 and 30 nm for the micelles and 250 to 350 nm for the vesicles, which is promising for the potential application of modified HES vesicles in the areas of drug/protein encapsulation and drug delivery [42].

9.3.3 Chemical Modification of Cross-Linked Starch Nanoparticles

It is advantageous to prepare chemically modified StNP using nanosized cross-linked starch as the precursor. Needless to say, various techniques have been developed for the preparation of cross-linked StNP precursors. Emulsion crosslinking techniques, especially *w/o* emulsions with well-defined and simple preparation steps under mild processing conditions [29], are effective methods for the modification of StNP. Shi et al. [29] explored a convenient new synthetic route using high pressure homogenization combined with a *w/o* mini-emulsion cross-linking technique to prepare sodium trimetaphosphate (STMP) cross-linked StNP. Some literature focuses on the chemical modification of cross-linked StNP [67, 68].

StNP were prepared with *w/o* microemulsions in which starch was hydrolyzed in a boiling water bath until the solution was clear. $POCl_3$ was used as the cross-linking agent [67]. Poly-L-lysine (PLL) was linked to the surface of StNP to prepare a novel nonvirus nanoparticle gene carrier with a nucleus-shell structure, which had stronger adherence to cells and lower cell toxicity, presenting prospects as a gene carrier [67]. Folate modified with NH_2-terminated poly(ethylene glycol) (FA-PEG) was conjugated to the surface of StNP prepared by *w/o* microemulsions. The obtained folate-conjugated StNP (FA-PEG/StNP) can carry and sustained-release the anticancer drug DOX. Consequently, the FA-PEG-conjugated StNP may be quite valuable as targeted carriers of anticancer drugs [68].

9.3.4 Chemical Modification of Starch Nanocolloids

Starch colloids can be prepared using high pressure homogenization to disperse starch particles in water and reduce their size from micrometer to nanometer [40]. The structure and thermal stability of starch colloids were unchanged, due to the unchanging nature of starch itself. Viscosity of the colloids increased with an increased number of homogenization passes, accompanied by a decreased particle size, narrower particle size distribution (PSD), and an increased absolute zeta potential, indicative of the formation of a suspension or stable gel composed of nanoparticles. PSD and particle size were key factors influencing the properties of the colloid systems. Properties of viscosity were greatly altered between 10 and 12 passes, suggesting a critical sol–gel transition [40]. The chemical modification of nanocolloids deserves wider and further study as the literature on this is scarce.

9.4 ROLE OF CHEMICAL MODIFICATION IN THE APPLICATION OF STARCH NANOPARTICLES

StNP have great potential to reinforce and install special functions in polymer matrices to achieve high performance in the resulting composites. Nonetheless, the distribution of the StNP nanofillers in the polymer matrix, as well as functional development based on chemical modification of StNP, dictate the ultimate performance of the composites. Special attention has therefore been paid to preserve and protect the unique/excellent properties, such as well-dispersed small particles, large surface areas, and biodegradability, of StNP in environmentally friendly nanocomposites.

9.4.1 Compatibilization at the Filler–Matrix Interface in Nanocomposites

Distribution of StNP in the matrix has been proven to have a significant effect on the enhancement of the nanocomposite's mechanical properties. The addition of starch nanocrystals may also induce a decrease in permeability to both water vapor and oxygen and an increase in the tortuosity of the diffusion path, ultimately affecting the barrier properties of the nanocomposites [60]. It has been reported that in some matrices StNP dispersed well, leading to good properties in the nanocomposites. For example, starch nanocrystal filler was introduced as the reinforcing phase in a natural rubber latex matrix [21, 60, 69]. Good mechanical properties resulted because the starch nanocrystals were evenly distributed within the natural rubber matrix, and the crystalline structure of the nanocomposites was not destroyed during preparation, as verified by wide-angle X-ray diffraction analysis [21, 60]. A similar conclusion has been reached in research on soy protein isolate (SPI)-based nanocomposites enhanced by pea starch nanocrystals [25]. In addition, the platelet-like starch nanocrystals have been used as reinforcing nanofillers for biomaterials. For example, starch nanocrystals were incorporated into alginate microsphere drug carriers [70] and supramolecular hydrogels based on cyclodextrin/polymer inclusion [71]. The results showed that the starch nanocrystals not only reinforced the materials, but also produced more predominant sustained-release profiles.

Interactions between the starch nanofillers and the polymer matrix also make a great difference. It is interesting to note that in toluene swelling of the natural rubber nanocomposites decreased with the introduction of starch nanocrystals, while in water the swelling increased. The decreasing swelling in toluene was attributed to favorable interactions between the natural rubber matrix and starch nanofillers and to the formation of a three-dimensional starch nanocrystal network through hydrogen linkages [60]. On the other hand, the formation of a continuous polar network of starch nanocrystals in a natural rubber matrix seemed to favor the swelling in water [60]. Waxy maize starch-based nanocomposites, with waxy maize starch nanocrystals as reinforcing agents, showed good miscibility and strong interfacial adhesion between the starch nanocrystal filler and the starch matrix. The clear increase in the mechanical properties of the nanocomposite was attributed to the similarity of the chemical structures of the matrix and nanofiller [26]. Moreover, poly(butylene succinate) (PBS)-based nanocomposites incorporated with platelet-like starch nanocrystal at a loading level of 5 wt% showed simultaneous enhancement of strength and elongation compared with neat PBS because of nucleation of the nanocrystals, formation of a percolating network, and interaction between the filler and the matrix [72].

Although the use of starch nanocrystals as filler in some polymeric matrices has lead to a desired reinforcing effect [1, 22, 28, 73], the relatively poor compatibility between the starch nanofillers and the matrix has been reported to restrict the development of nanocomposites using StNP as filler. In the studies of polyvinyl alcohol (PVOH) based nanocomposites using pea starch nanocrystals as a reinforcing filler [74, 75], the tensile strength and elongation at break of PVOH-based nanocomposites filled with starch nanocrystals decreased rapidly with increasing starch content (up to 40 wt%). This property deterioration was attributed to the relatively poor compatibility between starch and PVOH [74, 75]. As a result, surface modification may become a necessary and effective way to improve compatibility at the nanofiller /matrix interface in nanocomposites.

When modified by small molecules (such as acetic anhydride, ASA, and isocyanate), StNP may show different polarity or solubility. This may lead to good dispersion of the fillers in the polymer matrix, and hence change the original filler–filler and filler–matrix interactions. Consequently, the microstructure of such nanocomposites would appear quite different from their unmodified StNP-filled counterparts. In particular, the covalent linkage between the filler and the matrix in the near-perfect stress transfer at the interface will result in excellent mechanical properties of the nanocomposites. However, it should be noted that although some chemical modifications, such as ASA or PI, of starch nanocrystals improve the adhesion between the filler and the matrix, they also partially or totally destroy the nanocomposite's three-dimensional network, leading to a strong decrease in the mechanical performance of the nanocomposites [21, 60]. These results were also proven in the study of natural rubber nanocomposites reinforced with chemically modified crab shell chitin nanowhiskers [5, 76]. Moreover, when starch nanocrystals were modified with ASA or PI, the swelling rate of natural rubber/modified starch nanocrystals was systematically higher than that of natural rubber/unmodified nanocrystals. This could be ascribed to lower interactions between modified nanocrystals, higher interactions between the filler and the matrix, and a higher affinity between modified nanocrystals and toluene [60].

Polymer chains were grafted onto the StNP to study compatibilization at the filler–matrix interface in nanocomposites. The polymer chains on the surface of the StNP entangled and interacted with the polymer matrix to form a co-continuous structure, resulting in the close proximity between nanofillers and matrix. Interfacial interaction among the starch nanocrystals, grafted polymer chains, and the polymer matrix is also a crucial factor affecting the structure and properties of composites [62, 63]. Polystyrene was essentially grafted onto the surface of starch nanocrystals by Song et al. [64]. X-ray diffraction (XRD) patterns confirmed that the crystalline structure of the starch nanocrystal was not changed after grafting, and the grafted starch nanocrystals showed excellent amphiphilic characteristics, such as good dispersion both in polar and in nonpolar solvents, attributed to the different conformations of the hydrophobic polystyrene side chains and the hydrophilic starch backbone when in polar or nonpolar solvent [64].

9.4.2 Function Development Based on Chemical Modification of Starch Nanoparticles

Chemical modification of StNP can not only improve the compatibility of nanoparticles in the polymer matrix to enhance the mechanical properties of the nanocomposites, but can also develop various unique functions that may broaden the application for nanocomposites.

As reported, chemically modified StNP can be used as cross-linkers in many nanocomposites. For example, HMDI-modified StNP were used as cross-linkers in the synthesis of poly(propylene glycol) (PPG)-based polyurethanes. Compared with polyurethane cross-linked with unhydrolyzed waxy starch, the highly cross-linked and filled nanocomposites showed higher conductivity at all temperatures. The considerable effect of temperature on conductivity indicates that the nanocomposites cross-linked with HMDI-modified StNP probably exhibit thermoresponsive electrical conductivity behavior [38]. AAS nanospheres have also been used as a cross-linker, polymerized with acrylamide to produce hydrogels with high mechanical strength [43].

Chemically modified StNP are widely used in the pharmaceutical sector, especially in drug delivery systems [29, 67, 68, 70, 71, 77, 78]. Modified StNP show extraordinary characteristics such as amphiphilicity [64], controllable size [68], and targeted modification [68], which help to encapsulate and deliver hydrophilic drugs to targeted spots, minimize the undesired release of the drug (initial burst), improve the efficacy of drug delivery, prolong the residence time of drugs, and reduce drug-related side effects [29]. Xiao et al. [68] prepared StNP conjugated with folate modified with NH_2-terminated poly(ethylene glycol) (FA-PEG) to form a tumor-targeted drug delivery vector. Long chain chemical modification on the surface of StNP can be used to make binding sites for active agents in drug delivery systems or for toxins in purification and treatment systems [20, 28, 79]. As mentioned previously, a nucleus-shell PLL–StNP can be used as a nonvirus nanoparticle gene carrier because of the many advantages of PLL–StNP, including its ability to accept a large DNA load, high transfection efficiency, low cell toxicity, and biodegradability [67]. The fatty acid-grafted StNP was also found to be a good vehicle for controlled oral drug delivery [18]. Modified starch nanospheres have also been reported to have the ability to convert poorly soluble, poorly absorbed drugs into better deliverable drugs that are site specific, and to protect the drugs from various body fluids (enzyme action), which can degrade them during targeting [77]. The pH-sensitive nanogels based on starch are also used for drug release [80].

Chemically modified StNP have also been used in many other areas. Stearate-modified starch nanoplatelets can be employed as adsorbents for the removal of a wide range of dissolved aromatic compounds from water [81], and PLL–StNP, which was reported to have stronger water-absorbing and water-retaining properties, will hopefully be applied to water-saving agricultural production [67].

9.5 CONCLUSIONS AND PROSPECTS

StNP can be chemically modified in many ways by various small molecules, monomers, or polymers, which may offer the nanoparticles some outstanding characteristics to broaden their areas of application. The small molecules used to modify StNP provide special functional groups to help the nanoparticles distribute better in the polymer matrix or to interact more closely with the polymer matrix. The monomers and polymers grafted onto the surface of the StNP by the "graft onto" or "graft from" methods usually twist tightly together with the polymer matrix forming a co-continuous structure, and as a result achieve a near-perfect stress transfer between the matrix and modified nanoparticles. A polymer matrix phase may also form directly at the nanoparticle surface, which may improve the interplay between the modified nanofillers and the matrix, thanks to strong covalent linkages.

Besides enhancing the mechanical properties of the nanoparticles, chemical modification of StNP also promotes functional development. With chemical modification, nanomaterials with modified StNP can potentially be used in medicine as excellent vehicles in drug delivery systems. Modification of the StNP also makes it possible to use them as cross-linkers, adsorbents, and water-retaining agents, which broadens the application of StNP into the areas of sewage disposal and agriculture.

The majority of literature discusses the preparation of StNP by acid hydrolysis. Literature concerning other methods used for the preparation of StNP is scarce, which may be a potential field to be developed. StNP, as one kind of polysaccharide that is environmentally friendly, nontoxic, abundant in resources, low in price, and biodegradable, should be applied more broadly in medical areas and at the same time expanded into other areas, such as packing material, chemicals, and environmental remediation, to meet or exceed regulatory requirements, society's expectations, and consumer demands in terms of sustainability. Glycerol plasticized starch-based nanocomposites with CASN are reported to have potential applications in medicine, agriculture, drug release, and in packaging for use as edible films, food packaging, and one-off packaging [27]. Novel processing technology should be sought in order to accelerate the application of starch nanocrystals in nanomaterials. StNP can be introduced into the biomaterials field, where, similar to cellulose nanocrystals [82], they have potential for use in bioimaging applications with fluorescent labeling modifications. Moreover, StNP, nanospheres, and nanogels have potential for use as building blocks for the nanoscale construction of sensors, tissues, mechanical devices, and so on [32].

Chemical modifications can change filler–filler and filler–matrix interactions and hence affect the compatibility of the StNP and polymer matrices, and even form new microstructures in nanocomposites. Meanwhile, chemically modified StNP still contribute to unique functions. These positive factors may facilitate the extension of starch utilization in the fields of high performance and functional materials. As a result, chemical modification of StNP is worthy of extended and more in-depth research. First of all, the design of a modified molecular structure for StNP is essential; it is key for achieving an optimal compatibilization effect, realizing controlled assembly, and exploring functional applications. Second, efficient synthesis methods are crucial for the realization of the intended structures. In this case, modest reaction conditions (mimicking those of living biological systems) that are suitable for preservation of the original structure of starch nanocrystals, as well as capable of resulting in sufficient polymerization on the surface of StNP need to be established. On the whole, in-depth basic research on the chemical modification of StNP promises many valuable applied products, and a mass of high performance, high value-added biomaterials, packing materials, and environmental remediation materials is expected.

REFERENCES

[1] Xu, Y., Ding, W.Q., Liu, J., Li, Y., Kennedy, J.F., Gu, Q., and Shao, S.X. (2010) Preparation and characterization of organic-soluble acetylated starch nanocrystals. *Carbohydrate Polymers*, *80*(4), 1078–1084.

[2] Pérez, S., Bertoft, E. (2010) The molecular structures of starch components and their contribution to the architecture of starch granules: A comprehensive review. *Starch/Stärke*, *62*(8), 389–420.

[3] Buléon, A., Colonna, P., Planchot, V., and Ball, S. (1998) Starch granules: Structure and biosynthesis. *International Journal of Biological Macromolecules*, *23*(2), 85–112.

[4] Preiss, J. (2009) Biochemistry and molecular biology of starch biosynthesis. In: *Starch: Chemistry and Technology*, 3rd edn, eds. J.N. BeMiller, R.L. Whistler. New York: Academic Press, pp. 83–147.

[5] Lin, N., Huang, J., Chang, P.R., Anderson, D.P., and Yu, J.H. (2011) Preparation, modification, and application of starch nanocrystals in nanomaterials: A review. *Journal of Nanomaterials*, *2011*, doi:10.1155/2011/573687.

[6] Paris, M., Bizot, H., Emery, J., Buzaré, J.Y., and Buléon, A. (1999) Crystallinity and structuring role of water in native and recrystallized starches by ^{13}C CP-MAS NMR spectroscopy. 1: Spectral decomposition. *Carbohydrate Polymers*, *39*(4), 327–339.

[7] Angellier, H., Choisnard, L., Molina-Boisseau, S., Ozil, P., and Dufresne, A. (2004) Optimization of the preparation of aqueous suspensions of waxy maize starch nanocrystals using a response surface methodology. *Biomacromolecules*, *5*(4), 1545–1551.

[8] Avérous, L., Halley, P.J. (2009) Biocomposites based on plasticized starch. *Biofuels, Bioproducts, and Biorefining*, *3*(3), 329–343.

[9] Gidley, M.J., Bociek, S.M. (1985) Molecular organization in starches: A carbon 13 CP/MAS NMR study. *Journal of the American Chemical Society*, *107*(24), 7040–7044.

[10] Jobling, S. (2004) Improving starch for food and industrial applications. *Current Opinion in Plant Biology*, *7*(2), 210–218.

[11] Song, D., Zhao, Y., Dong, C., and Deng, Y. (2009) Surface modification of cellulose fibers by starch grafting with crosslinkers. *Journal of Applied Polymer Science*, *113*(5), 3019–3026.

[12] Le Corre, D., Bras, J., and Dufresne, A. (2010) Starch nanoparticles: A review. *Biomacromolecules*, *11*(5), 1139–1153.

[13] Oates, C.G. (1997) Towards an understanding of starch granule structure and hydrolysis. *Trends in Food Science and Technology*, *8*(11), 375–382.

[14] Vandeputte, G.E., Delcour, J.A. (2004) From sucrose to starch granule to starch physical behaviour: A focus on rice starch. *Carbohydrate Polymers*, *58*(3), 245–266.

[15] Gallant, D.J., Bouchet, B., and Baldwin, P.M. (1997) Microscopy of starch: Evidence of a new level of granule organization. *Carbohydrate Polymers*, *32*(3–4), 177–191.

[16] Hizukuri, S. (1986) Polymodal distribution of the chain lengths of amylopectins, and its significance. *Carbohydrate Research*, *147*(2), 342–347.

[17] Jenkins, P.J., Donald, A.M. (1995) The influence of amylose on starch granule structure. *International Journal of Biological Macromolecules*, *17*(6), 315–321.

[18] Simi, C.K., Abraham, T.E. (2007) Hydrophobic grafted and cross-linked starch nanoparticles for drug delivery. *Bioprocess and Biosystems Engineering*, *30*(3), 173–180.

[19] Bloembergen, S., Kappen, F., and Beelen, B., inventors (2005) EcosynthetixInc, assignee. US Patent 6,921,430 B2, 2005 Jul 26.

[20] Thielemans, W., Belgacem, M.N., and Dufresne, A. (2006) Starch nanocrystals with large chain surface modifications. *Langmuir*, *22*(10), 4804–4810.

[21] Angellier, H., Molina-Boisseau, S., and Dufresne, A. (2005) Mechanical properties of waxy maize starch nanocrystal reinforced natural rubber. *Macromolecules*, *38*(22), 9161–9170.

[22] Angellier, H., Putaux, J.L., Molina-Boisseau, S., Dupeyre, D., and Dufresne, A. (2005) Starch nanocrystal fillers in an acrylic polymer matrix. *Macromolecular Symposia*, *221*(1), 95–104.

[23] Kristo, E., Biliaderis, C.G. (2007) Physical properties of starch nanocrystal-reinforced pullulan films. *Carbohydrate Polymers*, *68*(1), 146–158.

[24] García, N.L., Ribba, L., Dufresne, A., Aranguren, M., and Goyanes, S. (2011) Effect of glycerol on the morphology of nanocomposites made from thermoplastic starch and starch nanocrystals. *Carbohydrate Polymers*, *84*(1), 203–210.

[25] Zheng, H., Ai, F., Chang, P.R., Huang, J., and Dufresne, A. (2009) Structure and properties ofstarch nanocrystal-reinforced soy protein plastics. *Polymer Composites*, *30*(4), 474–480.

[26] Angellier, H., Molina-Boisseau, S., Dole, P., and Dufresne, A. (2006) Thermoplastic starch-waxy maize starch nanocrystals nanocomposites. *Biomacromolecules*, *7*(2), 531–539.

[27] Ma, X.F., Jian, R.J., Chang, P.R., and Yu, J.G. (2008) Fabrication and characterization of citric acid-modified starch nanoparticles/plasticized-starch composites. *Biomacromolecules*, *9*(11), 3314–3320.

[28] Dufresne, A. (2008) Polysaccharide nanocrystal reinforced nanocomposites. *Canadian Journal of Chemistry*, *86*(6), 484–494.

[29] Shi, A.M., Li, D., Wang, L.J., Li, B.Z., and Adhikari, B. (2011) Preparation of starch-based nanoparticles through high-pressure homogenization and miniemulsion cross-linking: Influence of various process parameters on particle size and stability. *Carbohydrate Polymers*, *83*(4), 1604–1610.

[30] Namazi, H., Dadkhah, A. (2010) Convenient method for preparation of hydrophobically modified starch nanocrystals with using fatty acids. *Carbohydrate Polymers*, *79*(3), 731–737.

[31] Putaux, J.L., Molina-Boisseau, S., Momaur, T., and Dufresne, A. (2003) Platelet nanocrystals resulting from the disruption of waxy maize starch granules by acid hydrolysis. *Biomacromolecules*, *4*(5), 1198–1202.

[32] Chakraborty, S., Sahoo, B., Teraoka, I., Miller, L.M., and Gross, R.A. (2005) Enzyme-catalyzed regioselective modification of SNP. *Macromolecules*, *38*, 61(1.

[33] Cai, L.M., Shi, Y.C., Rong, L.X., and Hsiao, B.S. (2010) Debranching and crystallization of waxy maize starch in relation to enzyme digestibility. *Carbohydrate Polymers*, *81*(2), 385–393.

[34] An, H.J., Liang, H.B., Liu, Z.D., Yang, H.S., Liu, Q.D., and Wang, H.B. (2011) Nanostructures of debranched potato starch obtained by isoamylolysis. *Journal of Food Science*, *76*(1), N11–N14.

[35] Kim, J.Y., Lim, S.T. (2009) Preparation of nano-sized starch particles by complex formation with n-butanol. *Carbohydrate Polymers*, *76*(1), 110–116.

[36] Chin, S.K., Pang, S.C., and Tay, S.H. (2011) Size controlled synthesis of starch nanoparticles by a simple nanoprecipitationmethod. *Carbohydrate Polymers*, *86*(4), 1817–1819.

[37] Song, D., Thio, Y.S., and Deng, Y. (2011) Starch nanoparticle formation via reactive extrusion and related mechanism study. *Carbohydrate Polymers*, *85*(1), 208–214.

[38] Valodkar, M., Thakore, S. (2010) Isocyanate crosslinked reactive starch nanoparticles for thermo-responsive conducting applications. *Carbohydrate Research*, *345*(16), 2354–2360.

[39] Azizi Samir, M.A.S., Alloin, F., and Dufresne, A. (2005) Review of recent research into cellulosic whiskers, their properties and their application in nanocomposite field. *Biomacromolecules*, *6*(2), 612–626.

[40] Liu, D., Wu, Q., Chen, H., and Chang, P.R. (2009) Transitional properties of starch colloid with particle size reduction from micro- to nanometer. *Journal of Colloid and Interface Science*, *339*(1), 117–124.

[41] Siqueira, G., Bras, J., and Dufresne, A. (2009) Cellulose whiskers versus microfibrils: Influence of the nature of the nanoparticle and its surface functionalization on the thermal and mechanical properties of nanocomposites. *Biomacromolecules*, *10*(2), 425–432.

[42] Besheer, A., Hause, G., Kressler, J., and Mäder, K. (2007) Hydrophobically modified hydroxyethyl starch: Synthesis, characterization, and aqueous self-assembly into nano-sized polymeric micelles and vesicles. *Biomacromolecules*, *8*(2), 359–367.

[43] Tan, Y., Wang, P., Xu, K., Li, W., An, H., Li, L., Liu, C., and Dong, L. (2009) Designing starch-based nanospheres to make hydrogels with high mechanical strength. *Macromolecular Materials and Engineering*, *294*(12), 855–859.

[44] Han, J., Borjihan, G., Bai, R., Chen, X., and Jing, X. (2008) Synthesis and characterization of starch piperinic ester and its self-assembly of nanospheres. *Journal of Applied Polymer Science*, *108*(1), 523–528.

[45] LeCorre, D., Vahanian, E., Dufresne, A., and Bras, J. (2012) Enzymatic pretreatment for preparing starch nanocrystals. *Biomacromolecules*, *13*(1), 132–137.

[46] Gao, W., Lin, X., Lin, X., Ding, J., Huang, X., and Wu, H. (2011) Preparation of nano-sized flake carboxymethyl cassava starch under ultrasonic irradiation. *Carbohydrate Polymers*, *84*(4), 1413–1418.

[47] Qiu, W., Mai, K., and Zeng, H. (1999) Effect of macromolecular coupling agent on the property of PP/GF composites. *Journal of Applied Polymer Science*, *71*(10), 1537–1542.

[48] Vaz, C.M., Reis, R.L., and Cunha, A.M. (2002) Use of coupling agents to enhance the interfacial interactions in starch-EVOH/hydroxylapatite composites. *Biomaterials*, *23*(2), 629–635.

[49] Velasco, J.I., De Saja, J.A., and Martinez, A.B. (1996) Crystallization behavior of polypropylene filled with surface-modified talc. *Journal of Applied Polymer Science*, *61*(1), 125–132.

[50] Krysztafkiewicz, A., Domka, L. (1997) Surface-modified microporoustalcs as fillers of polymers and pigments. *Journal of Materials Chemistry*, *7*, 1655–1659.

[51] Chisholm, B.J., Moore, R.B., Barber, G., Khouri, F., Hempstead, A., Larsen, M., Olson, E., Kelley, J., Balch, G., and Caraher, J. (2002) Nanocomposites derived

from sulfonatedpoly(butylene terephthalate). *Macromolecules*, *35*(14), 5508–5516.

[52] Kim, J.T., Lee, D.Y., Oh, T.S., and Lee, D.H. (2003) Characteristics of nitrile-butadiene rubber layered silicate nanocomposites with silane coupling agent. *Journal of Applied Polymer Science*, *89*(10), 2633–2640.

[53] Zini, E., Scandola, M., and Gatenholm, P. (2003) Heterogeneous acylation of flax fibers: Reaction kinetics and surface properties. *Biomacromolecules*, *4*(3), 821–827.

[54] Fabbri, P., Champon, G., Castellano, M., Belgacem, M.N., and Gandini, A. (2004) Reactions of cellulose and wood superficial hydroxy groups with organometallic compounds. *Polymer International*, *53*(1), 7–11.

[55] Labet, M., Thielemans, W., and Dufresne, A. (2007) Polymer grafting onto starch nanocrystals. *Biomacromolecules*, *8*(9), 2916–2927.

[56] Belgacem, M.N., Gandini, A. (2009) Natural fibre-surface modification and characterisation. In: *Cellulose Fibre Reinforced Polymer Composites*, eds. T. Sabu, L. Pothan. Philadelphia: Old City Publishing, pp. 14–46. Chapter 2.

[57] Gandini, A., Belgacem, M.N. (2011) Physical and chemical methods of fiber surfacemodification. In: *Interface Engineering in Natural Fibre Composites for Maximum Performance*, ed. E. Zafeiropoulos. Cambridge, UK: Woodhead Publishing Ltd, pp. 3–42. Chapter 1.

[58] Belgacem, M.N., Gandini, A. (2005) The surface modification of cellulose fibres for use as reinforcingelements in composite materials. *Composite Interface*, *12*(1), 41–75.

[59] Angellier, H., Molina-Boisseau, S., Belgacem, M.N., and Dufresne, A. (2005) Surface chemical modification of waxy maize starch nanocrystals. *Langmuir*, *21*(6), 2425–2433.

[60] Angellier, H., Molina-Boisseau, S., Lebrun, L., and Dufresne, A. (2005) Processing and structural properties of waxy maize starch nanocrystals reinforced natural rubber. *Macromolecules*, *38*(9), 3783–3792.

[61] Namazi, H., Dadkhah, A. (2008) Surface modification of starch nanocrystals through ring-opening polymerization of ε-caprolactone and investigation of their microstructures. *Journal of Applied Polymer Science*, *110*(4), 2405–2412.

[62] Chang, P.R., Ai, F., Chen, Y., Dufresne, A., and Huang, J. (2009) Effects of starch nanocrystal-graft-polycaprolactone on mechanical properties of waterborne polyurethane-based nanocomposites. *Journal of Applied Polymer Science*, *111*(2), 619–627.

[63] Yu, J., Ai, F., Dufresne, A., Gao, S., Huang, J., and Chang, P.R. (2008) Structure and mechanical properties of poly(lactic acid) filled with (starch nanocrystal)-graft-poly(ε-caprolactone). *Macromolecular Materials and Engineering*, *293*(9), 763–770.

[64] Song, S., Wang, C., Pan, Z., and Wang, X. (2008) Preparation and characterization of amphiphilic starch nanocrystals. *Journal of Applied Polymer Science*, *107*(1), 418–422.

[65] Yu, D., Xiao, S., Tong, C., Chen, L., and Liu, X. (2007) Dialdehyde SNP: Preparation and application in drug carrier. *Chinese Science Bulletin*, *52*(21), 2913–2918.

[66] Tan, Y., Xu, K., Wang, P., Li, W., Sun, S., and Dong, L. (2010) High mechanical strength and rapid response rate of poly(N-isopropyl acrylamide) hydrogel cross-linked by starch-based nanospheres. *Soft Matter*, *6*, 1467–1471.

[67] Xiao, S., Liu, X., Tong, C., Liu, J., Tang, D., and Zhao, L. (2005) Studies of poly-L-lysine-starch nanoparticle preparation and its application as gene carrier. *Science in China. Series B, Chemistry*, *48*(2), 162–166.

[68] Xiao, S., Tong, C., Liu, X., Yu, D., Liu, Q., Xue, C., Tang, D., and Zhao, L. (2006) Preparation of folate-conjugated SNP and its application to tumor-targeted drug delivery vector. *Chinese Science Bulletin*, *51*(14), 1693–1697.

[69] Valodkar, M., Thakore, S. (2011) Organically modified nanosized starch derivatives as excellent reinforcing agents for bionanocomposites. *Carbohydrate Polymers*, *86*(3), 1244–1251.

[70] Lin, N., Huang, J., Chang, P.R., Feng, L.D., and Yu, J.H. (2011) Effect of polysaccharide nanocrystals on structure, properties, and drug release kinetics of alginate-based microspheres. *Colloids and Surfaces. B, Biointerfaces*, *85*, 270–279.

[71] Zhang, X.L., Huang, J., Chang, P.R., Li, J.L., Chen, Y.M., Wang, D.X., Yu, J.H., and Chen, J.H. (2010) Structure and properties of polysaccharide nanocrystal-doped supramolecular hydrogels based on cyclodextrin inclusion. *Polymer*, *51*(19), 4398–4407.

[72] Lin, N., Yu, J., Chang, P.R., Li, J., and Huang, J. (2011) Poly(butylene succinate)-based biocomposites filled with polysaccharide nanocrystals: Structure and properties. *Polymer Composites*, *32*(3), 472–482.

[73] Angellier, H., Molina-Boisseau, S., and Dufresne, A. (2006) Waxy maize starch nanocrystals as filler in natural rubber. *Macromolecular Symposia*, *233*(1), 132–136.

[74] Chen, Y., Cao, X., Chang, P.R., and Huneault, M.A. (2008) Comparative study on the films of poly(vinyl alcohol)/pea starch nanocrystals and poly(vinyl alcohol)/ native pea starch. *Carbohydrate Polymers*, *73*(1), 8–17.

[75] Tang, X., Alavi, S. (2011) Recent advances in starch, polyvinyl alcohol based polymer blends, nanocomposites and their biodegradability. *Carbohydrate Polymers*, *85*(1), 7–16.

[76] Nair, K.G., Dufresne, A., Gandini, A., and Belgacem, M.N. (2003) Crab shell chitin whiskers reinforced natural rubber nanocomposites. 3. Effect of chemical modification of chitin whiskers. *Biomacromolecules*, *4*(6), 1835–1842.

[77] Singh, A., Garg, G., and Sharma, P.K. (2010) Nanospheres: A novel approach for targeted drug delivery system. *International Journal of Pharmaceutical Sciences Review and Research*, *5*(3), 84–88. (Article-015).

[78] Saboktakin, M.R., Tabatabaie, R.M., Maharramov, A., and Ramazanov, M.A. (2011) Synthesis and *in vitro* evaluation of carboxymethyl starch-chitosan nanoparticles as drug delivery system to the colon. *International Journal of Biological Macromolecules*, *48*(3), 381–385.

[79] Dufresne, A. (2010) Processing of polymer nanocomposites reinforced with polysaccharide nanocrystals. *Molecules (Basel, Switzerland)*, *15*(6), 4111–4128.

[80] Saboktakin, M.R., Maharramov, A., and Ramazanov, M.A. (2009) pH-sensitive starch hydrogels via free radical graft copolymerization, synthesis and properties. *Carbohydrate Polymers*, *77*(3), 634–638.

[81] Alila, S., Aloulou, F., Thielemans, W., and Boufi, S. (2011) Sorption potential of modified nanocrystals for the removal of aromatic organic pollutant from aqueous solution. *Industrial Crops and Products*, *33*, 350–357.

[82] Dong, S., Roman, M. (2007) Fluorescently labeled cellulose nanocrystals for bio-imaging applications. *Journal of the American Chemical Society*, *129*(45), 13810–13811.

CHAPTER 10

Starch-Based Bionanocomposite: Processing Techniques

REKHA ROSE KOSHY, LALY A. POTHAN, and SABU THOMAS

10.1 INTRODUCTION

In recent years, the development of biodegradable materials from renewable resources with excellent mechanical and barrier properties has been an important research challenge for the plastics industry [1, 2]. Petroleum-based polymers are widely used in different fields due to their low density and high mechanical and physical properties. But petroleum is a fossil fuel that is estimated to last for another 50–60 years at the current rate of consumption. In addition, there are significant environmental concerns caused by the use of nondegradable petroleum-based polymers and composites. Most of the resins made from petroleum are nondegradable and in the composite form they are nonrecyclable. With double digit growth in the use of these materials in the past few years, their disposal has become a critical issue [3].

Therefore, extensive efforts are being made to develop biodegradable composites using renewable resources in an attempt to replace the nonbiodegradable synthetic polymers used for composites [4]. A number of biodegradable materials have been investigated for use as plastics. Some are naturally occurring polymeric materials (e.g., starches, proteins, cellulose, gums) [5]; some are produced by polymerizing organic molecules to form polymers (e.g., polyurethanes by polyols obtained from vegetable oils, polylactic acid by lactide monomer from 100% renewable resources like corn, sugar, and other starch-rich products) [6, 7]; and others are manufactured from petrochemicals (e.g., poly(butylene succinate)—PBS, biodegradable aliphatic polyesters—BAPs, poly(e-caprolactone)—PCL, and poly(vinyl alcohol)—PVA) [8]. The

Biopolymer Nanocomposites: Processing, Properties, and Applications, First Edition. Edited by Alain Dufresne, Sabu Thomas, and Laly A. Pothan.

processing properties, properties in finished products, and biodegradability of these different materials vary widely [9]; however, several commonly occurring factors that limit their industrial application have been identified. First, in some instances, their mechanical properties are relatively poor compared with many petroleum-based plastics due to the inherent lower stiffness and strength. Second, many are relatively sensitive to water, with some materials dissolving rapidly, or exhibiting a substantial decrease in mechanical properties when they absorb water, especially in moist environments [10]. Third, the present low level of production and high cost restrict them from a wide range of applications. Therefore, there is an urgent need to develop biodegradable plastics made from renewable resources to reduce the quantity of plastic wastes [6], but with properties comparable to petroleum-based plastics. An attractive option is nanocomposite materials based on materials of biological origin that are biodegradable materials, called bionanocomposites. They include bionanocomposites based on polysaccharides such as cellulose and starch [11–21], and bionanocomposites based on proteins such as dairy protein, soybean protein, and gelatin [22–24]. The use of these materials serves a number of important functions, such as extending the food shelf life, enhancing food quality because they can serve not only as barriers to moisture, water vapor, gases, and solutes, but also serve as carriers of some active substances, such as antioxidants and antimicrobials [25–27].

In the family of renewable source-based polymeric materials, starch is an inexpensive and abundant natural resource [1]. Starch has been considered as an attractive biodegradable filler because of its low cost, low density, nonabrasive nature, and biodegradability [28]. It is the second most abundant biomass material in nature. It is found in plant roots, stalks, crop seeds, and staple crops such as rice, corn, wheat, tapioca, and potato. The starch industry extracts and refines starches by wet grinding, sieving, and drying. Starch is either used as extracted from the plant and is called "native starch" or it undergoes one or more chemical modifications to reach specific properties and is called "modified starch" [29].

10.2 MICROSTRUCTURES OF STARCH

Chemically, starch is a polymeric carbohydrate consisting of anhydroglucose units linked together primarily through ά-d-(1→4) glucosidic bonds. Starch is a heterogeneous material containing two microstructures: linear (amylose) and branched (amylopectin).

Amylose is essentially a linear structure of ά-1,4 linked glucose units, and amylopectin is a highly branched structure of short ά-1,4 chains linked by ά-1,6 bonds. Depending on its source and the processing conditions employed during its extraction, the molecular weight of amylose is about 10^6, which is 10 times higher than conventional synthetic polymers.

Amylopectin, on the other hand, is a branched polymer and its molecular weight is much greater than amylose. The high molecular weight and branched structure of amylopectin reduce the mobility of the polymer chains, and interfere with any tendency for them to become oriented closely enough to permit significant levels of hydrogen bonding. Between the linear amylose and short-branched amylopectin, a long-branched structure has been detected, such as in tapioca starch. Physically, most native starches are semicrystalline, having a crystallinity of about 20–45% [2]. Amylose and the branching points of amylopectin form amorphous regions. The short-branched chains in the amylopectin are the main crystalline component in granular starch. The ratio of amylose/amylopectin depends on the source and age of the starch, and it can also be controlled by the extraction process employed. Starch granules also contain small amounts of lipids and proteins. Figure 10.1 shows the microstructure of starch.

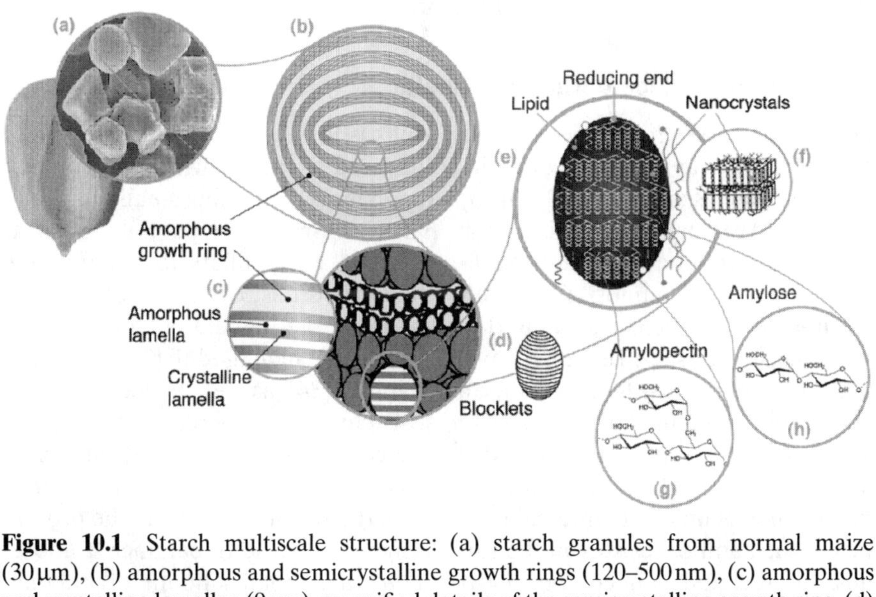

Figure 10.1 Starch multiscale structure: (a) starch granules from normal maize (30 µm), (b) amorphous and semicrystalline growth rings (120–500 nm), (c) amorphous and crystalline lamellae (9 nm), magnified details of the semicrystalline growth ring, (d) blocklets (20–50 nm) constituting a unit of the growth rings, (e) amylopectin double helices forming the crystalline lamellae of the blocklets, (f) nanocrystals: other representation of the crystalline lamellae called starch nanocrystals when separated by acid hydrolysis, (g) amylopectin's molecular structure, and (h) amylose's molecular structure (0.1–1 nm). Reprinted with permission from Reference [29]. Copy right 2010 American Chemical Society. See color insert.

10.3 STARCH-BASED BIONANOCOMPOSITE

Starch-based biodegradable polymers have great commercial potential for bioplastic, but some of the properties such as brittleness, low heat distortion temperature, high gas permeability, and low melt viscosity for further processing restrict their use in a wide range of applications. Therefore, modification of the biodegradable polymers through innovative technology is a challenge for material scientists [30]. Adding nano-reinforcement to pristine polymers for preparing nanocomposite has already proven to be an effective way to improve these properties concurrently [31]. Thus, green nanocomposites are being considered as the next-generation materials. A number of researchers have presented work in the field of starch-based bionanocomposites, which can be obtained by filling a thermoplastic starch (TPS) matrix with nanofillers such as layer silicates, carbon nanotubes, carbon black, cellulose, and starch nanocrystals. Montmorillonite (MMT) [32, 33] and kaolinite [34] are the usual layer silicates used in starch-based bionanocomposites. Cellulose nanocrystals or whiskers have been used to reinforce starchy material [31, 35, 36]. Starch can also be used as a filler to produce reinforced plastics [37, 38]. Usually, a small amount of starch (6–30%) is compounded with synthetic polymer, increasing the biodegradability of the product.

10.3.1 Starch Silicate Nanocomposites

One of the most promising nanocomposites is formed from organic polymer and inorganic clay minerals consisting of layered silicates. This type of polymer nanocomposites are a class of reinforced polymers containing small quantities (1–5 wt%) of nanometric-sized clay particles. Smectite-type clays, such as hectorite, synthetic mica, and MMTare employed as fillers to enhance the properties of the composites. The functional properties of the nanocomposites are improved markedly compared with those of the unfilled polymer or conventional composites. These improvements include high moduli [39], increased tensile strength (TS) [40], and thermal stability [34], as well as decreased gas permeability, flammability, and water absorbance, and increased biodegradability. Generally, polymer/layered silicate composites are divided into three main types: tactoid, intercalated, and exfoliated nanocomposites. In a tactoid, the polymer is unable to intercalate between the silicate sheets and the properties of the composites stay in the same range as the traditional microcomposites. Intercalated nanocomposites occur when a small amount of polymer moves into the gallery spacing between the silicate platelets. When the silicate layers are completely and uniformly dispersed in a continuous polymer matrix, an exfoliated or delaminated structure is formed [1]. The intercalated and exfoliated nanocomposites currently are of primary interest because their properties are significantly improved, even at low clay concentrations. Figure 10.2 represents three main types of composites based on layered silicate materials.

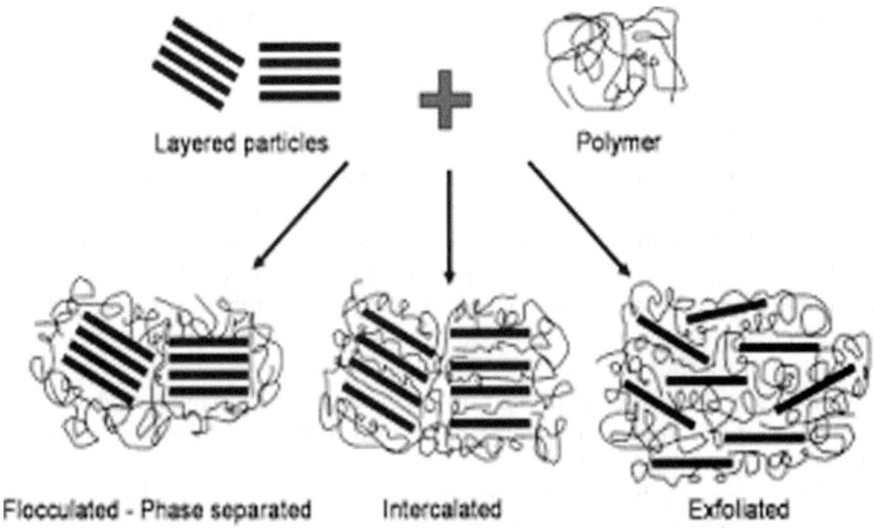

Figure 10.2 Illustration of different types of composite that can arise from the interaction between layered silicates and polymers. Reprinted with permission from Reference [29]. Copyright 2010 American Chemical Society.

10.3.2 Starch Cellulose Nanocomposites

Starch reinforced by cellulose is a typical example of natural polymer nanocomposites. These materials have the advantages of being renewable, biodegradable, abundantly available, and inexpensive, and as such they have attracted great attention in the last two decades. The performance improvements of the starch cellulose nanocomposites may be ascribed to the chemical similarities between starch and cellulose and the hydrogen bonding interactions existing in filler/matrix [41]. The reinforcing potential of wheat straw nanofibers obtained from agro-residues in a starch-based thermoplastic polymer was investigated by Alemdar and Sain, and it was found that the TS and modulus of the nanocomposite films increased linearly with the increase in nanofiber content. The TS of the nanocomposite film with 10% nanofibers loading (TPS-10% nanofiber [NF]) showed a 73% increase compared with the pure TPS. Also significant increase in the modulus with respect to the unfilled TPS was obtained as a function of the nanofiber content [42]. Enhanced properties were obtained by using various natural cellulosic fibers, such as sisal, cotton, bamboo, jute, straw, kenaf, and wood [43]. Mathew and Dufresne [44] prepared plasticized starch (PS)/tunicin whiskers nanocomposites, in which the cellulose whiskers consisted of slender parallelepiped rods with a high aspect ratio. Microfibrillated cellulose(MFC) nanofiber is combined with an almost viscous polysaccharide matrix in the form of a 50/50 amylopectin–glycerol blend [45]. Waxy maize starch nanocrystals obtained by hydrolysis of native granules are also used as the reinforcing filler in a thermoplastic waxy maize starch matrix plasticized with glycerol [37].

10.3.3 Starch Synthetic Polymer-Based Nanocomposite

Starch is usually added as filler in biodegradable polymers such as aliphatic polyesters, polyvinyl alcohol (PVA), and other biopolymers. The commonly used polyesters are poly(β-hydroxyalkanoates) (PHA), obtained by microbial synthesis, and polylactide (PLA) or poly(ε-caprolactone) (PCL), derived from chemical polymerization. The goal of blending completely degradable polyester with low cost starch is to improve its cost competitiveness while maintaining other properties at an acceptable level [46].

The main challenge in fabrication of these polymer nanocomposites for structural applications is obtaining a uniform dispersion of nanoparticles in the polymer matrix. Good dispersion for nanoparticles in polymer composite materials is extremely difficult to achieve because nanoparticles tend to aggregate during fabrication. The degree to which nanoparticles can be homogeneously dispersed in the polymer matrix significantly influences the thermal, mechanical, and optoelectronic properties of the material. Researchers have used several techniques for dispersing nanoparticles including: (1) mechanical agitation, such as ball milling or magnetic stirring, (2) ultrasonic vibration, (3) shear mixing, (4) noncontact mixing, and (5) using the dispersing agent. The curing process is another important factor in improving the mechanical, thermal, electrical, and optical properties. In polymer composite fabrication, the curing (polymerization) step is a crucial and time-consuming process. The epoxy-based thermoset polymers are generally cured using conventional heating. Several alternate curing methods have been tested for epoxy resin processing and they include ultraviolet (UV) rays, gamma rays, electron beam, and microwave heating [47].

10.4 CONVENTIONAL NANOCOMPOSITES MANUFACTURING TECHNIQUES

Two principal factors cause the properties of nanomaterials to differ significantly from other materials: increased relative surface area and quantum effects [48]. Some nanocomposites may show properties predominated by the interfacial interactions and others may exhibit quantum effects associated with nanodimensional structures [49]. The choice of fabrication method is very important for manufacturing nano-phased structural polymer composite materials. Some of the widely used methods for manufacturing conventional composite parts are wet lay-up, pultrusion, resin transfer molding (RTM), vacuum-assisted resin transfer molding (VARTM), autoclave processing, resin film infusion (RFI), prepreg method, filament winding, and fiber placement technology, among other methods [50]. Wet lay-up is a simple method compared with other composite manufacturing methods. It allows the resin to be applied only in the mold, but the mechanical properties of the product are poor due to voids and the final product is nonuniform. Pultrusion is a low-cost continuous process with a high production rate. But near the die assembly, the prepreg or materials accumulate and can create a jam. But the fibers in pultruded material are generally well aligned [51]; it helps to reduce fiber misalignment in the composite through optimization of manufacturing process variables, such as pull-speed, preformer temperature, and nanoparticle alignment and/or dispersion. The RTM is a closed mold operation. In this process, resin flow and fiber wet-out are critical issues; resin flows in the plane as well as in the transverse directions of the preform. Fiber wet-out depends on the fiber architecture and permeability of the perform. The VARTM is an adaptation of the RTM which is a widely used single-sided tooling process in which open molds can be used to make the parts using vacuum [50]. This process has certain advantages: (1) it requires a relatively low cost for high volume production; (2) very large and complex parts are possible with improved surface finish, higher fiber volume fraction than hand lay-up; and (3) as it is a closed system, it reduces environmental concerns more than hand lay-up processes.

Autoclave processing is a promising technique to process and manufacture different complex shapes of high quality composite structures [50]. It has the ability to process both thermoset and thermoplastic composites with uniform thickness and minimum porosity. Resin film infusion (RFI) is similar to RTM where a thin film or sheet of solid resin is laid into the mold and preform is laid on top of the resin film under heat and pressure. Resin-impregnated unidirectional or woven fabric (partially cured) is used in the prepreg method with vacuum bagging and autoclave processing. In filament winding, resin-impregnated fibers are wrapped over a mandrel at the same or different winding angles to form a part. Complicated cylindrical parts, pressure vessels, fuel and water tanks for storage and transportation, and pipes can be manufactured by this method. The viscosity-related problems of the resin systems can be eliminated in this technique [52].

10.5 PROCESSING TECHNIQUES OF STARCH-BASED BIONANOCOMPOSITES

The preparative methods are divided into three main groups according to the starting materials and processing techniques: (1) solution casting method, (2) melt intercalation method, and (3) *in situ* polymerization [53, 54].

10.5.1 Solution Casting Method

In the solution casting method, polymer solution is heated with layered nanofillers to form nanocomposites. This technique is mostly used in water-soluble polymers to produce intercalated nanocomposites. The main factors that help in the formation of nanocomposites are temperature and swelling of nanofillers. Temperature increases the movement of polymer chains, which results in the intercalated nanostructures [55].

10.5.1.1 Preparation of Glycerol Plasticized Pea Starch/Nano-ZnO Composites Chang et al. [56] prepared glycerol plasticized pea starch (GPS) zinc oxide bionanocomposite using the solution cast method. Nano-ZnO particles were dispersed in a solution of distilled water (100 mL) and glycerol (1.5 g) and ultrasonicated for 1 hour before addition of 5 g pea starch. Nano-ZnO was produced by the following method. 1.5 g of soluble starch was added to 250 mL distilled water. The mixture was heated at 90°C for approximately 10 minutes with constant stirring for the complete dissolution of soluble starch. The solution was then cooled to room temperature. Zinc nitrate hexahydrate, 7.44 g (0.025 mol), was added to the solution and then 250 mL of sodium hydroxide solution (0.2 mol/L) was added dropwise with constant stirring. The transparent solution became milky white with no precipitation. The reaction continued for 2 hours after addition of NaOH. The solution was then centrifuged at 8000 rpm for 20 minutes, and the settled nano-ZnO was washed three times using distilled water to remove the salt and the excess soluble starch. The nano-ZnO was dried at 80°C for 3 hours to complete conversion of the remaining ZnOH to ZnO. The nano-ZnO filler loading level (0 wt%, 0.5 wt%, 1 wt%, 2 wt%, 3 wt%, or 4 wt%) was based on the amount of pea starch. The mixture was heated at 90°C for 0.5 hour with constant stirring in order to plasticize the pea starch. To obtain the GPS/nano-ZnO composite films, the mixture was cast into a dish and placed in an air-circulating oven at 50°C until dry (about 6–8 hours). Fourier transform infrared (FT-IR) spectroscopy, thermogravimetric analysis (TGA), and transmission electron microscopy (TEM) showed that in GPS/nano-ZnO nanocomposites, loading a low level of nano-ZnO particles improved the pasting viscosity, storage modulus, and glass transition temperature (Tg). When the nano-ZnO content varied from 0 wt% to 4 wt%, the tensile yield strength and Young's modulus increased from 3.94 to 10.80 MPa and from 49.80 to 137 MPa, respectively. The water vapor permeability (WVP) decreased from 4.76×10^{-10} to 2.18×10^{-10}/gm/s/Pa. The improvement in these properties may be attributed to the interaction between the nano-ZnO filler and GPS matrix.

10.5.1.2 Preparation of Starch/Poly(Vinyl Alcohol)/Sodium Montmorillonite Nanocomposite Samer S. Ali et al. [32] prepared starch/poly(vinyl alcohol)/sodium MMTnanocomposite films by the film casting method. A solution was prepared by mixing four parts (by weight) starch/poly(vinyl alcohol) (PVOH)/Na +MMT/glycerol with 96 parts of water and then heating this mixture at 95°C for 30 minutes with constant stirring. Glycerol was used as a plasticizer at a concentration of 30% (polymer basis) in all experiments. The heated solution was cooled to 55°C, and equal amounts (60 g) were poured in 150 mm–15 mm petri dishes. The water was allowed to evaporate by drying

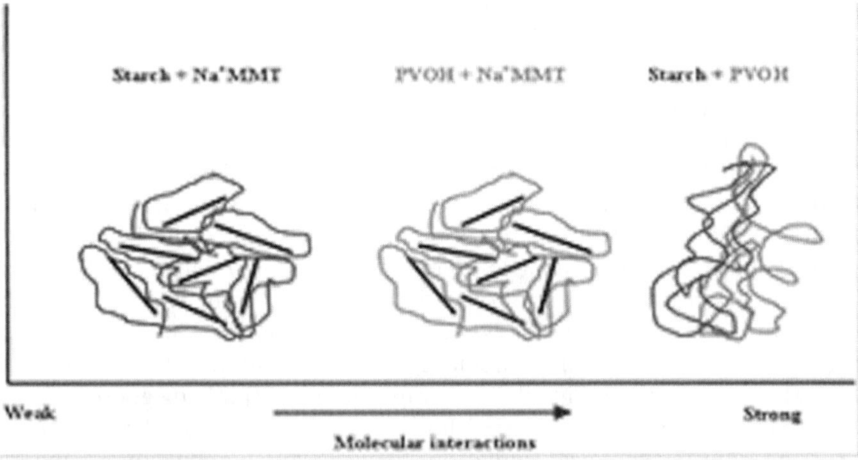

Figure 10.3 Schematic presentation of preferred interactions between different components. Reprinted with permission from Reference [32]. Copyright 2010 American Chemical Society.

for 24–36 hours at room temperature, and the resulting films were peeled off and stored in airtight bags at room temperature for further tests. TS and elongation at break (E%) of the films ranged from 11.60 to 22.35 MPa and 28.93–211.40%, respectively, while WVP ranged from 0.718 to 1.4303 mm/kPa h m^2. In general, an increase in Na+MMT content (0–20%) enhanced TS and decreased E% and WVP. Use of higher molecular weight PVOH increased both TS and E% and also decreased WVP. Mechanical properties were negatively affected, but water vapor barrier properties improved with increasing starch content (0–80%). X-ray diffraction (XRD) and TEM showed that an increase in the Na+MMT level from 5% to 20% led to a steady decrease in Tg. This was found because interactions between starch and PVOH were strongest, followed by polymer/clay interactions. The schematic presentation of preferred interactions between different components is shown in Figure 10.3.

10.5.1.3 Preparation of Starch/Flax Cellulose Nanocomposites X. Cao et al. [41] prepared starch-based nanocomposites reinforced with flax cellulose nanocrystals(FCN) by casting solution process. Starch, glycerol, cellulose nanocrystallites suspension, and distilled water were mixed together in order to obtain nanocomposite films with a homogeneous dispersion. The glycerol content was fixed at 36 wt% based on the dry starch matrix. Then this mixture was charged into a bottom flask equipped with a stirrer and heated at 100°C for 30 minutes for the starch to be gelatinized. After cooling down to about 70°C, the resulting paste was degassed under vacuum to remove the remaining air and cast in a square polystyrene petri dish, followed by drying in a vent oven at 50°C. The effects of FCNs loading on the morphology, thermal behavior, mechanical properties, and water sensitivity of the films were investigated by means of wide-angle X-ray diffraction (WAXRD), differential scanning calorimetry (DSC), tensile testing, and water absorption testing. Scanning electron microscopy (SEM) photographs of the failure surfaces clearly demonstrated a homogeneous dispersion of FCNs within the PS matrix and strong interfacial adherence between matrix and fillers, which led to an increase of Tg ascribed to the starch molecular chains in the starch-rich phase. In particular, these nanocomposite films exhibited a significant increase in TS and Young's modulus from 3.9 to 11.9 MPa and from 31.9 to 498.2 MPa, respectively, with increasing FCNs content from 0 wt% to 30 wt%. No large aggregates were formed and a homogeneous distribution of the FCNs in the PS matrix were observed in all nanocomposites films, implying good adhesion between the fillers and the matrix. This is attributed to the good compatibility resulting from the chemical similarities between starch and cellulose and the hydrogen bonding interactions existing in the interface between fillers and matrix. Such an even and uniform distribution of the fillers in the matrix could play an important role in improving the mechanical performance of the resulting nanocomposite film. The atomic force microscopy (AFM) image of a dilute suspension of FCNs shows that the suspension contains flax cellulose fragments consisting of both individual and aggregated nanocrystals. These fragments display slender rods and have a broad distribution in size, which have lengths (L) ranging from 100 to 500 nm and diameters (D) ranging from 10 to 30 nm.

10.5.1.4 Preparation of Green Composites Reinforced with Hemp Nanocrystals in Plasticized Starch Cao et al. [12] prepared nanocomposite films from a mixed suspension of hemp cellulose nanocrystals (HCNs) and TPS, or PS, by the casting and evaporating method. Starch, glycerol, CN suspension, and distilled water were mixed together to obtain a nanocomposite solution with a homogeneous dispersion. The glycerol content was fixed at 36 wt% on the basis of the dry starch matrix. Then, this mixture was charged into a bottom flask equipped with a stirrer and heated at 100°C for 30 minutes for the starch to be gelatinized. After it was cooled to about 708°C, the resulting paste was degassed *in vacuo* to remove the remaining air and cast in a polystyrene square petri dish; this was followed by drying in a vent oven at 50°C. By changing the content of HCNs over the range 5 wt%, 10 wt%, 15 wt%, 20 wt%, 25 wt%, and 30 wt%, they prepared a series of PS/HCN nanocomposites films with a thickness of about 0.3 mm and coded them as PS/HCN-5, PS/HCN-10, PS/HCN-15, PS/HCN-20, PS/HCN-25, and PS/HCN 30, respectively, in which the HCN content was expressed on a water-free PS matrix. As a control measure, a pure PS film without the addition of HCNs was obtained with the same fabrication process. Before the various characterizations, the resulting films were kept in a conditioning desiccator at 43% relative humidity (RH) for 1 week at room temperature to ensure the equilibration of the water

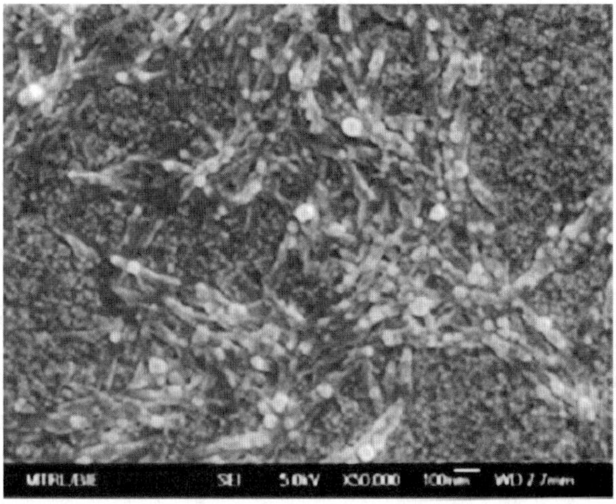

Figure 10.4 FE-SEM image of the HCNs dried on gold-coated wafers. Reprinted with permission from Reference [12]. Copyright Wiley Publications.

content in the films. The field emission scanning electron microscopy (FE-SEM) image of a dilute suspension of the HCNs is shown in Figure 10.4. The HCNs were very clearly identified from the background, although they were not dispersed individually because of the strong interactions among the nanocrystals themselves. An examination of the fractured surfaces after the tensile testing of the PS/HCN nanocomposites was also carried out with FE-SEM. A homogeneous distribution of the HCNs in the PS matrix was observed in all of the nanocomposites, which implied that the strong interactions among HCNs themselves were partly destroyed and a new strong adhesion between fillers and matrix was formed. Improved compatibility between HCNs and starch was ascribed to the chemical similarities in starch and cellulose, nanometer-scale size effect from the HCNs, and hydrogen bonding interactions between the fillers and matrix. This homogenous dispersion resulted in an increase in the Tg, the TS, and Young's modulus, from 3.9 to 11.5 MPa and from 31.9 to 823.9 MPa, respectively, and decrease in the water sensitivity of the final composite material.

10.5.1.5 Preparation of Thermoplastic Starch Nanocomposites Reinforced with Cellulose Nanofibers Extracted from Wheat Straw TPS nanocomposites reinforced with cellulose nanofibers (CNFs) extracted from wheat straw was prepared by Kaushik et al. [30] using the solution cast method. CNFs were dispersed in distilled water and sonicated for almost 3 hours. Maize starch was added with 30% glycerol and shear mixed for 10 minutes by using a Fluko FA25 homogenizer. Dispersed CNFs were added to the starch and glycerol mixer, and further shear mixed for 20 minutes. The starch + glycerol + CNFs mixture was continuously stirred using a mechanical stirrer and heated at 75 ± 3°C. After the solution became viscous, it was poured onto levelled Borosil glass petri dishes and kept at around 37°C for 2 days until it was completely dry. Solution cast films of TPS cellulose nanocomposites were made with 5%, 10%, and 15% nanofibers (as per dry weight of nanocomposites). The nanocomposites films were analyzed in terms of SEM, WAXRD, TGA, DSC, and mechanical and barrier properties. WAXRD profile for TPS/cellulose nanocomposites showed that the crystalline nature of native starch disappeared during processing. But the increase in the cellulose nanofibril content resulted in the preservation of the crystallinity of starch nanocrystals during processing method. The results of TGA and DSC experiments indicated an interaction between the fiber and the plasticizer, causing the reduction in onset of degradation temperatures and a reduction in water sorption compared with the pure matrix. Mechanical properties including dynamic mechanical analysis (DMA) and tensile modulus showed marked improvement with addition of CNF. Figure 10.5 shows SEM pictures of starch cellulose nanocomposites with 10% cellulose nanofibrils.

10.5.2 Melt Intercalation Technique

Recently, the melt intercalation technique has become the standard for the preparation of polymer layered silicate (PLS)–nanocomposites [1]. This process involves annealing a mixture of the polymer and layered nanofiller

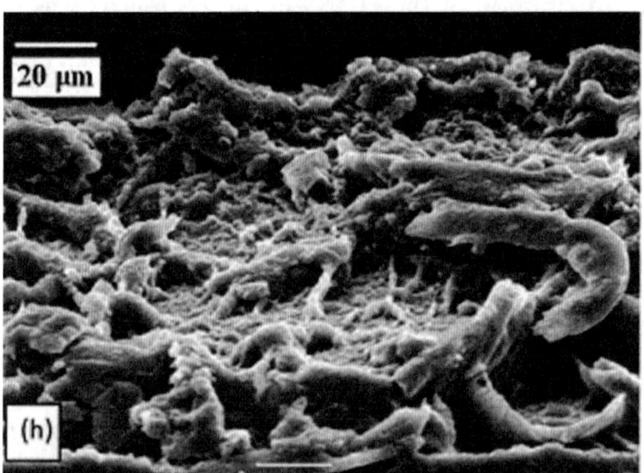

Figure 10.5 SEM images of starch cellulose nanofibril nanocomposites with 10% cellulose nanofibrils composite. Reprinted with permission from Reference [30]. Copyright Elsevier.

above the softening point of the polymer, statically or under shear. While annealing, the polymer chains diffuse from the bulk polymer melt into the galleries between the nanofiller layers. The filler platelets are sheared during compounding and kneading in a compounder and thus are well mixed with the polymer phase.A range of nanocomposites with structures from intercalated to exfoliate can be obtained, depending on the degree of penetration of the polymer chains into the nanofiller galleries [57].

Direct melt intercalation has many advantages over solution intercalation. During polymer intercalation from solution, a relatively large number of solvent molecules have to be desorbed from the host to accommodate the incoming polymer chains. The desorbed solvent molecules gain one translational degree of freedom, and the resulting entropic gain compensates for the decrease in conformational entropy of the confined polymer chains. Also, direct melt intercalation is highly specific for the polymer, leading to new hybrids that were previously inaccessible. In addition, the absence of a solvent makes direct melt intercalation an environmentally sound and an economically favorable method for industries from a waste perspective [6]. Extrusion has been commonly used in the preparation of polymer nanocomposite through melt intercalation processes [58].

10.5.2.1 Preparation of Thermoplastic Starch/Clay Hybrids Biodegradable TPS/clay hybrids were prepared by Park et al. [59] by melt intercalation. Three organically modified MMT with different ammonium cations and one unmodified Na+ MMT (Cloisite Na+; Southern Clay Products, Austin, TX) were used. TPS and clays were dried under vacuum at 80°C for at least 24 hours before use. TPS was mixed with clays in a Haake Rheocoder 600 roller mixer for 20 minutes. The contents of clays were fixed at 5 wt%. The rotor speed was 50 rpm and the temperature was set at 110°C. The mixed product was injection molded using a Mini-MAX molder to obtain dog bone-shaped specimens for characterization and property measurements. The blends after preparation were placed in tightly sealed polyethylene bags to prevent moisture absorption. From the XRD patterns and TEM observations, it is clearly established that the nanostructure of polymer/clay hybrids depends on the compatibility and interaction among the polymer, the silicate layers, and the nature of surfactant used for the modification of silicate layers. Due to the strong polar interactions between a small amount of the polar hydroxyl group of water in the TPS chain and the silicate layer of the pristine MMT, the polymer chains were intercalated into the silicate layers of pristine MMT and formed an intercalated TPS/MMT nanocomposite. It was found that Cloisite Na+ had the best dispersion in the TPS matrix. The TPS/Cloisite Na+ hybrid showed an intercalation of TPS in the silicate layer due to the matching of the surface polarity and interactions of the Cloisite Na+ and the TPS, which gives higher TS and better barrier properties to water vapor than the other TPS/organoclay hybrids as well as the pristine TPS. It was found that the dynamic mechanical properties of the TPS/clay hybrids were also affected by the polar interactions.

10.5.2.2 Preparation of Thermoplastic Starch–Clay Nanocomposites Chen et al. [16] prepared nanocomposites of TPS–clay with four types of clay by melt processing. The clay chosen were natural sodium MMT, natural hectorite, hectorite modified with 2 methyl, 2 hydrogenated tallow quaternary ammonium chloride, and kaolinite. TPS was prepared by melt processing of starch and glycerol at a ratio of 10:3 by mass. To prepare the TPS–clay composites, pre-weighed amounts of clay were gradually added to the polymer melt on the twin roll mill. To ensure lateral mixing, the mixture was stripped from the rolls and re-fed at least five times, which took approximately 15 minutes. It was found that in all composites the presence of clay increased both Young's and shear moduli, and the MMTand unmodified hectorite provided a significantly larger increase than kaolinite, which formed conventional composites with TPS. TEM showed that the untreated hectorite nanocomposites were partially exfoliated while the TPS-treated hectorite composites were conventional. Figure 10.6 shows TEM micrographs of (a) treated hectorite and (b) TPS-treated hectorite composites containing 3 wt% clay platelets.

10.5.2.3 Preparation of Thermoplastic Starch/Zein Blends Corradini et al. [60] prepared blends of starch and zein plasticized with glycerol by melting processing in an intensive batch mixer connected to a torque rheometer at 160°C. TPS blends with zein were prepared with starch/zein w.w^{-1} ratios

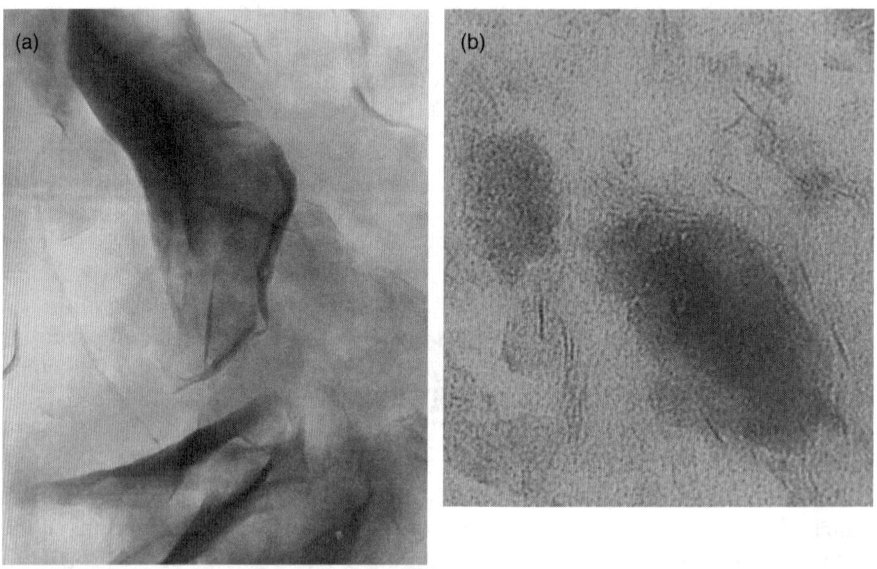

Figure 10.6 (a) TEM micrographs of treated hectorite. (b) TEM micrographs of TPS-treated hectorite composites containing 3 wt% clay platelets. Reprinted with permission from Reference [16]. Copyright Elsevier.

of 100/0, 80/20, 50/50, 20/80, and 0/100. The content of glycerol was 22% by weight of the total weight of polymers (starch and zein), on a dry basis. Starch, zein, and glycerol were weighed separately, pre-mixed in a beaker and the resulting mixtures were then processed at 160°C in a Haake Rheomix 600 batch mixer connected to a torque rheometer, with the rotors operating at 50 rpm. The mixing time used for all preparations was 6 minutes, since higher mixing times lead to starch degradation. The materials thus obtained were pressed for 5 minutes at 160°C with a force of 0.2 MPa, to produce 150 × 120 × 2.5 mm molded sheets. These sheets were conditioned for 4 weeks at 22 ± 3°C at a relative humidity (RH) of 52 ± 2%, prior to characterization. Glycerol losses were negligible, since the processing temperature was kept under the glycerol boiling point and no exudation was observed during processing. The sheet was characterized by SEM, DSC, WAXRD, and water absorption experiments. The blends were immiscible, showing two distinct phases of starch and zein. The SEM of the blends also shows interfacial voids, indicating poor adhesion between starch and zein phases. The water uptake at equilibrium and its diffusion coefficient were determined. The water uptake at equilibrium decreased with increasing zein content. The diffusion coefficient fell sharply on addition of 20% zein and remained constant as zein content was increased. No appreciable effect of zein on starch crystallization was observed by X ray diffraction. The use of zein in TPS compositions causes a decrease in the water sensitivity of these materials and lowers its melt viscosity during processing making zein a suitable and very promising component in TPS compositions.

10.5.2.4 Preparation of Starch Based Biodegradable Poly(Butylene Adipate-Co-Terephthalate) Nanocomposites Mohanty and Nayak [61] prepared Starch based biodegradable poly(butylene adipate-co-terephthalate) (PBAT) and organically modified nanoclays using melt intercalation technique. Two different organically modified nanoclays Cloisite C20 A and Cloisite C30B at various wt% have been used for fabrication of nanocomposites. Starch was gelatinized to prepare TPS for increasing the compatibility with the PBAT matrix. Subsequently, films of PBAT/TPS blends at various TPS content (10 wt%, 20 wt%, 30 wt%, 40 wt%) and PBAT/TPS organoclay biodegradable blend nanocomposites at different wt% of nanoclays were prepared using the solvent casting method. The interfacial region between the biodegradable polymer matrix and organoclay were also modified with grafting of maleic anhydride with PBAT chains during melt blending. This was done through a two-stage reactive extrusion process using free radical initiator inorder to provide the reactive functional group to chemically react with the hydroxyl group of starch as well as nanoclays. Mechanical test revealed an increase in tensile modulus and elongation at break with the incorporation of 30wt% TPS and C30B nanoclay to the tune of 44.45% and 776.9% as compared with the PBAT matrix. The morphology of PBAT/TPS 30%/C30B3% biodegradable blend nanocomposites studied using WAXRD and SEM indicated that intercalation occurred

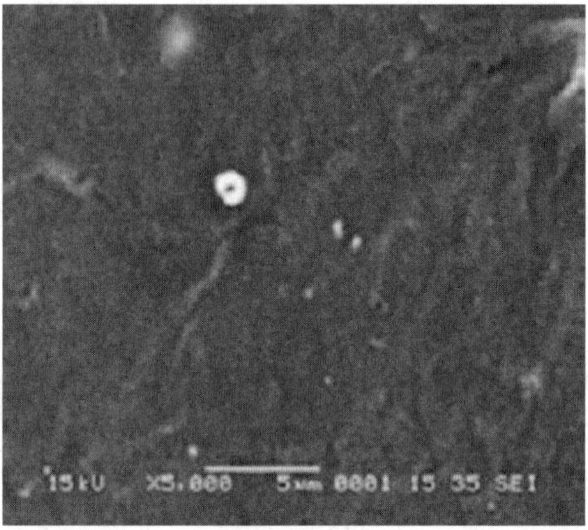

Figure 10.7 SEM micrograph of PBAT/TPS nanocomposite. Reprinted with permission from Reference [61]. Copyright Taylor & Francis.

along with an improved dispersion of TPS within PBAT with the inclusion of C30B. Figure 10.7 shows the SEM micrograph of PBAT/TPS nanocomposite.

10.5.2.5 Preparation of Thermoplastic Corn Starch Reinforced with Cotton Cellulose Nanofibers Teixeira et al. [62] prepared thermoplastic corn starch reinforced with cotton cellulose nanofibers by conventional extrusion process. For the preparation of nanocomposites, the dry cotton cellulose nanofibers (CCN) was first redispersed in water using a sonificator for 5 minutes, and it was composed of finely dispersed nanofibers in distinct concentrations 0.5 wt%, 1.5 wt%, 2.5 wt%, 3.5 wt%, and 5 wt% (dry starch basis). The resulting suspension was added to the system starch, glycerol, stearic acid, and citric acid. The resulting mixture was manually homogenized, and the nanocomposites were processed in a twin-screw extruder through the use of a temperature profile between 140°C and 160°C. The films were obtained by hot pressing process at 160°C, under a force of 10 metric tons and for 5 minutes after a prefusion of 5 minutes. Scanning transmission electron microscopy (STEM) observations showed some degree of agglomeration, especially for nanocomposites with CCN content of 2.5 wt% and above. XRD patterns recorded a small decrease of crystallinity. The crystallinity reduction was attributed to the decrease in the rearrangement resistance of the starch chains due to the transcrystallization of amylopectin in the nanofiber surface. This is a consequence of the coating of cellulose nanofibers by glycerol plasticizer.

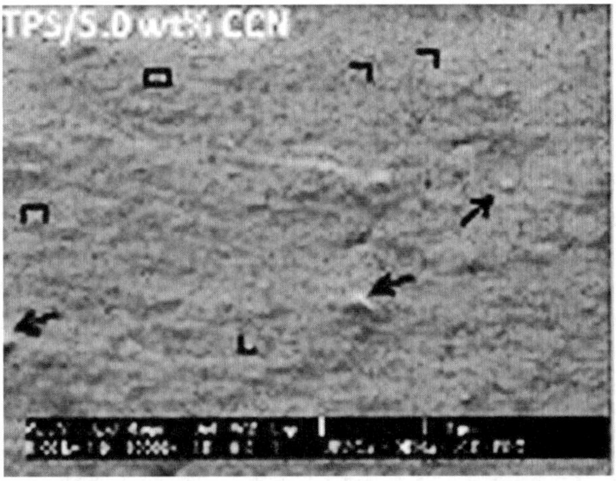

Figure 10.8 Micrograph of FE-SEM for thermoplastic corn starch and cotton cellulose nanofiber composite. Reprinted with permission from Reference [62]. Copyright John Wiley and Sons.

This phenomenon hinders the stress transfer in the filler–matrix interface, compromising the reinforcement effect of the nanocellulose filler. It was also found that in comparison with neat TPS, both elastic modulus and TSwere improved to up to 48.5% and 150%, respectively, for the composition with 5 wt% CCN, indicating a synergistic effect of the nanofibers on the TPS matrix. Figure 10.8 shows FE-SEM of thermoplastic corn starch and cotton cellulose nanofiber.

10.5.3 *In Situ* Intercalative Polymerization

In this method, the nanofiller is swollen within the liquid monomer or a monomer solution so the polymer formation can occur between the intercalated sheets. Polymerization can be initiated either by heat or radiation, by the diffusion of a suitable initiator, or by an organic initiator or catalyst fixed through cation exchange inside the interlayer before the swelling step [1]. At first this approach was successfully applied in manufacturing of nylon–MMT nanocomposite, and later it was extended to other thermoplastics [63]. Messermith and Giannelis [64] have modified MMT by bis(2-hydroxyethyl) methyl hydrogenated tallow alkyl ammonium cation, and they found the modified clay dispersed readily in diglycidyl ether of bisphenol A (DGEBA) by using this process. One obvious advantage of *in situ* polymerization is the tethering effect, which enables the nanoclay's surface organic chemical, such as 12-aminododecanoic acid (ADA), to link with nylon-6 polymer chains during polymerization. Polyethylene/layered silicate nanocomposites have also been prepared by *in situ* intercalative polymerization of ethylene using the so-called polymerization filling technique [65]. In this case the polymerizing catalyst is fixed on the clay surface and the polymerization is carried out using the modified clay as the catalyst. Pristine MMT and hectorite were first treated with trimethyl aluminum-depleted methyl aluminoxane, before being contacted by a Ti-based constrained geometry catalyst. The nanocomposite was formed by addition and polymerization of ethylene. In the absence of a chain transfer agent, ultra-high molecular weight (HME) polyethylene was produced. The tensile properties of the nanocomposite was poor and essentially independent of the nature and content of the silicate. Upon hydrogen addition, the molecular weight of polyethylene was decreased with corresponding increase in mechanical properties. The formation of exfoliated nanocomposite was confirmed XRD and TEM.

10.5.3.1 Preparation of Starch/Nanosilicate Layer/Polycaprolactone Composites Namazi and Mosadegh [66] prepared starch-g-polycaprolactone (PCL) nanocomposites with graft polymerization through *in situ* ring opening polymerization of ε-caprolactone in the presence of starch, $Sn(Oct)_2$ (Tin(II) 2-ethyl hexanoate) as an initiator/catalyst and silicate layers. The silicates used were MMT and kaolin. In fact, the related composites were prepared in solution method, bulk polymerization, and *in situ* polymerization methods with introducing the mineral clay. Starch–MMT/PCL and starch–kaolin/PCL nanocomposites were prepared by bulk polymerization technique (Fig. 10.9). The starch–MMT composite (5 wt%) (starch–kaolin composite) was powdered and dried. Three grams of composite was added to a flask containing caprolactone monomer (3 g), then 0.4 wt% $Sn(Oct)_2$ (initiator) was added to the flask with continuous stirring. Polymerization was performed at 100°C for 20 hours under nitrogen atmosphere. Polymerization was stopped by fast cooling at room temperature.

Starch-g-PCL nanocomposite was prepared through *in situ* polymerization method as follows.

The 5 wt% organoclay Cloisite 15A (C15A) was swelled and dispersed in 3 g ε-caprolactone for 18 hours under nitrogen atmosphere, then 3 g starch was

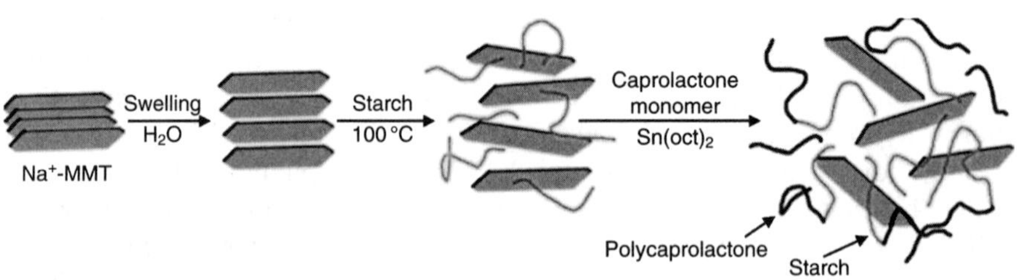

Figure 10.9 Preparation route of starch/MMT/PCL nanocomposite. Reprinted with permission from Reference [66]. Copyright Springer.

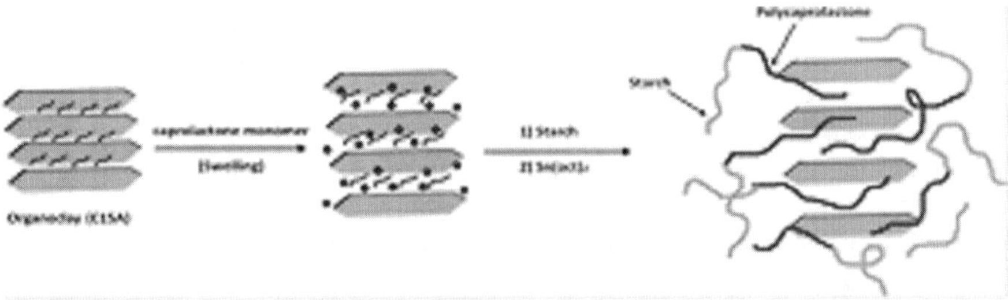

Figure 10.10 Preparation route of starch/C15A/PCL nanocomposite. Reprinted with permission from Reference [66]. Copyright Springer.

added and left for 2.5 hours with continuous stirring, then 0.4 wt% $Sn(Oct)_2$ was introduced via a conditioned syringe at 100°C temperature and left for 20 hours. Polymerization was stopped by fast cooling at room temperature. Figure 10.10 shows the method of preparation for starch/C15A/PCL nanocomposite. The effect of kind of clay on *d*-spacing of silicate layers was investigated and the obtained nanocomposites were analyzed using XRD. The obtained compounds were characterized by FT-IR. The XRD results showed that the biopolymer starch with hydrophilic property is compatible with MMT. The basal peak in MMT ($2\theta = 7.13$, *d*-spacing = 12.38 Å) broadened and shifted to a low angle in starch/MMT samples, which is an indication of intercalation/exfoliation state. SEM images showed the change in morphology of granules of starch in composite samples. Granules were destroyed due to gelatinization affected by water, temperature, and polymerization reaction on the surface of starch. From DSC thermograms of the nanocomposite, it was possible to see that the melting (Tm) of crystalline increased, except the starch/MMT sample. Decrease in Tm with addition of MMT, around 125°C, was attributed to the reduced crystalline size and presence of crystal imperfections due to compatibility of the MMT with starch, which suppressed the crystallization where the silicate platelet prevented reorganization of the amylase chains. Furthermore, the presence of grafted and blended PCL in starch/MMT/PCL, starch/kaolin/PCL, and starch/C15A/PCL samples caused the shifting of the endothermic peak to higher temperatures.

10.6 CONCLUSION

Various literature surveys show that nanocomposites with superior properties have been successfully made from starch matrix blends in the presence of nanoparticles through different processing techniques. Nanoparticles could be finely dispersed into polymer matrix materials if good processing conditions are employed and the addition of nanoparticles at a lower level (less than 5%

wt) could significantly improve the mechanical, physical, and thermal properties of nanocomposites, such as TS, Young's modulus, elongation, and water resistance. Starch-based nanocomposites are gaining interest due to their characteristics, biodegradability in soil and water, and low cost. There are many opportunities for starch-based nanocomposite and therefore extensive research is being performed all over the world with regard to the preparation and determining applications for starch-based nanocomposites.

REFERENCES

[1] Zhao, R., Torley, P., and Halley, P. (2008) Emerging biodegradable materials: Starch- and protein-based bio-nanocomposites. *Journal of Materials Science*, *43*, 3058–3071.

[2] Ahmed, J., Tiwari, B.K., Imam, S.H., and Rao, M.A., eds. (2012) *Starch-Based Polymeric Materials and Nanocomposites: Chemistry, Processing, and Applications*. Boca Raton, FL: CRC Press.

[3] Huang, X., Netravali, A. (2009) Biodegradable green composites made using bamboo micro/nano-fibrils and chemically modified soy protein resin. *Composites Science and Technology*, *69*, 1009–1015.

[4] Reddy, N., Yang, Y. (2011) Completely biodegradable soyprotein–jute biocomposites developed using water without any chemicals as plasticizer. *Industrial Crops and Products.*, *33*, 35–41.

[5] Savadekar, N.R., Mhaske, S.T. (2012) Synthesis of nano cellulose fibers and effect on thermoplastics starch based films. *Carbohydrate Polymers*, *89*, 146–151.

[6] Shen, L., Worrell, E., and Patel, M. (2010) Present and future development in plastics from biomass. *Biofuels, Bioproducts and Biorefining*, *4*, 25–40.

[7] Mallapragada, S.K., Narasimhan, B. (2006) *Handbook of Biodegradable Polymeric Materials and Applications*, Vol. 1–6. New York: American Chemical Society.

[8] Ray, S.S., Bousmina, M. (2005) Biodegradable polymers and their layered silicate nanocomposites: In greening the 21st century materials world. *Progress in Materials Science*, *50*, 962.

[9] Visakh, P.M., Mathew, A.P., Oksman, K., and Thomas, S. (2012) Starch based bionanocomposites. Processing and properties. In: *Polysaccharide Building Blocks: A Sustainable Approach to the Development of Renewable Biomaterials*, 1st ed, eds. Y. Habibi, L.A. Lucia. NewJersy: Wiley

[10] Debeaufort, F., Quezada-Gallo, J.A. (2010) Edible films and coatings: Tomorrow's packagings: A review. *Critical Reviews in Food Science and Nutrition*, *38*, 299–313.

[11] Siró, I., Plackett, D. (2010) Microfibrillated cellulose and new nanocomposite materials: A review. *Cellulose (London, England)*, *17*, 459–494.

[12] Cao, X., Chen, Y., Chang, P.R., Stumborg, M., and Huneault, M.A. (2008) Green composites reinforced with hemp nanocrystals in plasticized starch. *Journal of Applied Polymer Science*, *109*, 3804–3810.

[13] Woehl, M.A., Canestraro, C.D., Mikowski, A., Sierakowski, M.R., Ramos, L.P., and Wypych, F. (2010) Bionanocomposites of thermoplastic starch reinforced with

bacterial cellulose nanofibres: Effect of enzymatic treatment on mechanical properties. *Carbohydrate Polymers*, *80*, 866–873.

[14] Thomas, S., Paul, S.A., Pothan, L.A., and Deepa, B. (2011) Natural fibres. structure, properties and applications. In: *Cellulose Fibers: Bio- and Nano-Polymer Composites*, eds. S. Kalia, B.S. Keith, and I. Kaur. New York: Springer, pp. 3–37.

[15] Spence, K., Habibi, Y., and Dufresne, A. (2011) Nanocellulose-based composites. In: *Cellulose Fibers: Bio- and Nano-Polymer Composites*, eds. S. Kalia, B.S. Keith, and I. Kaur. New York: Springer, pp. 180–187.

[16] Chen, B., Evans, J.R.G. (2005) Thermoplastic starch–clay nanocomposites and their characteristics. *Carbohydrate Polymers*, *61*, 455–463.

[17] Goffin, A.L., Raquez, J.M., Duquesne, E., Siqueira, G., Habibi, Y., Dufresne, A., and Dubois, P.H. (2011) Poly(ε-caprolactone) based nanocomposites reinforced by surface-grafted cellulose nanowhiskers via extrusion processing: Morphology, rheology, and thermo-mechanical properties. *Polymers*, *52*, 1532–1538.

[18] Deepa, B., Abraham, E., Cherian, B.M., and Pothan, L.A. (2011) Structure, morphology and thermal characteristics of banana nano fibers obtained by steam explosion. *Bioresource Technology*, *102*, 1988–1997.

[19] Samir, S.A., Alloin, F., and Dufresne, A. (2005) Review of recent research into cellulosic whiskers, their properties and their application in nanocomposite field. *Biomacromolecules*, *6*, 612–626.

[20] Siqueira, G., Bras, J., and Dufresne, A. (2010) Cellulosic bionanocomposites: A review of preparation, properties and applications. *Polymers*, *2*, 728–765.

[21] Liu, H., Xiea, F., Yua, L., Chena, L., and Li, L. (2009) Thermal processing of starch-based polymers. *Progress in Polymer Science*, *34*, 1348–1368.

[22] Kumar, P., Sandeep, K.P., Alavi, S., Truong, V.D., and Gorga, R.E. (2010) Effect of type and content of modified montmorillonite on the structure and properties of bio-nanocomposite films based on soy protein isolate and montmorillonite. *Journal of Food Science*, *75*, 46–56.

[23] Dean, K., Yu, L., and Smith, R., eds. (2005) *Biodegradable Polymers for Industrial Application*. Boca Raton, FL: CRC Press.

[24] Kumar, P., Sandeep, K.P., Alavi, S., and Truong, V.D. (2011) A review of experimental and modeling techniques to determine properties of biopolymer-based nanocomposites. *Journal of Food Science*, *76*, 2–14.

[25] Rhim, J.W., Ng, P.K.W. (2007) Natural biopolymer-based nanocomposite films for packaging applications. *Critical Reviews in Food Science and Nutrition*, *47*, 411–433.

[26] Mensitieri, G., Maio, E.D., Buonocore, G.G., Nedi, I., Oliviero, M., Sansone, L., and Iannace, S. (2011) Processing and shelf life issues of selected food packaging materials and structures from renewable resources. *Trends in Food Science & Technology*, *22*, 72–80.

[27] Nussinovitch, A. (2009) Biopolymer films and composite coatings. In: *Modern Biopolymer Science*, 1st ed, eds. S. Kasapis, I.T. Norton, and J.B. Ubbink. London: Elsevier, pp. 295–300.

[28] De Azeredo, H.M.C. (2009) Nanocomposites for food packaging applications. *Food Research International (Ottawa, Ont.)*, *42*, 1240–1253.

[29] Corre, D.L., Bras, J., and Dufresne, A. (2010) Starch nanoparticles: A review. *Biomacromolecules*, *11*, 1139–1153.

[30] Kaushik, A., Singh, M., and Verma, G. (2010) Green nanocomposites based on thermoplastic starch and steam exploded cellulose nanofibrils from wheat straw. *Carbohydrate Polymers*, *82*, 337–345.

[31] Ray, S.S., Okamoto, M. (2003) Polymer/layered silicate nanocomposites: A review from preparation to processing. *Progress in Polymer Science*, *28*, 1539–1641.

[32] Ali, S., Tang, X., Alavi, S., and Faubion, F. (2011) Structure and physical properties of starch/poly vinyl alcohol/sodium montmorillonite nanocomposite films. *Journal of Agricultural and Food Chemistry*, *59*, 12384–12395.

[33] Ma, X., Yu, J., and Wang, N. (2007) Production of thermoplastic starch/MMT-sorbitol nanocomposites by dual-melt extrusion processing. *Macromolecular Materials and Engineering*, *292*, 723–728.

[34] Arora, A., Padua, G.W. (2010) Review: Nanocomposites in food packaging. *Journal of Food Science*, *75*, 43–49.

[35] Dufresne, A., Dupeyre, D., and Vignon, M.R. (2000) Cellulose microfibrils from potato tuber cells: Processing and characterization of starch–cellulose microfibril composites. *Journal of Applied Polymer Science*, *76*, 2080–2092.

[36] Ma, X., Chang, P.R., and Yu, J. (2008) Properties of biodegradable thermoplastic pea starch/carboxymethyl cellulose and pea starch/microcrystalline cellulose composites. *Carbohydrate Polymers*, *72*, 369–375.

[37] Angellier, H., Boisseau, S.M., and Dufresne, A. (2006) Waxy maize starch nanocrystals as filler in natural rubber. *Macromolecular Symposia*, *233*, 132–136.

[38] Dufresne, A., Cavaillé, J.Y., and Helbert, W. (1996) New nanocomposite materials: Microcrystalline starch reinforced thermoplastic. *Macromolecules*, *29*, 7624–7626.

[39] Lim, Y.T., Park, O. (2000) Rheological evidence for the microstructure of intercalated polymer/layered silicate nanocomposites. *Macromolecular Rapid Communications*, *21*, 231–235.

[40] Ren, P., Shen, T., and Wang, F. (2009) Study on biodegradable starch/OMMT nanocomposites for packaging applications. *Journal of Polymers and the Environment*, 203–207.

[41] Cao, X., Chen, Y., Chang, P.R., Muir, A.D., and Falk, G. (2008) Starch-based nanocomposites reinforced with flax cellulose nanocrystals. *eXPRESS Polymer Letters*, *2*, 502–510.

[42] Alemdar, A., Sain, M. (2008) Biocomposites from wheat straw nanofibers: Morphology, thermal and mechanical properties. *Composites Science and Technology*, *68*, 557–565.

[43] Kamel, S. (2007) Nanotechnology and its applications in lignocellulosic composites, a mini review. *eXPRESS Polymer Letters*, *1*, 546–575.

[44] Mathew, A.P., Dufresne, A. (2008) Morphological investigation of nanocomposites from sorbitol plasticized starch and tunicin whiskers. *Biomacromolecules*, *3*, 609–617.

[45] Svagan, A.J., Samir, A.S.A., and Berglund, L.A. (2007) Biomimetic polysaccharide nanocomposites of high cellulose content and high toughness. *Biomacromolecules*, *8*, 2556–2563.

[46] Lu, D.R., Xiao, C.M., and Xue, S.J. (2009) Starch-based completely biodegradable polymer materials. *eXPRESS Polymer Letters*, *3*, 366–375.

[47] Bogdal, D., Penczek, P., Pielichowski, J., and Prociak, A. (2003) Synthesis and characterization of epoxy resins prepared under microwave irradiation. *Advances in Polymer Science*, *163*, 193–263.

[48] Hussain, F., Hojjati, M. (2006) Polymer-matrix nanocomposites, processing, manufacturing, and application: An overview. *Journal of Composite Materials*, *40*, 6–8.

[49] Christopher, O.O., Lerner, M. (2006) Nanocomposites and intercalation compounds. In *Encyclopedia of Physical Science and Technology*, Vol. 10. San Diego, CA: Academic Press (reproduced in part with permission from Elsevier).

[50] Mazumder, S.K. (2002) *Composites Manufacturing, Materials, Product and Process Engineering*. Boca Raton, FL: CRC Press.

[51] Roy, S., Hussain, F., Lu, H., and Narasimhan, K. Characterization and Modeling of Strength Enhancement Mechanism in Polymer Clay Nanocomposites. *6th AIAA Conference on Structures, Structural Dynamics, and Materials*. 2005 April 18–21; Texas.

[52] Mahfuz, H., Baseer, M.A., and Zeelani, S. (2005) Fabrication, characterization and mechanical properties of nano phased carbon prepeg laminates. *SAMPE Journal*, *41*, 2–3.

[53] Chen, B., Evans, J.R.G., Greenwell, H.C., Boulet, P., Coveney, P.V., Bowden, A.A., and Whiting, A. (2008) A critical appraisal of polymer-clay nanocomposites. *Chemical Society Reviews*, *37*, 568–594.

[54] Oriakhi, C. (1998) Nano sandwiches. *Chemistry in Britain*, *34*, 59–62.

[55] Ali, S.S. Fundamental interactions and physical properties of starch, poly vinyl alcohol and montmorillonite clay based nanocomposites prepared using solution mixing and melt extrusion.[Dissertation] Department of Grain Science and Industry; 2010. Available from K-State Research Exchange.

[56] Ma, X., Chang, P.R., Yang, J., and Yu, J. (2009) Preparation and properties of glycerol plasticized-pea starch/zinc oxide-starch bionanocomposites. *Carbohydrate Polymers*, *75*, 472–478.

[57] Mittal, V. (2011) *Nanocomposites with Biodegradable Polymers*. Oxford, UK: Oxford University press.

[58] Bhattacharya, S.N., Kamal, M.R., and Gupta, R.K., eds. (2008) *Polymer Nanocomposite: Theory and Practice*. Munich Germany: Hanser.

[59] Park, H.M., Li, X., Jin, C.Z., Park, C.Y., Cho, W.J., and Ha, C.S. (2002) Preparation and properties of biodegradable thermoplastic starch/clay hybrids. *Macromolecular Materials and Engineering*, *287*, 553–558.

[60] Corradini, E., Carvalho, A.J.F., Curvelo, A.A.S., Agnelli, J.A.M., and Mattoso, L.H.C. (2007) Preparation and characterization of thermoplastic starch/zein blends. *Materials Research 10*, 227–231.

[61] Mohanty, S., Nayak, S.K. (2009) Starch based biodegradable PBAT nanocomposites. *International Journal of Plastics Technology*, *13*, 163–185.

[62] Teixeira, E.M., Lotti, C., Corre, A., Teodoro, K.B.R., Marconcini, J.M.M., and Mattoso, L.H.C. (2011) Thermoplastic corn starch reinforced with cotton cellulose nanofibers. *Journal of Applied Polymer Science*, *120*, 2428–2433.

[63] Kozima, Y., Usuki, A., Kawasumi, M., Okada, A., Fukushima, Y., Kurauchi, T., and Kamigaito, O.J. (1993) Mechanical properties of nylon-6 clay hybrid. *Journal of Material Research*, *8*, 1185–1189.

[64] Yu, Z., Yang, M., Zhang, Q., Zhao, C., and Mai, Y.W. (2003) Dispersion and distribution of organically modified montmorillonite in nylon-66 matrix. *Journal of Polymer Science. Part B: Polymer Physics*, *41*, 1234–1243.

[65] Alexandre, M., Dubois, P., Sun, T., Jarces, J.M., and Jerome, R. (2002) Polyethylene layered silicate nanocomposite prepared by polymerization filling technique. Synthesis and mechanical properties. *Polymer*, *43*, 2123–2132.

[66] Namazi, H., Mosadegh, M. (2011) Preparation and properties of starch/nanosilicate layer/polycaprolactone composites. *Journal of Polymers and the Environment*, *19*, 980–987.

CHAPTER 11

Morphological and Thermal Investigations of Starch-Based Nanocomposites

PETER R. CHANG, JIN HUANG, QING HUANG, and DEBBIE P. ANDERSON

11.1 INTRODUCTION

As a natural biopolymer, starch is abundant, renewable, inexpensive, biodegradable, environmentally friendly, and easy to chemically modify, making it one of the most attractive and promising bioresource materials [1, 2]. Unfortunately, contemporary research indicates that unsatisfactory reinforcement by starch [3–8] and undesirable properties (brittleness, low heat distortion temperature, high gas permeability, high viscosity, and poor melt properties) restrict its use in a wide range of applications [9–12]. Over the past few decades, the possibility of adding nano-reinforcement to pristine polymers as load bearing constituents has been explored and has produced encouraging results [13–16]. Starch-based nanocomposites are now considered to be next-generation "green" materials. There are plenty of studies reporting on the preparation of starch-based nanocomposites using layered silicates [17–23], carbon nanoparticles [24–32], polysaccharide nanocrystals [33–42], and other nanofillers such as SiO_2 [43–45], TiO_2 [46], hydroxyapatite (HAp) [47], ZnO [48], CdS, Ag and Ag_2S, α-zirconium phosphate (α-ZrP) [49], and layered double hydroxide [50]. The properties of the resulting starch-based nanocomposites vary with the different sizes, shapes, and surface chemistries of the nanofillers used. Nano-reinforced starch-based nanocomposites generally exhibit enhanced mechanical and thermal properties when nanofillers are well

Biopolymer Nanocomposites: Processing, Properties, and Applications, First Edition. Edited by Alain Dufresne, Sabu Thomas, and Laly A. Pothan.

dispersed, while the nature of the matrix and/or nanofiller contributes to its biological properties [51, 52].

Furthermore, starch-based nanocomposites made with different nanofillers may exhibit various beneficial properties. Starch nanocomposite with graphene oxide (GO), for example, could effectively protect against ultraviolet (UV) light, leading to its application as a UV-shielding packing material [27]. Glycerol plasticized starch/multiwalled carbon nanotube (MWCNT) nanocomposite exhibits electrical conductivity and promises a wide variety of potential applications, such as antistatic plastics, biosensors, artificial muscles, corrosion protection, electronic shielding, environmentally sensitive membranes, and solar materials [53]. Silica has been introduced into soluble potato starch using a sol–gel synthesis method. The resulting nanocomposite exhibits an excellent ability to bind cadmium from aqueous solution, leading to potential use in water remediation. It also shows luminescence in the UV-Vis region and could therefore have possible applications as sensors and optical data storage devices [44].

The technologies used to manufacture starch-based nanocomposites are broadly classified into three main categories [13, 54–58]: solution intercalation, *in situ* intercalative polymerization, and melt intercalation. Solution intercalation is based on a solvent system in which soluble biopolymer chains intercalate and replace the solvent molecules that were used to swell the nanofiller. In *in situ* intercalative polymerization, the nanofillers are swollen within a liquid monomer or a monomer solution so that polymer formation occurs between the intercalated sheets. In melt intercalation processing, the polymer chains diffuse between aggregated nanofillers to produce a nanostructured system. Compared with solution intercalation, melt intercalation has many advantages, including the absence of organic solvents and compatibility with extrusion and injection molding techniques, which help to make this the standard method [56, 57].

The so-called nano-effect forms when nanoparticles with a high specific surface area and surface energy physically interact with the polymer matrix, causing the same additive amount of well-distributed nanoparticles to have a higher contact area with the matrix. This leads to a series of effects including altered crystal structure and order in the crystalline regions of the polymer matrix, changes in organization of the multiphase structure, disruption and/or formation of intermolecular bonds, and relaxation of the polymer chain due to steric hindrance of the rigid nanoparticles. Although the high surface energy of nanoparticles makes them extremely reactive, most systems undergo severe aggregation [51]; thus, research on the distribution of nanofillers and their interactions with the matrix is significant.

This chapter details morphologies and thermal behaviors of starch-based nanocomposites, as well as their application. The preparation methods, crystalline structures, and reinforcement mechanisms of various nanofillers in starch-based nanocomposites are recognized and discussed; and the prospects and outlook for application in nanomaterials and biomaterials are presented.

11.2 MORPHOLOGIES OF STARCH-BASED NANOCOMPOSITES

11.2.1 Factors Affecting Nanofiller Distribution

Distribution of the nanofiller in the matrix has a great effect on interactions between the nanofiller and matrix, as well as on the mechanical properties of the nanocomposites, as shown in the literature [26, 59]. Dispersibility of nanofillers depends on factors such as shape, size, concentration in the starch matrix, functional groups on the nanofiller surface, and the mechanism of interactions between nanofiller and matrix. As a result, the dispersibility of different nanofillers needs to be discussed in detail.

11.2.1.1 Layered Silicates Layered silicates are a family in which clays with varying structures, textures, and morphologies can be found. Layered silicates present three levels of organization depending on the scale of observation: (1) the layer, (2) the primary particle, and (3) the aggregate [60]. The most intensive research has been focused on using layered silicates as the reinforcing phase due to their availability and versatility [13, 61]. Researchers concentrate particularly on montmorillonite (MMT) since its homogeneous dispersion in nanosheets of a continuous polymer matrix may greatly improve the properties of polymers. Meanwhile, besides improving the intrinsic properties of a polymer, specific new properties, which cannot normally be derived from a component in the composites, may also be installed as a contribution of homogeneously dispersed MMT nanosheets in a continuous polymer matrix [62–69]. In previous studies, nanoclays (natural and chemically modified) have been incorporated into plasticized starches such as wheat [70], potato [71], corn [72], and waxy [73] and high amylose maize [74], with the nanoclay content varying from 0 wt% to 10 wt% [56]. Plasticizers such as glycerol [75], sorbitol [76], citric acid [77], urea [78], formamide [79], N-(2-hydroxyethyl) formamide [80], and ethanolamine [81] are widely used in the preparation of starch/clay nanocomposites.

The structure of layered silicates blended with polymers has been extensively discussed [61, 82–88]. Vaia et al. [83] proposed that the intercalated chains adopt a more disordered, liquid-like structure as the chain length or the interlayer packing density decreases. This was subsequently confirmed by Hackett et al. using molecular dynamic simulations [84]. With the increase in chain length, the interlayer structure appears to evolve in a stepwise fashion, from a disordered to a more ordered monolayer and then "jumping" to a disordered pseudo-bilayer [87, 88]. Intercalation occurs when a small amount of polymer moves into the gallery spacing between the clay platelets. Exfoliation occurs when silicate layers are substantially expanded and are no longer parallel [87] (seen in Fig. 11.1).

The structure of the nanoclay will have a major effect on its dispersibility. It is generally believed that the best performance is achieved with exfoliated structures [86] in a well-dispersed nanocomposite consisting of exfoliated

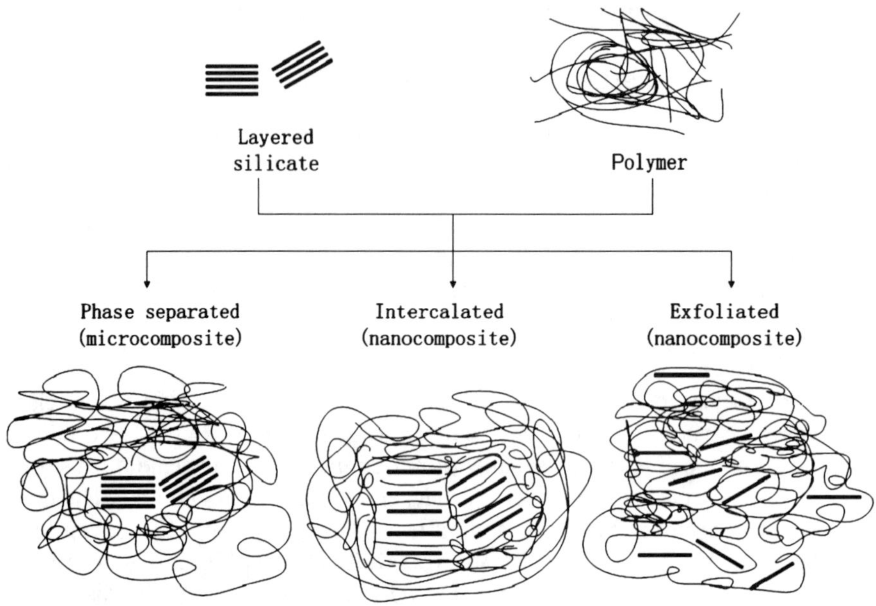

Figure 11.1 Scheme of different types of composite arising from the interaction of layered silicates and polymers: (a) phase separated microcomposite, (b) intercalated nanocomposite, and (c) exfoliated nanocomposite.

platelets distributed homogeneously throughout the polymer matrix forms. This has been attested to by Liu et al. [89].

Surface modification of the nanofiller can significantly influence its degree of dispersion [87, 90, 91]. MMT/cassava starch composite films were prepared by Kampeerapappun et al. [92] using the casting method. In the starch film containing pristine MMT, populations of large aggregates were common, whereas particles of a much smaller size were found scattered in the starch film containing chitosan-treated MMT. Typically there is no attractive force between starch and MMT resulting in poor distribution of MMT particles in the starch matrix. These results show that the chitosan, acting as a compatibilizing agent due to its hydrophilicity and its hydrogen bonding with the starch matrix, promoted both disaggregation of the MMT and a more homogeneous dispersion in the starch matrix [92].

Various characteristics of starch, such as the amount of amylose leached from the granule, the shape and swelling of granules, entanglement in the matrix of amylose and amylopectin molecules, and interactions between components [93], are reported to impact the dispersion of clay. For high amylose starches, a high degree of chain–chain interaction allows the formation of highly entangled networks [94], which may not favor dispersion and exfoliation of the clay silicate layers. Conversely, in waxy starches only the short outer chains participate in the formation of weak entangled structures [93], which may favor diffusion of these structures within the silicate galleries and, therefore, formation of partially exfoliated structures. The fact that branched polymers are more able to intercalate into the clay galleries than linear polymers [95, 96] is explained by the self-consistent field theory [58]. The effect of plasticizer loading level on the interlayer space of MMT in starch/MMT nanocomposites was investigated by Liu et al. [97]. The presence of plasticizer improved expansion of the MMT *d*-spacing in starch-based nanocomposites, with 15% glycerol resulting in the largest *d*-spacing. The interaction between plasticizer/plasticizer and plasticizer/starch seemed to resume its dominance when the plasticizer loading level was increased [97].

Mondragón et al. [58] used three types of maize starch with different amounts of amylose and amylopectin to prepare plasticized starch (PS)/MMT nanocomposite films. Waxy maize starch/MMT nanocomposites (WMNC), normal starch/MMT nanocomposites (NMNC), and high amylose maize starch/MMT nanocomposites (HAMNC) were prepared using the casting technique. WMNC could not be observed with the transmission electron microscope (TEM) because those embedded films could not be sectioned. However, at a relatively low filler content (2.0 wt%), NMNC and HAMNC presented mostly well-dispersed exfoliated structures in the TEM images. At a higher content (5.0 wt%) intercalated structures and poorly dispersed MMT aggregates were clearly visible. Dispersion of MMT in the PS matrices was corroborated by X-ray diffraction (XRD) patterns. According to XRD, dispersion of silicates in WMNC was not affected by increasing the MMT content; however, the small characteristic MMT diffraction peak in NMNC and HAMNC with 2.0 wt% MMT content increased, and the full-width-at-half-maximum decreased on further addition of MMT (5.0 wt%), indicating mainly intercalated structures [98]. These results indicate that dispersion of the silicates was affected by the type of starch and the MMT content.

In order to produce starch-based nanocomposites with better properties or special characteristics, many other polymers may be co-blended with starch, for example, starch/poly(lactic acid) (PLA) [99], starch/poly(caprolactone) (PCL) [100], and starch/poly(vinyl alcohol) (PVA) [101, 102]. While the nanoclays tend to stay in the thermoplastic starch (TPS), additional materials may affect the size of the TPS phase rather than the distribution of the nanoclay. This phenomenon was studied in detail by Arroyo et al. [103]. TPS and PLA or maleic anhydride grafted polylactic acid (PLAg) were compounded with natural MMT using a twin-screw extrusion process. In all the TPS/PLAg blends studied, large clay aggregates appeared with a few MMT stacks (formed by three or fewer clay layers) surrounding them, indicating that the PLAg did not have a significant effect on MMT dispersion. To verify this assumption, additional experiments were carried out where MMT was introduced into the PLA phase before mixing with TPS rather than in the starch–water suspension. Based on the images from atomic force microscopy (AFM) and TEM, it was concluded that MMT had migrated from the PLA phase to the TPS/PLA (or

PLAg) interface or had crossed the interface into the TPS phase due to the strong affinity between natural MMT and the hydrophilic TPS phase [103].

11.2.1.2 Carbon Nanoparticles Carbon nanoparticles are considered to be ideal fillers for reinforcing polymer matrices to achieve high performance and special functions. Various carbon nanoparticles, such as carbon nanotubes (both single- and multiwalled [SWCNT and MWCNT, respectively]) [24–26, 29], graphite (including colloidal and crystalline graphite [CGR and GR, respectively]) [25, 104], and fullerene C60 [25], have been used as nanofillers. There are also many derivatives of carbon nanoparticles, for example, GO [27], used as nanofillers that have achieved excellent properties. It is well known that effective utilization of nanofillers in composite applications, without destroying the original organization of the polymer matrix, depends strongly on the ability to homogeneously disperse the nanofiller throughout the matrix [26].

Many factors may affect the dispersibility of carbon nanoparticles. The nonreactive surface of carbon nanoparticles makes them unable to disperse well in nanocomposites because of a lack of adhesion between the carbon nanoparticles and the matrix. In the past decade, much effort has been made to improve dispersibility of carbon nanoparticles including the use of surfactants [53], oxidation [27], and chemical functionalization of the carbon nanoparticle's surface [26]. Introducing carboxylic, carbonyl, and hydroxyl groups to MWCNTs by refluxing with nitric acid or mixed acids can greatly enhance their interaction with the matrix. The distribution of carbon nanoparticles is also closely correlated with the shape and size of the added carbon nanoparticles, the concentration of carbon nanoparticles in the starch matrix, and the interaction between carbon nanoparticles and the matrix.

Oxidation is a good method for enhancing the distribution of graphite. GO is hydrophilic and forms strong physical interactions with the polymer due to its various oxygenated groups, including hydroxyls, epoxides, carbonyls, and carboxyls. GO sheets can also be intercalated or exfoliated by small molecules or polymers [105, 106]. Li et al. [27] prepared glycerol plasticized pea starch/graphene oxide (GPS/GO) nanocomposite films with different GO loading levels using a solution casting method. The TEM and AFM images shown in Figure 11.2 suggest that compatibility between GO and starch was good and that the GO filler was dispersed uniformly in the starch matrix. This uniform dispersion could be attributed to plenty of oxygen functional groups on the GO surface, which easily formed strong hydrogen bonds with the starch molecules. However, when the GO loading level was 5.0 wt%, blisters and depressions could be observed on the surface of the GPS/GO nanocomposite film (Fig. 11.2d), resulting in a higher roughness ($Ra = 14.009\,\text{nm}$, $Rq = 17.926\,\text{nm}$). This difference in microstructure was presumably due to poor dispersion and aggregation of GO at high loading levels [27].

The influence of the concentration of carbon nanoparticles on distribution was also discussed by Cao et al. [26] when MWCNTs were used as a

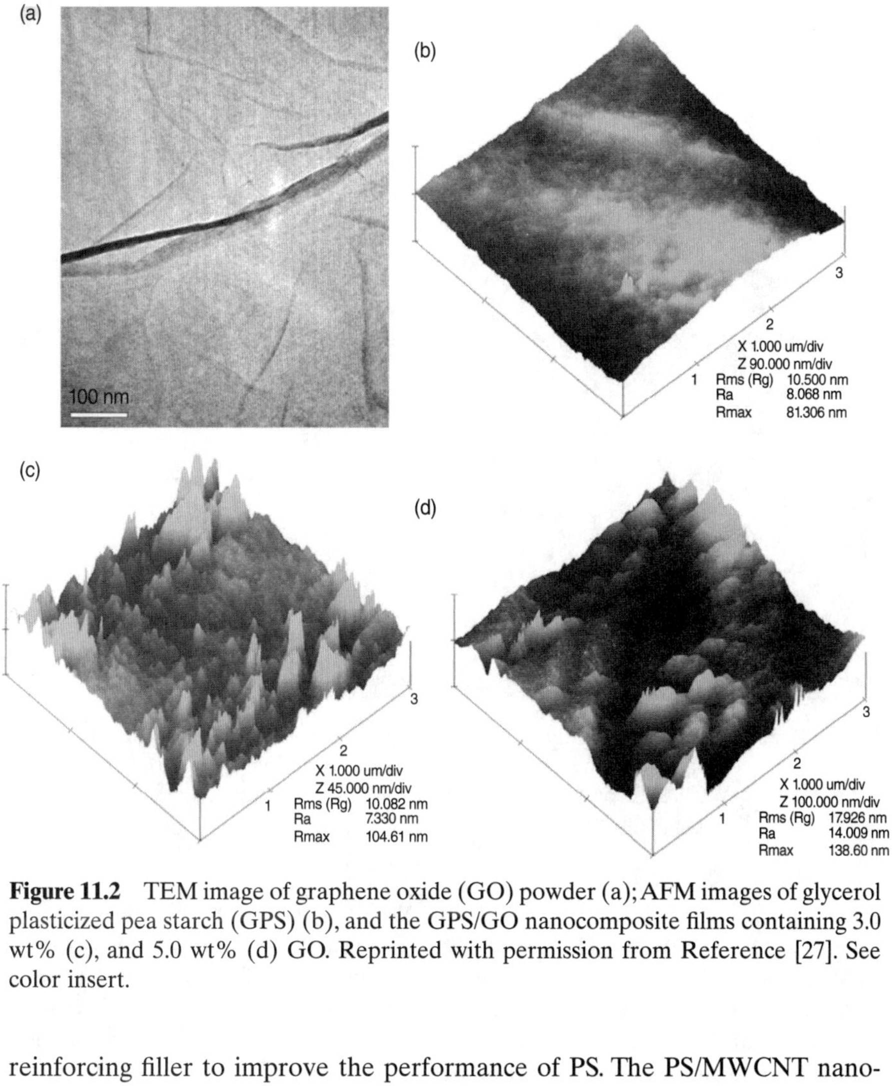

Figure 11.2 TEM image of graphene oxide (GO) powder (a); AFM images of glycerol plasticized pea starch (GPS) (b), and the GPS/GO nanocomposite films containing 3.0 wt% (c), and 5.0 wt% (d) GO. Reprinted with permission from Reference [27]. See color insert.

reinforcing filler to improve the performance of PS. The PS/MWCNT nanocomposites were successfully prepared using a simple method of solution casting and evaporation. As seen in the SEM images in Figure 11.3, no obvious large aggregates of MWCNTs were present on the fracture surfaces when the MWCNT content was lower than 1.0 wt%, suggesting that MWCNTs were homogenously dispersed in the PS matrix. At higher MWCNT levels (2.0 wt% and 3.0 wt%), large aggregates of MWCNTs were easily observed at the same magnification. This phenomenon attests to the fact that the MWCNT concentration makes a big difference, which is in accordance with previous conclusions.

The shape of carbon nanoparticles and the interaction between nanofiller and matrix also significantly influence the distribution of nanofiller in the

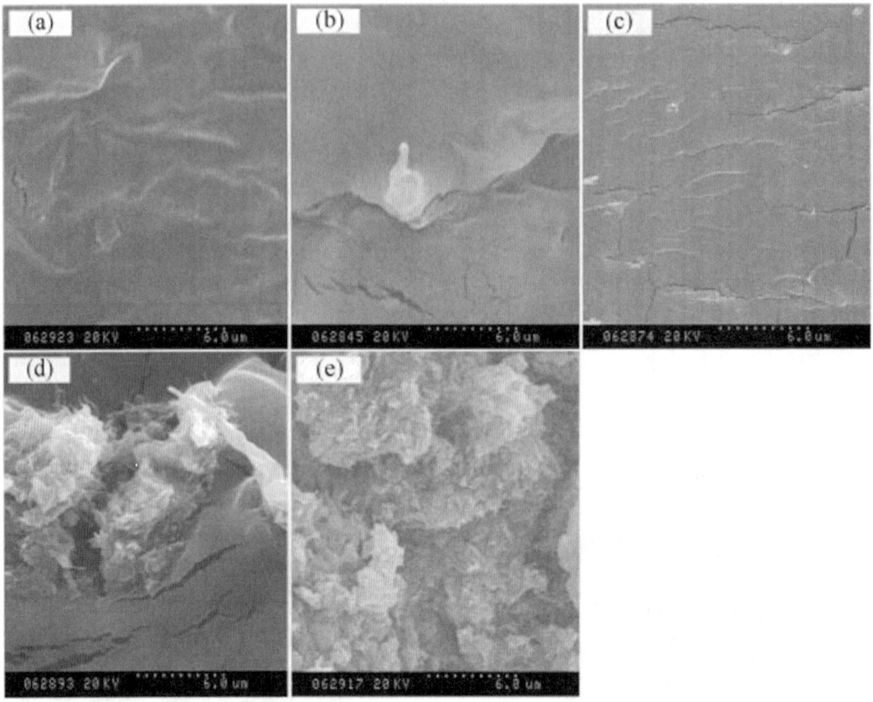

Figure 11.3 SEM images of failure surfaces of plasticized starch (PS) film (a) and PS/MWCNT nanocomposite films containing 0.2 wt% (b), 1.0 wt% (c), 2.0 wt% (d), and 3.0 wt% (e) MWCNT. Reprinted with permission from Reference [26].

matrix [25]. Lii et al. [25] reported on SWCNTs that were blended into an aqueous potato amylose (PA) solution to form a suspension, which was then allowed to separate into a homogenous suspension of nanotubes with the larger SWCNT bundles precipitated on the bottom of the container. CGR and GR were similarly wetted in PA solution, whereas fullerene C60 remained nonwetted. Differences in the affinity of these four forms of carbon nanoparticles (CGR, GR, fullerene C60, and SWCNTs) to PA strongly suggests that PA–GR(CGR) and PA–SWCNTs interactions may be specific and require multipoint interactions between the surface of a carbon species and PA. In this case, it is not possible for the spherical structure and the relatively small size of zero-dimensional fullerene nanoparticles to form multipoint interactions between filler and matrix.

11.2.1.3 Polysaccharide Nanocrystals Nanocrystals from natural polysaccharides, such as cellulose nanocrystals [2, 33, 35, 107–112], starch nanocrystals [40–42, 113–115], and chitin nanocrystals (CNC) [116–118], are potential reinforcements where the nanosized fillers impart enhanced mechanical and barrier properties. The main advantages of these natural nanofillers are their renewable nature, availability, high specific strength, nonabrasive nature (that allows easier processing even at high loading levels), biodegradability, and their relatively reactive surfaces, which can be modified accordingly [119–121]. It has been reported [121] that the outstanding reinforcement effects of these natural nanofillers in nanocomposites are defined mainly by the nature of the polymer–particle interface [122, 123] and/or the strength of filler–filler interactions leading to the establishment of a particle percolating network [124–126]. The interface is related to the distribution of the nanofiller, so it is therefore very important to pay attention to the dispersibility of nanofiller in the matrix.

Cellulose nanocrystals can be extracted from plant sources, such as cotton, flax, hemp, jute, sisal, and wood fibers, through mechanical separation or chemical and/or enzymatic fiber pre-treatments [127]. Compared with inorganic reinforcing fillers, cellulose nanocrystals display outstanding advantages due to their high surface area, high strength, unique morphology, low weight, and biodegradability, as well as their good mechanical response to stress [59, 127]. In order to prevent cellulose nanocrystals from aggregating, and to improve dispersion and compatibility in the matrix material, modification of the cellulose nanocrystal surface is required [59]. Some methods of cellulose nanocrystal surface modification include graft polymerization, silylation, and the use of surfactants. Dispersion in organic solvent without modification has also been reported. Chen et al. [38] explored the use of both pea hull fiber (PHF) and PHF-derived cellulose nanocrystals (PHFCN) (isolated from PHF by acid hydrolysis) as filler in pea starch (PSt)-based biocomposite films. The original PHF exhibited an irregular shape and various sizes, while the PHFCN were needle- or rod-like with one or two sharpened ends. As shown in scanning electron microscope (SEM) photographs (Fig. 11.4), the cross sections of the PSt/PHFCN nanocomposites were much smoother and more compact than the corresponding PSt/PHF films, although both were rougher than that of the PSt film. The obvious morphological differences between the PSt/PHF and PSt/PHFCN films were due to the differences in structure and performance

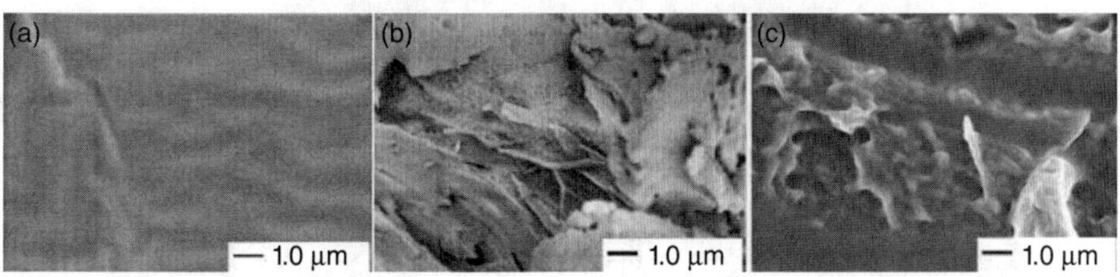

Figure 11.4 SEM photographs of cross sections of pea starch (PSt) (a), PSt/PHF biocomposite films containing 10 wt% of pea hull fiber (PHF) (b), and PSt/PHFNW nanocomposite film containing 10 wt% of PHF derived nanowhiskers (PHFNW) (c). Reprinted with permission from Reference [38].

between PHF and PHFCN. Compared with PHF, PHFCN had a higher aspect ratio and smaller size. It also dispersed more homogenously within the PSt matrix, resulting in more chances to interact with PSt, and thus forming stronger interactions and adhesions to the PSt and filler interfaces [38].

There are also many kinds of starch nanocrystals that can be used as reinforcements in the starch matrix. It has been reported that the distribution of starch nanocrystal in the matrix is related not only to the shape, size, and concentration of the nanofiller, but also to the structure of the starch matrix. García et al. [42] used the same waxy maize starch to prepare starch nanocrystals and glycerol plasticized and unplasticized matrices for nanocomposites. Figure 11.5 shows micrographs of the cryofractured surfaces of the glycerol

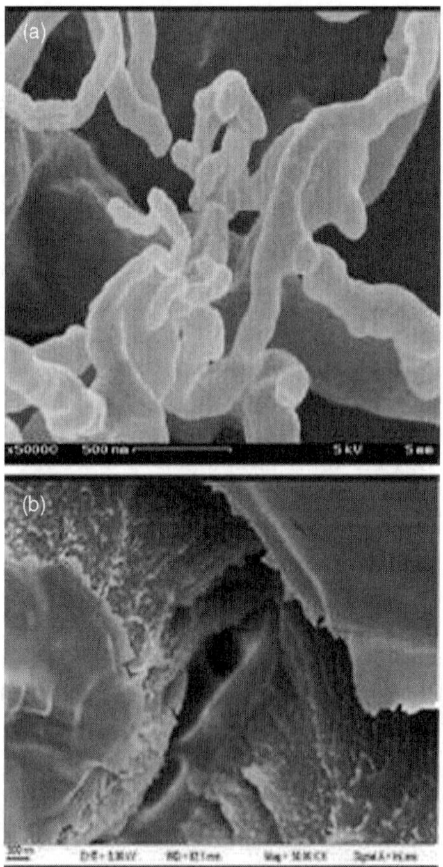

Figure 11.5 SEM photographs of fractured surfaces of (a) glycerol plasticized composite film, and (b) unplasticized composite film containing 2.5 wt% starch nanocrystals. Reprinted with permission from Reference [42].

plasticized and unplasticized composite films containing 2.5 wt% starch nanocrystals. The unplasticized sample displayed a rather smooth homogeneous surface associated with a brittle fracture, while an obviously rougher, nanometric fibrillar structure was observed on the plasticized sample. The high glycerol content probably increased phase separation and was responsible for higher interaction between the glycerol and starch nanocrystals, which was in accordance with the literature [128].

Chitin is also a biopolymer that is abundant in nature. Under hydrolytic conditions of boiling HCl and vigorous stirring, CNC of slender parallelepiped rods have been successfully prepared from different sources of chitin such as crab shells, squid pens, and *Riftia pachyptila* tubeworms. By treating chitin with consecutive acid hydrolysis and mechanical ultrasonication treatments, Chang et al. [120] prepared unique CNC of about 50–100 nm that exhibited lower crystallinity than conventional CNC. All-natural nanocomposites were prepared by combining glycerol plasticized potato starch (GPSt) with CNC using casting and evaporation. SEM showed that at low concentrations (2 wt%) most CNC were dispersed uniformly in the GPSt matrix without obvious aggregation, which is ascribed to strong interactions between the similar polysaccharide structures of the CNC and potato starch in the GPSt matrix. However, conglomeration of CNC was obvious at a high filler content (5 wt%) [120]. Again, this fact indicates that the nanofiller content plays a very significant role in the dispersibility of the nanofiller.

11.2.1.4 Other Nanofillers There are plenty of studies to indicate that various nanoparticles added into the starch matrix can achieve outstanding properties. Many types of nanofillers (SiO_2 [43–45], TiO_2 [46], HAp [47], ZnO [48], CdS, Ag and Ag_2S, α-zirconium phosphate (α-ZrP) [49], and layered double hydroxide [50], etc.) are added to the starch matrix for their excellent properties such as small size, large specific surface area, high surface energy, unsaturated chemical bonds, and surface hydroxyl groups.

Nanosilicon dioxide used as nanofiller in the starch matrix may achieve excellent performance if it is well distributed. Tang et al. [43, 45] prepared starch/poly(vinyl alcohol) (PVA)/nanosilicon dioxide (nano-SiO_2) biodegradable blend films, with various nano-SiO_2 concentrations, using a solution casting method and a coating method. SEM analysis predicted that there was good miscibility between the nano-SiO_2 and starch/PVA in the blends when the nano-SiO_2 content was 3 wt%, while at higher levels (5 wt%) aggregated particles appeared. This fact indicates that the concentration of nanofiller in the matrix is highly significant for its distribution. Nano-ZnO particles at a relatively low concentration (1 wt%) were also distributed evenly in the glycerol plasticized pea starch (GPS) matrix. At a higher content (3 wt%), a few agglomerates of nano-ZnO were observed [48]. An experiment carried out by Wu et al. [49], in which glycerol plasticized pea starch/α-ZrP (GPS/ZrP) nanocomposite films with different α-ZrP loading levels were prepared using a casting and solvent evaporation method, attests to this conclusion. When the α-ZrP content in the nanocomposite films was relatively low, even with progressive increases in the content, α-ZrP dispersed uniformly in the starch matrix during fabrication. The fracture surfaces of the GPS/ZrP films with a relatively high α-ZrP content became rougher, indicating a certain degree of phase separation between starch and α-ZrP, resulting from poor dispersion and aggregation of α-ZrP nanoparticles [49].

11.2.2 Fracture Morphologies

Fracture morphologies reflect information (such as nanofiller distribution, interface structure, and interaction between nanofiller and matrix) that is closely correlated to the structures of the nanofiller and its surface. Research into the effects of nanofiller structure, as well as that of its surface characteristics, will contribute to the regulation of interface structure and improved interfacial miscibility, and hence help to enhance the properties of nanocomposites.

11.2.2.1 Effects of Nanofiller Structure Nanofillers with different structures (spherical, layered, tubular, rod, tridimensional network structure, etc.) added to the starch matrix result in different effects that induce different fracture morphologies.

Nanocomposites filled with layered fillers, which are obtained almost exclusively by intercalation of the polymer inside the galleries of layered host crystals, always result in unique interface structures. There is a wide variety of both synthetic and natural crystalline fillers that are able, under specific conditions, to intercalate a polymer. Table 11.1 [13, 50, 60–103, 129–136] presents a nonexhaustive list of possible layered host crystals [88]. Interaction between the layered structure and the starch matrix has been discussed previously and

TABLE 11.1 Example of Layered Nanofillers Susceptible to Formation of an Intercalated Structure in a Polymer Matrix

Chemical Nature	Examples
Layered silicates and their modified products	Montmorillonite (MMT), Cloisite 25a, kaolinite, mica, saponite, fluoromica, fluorohectorite, vermiculite, hectorite, magadiite, chitosan-treated MMT [13, 50, 60–103]
Carbon-based nanomaterials	Graphite [29], graphite oxide [130, 131]
Metal-based nanomaterials	Metal chalcogenides, such as $(PbS)_{1.18}(TiS_2)_2$ [132] and MoS_2 [133]; metal phosphates, such as $Zr(HPO_4)$ [134]; layered double hydroxides [47], such as $Mg_6Al_2(OH)_{16}CO_3 \cdot nH_2O$ [135], and $Zn_6Al_2(OH)_{16}CO_3 \cdot nH_2O$ [136]

is shown in Figure 11.1. In intercalated nanocomposites, the original layered structure was virtually unchanged except for an increase in the interlayer distance (*d*-spacing) as the polymer chains penetrated the interlayers. In exfoliated nanocomposites the layered structure was totally destroyed, meaning the layers were delaminated and separated from one another [137]. Plasticizers used in extrusion and hot-pressing of starch compete with starch molecules to enter into the interlayer of the nanofiller [21]. Because of their smaller size, the addition of plasticizers such as glycerol may inhibit intercalation and exfoliation of the layered nanofillers in the starch matrix [137] in spite of increasing interlayer space [97]. Exfoliated layered particles, such as clays, mica, or graphite has been used to provide very effective reinforcement of polymers at loading levels much lower than required in the case of solid particles, such as carbon black and silica [138]. Other properties including increased solvent resistance and reduced permeability and flammability can also be substantially improved [139]. Chung et al. [137] prepared starch-clay nanocomposites by adding a dilute clay dispersion to a starch solution followed by coprecipitation in ethanol. Usually, it is difficult to obtain either an intercalated or exfoliated structure in systems of starch-clay nanocomposites because the starch, as a neutral polymer, lacks the driving force required to replace sodium ions in the interlayer through ion exchange, and high molecular weight inhibits its penetration into interlayers. After precipitation in ethanol, however, TEM observation showed the clay layers were exfoliated and hence trapped by starch molecules. XRD patterns further verified that glycerol added after mixing starch and clay, did not penetrate into the clay galleries to expand the interlayer space. A homogeneous and smooth fracture morphology similar to the fracture surface of unfilled starch film was observed by SEM on the starch-based nanocomposite film filled with 5 wt% MMT. This indicated that the starch granules were completely disrupted and the exfoliated clay was uniformly dispersed in the polymer matrix. Good dispersion of layered clay filler in nanocomposites was achieved by mixing in dilute aqueous solutions and subsequent plasticization to prevent interference from plasticizer penetrating into the clay interlayer [137].

Some nanofillers exhibit a network structure that may play a significant role in improving the mechanical properties of nanocomposites. Nano-SiO_2 is an amorphous powder that has a tridimensional net molecular structure. It deviates from a stable silicon–oxygen structure because of a lack of oxygen on its surface [43, 45]. When added to the matrix, nano-SiO_2 forms an interpenetrating network structure with the starch hybrid materials, such as the starch/PVA blend, by combining the stiff structure of nano-SiO_2 with the flexible structure of starch [43, 45]. This may greatly enhance the mechanical properties of the nanocomposites. It has also been reported [140] that bacterial cellulose (BC) is characterized by an ultrafine network structure. Wan et al. [141] incorporated BC nanofibers into starch plasticized with glycerol via a solution impregnation method. The typical tensile fracture surface of BC/starch nanocomposites is displayed in Figure 11.6. The BC/starch nanocomposites have a layered

Figure 11.6 Scanning electron micrograph of the tensile fracture surface of bacterial cellulose (BC)/starch nanocomposite. Reprinted with permission from Reference [141].

structure, which is similar to the morphology of the microfibrillated cellulose–melamine formaldehyde composites as reported [142]. This layered structure is characteristic of the BC network as the difference in the cellulose network structure is usually observed between the upper surface layer (exposed to air) and the lower layer (exposed to culture medium during cultivation) [140]. It is worth noting that, as seen in the BC nanofiber ends in Figure 11.6, the length of nanofiber pulled out of the nanocomposites at the fracture is quite modest, indicating good fiber–matrix bonding [141]. This may be attributed to the formation of a unique network structure.

Nanocomposite films of flax cellulose nanocrystals (FCNs) and PS were obtained using the casting method as reported by Cao et al. [34]. Field emission SEM (FE-SEM) of the fractured films revealed no large aggregates and a homogeneous distribution of the FCNs in the PS matrix in all nanocomposite films. This is an indication of good adhesion between filler and matrix, and is attributed to the good compatibility resulting from both the chemical similarities between starch and cellulose, and the hydrogen bond interactions at the filler–matrix interface [34]. The similar structures of the nanofiller and matrix favor the development of enhanced properties in nanocomposites. This conclusion was confirmed in GPSt nanocomposites that used CNC as the nanofiller [120], where, because of the strong interaction between the similar polysaccharide structures of CNC and potato starch, most of the CNC dispersed uniformly in the GPSt matrix without obvious aggregation.

The loading level of nanofillers distributed in the amorphous area of the starch matrix may also make an obvious difference in the fracture surfaces.

When very small quantities of MWCNTs (0.027 wt% and 0.055 wt%) were introduced into the tapioca starch matrix, a well-developed vein pattern was observed in the cryogenic fracture surfaces of the starch-based nanocomposites, which was similar to the image in Figure 11.3c. This suggests that the residual amorphous matrix determined the deformation behavior of the material, and the space between veins was found to become thinner and denser with increased filler content [143].

11.2.2.2 Effects of Nanofiller Surface Structure It is well known that interfacial interaction is crucial for improving polymer properties, and, as a result, the nanofiller surface structure plays an important role in producing nanocomposites with excellent properties. Many types of nanofiller stand out from other materials because of their surface structures. Nano-SiO_2, for example, is widely used because of its surface hydroxyl groups and unsaturated chemical bonds, its large specific surface area, and its high surface energy [43, 45].

There are many cases in which nanofillers are treated with acids or modified by other materials containing the needed functional groups in order to obtain the required surface structure. Graphene and carbon nanotubes are treated with $NaNO_3$ and H_2SO_4 to make hydrophilic GO with oxygen functional groups such as hydroxyls, epoxides, carbonyls, and carboxyls [25, 27]. MMT has been modified by plasticizer [144] or organic reagents such as tetraalkylammonium cation [145], methyl tallow bis-2-hydroxyethyl ammonium cation [146], and/or *N*-(2-hydroxyethyl) formamide [80] to improve its compatibility with the starch matrix. Polysaccharide nanocrystals prepared conventionally using the H_2SO_4 method always have sulfate groups on their surfaces. Cellulose nanocrystals can be further modified by chemical reactions of cationization, esterification, silylation, polymer grafting, and so on, which may introduce stable negative or positive electrostatic charges onto the surface of cellulose nanocrystals. This will allow for better dispersion and/or tune the surface energy characteristics of cellulose nanocrystals to improve compatibility, especially when used in conjunction with nonpolar or hydrophobic matrices in nanocomposites [108]. Starch nanocrystals were reported to have been modified by isocyanate groups (–NCO) as well as long chain polymers based on the "graft onto" and "graft from" strategies [114].

Yu et al. [51] reported on the preparation of a nanocomposite by casting using ZnO nanoparticles stabilized by carboxymethylcellulose sodium (CMC) as filler in a glycerol plasticized pea starch (GPS) matrix. Using SEM, the author found that agglomerations of ZnO-CMC appeared in the GPS/ZnO-CMC composites with higher ZnO-CMC content. In addition, the surfaces of the ZnO-CMC particles appeared to be covered by GPS, which was attributed to a strong interaction between the hydrophilic surface of the CMC in the ZnO-CMC filler and the GPS matrix [51].

As previously mentioned [26], when nanocomposites failed, most of the MWCNTs broke rather than pull out wholly from the matrix. The ends of the broken MWCNTs can be seen in SEM images of the failure surfaces of PS/MWCNT nanocomposite films with relatively high MWCNT contents (2.0 and 3.0 wt%), especially at the highest magnification (Fig. 11.3). This phenomenon indicates a strong interfacial adhesion between the MWCNTs and the PS matrix. Besides influencing the structure of MWCNTs, it has been proven that acid treatment incorporated polar groups with the MWCNTs, improved the hydrophilicity of MWCNTs, and reduced agglomerations [147]. These factors contribute greatly to the enhanced hydrogen bonding interactions between acid-treated MWCNTs and PS; they increase the compatibility between MWCNTs and PS, and result in improved mechanical performance of the PS/MWCNT nanocomposite films [26].

11.2.3 Crystalline Structure of the Starch Matrix

11.2.3.1 Factors Affecting Crystalline Structure The crystallization of a polymer directly affects its processability and performance in the resultant materials. Crystallization behavior, including nuclear crystallization and retrogradation, is common in starch-based materials. It has also been attested [148] that, to a certain extent, the type of starch impacts the crystalline structure. For example, in high amylose starch the fractional association of amylopectin and amylose was significantly reduced, allowing only amylose–amylose interaction in the presence of water. There is much literature that focuses on the mechanisms of crystalline structure.

It has been reported that the type of nanofiller used has an effect on the crystalline behavior of the starch matrix. Huang et al. [149] prepared MMT-reinforced thermoplastic starch composites (MTPSC) with MMT and glycerol plasticized thermoplastic starch (GTPS) using a melt extrusion method. Crystallization of natural corn starch is of the A-type. In the process of making GTPS, most of the crystallization of the original starch was destroyed due to mechanical and thermal effects, and a small amount of V_H crystallization was present [150], as seen in Figure 11.7A. Figure 11.7 shows the XRD patterns of MTPSC containing different concentrations of MMT stored at relative humidity (RH) = 50% for 60 days, and MTPSC containing 20 wt% MMT stored at RH = 50% for 0, 30, 60, and 90 days. As shown in Figure 11.7A, an obvious V_H-type crystallization peak and an A-type crystallization peak from the original starch were present after the GTPS had been stored at RH = 50% for 60 days. No obvious crystallization peak appeared in the MTPSC containing MMT under the same conditions. In Figure 11.7B, no obvious crystallization peak appears in the XRD patterns, which indicates that the introduction of MMT restrained the formation of hydrogen bonds between the starch molecules, and inhibited the crystallization of GTPS [149].

When organo-modified MMT clay was added into the starch matrix, the situation was exactly the opposite. Chaudhary prepared a nanocomposite with high amylose starch and the organo-modified MMT clay, Cloisite 25a (C25a) [151]. The starch matrix developed nuclear crystallization when the C25a clay

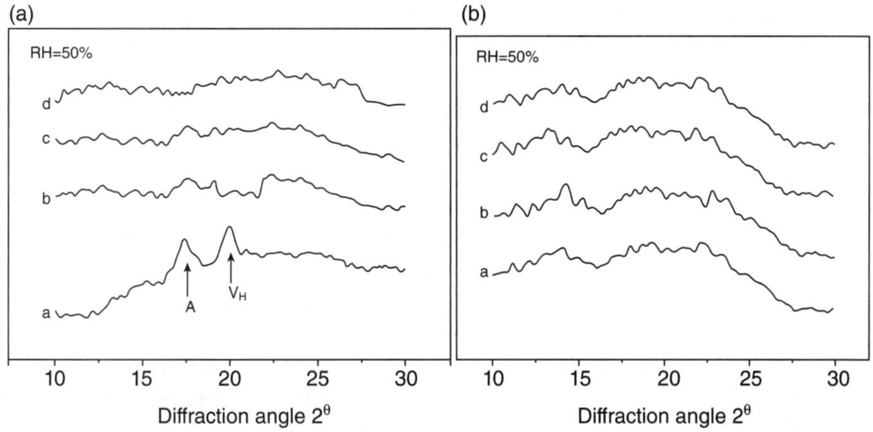

Figure 11.7 X-ray diffraction patterns of (A) starch-based nanocomposites stored at RH=50% for 60 days containing: 10 wt% MMT (b), 20 wt% MMT (c) and 30 wt% MMT (d), as well as glycerol plasticized thermoplastic starch (GTPS) (a) for comparison; and (B) the starch-based nanocomposites containing 20 wt% MMT stored at RH=50% for 0 day (a), 30 days (b), 60 days (c), and 90 days (d). Reprinted with permission from Reference [149].

particles acted as a nucleating agent for crystallization of the matrix. The linear growth rate and overall crystallization rate were not significantly influenced by the presence of clay. A unique mechanism that explains the higher crystallization rate, along with the morphology and developed internal structure, is that a sandwich structure in which each silicate layer was strongly covered by polymer crystals made the system very rigid [13].

Matrix crystallinity was also affected by the concentration of the nanofiller added. In the starch-based nanocomposites filled with C25a prepared by Chaudhary [151] as described above, it was found that some degree of intercalation/exfoliation appeared gradually with the addition of clay, shown as diffraction peaks shifting to lower 2θ values in the XRD patterns. The presence of water in the system allows starch nanocomposites greater ability to achieve exfoliation than their polyolefin counterparts at higher clay loading levels. This was corroborated by SEM in which the starch nanoclay system showed significant compatibility. After the total crystallinity decreased slightly when the clay content was lowest (i.e., 1.5 wt%), it increased as the clay content increased to greater than 2.5 wt%. On the other hand, the rigid amorphous fraction increased continuously with increasing clay content. The changes in total crystallinity and the rigid amorphous phase synergistically contributed to enhanced resistance to initial deformation [151].

11.2.3.2 Effects of Retrogradation Retrogradation or recrystallization of amylose in the starch polymer is a significant limitation for starch-based nanocomposites, leading to increased brittleness and loss of mechanical strength [42, 148, 149, 152]. The crystalline structure resulting from spontaneous retrogradation of starch molecules after melting or gelatinization has often been detected in food and thermoplastic materials [153]. Smits et al. [154] found that retrogradation occurred in water plasticized starch, and Ma and Yu [155] found that glycerol plasticized starch presented an increased crystallinity after storing at RH = 0% and at RH = 50% for 25 days. Alkyl ammonium salt modifier in hectorite may inhibit the occurrence of recrystallization. Calvert [156] attributed the problem of cracking to evaporation of the plasticizer, and Ma and Yu [155] recently reported that starch plasticized with urea did not present a problem with retrogradation [19].

Based on the above facts, it has been concluded that retrogradation of PS is greatly dependent on the additive's ability to form hydrogen bonds with starch molecules. The stronger the hydrogen bond between the starch and additive, the more difficult it is for starch to recrystallize during the storage period [19, 26, 149]. As a result, the water content, type of starch [157], type of plasticizer [19, 42, 149], and the type and content of the nanofiller used [158] may greatly impact the crystalline structure.

MWCNTs were added at different concentrations to PS to investigate the effect of nanofillers on retrogradation using DSC [26]. It was observed that the heat of fusion (ΔH_m) decreased as more MWCNTs were incorporated. This demonstrated that the carboxylic, carbonyl, and hydroxyl groups on acid-treated MWCNTs formed relatively strong hydrogen bonds between the PS and MWCNTs, which suppressed starch retrogradation. Furthermore, there was no shoulder peak on the melt endotherm on any of the differential scanning calorimetry (DSC) curves, indicating that there was no new crystallinity. This was also supported by wide-angle X-ray diffraction (WAXRD) where no additional peaks were detected, unlike the extra peak at $2\theta = 21.15°$ in glycerol PS-based nanocomposites filled with tunicin whiskers (a kind of cellulose nanocrystal) as reported by Anglès and Dufresne [107]. This extra peak probably resulted from accumulation of plasticizers at the starch–nanofiller interface and from formation of a glycerol–starch *V* structure. It can be concluded that incorporation of MWCNTs did not have a significant effect on the crystalline structure of the PS matrix, although retrogradation was apparently reduced [26].

Ma et al. [158] prepared composites based on TPS and winceyette fiber with various weight ratios using urea and formamide as plasticizer. They investigated the crystalline characteristics after storage at 50% RH for two months. As reported [159, 160], the addition of nanofiller improved the crystallinity of the starch matrix in systems with glycerol as plasticizer; however, when the fiber content level was below 20 wt% the TPS matrix, plasticized with a mixture of urea and formamide, did not show the obvious crystallinity peaks assigned to starch. This suggests that retrogradation of starch was inhibited by the fiber at a lower loading level. On the other hand, when TPS was filled with greater than 20 wt% fiber, the small crystallinity peaks assigned to starch could be identified, indicating that the inhibition to starch retrogradation was absent

[158]. Thus, it can be concluded that retrogradation of the starch matrix can be effectively inhibited by using given plasticizers and/or controlling the nano-filler content. This information may be used to solve the problems caused by starch retrogradation, that is, brittleness and loss of mechanical strength, and to contribute to the development and application of starch-based materials.

11.3 THERMAL BEHAVIOR OF STARCH-BASED NANOCOMPOSITES

The main components of thermal behavior of starch-based nanocomposites include glass transition and thermal decomposition. Glass transition is often characterized by dynamic mechanical analysis (DMA), dynamic mechanical thermal analysis (DMTA), and DSC, with which the glass transition temperature (Tg) can be determined. Thermal decomposition is determined using thermogravimetric analysis (TGA) and derivative thermogravimetric analysis (DTG).

11.3.1 Glass Transition of Plasticized Starch-Based Nanocomposites

11.3.1.1 Shift of Glass Transition Temperature The Tg is the critical temperature at which a material changes its behavior from "glassy" (hard and brittle) to "rubbery" (elastic and flexible). Thermal behavior was studied by Cao et al. [26], who analyzed the Tg of MWCNT-enforced starch nanocomposites. The DSC curves showed that by incorporating 0–3 wt% MWCNT filler into a PS matrix, the Tg for the starch-rich phase shifted from 16.5°C to 25.3°C. The dependence of Tg on the MWCNT content may be attributed to the occurrence of intermolecular interactions between the starch and the MWCNTs, which reduce the flexibility of the starch molecular chains in contact with the MWCNT surface [26].

DMTA was used to investigate the change in Tg when CNC was introduced into a GPSt matrix [120]. Figure 11.8 shows the curves for loss factor (tan delta) as a function of temperature for GPSt/CNC composites. The curve for GPSt revealed two thermal transitions that correspond to the two separate phases in GPSt [161]. The upper transition (19.6°C) assigned to a starch-rich phase is regarded as the Tg of GPSt; the lower transition (–57.2°C) was due to a starch-poor phase. In the GPSt/CNC composites, both the upper and lower transitions shifted to higher temperatures, which indicated that both the starch-rich and starch-poor phases interacted well with CNC. CNC improved the intermolecular interaction of GPSt in the starch-rich phase by bringing adjacent chains of starch closer, restraining chain mobility, and reducing the free volume [162], thereby raising the Tg of nanocomposites as the CNC content increased. It was noted that a double peak appeared at 31.6°C and 46.4°C in GPSt/CNC composites with a CNC content of 5 wt% in which CNC aggregation was obvious when observed by SEM. Aggregation of CNC resulted in an uneven fractured surface of the GPSt matrix and dispersion of the CNC-filler phase. The higher

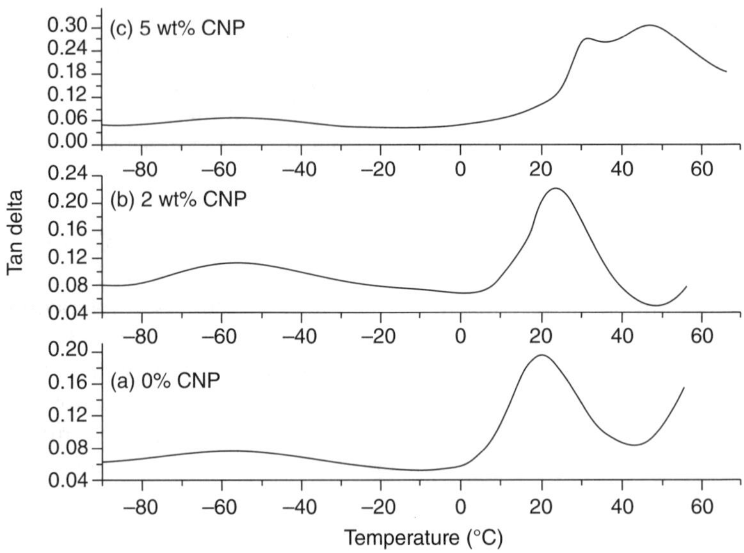

Figure 11.8 Dependence of Tan delta upon temperature for glycerol plasticized potato starch (GPSt)-based nanocomposites with different chitin nanoparticle (CNP) contents. Reprinted with permission from Reference [120].

transition temperature (46.4°C) may have been due to a CNC-rich phase, while the lower transition (31.6°C) to a CNC-poor phase [120].

11.3.1.2 Factors Affecting Glass Transition It can be concluded that with the addition of nanofillers, the Tg of starch-based nanocomposites tends to shift to a higher temperature as compared to the neat matrix [26, 60, 120] because interactions between the nanofiller and the starch matrix can hinder polymer chain mobility. Since Tg is closely correlated to the mobility of the polymer, the factors that restrict polymer mobility may affect the glass transition—the greater the resistance, the higher Tg. This trend is likely related to the dispersion state of the nanofiller, which can restrict the movement of starch chains. Another possible explanation is linked to the structure of the nanofiller. Xu et al. [163] attributed the increase in Tg to the formation of an intercalated structure. Surface structures such as hydrogen bonds established between the -OH groups of the nanofiller and those of the starch chains may reduce the starch chain mobility and affect the glass transition.

11.3.2 Thermal Decomposition Behavior of Starch-Based Nanocomposites

The thermal decomposition temperature is the temperature at the maximum rate of mass loss, that is, the peak temperature shown in DTG curves. The

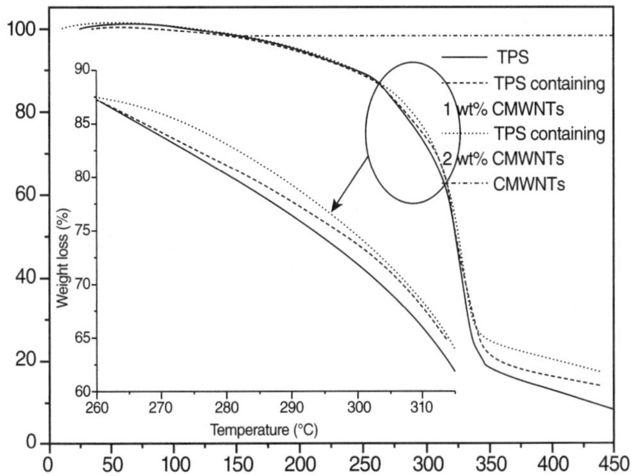

Figure 11.9 Thermogravimetric curves for thermoplastic starch (TPS), carboxylate multiwalled carbon nanotubes (CMWNTs), and TPS/CMWNT nanocomposites containing 1 wt% and 2 wt% CMWNTs. Reprinted with permission from Reference [164].

decomposition temperature of soluble starch was 310.1°C, but changed with the addition of nanofiller. It was reported [164] that carboxylated multiwalled carbon nanotubes (CMWNTs) could increase the thermal stability of the TPS matrix. TGA curves (seen in Fig. 11.9) were recorded over the temperature range of 25 to 450°C to evaluate the thermal stability of TPS and TPS/CMWNT nanocomposites. According to the TGA curves, the first weight-loss stage of TPS and TPS/CMWNT nanocomposites, ascribed mainly to water evaporation, occurred at about 100°C; the second stage of weight loss was the decomposition of starch and the removal of glycerol [165]. The third stage, over a range of 260–315°C, showed that the rate of weight loss gradually decreased with an increase in CMWNT content. Compared with TPS, the CMWNTs were more stable, and the weight loss due to the decomposition of free carboxyl groups on the surface of CMWNTs was only 2 wt%. The effect of CMWNTs on the thermal stability of the nanocomposites may be explained by a barrier function of the CMWNTs that hindered diffusion of the degradation products from the bulk of the polymer to the gas phase [166]. Additionally, carboxyl groups on the surface of CMWNTs may improve interfacial interactions between CMWNTs and the TPS matrix, and therefore increase the activation energy of degradation [164].

Generally, the incorporation of clay into the polymer matrix was found to enhance thermal stability by acting as an insulator and a barrier to the mass transport of volatile products generated during decomposition [149]. Compared with organic materials, inorganic materials had better thermal stability and thermal resistant properties; therefore, the introduction of inorganic particles should improve the thermal stability of organic polymer materials, as reported by Huang et al. [149].

As regards the starch nanocrystals-filled nanocomposites, a wide peak around 113°C in the DTG curves that was associated with the maximum rate of water loss in the neat starch film disappeared in the nanocomposites. This result was attributed to the presence of hydrogen bonding between glycerol and starch nanocrystals in the composite, and to the reduced phase separation observed by SEM. Moreover, in the first step of decomposition the nanocrystal filled starch-based composite film showed higher water content than the neat starch matrix. This indicated an increased availability of hydroxyl groups ready for the formation of hydrogen bonds with the absorbed water [115].

11.4 CONCLUSIONS AND PROSPECTS

Starch-based nanocomposites are environmentally friendly materials based on renewable agricultural resources that show great promise for potential applications as high performance biodegradable materials. In this chapter, morphologies and thermal behavior of starch-based nanocomposites are discussed to provide insight into the internal structure of these nanocomposites. All the properties and thermal behaviors have a close correlation with the structure of the starch-based nanocomposites, especially the structure of the interface between the starch matrix and the nanofiller. It is essential to establish the complex structure–property relationships in various nanocomposites, and to design the optimal structure based on these relationships to obtain nanocomposites with excellent properties. Various properties of nanofillers, including shape, size, concentration in the starch matrix, functional groups on the nanofiller surface, and the mechanisms of interactions between the nanofiller and the matrix, must be altered to obtain the desired nanocomposite properties.

From this chapter, we can conclude that external performance is closely tied to the internal structure. The distribution of nanofiller can be regulated by choosing different kinds of nanofillers with the desired characteristics (shape, size), changing the amount of filler added to the starch matrix, and by chemical modification. Fracture morphology vividly reflects the interaction between the nanofiller and the starch matrix, which is affected by the structure of both. The crystalline structure may be influenced by nuclear crystallization and retrogradation. Retrogradation can be controlled to some extent, as it is dependent on the abilities of the additives to form hydrogen bonds with starch molecules. The thermal decomposition behavior may be affected by many factors including nanofiller structure, concentration of the nanofiller in the nanocomposite, the type of starch, and the methods used for preparation of the nanocomposites. Tg of the starch-based nanocomposites tends to be higher with the addition of nanofiller because interactions between the nanofiller and starch matrix restrict the mobility of the starch chain.

Starch-based nanocomposites are a valid answer for production of low cost, highly competitive, environmentally friendly materials if the parameters and their interactions identified above are optimized. Through a concerted effort by academia, industry, government, and consumers, starch-based nanocomposites with outstanding performance can be successfully developed and commercialized for a wide range of applications, such as packaging, agricultural devices, or biomedical applications.

REFERENCES

[1] Mohanty, A.K., Misra, M., and Hinrichsen, G. (2000) Biofibres, biodegradable polymers and biocomposites: An overview. *Macromolecular Materials and Engineering*, *276/277*(1), 1–24.

[2] Mathew, A.P., Dufresne, A. (2002) Morphological investigation of nanocomposites from sorbitol plasticized starch and tunicin whiskers. *Biomacromolecules*, *3*(3), 609–617.

[3] Buchanan, R.A., Weislogel, O.E., Russell, C.R., and Rist, C.E. (1968) Starch in rubber. Zinc starch xanthate in latex masterbatching. *IEC Product Research and Development*, *7*(2), 155–158.

[4] Buchanan, R.A., Kwolek, W.F., Katz, H.C., and Russell, C.R. (1971) Starch in rubber. Influence of starch type and concomitant variables in reinforcement of styrene-butadiene rubbers. *Starch/Stärke*, *23*(10), 350–359.

[5] Buchanan, R.A., Katz, H.C., Russell, C.R., and Rist, C.E. (1971) Starch in rubber. Powered elastomers from starch-encased latex particles. *Rubber Journal*, *153*(10), 28–35.

[6] Buchanan, R.A. (1974) Starch xanthide styrene-butadiene rubbers. Effect of prolonged water immersion. *Starch/Stärke*, *26*(5), 165–172.

[7] Buchanan, R.A., Doane, W.M., Russel, C.R., and Kwolek, W.F. (1975) Starch xanthide reinforced styrene-butadiene rubber: Compounding to reduce water sensitivity. *Journal of Elastomers and Plastics*, *7*(2), 95–113.

[8] Abbott, T.P., Doane, W.M., and Russell, C.R. (1973) Starch xanthide-encased rubbers. *Rubber Age*, *105*(8), 43–49.

[9] Cao, X.D., Chen, Y., Chang, P.R., Stumborg, M., and Huneault, M.A. (2008) Green composites reinforced with hemp nanocrystals in plasticized starch. *Journal of Applied Polymer Science*, *109*(6), 3804–3810.

[10] Bhattacharya, M. (1998) Stress relaxation of starch/synthetic polymer blends. *Journal of Materials Science*, *33*(16), 4131–4139.

[11] Shogren, R. (1997) Water vapor permeability of biodegradable polymers. *Journal of Environmental Polymer Degradation*, *5*(2), 91–95.

[12] Majdzadeh-Ardakani, K., Nazari, B. (2010) Improving the mechanical properties of thermoplastic starch/poly(vinyl alcohol)/clay nanocomposites. *Composites Science and Technology*, *70*(10), 1557–1563.

[13] Ray, S.S., Okamoto, M. (2003) Polymer/layered silicate nanocomposites: A review from preparation to processing. *Progress in Polymer Science*, *28*(11), 1539–1641.

[14] Sain, M.M., Kokta, B.V., and Imbert, C. (1994) Structure-property relationships of wood fiber-filled polypropylene composites. *Polymer-Plastics Technology and Engineering*, *33*(1), 89–104.

[15] Gassan, J., Bledzki, A.K. (1997) The influence of fiber-surface treatment on the mechanical properties of jute-polypropylene composites. *Composites Part A: Applied Science and Manufacturing*, *28*(12), 1001–1005.

[16] Hornsby, P.R., Hinrichsen, E., and Tarverdi, K. (1997) Preparation and properties of polypropylene composites reinforced with wheat and flax straw fibres: Part II analysis of composite microstructure and mechanical properties. *Journal of Materials Science*, *32*(4), 1009–1015.

[17] Kvien, I., Sugiyama, J., Votrubec, M., and Oksman, K. (2007) Characterization of starch based nanocomposites. *Journal of Materials Science*, *42*(19), 8163–8171.

[18] Liao, H.T., Wu, C.S. (2005) Synthesis and characterization of polyethylene-octene elastomer/clay/biodegradable starch nanocomposites. *Journal of Applied Polymer Science*, *97*(1), 397–404.

[19] Chen, B., Evans, J.R.G. (2005) Thermoplastic starch-clay nanocomposites and their characteristics. *Carbohydrate Polymers*, *61*(4), 455–463.

[20] Chen, M., Chen, B., and Evans, J.R.G. (2005) Novel thermoplastic starch-clay nanocomposite foams. *Nanotechnology*, *16*(10), 2334–2337.

[21] Pandey, J.K., Singh, R.P. (2005) Green nanocomposites from renewable resources: Effect of plasticizer on the structure and material properties of clay-filled starch. *Starch/Stärke*, *57*(1), 8–15.

[22] Namazi, H., Mosadegh, M., and Dadkhah, A. (2009) New intercalated layer silicate nanocomposites based on synthesized starch-g-PCL prepared via solution intercalation and *in situ* polymerization methods: As a comparative study. *Carbohydrate Polymers*, *75*(4), 665–669.

[23] Wilhelm, H.M., Sierakowski, M.R., Souza, G.P., and Wypych, F. (2003) Starch films reinforced with mineral clay. *Carbohydrate Polymers*, *52*(2), 101–110.

[24] Chang, P.R., Zheng, P.W., Liu, B.X., Anderson, D.P., Yu, J.G., and Ma, X.F. (2011) Characterization of magnetic soluble starch-functionalized carbon nanotubes and its application for the adsorption of the dyes. *Journal of Hazardous Materials*, *186*(2–3), 2144–2150.

[25] Lii, C.Y., Stobinski, L., Tomasik, P., and Liao, C.D. (2003) Single-walled carbon nanotube–potato amylose complex. *Carbohydrate Polymers*, *51*(1), 93–98.

[26] Cao, X.D., Chen, Y., Chang, P.R., and Huneault, M.A. (2007) Preparation and properties of plasticized starch/multiwalled carbon nanotubes composites. *Journal of Applied Polymer Science*, *106*(2), 1431–1437.

[27] Li, R., Liu, C.H., and Ma, J. (2011) Studies on the properties of graphene oxide-reinforced starch biocomposites. *Carbohydrate Polymers*, *84*(1), 631–637.

[28] Qi, Q., Wu, Y.P., Tian, M., Liang, G.H., Zhang, L.Q., and Ma, J. (2006) Modification of starch for high performance elastomer. *Polymer*, *47*(11), 3896–3903.

[29] Zhao, Y.C., Yang, X.L., Zhan, L., Ou, S.J., and Tian, J.N. (2011) High electrocatalytic activity of PtRu nanoparticles supported on starch-functionalized multi-walled carbon nanotubes for ethanol oxidation. *Journal of Materials Chemistry*, *21*(12), 4257–4263.

[30] Andrews, R., Weisenberger, M.C. (2004) Carbon nanotube polymer composites. *Current Opinion in Solid State and Materials Science*, *8*(1), 31–37.

[31] Matsui, H., Karuppuchamy, S., Yamaguchi, J., and Yoshihara, M. (2007) Electronic behavior of calcined materials obtained from SnO_2 hydrosol/starch composite materials. *Journal of Photochemistry and Photobiology A: Chemistry*, *189*(2–3), 280–285.

[32] Bai, J. (2003) Evidence of the reinforcement role of chemical vapour deposition multi-walled carbon nanotubes in a polymer matrix. *Carbon*, *41*(6), 1325–1328.

[33] Mathew, A.P., Thielemans, W., and Dufresne, A. (2008) Mechanical properties of nanocomposites from sorbitol plasticized starch and tunicin whiskers. *Journal of Applied Polymer Science*, *109*(6), 4065–4074.

[34] Cao, X., Chen, Y., Chang, P.R., Muir, A.D., and Falk, G. (2008) Starch-based nanocomposites reinforced with flax cellulose nanocrystals. *eXPRESS Polymer Letters*, *2*(7), 502–510.

[35] Anglès, M.N., Dufresne, A. (2001) Plasticized starch/tunicin whiskers nanocomposite materials. 2. Mechanical behavior. *Macromolecules*, *34*(9), 2921–2931.

[36] Chen, Y., Liu, C.H., Chang, P.R., Cao, X.D., and Anderson, D.P. (2009) Bionanocomposites based on pea starch and cellulose nanowhiskers hydrolyzed from pea hull fibre: Effect of hydrolysis time. *Carbohydrate Polymers*, *76*(4), 607–615.

[37] Alemdar, A., Sain, M. (2008) Biocomposites from wheat straw nanofibers: Morphology, thermal and mechanical properties. *Composites Science and Technology*, *68*(2), 557–565.

[38] Chen, Y., Liu, C.H., Chang, P.R., Anderson, D.P., and Huneault, M.A. (2009) Pea starch-based composite films with pea hull fibers and pea hull fiber-derived nanowhiskers. *Polymer Engineering and Science*, *49*(2), 369–378.

[39] Eichhorn, S.J. (2011) Cellulose nanowhiskers: Promising materials for advanced applications. *Soft Matter*, *7*, 303–315.

[40] Angellier, H., Molina-Boisseau, S., Dole, P., and Dufresne, A. (2006) Thermoplastic starch-waxy maize starch nanocrystals nanocomposites. *Biomacromolecules*, *7*(2), 531–539.

[41] Viguié, J., Molina-Boisseau, S., and Dufresne, A. (2007) Processing and characterization of waxy maize starch films plasticized by sorbitol and reinforced with starch nanocrystals. *Macromolecular Bioscience*, *7*(11), 1206–1216.

[42] García, N.L., Ribba, L., Dufresne, A., Aranguren, M., and Goyanes, S. (2011) Effect of glycerol on the morphology of nanocomposites made from thermoplastic starch and starch nanocrystals. *Carbohydrate Polymers*, *84*(1), 203–210.

[43] Tang, H.L., Xiong, H.G., Tang, S.W., and Zou, P. (2009) A starch-based biodegradable film modified by nano silicon dioxide. *Journal of Applied Polymer Science*, *113*(1), 34–40.

[44] Singh, V., Singh, S.K., Pandey, S., and Kumar, P. (2011) Sol-gel synthesis and characterization of adsorbent and photoluminescent nanocomposites of starch and silica. *Journal of Non-Crystalline Solids*, *357*(1), 194–201.

[45] Xiong, H.G., Tang, S.W., Tang, H.L., and Zou, P. (2008) The structure and properties of a starch-based biodegradable film. *Carbohydrate Polymers*, *71*(2), 263–268.

[46] Yun, Y.H., Hwang, K.J., Wee, Y.J., and Yoon, S.D. (2011) Synthesis, physical properties, and characterization of starch-based blend films by adding nano-sized TiO_2/poly(methyl metacrylate-co-acrylamide). *Journal of Applied Polymer Science*, *120*(3), 1850–1858.

[47] Sadjadi, M.S., Meskinfam, M., Sadeghi, B., Jazdarreh, H., and Zare, K. (2010) *In situ* biomimetic synthesis, characterization and in vitro investigation of bone-like nanohydroxyapatite in starch matrix. *Materials Chemistry and Physics*, *124*(1), 217–222.

[48] Ma, X.F., Chang, P.R., Yang, J.W., and Yu, J.G. (2009) Preparation and properties of glycerol plasticized-pea starch/zinc oxide-starch bionanocomposites. *Carbohydrate Polymers*, *75*(3), 472–478.

[49] Wu, H.X., Liu, C.H., Chen, J.G., Chang, P.R., Chen, Y., and Anderson, D.P. (2009) Structure and properties of starch/α-zirconium phosphate nanocomposite films. *Carbohydrate Polymers*, *77*(2), 358–364.

[50] Wilhelm, H.M., Sierakowski, M.R., Souza, G.P., and Wypych, F. (2003) The influence of layered compounds on the properties of starch/layered compound composites. *Polymer International*, *52*(6), 1035–1044.

[51] Yu, J.G., Yang, J.W., Liu, B.X., and Ma, X.F. (2009) Preparation and characterization of glycerol plasticized-pea starch/ZnO-carboxymethylcellulose sodium nanocomposites. *Bioresource Technology*, *100*(11), 2832–2841.

[52] Kaushik, A., Singh, M., and Verma, G. (2010) Green nanocomposites based on thermoplastic starch and steam exploded cellulose nanofibrils from wheat straw. *Carbohydrate Polymers*, *82*(2), 337–345.

[53] Ma, X.F., Yu, J.G., and Wang, N. (2008) Glycerol plasticized-starch/multiwall carbon nanotube composites for electroactive polymers. *Composites Science and Technology*, *68*, 268–273.

[54] Giannelis, E.P. (1996) Polymer layered silicate nanocomposites. *Advanced Materials*, *8*(1), 29–35.

[55] Ray, S.S., Bousmina, M. (2005) Biodegradable polymers and their layered silicate nanocomposites: In greening the 21st century materials world. *Progress in Materials Science*, *50*(8), 962–1079.

[56] Xie, F.W., Halley, P.J., and Avérous, L. (2011) Bio-nanocomposites based on starch. In: *Nanocomposites with Biodegradable Polymers: Synthesis, Properties and Future Perspectives*, ed. V. Mittal. New York: Oxford University Press, pp. 234–253.

[57] Zhao, R.X., Torley, P., and Halley, P.J. (2008) Emerging biodegradable materials: Starch- and protein-based bio-nanocomposites. *Journal of Materials Science*, *43*(9), 3058–3071.

[58] Mondragón, M., Mancilla, J.E., and Rodríguez-González, F.J. (2008) Nanocomposites from plasticized high-amylopectin, normal and high-amylose maize starches. *Polymer Engineering and Science*, *48*(7), 1261–1267.

[59] Habibi, Y., Goffin, A.L., Schiltz, N., Duquesne, E., Dubois, P., and Dufresne, A. (2008) Bionanocomposites based on poly(ε-caprolactone)-grafted cellulose nanocrystals by ring-opening polymerization. *Journal of Materials Chemistry*, *18*(41), 5002–5010.

[60] Chivrac, F., Pollet, E., and Avérous, L. (2009) Progress in nano-biocomposites based on polysaccharides and nanoclays. *Materials Science and Engineering R: Reports*, *67*(1), 1–17.

[61] Chivrac, F., Gueguen, O., Pollet, E., Ahzi, S., Makradi, A., and Avérous, L. (2008) Micromechanical modeling and characterization of the effective properties in starch-based nano-biocomposites. *Acta Biomaterialia*, *4*(6), 1707–1714.

[62] Lu, Y.F., Yang, Y., Sellinger, A., Lu, M.C., Huang, J.M., Fan, H.Y., Haddad, R., Lopez, G., Burns, A.R., Sasaki, D.Y., Shelnutt, J., and Brinker, C.J. (2001) Self-assembly of mesoscopically ordered chromatic polydiacetylene/silica nanocomposites. *Nature*, *410*, 913–917.

[63] Merkel, T.C., Freeman, B.D., Spontak, R.J., He, Z., Pinnau, I., Meakin, P., and Hill, A.J. (2002) Ultrapermeable, reverse-selective nanocomposite membranes. *Science*, *296*(5567), 519–522.

[64] Gross, R.A., Kalra, B. (2002) Biodegradable polymers for the environment. *Science*, *297*(5582), 803–807.

[65] Usuki, A., Hasegawa, N., Kadoura, H., and Okamoto, T. (2001) Three-dimensional observation of structure and morphology in nylon-6/clay nanocomposite. *Nano Letters*, *1*(5), 271–272.

[66] Isik, I., Yilmazer, U., and Bayram, G. (2003) Impact modified epoxy/montmorillonite nanocomposites: Synthesis and characterization. *Polymer*, *44*(20), 6371–6377.

[67] Lee, K.M., Han, C.D. (2003) Effect of hydrogen bonding on the rheology of polycarbonate/organoclay nanocomposites. *Polymer*, *44*(16), 4573–4588.

[68] Koerner, H., Jacobs, J.D., Tomlin, D.W., Busbee, J.D., and Vaia, R.A. (2004) Tuning polymer nanocomposite morphology: AC electric field manipulation of epoxy-montmorillonite (clay) suspensions. *Advanced Materials*, *16*(4), 297–302.

[69] Huang, M.F., Yu, J.G., Ma, X.F., and Jin, P. (2005) High performance biodegradable thermoplastic starch-EMMT nanoplastics. *Polymer*, *46*(9), 3157–3162.

[70] Majdzadeh-Ardakani, K., Navarchian, A.H., and Sadeghi, F. (2010) Optimization of mechanical properties of thermoplastic starch/clay nanocomposites. *Carbohydrate Polymers*, *79*(3), 547–554.

[71] Avella, M., De Vliger, J.J., Errico, M.E., Fischer, S., Vacca, P., and Volpe, M.G. (2005) Biodegradable starch/clay nanocomposite films for food packaging applications. *Food Chemistry*, *93*(3), 467–474.

[72] Zhang, Q.X., Yu, Z.Z., Xie, X.L., Naito, K., and Kagawa, Y. (2007) Preparation and crystalline morphology of biodegradable starch/clay nanocomposites. *Polymer*, *48*(24), 7193–7200.

[73] Chiou, B.S., Yee, E., Glenn, G.M., and Orts, W.J. (2005) Rheology of starch-clay nanocomposites. *Carbohydrate Polymers*, *59*(4), 467–475.

[74] Dean, K., Yu, L., and Wu, D.Y. (2007) Preparation and characterization of melt-extruded thermoplastic starch/clay nanocomposites. *Composites Science and Technology*, *67*(3–4), 413–421.

[75] Wang, X., Zhang, X., Liu, H., and Wang, N. (2009) Impact of pre-processing of montmorillonite on the properties of melt-extruded thermoplastic starch/montmorillonite nanocomposites. *Starch/Stärke*, *61*(9), 489–494.

[76] Chivrac, F., Pollet, E., Dole, P., and Avérous, L. (2010) Starch-based nano-biocomposites: Plasticizer impact on the montmorillonite exfoliation process. *Carbohydrate Polymers*, *79*(4), 941–947.

[77] Wang, N., Zhang, X., Han, N., and Bai, S. (2009) Effect of citric acid and processing on the performance of thermoplastic starch/montmorillonite nanocomposites. *Carbohydrate Polymers*, *76*(1), 68–73.

[78] Huang, M., Yu, J. (2006) Structure and properties of thermoplastic corn starch/montmorillonite biodegradable composites. *Journal of Applied Polymer Science*, *99*(1), 170–176.

[79] Tang, X., Alavi, S., and Herald, T.J. (2008) Effects of plasticizers on the structure and properties of starch/clay nanocomposites films. *Carbohydrate Polymers*, *74*(3), 552–558.

[80] Dai, H.G., Chang, P.R., Geng, F.Y., Yu, J.G., and Ma, X.F. (2009) Preparation and properties of thermoplastic starch/montmorillonite nanocomposite using N-(2-hydroxyethyl) formamide as a new additive. *Journal of Polymers and the Environment*, *17*(4), 225–232.

[81] Zeppa, C., Gouanvé, F., and Espuche, E. (2009) Effect of a plasticizer on the structure of biodegradable starch/clay nanocomposites: Thermal, water-sorption, and oxygen-barrier properties. *Journal of Applied Polymer Science*, *112*(4), 2044–2056.

[82] Lagaly, G. (1986) Interaction of alkylamines with different types of layered compounds. *Solid State Ionics*, *22*(1), 43–51.

[83] Vaia, R.A., Teukolsky, R.K., and Giannelis, E.P. (1994) Interlayer structure and molecular environment of alkylammonium layered silicates. *Chemistry of Materials*, *6*(7), 1017–1022.

[84] Hackett, E., Manias, E., and Giannelis, E.P. (1998) Molecular dynamics simulations of organically modified layered silicates. *Journal of Chemical Physics*, *108*(17), 7410–7415.

[85] Vaia, R.A., Giannelis, E.P. (1997) Polymer melt intercalation in organically modified layered silicates: Model predictions and experiment. *Macromolecules*, *30*(25), 8000–8009.

[86] Pollet, E., Delcourt, C., Alexandre, M., and Dubois, P. (2004) Organic-inorganic nanohybrids obtained by sequential copolymerization of ε-caprolactone and L, L-lactide from activated clay surface. *Macromolecular Chemistry and Physics*, *205*(16), 2235–2244.

[87] McGlashan, S.A., Halley, P.J. (2003) Preparation and characterisation of biodegradable starch-based nanocomposite materials. *Polymer International*, *52*(11), 1767–1773.

[88] Alexandre, M., Dubois, P. (2000) Polymer-layered silicate nanocomposites: Preparation, properties and uses of a new class of materials. *Materials Science and Engineering, R: Reports*, *28*(1–2), 1–63.

[89] Liu, H.H., Chaudhary, D., Yusa, S., and Tadé, M.O. (2011) Preparation and characterization of sorbitol modified nanoclay with high amylose bionanocomposites. *Carbohydrate Polymers*, *85*(1), 97–104.

[90] Dennis, H.R., Hunter, D.L., Chang, D., Kim, S., White, J.L., Cho, J.W., and Paul, D.R. (2001) Effect of melt processing conditions on the extent of exfoliation in organoclay-based nanocomposites. *Polymer*, *42*(23), 9513–9522.

[91] Messersmith, P.B., Giannelis, E.P. (1995) Synthesis and barrier properties of poly(ε-caprolactone)-layered silicate nanocomposites. *Journal of Polymer Science Part A: Polymer Chemistry*, *33*(7), 1047–1057.

[92] Kampeerapappun, P., Aht-ong, D., Pentrakoon, D., and Srikulkit, K. (2007) Preparation of cassava starch/montmorillonite composite film. *Carbohydrate Polymers*, *67*(2), 155–163.

[93] Van Soest, J.J.G., DeWit, D., and Vliegenthart, J.F.G. (1996) Mechanical properties of thermoplastic waxy maize starch. *Journal of Applied Polymer Science*, *61*(11), 1927–1937.

[94] Van Soest, J.J.G., Borger, D.B. (1997) Structure and properties of compression-molded thermoplastic starch materials from normal and high-amylose maize starches. *Journal of Applied Polymer Science*, *64*(4), 631–644.

[95] Singh, C., Balazs, A.C. (2000) Effect of polymer architecture on the miscibility of polymer/clay mixtures. *Polymer International*, *49*(5), 469–471.

[96] Robello, D.R., Yamaguchi, N., Blanton, T., and Barnes, C. (2004) Spontaneous formation of an exfoliated polystyrene-clay nanocomposite using a star-shaped polymer. *Journal of the American Chemical Society*, *126*(26), 8118–8119.

[97] Liu, H.H., Chaudhary, D., Yusa, S., and Tadé, M.O. (2011) Glycerol/starch/Na^+-montmorillonite nanocomposites: A XRD, FTIR, DSC and ^1H NMR study. *Carbohydrate Polymers*, *83*(4), 1591–1597.

[98] Wu, Y.P., Ma, Y., Wang, Y.Q., and Zhang, L.Q. (2004) Effects of characteristics of rubber, mixing and vulcanization on the structure and properties of rubber/clay nanocomposites by melt blending. *Macromolecular Materials and Engineering*, *289*(10), 890–894.

[99] Lee, S.Y., Hanna, M.A. (2009) Tapioca starch-poly(lactic acid)-Cloisite 30B nanocomposite foams. *Polymer Composites*, *30*(5), 665–672.

[100] Pérez, C.J., Alvarez, V.A., and Vázquez, A. (2008) Creep behaviour of layered silicate/starch-polycaprolactone blends nanocomposites. *Materials Science and Engineering: A*, *480*(1–2), 259–265.

[101] Dean, K.M., Do, M.D., Petinakis, E., and Yu, L. (2008) Key interaction in biodegradable thermoplastic starch/poly(vinyl alcohol)/montmorillonite micro- and nanocomposites. *Composites Science and Technology*, *68*(6), 1453–1462.

[102] Spiridon, I., Popescu, M.C., Bodârlău, R., and Vasile, C. (2008) Enzymatic degradation of some nanocomposites of poly(vinyl alcohol) with starch. *Polymer Degradation and Stability*, *93*(10), 1884–1890.

[103] Arroyo, O.H., Huneault, M.A., Favis, B.D., and Bureau, M.N. (2010) Processing and properties of PLA/thermoplastic starch/montmorillonite nanocomposites. *Polymer Composites*, *31*(1), 114–127.

[104] Jiang, L., Shen, X.P., Wu, J.L., and Shen, K.C. (2010) Preparation and characterization of graphene/poly(vinyl alcohol) nanocomposites. *Journal of Applied Polymer Science*, *118*(1), 275–279.

[105] Liu, P.G., Gong, K.C., Xiao, P., and Xiao, M. (2000) Preparation and characterization of poly(vinyl acetate)-intercalated graphite oxide composite. *Journal of Materials Chemistry*, *10*, 933–935.

[106] Matsuo, Y., Tahara, K., and Sugie, Y. (1997) Structure and thermal properties of poly (ethylene oxide)-intercalated graphite oxide. *Carbon*, *35*(1), 113–120.

[107] Anglès, M.N., Dufresne, A. (2000) Plasticized starch/tunicin whiskers nanocomposites: 1. Structural analysis. *Macromolecules*, *33*(22), 8344–8353.

[108] Habibi, Y., Lucia, L.A., and Rojas, O.J. (2010) Cellulose nanocrystals: Chemistry, self-assembly, and applications. *Chemical Reviews*, *110*(6), 3479–3500.

[109] Chang, P.R., Jian, R., Zheng, P., Yu, J., and Ma, X. (2010) Preparation and properties of glycerol plasticized-starch (GPS)/cellulose nanoparticle (CN) composites. *Carbohydrate Polymers*, *79*(2), 301–305.

[110] Teixeira, E.D.M., Lotti, C., Corrêa, A.C., Teodoro, K.B.R., Marconcini, J.M., and Mattoso, L.H.C. (2011) Thermoplastic corn starch reinforced with cotton cellulose nanofibers. *Journal of Applied Polymer Science*, *120*(4), 2428–2433.

[111] Wang, Y.X., Chang, C.Y., and Zhang, L.N. (2010) Effects of freezing/thawing cycles and cellulose nanowhiskers on structure and properties of biocompatible starch/PVA sponges. *Macromolecular Materials and Engineering*, *295*(2), 137–145.

[112] Teixeira, E.D.M., Pasquini, D., Curvelo, A.A.S., Corradini, E., Belgacem, M.N., and Dufresne, A. (2009) Cassava bagasse cellulose nanofibrils reinforced thermoplastic cassava starch. *Carbohydrate Polymers*, *78*(3), 422–431.

[113] Le Corre, D., Bras, J., and Dufresne, A. (2010) Starch nanoparticles: A review. *Biomacromolecules*, *11*(5), 1139–1153.

[114] Lin, N., Huang, J., Chang, P.R., Anderson, D.P., and Yu, J.H. (2011) Preparation, modification, and application of starch nanocrystals in nanomaterials: A review. *Journal of Nanomaterials*. Article ID 573687, 13 pages.

[115] García, N.L., Ribba, L., Dufresne, A., Aranguren, M.I., and Goyanes, S. (2009) Physico-mechanical properties of biodegradable starch nanocomposites. *Macromolecular Materials and Engineering*, *294*(3), 169–177.

[116] Nair, K.G., Dufresne, A., Gandini, A., and Belgacem, M.N. (2003) Crab shell chitin whiskers reinforced natural rubber nanocomposites. 3. Effect of chemical modification of chitin whiskers. *Biomacromolecules*, *4*(6), 1835–1842.

[117] Zeng, J.B., He, Y.S., Li, S.L., and Wang, Y.Z. (2012) Chitin whiskers: An overview. *Biomacromolecules*, *13*(1), 1–11.

[118] Wang, J., Wang, Z.B., Li, J., Wang, B.J., Liu, J., Chen, P., Miao, M.H., and Gu, Q. (2012) Chitin nanocrystals grafted with poly(3-hydroxybutyrate-co-3-hydroxyvalerate) and their effects on thermal behavior of PHBV. *Carbohydrate Polymers*, *87*(1), 784–789.

[119] Dufresne, A. (2008) Polysaccharide nanocrystal reinforced nanocomposites. *Canadian Journal of Chemistry*, *86*(6), 484–494.

[120] Chang, P.R., Jian, R.J., Yu, J.G., and Ma, X.F. (2010) Starch-based composites reinforced with novel chitin nanoparticles. *Carbohydrate Polymers*, *80*(2), 420–425.

[121] Kristo, E., Biliaderis, C.G. (2007) Physical properties of starch nanocrystal-reinforced pullulan films. *Carbohydrate Polymers*, *68*(1), 146–158.

[122] Dionne, P.J., Ozisik, R., and Picu, C.R. (2005) Structure and dynamics of polyethylene nanocomposites. *Macromolecules*, *38*(22), 9351–9358.

[123] Smith, J.S., Bedrov, D., and Smith, G.D. (2003) A molecular dynamics simulation study of nanoparticle interactions in a model polymer-nanoparticle composite. *Composites Science and Technology*, *63*(11), 1599–1605.

[124] Dubief, D., Samain, E., and Dufresne, A. (1999) Polysaccharide microcrystals reinforced amorphous poly(β-hydroxyoctanoate) nanocomposite materials. *Macromolecules*, *32*(18), 5765–5771.

[125] Dufresne, A., Cavaillé, J.Y. (1998) Clustering and percolation effects in microcrystalline starch-reinforced thermoplastic. *Journal of Polymer Science Part B: Polymer Physics*, *36*(12), 2211–2224.

[126] Morin, A., Dufresne, A. (2002) Nanocomposites of chitin whiskers from *Riftia* tubes and poly(caprolactone). *Macromolecules*, *35*(6), 2190–2199.

[127] Siró, I., Plackett, D. (2010) Microfibrillated cellulose and new nanocomposite materials: A review. *Cellulose*, *17*(3), 459–494.

[128] Angellier-Coussy, H., Putaux, J.L., Molina-Boisseau, S., Dufresne, A., Bertoft, E., and Perez, S. (2009) The molecular structure of waxy maize starch nanocrystals. *Carbohydrate Research*, *344*(12), 1558–1566.

[129] Shioyama, H. (1997) Polymerization of isoprene and styrene in the interlayer spacing of graphite. *Carbon*, *35*(10–11), 1664–1665.

[130] Matsuo, Y., Tahara, K., and Sugie, Y. (1996) Synthesis of poly(ethylene oxide)-intercalated graphite oxide. *Carbon*, *34*(5), 672–674.

[131] Wang, G.X., Wang, B., Park, J., Yang, J., Shen, X.P., and Yao, J.N. (2009) Synthesis of enhanced hydrophilic and hydrophobic graphene oxide nanosheets by a solvothermal method. *Carbon*, *47*(1), 68–72.

[132] Hernan, L., Morales, J., and Santos, J. (1998) Synthesis and characterization of poly(ethylene oxide) nanocomposites of misfit layer chacogenides. *Journal of Solid State Chemistry*, *141*(2), 323–329.

[133] Harris, D.J., Bonagamba, T.J., and Schmidt-Rohr, K. (1999) Conformation of poly(ethylene oxide) intercalated in clay and MoS_2 studied by two-dimensional double-quantum NMR. *Macromolecules*, *32*(20), 6718–6724.

[134] Ding, Y., Jones, D.J., Maireles-Torres, P., and Roziere, J. (1995) Two-dimensional nanocomposites: Alternating inorganic-organic polymer layers in zirconium phosphate. *Chemistry of Materials*, *7*(3), 562–571.

[135] Wilson Jr., O.C., Olorunyolemi, T., Jaworski, A., Borum, L., Young, D., Siriwat, A., Dickens, E., Oriakhi, C., and Lerner, M. (1999) Surface and interfacial properties of polymer-intercalated layered double hydroxide nanocomposites. *Applied Clay Science*, *15*(1–2), 265–279.

[136] Oriakhi, C.O., Farr, I.V., and Lerner, M.M. (1997) Thermal characterization of poly(styrene sulfonate)/layered double hydroxide nanocomposites. *Clays and Clay Minerals*, *45*(2), 194–202.

[137] Chung, Y.L., Ansari, S., Estevez, L., Hayrapetyan, S., Giannelis, E.P., and Lai, H.M. (2010) Preparation and properties of biodegradable starch-clay nanocomposites. *Carbohydrate Polymers*, *79*(2), 391–396.

[138] Vu, Y.T., Mark, J.E., Pham, L.H., and Engelhardt, M. (2001) Clay nanolayer reinforcement of cis-1,4-polyisoprene and epoxidized natural rubber. *Journal of Applied Polymer Science*, *82*(6), 1391–1403.

[139] Mark, J.E. (2006) Some novel polymeric nanocomposites. *Accounts of Chemical Research*, *39*(12), 881–888.

[140] Klemm, D., Schumann, D., Udhardt, U., and Marsch, S. (2001) Bacterial synthesized cellulose-artificial blood vessels for microsurgery. *Progress in Polymer Science*, *26*(9), 1561–1603.

[141] Wan, Y.Z., Luo, H.L., He, F., Liang, H., Huang, Y., and Li, X.L. (2009) Mechanical, moisture absorption, and biodegradation behaviours of bacterial cellulose

fibre-reinforced starch biocomposites. *Composites Science and Technology*, *69*(7–8), 1212–1217.

[142] Henriksson, M., Berglund, L.A. (2007) Structure and properties of cellulose nanocomposites films containing melamine formaldehyde. *Journal of Applied Polymer Science*, *106*(4), 2817–2824.

[143] Famá, L.M., Pettarin, V., Goyanes, S.N., and Bernal, C.R. (2011) Starch/multi-walled carbon nanotubes composites with improved mechanical properties. *Carbohydrate Polymers*, *83*(3), 1226–1231.

[144] Cyras, V.P., Manfredi, L.B., Ton-That, M.T., and Vazquez, A. (2008) Physical and mechanical properties of thermoplastic starch/montmorillonite nanocomposites films. *Carbohydrate Polymers*, *73*(1), 55–63.

[145] Doh, J.G., Cho, I. (1998) Synthesis and properties of polystyrene-organoammonium montmorillonite hybrid. *Polymer Bulletin*, *41*(5), 511–518.

[146] Park, H.M., Lee, W.K., Park, C.Y., Cho, W.J., and Ha, C.S. (2003) Environmentally friendly polymer hybrids. Part I: Mechanical, thermal, and barrier properties of thermoplastic starch/clay nanocomposites. *Journal of Materials Science*, *38*(5), 909–915.

[147] Ratnayake, W.S., Hoover, R., and Warkentin, T. (2002) Pea starch: Composition, structure and properties—A review. *Starch/Stärke*, *54*(6), 217–234.

[148] Van Soest, J.J.G., Knooren, N. (1997) Influence of glycerol and water content on the structure and properties of extruded starch plastic sheets during aging. *Journal of Applied Polymer Science*, *64*(7), 1411–1422.

[149] Huang, M.F., Yu, J.G., and Ma, X.F. (2004) Studies on the properties of montmorillonite-reinforced thermoplastic starch composites. *Polymer*, *45*(20), 7017–7023.

[150] Van Soest, J.J.G., Hulleman, S.H.D., De Wit, D., and Vliegenthart, J.F.G. (1996) Crystallinity in starch bioplastics. *Industrial Crops and Products*, *5*(1), 11–22.

[151] Chaudhary, D.S. (2008) Understanding amylose crystallinity in starch-clay nanocomposites. *Journal of Polymer Science Part B: Polymer Physics*, *46*(10), 979–987.

[152] Imberty, A., Chanzy, H., Pérez, S.J., Belèon, A., and Tran, V. (1988) The double-helical nature of the crystalline part of A-starch. *Journal of Molecular Biology*, *201*(2), 365–378.

[153] Souza Rosa, R.C.R., Andrade, C.T. (2004) Effect of chitin addition on injection-molded thermoplastic corn starch. *Journal of Applied Polymer Science*, *92*(4), 2706–2713.

[154] Smits, A.L.M., Ruhnau, F.C., Vliegenthart, J.F.G., and Van Soest, J.J.G. (1998) Ageing of starch based systems as observed with FT-IR and solid state NMR spectroscopy. *Starch/Stärke*, *50*(11–12), 478–483.

[155] Ma, X.F., Yu, J.G. (2004) The plasticizers containing amide groups for thermoplastic starch. *Carbohydrate Polymers*, *57*(2), 197–203.

[156] Calvert, P. (1997) Biopolymers: The structure of starch. *Nature*, *389*(6649), 338–339.

[157] García, N.L., Famá, L., Dufresne, A., Aranguren, M., and Goyanes, S. (2009) A comparison between the physico-chemical properties of tuber and cereal starches. *Food Research International*, *42*(8), 976–982.

[158] Ma, X.F., Yu, J.G., and Kennedy, J.F. (2005) Studies on the properties of natural fibers-reinforced thermoplastic starch composites. *Carbohydrate Polymers*, *62*(1), 19–24.

[159] Liu, D.G., Zhong, T.H., Chang, P.R., Li, K.F., and Wu, Q.L. (2010) Starch composites reinforced by bamboo cellulosic crystals. *Bioresource Technology*, *101*(7), 2529–2536.

[160] Amash, A., Zugenmaier, P. (2000) Morphology and properties of isotropic and oriented samples of cellulose fibre-polypropylene composites. *Polymer*, *41*(4), 1589–1596.

[161] Ma, X.F., Chang, P.R., Yu, J.G., and Stumborg, M. (2009) Properties of biodegradable citric acid-modified granular starch/thermoplastic pea starch composites. *Carbohydrate Polymers*, *75*(1), 1–8.

[162] Yu, J.G., Wang, N., and Ma, X.F. (2008) Fabrication and characterization of poly(lactic acid)/acetyl tributyl citrate/carbon black as conductive polymer composites. *Biomacromolecules*, *9*(3), 1050–1057.

[163] Xu, Y., Zhou, J., and Hanna, M.A. (2005) Melt-intercalated starch acetate nanocomposite foams as affected by type of organoclay. *Cereal Chemistry*, *82*(1), 105–110.

[164] Liu, Z.J., Zhao, L., Chen, M.N., and Yu, J.G. (2011) Effect of carboxylate multiwalled carbon nanotubes on the performance of thermoplastic starch nanocomposites. *Carbohydrate Polymers*, *83*(2), 447–451.

[165] Wang, N., Yu, J.G., Chang, P.R., and Ma, X.F. (2008) Influence of formamide and water on the properties of thermoplastic starch/poly(lactic acid) blends. *Carbohydrate Polymers*, *71*(1), 109–118.

[166] Chatterjee, A., Deopura, B.L. (2006) Thermal stability of polypropylene/carbon nanofiber composite. *Journal of Applied Polymer Science*, *100*(5), 3574–3578.

CHAPTER 12

Mechanical Properties of Starch-Based Nanocomposites

HÉLÈNE ANGELLIER-COUSSY and ALAIN DUFRESNE

12.1 INTRODUCTION

Starch is a natural, renewable, biodegradable polymer produced by many plants as a source of storage energy. It is the second most abundant biomass material in nature. The starch industry extracts and refines starches by wet grinding, sieving, and drying. Starch is used as extracted from the plant and is called "native starch," or it undergoes one or more chemical modifications to reach specific properties and is then called "modified starch." Worldwide the main sources of starch are maize, wheat, potatoes, and cassava, from which tapioca starch is derived. In 2000, the world starch market was estimated to be 48.5 million tons, including native and modified starches. The value of the output is worth €15 billion per year, which explains why industrialists and researchers are seeking new properties and high value applications.

Among naturally biodegradable polymers using renewable resources, starch is probably the most promising material. It is a versatile biopolymer with huge potential for use in the nonfood industries due to its high availability at low prices and nondependence on fossil sources. The first generation of starch-based materials consisted of the physical incorporation of starch granules or derivatives as functional additives to, and fillers for, synthetic polymers during processing. However, the mechanical properties of ensuing materials are generally reduced by incorporation of starch due to its immiscibility with most synthetic polymers at the molecular level. Starch is not truly thermoplastic but can be converted into a continuous polymeric entangled phase by mixing with enough water or nonaqueous plasticizer (generally polyols, such as glycerol).

Biopolymer Nanocomposites: Processing, Properties, and Applications, First Edition. Edited by Alain Dufresne, Sabu Thomas, and Laly A. Pothan.

The resulting material, corresponding to the second generation of starch-based materials, can be manufactured using technology already developed for the production of synthetic plastics, thus representing a minor investment. However, the hydrophilic nature of thermoplastic starches makes them susceptible to moisture attack and resultant changes in dimensional stability and mechanical properties. In addition, retrogradation and crystallization of the mobile starch chains lead to an undesired change in thermomechanical properties. Thermoplastic starch can be used as matrix and its properties can be tailored by adding different fillers, including nanosized particles. Nanoparticles allow not only enhancing mechanical properties but also physical properties such as permeability or fire retardancy. Their properties depend on the nature and effectiveness of interactions at the interfacial region, that is, on both the surface area and the dispersion of the particles.

The semicrystalline nature of starch can be exploited to extract nanocrystals as for cellulose or chitin, by a selective hydrolysis process. Contrarily to cellulose or chitin needle-like nanocrystals, starch nanocrystals occur as platelet-like nanoparticles. They can be used as nanoscale filler in a polymeric matrix.

This chapter reports the mechanical properties of starch-based nanocomposite materials, considering that the starchy material can be used either as the continuous polymeric phase (matrix), or the nanodispersed phase (filler), or both. The first part of the chapter will discuss the reinforcing effect of the different fillers used to improve the mechanical properties of thermoplastic starch, that is, inorganic clays, carbon nanotubes, and polysaccharide-based (cellulose, starch) nanoparticles. The second section will be devoted to the reinforcing effect of starch nanocrystals, as well as the mechanisms proposed to explain it.

12.2 STARCH AS MATRIX IN NANOCOMPOSITES

Starch is one of the most studied natural and renewable polymers with many promising applications in biodegradable packaging. During the last decade, starch-based nanocomposites with the addition of low quantities of inorganic or organic fillers have given rise to significant improvements in physical, mechanical, and barrier properties. This section will focus on the mechanical properties of starch-based nanocomposites considering starch as matrix. It will discuss some examples of starch nanocomposites reinforced with inorganic clays or carbon nanotubes, and polysaccharide-based (cellulose, starch) fillers, and attempt to give a critical comparison of the different reinforcements.

12.2.1 Mechanical Properties of Starch-Based Materials

12.2.1.1 Botanical Origin: Amylose/Amylopectin Ratio The botanical origin, especially the amylose/amylopectin ratio, is known to influence the mechanical properties of starch. The idea generally spread is that amylose-rich materials display higher strength and lower strain at break than amylopectin-rich materials [1]. However, this affirmation should be nuanced due to some opposite results described in the literature. For example, Myllärinen et al. [2] showed that above 20 wt% glycerol the strain at break was more important for amylose-rich films. In fact, two behaviors can be reported according to the plasticizer content [1, 2]. (1) For low plasticizer contents, an increase in the amylose/amylopectin ratio would improve both the strength and the strain at break because of the linear structure of amylose chains. (2) For high plasticizer contents, an increase in the amylose content results in a mechanical reinforcement (the strength increases and the strain decreases). This last tendency is related to fast amylose crystallization in comparison with slower amylopectin retrogradation. In any case, it can be concluded that the amylose/amylopectin ratio is not a determining parameter for the improvement of ultimate mechanical properties [3].

12.2.1.2 Plasticization Starch is most often plasticized with water and polyols such as glycerol. For example, a glycerol content of 25% allows decreasing the glass transition temperature of potato starch-based materials stored at 57% relative humidty (RH) from 285°C to −10°C [4, 5]. Depending on the plasticizer content, two mechanical behaviors are evidenced whatever the starch origin. For a low plasticizer content, mechanical properties present similarities to the antiplasticization effect generally observed in synthetic polymers; that is, both the stress and strain at break decrease when the plasticizer concentration increases to up to 12 wt% [5]. This phenomenon would correspond to a local macromolecular motion limitation due to strong interactions between starch chains and plasticizer resulting in a reinforcing hydrogen network [5–7]. For higher amounts of plasticizer, mechanical properties are controlled by the glass transition, which involves the molecular motions. The classical plasticization effect is characterized by an increase in the strain at break and a decrease in both the strength and Young's modulus [5, 8]. Plasticizers (water and glycerol) play the role of internal lubricants, preventing the rigidification of noncrystallized macromolecular starch chains at ambient temperature [5]. The plasticization effect is the phenomenon that governs the mechanical properties of unaged systems. Currently, all formulations of thermoplastic starch materials require the use of very large amounts of plasticizer (roughly 20 wt%) to avoid antiplasticization events. As a result, starch-based materials have globally poor mechanical properties due to the relaxation of the polymer network in the presence of plasticizer.

In the case of plasticized starch/glycerol systems, dynamic mechanical analysis (DMA) highlights the presence of two relaxation phenomena associated with a two-step storage modulus drop. It is characterized by two peaks on loss modulus and tan δ curves, around −40°C and 40°C (25 wt% of glycerol and 15 wt% of water) [5], revealing the heterogeneity of the system and a phase separation. The first peak is associated with the relaxation of glycerol-rich domains, and the second one with starch-rich domains [9–11].

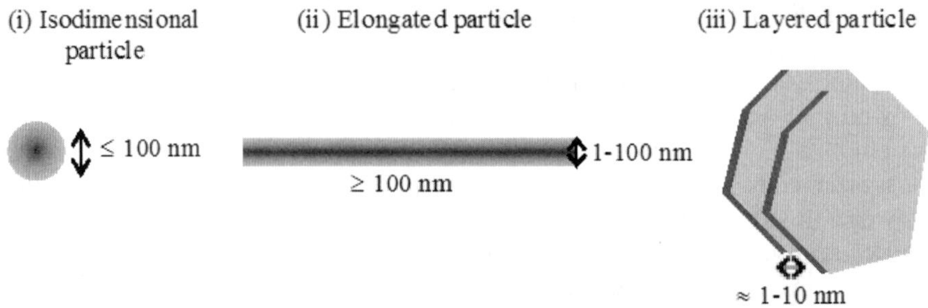

Figure 12.1 Schematic representation of the three categories of fillers.

12.2.1.3 Aging Physical aging, that is, retrogradation, induces structural changes without chemical modification. This phenomenon is slow because it generally occurs at temperatures below the glass transition temperture of starch-based materials (T_g) for which the chain mobility is reduced. It only occurs at room temperature for plasticized systems displaying a T_g below room temperature. Starch-based materials in the rubbery state are the site of recrystallization, which takes place in two steps: (1) a fast recrystallization step during film processing where gelatinized starch recrystallizes according to the B-type, and (2) a slow recrystallization upon film storage. Mechanical properties of starch-based materials are directly affected by physical aging [12–14]. It inevitably leads to higher strength and Young's modulus [12] and lower strain at break [13, 15] whatever the starch origin. Crystallites play the role of physical cross-linking bonds, resulting in stress concentration and thus making more brittle the material [16]. Therefore, the mechanical properties of starch-based materials are very sensitive to the ambient conditions and moisture content, which are difficult to control.

12.2.2 Fillers

Fillers can be classified into three major categories according to their particle size and shape (Fig. 12.1). (i) *Isodimensional particles* have the same size in all directions and their aspect ratio is usually close to unity (examples include spherical silica, carbon black, and fullerenes). This type of particles has not yet been used as fillers in a starch matrix, probably due to their moderate reinforcing effect. (ii) *Elongated particles* consist of fibrils with diameters ranging between 1 and 100 nm and lengths of several hundred of nanometers. They included cellulose and chitin nanofibrils and carbon nanotubes. These nanocomposites are known to present better mechanical properties due to the higher particle aspect ratio. (iii) The third category is composed of *layered particles* such as layered silicates, but also starch nanocrystals. They are characterized by one dimension, often the thickness, ranging from several angstroms to several nanometers. The most common fillers used as reinforcing agents in a starch matrix are layered silicates and carbon nanotubes, which are both inorganic fillers, and polysaccharide-based (cellulose, chitin, starch) nanocrystals, which are organic fillers.

12.2.2.1 Inorganic Fillers

Clays Clays, also called layered silicates, are constituted of colloidal fragments of primary silicates made up of tetrahedral and octahedral sheets of small cations, such as aluminum or magnesium coordinated by oxygen atoms. One important feature of clays is that the space between layers contains hydrated cations such as Na^+ or K^+ that can undergo exchange reaction with organic as well as inorganic cation. Smectites are among the most used layered silicates for the preparation of nanocomposites. They belong to the phyllosilicate family and display a perfect crystalline structure consisting of two-dimensional layers where a central octahedral sheet of alumina or magnesia is linked with two external silica tetrahedrons [17]. The layer thickness is around 1 nm and the lateral dimensions of these layers vary from 30 nm to several micrometers [17]. Smectites include natural clays such as montmorillonite (MMT), hectorite, and kaolinite. The limiting aspect of the use of such nanofillers is the difficulty to correctly disperse them within a polymeric matrix due to their preferred face-to-face stacking in agglomerated tactoïds. Cation exchange reactions in clays are very important to impart functionality and compatibility with polymers. Different types of nanocomposites from polymers and clays can be obtained depending on the dispersion state of clays: a conventional microcomposite structure, an intercalated nanocomposite structure, or an exfoliated nanocomposite structure (Fig. 12.2). The achievement of a well-dispersed nanocomposite structure is required to improve mechanical properties of the polymer matrix.

Carbon Nanotubes Carbon nanotubes are molecular-scale tubes formed by rolled-up graphene sheets with outstanding mechanical and electrical properties. There are two main types of nanotubes available today. Single-walled carbon nanotubes (SWCNT) consist of a single sheet of graphene rolled seamlessly to form a cylinder with diameters of the order of 1 nm and lengths of up to centimeters (Fig. 12.3a,b). Due to their high flexibility and high surface energy, SWNTs tend to aggregate into large bundles, which are extremely difficult to separate, thus restricting the use of SWNTs for real applications [18]. SWCNTs have not yet been used as reinforcing fillers in a starch matrix. Multiwalled carbon nanotubes (MWCNT) consist of an array of such cylinders formed concentrically and separated by 0.36 nm, similar to the basal plane separation in graphite. MWCNTs can have outer diameters of about 30 nm and lengths of 1–100 micrometers [18] (Fig. 12.3c,d). They display high moduli (1 TPa) along their length direction, as well as high bending moduli [19]. The effective utilization of MWCNTs in composite applications depends strongly on the ability to disperse them homogeneously within the polymer matrix

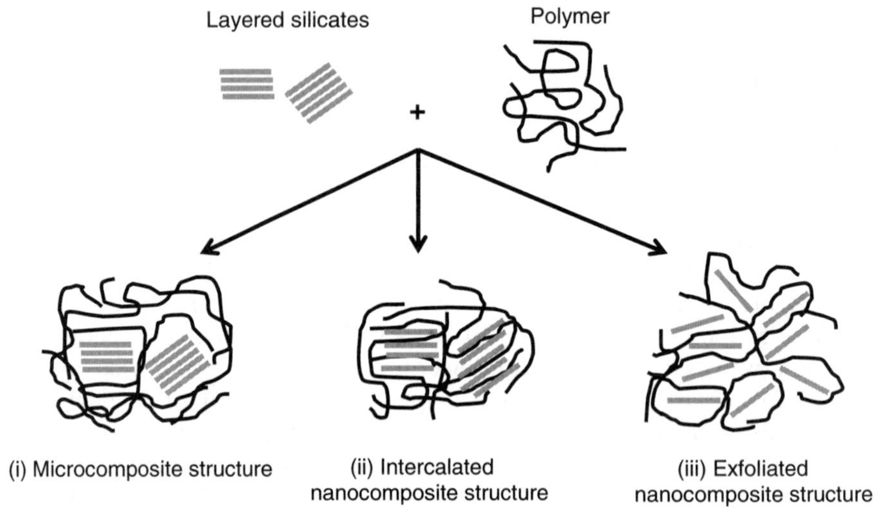

Figure 12.2 Different kinds of nanocomposites from polymer and layered silicates, depending on the dispersion state of particles.

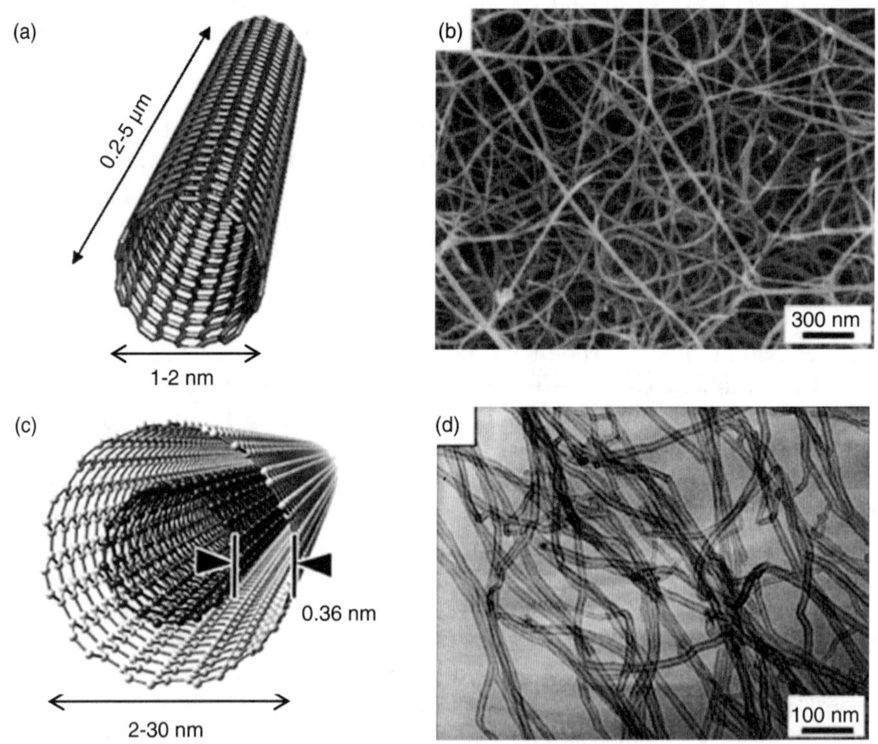

Figure 12.3 (a) Conceptual diagram (adapted from Reilly [79]) and (b) scanning electron microscopy picture of SWCNTs [80]; (c) conceptual diagram (adapted from Reference [79]) and (d) TEM picture [80] of MWCNTs.

without destroying their integrity [19]. To improve their dispersibility, their surface can be chemically functionalized to optimize the interfacial interaction between carbon nanotubes and the polymeric matrix. To solve this problem, both covalent and noncovalent functional modifications have been widely studied [18].

12.2.2.2 Organic Fillers Polysaccharides (cellulose, chitin, starch) are among the most promising sources for the production of nanoparticles with various different morphologies and geometrical characteristics providing a wide range of nanoparticle properties. As layered silicates, polysaccharides offer significant potential for surface modification using well-established carbohydrate chemistry, allowing researchers to tailor the surface functionality of the nanoparticles.

Cellulose Nanocrystals Aqueous suspensions of cellulose nanocrystals can be prepared by acid hydrolysis of native substrates. The object of this treatment is to dissolve away regions of low lateral order so that the water-insoluble, highly crystalline residue may be converted into a stable suspension. The resulting nanocrystals occur as rod-like particles or whiskers, whose dimensions depend on the nature of the substrate and preparation conditions. They are typically a few hundred nanometers long and in the order of few nanometers in diameter [20] (Table 12.1). The aspect ratio of these whiskers is defined as the ratio of the length to the diameter.

Starch Nanocrystals Starch can also be used as a source for the production of nanocrystals. Starch nanocrystals are obtained from mild acid hydrolysis of native starch granules using hydrochloric or sulfuric acid. The use of sulfuric acid leads to more stable aqueous suspensions due to resulting negatively charged surfaces [21]. Starch nanocrystals correspond to the highly crystalline

TABLE 12.1 Geometrical Characteristics of Rod-Like Cellulose Nanofibers: Length (L) and Diameter (D)

Source of Cellulose	L (nm)	D/W (nm)	Ref.
Potato pulp	>1000	2–4	[63, 64]
Tunicate	500–2000	10	[7, 66]
Cottonseed linter	170–490	40–60	[69]
Cotton	100–300	5–10	Data from Reference [81] used in Reference [65]
Wheat straw	>1000	30–40	[70, 71]
Hemp	Not estimated	20–40	[59]
Pea hull	240–400	7–12	[72]
Cassava bagasse	360–1700	2–11	[68]

water-insoluble lamellae of the native starch granules. Three main botanical origins were used for that purpose: waxy maize [21–30], pea [31–35], and potato [36, 37]. Starch nanocrystals generally appear as aggregates of parallelepiped platelets exhibiting 60–65° acute angles. The botanical origin of starch, namely the crystallinity type and the relative proportion of amylose and amylopectin, determines the structure and the morphology of starch nanocrystals. The conditions of acid hydrolysis also affect the yield and the final dimensions of starch nanocrystals. The dimensions of starch nanocrystals have been characterized by image analysis of transmission electron microscopy (TEM) pictures, with a length ranging from 20 to 150 nm, width from 10 to 30 nm, and thickness from 5 to 10 nm, depending on the botanical origin [38]. It has been recently reported that at any time, both microscaled and nanoscaled particles can be found and coexist in suspensions [39]. Cross-flow filtration was proved to be an efficient continuous operation for separating the nanoparticles from the bulk suspension [40].

Starch nanocrystals, as cellulose whiskers, possess a highly reactive surface covered with hydroxyl groups, which provides the possibility of modification via a chemical reaction strategy. The purpose of the chemical modification is to contribute to specific functions and to expand the applications of starch nanocrystals. There are three strategies for chemical modification of starch nanocrystals: modification by chemical reaction with small molecules, grafting of polymer chains using coupling agents (i.e., the "grafting onto" strategy), and grafting of polymer chains by polymerization of a monomer (also called the "grafting from" strategy) [38]. The effect of surface chemical modification on mechanical properties of starch nanocrystals reinforced nanocomposites will be discussed in Section 12.3.2.

12.2.3 Mechanical Properties of Nanocomposites Based on Starch and Inorganic Reinforcements

Clays are the inorganic nanoparticles the most largely used as reinforcing filler in starch-based matrices with first papers published in 2002 and already more than tens publications. To reach exfoliation of layered silicates, different nanofillers (natural or organically modified), and processing routes (thermomechanical process or casting) were developed. Most studies are devoted to the introduction of unmodified clays, especially MMT [41–52], and to a less extent, of hectorite [11, 44, 53], laponite [47], or kaolinite [44, 53]. Thanks to the hydrophilic nature of both starch and unmodified clays, this type of nanofiller was expected to give an enhanced nanodispersion state. However, at best, only an intercalated nanocomposite structure with the presence of aggregates was obtained, as shown by X-ray diffraction (XRD) and TEM analyses [42, 44–46, 48–50, 52]. It was demonstrated that the processing conditions and the plasticizer content strongly affect the dispersion state of layered silicates and should be thus optimized for each starch/clay couple [45, 54]. Furthermore, it was shown that the hydrophilic plasticizer (often glycerol) could disturb the clay exfoliation process due to hydrogen bonds established between plasticizer and clay platelets [43, 53]. Park et al. [41, 42] were the first to report the reinforcing effect of unmodified MMT in a potato starch/glycerol-based matrix, as demonstrated by DMA and tensile tests. They noted an increase of the storage modulus for filler contents higher than 2.5 wt% over the entire temperature span, with a more pronounced reinforcing effect at higher temperatures where the molecular relaxation of the starch-rich phase occurs. At the same time, a shift of the relaxation temperature for both glycerol-rich and starch-rich phases toward higher temperatures (from –64°C to –53°C/–32°C and from 7°C to 12°C/20°C for systems filled with 5 wt% or 10 wt% of unmodified MMT) was noted, indicating that MMT strongly influences the chain mobility of the starch chains [41, 42]. These results were further confirmed for starch/glycerol-based materials reinforced with hectorite or kaolinite [53], potato starch/sorbitol-based materials reinforced with unmodified laponite [47], and wheat starch/glycerol-based films filled with unmodified MMT [51]. The reinforcing effect of unmodified clays was ascribed to several parameters including the creation of an intercalated (not exfoliated) nanocomposite structure, as demonstrated by X-ray diffraction (XRD) and TEM analysis, the creation of a three-dimensional network of interconnected silicate layers, strengthening the material through mechanical percolation [42, 47], the film contraction resulting from the loss of moisture during the heating process [11], and also the increase of crystallinity of the clay filler itself [11]. Kvien et al. [47] also concluded that the plasticizer was not present in the starch/reinforcements interfacial zone. This was confirmed by Chivrac et al. [51], who reported that TEM and DMA clearly demonstrated that clay platelets were preferentially located in the domains rich in polysaccharide chains as a result of phase separation. Tensile tests confirmed the reinforcing effect of unmodified clays (Table 12.2). It was unanimously observed that with the increasing of clay content up to a certain value (15–20 wt%), the stress at break and Young's modulus of composites are improved, while the strain at break decreases, whatever the type of clay and the botanical origin of starch [43, 48–50, 52]. This was ascribed to hydrogen bonds between unmodified natural clays and starch [43]. Tang et al. [49] showed that the reinforcing effect of a given nanoparticle was rather similar whatever the botanical origin of the starch matrix considered, that is, corn, wheat, or potato.

It is worth noting that the reinforcing effect of layered silicates was rather moderate and sometimes inexistent due to a low dispersion level of nanoparticles within the starch-based matrix. To promote the clay exfoliation process, some work has been thus devoted to the surface modification of layered silicates. Already in the earliest study, it was demonstrated that the introduction of clays organically modified with hydrophobic alkylammoniums led to the formation of conventional microcomposites due to the lack of compatibility between both components, as evidenced by X-ray analysis, resulting in very poor mechanical properties, even lower than the neat matrix [41, 44, 45]. Among the hydrophobic alkylammoniums studied, we can report either dimethyl benzyl hydrogenated-tallow [55] or dimethyl dehydrogenated tallow ammoni-

TABLE 12.2 Relative Tensile Test Characteristics (Nominal Young's Modulus, E_R; Nominal Stress at Break, $\sigma_{b,R}$; Nominal Strain at Break, $\varepsilon_{b,R}$) for Starch-Layered Silicates Nanocomposites

Starch Matrix	RH (%)	Layered Silicates	Filler Content (wt%)	E_R	$\sigma_{b,R}$	$\varepsilon_{b,R}$	Year	Ref.
Potato starch/20 wt% water + 30 wt% glycerol	43	Na-MMT (Cloisite Na^+)	5	–	1.27	1.21	2002	[41]
		OMMT (Cloisite 30B)	5	–	1.07	0.95		
		OMMT (Cloisite 6A)	5	–	0.82	0.74		
		OMMT (Cloisite 10A)	5	–	0.96	0.81		
Corn starch/25 wt% glycerol	39	Na-MMT (Cloisite Na^+)	5	–	1.3	0.75	2004	[43]
			30	5.4	5.0	0.2		
Corn starch/20 wt% glycerol	5	Na-MMT	5	1.04	–	0.6	2005	[45]
Cassava starch/17 wt% glycerol	50	Na-MMT	5	1.06	1.04	0.88	2007	[48]
			10	1.06	1.04	0.70		
			15	1.06	0.71	0.33		
		OMMT (chitosan-20 wt%)	5	1.31	1.83	0.53		
			10	1.42	2	0.53		
			15	1.09	0.83	0.40		

Wheat starch/23 wt% water + 23 wt% glycerol	57	Na-MMT (Dellite LVF)	3	1.09	–	–	2008	[58]
			6	1.15	–	–		
		OMMT (cationic starch)	3	1.25	–	–		
			6	1.45	–	–		
Corn starch/19 wt% water + 15 wt% glycerol	50	Na-MMT	3	–	1.17	1.20	2008	[49]
			6	–	1.30	0.84		
			9	–	1.65	0.92		
Wheat starch/19 wt% water + 15 wt% glycerol	50	Na-MMT	9	–	1.51	0.83		
Potato starch/19 wt% water + 15 wt% glycerol	50	Na-MMT	9	–	1.52	1.10		
Potato starch/30 wt% glycerol	90	Na-MMT (Cloisite Na^+)	2	1	0.94	0.88	2008	[50]
			3	5.05	0.97	0.60		
			5	6.56	1.57	0.75		
Potato starch/37.5 wt% glycerol	3 wt% of water content	Na-MMT	1	1.16	1.09	0.88	2009	[52]
			3	1.75	1.41	0.73		
			6	2.55	1.90	0.37		

Na-MMT, unmodified montmorillonite; OMMT, organically modified MMT.

ums [44]. In the case of more hydrophilic surfactants, that is, alkylammoniums containing hydroxyl groups such as Cloisite 30B, the dispersion state was better due to hydrogen bonds established between starch and clay surfactant, leading to higher mechanical properties, but still lower than those obtained for unmodified clays [41, 42, 56]. Recently, authors have focused their attention on the use of compatibilizers as MMT organo-modifier to enhance the MMT platelets exfoliation [57], including chitosan [48] and cationic starch [58]. TEM and XRD analyses confirmed the achievement of a well-exfoliated structure, ascribed to the preferential interactions established between the hydroxyl groups of the polysaccharide chains. It results in an enhanced reinforcing effect of clays as compared to unmodified clays (Table 12.2).

We can conclude that the dispersion state of layered silicates, which is strongly affected by the clay surface polarity and the clay/starch interactions, and of course the filler content are the two key parameters governing the mechanical properties of starch/clay nanocomposites. On the one hand, all studies clearly demonstrated that the achievement of a well-exfoliated nanocomposite structure is desirable to significantly improve the mechanical properties. On the other hand, significant improvement in mechanical properties is generally observed only for filler contents higher than 3–5 wt% [50].

Cao et al. [19] first attempted to use multiwalled carbon nanotubes (MWCNT) as a reinforcing filler in a pea starch-based matrix. To improve the hydrophilicity of MWCNTs and thus the compatibility between MWCNTs and starch, carboxylic acid groups were introduced onto the tips and defect sites of carbon nanotube surfaces by refluxing in a mixture of concentrated acid (sulfuric and nitric). Whatever the filler content (up to 3 wt% in the study), tensile properties of nanocomposites were better than the neat matrix. For a filler content of 3 wt%, a significant improvement of a factor of 1.65 was noted for the tensile strength (from 2.85 to 4.73 MPa) and of a factor 1.89 for Young's modulus (from 20.74 to 39.18 MPa). It is worth noting that the increase in strength did not occur at the expense of the elongation at break (Table 12.3). At the same time, the T_g value ascribed to the starch-rich phase increased with an increase in MWCNT content (from 16.5°C to 25.3°C for 3 wt%), indicating a good adhesion between MWCNTs and the starch matrix, as confirmed by scanning electron microscopy (SEM) observations. The improvement of mechanical properties was explained by the establishment of hydrogen bonds between MWCNTs and starch chains [59]. In a similar study, that is, corn thermoplastic starch reinforced with mixed acid treated MWCNTs, the improvement of the tensile strength was only noted for MWCNTs contents up to 1.5 wt% (from 4.5 to 7.7 MPa). This was attributed to the hydrogen bonding interactions between MWCNTS and starch chains, as already reported by Cao et al. [59]. Then, the decrease in the tensile strength reported for filler content above 1.5 wt%, which corresponds to the electrical percolation threshold, was ascribed to the deterioration of the plasticization and the destruction of the continuity of the matrix [60]. However, contrarily to Cao et al. [59], a decrease in elongation at break was observed whatever the filler content

TABLE 12.3 Relative Tensile Test Characteristics (Nominal Young's Modulus, E_R; Nominal Stress at Break, $\sigma_{b,R}$; Nominal Strain at Break, $\varepsilon_{b,R}$) for Starch–Carbon Nanotubes Nanocomposites

Starch Matrix	RH (%)	Carbon Nanotubes	Filler Content (wt%)	E_R	$\sigma_{b,R}$	$\varepsilon_{b,R}$	Year	Ref.
Pea starch/36 wt% glycerol	43	Mixed acid-treated MWCNT	0.1	1.16	1.06	1	2007	[59]
			0.5	1.19	1.24	1.25		
			1	1.34	1.43	1.41		
			2	1.51	1.50	1.28		
			3	1.89	1.65	1.08		
Corn starch/23 wt% glycerol	50	Nitric acid + sodium dodecylsulfate-treated MWCNT	0.95	1.08	1.14	0.79	2008	[61]
			1.9	1.22	1.27	0.82		
			2.85	1.31	1.41	0.71		
			3.8	1.38	1.45	0.63		
			4.75	1.42	1.65	0.57		
Corn starch/23 wt% glycerol	50	Mixed acid-treated MWCNT	0.5	–	1.28	0.94	2011	[60]
			1	–	1.57	0.86		
			1.5	–	1.71	0.72		
			2	–	1.6	0.62		
			3	–	1.3	0.51		
Tapioca starch/35 wt% of glycerol	57	MWCNT wrapped with starch–iodine complex	0.027	1.44	1.27	1.02	2011	[62]
			0.055	1.68	1.36	1.12		

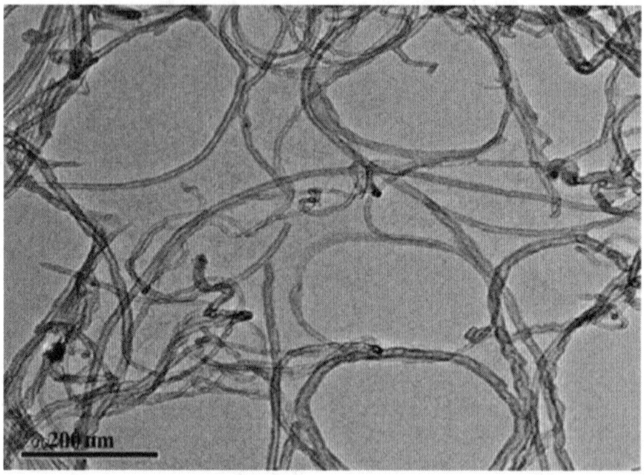

Figure 12.4 TEM micrograph of corn plasticized starch filled with 2.85 wt% MWCNTs [61].

(Table 12.3) [60]. This difference in behavior may be explained by the difference in botanical origin of the starch matrix. In addition to a nitric acid treatment of MWCNTs, Ma et al. [61] used sodium dodecylsulfate as surfactant to prepare corn starch/MWCNT composites. They only found a more modest increase in Young's modulus and tensile strength together with a slight decrease in elongation at break (Table 12.3), in spite of a rather good filler dispersion state and interfacial adhesion between the two components, which were highlighted by TEM and SEM observations (Fig. 12.4). This can be explained by the huge outer diameter and smooth surface of MWCNTs that may cause the compatibility between both components of the composites to deteriorate [60]. Very recently, Fama et al. [62] proposed a noncovalent functionalization to improve the compatibility between MWCNTs and starch. The treatment consists of wrapping MWCNTs with a starch–iodine complex, containing the same starch as the matrix, tapioca in the present case [62]. They suggested that iodine induces an initial preorganization of amylose into a helical conformation that makes its hydrophobic cavity accessible to carbon nanotubes. This treatment was highly effective to obtain very well-dispersed nanotubes and strong adhesion between the phases. It seems to be a very promising route to enhance mechanical properties of thermoplastic starch since very small amounts of MWCNTs, only 0.055 wt%, allow improving Young's modulus by a factor of 1.68, the tensile strength by a factor of 1.36, and the elongation at break by a factor of 1.12 (Table 12.3).

12.2.4 Mechanical Properties of Nanocomposites Based on Starch and Organic Reinforcements

Dufresne and Vignon [63] were the first to report the reinforcing effect of cellulose microfibrils, which were obtained from potato tuber cells, in a glycerol plasticized potato starch-based matrix (Table 12.4). Above T_g, a great increase in the storage modulus (more than 2 decades) and in the thermal stability was observed even with a filler content of as low as 5 wt%. With increasing filler content, no marked evolution of the rubbery modulus was noted, but the terminal zone occurred at higher temperatures [63]. The reinforcing effect of such cellulose microfibrils was further confirmed via tensile tests by the same authors [64] and also later by Orts et al. [65] for cotton-derived microfibrils (Table 12.4).

Dufresne et al. [64] first highlighted that mechanical properties of highly plasticized materials were strongly dependent on relative humidity conditions. The high dependence of mechanical properties of starch/cellulose-based nanocomposite materials on plasticizer and moisture content was greater than glycerol [7] and sorbitol [66] plasticized waxy maize starch reinforced with tunicin whiskers.

In the case of glycerol plasticized systems, Anglès and Dufresne [7] reported that the reinforcing effect of tunicin whiskers was very low as compared to experimental data obtained for other matrices such as poly(S-*co*-BuA) [67]. The reinforcing effect of cellulose whiskers is generally ascribed to the presence of strong interactions between whiskers such as hydrogen bonds, which lead to the formation of a rigid network governed by the percolation threshold, besides the geometry and stiffness of the whiskers [20, 67]. Thus, it was deduced that plasticizers (glycerol and water) probably hindered the formation of such hydrogen-bonded cellulose network within the starch-based matrix due to an accumulation of plasticizer in the cellulose/amylopectin interfacial zone [7, 9]. This phenomenon was also reported for glycerol/sorbitol plasticized cassava starch/cassava bagasse cellulose nanofibers systems [68] (Table 12.4). However, the reinforcing effect of tunicin whiskers in sorbitol plasticized waxy maize starch was higher than in corresponding glycerol plasticized systems (Table 12.4) due to the absence of transcrystallization of sorbitol at the cellulose/amylopectin interface, as proved by DSC and WAXS studies [66]. Thus, sorbitol seems to be more adequate to be used as plasticizer.

The relatively low reinforcing effect of tunicin whiskers observed by Anglès and Dufresne [7] disagreed with previous experimental data reported for glycerol plasticized systems reinforced with cellulose microfibrils [63, 64]. This was explained by the difference in flexibility between the two kinds of fillers [7]. Cellulose whiskers are stiff and straight well-defined objects, whereas cellulose microfibrils are flexible and hairy fibers. Therefore, in composites filled with cellulose microfibrils, the reinforcing effect occurs most probably from both the formation of the hydrogen-bonded cellulose network and from a tangling effect. In composites filled with cellulose whiskers, only the former phenomenon is likely occurring.

The low reinforcing effect of cellulose whiskers in glycerol plasticized systems was not unanimously reported. For example, in the case of glycerol plasticized wheat starch/cottonseed linter whiskers [69], potato starch [70] or

TABLE 12.4 Relative Rubbery Storage Tensile Modulus (E'_R) and Relative Tensile Test Characteristics (Nominal Young's Modulus, E; Nominal Stress at Break, $\sigma_{,R}$; Nominal Strain at Break, $\varepsilon_{b,R}$) for Starch–Cellulose Whiskers Nanocomposites

Starch Matrix	Cellulose Origin	RH (%)	Filler Content (wt%)	E'_R	E_R	$\sigma_{b,R}$	$\varepsilon_{b,R}$	Year	Ref.
Potato starch/30 wt% glycerol	Potato pulp	0	5	125	–	–	–	1998	[63]
			10	150	–	–	–		
Potato starch/30 wt% glycerol	Potato pulp	58	7.5	–	18	–	–	2000	[64]
			15	–	25	–	–		
			23	–	40	–	–		
			31	–	50	–	–		
			39	–	60	–	–		
Waxy maize starch/33 wt% glycerol	Tunicate	58	3.2	0.3	1.5	8	1.6	2001	[7]
			6.2	0.6	1.5	8	1.3		
			16.7	1	2.5	8	1.2		
			25	4.5	6	12	1		
Waxy maize starch/33 wt% sorbitol	Tunicate	58	5	1	1.2	1.8	1	2008	[66]
			10	2.5	3.6	4	1		
			15	2.5	5.3	4.3	1		
			25	3.5	7.2	10.3	1		

Wheat starch/42 wt% glycerol	Cottonseed linter	50	2.5	2.2	1.6	1.2	0.9	2005	[69]
			5	3	2.5	1.7	0.85		
			10	5.2	4.3	2.1	0.8		
			15	7.2	5.8	2.5	0.7		
			20	10	8.3	2.6	0.65		
			30	15	10	3.1	0.45		
Wheat starch	Cotton (from Whatman filter paper)	50	2.1	–	3.6	–	1.4	2005	[65]
			5	–	6.7	–	3.1		
			10.3	–	9	–	3.3		
Modified potato starch	Wheat straw	50	2	1.55	1.4	1.2	–	2008	[70]
			5	1.8	1.75	1.35	–		
			10	2.75	2.45	1.75	–		
Pea starch/36 wt% glycerol	Hemp	43	5	–	1.1	1.15	0.85	2008	[59]
			10	–	3.5	1.55	0.75		
			15	–	5.3	1.75	0.45		
			20	–	7.6	1.9	0.3		
			30	–	25.8	2.95	0.1		
Pea starch/30 wt% glycerol	Pea hull	43	10	–	–	1.95	2.1	2009	[72]
Cassava starch/30 wt% glycerol	Cassava bagasse	53	5	1.1	1.4	1.4	2.6	2009	[68]
			10	2.1	1.5	1.6	2.6		
			20	2.1	1.6	1.4	1.8		
Maize starch/30 wt% glycerol	Wheat straw	75	5	1.05	1.4	1.1	–	2010	[71]
			10	2.1	2.3	1.2	–		
			15	3.4	2.95	1.4	–		

maize starch [71]/wheat straw nanofibers, or pea starch/hemp nanofibers [59] nanocomposite materials, Young's modulus, the tensile strength, and the storage modulus at the rubbery state were significantly improved with increasing filler content (Table 12.4). These results indicated an efficient stress transfer from matrix to filler [70], which was attributed to strong interfacial interactions between the cellulose crystallite surface and the starch matrix [59, 69], and also to the formation of a rigid network of nanofibers connected by hydrogen bonds [71]. The shift of T_g to higher temperatures (about 30–40°C higher in the case of glycerol plasticized potato starch/wheat straw nanofibers systems [70]) confirmed this hypothesis [70, 71]. For some systems, the elongation at break decreased at the same time [59, 69]. However, the reinforcing effect of cellulose whiskers is not necessarily accompanied by a decrease in the elongation at break, as observed for different glycerol plasticized systems [65, 66, 71, 72] (Table 12.4).

The reinforcing effect of starch nanocrystals will be discussed in the next section.

12.3 STARCH AS FILLER IN NANOCOMPOSITES

As reported by Le Corre et al. [73], starch nanocrystals were first used as reinforcements in elastomer-based matrices such as poly(styrene-co-butyl acrylate) [21, 36] and natural rubber (NR) [22, 23, 25, 29]. Increasing number of investigations related to starch nanocrystals was noted in recent literature (Fig. 12.5). Since 2006, most work has been oriented toward the use of bio-based polymers such as waterborne polyurethane (WPU) [34, 37], starch [26, 27, 74, 75], pullulan [28], poly(lactic acid (PLA) [31], poly(vinyl alcohol) (PVA) [32], poly(caprolactone) (PCL) [30], poly(butylene succinate) (PBS) [35], and soy protein isolate (SPI) [33]. The reinforcing effect of starch nanocrystals has been evaluated both in the linear range using DMA and in the nonlinear range through tensile tests. This section discusses the reinforcing effect of starch nanocrystals and the influence of surface modification on this reinforcing effect, as well as the mechanisms proposed to explain it.

12.3.1 Reinforcing Effect of Starch Nanocrystals

Dufresne and Cavaillé [36] were the first to report the reinforcing effect of starch nanocrystals in an elastomer-based matrix, especially in the rubbery state as demonstrated by DMA experiments. For instance, the relaxed modulus at the rubbery plateau of a poly(styrene-co-butyl acrylate)-based film containing 30 wt% of potato starch nanocrystals was 100 times higher than that of the matrix (Table 12.5) [36]. In the case of an NR matrix, Angellier et al. [22] reported an increase in the relaxed storage modulus of 10%, 75%, and 200% for 10 wt%, 20 wt%, and 30 wt% of waxy maize starch nanocrystals, respectively. In both cases, no increase of the main relaxation temperature associated

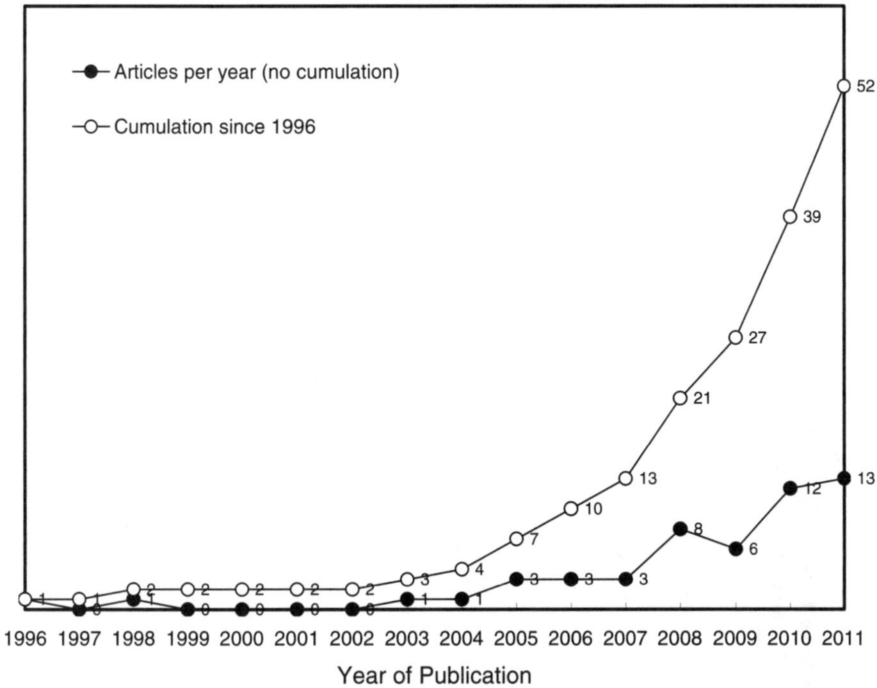

Figure 12.5 Evolution of the number of articles dealing with SNC (updated September 1, 2011).

with T_g was observed, which was an indication of poor adhesion between the filler and the elastomer-based matrix [22, 36]. Most authors further confirmed the reinforcing effect of starch nanocrystals for other types of matrices, such as thermoplastic starch [26, 27, 74], pullulan [28] and SPI [33]. In the case of thermoplastic starch either plasticized with glycerol [26] or sorbitol [27], the increase in filler content led to a shift of the main relaxation process toward higher temperatures (from 3°C for the unfilled matrix plasticized with 25 wt% of glycerol to 39.8°C for the same material filled with 15 wt% starch nanocrystals) due to a reduction in the mobility of amylopectin chains [26]. This demonstrates the presence of interactions between the filler and the matrix due to hydrogen bonding forces. Relaxation temperature increases with increasing filler content, which was attributed to filler/matrix interactions, was also evident in pullulan [28] and waterbone polyurethane matrices [37]. However, in the case of WPU matrix, no reinforcing effect of starch nanocrystals was evidenced by DMA experiments [37].

The reinforcing effect of starch nanocrystals was also evidenced by tensile tests in the nonlinear range. Relative values of Young's modulus, strength (or stress at break), and strain at break reported by different authors are listed in

TABLE 12.5 Relative Rubbery Storage Tensile Modulus (E'_R) and Relative Tensile Test Characteristics (Nominal Young's Modulus, E; Nominal Stress at Break, $\sigma_{,R}$; Nominal Strain at Break, $\varepsilon_{b,R}$) for Starch Nanocrystal-Filled Nanocomposites

Matrix	Starch Origin	RH (%)	Filler Content (wt%)	E'_R	E_R	$\sigma_{b,R}$	$\varepsilon_{b,R}$	Year	Ref.
Poly(styrene-co-butyl acrylate)	Potato starch	0	5	2	–	–	–	1998	[36]
			10	4	–	–	–		
			20	22	–	–	–		
			30	100	–	–	–		
			40	400	–	–	–		
			60	1000	–	–	–		
Poly(styrene-co-butyl acrylate)	Waxy maize	0	10	2	–	–	–	2005	[21]
			30	5	–	–	–		
			50	100	–	–	–		
Natural rubber	Waxy maize	0	2	–	1.5	1.2	0.95	2005	[22]
			5	2	2.5	1.9	0.91		
			10	10	7.5	3.0	0.90		
			15	–	19.2	3.1	0.81		
			20	75	40.8	3.9	0.76		
			25	–	52.8	3.2	0.67		
			30	200	119.7	2.4	0.49		
Thermoplastic waxy maize starch/25 wt% of glycerol	Waxy maize	43	2	–	4.8	3.3	0.41	2006	[26]
			5	1.7	7.3	3.6	0.32		
			10	4.7	7.5	4.2	0.19		
			15	7.2	22	9.8	0.07		
Thermoplastic waxy maize starch/25 wt% of sorbitol	Waxy maize	43	5	1.5	2.1	2.6	0.90	2007	[27]
			10	2.0	2.2	3.8	0.92		
			15	2.5	2.7	4.2	0.65		

Pullulan/30 wt% of sorbitol	Waxy maize	53	3	2	2.3	1.5	0.58	2007	[28]
			6	–	2.5	1.7	0.50		
			15	9	7	2.6	0.27		
			20	–	8.6	2.9	0.21		
			40	–	15.5	5.1	0.06		
Waterbone polyurethane	Potato starch	–	1	1	1.1	2.4	1.4	2008	[37]
			2	0.93	1.1	2.5	1.7		
			4	0.2	1.1	3	1.4		
			5	0.2	1.9	4.7	1.2		
			8	1.1	1.7	3.4	1.1		
Poly(vinyl alcohol)	Pea	43	5	–	–	1.1	1.04	2008	[32]
			10	–	–	1.1	1.03		
			15	–	–	0.9	1		
			20	–	–	0.8	0.9		
			30	–	–	0.7	0.8		
			40	–	–	0.6	0.75		
Thermoplastic cassava starch/33 wt% of glycerol	Waxy maize	43	2.5	3.9	–	–	–		[74]
Soy protein isolate/30 wt%glycerol	Pea	0	1		2.4	1.4	0.35	2009	[33]
			2		2.9	1.5	0.3		
			4	–	2.5	1.3	0.2		
			8		2.3	1.25	0.4		
			12	–	2	1	0.25		
			16		1.9	0.9	0.07		
Poly(butylene succinate)	Pea	0	5	–	1.02	1.03	1.2	2011	[35]
			10	–	1.12	0.85	1.1		
			15	–	1.15	0.55	0.6		
			20	–	1.2	0.46	0.3		

Table 12.5 for comparison. For most authors, the introduction of starch nanocrystals resulted in the increase of both the relaxed storage modulus and the stress at break, with a decrease in the strain at break [22, 26–28, 33]. In the case of NR, considering the ultimate properties, a good compromise between the increase in the strength and the decrease of the elongation at break seemed to be reached for a filler content around 20 wt% [22]. The tensile modulus increased nearly exponentially with filler content, from 0.64 MPa for the neat matrix up to 77.8 MPa for nanocomposite films filled with 30 wt%. These values are lower than the storage tensile modulus values reported from DMA experiments (respectively 1.28 and 257 MPa) at room temperature. Given that in DMA the adhesion between the filler and the matrix is less involved than in tensile tests because of weaker stresses applied, this result therefore confirms a lack of intimate adhesion between both components of the composite structure [22]. For thermoplastic starch, the particularly drastic decrease in the strain at break may be explained at least partially by the filler/matrix interactions, which were evidenced by DMA tests [26–28]. In the case of a protein-based matrix, a filler content of 2 wt% seemed to be the optimal content, with the highest strength and Young's modulus [33]. With increasing filler content, the strength and Young's modulus gradually decreased, but almost all nanocomposites maintained higher resistance and rigidity than the neat matrix (Table 12.5). At the same time, the addition of pea starch nanocrystals caused the decrease of the elongation at break. The reinforcement for low-starch nanocrystal loading levels was most likely attributed to the uniform distribution of starch nanocrystals as a stress-concentrated point in the SPI matrix. However, with an increase of starch nanocrystals content, the starch nanocrystals easily self-aggregated as large domains, which decreased the effective active starch nanocrystal surface for interacting with the SPI matrix [33]. The reinforcing effect of pea starch nanocrystals in a WPU matrix was quite singular [37] and might be explained by the range of studied filler contents (up to 8 wt% only). Compared with the neat WPU matrix, all nanocomposites showed a simultaneous enhancement in strength, Young's modulus, and elongation at break. However, a decrease in elongation at break was noted for filler contents higher than 2 wt%, whereas a decrease in Young's modulus and strength was noted for filler contents higher than 5 wt%. This was also attributed to the uniform dispersion of starch nanocrystals in nanoscale as well as to interfacial interactions between starch nanocrystals and the WPU matrix. With an increase in the loading level, XRD analysis confirmed that the self-aggregation of starch nanocrystals leads to a break in the original structure and interactions, and even causes severe microphase separation between pea starch nanocrystals and WPU matrix [37]. A little but significant simultaneous enhancement of the strength, toughness, and rigidity was also observed for PBS filled with pea starch nanocrystal loadings of up to 5 wt% [35]. For filler contents higher than 5 wt%, only Young's modulus was improved (Table 12.5). As already mentioned for SPI and WPU matrices, the incorporation of large amounts of nanofiller (above the percolation threshold) may lead to self-aggregation and separation between starch nanocrystal nanophase and PBS matrix, decreasing the effective active nanofiller surface available for interaction with the PBS molecules [35]. Finally, Chen et al. [32] attempted to compare the reinforcing effect of pea starch nanocrystals and native pea granules in a PVA matrix. Even if the reinforcing effect of starch nanocrystals was clearly better than the one of the native granules, it remained very poor since only a very little improvement of strength and strain at break (10% at best) was noted for filler contents lower than 10 wt%. This was explained by the poor compatibility of both components [32].

Globally, waxy maize starch nanocrystals showed a better reinforcing effect than pea starch nanocrystals, probably due to the higher size aspect ratio of the former (Table 12.5). Furthermore, the reinforcing effect of waxy maize starch nanocrystals is more significant in thermoplastic starch than in NR due to stronger interactions between the filler and the amylopectin chains. It is also possible that the reinforcing effect of starch nanocrystals in thermoplastic may be favored by an eventual crystallization at the filler/matrix interface [26].

12.3.2 Effect of Starch Nanocrystal Surface Chemical Modification

The first attempt to chemically modify the surface of starch nanocrystals consisted in grafting small molecules using the hydroxyl groups from the nanoparticle surface to broaden the number of possible compatible polymeric matrices. Indeed, the chemical modification can be used for transforming the polar hydroxyl group sitting at the surface of starch nanocrystals in moieties capable of enhancing interactions with nonpolar polymers. The two mechanisms chosen were esterification with the anhydride function of alkenyl succinic anhydride (ASA) and the formation of urethane using an isocyanate such as phenylisocyanate (PI) [24]. Modified starch nanocrystals were introduced in a NR matrix using a casting/evaporation technique involving toluene as the liquid phase. Their reinforcing effect was compared to the one reported for unmodified nanoparticles filled systems [22]. DMA experiments demonstrated that the reinforcing effect of chemically modified starch nanocrystals in the rubbery state of the NR matrix was considerably reduced, as well as the thermal stability of the materials. Tensile tests confirmed that the reinforcing effect of starch nanocrystals displayed through the tensile modulus was much lower for modified nanoparticles than for unmodified ones. The ultimate properties (strength and strain at break) were also significantly affected. This was ascribed to hindered interactions between chemically modified particles resulting from their coating with the grafting agent [22].

Although modification by chemical reaction of small molecules on starch nanocrystals improved the adhesion between the filler and the matrix, it was demonstrated that it resulted in a sharp decrease in mechanical performance owing to the partial or total destruction of the three-dimensional network in the nanocomposites [22, 23]. In this context, the strategy of "grafting onto" polymer chains was then applied to starch nanocrystals with various polymers such as poly(ethylene glycol) monobuthyl ether (PEGME) [76], poly(propylene glyceol) monobutyl ether (PPGBE) [77], poly(tetrahydrofuran) (PTHF) [77], polycaprolactone (PCL) [30, 77] and styrene [78]. However, there are still shortcomings such as low controllability and grafting efficiency, as well as difficulty in grafting long-chain polymers [35]. To our knowledge, only the work of Habibi and Dufresne [30] has been yet published concerning the incorporation of such chemically modified starch nanocrystals within a polymer matrix, PCL in that case, with the objective of modulating mechanical properties of the resulting nanocomposite materials. They demonstrated that nanocomposite materials processed from modified nanoparticles display both a high modulus and stress at break as well as an improved ductility as compared to unmodified nanocrystals-based composites, due to an intimate adhesion between the filler and the polymeric matrix and a better dispersion of the filler within the matrix [30].

The "grafting from" strategy has been recently proposed. By this method, the polymerization of monomers and chain propagation of polymers are induced, and ultimately the polymer chains are coupled with the starch nanocrystals. In the study of Yu et al. [31], poly(ε-caprolactone) was grafted onto the surface of pea starch nanocrystals via microwave-assisted ring-opening polymerization based on "grafting from" strategy, and the resultant nanoparticles were used as filler in poly(lactic acid) (PLA) as matrix. Authors reported that the process of grafting PCL chains destroyed the starch nanocrystals aggregates and, hence, produced nano-objects with a relatively uniform size. It resulted in a simultaneous enhancement in strength (from about 42 to 58 MPa) and elongation at break (from about 10% to 135%) together with a decrease in Young modulus (from about 2000 to 1500 MPa) for a filler content of 5 wt%. However, when the loading level was higher (≥10 wt%) ultimate properties were also dramatically decreased. This was attributed to the introduction of the PCL component, which provided flexibility to the nanocomposites, as well as the formation of an interfacial layer, which facilitated the stress transfer to rigid starch nanocrystals [31]. In continuation, PCL-grafted pea starch nanocrystals were incorporated into a WPU matrix [34]. Authors also noted that the lowest loading level of 5 wt% resulted in a maximum tensile strength (from about 31 to 40 MPa), but contrarily to PLA, the elongation at break remainedstable up to a filler content of 25 wt% while Young modulus increased strongly, with a maximum value for a filler content of 20 wt% (increase from about 1 up to 220 MPa for a filler content of 20 wt%) [34]. This enhancement was attributed to the uniform dispersion of grafted starch nanocrystals because of their nanoscale size, the increased entanglements mediated with grafted PCL chains, and the reinforcing function of rigid starch nanocrystals. With an increase of the filler content, the rubbery PCL component increased, inducing the self-aggregation of starch nanocrystals as crystalline domains, which impeded improvement in tensile strength and elongation at break, but significantly enhanced Young modulus [34].

12.3.3 Reinforcing Mechanisms

The reinforcing effect of starch nanocrystals is generally ascribed to the formation of a hydrogen bonded percolating filler network above a given starch content corresponding to the percolation threshold. This percolation phenomenon was evidenced in the case of NR reinforced with waxy maize starch nanocrystals from swelling experiments (toluene and water uptake) for which changing behavior was reported above a certain filler concentration [23]. In that case, the percolation threshold was estimated at 10 wt% [22]. The percolation mechanism has been verified by successive tensile experiments, which consisted of stretching the material to a given elongation, releasing the force, stretching again to a higher elongation, and repeating the process until the break of the material [20, 23]. It was shown that for the unfilled NR matrix, the tensile modulus continuously increased during successive cycles due to the strain-induced crystallization of NR, whereas for highly filled nanocomposite materials the tensile modulus decreased during the five first cycles [23]. This was ascribed to the progressive disruption of the continuous starch nanocrystals network. The establishment of a percolating network was further confirmed by the same authors. They showed that any deterioration of these filler–filler interactions, for example induced by the surface chemical modification of starch nanocrystals, results in a dramatic decrease of the mechanical performances of the ensuing composites [22, 23].

Recently, new highlights have been given concerning the reinforcing mechanisms of starch nanocrystals in the NR matrix [29]. The presence of both Mullins and Payne effects was demonstrated, even for low filler contents, suggesting that the strain dependence of the viscoelastic properties of filler nonvulcanized NR could not be only related to the development of filler–filler interactions [29]. Two phenomenological models were used to predict the Payne effect and thus discriminate the possible origins of the nonlinear viscoelastic behavior: (1) the Kraus model considering that filler–filler interactions are preponderant, and (2) the Maier and Göritz model, which is based on matrix–filler interactions [29]. The use of the Kraus model lied with the confirmation of the formation of a percolating network for filler contents higher than 6.7 vol% (10 wt%) [29]. However, the Kraus model was unable to predict the dissipative properties of the different nanocomposites. The use of the Maier and Goritz model allowed demonstrating that nonlinear viscoelastic properties were governed by phenomena of adsorption and desorption of NR chains on the filler surface, even if the formation of a percolating network for filler contents >6.7 vol % (10 wt%) was evidenced [29].

12.4 CONCLUSION

This chapter highlighted the promising use of starch as raw material for the production of biodegradable nanocomposite materials, either as matrix or as reinforcement in the form of nanocrystals. The development of nanocomposite structures by the introduction of either inorganic, such as layered silicates or carbon nanotubes, or organic nanofillers, such as cellulose or starch nanocrystals, within starch matrices allowed obtaining materials with enhanced mechanical properties. It is worth noting that the comparison of the reinforcing effect of the different fillers is quite difficult because of botanical sources of starch, types and contents of plasticizer, storage conditions, and processing methods, which differ according to the studies. However, some general tendencies can be given. Globally, organic fillers have a better reinforcing effect than inorganic fillers. This may be ascribed to the natural affinity between polysaccharide nanocrystals and starch matrices, allowing the establishment of hydrogen bonds between both components and a good dispersion of fillers within the matrix. In the case of inorganic fillers, some additional efforts including surface chemical modification should be carried out to improve their dispersion within the matrix and the adhesion between both components. Starch nanocrystals also appear as promising candidates for improving mechanical properties of polymer matrices in light of their platelet-like shape. The possibility of surface chemical modification of starch nanocrystals to improve the affinity between the filler and the polymer matrix opens large windows of applications. In conclusion, effectiveness of interactions at the interfacial region between the reinforcements and the polymer matrix continues to be a challenge to expand the use and commercialization of biodegradable nanocomposites based on starch.

REFERENCES

[1] Hulleman, S.H., Janssen, F.H.P., and Feil, H. (1998) The role of water during plasticization of native starches. *Polymer*, *39*(10), 2043–2048.

[2] Myllärinen, P., Partanen, R., Seppälä, J., and Forssell, P. (2002) Effect of glycerol on behaviour of amylose and amylopectin films. *Carbohydrate Polymers*, *50*(4), 355–361.

[3] Follain, N., Joly, C., Dole, P., and Bliard, C. (2005) Mechanical properties of starch-based materials. I. Short review and complementary experimental analysis. *Journal of Applied Polymer Science*, *97*, 1783–1794.

[4] Lourdin, D., Coignard, L., Bizot, H., and Colonna, P. (1997) Influence of equilibrium relative humidity and plasticizer concentration on the water content and glass transition of starch materials. *Polymer*, *38*(21), 5401–5406.

[5] Lourdin, D., Bizot, H., and Colonna, P. (1997) Antiplastization in starch-glycerol films? *Journal of Applied Polymer Science*, *63*(8), 1047–1053.

[6] Benczédi, D., Tomka, I., and Escher, F. (1998) Thermodynamics of amorphous starch-water systems. 1. Volume fluctuations. *Macromolecules*, *31*(9), 3055–3061.

[7] Anglès, M.N., Dufresne, A. (2001) Plasticized starch/tunicin whiskers nanocompositematerilas. 2. Mechanical behavior. *Macromolecules*, *34*, 2921–2931.

[8] Poutanen, K., Forssell, P. (1996) Modification of starch properties with plasticizers. *TRIP*, *4*(4), 128–131.

[9] Anglès, M.N., Dufresne, A. (2000) Plasticized starch/tunicin whiskers nanocomposites. 1. Structural analysis. *Macromolecules*, *33*(22), 8344–8353.

[10] Stading, M., Rindlav-Westling, A., and Gatenholm, P. (2001) Humidity-induced transitions in amylose and amylopectin films. *Carbohydrate Polymers*, *45*(3), 209–217.

[11] Wilhelm, H.M., Sierakowski, M.R., Souza, G., and Wypych, F. (2003) Starch films reinforced with mineral clay. *Carbohydrate Polymers*, *52*(2), 101–110.

[12] Van Soest, J.J.G., Hulleman, S.H.D., De Wit, D., and Vliegenthart, J.F.G. (1996) Changes in the mechanical properties of thermoplastic potato starch in relation with changes in B-type crystallinity. *Carbohydrate Polymers*, *29*(3), 225–232.

[13] Van Soest, J.J.G., De Wit, D., and Vliegenthart, J.F.G. (1996) Mechanical properties of thermoplastic waxy maize starch. *Journal of Applied Polymer Science*, *61*(11), 1927–1937.

[14] Kuutti, L., Peltonen, J., Myllärinen, P., Teleman, O., and Forssell, P. (1997) AFM in studies of thermoplastic starches during ageing. *Carbohydrate Polymers*, *37*(1), 7–12.

[15] Forssell, P.M., Hulleman, S.H.D., Myllaärinen, P.J., Moates, G.K., and Parker, R. (1999) Ageing of rubbery thermoplastic barley and oat starches. *Carbohydrate Polymers*, *39*(1), 43–51.

[16] Van Soest, J.J.G., Knooren, N. (1997) Influence of glycerol and water content on the structure and properties of extruded starch plastic sheets. *Journal of Applied Polymer Science*, *64*(7), 1411–1422.

[17] Alexandre, M., Dubois, P. (2000) Polymer/layered silicate nanocomposites. Preparation, properties and uses of a new class of materials. *Materials Science and Engineering*, *28*, 1–63.

[18] Coleman, J.N., Khan, U., Blau, W.J., and Gun'ko, Y.K. (2006) Small but strong: A review of the mechanical properties of carbon nanotube-polymer composites. *Carbon*, *44*, 1624–1652.

[19] Cao, X., Chen, Y., Chang, P.R., and Huneault, M.A. (2007) Preparation and properties of plasticized starch/multiwalled carbon nanotubes composites. *Journal of Applied Polymer Science*, *106*, 1431–1437.

[20] Dufresne, A. (2008) Polysaccharide nanocrystal reinforced nanocomposites. *Canadian Journal of Chemistry*, *86*(6), 484–494.

[21] Angellier, H., Putaux, J.L., Molina-Boisseau, S., Dupeyre, D., and Dufresne, A. (2005) Starch nanocrystal fillers in an acrylic polymer matrix. *Macromolecular Symposia*, *221*, 95–104.

[22] Angellier, H., Molina-Boisseau, S., and Dufresne, A. (2005) Mechanical properties of waxy maize starch nanocrystals reinforced natural rubber. *Macromolecules*, *38*, 9161–9170.

[23] Angellier, H., Molina-Boisseau, S., Lebrun, L., and Dufresne, A. (2005) Processing and structural properties of waxy maize starch nanocrystals reinforced natural rubber. *Macromolecules*, *38*(9), 3783–3792.

[24] Angellier, H., Molina-Boisseau, S., Belgacem, M.N., and Dufresne, A. (2005) Surface chemical modification of waxy maize starch nanocrystals. *Langmuir*, *21*, 2425–2433.

[25] Angellier, H., Molina-Boisseau, S., and Dufresne, A. (2006) Waxy maize starch nanocrystals as filler in natural rubber. *Macromolecular Symposia*, *233*, 132–136.

[26] Angellier, H., Molina-Boisseau, S., Dole, P., and Dufresne, A. (2006) Thermoplastic starch-waxy maize starch nanocrystals nanocomposites. *Biomacromolecules*, *7*, 531–539.

[27] Viguié, J., Molina-Boisseau, S., and Dufresne, A. (2007) Processing and characterization of waxy maize starch films plasticized by sorbitol and reinforced with starch nanocrystals. *Macromolecular Bioscience*, *7*, 1206–1216.

[28] Kristo, E., Biliaderis, C.G. (2007) Physical properties of starch nanocrystal-reinforced pullulan films. *Carbohydrate Polymers*, *68*, 146–158.

[29] Mélé, P., Angellier-Coussy, H., Molina-Boisseau, S., and Dufresne, A. (2011) Reinforcing mechanisms of starch nanocrystals in a non-vulcanized natural rubber matrix. *Biomacromolecules*, *12*, 1487–1493.

[30] Habibi, Y., Dufresne, A. (2008) Highly filled bionancomposites from functionalized polysaccharide nanocrystals. *Biomacromolecules*, *9*, 1974–1980.

[31] Yu, J., Ai, F., Dufresne, A., Gao, S., Huang, J., and Chang, P.R. (2008) Structure and mechanical properties of poly(lactic acid) filled with (starch nanocrystal)-graft-poly(ε-caprolactone). *Macromolecular Materials and Engineering*, *293*, 763–770.

[32] Chen, Y., Cao, X., Chang, P.R., and Huneault, M.A. (2008) Comparative study on the films of poly(vinyl alcohol)/pea starch nanocrystals and poly(vinyl alcohol)/native pea starch. *Carbohydrate Polymers*, *73*, 8–17.

[33] Zheng, H., Ai, F., Chang, P.R., Huang, J., and Dufresne, A. (2009) Structure and properties of starch nanocrystal-reinforced soy protein plastics. *Polymer Composites*, *30*(4), 474–480.

[34] Chang, P.R., Ai, F., Chen, Y., Dufresne, A., and Huang, J. (2009) Effects of starch nanocrystal-graft-polycaprolactone on mechanical properties of waterbone polyurethane-based nanocomposites. *Journal of Applied Polymer Science*, *111*, 619–627.

[35] Lin, N., Yu, J., Chang, P.R., Li, J., and Huang, J. (2011) Poly(butylene succinate)-based biocomposites filled with polysaccharide nanocrystals: Structure and properties. *Polymer composites*, *32*(3), 472–482.

[36] Dufresne, A., Cavaillé, J.-Y. (1998) Clustering and percolation effects in microcrystalline starch-reinforced thermoplastic. *Journal of Polymer Science Part B: Polymer Physics*, *36*, 2211–2224.

[37] Chen, G., Wei, M., Chen, J., Huang, J., Dufresne, A., and Chang, P.R. (2008) Simultaneous reinforcing and toughening: New nanocomposites of waterbone polyurethane filled with low loading level of starch nanocrystals. *Polymer*, *49*, 1860–1870.

[38] Lin, N., Huang, J., Chang, P.R., Anderson, D.P., and Yu, J. (2011) Preparation, modification and application of starch nanocrystals in nanomaterials: A review. *Journal of Nanomaterials.* doi: 10.1155/2011/573687.

[39] LeCorre, D., Bras, J., and Dufresne, A. (2011) Evidence of micro and nano-scaled particles during starch nanocrystals production and their isolation. *Biomacromolecules*, *12*, 3039–3046.

[40] LeCorre, D., Bras, J., and Dufresne, A. (2011) Ceramic membrane filtration for isolating starch nanocrystals. *Carbohydrate Polymers*, *86*, 1565–1572.

[41] Park, H.M., Li, X., Jin, C.Z., Park, C.Y., Cho, W.J., and Ha, C.S. (2002) Preparation and properties of biodegradable thermoplastic starch/clay hydrids. *Macromolecular Materials and Engineering*, *287*(8), 553–558.

[42] Park, H.M., Lee, W.K., Park, C.Y., Cho, W.J., and Ha, C.S. (2003) Environmentally friendly polymer hybrids. Part I. Mechanical, thermal, and barrier properties of thermoplastic starch/clay nanocomposites. *Journal of Materials Science*, *38*, 909–915.

[43] Huang, M.F., Yu, J.G., and Ma, X.F. (2004) Studies on the properties of montmorillonite-reinforced thermoplastic starch composites. *Polymer*, *45*, 7017–7023.

[44] Chen, B., Evans, J.R.G. (2005) Thermoplastic starch-clay nanocomposites and their characteristics. *Carbohydrate Polymers*, *61*, 455–463.

[45] Pandey, J.K., Singh, R.P. (2005) Green nanocomposites from renewable resources: Effect of plasticizer on the structure and material properties of clay-filled starch. *Starch/Stärke*, *57*(1), 8–15.

[46] Bagdi, K., Müller, P., and Pukanszky, B. (2006) Thermoplastic starch/layered silicate composites: Structure, interaction, properties. *Composite Interfaces*, *13*(1), 1–17.

[47] Kvien, I., Sugiyama, J., Votrubec, M., and Oksman, K. (2007) Characterization of starch based nanocomposites. *Journal of Materials Science*, *42*, 8163–8171.

[48] Kampeerapappun, P., Aht-ong, D., Pentrakoon, D., and Srikulkit, K. (2007) Preparation of cassava starch/montmorillonite composite film. *Carbohydrate Polymers*, *67*, 155–163.

[49] Tang, X., Alavi, S., and Herald, T.J. (2008) Barrier and mechanical properties of starch-clay nanocomposite films. *Cereal Chemistry*, *85*(3), 433–439.

[50] Cyras, V.P., Manfredi, L.B., Ton-That, M.T., and Vasquez, A. (2008) Physical and mechanical properties of thermoplastic starch/montmorillonitenanocomposite films. *Carbohydrate Polymers*, *73*, 55–63.

[51] Chivrac, F., Angellier-Coussy, H., Guillard, V., Pollet, E., and Averous, L. (2010) How does water diffuse in starch/montmorillonite nano-biocomposite materials? *Carbohydrate Polymers*, *82*, 128–135.

[52] Maksimov, R.D., Lagzdins, A., Lilichenko, N., and Plume, E. (2009) Mechanical properties and water vapour permeability of starch/montmorillonite nanocomposites. *Polymer Engineering and Science*, *49*(12), 2421–2429.

[53] Wilhelm, H.M., Sierakowski, M.R., Souza, G.P., and Wypych, F. (2003) The influence of layered compounds on the properties of starch/layered compound composites. *Polymer International*, *52*(6), 1035–1044.

[54] Dean, K., Yu, L., and Wu, D.W. (2007) Preparation and characterization of melt-extruded thermoplastic starch/clay nanocomposites. *Composites Science and Technology*, *67*, 413–421.

[55] Ren, P., Shen, T., Wang, F., Wang, X., and Zhang, Z. (2009) Study on biodegradable starch/OMMT nanocomposites for packaging applications. *Journal of Polymers and the Environment*, *17*, 203–207.

[56] Majdzadeh-Ardakani, K., Navarchian, A.H., and Sadeghi, F. (2010) Optimization of mechanical properties of thermoplastic starch/clay nanocomposites. *Carbohydrate Polymers*, *79*, 547–554.

[57] Chivrac, F., Pollet, E., and Avérous, L. (2009) Progress in nano-biocomposites based on polysaccharides and nanoclays. *Materials Science and Engineering R*, *67*, 1–17.

[58] Chivrac, F., Gueguen, O., Pollet, E., Ahzi, S., Makradi, A., and Averous, L. (2008) Micromechanical modelling and characterization of the effective properties in starch-based nano-biocomposites. *Acta Biomaterialia*, *4*, 1707–1714.

[59] Cao, X., Chen, Y., Chang, P.R., Stumborg, M., and Huneault, M.A. (2008) Green composites reinforced with hemp nanocrystals in plasticized starch. *Journal of Applied Polymer Science*, *109*, 3804–3810.

[60] Liu, Z., Zhao, L., Chen, M., and Yu, J. (2011) Effect of carboxylate multi-walled carbon nanotubes on the performance of thermoplastic starch nanocomposites. *Carbohydrate Polymers*, *83*, 447–451.

[61] Ma, X., Yu, J., and wang, N. (2008) Glycerol plasticized-starch/multiwall carbon nanotube composites for electroactive polymers. *Composites Science and Technology*, *68*, 268–273.

[62] Fama, L.M., Pettarin, V., Goyanes, S.N., and Bernal, C.R. (2011) Starch/multi-walled carbon nanotubes composites with improved mechanical properties. *Carbohydrate Polymers*, *83*, 1226–1231.

[63] Dufresne, A., Vignon, M. (1998) Improvement of starch film performances using cellulose microfibrils. *Macromolecules*, *31*(8), 2693–2696.

[64] Dufresne, A., Dupeyre, D., and Vignon, M.R. (2000) Cellulose microfibrils from potato tuber cells: Processing and characterization of starch-cellulose microfibril composites. *Journal of Applied Polymer Science*, *76*, 2080–2092.

[65] Orts, W.J., Shey, J., Imam, S.H., Glenn, G.M., Guttman, M.E., and Revol, J.F. (2005) Application of cellulose microfibrils in polymer nanocomposites. *Journal of Polymers and the Environment*, *13*(4), 301–306.

[66] Mathew, A.P., Thielemans, W., and Dufresne, A. (2008) Mechanical properties of nanocomposites from sorbitol and tunicin whiskers. *Journal of Applied Polymer Science*, *109*, 4065–4074.

[67] Favier, V., Canova, G.R., Cavaillé, J.Y., Chanzy, H., Dufresne, A., and Gauthier, C. (1995) Nanocomposite materials from latex and cellulose whiskers. *Polymers for Advanced Technologies*, *6*(5), 351–355.

[68] Teixeira, E., Pasquini, D., Curvelo, A.A.S., Corradini, E., Belgacem, M.N., and Dufresne, A. (2009) Cassava bagasse cellulose nanofibrils reinforced thermoplastic cassava starch. *Carbohydrate Polymers*, *78*, 422–431.

[69] Lu, Y.S., Weng, L.H., and Cao, X.D. (2005) Biocomposites of plasticized starch reinforced with cellulose crystallites from cottonseed linter. *Macromolecular Bioscience*, *5*, 1101–1107.

[70] Alemdar, A., Sain, M. (2008) Biocomposites from wheat straw nanofibers: Morphology, thermal and mechanical properties. *Composites Science and Technology*, *68*, 557–565.

[71] Kaushik, A., Singh, M., and Verma, G. (2010) Green nanocomposites based on thermoplastic starch and steam exploded cellulose nanofibrils from wheat straw. *Carbohydrate Polymers*, *82*, 337–345.

[72] Chen, Y., Liu, C., Chang, P.R., Cao, X., and Anderson, D.P. (2009) Bionanocomposites based on pea starch and cellulose nanowhiskers hydrolyzed from pea hull fibre: Effect of hydrolysis time. *Carbohydrate Polymers*, *76*, 607–615.

[73] Le Corre, D., Bras, J., and Dufresne, A. (2010) Starch nanoparticles: A review. *Biomacromolecules*, *11*, 1139–1153.

[74] Garcia, N.L., Ribba, L., Dufresne, A., Aranguren, M., and Goyanes, S. (2009) Physico-mechanicalproperties of biodegradable starch nanocomposites. *Macromolecular Materials and Engineering*, *294*, 169–177.

[75] Garcia, N.L., Ribba, L., Dufresne, A., Aranguren, M., and Goyanes, S. (2011) Effect of glycerol on the morphology of nanocomposites made from thermoplastic starch and starch nanocrystals. *Carbohydrate Polymers*, *84*, 203–210.

[76] Thielemans, W., Belgacem, M.N., and Dusfresne, A. (2006) Starch nanocrystals with large chain surface modifications. *Langmuir*, *22*(10), 4804–4810.

[77] Labet, M., Thielemans, W., and Dufresne, A. (2007) Polymer grafting onto starch nanocrystals. *Biomacromolecules*, *8*(9), 2916–2927.

[78] Song, S., Wang, C., Pan, Z., and Wang, X. (2008) Preparation and characterization of amphiphilic starch nanocrystals. *Journal of Applied Polymer Science*, *107*, 418–422.

[79] Reilly, R.M. (2007) Carbon nanotubes: Potential benefits and risks of nanotechnology in nuclear medicine. *Journal of Nuclear Medicine*, *48*(7), 1039–1042.

[80] Harris, P.J.F. (2004) Carbon nanotube composites. *International Materials Reviews*, *49*(1), 31–43.

[81] Ebeling, T., Paillet, M., Borsali, R., Diat, O., Dufresne, A., Cavaillé, J.Y., and Chanzy, H. (1999) Shear-induced orientation phenomena in suspensions of cellulose microcrystals, revealed by small angle X-ray scattering. *Langmuir*, *15*(19), 6123–6126.

CHAPTER 13

Applications of Starch Nanoparticles and Starch-Based Bionanocomposites

SIJI K. MARY, LALY A. POTHAN, and SABU THOMAS

13.1 INTRODUCTION

Nanoscience and nanotechnology have already been applied in various fields, such as computer electronics, communication, energy production, medicine, and food processing and packaging. The nanoscale devices are often manufactured with the view of imitating the nanodevices found in nature, which include proteins, DNA, membranes, and other natural biomolecules. In today's world, food materials are often considered not only a source of nutrients but also as having to contribute to the health of consumers [1, 2].

Synthetic polymeric materials have been widely used in all areas of human activity. The disposal of a high volume of plastics, which take a long time to decompose, poses a huge environmental problem. With increased environmental awareness in society, there is growing recognition of the need to reduce the quantity of plastic materials discarded. Therefore, intense research has been carried out in the areas of engineering and biological sciences to develop polymeric materials that are easily disposable but not environmentally harmful, particularly for use in the packaging industry. Starches are considered one of the most promising available natural polymer candidates for the development of biodegradable materials [3].

Natural resources such as starch, cellulose, and chitin are used as both the polymer matrix and the nanoscale reinforcement agent. Most of the nanoparticles used traditionally belong to the group of colloids. Owing to the

Biopolymer Nanocomposites: Processing, Properties, and Applications, First Edition. Edited by Alain Dufresne, Sabu Thomas, and Laly A. Pothan.

higher surface area of nanoparticles per mass unit, they are expected to be more biologically active than larger sized particles of the same chemical composition [4].

Starch is probably the most promising for use as a biodegradable polymer using renewable resources. A versatile biopolymer, it has great potential, low price, and abundant availability, not dependent on fossil sources, for use in the nonfood industries. The most commercially important sources of starch are corn, wheat, rice, potatoes, tapioca, and peas. Starch has received considerable attention during the past three decades as a biodegradable thermoplastic polymer or as biodegradable particulate filler. Although it is not truly thermoplastic, but by mixing with sufficient water or nonaqueous plasticizer (generally polyols, such as glycerol), it can be converted into a continuous polymeric entangled phase. The resulting material can be manufactured using technology already developed for the production of synthetic plastic [5]. Great progress has been achieved in the development of biodegradable products based on polysaccharides such as cellulose, chitin, and starch, which are widely available in nature and constitute an important class of biopolymers. Among them, starch is a renewable carbohydrate polymer procurable at low cost from a great variety of crops. Native starch commonly exists in a granular structure, which can be processed into plasticized starch materials [6]. The starch industry extracts and refines starches by wet grinding, sieving, and drying. Starch is used as extracted from the plant and is called "native starch," or it undergoes one or more chemical modifications to reach specific properties and is called "modified starch."

Nanofillers have strong reinforcing effects in polymers. However, for decades, studies have been conducted with nonrenewable inorganic fillers and a petroleum-based matrix. Increasing environmental concerns have led to developing new flexible barrier biobased packaging and investigating the potential uses of renewable resources for such an application. However, the usage of these materials has been limited by their poor performance, such as brittleness and poor gas and moisture barrier. Polysaccharides are good candidates for renewable nanofillers because they have partly crystalline structures conferring interesting properties [7].

The predominant model for starch is a concentric semicrystalline multiscale structure that allows the production of new nanoelements: (1) starch nanocrystals resulting from the disruption of amorphous domains from semicrystalline granules by acid hydrolysis and (2) starch nanoparticles produced from gelatinized starch. Recent studies have shown that starch nanoparticles could be used as fillers to improve mechanical and barrier properties of biocomposites. Their use for industrial packaging, with the ongoing search for innovative solutions for efficient and sustainable systems, is being investigated. Therefore, recently, starch nanoparticles have been the focus of an exponentially increasing number of works devoted to developing biocomposites by blending starch nanoparticles with different biopolymeric matrices [7].

13.2 APPLICATIONS OF STARCH NANOPARTICLE

A large number of organic and inorganic materials and lignocellulosic fibers have been tested as potential sources of nanoparticles and nanofibers to be used as nanofillers. These materials have the advantage of being renewable, which means they have a much smaller environmental impact than their synthetic equivalents. The field of reinforced polymer materials with polysaccharide nanofillers has seen rapid advances and gained considerable interest in the last decade owing to the renewable character, high mechanical properties, low density, and diversity of the sources [8]. Starch nanoparticles are receiving attention due to the abundant availability of starch and its low cost, renewability, biocompatibility, biodegradability, and nontoxicity. Starch nanocrystals obtained by acid hydrolysis of starch have been used as fillers in natural and synthetic polymeric matrices and as reinforcing agents [8]. Nanoparticles (fillers) enhance not only mechanical properties, but also physical properties, such as permeability or fire retardancy. Their properties depend on the nature and effectiveness of interactions at the interfacial region, that is, on both the surface area and the dispersion of the particles [7]. The choice of the matrix depends on several parameters such as the application, the compatibility between components, the process, and the cost. Starch is the second most studied organic material for producing nanocrystals. Among the components of bionanocomposites, the nanometer-sized biofillers from biomass show unique advantages over traditional inorganic nanoparticles by virtue of their biodegradability and biocompatibility.

13.2.1 Starch Nanofillers Used in Natural Polymer Matrices

Carbon black (manufactured by burning oil or natural gas in controlled conditions) is the most important reinforcing agent used in the rubber industry. The attempt to replace carbon black in rubber compounds with other reinforcing agents has been the main focus of researchers in the past decade. This is because carbon black causes pollution and gives a black color to the rubber. Kaolin and silica were commonly used as reinforcing agents, but their reinforcing properties are lower than those obtained with carbon black. Due to the ability of the clay silicate layers to disperse into polymer rubber matrices, a variety of clays have been used to obtain unusual nanocomposites. Clay minerals such as montmorillonite and organoclays seem to be a potential substitute to carbon black. Recently, Novamont (Novara, Italy), working in partnership with Goodyear Tire and Rubber, has developed tires using nanoparticles derived from cornstarch, partially replacing the conventional carbon black and silica used in making tires [8]. Both native starch and starch nanoparticles are found to induce excellent reinforcement in natural rubber. The main hurdle for the use of starch as a reinforcing phase is its hydrophilicity. Hence, there is a growing interest in organically modified derivatives of polysaccharides for different applications. Starch nanocrystals obtained by acid hydrolysis of potato and waxy maize starch granules have been used as a filler in a ynthetic polymeric matrix and appear to be an interesting reinforcing agent. Nanocomposite materials were obtained using a latex of natural rubber as the matrix and an aqueous suspension of waxy maize starch nanocrystals as the reinforcing phase. Starch nanocrystals were obtained after sulfuric acid hydrolysis of waxy maize starch granules. They consisted of crystalline platelets 6–8 nm thick, 40–60 nm long, and 15–30 nm wide. After mixing of the latex and the starch nanocrystals, the resulting aqueous suspension was cast and evaporated [9].

Chang et al. prepared bionanocomposites of citric acid modified starch nanoparticle (CASN)–glycerol plasticized pea starch (GPS) by using CASN as the filler and GPS as the matrix. They synthesized CASN by cross-linking starch in starch nanoparticles with citric acid (CA). Most of CASN ranged in size from about 50 to 100 nm. CASN was in the amorphous phase, which could not be gelatinized in hot water even at a high temperature (95°C), because of the cross-linking. The nanocomposites were prepared using CASN as the filler in the GPS matrix by the casting process. Scanning electron microscopy (SEM) revealed that CASN was dispersed evenly in the GPS matrix. The introduction of CASN could improve the storage modulus and the glass transition temperature of CASN/GPS composites. The work is finally focused on processing and characterization of CASN/GPS nanocomposites in terms of the morphology, mechanical properties, dynamic mechanical thermal analysis, and water vapor permeability. CASN was dispersed evenly in the GPS matrix and remained at their original size of CASN without obvious aggregation. As the filler of the GPS matrix, CASN had an obvious reinforcement effect on the tensile strength and Young's modulus, which was attributed to the strong interaction because of chemical similarities between CASN and pea starch in the GPS matrix. The introduction of CASN could improve the storage modulus and the glass transition temperature of the composites and water vapor barrier in comparison with pure GPS. These starch (starch derivatives/GPS) nanocomposites will have potential application in the medical, agriculture, drug release, and packaging fields, such as edible films, food packaging, and one-off packaging. In addition, CASN could be also used as the filler for the matrix of other natural polysaccharides (agar, alginate, and chitin) [6].

In the processing of starch nanoparticles (SN), native granular corn starch (Fig. 13.1a) was gelatinized in water and formed a starch paste. The preparation of SN in water by dropwise addition of ethanol resulted in the nanoparticles (Fig. 13.1b). Most of SN ranged in size from 50 to 300 nm. It was known that the stability of the nanoparticle suspensions was an important prerequisite for preparing these novel nanoparticles. The interaction of hydrogen bonds between SN and starch paste plays an important role in the stability of the precipitated SN in the suspension. The reaction of SN and CA seemed to decrease the aggregation of SN and large particles of SN were reduced, as

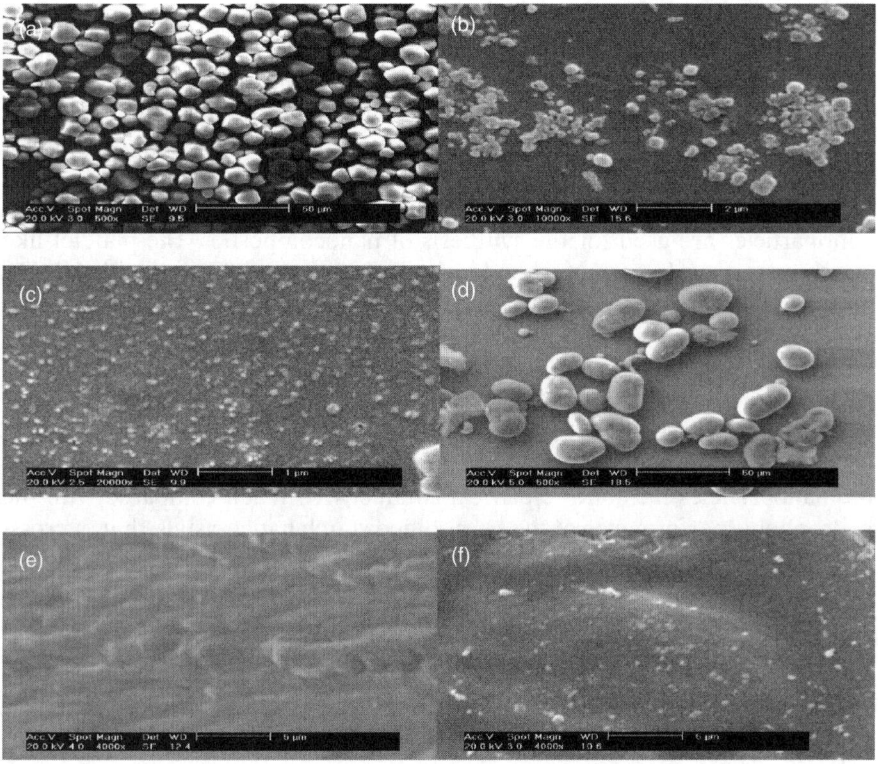

Figure 13.1 SEM micrograph of corn starch (×500, a), SN (×10,000, b), CASN (×20,000, c), pea starch (×500, d) powders, the fragile fractured surface of CASN/GPS composites containing 0 wt% CASN (×4000, e), and 3 wt% CASN (×4000, f). Reprinted with permission from Reference [6]. Copyright 2008 American Chemical Society.

shown in Figure 13.1c. Most of CASN ranged in size from about 50 to 100 nm. As shown in Figure 13.1d,e, no residual granular structure of pea starch was observed in the continuous GPS phase. At high temperature, water and glycerol are known to physically break up the granules of pea starch and disrupt intermolecular and starch intramolecular hydrogen bonds and make the native starch plastic. CASN was dispersed well in the GPS matrix and remained at their original size without obvious aggregation, which was attributed to the strong interaction because of chemical similarities between CASN and pea starch in the GPS matrix (Fig. 13.1f) [6].

13.2.2 Starch Nanofillers Used in Synthetic Polymer Matrix

Chen et al. studied the effects of low levels of starch nanocrystals in modifying waterborne polyurethane (WPU). WPU is regarded as a nontoxic, nonflammable, and biodegradable material. Environmentally friendly WPU can be applied to leather and textile finishing, floor coverings, adhesives, pressure-sensitive adhesives, and so on. Starch, soy protein, and lignin have been incorporated into WPU for the purposes of reducing cost, improving biodegradability, and enhancing mechanical performance. Synergistic reinforcement of WPU by both starch nanocrystals and cellulose whiskers is one interesting and innovative approach [10, 11].

1,4-Hexamethylene diisocyanate (HMDI)-cross-linked insoluble starch nanoparticles are used for the synthesis of nanocomposites. The platelet-like starch nanocrystals become pseudospherical after modification with HMDI, and the size increases or decreases depending on diisocyanate concentration compared with the ungrafted particles, as revealed by transmission electron microscopy (TEM) results. Because iisocyanates are highly reactive compounds, reactions take place easily under mild ambient conditions. Among various isocyanates, HMDI has a flexible structure with an aliphatic chain of six methylene groups. The high degree of cross-linking and effective filling by the nanoparticles facilitates electrical conductivity. When compared with the hydrophobic performance of the unmodified starch nanocrystals, that of cross-linked starch nanocrystals significantly increased. X-ray diffraction reveals that the crystalline structure of modified starch nanocrystals was preserved. The resulting hydrophobic starch nanoparticles are versatile precursors to the development of nanocomposites (Fig. 13.2).

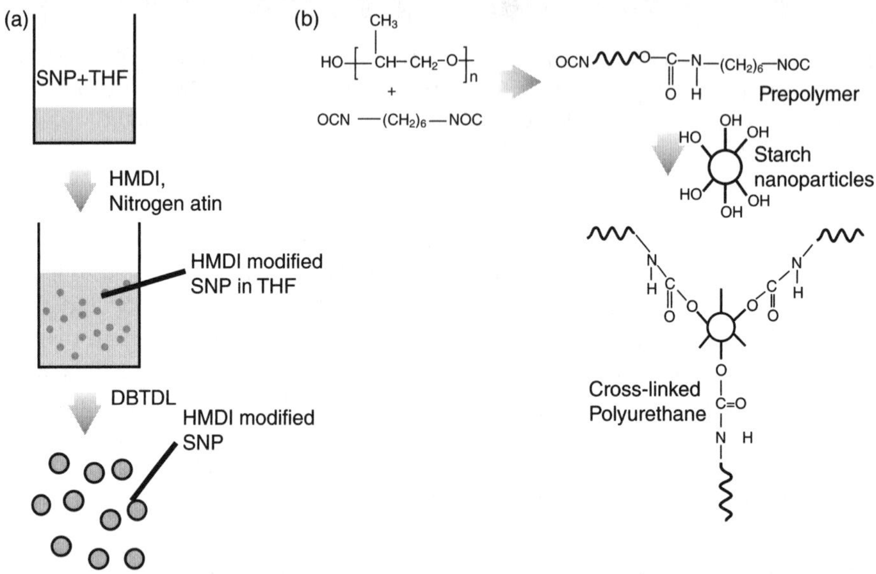

Figure 13.2 An illustration of formation of (a) HMDI-modified starch nanoparticles and (b) photoplethysmography-HMDI-modified starch nanoparticles polyurethane nanocomposite. For simplicity only 3 –OH groups are shown to have reacted. Reprinted with permission from Reference [12].

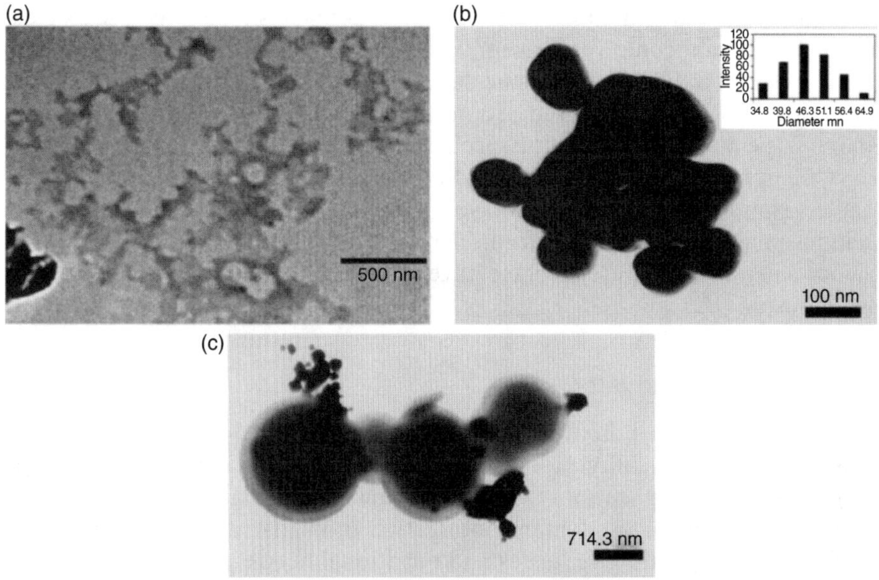

Figure 13.3 TEM images of (a) starch nanocrystals and (b, c) HMDI-modified starch nanoparticles. Particle size distribution of HMDI-modified starch nanoparticles (inset of b). Reprinted with permission from Reference [12].

Figure 13.3 shows TEM images of nanocrystals before (Fig. 13.3a) and after modification (Fig. 13.3b) [12].

13.3 APPLICATIONS OF STARCH-BASED COMPOSITES

Bionanocomposites are materials with a biological origin that are biodegradable. These materials serve a number of important functions, such as extending food shelf life and enhancing food quality, because they not only are barriers to moisture, water vapor, gases, and solutes, but also are carriers of some active substances, such as antioxidants and antimicrobials.

In recent years the development of biodegradable materials from renewable resources with excellent mechanical and barrier properties has been an important research challenge for the plastics industry. The main types of plastics that are currently used in all applications are derived from nonrenewable petroleum resources, and result in nonbiodegradable plastic materials. Thus, the volume of plastics discarded annually creates a substantial waste disposal issue, and alternative methods to reduce the quantity of persistent plastic wastes are required. One approach is to make materials from biodegradable materials that can be disposed of through composting or feedstock recycling.

A number of biodegradable materials have been investigated for use as plastics [13]. Bionanocomposites are mixtures of polymers with nanosized inorganic or organic fillers with particular size, geometry, and surface chemistry properties. The polymers used are normally hydrocolloids, such as proteins, starches, pectins, and other polysaccharides. Various inorganic nanoparticles have been recognized as possible additives to enhance the polymer performance. Nanofillers include solid layered clays, synthetic polymer nanofibers, cellulose nanowhiskers, and carbon nanotubes. Up to now only layered inorganic solids such as layered silicate have attracted the attention of the packaging industry.

13.3.1 In Food Industry

Nanotechnology has potential applications in all aspects of the food industry, including storage, quality monitoring, food processing, and food packaging. Nanotechnology applications in the food industry range from intelligent packaging to creation of on-demand interactive food that allows consumers to modify food, depending on tastes and nutritional needs.

Packaging plays a variety of important roles in the food industry. The major role of packaging is to protect food from spoilage by microbial contamination, physical damage, or biochemical reactions. Packaging also provides ease in handling, storage efficiency, attractiveness, and product information for food. The ideal food packaging material serves all of these purposes and is also cost-efficient. Different types of materials, including plastics, cardboard, and metal, are used for food packaging depending on specific needs. The use of plastics in food packaging, including films, is common and increasing because of low cost and functional advantages over other materials [14–16]. Although plastics are one of the cheapest sources available for food packaging, their reliance on petroleum and their long-term impact on the environment have spurred research in recent years on alternative packaging based on renewable and biodegradable materials. Starch is one such inexpensive, abundantly available, and renewable material that can be used for making biodegradable packaging films. Bionanocomposites are hybrid nanostructured materials with improved mechanical, thermal, and gas barrier properties. The use of bionanocomposites for food packaging not only protects the food and increases its shelf life but can also be considered a more environmentally friendly solution because it reduces the requirement to use plastics as packaging materials (Fig. 13.4). Most of the traditional packaging materials are made from nondegradable materials, which increase environmental pollution [17, 18].

To improve the properties of starch-based films, researchers have blended starch with other polymers such as polyhydroxyalkanoates, poly(lactic acid) (PLA), and poly(vinyl alcohol) (PVOH) [17, 18]. An innovative approach to improve the mechanical and barrier properties of polymer films is to produce nanocomposites by adding nanoscale particles. Nanocomposites of starch, PVOH, and sodium montmorillonite (Na^{+}MMT) were produced by solution

Figure 13.4 Toast packaged with a biodegradable film based on cassava starch formulated with glycerol, sucrose, and inverted sugar as plasticizers. Reprinted with permission from Reference [27]. See color insert.

mixing and cast into films. Clay minerals are a diverse class of layered silicates that have been used for producing nanocomposites with polymers. Starch is compatible with Na^+MMT due to the interaction between its polar hydroxyl groups and the Na^+ ions of nanoclay. This results in well-intercalated or exfoliated nanocomposites, which help in improving the mechanical and barrier properties of starch films. PVOH is also highly compatible with Na^+MMT, and films made from these nanocomposites exhibit better mechanical and barrier properties than films from PVOH alone [19–22].

Starch and its derivatives are important natural polymers that could be further improved through nanotechnology approaches. After extrusion, the starch is typically converted to a thermoplastic material with low mechanical resistance and poor protection against oxygen and moisture. Park et al. prepared hybrids of thermoplastic starch (TPS) with nanoclay and investigated the resulting properties. They found that the strong interaction between the TPS and the nanoclay improved tensile strength and lowered water vapor permeability compared with native TPS matrix [22–25].

Cassava starch has been extensively used to produce biodegradable films, and these carbohydrates are promising materials in this regard. Films developed from starch are described as isotropic, odorless, tasteless, colorless, nontoxic, and biologically degradable [26, 27]. The association of cassava starch with plasticizers as glycerol, sucrose, and inverted sugar can promote alterations in the films, justifying the study of these additives to develop a potential and ecological alternative to the synthetic packaging of several food products. Tadini et al. studied the influence of glycerol and nanoclay particles on tensile strength and percent elongation at break, barrier properties (water vapor permeability and oxygen permeability coefficient), and glass transition temperature of biodegradable films based on cassava starch. The results establish that films based on plasticized cassava starch reinforced with clay nanoparticles can be considered as an interesting biodegradable alternative packaging material [27].

As a natural biopolymer, besides its biodegradable character, starch would be a promising alternative for the development of new food packaging materials because of its attractive combination of availability and price [27].

13.3.2 Biomedical Applications

Studies with biodegradable starch-based polymers have recently demonstrated that these materials have a range of properties that make them suitable for use in several biomedical applications, ranging from bone plates and screws to drug delivery carriers and tissue engineering scaffolds [28]. The development of new processing technique and the reinforcement with various fillers results in materials with mechanical properties matching those of bone. The performance of a medical device is controlled by two sets of characteristics, those that determine the ability of a device to perform the appropriate and specific function and those that determine the compatibility of the material within the body: biofunctionality and biocompatibility. As such, the approach in the assessment of material biocompatibility encompasses the evaluation of the effects of physiological environments on materials and the materials' effects on the environment. Marques et al. [28] studied the biocompatibility of starch-based polymers. The biocompatibility of two different blends of corn starch, starch/ethylene vinyl alcohol (SEVA-C) and starch/cellulose acetate (SCA), and their respective composites with hydroxyapatite (HA) was assessed by cytotoxicity and cell adhesion tests. Both types of starch-based polymers exhibit a cytocompatibility that might allow for their use as biomaterials. SEVA-C blends were found to be the less cytotoxic for the tested cell line, although cells adhere better to SCA surface. The cytotoxicity test also revealed that SCA and SEVA-C composites have a similar response to the one obtained for SCA polymer [28–30].

Biocompatible and biodegradable drug delivery systems can preferably be made of naturally occurring polymers such as chitosan, gelatin, polysaccharides, and silk fibroin. The major limitation of current anticancer drugs is their toxicity and lack of specificity [31]. In an attempt to overcome these problems, endocytosable carriers such as microsphere, liposomes, and nanoparticles have been suggested to make the treatment more effective. Chemotherapy with

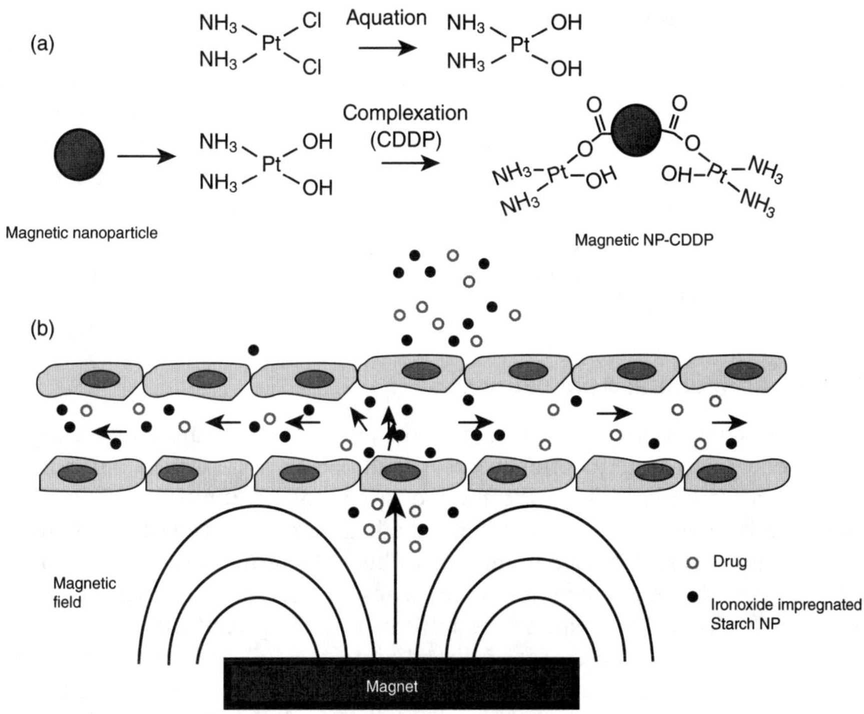

Figure 13.5 Schematic presentation of (a) binding of cisplatin drug molecules to the iron oxide-impregnated starch nanoparticles and (b) possible targeting of cells by released drug due to applied magnetic field. Reprinted with permission from Reference [31].

cisplatin is associated with various secondary effects, such as anemia, nausea, vomiting, neurotoxicity, and nephrotoxicity. Due to these side effects, alternative methods of administering cisplatin are needed. Starch has been selected as the raw material to prepare the drug carrier because of its biocompatibility, biodegradability, and inert nature. Bajpai et al. [31] synthesized and characterized cisplatin containing iron oxide-impregnated starch nanoparticles as a possible and potential drug carrier for magnetically mediated targeted drug delivery (see Fig. 13.5a,b). Figure 13.5a illustrates the binding of cisplatin molecules to the iron oxide-impregnated starch nanoparticles, whereas Figure 13.5b visualizes its possible applications in targeting infected cells by applying external magnetic field to enable the release of the drug from the drug-loaded nanoparticles.

Iron oxide-impregnated starch nanoparticles were prepared by the emulsion cross-linking method, which effectively delivers the antitumor drug cisplatin in the presence and absence of magnetic field via a diffusion-controlled pathway. The structural characterization of prepared nanoparticles by Fourier transform infrared (FT-IR) spectral analysis confirms the presence of functional groups of starch, iron oxide-impregnated starch nanoparticles, cisplatin-loaded nanoparticles, and pure drug. The influence of factors such as chemical composition of nanoparticles, pH, temperature of the release media, and applied magnetic field was investigated for the release profiles of the drug. The prepared nanoparticles could provide a possible pathway for targeted and controlled delivery of anticancer drugs, minimizing side effects and achieving higher efficacy [31–33].

13.3.3 Agricultural Applications

The world consumption of plastic materials in agriculture amounts yearly to 6.5 million tons, used to improve crop cultivation and protect agricultural products after harvesting, in the form of greenhouses, tunnels, mulch, silage films, and bale wraps. Therefore, there is a huge amount of plastics discarded into the environment, buried in the soil, or burnt by farmers, releasing harmful substances with the associated negative consequences to the environment. A solution to this problem can be the introduction in agriculture of biodegradable films, which can be disposed directly into the soil or into a composting system at the end of their lifetime. When biodegradable raw materials are mixed with some nanograde additives such as TiO_2, layered silicate, and MMT, bionanocomposites or films result in many advantages:

- biodegradation in the soil due to the action of microorganisms such as bacteria, fungi, and algae
- degradation by sunlight and water
- carriers for pesticide resistance
- controlled release of active substances, such as insecticides
- controlled degradation, such as biopolymer/TiO_2 nanofilms
- enhanced durability and optical properties, such as ultraviolet, visible, infrared
- annually renewable resources.

Up to now, degradable or biodegradable materials have been extensively tested in field trials in different countries and for different crops. Successful results have been obtained such as in maize, melon, strawberries, and cotton mulch film applications. The use of biodegradable materials in agriculture can promote sustainable and environmentally friendly cultivation, reducing the contamination of the soil, enhancing the protection of the landscape in rural areas against pollution, and increasing the use of renewable nonoil raw materials such as starch. At the end of their lifetimes, these biodegradable materials can be safely buried in the soil for final composting [13].

13.4 CONCLUSION

Nanocomposites and nanoparticles from starch are a new generation of polymers emerging into every aspect of our lives. They show great promise for potential applications as high performance biodegradable materials, which are entirely new types of materials based on plants, animals, and other natural products. Starch-based bionanocomposites have various applications in fields such as the food, biomedical, and agricultural industries. We believe that the next generation of packaging materials will be to fit the requirements of preserving fruits, vegetables, beverages, wines, chocolate, and other foods. By adding appropriate nanoparticles, it will be possible to produce packages with stronger mechanical, barrier, and thermal performance. Regarding food safety, nanostructured materials will prevent the invasion of bacteria and microorganisms. Embedded nanosensors in the packaging will alert the consumer if a food has gone bad. Biofunctionality and biocompatibility of biodegradable starch-based polymers have recently demonstrated that these materials have a range of properties that make them suitable for use in biomedical applications as well. In agriculture, the use of biodegradable materials can promote sustainable and environmentally friendly cultivation and reduce the contamination of the soil and pollution of landscape in rural areas. For renewable polymer-based bionanocomposites to meet a wide range of applications, bionanocomposite formulation must be further researched and modified so that mechanical and other properties can be easily manipulated, depending on the end-users' requirements. In all, natural polymer-based film materials originating in controlled bionanocomposites pave the way to a much broader range of applications and open a new dimension for plastics and composites in the future.

REFERENCES

[1] Sanguansri, P., Augustin, M.A. (2006) Nanoscale materials development—A food industry perspective. *Trends in Food Science & Technology*, *17*, 547–556.

[2] Fendler, J.H., Korean, J.H. (2001) Colloid chemical approach to nanotechnology. *Journal of Chemical Engineering*, *18*, 1–13.

[3] Ramírez, M.G.L., Satyanarayana, K.G., Iwakiri, S., Muniz, G.B., Tanobe, V., and Sahagun, T.S.F. (2011) Study of the properties of biocomposites. Part I. Cassava starch-green coir fibers from Brazil. *Carbohydrate Polymers*, *86*, 1712–1722.

[4] German, J.B., Smilowitz, J.T., and Zivkovic, A.M. (2006) Lipoproteins: When size really matters. *Current Opinion in Colloid & Interface Science*, *11*, 171–183.

[5] Angellier, H., Molina-Boisseau, S., Dole, P., and Dufresne, A. (2006) Thermoplastic starch-waxy maize starch nanocrystals nanocomposites. *Biomacromolecules*, *7*, 531–539.

[6] Xiaofei, M., Jian, R., Chang, P.R., and Yu, J. (2008) Fabrication and characterization of citric acid- modified starch nanoparticles/plasticized-starch composites. *Biomacromolecules.*, *9*, 3314–3320.

[7] Corre, D.L., Bras, J., and Dufresne, A. (2010) Starch nanoparticles: A review. *Biomacromolecules*, *11*, 1139–1153.

[8] Valodkar, M., Thakore, S. (2011) Organically modified nanosized starch derivatives as excellent reinforcing agents for bionanocomposites. *Carbohydrate Polymers*, *86*, 1244–1251.

[9] Angellier, H., Molina-Boisseau, S., Lebrun, L., and Dufresne, A. (2011) Processing and structural properties of waxy maize starch nanocrystals reinforced natural rubber. *Biomacromolecules*, *38*, 3783–3792.

[10] Zou, J., Zhang, F., Huang, J., Chang, P.R., Su, Z., and Yu, J. (2011) Effects of starch nanocrystals on structure and properties of waterborne polyurethane-based composites. *Carbohydrate Polymers*, *85*, 824–831.

[11] Wicks, Z.W., Wicks, D.A., and Rosthauser, J.W. (2002) Two package waterborneurethane systems. *Progress in Organic Coatings*, *44*, 161–183.

[12] Valodkar, M., Thakore, S. (2010) Isocyanate crosslinked reactive starch nanoparticles for thermo-responsive conducting applications. *Carbohydrate Research*, *345*, 2354–2360.

[13] Zhao, R., Torley, P., and Halley, P.J. (2008) Emerging biodegradable materials: Starch- and protein-based bio- nanocomposites. *Journal of Materials Science*, *43*, 3058–3071.

[14] Marsh, K., Bugusu, B. (2007) Food packaging: Roles, materials, and environmental issues. *Journal of Food Science*, *72*, R39–R55.

[15] Robertson, G.L. (2006) *Food Packaging: Principles and Practice*, 2nd ed. BocaRaton, FL: CRC Press.

[16] Nielsen, L.E., Landel, R.F. (1994) *Mechanical Properties of Polymers and Composites*, 2nd ed. New York: Marcel Dekker, Inc.

[17] Yang, S., Huang, C. (2008) Plasma treatment for enhancing mechanical and thermal properties of biodegradablePVA/starch blends. *Journal of Applied Polymer Science*, *109*, 2452–2459.

[18] Ramaraj, B. (2007) Crosslinked poly(vinyl alcohol) and starch composite films: Study of their physicomechanical, thermal, and swelling properties. *Journal of Applied Polymer Science*, *103*, 1127–1132.

[19] Tang, X., Alavi, S., and Herald, T. (2008) Barrier and mechanical properties of starch-clay nanocomposite films. *Journal of Cereal Chemistry*, *85*, 433–439.

[20] Dean, K., Yu, L., and Wu, D.Y. (2007) Preparation and characterization of melt-extruded thermoplastic starch/clay nanocomposites. *Composites Science and Technology*, *67*, 413–421.

[21] Ali, S.S., Tang, X., Alavi, S., and Faubion, J. (2011) Structure and physical properties of starch/poly vinyl alcohol/sodium montmorillonite nanocomposite films. *Journal of Agricultural and Food Chemistry*, *59*, 12384–12395.

[22] Sozer, N., Kokin, J.L. (2009) Nanotechnology and its applications in the food sector. *Trends in Biotechnology*, *27*, 82–89.

[23] Chen, B., Evans, J.R.G. (2005) Thermoplastic starch-clay nanocomposites and their characteristics. *Carbohydrate Polymers*, *61*, 455–463.

[24] De Carvalho, A.J.F., Curvelo, A.A.S., and Agnelli, J.A.M. (2001) A first insight on composites of thermoplastic starch and kaolin. *Carbohydrate Polymers*, *45*, 189–194.

[25] McGlashan, S.A., Halley, P.J. (2003) Preparation and characterization of biodegradable starch-based nanocomposite materials. *Polymer International*, *52*, 1767–1773.

[26] Flores, S., Fama, L., Rojas, A.M., Goyanes, S., and Gerschenson, L. (2007) Physical properties of tapioca-starch edible films: Influence of filmmaking and potassium-sorbate. *Food Research International*, *40*, 257–265.

[27] Souza, A.C., Benze, R., Ferrão, E.S., Ditchfield, C., Coelho, A.C.V., and Tadini, C.C. (2012) Cassava starch biodegradable films: Influence of glycerol and clay nanoparticles content on tensile and barrier properties and glass transition temperature. *Food Science and Technology*, *46*, 110–117.

[28] Marquesa, A.P., Reis, R.L., and Hunt, J.A. (2002) The biocompatibility of novel starch-based polymers and composites: *In vitro* studies. *Biomaterials*, *23*, 1471–1478.

[29] Weinheim (1992) *Medical, Dental Materials, J Mater Sci Technol—A Comprehensive Treatment*. New York, Basel, Cambridge: VCH.

[30] Williams, D.F. (1986) On the biocompatibility of high technology materials. *Materials Research Society Symposia Proceedings*, *55*, 117–125.

[31] Likhitkar, S., Bajpai, A.K. (2012) Magnetically controlled release of cisplatin from superparamagnetic starch nanoparticles. *Carbohydrate Polymers*, *87*, 300–308.

[32] Bertholon, I., Hommel, H., Labarre, D., and Vauthier, C. (2006) Properties of polysaccharides grafted on nanoparticles investigated by EPR. *Langmuir*, *22*, 5485–5490.

[33] Sangeetha, S., Harish, G.L., and Samanta, M.K. (2010) Chitosan-based nanospheres as drug delivery system for cytarabine. *International Journal of Pharmaceutical and Biological Sciences*, *2*, 61–64.

CHAPTER 14

Preparation of Nanofibrillated Cellulose and Cellulose Whiskers

DAVID PLACKETT and MARCO IOTTI

14.1 INTRODUCTION

14.1.1 Cellulose

As the main structural component in the cell walls of all plants, cellulose is a fibrous, tough, water-insoluble substance and is widely considered to be the most abundant bioderived polymer on the planet. From an industrial perspective, cellulose has of course been the mainstay of the wood products and pulp and paper industries for the best part of the past 100 years. Estimates indicate that between 10^{10} and 10^{11} tons of cellulose are produced each year on Earth, but of this amount only about 6×10^9 tons are used annually by industry to generate pulp, paper, textiles, chemicals, and other materials [1]. In addition to its role in the plant kingdom, cellulose is produced by algae and fungi but, as far as is known, in only one animal species, the marine tunicates [2]. In plants, cellulose can occur in the stem, leaves, or fruit, depending on the species.

Cellulose is a linear homopolysaccharide of β-1,4-linked anhydro-D-glucose units with a degree of polymerization (DP) of approximately 10,000 but with a value of 15,000 reported for native cellulose in cotton. The chemical structure of cellulose is shown in Figure 14.1 and consists of cellobiose dimer units as the repeating segment. The defining feature of the cellulose structure is the availability of three hydroxyl groups per anhydroglucose unit. In particular, these groups participate in extensive hydrogen bonding, which contributes to the multiscale nature of cellulose (Fig. 14.2), its crystalline and amorphous structure, and its highly cohesive nature [3].

Biopolymer Nanocomposites: Processing, Properties, and Applications, First Edition. Edited by Alain Dufresne, Sabu Thomas, and Laly A. Pothan.

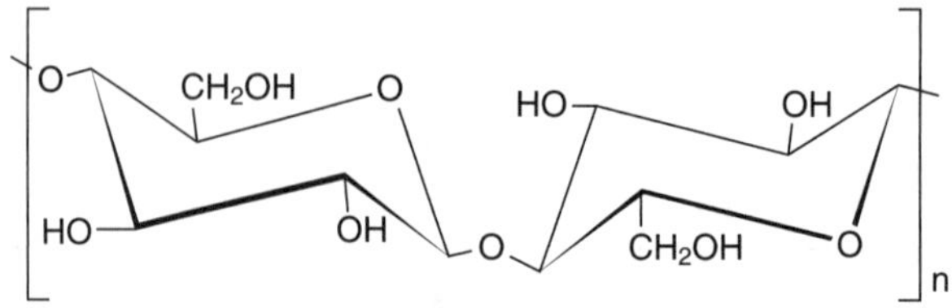

Figure 14.1 The cellobiose repeating unit of cellulose.

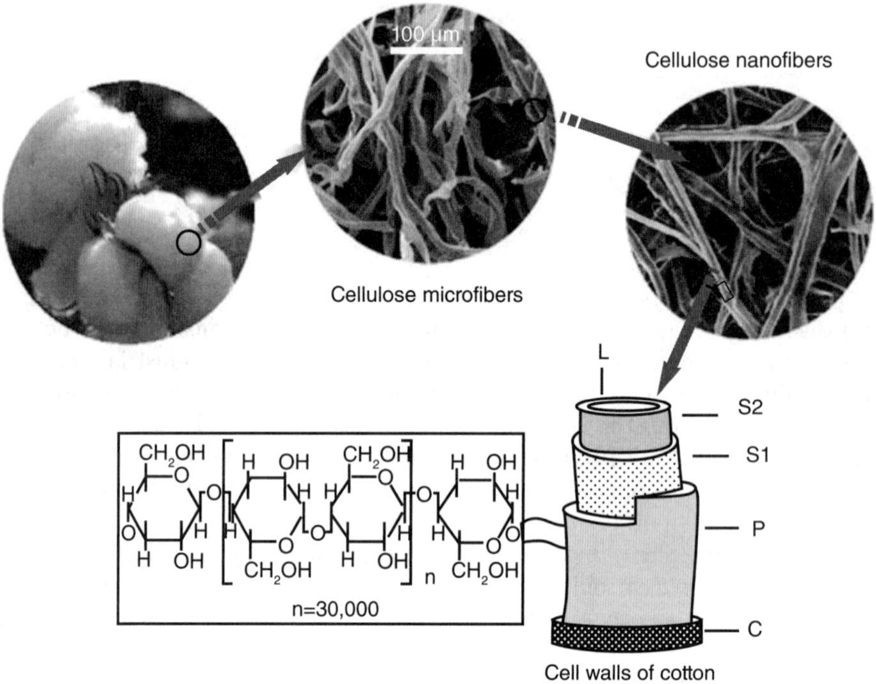

Figure 14.2 The multiscale nature of cellulose as illustrated for cotton (sourced from Reference [3]).

Cellulose exists as six different polymorphs that can be interconverted: cellulose I, II, III_I and III_{II}, IV_I, and IV_{II} [4–7] (Fig. 14.3). Cellulose I, the native form found in nature, occurs in two allomorphs, I_α and I_β. Proof of the polymorphy of cellulose comes from nuclear magnetic resonance (NMR) spectroscopy, infrared spectroscopy, and diffraction studies [8, 9]. Cellulose II is the most stable crystalline form and arises after recrystallization or mercerization with sodium hydroxide. Cellulose II has antiparallel packing of the polymer chains, whereas the chains are packed in parallel fashion in cellulose

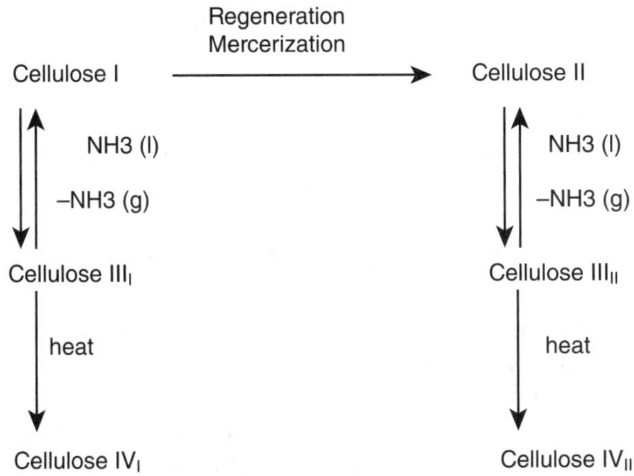

Figure 14.3 Cellulose polymorphs.

I. Cellulose III$_I$ and III$_{II}$ are obtained in a reversible process by liquid ammonia treatment of celluloses I and II, respectively, followed by evaporation of excess ammonia, but certain amines can also be used for this process. Cellulose polymorphs IV$_1$ and IV$_{11}$ are derived by heating celluloses III$_I$ and III$_{II}$, respectively, up to 206°C in glycerol [10, 11]. For the purposes of this chapter, native cellulose I can be considered the key source of nanocellulose.

14.1.2 Nanocellulose

14.1.2.1 Background The microfibrils of cellulose formed during biosynthesis are 2–20 nm in diameter, depending on the source, and can be several micrometers in length. Each microfibril consists of crystalline domains interspersed with disordered amorphous regions [12]. Materials that fall into the general category of nanocelluloses are obtained either by mechanical treatments to obtain nanofibrillated cellulose (NFC) or by acid hydrolysis to remove the amorphous segments and leave highly crystalline cellulose whiskers, otherwise known as cellulose nanocrystals, nanorods, or nanowhiskers [13–15]. A third type of nanocellulose is bacterial cellulose, composed of cellulose nanofibers secreted extracellularly by certain bacteria. These bacteria are usually strains of *Gluconacetobacter*, and, for example, *Gluconacetobacter xylinus* secretes cellulose nanofibers having widths of ~3.5 nm [16]. The unique nanofibrillar structure of bacterial cellulose confers excellent physical and mechanical properties such as high porosity, high elastic modulus, and high crystallinity [17–19].

As discussed elsewhere in this book, and as has been previously reviewed [15, 20], nanocellulose has long been recognized for its potential in a wide variety of applications including paper and paperboard, composites, hygiene and absorbent products, food additives, emulsifiers and dispersants, oil recovery, and cosmetic, medical, and pharmaceutical uses. This chapter deals specifically with the preparation and characterization of NFC or cellulose whiskers and the current state of their commercial development. The post-treatment of nanocellulose for specific applications and a review of these applications are topics that are covered elsewhere in this book.

14.1.2.2 Nanofibrillated Cellulose NFC, otherwise referred to as microfibrillated cellulose (MFC), was first prepared by Turbak and colleagues at the ITT Rayonier laboratories in the United States in the late 1970s. In this earliest work, a Gaulin-type milk homogenizer was used to process wood pulp at high temperatures and pressures into a gel-like material, which was obtained after ejection from the homogenizer against a hard surface. The Gaulin homogenizer was first demonstrated at a dairy exhibition in Paris in 1900 and the equivalent is now manufactured, on the basis of more than 100 years' know-how, by SPX Corporation in Germany and the United States. The first publications on the preparation of MFC were those of Turbak and coworkers at ITT Rayonier in the early 1980s, when the term microfibrillated cellulose first came into use [21–23].

NFC, as it is now described, is composed of expanded high-volume cellulose, moderately degraded and greatly expanded in surface area, obtained by a homogenization process. NFC consists of aggregates of cellulose microfibrils and has fibril diameters in the range of 20–60 nm. Since cellulose microfibrils are 2–10 nm in diameter and are several tens of micrometers in length, the individual elements of which NFC is comprised consist of between 10 and 50 microfibrils. NFC exhibits gel-like properties in water with pseudoplastic and thixotropic properties and can be cast into transparent films [24] (Fig. 14.4).

14.1.2.3 Cellulose Whiskers Cellulose whiskers are obtained by acid hydrolysis of various forms of cellulose. The first report outlining the preparation of such cellulosic material was published by Rånby and Ribi [25]. These initial investigations, involving sulfuric acid treatments, were inspired by the work of Nickerson and Habrle, who noted that exposure of cellulose to boiling acidic solutions resulted in cellulose degradation, but only up to a certain limit [26]. Transmission electron microscopy (TEM) studies showed that the colloidal suspensions after acidic treatment of cellulose could be dried to produce aggregates of needle-shaped particles. Further analysis using electron diffraction showed that these particles had the same crystalline structure as the original fibers. At the same time, Battista developed the hydrochloric acid-assisted degradation of cellulose fibers from high-quality wood pulps, followed by sonication treatment, which led eventually to the production and commercialization of microcrystalline cellulose (MCC) [27]. According to the United Nations Food and Agriculture Organization (FAO), MCC is defined as purified, partially depolymerized cellulose prepared by treating α-cellulose, obtained as a pulp from fibrous plant material, with mineral acids. The degree

Figure 14.4 NFC film and 0.4% consistency NFC gel produced at the Paper and Fiber Research Institute (PFI), Trondheim, Norway, being handled by Kristin Syverud (left) and Ingebjorg Leirset (right). The photograph was taken by Oddbjørn Svarlien and kindly supplied by Kristin Syverud. See color insert.

of polymerization is typically less than 400 and not more than 10% of the material has a particle size of less than 5 μm.

14.2 PREPARATION OF NANOFIBRILLATED CELLULOSE

14.2.1 Mechanical Treatments

The Gaulin homogenizer first used to generate NFC gels by Turbak and coworkers is still in use today to manufacture various forms of NFC. In this process, cellulose slurries are pumped at high pressure and fed through a spring-loaded valve assembly. The valve opens and closes in rapid succession and the fibers are thereby subjected to large changes in pressure with high shearing forces. This procedure results in a high degree of cellulose fiber fibrillation.

An alternative to the Gaulin homogenizer is the microfluidizer, in which wood pulp for example is passed through z-shaped channels of 200–400 μm diameter at high pressure (e.g., ~2000 bar) (Fig. 14.5). The resulting shear rate is very high and produces very thin cellulose nanofibers. The microfluidizing process has become popular because of the greater uniformity of the fibril

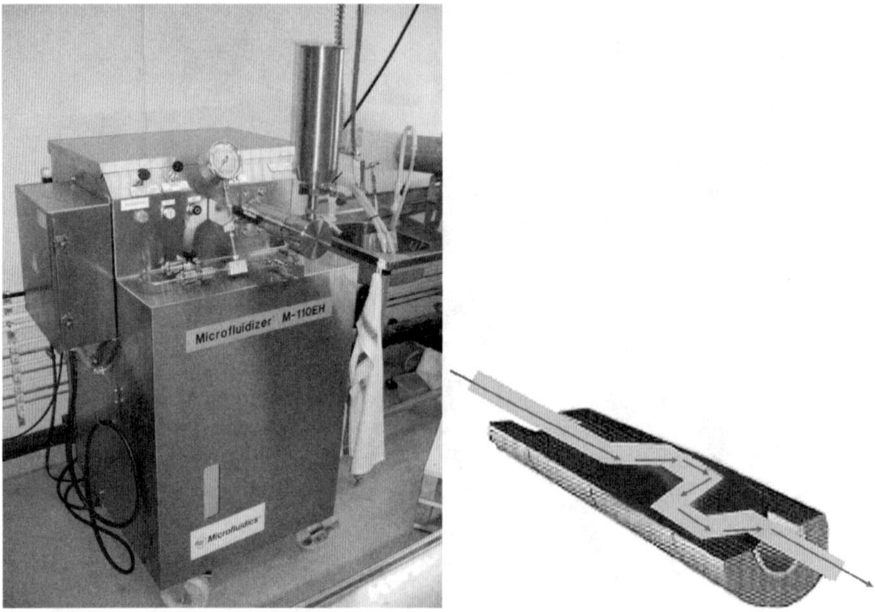

Figure 14.5 A microfluidizer used for production of nanofibrillated cellulose at Innventia AB, Stockholm, with a schematic (right) of the pulp flow in high shear through the homogenizer chamber containing 100- to 400-micron-wide channels. Photograph and schematic were kindly supplied by Mikael Ankerfors, Innventia AB.

dimensions, but the use of different-sized chambers has remained necessary in order to obtain the required degree of fibrillation. Unfortunately, without further modification, such processes have typically required 10 or even as many as 30 passes of the raw fibers through the microfluidizer with excessively high energy consumption, reaching values as high as 30,000 kWh/ton or more [28, 29]. As noted in the next section, considerable progress has been made by adopting cellulose pretreatments, which have significantly reduced the energy required for NFC production.

Grinding processes have also been used to produce NFC and Masuko Sangyo Co., Ltd. (Kawaguchi City, Japan) was the first company to build grinding equipment specifically for NFC production. The principle of the grinding process is the breakdown of the plant cell wall structure by shearing forces as the wood pulp passes between a static grindstone and a rotating grindstone revolving at ~1500 rpm. Iwamoto et al. fibrillated pulp from radiata pine (*Pinus radiata*) fibers and noted changes in morphology as a function of the number of passes [30]. Five passes through the grinder was found to be optimum since no changes in fibril morphology were noted with more passes and a mixture of micro- and nanosized fibers was generated when fewer passes were used.

The same researchers noted that the grinding process resulted in a decrease in fibril length, which would be undesirable for some applications. Unfortunately, as noted by a number of authors, the physical nature of NFC gels with intertwined fibrils makes accurate measurement of fibril length a significant challenge.

Uetani and Yano recently used a new approach to manufacture NFC, involving use of a Waring blender [31]. Wood fiber suspensions at concentrations varying from 0.1 wt% to 1.5 wt% were processed at speeds ranging from 5,000 to 37,000 rpm, but the resulting gels were not very homogeneous and not all fibers were converted to NFC. However, the gels produced after blending for 30 minutes showed the same degree of fibrillation with less damage than those obtained after one pass in a grinder at 1500 rpm.

A further approach to NFC production involves cryocrushing by applying high shear forces to pulp that has been exposed to liquid nitrogen [32]. The principle of cryocrushing is based on the formation of ice crystals within the plant cells, which, when exposed to mechanical crushing, cut the cell walls and release wall fragments [33]. Alemdar and Sain used cryocrushing to obtain NFC from wheat straw and soy hulls, and, in this case, fibrils had diameters of 30–40 nm and lengths of several micrometers [34]. Similarly, flax, hemp, and rutabaga fibers were cryocrushed to fibrils of 5–80 nm diameter and soybean stock was converted to NFC with fibril diameters of 50–100 nm [33, 35, 36].

As indicated in Table 14.1, reports of energy consumption for production of NFC by homogenization have been as high as 70,000 kWh/ton. This factor has been significant in terms of delaying the development and uptake of NFC

TABLE 14.1 Approximate Energy Requirements for NFC Production from Various Studies as Reported in References [24, 38]

Equipment	Pretreatment	Raw Material	Energy Requirement (KWh/tonne)
Homogenizer	None	Bleached kraft pulp	12,000–70,000
Homogenizer	None	Bleached sulfite pulp	27,000
Homogenizer	Carboxymethylation (Degree of substitution (DS) = 0.1)	Bleached kraft/ sulfite pulp	500
Homogenizer	Enzymatic/refining	Bleached sulfite pulp	1,500
Homogenizer	Valley beater	Bleached/ unbleached hardwood pulp	21,900
Microgrinding	None	Bleached/ unbleached hardwood pulp	1,550
Microfluidizer	Valley beater	Bleached/ unbleached hardwood pulp	2,290

technology. In a recent study, Zimmerman et al. reported that a microfluidizer required 8.5 kWh (manufacturer's data) at a processing pressure of 1500 bar [37]. Since 10 L of cellulose pulp at 1–2 wt% needed about 15 minutes to pass once through the microfluidizer, the most important parameter was deemed to be the number of passes. Spence et al. examined a number of properties of NFC from bleached and unbleached hardwood pulp samples and the energy consumption during production as a function of the manufacturing process [38]. These researchers discovered that microfluidization with refining pretreatment or micro-grinding processes used less energy than homogenization and also produced cast NFC films with better physical, optical, and water interaction properties.

The need to develop NFC processes involving much less energy consumption is naturally of considerable interest. As a result, alternative processing methods (e.g., extrusion) have been studied. Similarly, production of NFC through use of the Cavitron® (Cavitron v. Hagen & Funke GmbH, Sprockhövel, Germany) has been investigated by the Paper Technical Center (PTS) in Germany as part of the EU SUNPAP project (http://sunpap.vtt.fi), and the GEA Niro Soavi Ariete NS3075H developed by the French Technical Center of Paper (CTP) has been proposed. However, few results from these new methods have been reported so far. Another strategy that has attracted some attention is the fractionation of partially ground *Pinus radiata* NFC, as reported by Abe et al. [39]. In this approach, wood powder sieved to pass through a 60-mesh screen was sequentially extracted with solvents to remove lignin and hemicelluloses and finally a 1 wt% slurry was passed through a Masuko grinder. The output consisted of cellulose nanofiber suspensions with uniform nanofiber widths of 15 nm.

14.2.2 Pretreatments

A number of methods have been suggested for pretreatment of plant or wood fibers with the objective of reducing energy consumption in NFC production. Generally, the methods involve one of three different principles: (1) interfering with hydrogen bonding between fibrils, (2) adding electrostatic charges, or (3) decreasing the DP or amorphous linkages between fibrils.

The use of combined refining and enzymatic pretreatments for NFC production has proven effective as a means of reducing energy consumption. Researchers have produced high aspect ratio NFC with good nanofibril diameter uniformity through a combination of enzymatic hydrolysis with mechanical shearing and high pressure homogenization [40–42]. Pääkkö et al. used an endoglucanase to treat fibers after refining and before passage of the pulp through a microfluidizer [40]. Enzymatic hydrolysis has the advantage that it is selective toward the amorphous segments in cellulose, thereby making mechanical disintegration easier. From a practical perspective, an important benefit of enzymatic pretreatment is reduced frequency of channel blocking in the microfluidizer. In another study, Henriksson et al. used a C-type endoglucanase, carried out mechanical treatments using a Gaulin homogenizer, and compared the resulting NFC with that produced by other hydrolytic pretreatments or no treatment of the incoming pulp [42]. In addition to facilitating wood pulp fiber disintegration through enhanced swelling in water, this enzymatic approach resulted in shorter fiber length and increased the amount of fine material relative to an acid hydrolysis pretreatment. Engström et al. undertook further investigations on enzymatic pretreatment and the effect on cellulose fiber reactivity [43]. This research and other investigations led the way to a scaled-up process at Innventia AB in Stockholm, which was announced by press release in 2011 and is discussed further at the end of this chapter. Recent work by Siqueira et al. analyzed the impact of enzymatic treatment on the final properties of NFC and the effect on the mechanical properties of NFC–rubber composite materials [44, 45]. In the first of these studies, the authors noted that enzymatic pretreatment allowed production of a wide variety of cellulose nanoparticles and that it was essential to report material output in terms of the pretreatment that was used [44]. In later work, it was shown how enzymatic pretreatment of sisal fibers could be used to adjust the properties of composites based on natural rubber as a model matrix [45].

The use of (2,2,6,6 tetramethyl piperidin-1-yl)oxyl (TEMPO)-mediated oxidation as a method for fiber pretreatment has been widely reported. This procedure allows carboxylate and aldehyde groups to be introduced into solid native cellulose under mild aqueous conditions. Furthermore, compared with repeated high pressure homogenization, TEMPO-mediated oxidation significantly decreases the required energy consumption. The principle behind TEMPO oxidation is the addition of NaOCl to aqueous cellulose suspensions in the presence of catalytic quantities of TEMPO and NaBr at pH 10–11 and room temperature. The process is shown schematically in Figure 14.6. The outcome is the selective conversion of C6 primary hydroxyl groups on cellulose to aldehyde and then carboxylate groups. As a consequence, the presence of ionized carboxylate groups on the fibril surfaces leads to repulsive forces, which cause improved separation. In general, increase in concentration of NaOCl leads to higher levels of carboxylation, but also to a requirement for longer oxidation times.

Application of TEMPO oxidation to nanocellulose production was primarily developed by Isogai and coworkers and has been used to process various different forms of cellulosic biomass [46–48]. Oxidation efficiencies vary from 62% to 96%, depending on the type of starting plant material. An alternative approach to TEMPO oxidation involves the use of NaOCl in combination with NaO_2Cl instead of NaBr at pH 5–7 [49–51]. In contrast to the first method, in which some residual aldehyde groups are present on the cellulose nanofibrils after pretreatment, the combined TEMPO/NaOCl/NaO_2Cl method at pH 5–7 leaves no residual aldehyde groups. However, the carboxylate content is lower in this case and the optimum reaction time and temperature are increased. A discussion of the difference between the two approaches to TEMPO oxidation is included in a recent review article [52]. A representative process for producing TEMPO-oxidized cellulose is shown schematically in Figure 14.7.

Isogai et al. investigated a TEMPO electro-mediated oxidation, using either TEMPO at pH 10 or 4-acetamido TEMPO at pH 6.8 in buffer solution [50].

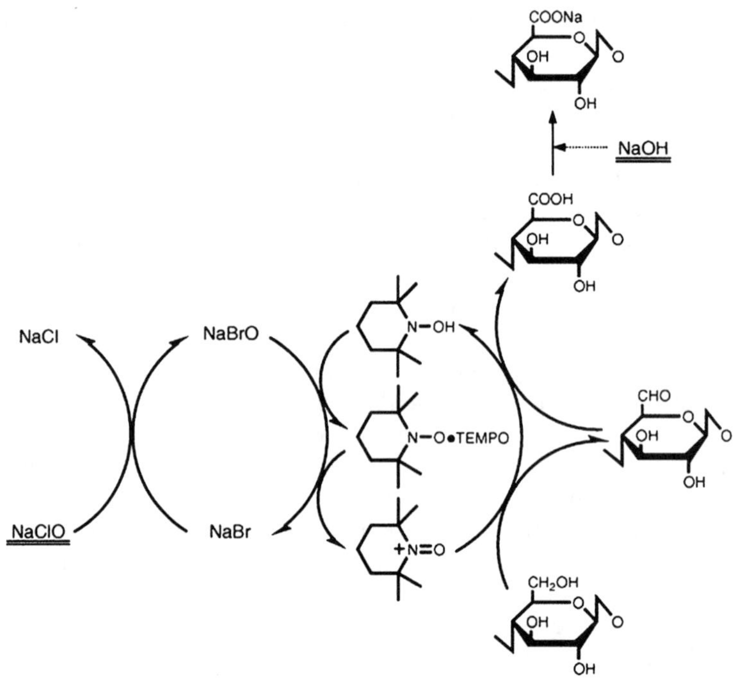

Figure 14.6 Schematic of TEMPO oxidation of cellulose (adapted from Reference [52]).

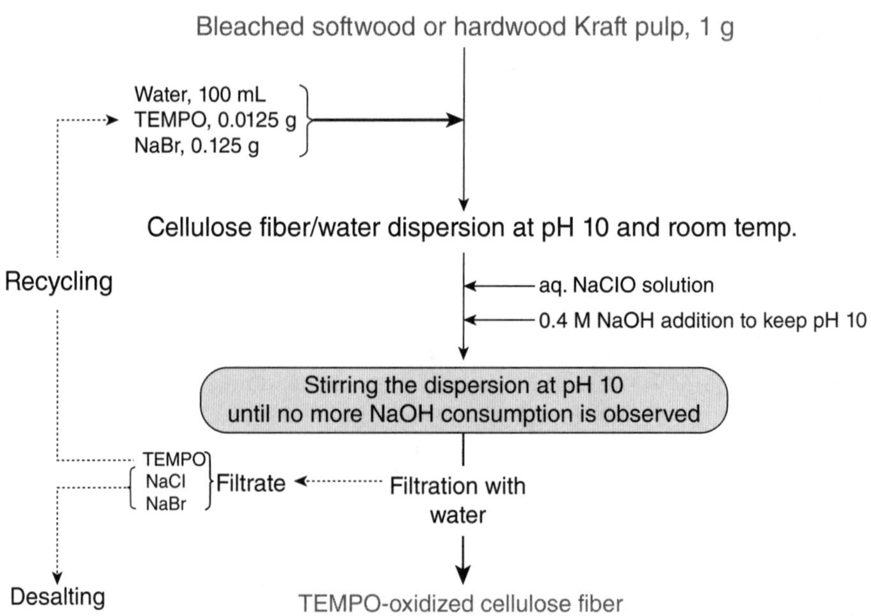

Figure 14.7 Representative process for preparation of TEMPO-oxidized cellulose by the TEMPO/NaBr/NaOCl system at pH 10 (sourced from Reference [52]).

Although requiring longer oxidation times, the electro-mediated reaction provided a yield from softwood kraft pulp in excess of 80% while retaining the characteristics of conventional TEMPO-oxidized NFC. TEMPO oxidation does, however, suffer from the high cost of the chemical as well as the continued need for optimized techniques for TEMPO recovery.

Following TEMPO oxidation, the resulting suspensions are usually subjected to mechanical treatment by blending or by the use of an Ultra Turrax (IKA, Wilmington, NC) mixer. Centrifugation is normally used to separate the incompletely fibrillated NFC. As also reported, blending can be replaced by sonication as a means of isolating TEMPO-oxidized pulp, and the length of sonication time can have a significant influence on nanofibril dimensions [53, 54]. Sonication has often been used in combination with mechanical treatments to enhance NFC production.

Pretreatment of fibers through carboxymethylation is also an established method for NFC production. In this case, as with TEMPO oxidation, the formation of carboxylate groups on the surface of the NFC introduces repulsive charges, which assists in the process of nanofibril separation [55–58]. Treatment using sodium salts ensures that the incoming wood pulp swells as much as possible. Swollen pulps have reduced fiber cell wall cohesion and therefore should be easier to delaminate. This provides a demonstrated benefit in terms of energy consumption as indicated in Table 14.1.

14.3 CHARACTERIZATION OF NANOFIBRILLATED CELLULOSE

14.3.1 Morphology and Dimensions

The characterization techniques used in studies on NFC production have included field emission gun-scanning electron microscopy (FEG-SEM), TEM, and atomic force microscopy (AFM). Use of these methods has allowed cellulose nanofibril dimensions to be determined as summarized in Table 14.2 [1, 7, 59, 60]. Typically, homogenization provides nanofibrils that are 10–40 nm in width and several micrometers in length. Aulin et al. applied the AFM method to measure the dimensions of carboxymethylated NFC produced using a microfluidizer [61]. In this case, nanofibril widths were 10–15 nm and 10–30 nm for pretreated and non-pretreated NFC, respectively. Pääkkö et al. applied

TABLE 14.2 Dimensions of Nanofibrillated Cellulose from Various Studies (Sourced from References [1, 7, 59, 60])

Cellulose Type	Diameter (nm)	Length (nm)	Aspect Ratio
Nanofibrillated cellulose	10–40	>10,000	>1,000
Cellulose whiskers	2–20	100–600	10–100
Bacterial cellulose	5–6	1,000–9,000	160–1,800
Microcrystalline cellulose	>1,000	>1,000	~1

enzymatic pretreatment to bleached softwood sulfite pulp and produced NFC with widths in the range of 20–30 nm as measured by AFM. These findings were consistent with widths determined by cross-polarization magic-angle spinning (CP-MAS) ^{13}C NMR spectroscopy, and TEM [40]. Using the NMR method, lateral fibril dimensions and lateral fibril aggregate dimensions can be determined in cellulose I samples. For this purpose, the C4 region of the NMR spectrum is subjected to spectral fitting as has been described previously [62, 63]. Grinding is reported to generate NFC elements that are 20–50 nm in width and over 1 μm in length [30]. As mentioned previously, Abe et al. produced NFC with a uniform width of 15 nm by means of grinding [39]. In a recent study, Uetani and Yano processed wood pulp with a blender to obtain NFC with a uniform diameter of 15–20 nm [31]. However, TEMPO oxidation followed by the same mechanical treatment and a centrifugation step to remove larger NFC or fibers still present in the suspension resulted in nanofibrils with diameters as low as 3–5 nm [46, 53].

The morphology and dimensions of NFC show variations according to pretreatment, mechanical treatment, and raw material source. The incoming fibers naturally vary in respect to length, microfibril angle, amount of noncellulosic residues (e.g., lignin, hemicelluloses), and quality (e.g., as indicated by fiber defects) according to source and to whether the biomass is a primary stream or a waste stream. As noted by Alemdar and Sain, NFC from wheat straw and soy hulls had diameters of 10–80 nm and 20–120 nm, respectively, and lengths of several micrometers [34]. Other researchers processed a variety of agricultural residues and obtained a wide range of NFC diameters according to the particular raw material [44, 45, 64–66]. The picture is also complicated by variations within a particular crop as well as variations in terms of processing method. General conclusions about NFC processing can therefore be hard to draw.

An interesting aspect is the influence of lignin on NFC properties. This question has been studied by Spence et al., who discovered that unbleached softwood pulp with lignin content of ~9% gave a mean NFC width of 30.8 nm, while pulp with a lignin content of ~14% produced an NFC width of 34.4 nm [67]. Hemicelluloses also limit the association between cellulose nanofibers and hence can influence NFC diameter distribution [58]. As reported by Iwamoto et al., never-dried and once-dried pulps from Sitka spruce (*Picea sitchensis* Carr.) with different amounts of hemicelluloses were fibrillated using a grinding treatment. The degree of fibrillation was evaluated by SEM observation of the fibrillated pulps and light transmittance measurements of fibrillated pulp/acrylic resin composites. With a one-pass grinding treatment, the once-dried pulp with higher hemicellulose content was fibrillated into 10- to 20-nm-wide fibers as easily as the never-dried pulps, while the once-dried pulp with lower hemicellulose content was not fibrillated into uniform nanosized fibers. These findings confirm that hemicelluloses can inhibit the coalescence of microfibrils during drying and facilitate the nanofibrillation of once-dried pulp [58].

Of the various methods used to visualize NFC, it is recognized that FEG-SEM may overestimate the dimensions of cellulose nanofibrils due to the need to coat material surfaces with a conductive metal layer. With this in mind, Chinga-Carrasco and Syverud developed a new method that avoids the need for metal coating [68]. This new approach allows the cellulose nanofibril structure to be observed at high magnification (>50,000×) by applying a low acceleration voltage (<1 kV) and short working distance (<1 mm). Despite these advances, estimation of the length of NFC nanofibrils generally remains an approximation. However, in a recent notable development, Fukuzumi et al. applied an analysis of the storage and loss moduli of TEMPO-oxidized NFC and used the theory of linear viscoelasticity to obtain a calculated mean length of TEMPO-oxidized NFC of 2.2 μm [69].

14.3.2 Physical Properties of Nanofibrillated Cellulose

14.3.2.1 Degree of Polymerization and Strength Determination of the degree of polymerization (DP) of NFC has generally been carried out using a viscosity method and a cupriethylene diamine solution according to ISO Standard 5351 [70]. In this method, average DP is calculated using the Mark–Houwink equation, which relates intrinsic viscosity to molecular weight. Conversion of fibers to NFC is accompanied by a reduction in DP as revealed for example by the work of Zimmerman et al., in which case the value of 2249 for the initial pulp decreased to 825 in the resulting NFC [37]. Since DP and NFC film strength are correlated, the DP value might be useful in assessing the likely performance of NFC as a polymer reinforcing material, especially since techniques for measuring the strength properties of individual NFC elements are still to be optimized. Various estimates of the strength and modulus of cellulose nanofibers have, however, been made. For example, Yano and Nakahara [71] stated that the strength of cellulose microfibrils could be estimated to be at least 2 GPa based on the tensile strength of 1.7 GPa obtained for kraft pulp, for which 70–80% of the cellulose microfibrils are thought to be aligned in the fiber direction [72]. This estimate may be contrasted with the elastic modulus of a perfect crystal of native cellulose, which was previously calculated to be in the range of 130–250 GPa [73, 74]. This large variation may be related to the different sources of cellulose and consequent variations in crystal structure or to the use of different measuring systems. Recently, Iwamoto et al. used an AFM cantilever and three-point bending to determine an elastic modulus of 145 GPa for a single TEMPO-oxidized tunicate cellulose nanofiber [75]. It is worth noting that such values are comparable to those reported for high-strength synthetic aramid fibers [76].

14.3.2.2 Crystallinity As outlined in a recent review, X-ray diffraction in combination with either peak area or peak intensity measurements can be used to determine the crystallinity of cellulose [77]. Typically, the crystallinity index is calculated as a ratio of diffraction from the crystalline part to the diffraction from the whole sample. The crystallinity of NFC obtained from a variety of different raw materials has been reported in the literature. For example, Heux et al. found relatively low values of 30–40% for sugar beet pulp [78]. This contrasts with results obtained by Agoda-Tandjawa et al., who calculated the degree of crystallinity of NFC from beet pulp to be 13% and 11%, respectively, before and after mechanical treatment [79]. Other researchers determined NFC crystallinity to be as high as 90% when using sisal or *Luffa cylindrica* [44, 66]. Alemdar and Sain estimated the crystallinity of NFC from wheat straw and soy hulls to be 78% and 70%, respectively [34]. The crystallinity of NFC produced by grinding has been found to decrease as a function of the number of passes through the grinder [30], and this finding could be related to shear rate, since an initial crystallinity decrease of 5% during a 5-minute blending process compares with a decrease of about 10% as a result of one pass through a grinder. Overall, blender treatment may do less fiber damage than grinding. Homogenization may also do less damage to the cellulose crystalline structure than the grinding process. Saito and Isogai determined that crystallinity and crystallite size of cellulose I in the original native cellulose were unchanged by a TEMPO oxidation process and concluded that the majority of carboxylate and aldehyde groups are only present on the cellulose crystal surfaces and/or in amorphous disordered regions [47].

14.3.2.3 Surface Chemistry Conductometric titration has been used to determine the carboxylate and aldehyde group contents in TEMPO-oxidized NFC [80]. For example, using this approach, the combined carboxylate and aldehyde contents on cotton linters corresponded to roughly one C6-oxidized glucose unit per seven glucose units. In contrast, when using ramie or spruce as a starting material, the figure is one C6-oxidized glucose unit per five glucose units. Conductometric titration has also been applied to determine carboxymethylated NFC (~515 μeq/g) [81]. These conductometric titrations were based on a previously reported method [82].

14.4 PREPARATION OF CELLULOSE WHISKERS

14.4.1 Acid Hydrolysis

The preparation of cellulose whiskers or cellulose nanocrystals by acid hydrolysis is now an established laboratory process and has been applied to a wide variety of raw materials [83]. The various sources of biomass investigated for cellulose whisker production have included banana [84], sugar cane bagasse [85], mulberry [86], sugar beet pulp [87], wood [88], and cotton [89] among others (e.g., hemp, ramie, sisal, wheat straw) [12]. The typical appearance of cellulose whiskers when viewed using TEM is illustrated for the case of tunicate whiskers in Figure 14.8. In addition to these various types of biomass, there have also been studies aimed at using microcrystalline cellulose (MCC)

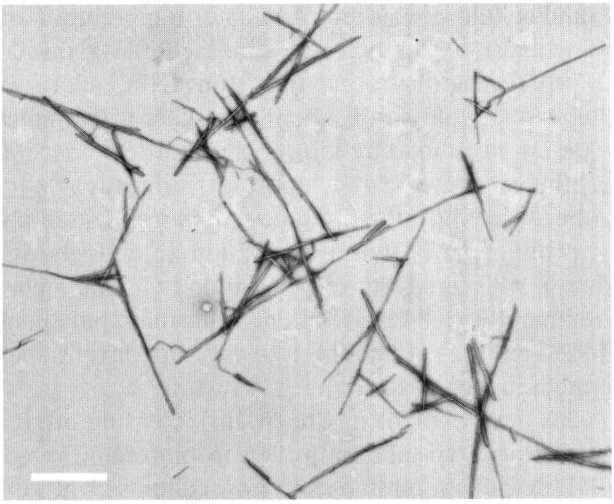

Figure 14.8 TEM image of cellulose whiskers isolated from tunicates (white scale bar = 1 micron). Image source: Jeff Capadona, Case Western Reserve University, Cleveland, OH.

as a convenient and pure starting material. Bondeson et al. in particular developed a rapid, high-yield optimized process to obtain a stable aqueous colloidal suspension of cellulose whiskers from MCC, assisted by the use of response surface methodology [90]. Cellulose whiskers measuring 200–400 nm in length were obtained by using 63.5 wt% sulfuric treatment of MCC over a period of 2 hours and with a yield of 30%.

New strategies and approaches to manufacturing cellulose whiskers continue to be in focus. If hydrolysis is carried out using hydrochloric acid instead of sulfuric acid, it seems that the cellulose whisker dispersion is limited and there is a tendency toward flocculation, which is not desirable for most end-applications [91]. The use of NFC as a starting material for cellulose whiskers has been discussed [87] and, more recently, Siqueira et al. applied enzymatic hydrolysis to obtain different types of cellulose nanoparticles. In this case, the new development was the use of cellulase enzymes for preparation of cellulose whiskers [44]. This enzymatic approach was motivated by the potential to obtain improvements in cellulose whisker size reproducibility, aspect ratio, and cost of preparation. Starting with a bleached sisal pulp, various combinations of mechanical shearing through a microfluidizer and enzymatic hydrolysis were studied. The mechanical shearing process alone resulted in a nonhomogeneous material with NFC diameters in the range of tens to hundreds of nanometers. Subsequently, hydrolysis of NFC with 55 wt% sulfuric acid at 55°C over 30 minutes produced a material with morphology intermediate between NFC and cellulose whiskers. At higher temperatures, more homogeneous cellulose whisker dispersions were obtained. Based on comparison with an earlier study [92] it appears that that using NFC as a starting material instead of cellulosic microfibers results in whiskers with similar morphology but may provide an advantage by allowing use of less concentrated sulfuric acid. However, the objective of the reported study was not to optimize the process but rather to confirm the influence of hydrolysis temperature on cellulose whisker production. Filson et al. applied an endoglucanase treatment to extract cellulose whiskers from recycled pulp [93]. The highest yields were obtained by treatment with 84 EGU of endoglucanase per 200 mg of recycled pulp at 50°C for 60 minutes of either microwave or conventional heating, with the former giving a higher yield.

Another recent development has been an increasing interest in cellulose whisker production as a specific output from biorefining processes [94–97]. Chirat et al. [94] provide an outline as to how cellulose whiskers may be produced from mixed softwood kraft pulp and also point out that the exact physical dimensions of the whiskers depend on factors such as source, hydrolysis conditions, and ionic strength. In their review, Duran et al. [95] discuss the general aspects of nanotechnology in the forest products industry, the synthesis of cellulose whiskers (referred to as nanocrystals) by various means, and applications of cellulose whiskers. In discussing cellulose whiskers as a co-product in bioethanol production, Duran et al. mention the work of a consortium at North Dakota State University in which an economic analysis of cellulose whisker production was made as a component in ethanol production from wheat straw. Using an ASPEN Plus-based process model (http://www.aspentech.com/core/aspen-plus.cfm), the production cost of ethanol including by-product credit was estimated at \$0.41 per liter. When production of cellulose whiskers was added to the base model, the cost of producing the whiskers from wheat straw was estimated to be \$1.25 per kg. This led to the conclusion that cellulose whisker production could enhance the economic feasibility of ethanol production from wheat straw [98]. In the report by Oksman et al. [96], residue from bioethanol production was examined as a low-cost source of raw material for cellulose whisker production. Bioresidue from a wood bioethanol pilot plant was used as a starting material and, after dewaxing, bleaching, and cleaning, cellulose whiskers were produced by mechanical treatments, including ultrasonication or high pressure homogenization, as well as chemical hydrolysis. Whiskers isolated using ultrasonication or high pressure homogenization had better thermal stability than whiskers isolated by acid hydrolysis. Although described as a "lignin residue," this material had a surprisingly high content of separable cellulose in crystalline form (~48%). Zhu et al. used commercial cellulase enzymes to fractionate amorphous cellulose from bleached kraft eucalyptus pulp and to generate crystalline cellulose which was suitable for mechanical homogenization [97]. Based on these reports, it seems likely that there will be many future investigations into nanocellulose as an added-value output from biorefining.

14.5 CHARACTERIZATION OF CELLULOSE WHISKERS

14.5.1 Morphology

Although similar in size to NFC, cellulose whiskers show much lower flexibility, as they do not contain amorphous regions. Nanocrystalline cellulose (NCC) crystals produced by acid hydrolysis may show different geometries, depending on their biological source [99–101]. The dimensions of cellulose whiskers also vary significantly depending upon the hydrolysis conditions. In general, cellulose whiskers are 5–70 nm in width and between 100 nm and several micrometers in length. These variations are partly due to differences in starting material and also to the diffusion-controlled nature of the acid hydrolysis treatment, which is in turn influenced by degree of cellulose crystallinity. Acid hydrolysis, often in combination with ultrasonication, can be followed up by separation techniques, such as differential centrifugation or ultracentrifugation, which may be used to obtain suspensions with more uniform cellulose whisker dimensions. A Finnish research group has found a relationship between pulp drying procedure before acid hydrolysis and cellulose whisker dimensions [102]. The recent study by Elazzouzi-Hafraoui et al. measured the dimensions of cellulose whiskers produced from cotton, MCC, and tunicate sources [99]. Using a range of techniques, these researchers found that tunicin gave the lowest dispersity in terms of lateral size, but the highest in respect to length. The authors rationalized this finding on the basis of a relatively low degree of lateral association of cellulose elementary fibrils or crystallites from tunicin, when compared with whiskers derived from plant materials. The lower length polydispersity in cellulose whiskers from plant materials was attributed to the presence of periodic domains susceptible to acid hydrolysis, resulting in a narrower distribution of extracted crystallite lengths.

The morphology of cellulose whiskers has usually been studied by means of TEM, SEM, FEG-SEM, or AFM, but scattering techniques such as small-angle neutron scattering or dynamic light scattering have also proven useful. A recent report has also highlighted the possibility of using cryo-TEM to examine cellulose whiskers without the agglomeration effect, which is prevalent when using other methods [99].

14.5.2 Physical Properties of Cellulose Whiskers

The Young's modulus of cellulose whiskers was estimated by Sakurada et al. at 130 GPa [73], but a value as high as 250 GPa has also been reported [74]. The latter value is close to the modulus of the perfect crystal of native cellulose. The strength properties of cellulose whiskers have seldom been reported on an experimental basis; however, Azizi Samir et al. reported the strength of cellulose whiskers to be 10 GPa [1]. Mechanical property data for cellulose whiskers are presented in Table 14.3 along with data calculated or estimated for the pure native cellulose crystal [70, 72, 103–109].

TABLE 14.3 Reported Mechanical Properties of Cellulose Whiskers and the Pure Native Cellulose Crystal from References [70, 72, 103–109]

Material	Technique	Modulus (GPa)
Tunicate whiskers	AFM bending stiffness	150 ± 28.8
Acid-hydrolyzed whiskers	Raman spectroscopy	50–100
Tunicate whiskers	Raman spectroscopy	143
Pure cellulose crystal	Theoretical estimates	100–160
Flax fibers	Inelastic X-ray scattering	220
Pure cellulose crystal	X-ray diffraction of plant fiber bundles	138

Cellulose whiskers obtained through the use of sulfuric acid have negatively charged sulfate groups on the fibril surfaces and are therefore easy to disperse in aqueous solvents due to electrostatic repulsion. A weaker charge distribution and therefore poorer dispersibility arises when hydrochloric acid is used as the hydrolysis medium. Recent developments have enabled cationic surface charges to be generated on cellulose nanowhiskers, albeit by an additional reaction with epoxypropyltrimethylammonium chloride, and not as a result of hydrolysis [110].

Marchessault and coworkers showed that rod-like cellulose crystallites could form birefringent and liquid crystal-like gels, reminiscent of spherulitic structures in polymers [111]. This liquid crystal gel formation occurs when cellulose whisker concentrations reach a high enough level to form a chiral nematic phase.

In the introduction to their paper on TEMPO-mediated oxidation of cellulose whiskers, Habibi et al. discussed the degree of cellulose polymerization expected after acid hydrolysis procedures [112]. After a rapid decrease in DP as a result of acid treatment, values reach a so-called level-off DP (LODP) that may remain constant during extended hydrolysis. The value of the LODP depends on the starting raw material, with values ranging from 250 for hydrolyzed cotton [27] to 140–200 for bleached wood pulp [113] and up to 600 for the highly crystalline *Valonia* cellulose [114].

14.6 COMMERCIAL DEVELOPMENT OF NANOCELLULOSES

14.6.1 Nanofibrillated Cellulose

This section summarizes the current commercial status of NFC and NCC. The development of this sector is proceeding rapidly and the companies and other organizations involved must be considered only in light of the current situation at the time of writing. Since nanocelluloses are in an early stage of commercialization, it is not easy to gather data, which in some cases are sensitive for the companies involved. In any case, the authors believe that a comprehensive study of the actual commercial situation of nanocelluloses is not presently available in the literature and so the summary presented here is considered necessary and potentially valuable, albeit that the commercialization of nanocellulose may evolve quite quickly in the coming years and the story will need to be updated.

As mentioned earlier, the manufacture of NFC was pioneered by Turbak and others in the United States in the late 1970s/early 1980s [21, 23]. The forcing of suspensions of wood-based cellulose fibers through a mechanical device, such as a high pressure homogenizer, was then shown to produce NFC. The major impediments for commercial success at this early stage were both economical and technical in nature. The very high energy consumption involved in the production of NFC as a result of the requirement for multiple-pass homogenization, together with the tendency of the homogenizers to become clogged, were found to be expensive economic and technical drawbacks.

In the early stages, interest in NFC was principally, if not totally, concentrated in the academic world. Discoveries in the following years were fundamental to creating an industrial interest in the field. In those years, it was discovered that it was easier to produce NFC from primary-wall materials (e.g., parenchyma cells from sugar beet and citrus fruits) than secondary-wall materials (e.g., wood), as the stabilization of these suspensions by glucuronic and galacturonic acid residues made them easier to delaminate [13, 115, 116]. As outlined earlier in this chapter, there has been a focus on energy-efficient production methods, whereby fibers are pretreated mechanically, chemically, and/or enzymatically by various methods before homogenization in order to decrease the total energy consumption. These various investigations have led to innovations in NFC manufacturing which are reflected in the patent literature. Just since the year 2000, this literature contains diverse concepts including, for example, a manufacturing process based on mechanical, chemical, or enzymatic means as well as a pigment in the dispersion [117], a similar process but incorporating mono-, di-, or oligosaccharides aimed at production of NFC as a shear-thinning additive [118], a method and a unit for producing a porous web from NFC suspension on a wire or wire mesh [119], the manufacturing of NFC pulps for use in paper products and composites [120], the use of an extruder to generate NFC with or without various additives [121], a process in which a pulp is first enzymatically treated, followed by mechanical treatment and then a repeat of this sequence to generate NFC in a more energy-efficient way [122], a method for producing NFC in which a cellulosic suspension and a solvent are pumped through an orifice and thereby subjected to a pressure drop [123], and the use of a hemicellulose-containing pulp subjected to refining and low-dosage enzymatic treatment followed by homogenization [41].

NFC has been particularly studied and developed in Europe and, especially in the north of the continent, the pulp and paper industry has identified NFC as an opportunity to add value and to stimulate this industrial sector. For many years, different organizations have been developing parallel competences and industrial processes that would allow rapid industrialization of NFC. In Sweden, the interests of Innventia AB in collaboration with KTH, the Royal Institute of Technology in Stockholm, and Chalmers University in Göteborg have focused more on chemical pulp, while in Norway, where mechanical pulping of wood is preferred, this type of process was selected by the Paper and Fiber Research Institute (PFI) and its research partner NTNU (the Norwegian University of Science and Technology). PFI and Innventia AB, part of the same group, presently claim a total NFC production capacity greater than 100 kg/day. Both these companies use homogenization techniques to produce NFC; however, production may differ in terms of the specific pulp pretreatments.

In Finland, another Nordic country with an important forestry sector, Stora Enso has made significant investment and in 2011 started up an NFC plant in the town of Imatra, in the east of the country. This investment in NFC technology, including the Imatra pre-commercial plant, is estimated to have cost approximately 10 million Euros. In November 2011, UPM, a Finnish-based company announced that it had started up pre-commercial production of fibrillar cellulose in Ottaniemi, which can be either micro- or nanoscale, and is currently developing new fibrillar cellulose applications in conjunction with VTT and Aalto University.

Another example of the interest in NFC in northern Europe is investment by the Norwegian company Borregaard AS as part of its approach to biorefining. In 2010, this company opened a plant with a capacity of 50–100 tonnes/year for production of Power Fiber, its commercial name for NFC, and had plans to install a manufacturing plant of 2000 tonnes/year in 2012–2013.

Still in Europe, Rettenmeier (Germany) and Cellucomp (Scotland) are producing and commercializing fibrillar cellulose. The interests of Rettenmeier seem more focused on using NFC as a binder, while Cellucomp is notable for having developed a process to obtain NFC from root vegetables.

There is also interest in NFC production and commercialization outside of Europe, as for example illustrated by the Japanese corporation Daicel Co., Ltd, which sells NFC under the trade name Celish K1. In Canada, GreenCore Composites Inc. is using NFC in combination with thermoplastics in its commercial resin.

14.6.2 Cellulose Whiskers

While Europe has been focusing more on NFC production, on the other side of the Atlantic, American and Canadian researchers have tended to concentrate more on cellulose whisker production, generally referred to industrially as NCC. Since the process for generating NCC is quite different from that for NFC, the size of the required plant and the corresponding investment are also different.

Canada has taken the lead in the production of NCC. On January 26, 2012, CelluForce officially inaugurated the world's first NCC demonstration plant at the Domtar pulp and paper mill site in Windsor, Québec. CelluForce is now ramping up its production of NCC and had a production target of 1000 kg (1 metric ton) per day in 2012. The company is trying to integrate NCC into the manufacturing process for various products and, with this objective, collaboration agreements are in effect between CelluForce and 15 companies based in Canada, the United States, Europe, and Asia. These companies represent four main industrial sectors: paints and coatings, films and barriers, textiles, and composites. In order to realize this project, the Canadian and Québec governments made significant contributions to the financing of the \$36 million plant, with \$23.2 million coming from Natural Resources Canada (Pulp and Paper Green Transformation Program and Transformative Technologies Program) and \$10.2 million from Québec's Natural Resources and Wildlife Department. CelluForce is a joint venture between FP Innovations and Domtar. ArboraNano is a Canadian research network that has been working in partnership with CelluForce to develop NCC applications.

14.7 SUMMARY

This chapter has provided an overview of past research efforts directed toward the production of nanocellulose with a particular focus on NFC and cellulose whiskers. Until quite recently, this research was very largely carried out on a small scale in various universities and institutes, typically by generating 1–2 kg batches of nanocellulose at a time, with low dry matter content. However, the situation has now changed and, as also outlined here, pilot-scale industrial facilities are coming on stream. These commercial developments will alter the picture significantly in terms of technology uptake since potential users will now be able to receive nanocellulose suspensions in quantities of tens or hundreds of kilograms rather than the smaller batches available heretofore. Although not covered as a topic within this chapter, at least two significant technical challenges still remain before NFC or cellulose whiskers can be fully adopted into a range of commercial applications. First, through hydrogen bonding, both forms of nanocellulose have a strong tendency to agglomerate during drying and consequently the key nanoscale characteristics are lost. Improved processes for drying nanocellulose that could alleviate this difficulty are a continuing subject of investigation. A successful drying process would allow nanocellulose to be transported in the dry or semidry state with very significant cost benefits. Second, chemical modification of nanocellulose is required in order to allow its effective blending with thermoplastics for new nanocomposite materials. This topic has been widely explored but, as yet, there are no processes that have been optimized in the context of rapid commercial adoption. It is therefore very likely that solutions to these two key challenges and the associated technology transfer will continue to stimulate new research and development in the coming years.

14.8 FURTHER READING

The references in this chapter hopefully include many of the key publications related to NFC or cellulose whisker production. This not an easy task given that publications relating to nanocellulose are now appearing at an estimated rate of several per day, which will likely increase further in the coming months and years. Considering articles included in the list of references that could be included in further reading, one might particularly suggest the article by Klemm et al. from 2009 [14], which provides a very interesting discussion regarding bacterial cellulose and its uses, particularly in medicine. Concerning the challenge of drying nanocellulose formulations, it is interesting to note that a research group at EMPA in Zurich, Switzerland, has just announced a new method for drying nanocellulose, which also allows its effective redispersion on the nanoscale back into aqueous suspension. This development builds on earlier research undertaken by Eyholzer and others [124]. Other reports [125, 126] on processes for drying nanocellulose also make useful further reading. Addressing the other key challenge in terms of nanocellulose use, there is a vast and expanding literature addressing chemical modification of nanocellulose for specific applications, ranging from reinforcement of thermoplastics for bulk uses such as food packaging through to higher value end products in medicine and drug delivery, which is best accessed through state-of-the-art online searching.

REFERENCES

[1] Azizi Samir, M.A.S., Alloin, F., and Dufresne, A. (2005) Review of recent research into cellulosic whiskers, their properties and their application in nanocomposite field. *Biomacromolecules*, *6*, 612–626.

[2] Kimura, S., Itoh, T. (2007) Biogenesis and function of cellulose in the tunicates. In: *Cellulose: Molecular and Structural Biology*, eds. R.M. Brown Jr., I.M. Saxena. New York: Springer, pp. 217–236, Chapter 13.

[3] Zhou, W., Shao, Z.P., Ran, R., Jin, W.Q., and Xu, N.P. (2008) Functional nanocomposite oxides synthesized by environmental-friendly auto-combustion within a micro-bioreactor. *Materials Research Bulletin*, *43*, 2248–2259.

[4] Marchessault, R.H., Sarko, A. (1967) X-ray structure of polysaccharides. In: *Advanced Carbohydrate Chemistry*, Vol. 22, ed. M.L. Wolfrom. New York: Academic Press, pp. 421–482.

[5] Walton, A.G., Blackwell, J. (1973) The polysaccharides. In: *Biopolymers*, Vol. 22, eds. A.G. Walton, J. Blackwell. New York: Academic Press, p. 468.

[6] Marchssault, R.H., Sundararajan, P.R. (1983) Cellulose. In: *Polysaccharides*, ed. G. Aspinall. New York: Academic Press, p. 11.

[7] O'Sullivan, A.C. (1997) Cellulose: The structure slowly unravels. *Cellulose*, *4*, 173–207.

[8] Blackwell, J., Marchessault, R.H. (1971) Infrared spectroscopy of cellulose. In: *Cellulose and Cellulose Derivatives*, eds. N. Bikales, L.E. Segal. New York: Wiley-Interscience.

[9] Blackwell, J. (1982) The macromolecular organization of cellulose and chitin. In: *Cellulose and Other Natural Polymer Systems*, ed. R.M. Brown Jr. New York: Plenum Press.

[10] Hess, K., Kissig, H. (1941) Zur Kenntnis der Hochtemperatur-Modifikation der Cellulose (Cellulose IV). *Zeitschrift für Physikalische Chemie B*, *49*, 235–239.

[11] Gardiner, E.S., Sarko, A. (1985) Packing analysis of carbohydrates and polysaccharides. 16. The crystal structures of celluloses IVI and IVII. *Canadian Journal of Chemistry*, *63*, 173–180.

[12] Habibi, Y., Lucia, L.A., and Rojas, A.J. (2010) Cellulose nanocrystals: Chemistry, self-assembly and applications. *Chemical Reviews*, *110*, 3479–3500.

[13] Klemm, D., Schumann, D., Kramer, F., Hessler, N., Hornung, M., Schmauder, H.-P., and Marsch, S. (2006) Nanocelluloses as innovative polymers in research and application. In: *Advances in Polymer Science (Polysaccharides II)*, ed. D. Klemm. Berlin, Heidelberg: Springer, 205, 49–96.

[14] Klemm, D., Schumann, D., Kramer, F., Hessler, N., Koth, D., and Sultanova, B. (2009) Nanocellulose materials—Different cellulose, different functionality. *Macromolecular Symposia*, *280*, 60–71.

[15] Siró, I., Plackett, D. (2010) Microfibrillated cellulose and new nanocomposite materials: A review. *Cellulose*, *17*, 459–494.

[16] Kose, R., Mitani, I., Kasai, W., and Kondo, T. (2011) "Nanocellulose" as a single nanofiber prepared from pellicle secreted by *Gluconacetobacter xylinus* using aqueous counter collision. *Biomacromolecules*, *12*, 716–720.

[17] Guhados, G., Wan, W., and Hutter, J.L. (2005) Measurement of the elastic modulus of single cellulose fibers using atomic force microscopy. *Langmuir*, *21*, 6642–6646.

[18] Saheb, D.N., Jog, J.P. (1999) Natural fiber polymer composites: A review. *Advances in Polymer Technology*, *18*, 351–363.

[19] Cjaza, W., Romanovicz, D., and Brown, R.M. (2004) Structural investigations of microbial cellulose produced in stationary and agitated culture. *Cellulose*, *11*, 403–411.

[20] Spence, K., Habibi, Y., and Dufresne, A. (2011) Nanocellulose-based composites. In: *Cellulose Fibers: Bio- and Nano-Polymer Composites*, eds. S. Kalia, B.S. Kaith, and I. Kaur. Berlin, Heidelberg: Springer-Verlag, pp. 179–214, Chapter 7.

[21] Herrick, F.W., Casebier, R.L., Hamilton, J.K., and Sandberg, K.R. (1983) Microfibrillated cellulose: Morphology and accessibility. *Journal of Applied Polymer Science: Applied Polymer Symposia*, *37*, 797–813.

[22] Turbak, A.F., Snyder, F.W., and Sandberg, K.R. (1983) Microfibrillated cellulose, a new cellulose product: Properties, uses and commercial potential. *Journal of Applied Polymer Science: Applied Polymer Symposia*, *37*, 815–827.

[23] Turbak, A.F., Snyder, F.W., (1985) Microfibrillated cellulose and process for producing it. Patent CH 648071 (A5).

[24] Klemm, D., Kramer, F., Moritz, S., Lindström, T., Ankerfors, M., Gray, D., and Dorris, A. (2011) Nanocelluloses: A new family of nature-based materials. *Angewandte Chemie International Edition*, *50*, 5438–5466.

[25] Rånby, B.G., Ribi, E. (1950) Über den feinbau des zellulose. *Experientia*, *6*, 12–14.

[26] Nickerson, R.F., Habrle, J.A. (1947) Cellulose intercrystalline structure—Study by hydrolytic methods. *Industrial and Engineering Chemistry*, *39*, 1507–1512.

[27] Battista, O.A. (1950) Hydrolysis and crystallization of cellulose. *Industrial and Engineering Chemistry*, *42*, 502–507.

[28] Nakagaito, A.N., Yano, H. (2004) The effect of morphological changes from pulp fiber towards nano-scale fibrillated cellulose on the mechanical properties of high-strength plant fiber based composites. *Applied Physics A: Materials Science and Processing*, *78*, 547–552.

[29] Lindström, T. (2007) Towards new perspectives in paper chemistry. *Das Papier IPW*, *10*, 32–36.

[30] Iwamoto, S., Nakagaito, A.N., and Yano, H. (2007) Nano-fibrillation of pulp fibers for the processing of transparent nanocomposites. *Applied Physics A: Materials Science and Processing*, *89*, 461–466.

[31] Uetani, K., Yano, H. (2011) Nanofibrillation of wood pulp using a high-speed blender. *Biomacromolecules*, *12*, 348–353.

[32] Chakraborty, A., Sain, M., and Kortschot, M. (2005) Cellulose microfibrils: A novel method of preparation using high shear refining and cryocrushing. *Holzforschung*, *59*, 102–107.

[33] Wang, B., Sain, M. (2007) Dispersion of soybean stock-based nanofiber in a plastic matrix. *Polymer International*, *56*, 538–546.

[34] Alemdar, A., Sain, M. (2008) Isolation and characterization of nanofibers from agricultural residues—Wheat straw and soy hulls. *Bioresource Technology*, *99*, 1664–1671.

[35] Bhatnagar, A., Sain, M. (2005) Processing of cellulose nanofiber-reinforced composites. *Journal of Reinforced Plastics and Composites*, *24*, 1259–1268.

[36] Wang, B., Sain, M. (2007) Isolation of nanofibers from soybean source and their reinforcing capability on synthetic polymers. *Composites Science and Technology*, *67*, 2521–2527.

[37] Zimmerman, T., Bordeanu, N., and Strub, E. (2010) Properties of nanofibrillated cellulose from different raw materials and its reinforcement potential. *Carbohydrate Polymers*, *79*, 1086–1093.

[38] Spence, K.L., Venditti, R.A., Rojas, O.J., Habibi, Y., and Pawlak, J.J. (2011) A comparative study of energy consumption and physical properties of microfibrillated cellulose produced by different processing methods. *Cellulose*, *18*, 1097–1111.

[39] Abe, K., Iwamoto, S., and Yano, H. (2007) Obtaining cellulose nanofibers with a uniform width of 15nm from wood. *Biomacromolecules*, *8*, 3276–3278.

[40] Pääkkö, M., Ankerfors, M., Kosonen, H., Nykänen, A., Ahola, S., Österberg, M., Ruokolainen, J., Laine, J., Larsson, P.T., Ikkala, O., and Lindström, T. (2007) Enzymatic hydrolysis combined with mechanical shearing and high-pressure

homogenization for nanoscale cellulose fibrils and strong gels. *Biomacromolecules*, *8*, 1934–1941.

[41] Lindström, T., Ankerfors, M., and Henriksson, G. (2007) Method for the manufacturing of microfibrillated cellulose. WO 2007091942.

[42] Henriksson, M., Henriksson, G., Berglund, L.A., and Lindström, T. (2007) An environmentally friendly method for enzyme-assisted preparation of microfibrillated cellulose (MFC) nanofibers. *European Polymer Journal*, *43*, 3434–3441.

[43] Engström, A.-C., Ek. M., and Henriksson, G. (2006) Improved accessibility and reactivity of dissolving pulp for the viscose process: Pretreatment with monocomponent endoglucanase. *Biomacromolecules*, *7*, 2017–2031.

[44] Siqueira, G., Tapin-Lingua, S., Bras, J., da Silva Perez, D., and Dufresne, A. (2010) Morphological investigation of nanoparticles obtained from combined mechanical shearing, and enzymatic and acid hydrolysis of sisal fibers. *Cellulose*, *17*, 1147–1158.

[45] Siqueira, G., Tapin-Lingua, S., Bras, J., da Silva Perez, D., and Dufresne, A. (2011) Mechanical properties of natural rubber nanocomposites reinforced with cellulosic nanoparticles obtained from combined mechanical shearing, and enzymatic and acid hydrolysis of sisal fibers. *Cellulose*, *18*, 57–65.

[46] Saito, T., Nishiyama, Y., Putaux, J.-L., Vignon, M., and Isogai, A. (2006) Homogeneous suspensions of individualized microfibrils for TEMPO-catalyzed oxidation of native celluloses. *Biomacromolecules*, *7*, 1687–1691.

[47] Saito, T., Isogai, A. (2007) Preparation of cellulose single microfibrils by TEMPO-mediated oxidation of native cellulose. *Cellulose Communications (Japan)*, *14*, 62–66.

[48] Saito, T., Kimura, T., Nishiyama, Y., and Isogai, A. (2007) Cellulose nanofibers prepared by TEMPO-mediated oxidation of native cellulose. *Biomacromolecules*, *8*, 2485–2491.

[49] Hirota, M., Tamura, N., Saito, T., Isogai, A. (2010) Water dispersion of cellulose II nanocrystals prepared by TEMPO-mediated oxidation of mercerized cellulose at pH 4.8. *Cellulose*, *17*, 279–288.

[50] Isogai, T., Saito, T., and Isogai, A. (2011) Wood cellulose nanofibrils prepared by TEMPO electro-mediated oxidation. *Cellulose*, *18*, 421–431.

[51] Saito, T., Hirota, M., Tamura, N., Kimura, S., Fukuzumi, H., Heux, L., and Isogai, A. (2009) Individualization of nano-sized plant cellulose fibrils by direct surface carboxylation using TEMPO catalyst under neutral conditions. *Biomacromolecules*, *10*, 1992–1996.

[52] Isogai, A., Saito, T., and Fukuzumi, H. (2011) TEMPO-oxidized cellulose nanofibers. *Nanoscale*, *3*, 71–85.

[53] Eichhorn, S.J., Dufresne, A., Aranguren, M., Marcovich, N.E., Capadona, J.R., Rowan, S.J., Weder, C., Thielemans, W., Roman, M., Renneckar, S., Gindl, W., Veigel, S., Keckes, J., Yano, H., Abe, K., Nogi, M., Nakagaito, A.N., Mangalam, A., Simonsen, J., Benight, A.S., Bismarck, A., Berglund, L.A., and Peijs, T. (2010) Review: Current international research into cellulose nanofibres and nanocomposites. *Journal of Materials Science*, *45*, 1–33.

[54] Johnson, R.K., Zink-Sharp, A. (2008) A new bio-based nanocomposite: Fibrillated TEMPO-oxidized celluloses in hydroxypropylcellulose matrix. *Cellulose*, *16*, 227–238.

[55] Walecka, J.A. (1956) An investigation of low degree of substitution carboxymethylcelluloses. *Tappi*, *39*, 458–463.

[56] Wågberg, L., Winter, L., Ödberg, L., and Lindström, T. (1987) On the charge stoichiometry upon adsorption of a cationic polyelectrolyte on cellulosic materials. *Colloids and Surfaces*, *27*, 163–173.

[57] Carlsson, G., Kolseth, P., and Lindström, T. (1983) Polyelectrolyte swelling behavior of chlorite delignified spruce wood fibers. *Wood Science and Technology*, *17*, 69–73.

[58] Iwamoto, S., Abe, K., and Yano, H. (2008) The effect of hemicelluloses on wood pulp nanofibrillation and nanofiber network characteristics. *Biomacromolecules*, *9*, 1022–1026.

[59] Tanem, B.S., Kvien, I., van Helvoort, A.T.J., and Oksman, K. (2006) Morphology of cellulose and its nanocomposites. In: *Cellulose Nanocomposites: Processing, Characterization, and Properties*, eds. K. Oksman, M. Sain. Washington, DC: American Chemical Society, *938*, 48–62.

[60] Hubbe, M.A., Rojas, O.J., Lucia, L.A., and Sain, M. (2008) Cellulosic nanocomposites: A review. *BioResources*, *3*, 929–980.

[61] Aulin, C., Ahola, S., Josefsson, P., Nishino, T., Hirose, Y., Österberg, M., and Wågberg, L. (2009) Nanoscale cellulose films with different crystallinities and mesostructures—Their surface properties and interaction with water. *Langmuir*, *25*, 7675–7685.

[62] Larsson, P., Wickholm, K., and Iversen, T. (1997) A CP/MAS ^{13}C NMR investigation of molecular ordering in celluloses. *Carbohydrate Research*, *302*, 19–25.

[63] Wickholm, K., Larsson, P., and Iversen, T. (1998) Assignment of non-crystalline forms in cellulose I by CP/MAS ^{13}C NMR spectroscopy. *Carbohydrate Research*, *312*, 123–129.

[64] Dufresne, A., Cavaillé, J.-Y., and Vignon, M.R. (1997) Mechanical behavior of sheets prepared from sugar beet cellulose microfibrils. *Journal of Applied Polymer Science*, *64*, 1185–1194.

[65] Leitner, J., Hinderstoisser, B., Wastyn, M., Keckes, J., and Gindl, W. (2007) Sugar beet cellulose nanofibril-reinforced composites. *Cellulose*, *14*, 419–425.

[66] Siqueira, G., Bras, J., and Dufresne, A. (2010) Luffa cylindrica as a lignocellulosic source of fiber, microfibrillated cellulose and cellulose nanocrystals. *BioResources*, *5*, 727–740.

[67] Spence, K.L., Venditti, R.A., Rojas, O.J., Habibi, Y., and Pawlak, J.J. (2010) The effect of chemical composition on microfibrillar cellulose films from wood pulps: Mechanical processing and physical properties. *Cellulose*, *17*, 835–848.

[68] Chinga-Carrasco, G., Syverud, K. (2010) Computer-assisted quantification of the multi-scale structure of films made of nanofibrillated cellulose. *Journal of Nanoparticle Research*, *12*, 841–851.

[69] Fukuzumi, H., Saito, T., Iwata, T., Kumamoto, Y., and Isogai, A. (2009) Transparent and high gas barrier films of cellulose nanofibers prepared by TEMPO-mediated oxidation. *Biomacromolecules*, *10*, 162–165.

[70] ISO Standard 5351 (2010) Pulps—Determination of limiting viscosity number in cupri-ethylenediamine solution. International Organization for Standardization, Geneva, Switzerland.

[71] Yano, H., Nakahara, S. (2004) Bio-composites produced from plant microfiber bundles with a nanometer unit web-like network. *Journal of Materials Science*, *39*, 1635–1638.

[72] Page, D.H., El-Hosseiny, F. (1983) Mechanical properties of single wood pulp fibres. Part VI. Fibril angle and the shape of the stress-strain curve. *Journal of Pulp and Paper Science*, *9*, 99–100.

[73] Sakurada, I., Nukushina, Y., and Ito, T. (1962) Experimental determination of the elastic modulus of crystalline regions in oriented polymers. *Journal of Polymer Science*, *57*, 651–660.

[74] Zimmerman, T., Pöhler, E., and Geiger, T. (2004) Cellulose fibrils for polymer reinforcement. *Advanced Engineering Materials*, *6*, 754–761.

[75] Iwamoto, S., Kai, W., Isogai, A., and Iwata, T. (2009) Elastic modulus of single cellulose microfibrils from tunicate measured by atomic force microscopy. *Biomacromolecules*, *10*, 2571–2576.

[76] Sperling, L.H. (1984) *Introduction to Physical Polymer Science*. New York: John Wiley & Sons Inc.

[77] Park, S., Baker, J.O., Himmel, M.E., Parilla, P.A., and Johnson, D.K. (2010) Cellulose crystallinity index: Measurement techniques and their impact on interpreting cellulase performance. *Biotechnology for Biofuels*, *3*, 10.

[78] Heux, L., Dinand, E., and Vignon, M.R. (1999) Structural aspects in ultrathin cellulose microfibrils followed by ^{13}C CP-MAS NMR. *Carbohydrate Polymers*, *40*, 115–124.

[79] Agoda-Tandjawa, G., Durand, S., Berot, C., Blassel, C., Gaillard, C., Garnier, C. and Doublier, J.-L. (2010) Rheological characterization of microfibrillated cellulose suspensions after freezing. *Carbohydrate Polymers*, *80*, 677–686.

[80] Saito, T., Isogai, A. (2004) TEMPO-mediated oxidation of native cellulose. The effect of oxidation conditions on chemical and crystal structures of the water-insoluble fractions. *Biomacromolecules*, *5*, 1983–1989.

[81] Wågberg, L., Decher, G., Norgren, M., Lindstrom, T., Ankerfors, M., and Axnäs, K. (2008) The build-up of polyelectrolyte multilayers of microfibrillated cellulose and cationic polyelectrolytes. *Langmuir*, *24*, 784–795.

[82] Katz, K., Beatson, R.P., and Scallan, A.M. (1984) The determination of strong and weak acidic groups in sulphite pulps. *Svensk Papperstidning*, *87*, R48–R53.

[83] Eichhorn, S.J. (2010) Cellulose nanowhiskers: Promising materials for advanced applications. *Soft Matter*, *7*, 303–315.

[84] Bolio-Lopez, G., Valadez-Gonzalez, A., Veleva, L., and Andreeva, A. (2011) Cellulose whiskers from agro-industrial banana wastes: Isolation and characterization. *Revista Mexicana de Ingeniería Química*, *2*, 291–299.

[85] De Morais Teixeira, E., Bondancia, T.J., Teodoro, K.B.R., Correa, A.C., Marconcini, J.M., and Mattoso, L.H.C. (2010) Sugarcane bagasse whiskers: Extraction and characterizations. *Industrial Crops and Products*, *33*, 63–66.

[86] Li, R., Cai, Y., Yao, J., Fei, J., Li, Y., and Feng, J. (2009) Cellulose whiskers extracted from mulberry: A novel biomass production. *Carbohydrate Polymers*, *76*, 94–99.

[87] Azizi Samir, M.A.S., Alloin, F., Paillet, M., and Dufresne, A. (2004) Tangling effect in fibrillated cellulose reinforced nanocomposites. *Macromolecules*, *37*, 4313–4316.

[88] Beck-Candenedo, S., Roman, M., and Gray, D. (2005) Effect of conditions on the properties behavior of wood cellulose nanocrystals suspensions. *Biomacromolecules*, *6*, 1048–1054.

[89] Weder, C., Capadona, J.R., Shanmuganathan, K., and Rowan, S.J. (2010) Bio-inspired mechanically adaptive nanocomposites. *Polymer Preprints (American Chemical Society, Division of Polymer Chemistry)*, *51*, 269–270.

[90] Bondeson, D., Mathew, A., and Oksman, K. (2006) Optimization of the isolation of nanocrystals from microcrystalline cellulose by acid hydrolysis. *Cellulose*, *13*, 171–180.

[91] Araki, J., Wada, M., Kuga, S., and Okano, T. (1998) Flow properties of microcrystalline cellulose suspension prepared by acid treatment of native cellulose. *Colloids and Surfaces A: Physicochemical and Engineering Aspects*, *142*, 75–82.

[92] Siqueira, G., Bras, J., and Dufresne, A. (2009) Cellulose whiskers versus microfibrils: Influence of the nature of the nanoparticle and its surface functionalization on the thermal and mechanical properties of nanocomposites. *Biomacromolecules*, *10*, 425–432.

[93] Filson, P.B., Dawson-Andoh, B., and Schwegler-Berry, D. (2009) Enzymatic-mediated production of cellulose nanocrystals from recycled pulp. *Green Chemistry*, *11*, 1808–1814.

[94] Chirat, C., Lachenal, D., and Dufresne, A. (2010) Biorefinery in a Kraft pulp mill. From bioethanol to cellulose nanocrystals. *Cellulose Chemistry and Technology*, *44*, 59–64.

[95] Duran, N., Lemes, A.P., Duran, M., Freer, J., and Baeza, J. (2011) A minireview of cellulose nanocrystals and its potential integration as co-product in bioethanol production. *Journal of the Chilean Chemical Society*, *56*, 672–677.

[96] Oksman, K., Etang, J.A., Mathew, A.P., and Jonoobi, M. (2011) Cellulose nanowhiskers separated from a bioresidue from wood bioethanol production. *Biomass and Bioenergy*, *35*, 146–152.

[97] Zhu, J.Y., Sabo, R., and Luo, X. (2011) Integrated production of nano-fibrillated cellulose and cellulosic biofuel (ethanol) by enzymatic fractionation of wood fibers. *Green Chemistry*, *13*, 1339–1344.

[98] Leistritz, F.L., Senechal, D.M., Stower, M.D., McDonald, W.F., Safron, C.M., and Hodur, N.M. (2006) Preliminary feasibility analysis for an integrated biomaterials and ethanol biorefinery using wheat straw feedstock. Agribusiness and Applied Economic report No. 590. http://ageconsearch.umn.edu/handle/23500 (accessed March 18, 2012).

[99] Elazzouzi-Hafraoui, S., Nishiyama, Y., Putaux, J.-L., Heux, L., Dubreuil, F., and Rochas, C. (2008) The shape and size distribution of crystalline nanoparticles prepared by acid hydrolysis of native cellulose. *Biomacromolecules*, *9*, 57–65.

[100] Itoh, T., Brown, R.M. (1988) Development of cellulose synthesizing complexes in *Boergesenia* and *Valonia*. *Protoplasma*, *144*, 160–169.

[101] Brown, R.M., Willison, J.H., and Richardson, C.L. (1976) Cellulose biosynthesis in *Acetobacter xylinum*: Visualization of the site of synthesis and direct measurement of the *in vivo* process. *Proceedings of the National Academy of Sciences of the United States of America*, *73*, 4565–4569.

[102] Kontturi, E., Vuorinen, T. (2009) Indirect evidence of supramolecular changes within cellulose microfibrils of chemical pulp fibers upon drying. *Cellulose*, *16*, 65–74.

[103] Rusli, R., Eichhorn, S.J. (2008) Determination of the stiffness of cellulose nanowhiskers and the fiber-matrix interface in a nanocomposite using Raman spectroscopy. *Applied Physics Letters*, *93*, 033111.

[104] Šturcová, A., Davies, G.R., and Eichhorn, S.J. (2005) Elastic modulus and stress transfer properties of tunicate cellulose whiskers. *Biomacromolecules*, *6*, 1055–1061.

[105] Tashiro, K., Kobayashi, M. (1991) Theoretical evaluation of three-dimensional elastic constants of native and regenerated celluloses: Role of hydrogen bonds. *Polymer*, *32*, 1516–1526.

[106] Eichhorn, S.J., Davies, G.R. (2006) Modelling the crystal deformation of native and regenerated cellulose. *Cellulose*, *13*, 291–307.

[107] Bergenstråhle, M., Berglund, L.A., and Mazeau, K. (2007) Thermal response in crystalline Iβ cellulose: A molecular dynamics study. *Journal of Physical Chemistry Part B*, *111*, 9138–9145.

[108] Nishino, T., Takano, K., and Nakamae, K. (1995) Elastic modulus of the crystalline regions of cellulose polymorphs. *Journal of Polymer Science Part B: Polymer Physics*, *33*, 1647–1651.

[109] Diddens, I., Murphy, B., Krisch, M., and Muller, M. (2008) Ansiotropic elastic properties of cellulose measured using inelastic X-ray scattering. *Macromolecules*, *41*, 9755–9759.

[110] Hasani, M., Cranston, E.D., Westman, G., and Gray, D.G. (2008) Cationic surface functionalization of cellulose nanocrystals. *Soft Matter*, *4*, 2238–2244.

[111] Marchessault, R.H., Morehead, F.F., and Walter, N.M. (1959) Liquid crystal systems from fibrillar polysaccharides. *Nature*, *184*, 632–633.

[112] Habibi, Y., Chanzy, H., and Vignon, M.R. (2006) TEMPO-mediated surface oxidation of cellulose whiskers. *Cellulose*, *13*, 679–687.

[113] Battista, O.A., Coppick, S., Howsmon, J.A., Morehead, F.F., and Sisson, W.A. (1956) Level-off degree of polymerization. Relation to polyphase structure of cellulose fibers. *Industrial and Engineering Chemistry*, *48*, 333–335.

[114] Kai, A. (1976) The fine structure of *Valonia* microfibril. Gel permeation chromatographic studies of *Valonia* cellulose. *Sen-I Gakkaishi*, *32*, T326–T334.

[115] Dinand, E., Chanzy, H., and Vignon, M.R. (1996) Parenchymal cell cellulose from sugar beet pulp. Preparation and properties. *Cellulose*, *3*, 183–188.

[116] Dinand, E., Vignon, M.R. (2001) Isolation and NMR characterisation of a (4-O-methyl-d-glucurono)-D-xylan from sugar beet pulp. *Carbohydrate Research*, *330*, 285–288.

[117] Heiskanen, I., Backfolk, K. (2011) A composition comprising microfibrillated cellulose and a process for the production of a composition. WO 2011141877.

[118] Heiskanen, I., Backfolk, K. (2011) A composition comprising microfibrillated cellulose and a process for production of a composition. WO 2011141876.

[119] Kajanto, I. (2011) A method and a system for making a fibre-containing product. WO 2011080386.

[120] Nuopponen, M., Paeivaelaeinen, T., Laukkanen, A., and Paltakari, J. (2011) Method for manufacturing nanofibrillated cellulose pulp and use of the pulp in paper manufacturing or in nanofibrillated cellulose composites. WO 2011064441.

[121] Heiskanen, I., Harlin, A., Backfolk, K., and Laitinen, R. (2011) Process for production of microfibrillated cellulose in an extruder and microfibrillated cellulose produced according to the process. WO 2011051882.

[122] Heiskanen, I., Backfolk, K., Vehvilaeinen, M., Kamppuri, T., and Nousiainen, P. (2011) Process for producing microfibrillated cellulose. WO 2011004301.

[123] Wichmann, J.-U., Haugen, O. (2010) Method for producing microfibrillated cellulose. EP 2196579.

[124] Eyholzer, C., Bordeanu, N., Lopez-Suevos, F., Rentsch, D., Zimmermann, T., and Oksman, K. (2010) Preparation and characterization of water-redispersible nanofibrillated cellulose in powder form. *Cellulose*, *17*, 19–30.

[125] Peng, Y., Gardner, D.J., and Han, Y. (2012) Drying cellulose nanofibrils: In search of a suitable method. *Cellulose*, *19*, 91–102.

[126] Voronova, M.I., Zakharov, A.G., Kuznetsov, O.Y., and Surov, O.V. (2012) The effect of drying technique for nanocellulose dispersions on properties of dried materials. *Materials Letters*, *68*, 164–167.

CHAPTER 15

Bacterial Cellulose

ELIANE TROVATTI

15.1 INTRODUCTION

Bacterial cellulose (BC) is a fascinating polysaccharide produced by bacteria in the form of a swollen gel-like membrane with the function of a protective structure. In nature, the BC membrane is secreted over the microorganism colony and acts as a protective layer against detrimental environmental conditions, playing multiple functions in its growth and survival. The BC membrane is able to prevent drying of natural substrates such as rotting fruit, retaining moisture close to the bacterial cells. Its low transparency to ultraviolet (UV) light provides protection against the damaging effects of the sun's radiation. BC can also prevent the colonization of the natural substrate by competitors organisms such as molds and other bacteria [1].

BC was described for the first time in 1886 by Brown [2]. In his work, Brown reported the discovery of the new exopolysaccharide and the strategies for isolation of the producing microorganism and identification of the polysaccharide. This discovery started from a study about microorganisms for vinegar fermentation, *Bacterium aceti*, when Brown found a new and peculiar acetic "ferment" known as the "vinegar plant" or "mother," which secreted a pelicule on the surface of the liquid media. The "ferment" differed greatly in appearance from any other form of *B. aceti* he had noticed, probably due to the fact that it was a distinct organism. To ascertain this hypothesis, and also to enable the studies of the chemical properties of the pelicule, he isolated a pure culture of the producing microorganism. When this new microorganism was cultivated in appropriate culture media, a jelly-like translucent membrane was formed on the surface of the liquid. The membrane rapidly increased until the whole surface was covered. After several sets of chemical tests, the membrane was

Biopolymer Nanocomposites: Processing, Properties, and Applications, First Edition. Edited by Alain Dufresne, Sabu Thomas, and Laly A. Pothan.

identified as cellulose. Regarding its ability to forming cellulose, the "vinegar plant" was denominated *Bacterium xylinum* (from Greek *xýlinalína*, literally "linen from wood"). Nowadays BC is widely consumed in several parts of the world as a healthy diet food (*nata de coco*).

Although it was first described in 1886, BC has attracted more attention in recent decades, and the increasing interest in this material is due to to its unique properties and applications.

15.2 PRODUCING MICROORGANISMS

Cellulose is synthesized by several living organisms, from prokaryotes to animals. Among the prokaryotes, BC is synthesized by bacteria belonging to the genus *Gluconacetobacter*, *Rhizobium*, *Agrobacterium*, *Escherichia* [3, 4], *Enterobacter* [5], and *Salmonella* [6], among others. The *Agrobacterium* is a plant pathogenic infectious microorganism that uses cellulose to attach its cells to the plant hosts cells [7, 8]. Similarly to *Agrobacterium*, the role of cellulose in *Rhizobium* species in the plant–microorganism attachment process, associated with atmospheric nitrogen fixing, is perhaps one of the most important symbiotic relationships of the biosphere [8, 9].

Eukaryotic organisms such fungi, amoebae, and trees produce cellulose as part of the basic structure of their cell walls. Trees and cotton plants are the main sources of cellulose for industrial purposes. Animals able to produce cellulose are the tunicates [10]. The cellulose from *Valoniaventricosa*, a green algae, is one of the most perfect crystalline forms of cellulose in nature [11]. However, the swollen cellulose in membranous form is synthesized only by bacteria and, in abundance, by bacteria belonging to the *Gluconacetobacter* genera.

The *Gluconacetobacter* genus belongs to the Acetobacteriaceae family. This family has undergone many taxonomic changes with respect to their genus status. Thus, the *Gluconacetobacter* genus was described for the first time in 1997 [12], and several new species were described in the last years within this genera [13]. The microorganisms belonging to the *Gluconacetobacter* genus are Gram negative, aerobic, nonpathogenic to humans and animals, rod-shaped, and occurring singly or, many times, in pairs. The optimum temperature for growth is around 30°C and the pH ranges from 3.5 to 6.0. In nature, *Gluconacetobacter* species are found in vinegar, tea fungus, sugarcane, mealy bug, flowers, and fruits in decomposition [1, 13].

There are several species within the *Gluconacetobacter* genus differentiated by phenotypic characteristics, mainly based on the consumption of substrate and metabolites. The production of cellulose can be observed in some strains such *G. europaeus*, *G. intermedius*, *G. oboediens* [13], *G. nataicola* [14], *G. sacchari* [15], and *G. xylinus* [13]. New cellulose-producing microorganisms have been isolated and described from food sources and wastes, in an attempt to obtain highly productive strains for industrial applications of BC [13, 15–18].

15.2.1 Isolation of Producing Microorganisms and Cellulose Identification

The isolation of the cellulose-producing strains begins with the search for a microorganism source. It is easily identified by the accumulation of cellulose on the substrate, such as in Kombucha tea [16], where a dense cellulose membrane can be found on the surface of the liquid. The second step is based on a simple microbiological approach for bacteria isolation, which starts with the inoculation of a sample of the microorganisms in a liquid culture media. After an adequate period of time for the microorganisms to grow, the liquid culture medium is diluted and spread on solid media in petri dish. A usual culture medium that can be used for this purpose was described by Hestrin and Schramm (HS), composed (in g/L) by 20.0 glucose, 5.0 peptone, 5.0 yeast extract, 1.15 citric acid, 2.7 phosphate disodium [19], and 20.0 agar–agar. Each isolated colony from solid media should be inoculated in assay tubes containing liquid culture media for BC production (HS). If a membrane of cellulose appears on the surface of the liquid, after an adequate period (5–10 days), the test confirms the cellulose production. The study of the morphology of the colony helps the identification of the cellulose-producing bacteria. Cellulose-producing colonies are circular, around 1 mm in diameter, transparent, and slightly yellow with a regular border after 3 days of incubation. Some species of cellulose-producing bacteria have colonies with a gelatinous aspect, as shown in Figure 15.1. The morphology of the colony also shows the presence of nonproducing cells, which is common in agitated culture.

The identification of cellulose can be performed by chemical or spectrophotometric techniques, after purification. The chemical method for cellulose identification involves hydrolysis via chemical or enzymatic reaction, generating glucose monomers, which can be detected by chromatographic methods. In addition, practical methods such as Fourier transform infrared (FT-IR), nuclear magnetic resonance (NMR), and X-ray diffractometry (XRD) are used for cellulose identification.

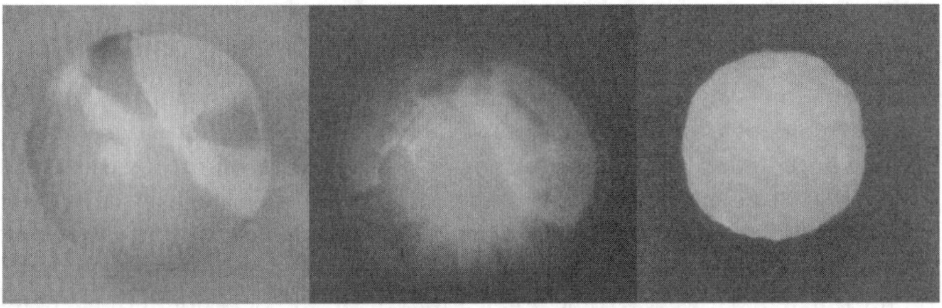

Figure 15.1 Morphological characteristic of the colonies of bacteria able to synthesize cellulose.

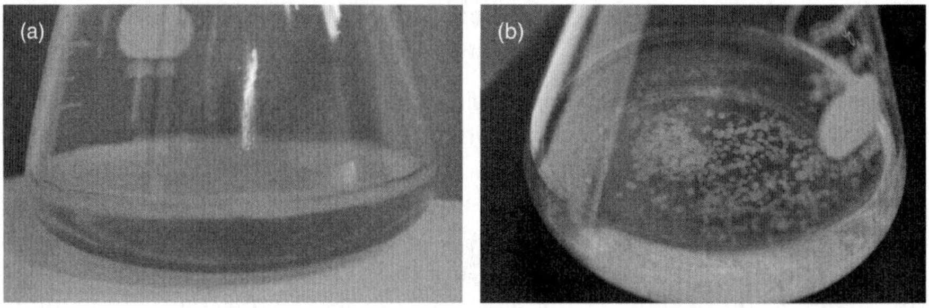

Figure 15.2 Bacterial cellulose membrane produced in static condition on the surface of culture media (a) and BC pellets produced under agitation (b).

In nature, cellulose-producing microorganisms are in symbiosis with other microorganisms, mostly yeasts, which are able to survive at low pH environments, typical of the *Acetobacter* family. For both industrial and research purposes, the strain should be pure, assuring no competition for nutrient sources and for reproducibility of the results.

15.2.2 Production of Bacterial Cellulose

BC is produced under adequate pH, temperature, oxygen rate, and nutrient sources. Changes in these parameters can affect the growth of the microorganisms and, consequently, the product formation and its properties. Concerning its form, BC can be obtained as a continuous membrane or in pellets (Fig. 15.2), according to the process used during the production phase, which can be static or agitated, respectively. Beyond the ideal pH and temperature, microorganisms also need adequate sources of nutrients such as carbon, nitrogen, oxygen, and mineral salts. Some properties of cellulose are dependent on the type of process and conditions used during the synthesis.

15.3 STATIC AND AGITATED CULTURES

The shape and structure of BC directs its application to several specific and highly technological fields. The industrial production of BC under shaking cultivation could be advantageous for some industrial purposes, since the production of pellets of cellulose avoids milling and saves energy. This is an economical alternative for BC production when compared with the high cost of conventional static cultivation, which is often not applicable to large-scale production [20]. However, some applications of BC are dependent of its membranous form. Then, the choice of the production process should be in agreement with the application of cellulose.

15.3.1 Shaking Cultivation

Perfect spheres, irregular granules, stellate, and fibrous cellulose forms can be obtained by tuning the parameters of cultivation in the agitated process, such as the speed of agitation, oxygen transfer rate, the initial number of viable cells, pH, temperature, shear stress, volume, and design of the reactor.

Several attempts to improve the yield via the aerated and agitated cultivation method focus on the design of specific reactors for BC production [20]. High agitation is required to provide a high oxygen transfer rate for the culture broth [21, 22]. In this way, several reactor types have been tested in an attempt to promote low shear and high oxygen transfer rates aiming to improve yield and reduce the noncellulose-producing mutants in shaking culture [22, 23].

BC produced via shaking culture exhibits some differences in the degree of polymerization and crystallinity, and Young's modulus is lower than that produced under static conditions [20]. The development of reactors for large-scale production of BC pellets is economically advantageous if large amounts of cellulose fibers are required for hydrolysis, as in the textile industry.

15.3.2 Static Process

The cellulose produced under static process results in a continuous membrane that assumes the shape of the container. It presents different properties compared with the cellulose produced via the shaking process. The cellulose membrane is synthesized in nature only by bacteria, mainly by those belonging to the *Gluconacetobacter* genus.

The static process for BC synthesis is normally performed in polypropylene trays where the bacterial strain is inoculated and incubated under ideal conditions until the desired thickness is reached. In the static process, the yield of cellulose is dependent on the ratio between the surface area and the volume of the container and is unaffected by the depth of the culture media [24, 25]. For instance, the concentration of BC produced in a 250-mL Erlenmeyer flask containing 50 mL of HS culture media after 96 hours under static cultivation at 30°C is around 2.7 g/L [15]. The production of BC in membranous form is advantageous compared with the pellet form regarding application and purification. The membrane synthesized by different species of microorganisms presents slight differences, mainly in terms of mechanical strength and water holding capacity.

Besides static and agitated cultivation, it is also possible to produce BC by a new and efficient semicontinuous process. It combines the advantages of static cultivation and continuous harvesting under steady-state cultivation. The cellulose is obtained in membranous form and retains the same properties of cellulose obtained from static cultivation [26]. The semicontinuous process seems to be the best way to scale up the BC production.

BC production, however, has not reached a significant scale for widespread application. This is probably due to the fact that BC has not received enough attention from the market as a unique material for many fields of application. The development of highly technological products based on BC could be cost-effective to attract sponsors and could be the way to optimize and scale up production.

15.4 CARBON AND NITROGEN SOURCES

The growth of the microorganism and cellulose production depends on the optimal combination of nutrient sources, including carbon, nitrogen, and mineral salts, for each specific strain. Large amounts of BC are produced from substrates such as glucose, fructose, mannitol, sucrose, and glycerol [19, 27, 28] by several cellulose-producing strains. The main nitrogen sources for these cellulose-producing microorganisms are peptone, yeast extract [19], and $(NH_4)_2SO_4$ [29, 30]. The mass of BC produced from the different culture media is related to the limitations of the substrate over the metabolic pathway of cellulose biosynthesis [27].

In general, the structure and morphological characteristics of BC are not drastically affected by the nutrient source. Some slight changes can be observed when BC is obtained using glycerol as substrate, particularly in the degree of crystallinity and water holding capacity. When glycerol is used as a substrate in shaking flask cultivation, the BC pellets present a denser and reticulated structure than those from the glucose medium [28]. This is probably due to the differences in crystallinity and average degree of polymerization [31]. However, X-ray diffractometry indicates the preservation of the same planar spacing in the crystalline regions [28].

15.5 PRODUCTION OF BACTERIAL CELLULOSE FROM FOOD AND AGRO-FORESTRY RESIDUES

Some abundant by-products and residues from the food and agro-forestry industries can be used as raw material for BC production, providing all the essential nutrients for the microorganisms. BC can be attained from potato waste water [32], grape juice, orange juice [33], Konjac powder [34], molasses [35], and other sources. Besides the high yield of cellulose produced from these raw materials, some properties such as the water content, thermal and mechanical properties, and the degree of crystallinity are highly conserved compared with the cellulose produced using the HS culture media [36]. The production of BC from media composed of waste, residues, or by-products can decrease costs and contribute to the environment. However, a closer analysis should be carried out concerning the composition of the residues and the possible contaminants affecting the BC, mainly, if it is applicable in biomedical fields. Several culture media prepared from dark residues have a dark or brown color, which requires additional steps of washing during the purification process. The excess of alkali washing can damage the surface of the membrane with loss of mass due to hydrolysis of cellulose fibers.

The conditions for production and purification of BC are of fundamental importance in the preparation of composite materials, as all the steps affect the quality of the final product.

15.6 PURIFICATION

Bacterial cellulose purification is highly advantageous compared with cellulose from plants. BC is produced as a high pure form of cellulose, is more environmentally friendly, and demands less energetic methods than cellulose from plants for purification.

Cellulose is a fibrous, tough, water-insoluble substance, which is found as a protective structure in plants, particularly in stalks, stems, trunks, and all woody portions [37]. In plants, fungi, and algae, cellulose is produced as part of the matrix of the cell wall. During cellulose extraction from plants, other elements that compose its basic structure, mainly lignin and hemicelluloses, are extracted with it. The industrial application of cellulose requires the separation of these compounds, performed through hard chemical and physical treatments that, many times, increases the costs of the purification process [3].

BC is produced as a highly swollen gel structure, embedded in the culture medium. The producing microorganisms are free in the media or attached to the surface of the cellulose fibers. Purification of this kind of cellulose is based on killing the microorganisms and removal of cell wastes and residues of the liquid culture media from the cellulose matrix.

Purification of BC constitutes a critical step for the quality of the cellulose and its derivatives. Purification can be performed by multiple washing with hot sodium hydroxide solution, followed by washing with water until pH neutralization [30]. Alternative treatments can be performed with hypochlorite, hydrochloric acid, chlorexedin, or a combination of chemical compounds (NaOH and hypochlorite) [30]. A suitable washing process should be chosen in agreement with the desired application of the cellulose. The contaminants usually found in BC (membranes or pellets) are bacteria, salts, and residual culture media. These contaminants change some properties of BC, such as the mechanical and thermal behavior. These residues can also influence the BC dry mass. The weighing of them with the BC leads to errors in determining the mass of the product, taking into account that they represent a relatively high mass in the dry membrane.

Sterilization with water vapor is a useful process for finishing the purification process of hydrated BC membranes for biomedical applications. Sterilization of dried BC membranes is usually performed through gamma irradiation. Both methods are efficient and do not change significantly the structure and properties of the material.

15.7 STRUCTURE OF BACTERIAL CELLULOSE

Cellulose is a long linear polymeric chain composed of D-glucose monomers connected by β-1,4 glucosidic bonds [8]. The molecular formula of cellulose produced by microorganisms is identical to cellulose derived from plants and other organisms.

The biosynthesis of cellulose by *Gluconacetobacter* bacteria is a complex process that involves the polymerization of glucose monomers and the cellulose secretion to the external environment.

The metabolic pathway for BC biosynthesis starts with the conversion of a substrate, usually a carbon source such as glucose, to uridinediphosphate glucose (UDP-glucose) [8], the cellulose precursor. UDP–glucose is the substrate for the enzyme cellulose synthase. UDP–glucose undergoes the catalytic action of the cellulose synthase, resulting in the formation of β-1,4 bonds between the glucose monomers [8, 38]. The enzymatic complex cellulose synthase is associated with pores on the surface of the outer membrane of the bacterial cell, where the polymerization takes place [39]. After polymerization, the cellulose is secreted by the pores. Each pore extrudes about 16 cellulose chains, or elementary fibrils (width of 1.5 nm), into the culture medium. The elementary fibrils aggregate to each other, possibly through hydrogen bonding, to form a tridimensional network with other elementary fibrils, promoting the formation of microfibrils [39]. The aggregation of microfibrils produces a ribbon assembly with a width of 40–60 nm [40–42] (Fig. 15.3).

The ribbon end curves over itself due to the viscosity of the medium. Combined with the bacteria displacement, such movement wraps the ribbon and creates fibers that are dragged by the bacteria through the culture media. The cross-passing of bacteria produces the entanglement of such fibers creating

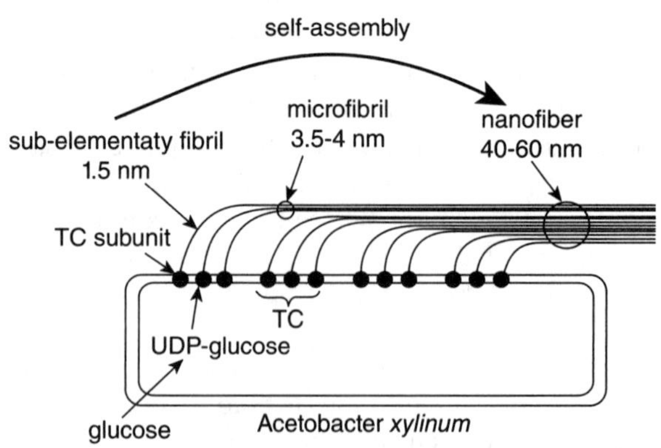

Figure 15.3 Schematic representation of formation of cellulose nanofiber on the side view of the bacteria. Reproduced with the permission from Reference [42].

the cellulose fibers. It is possible to observe the extrusion of the cellulose fibrils during the synthesis by bacteria [41].

The biosynthesis and the polymerization process confer to BC a nano- and microfibrillar tridimensional network structure. The BC structure is responsible for some properties such as high mechanical strength, high water content, high degree of crystallinity, and high purity [39, 43]. Besides these properties, the nano- and microfibrillar structure endows the wet BC membrane with a highly swollen gel aspect.

The BC membrane is formed on the surface of the culture media during the self-assembly of the cellulose molecules [39]. The wet membrane is composed of 0.5–1% wt in mass of pure cellulose and 99% wt or more in mass of water.

The nanofibrillar structure of BC under optical microscopy shows a semitransparent film characterized by great homogeneity with weakly marked fibrils distributed throughout the structure [44] (Fig. 15.4).

The structural characteristics of BC, related to their chemical groups, crystalline index, and morphological aspects, are well resolved by scanning electron microscopy (SEM), FT-IR, and XRD.

The tridimensional nano- and microfibrillar ultrastructure of BC and its details such as fiber width can be observed on the surface of a freeze-dried BC sample by SEM [27] (Fig. 15.5). The surface morphology of the oven-dried BC results in a collapsed network of nano- and microfibrils.

Infrared spectroscopy helps to clarify the cellulose chemical structure [45]. Cellulose absorbs the infrared radiation at specific energies, which are characteristic of their chemical groups (Fig. 15.6). Also, FT-IR is one of the most powerful tools to study hydrogen bonds and helps to identify [37] the intramolecular and intermolecular hydrogen bonds of cellulose and cellulose derivatives [45]. The FT-IR spectrum of BC reveals the typical profile, similar to plant cellulose. The absorption bands around 3300/cm are associated with the vibrations of the OH group [46]. The bands at 2890/cm are attributed to C–H

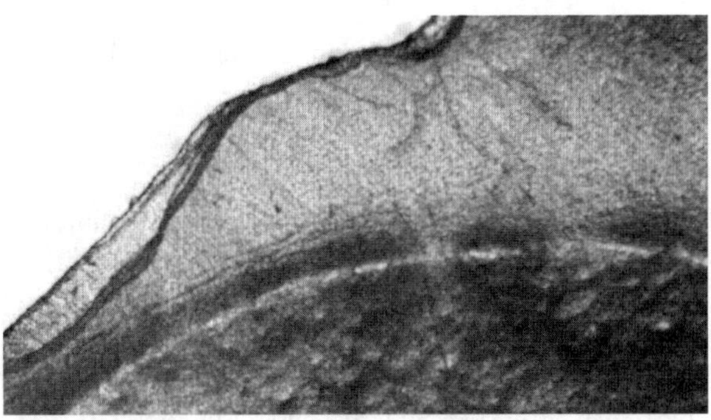

Figure 15.4 Morphological structure of bacterial cellulose at optical microscopy.

Figure 15.5 Scanning electron microscopy of freeze-dried bacterial cellulose.

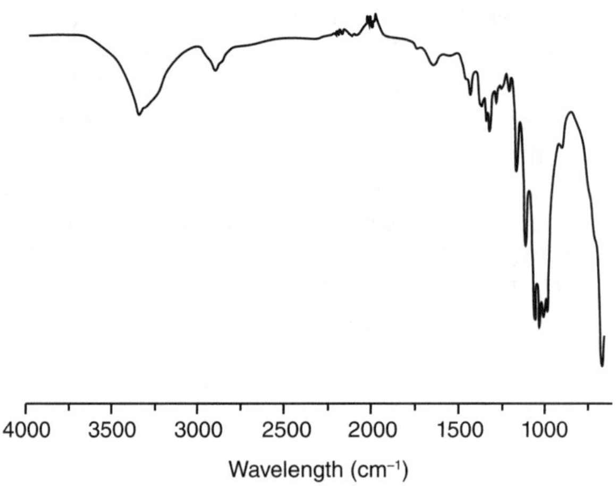

Figure 15.6 FT-IR spectrum of bacterial cellulose.

stretching. The bands at 1060 [47] and 1100/cm are associated with the vibration of C–O–C groups, typically reported for this material [48], and the bands at 1157/cm are due to the C–O antisymmetric bridge stretching. The band at 1642/cm originates from the H–O–H bending of the absorbed water [49, 50].

From the FT-IR spectra profile it is possible to infer the crystallinity degree of cellulose. The bands at 4000–2995/cm, 2900/cm, 1430/cm, 1375/cm, and 900/cm are especially sensitive to cellulose molecular order. The shape of these bands is related to the order or disorder in the polysaccharide phase morphology and can be related to the amount of crystalline versus amorphous fractions in cellulose. Sharpening of the characteristic cellulose bands suggests an increase in the crystallinity degree, and broadening of these bands suggests its decrease [51]. In spite of being a simple method, FT-IR only gives relative values, once the bands presents the contribution of both crystalline and amorphous phases [52].

Regarding the crystalline structure of cellulose, the allomorph cellulose I is the most abundant in nature, including BC. However, derivatives with different crystalline structures named cellulose II, III, and IV have been identified by XRD and solid-state ^{13}C NMR [37, 52]. The cellulose I is composed of two allomorphs, cellulose 1α and cellulose 1β, in different ratios for different cellulosic substrates [53].

Among several ways to determine the degree of crystallinity of cellulose (FT-IR, XRD, and NMR), the most useful is X-ray diffractometry.

In general, XRD of cellulose reveals four crystalline peaks at 2θ (14.4°, 16.7°, 22.6°, and 34.1°) diffraction angles, separated at 101, 10$\bar{1}$, 002, and 040 diffraction planes, for BC [15, 54] and other celluloses [55]. XRD spectra indicate that BC presents the typical crystalline profile of cellulose I (Fig. 15.7). The XRD peak height method [56] for crystallinity index determination is easy and rapid. This method is proposed for empirical determination of relative crystallinity of cellulose samples. However, the results gives an overestimation of the crystalinity index [52].

The most accurate method to obtain the crystallinity index of samples with high crystallinity degree (such as BC) is the XRD deconvolution method. It is based on the separation of the amorphous and crystalline contributions to the diffraction spectrum, requiring suitable software for peak deconvolution

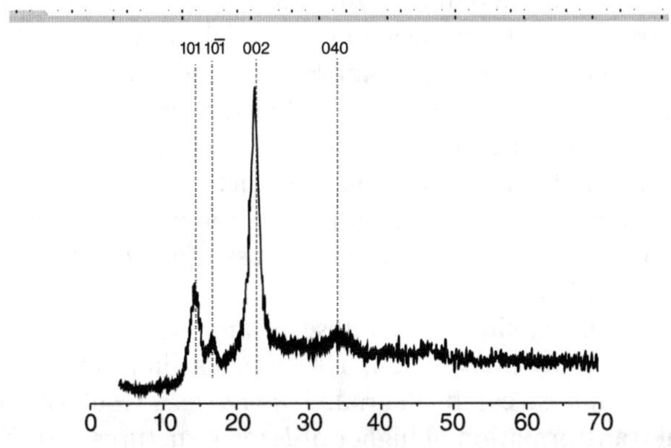

Figure 15.7 X-ray diffractogram of bacterial cellulose.

and curve-fitting. The total area corresponds to the area of the crystalline peaks added to the band of amorphous area [52].

The molecular structure of cellulose allows its interaction with chemically compatible substances through hydrogen bonds. The long chain of cellulose and its high degree of crystallinity are important to obtain composites or to use the cellulose as a matrix for composite materials.

15.8 PROPERTIES OF BACTERIAL CELLULOSE

The most important properties of BC are their mechanical strength, water content, high degree of crystallinity, and high purity [39].

The high water content of BC in the native state is a result of the synthesis, polymerization, and extrusion by bacteria. During cellulose synthesis, a ribbon-like cellulose assembly is formed, extruded in parallel to the long axis of the cell and twisted in a right-handed manner due to the movement of the microorganism in the liquid environment [57]. The microfibrils of cellulose are arranged in a tridimensional network of nanofibers, which allows water retention within the high hydrophilic thin porous structure. The wet or dry BC structure has a large number of pores, suitable for a wide range of applications.

The insolubility of cellulose in water and common organic solvents and its resistance to chemical and enzymatic hydrolysis are conferred by the conformation of the cellulose molecules and their supramolecular organization, in which the glucan chains are stabilized by hydrogen bonds inter- and intrachain [6, 58]. In addition, the linear chains of cellulose and the strong cohesion between the macromolecules are responsible for the high mechanical strength of BC [59].

The mechanical behavior of a single fiber of BC reveals a Young's modulus ranging from 78 GPa, when determined by atomic force microscopy (AFM) [60] to 114 GPa [61], when determined by Raman spectroscopy. The BC fibers behave mechanically as a homogeneous material, showing no dependence on their diameter [60]. Dried BC membranes present a Young's modulus of around 9.7 GPa, stress at break around 240 MPa, and strain at break around 2.6% [61]. Rheological measurements of BC fibers exhibit a reversible stress softening due to the buckling of fibers, destruction of its physical cross-links, and rearrangement of the network with formation of new cross-links. Also, the storage and loss modulus decrease with stress after a certain stress level is reached, in a reversible way [62].

The low biodegradability of cellulose in animal tissues is related to the absence of enzymes to cleave the β 1-4 linkages, limiting its degradation *in vivo*. However, cellulose can be degraded *in vivo* as a result of the susceptibility to physical transformation of higher ordered structures and tissue response [63, 64]. A systematic investigation about the biocompatibility of BC in subcutaneous implants in rats shows no macroscopic signs of inflammation such as redness, edema, or exudates, no histological signs of chronic inflammation, and no foreign body reaction. The high biocompatibility was complemented by the formation of new blood vessels around and growing into the BC implants [64].

The crystallinity degree of BC is around 80%, according to the culture conditions such as the media composition and the production process (static or agitated) [15, 35]. This parameter is in accord with the properties of the BC-based composites [65]. The ultrasonic processing in mild conditions is effective in changing the microfibrillar structure of BC, increasing its degree of crystallinity [66].

The cellulose produced by bacteria is excreted in a highly pure form, requiring a simple purification process to remove cells and residual culture media inside the membrane. Different from cellulose from plant, algae, and fungus, in which the purification is a complex process evolving several steps [67], the relatively ease, low cost, and efficient purification of BC has attracted the attention of industry in recent years [68]. The high purity of BC contributes to its high biocompatibility, increasing the interest in this material for biomedical and biological applications.

The thermal degradation profile of BC shows a single mass loss step with a maximum at 365°C (Fig. 15.8). The thermal character of BC can change the thermal stability of its derivative materials, changing both initial and maximum degradation temperatures. This behavior is dependent on the dispersion and compatibility between the compounds [69].

The properties of BC are essential in the field of cellulose-based composites, generating materials with improved properties such as increased mechanical strength and thermal stability; in addition, the transparency of the composite is not affected by low amounts of fibers.

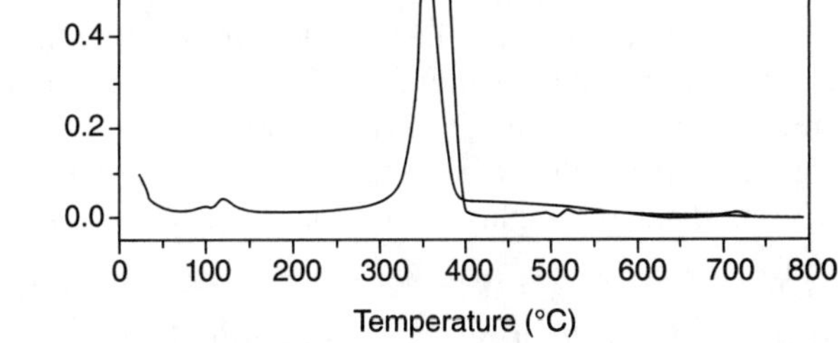

Figure 15.8 Thermogravimetric and differential curves for BC.

15.9 APPLICATIONS

15.9.1 Food Industry

BC is a rich source of food fibers. The growing market of food based on BC is represented by the increasing number of patents relating to it and its derivatives in the food industry [70]. The application of BC as a diet food is traditionally found in the Asian culture, as desserts made in the home or in small factories. The sweetened membrane is consumed directly, known as *nata de coco*. The fermented liquid containing green tea, sugar, and microbial metabolites, mainly acetic acid, is a traditional beverage used as a drink in several parts of the world, known as Kombucha tea. Kombucha tea can be prepared with 6 g/L of green tea and sucrose (100 g/L), followed by inoculation with a starter culture. The starter culture is a portion of tea originated from a previous Kombucha culture, containing the cellulose-producing microorganisms. The tea stays at 25°C during an ideal period of time, around 2 weeks. The microorganisms grow, acidify the liquid, and synthesize the cellulose on the interface air–liquid. After removal of the membrane, the acidified and sweetened liquid is used as a drink.

15.9.2 Bacterial Cellulose Composites

Interest in BC as a component (membrane or fiber) of composites is growing. The nanosized fibers are able to change the properties of the matrix, and, if the membranes are used as matrix, they are susceptible to changing their properties through the incorporation of other components to generate the composite [43].

Membrane modification can be made *in situ* if the modifying agent is introduced into the culture medium and the composite material is formed during the fiber assembly [43]. *In situ* modification sometimes generates more homogenous composites with improved properties, although it is not the best way for composite preparation. The compounds introduced in the culture media during the cellulose synthesis can disturb the growth of the microorganisms and decrease the cellulose synthesis, compromising the reproducibility of the method. Moreover, the modifying agent can be leached during the purification.

A traditional and efficient approach to prepare BC composites is *post-modification*, loading the BC membrane (swollen or dried) with the modifying agent [43]. Removal of part of the water from BC allows the material to absorb the same volume of other solution. If the volume and concentration are known, the composition of the product can be determined. BC-based composites have application in biomedical and highly technological fields, usually as functional materials.

15.9.2.1 Biomedical Applications BC has found application as wound dressing membranes and temporary skin substitutes for burns and chronic wounds. The topical application of BC membranes is highly effective, improving the healing process of wounds [25, 71] and promoting the re-epithelization process in human skin [72–77]. BC can be used as a blood vessel substitutes [78], tubes for vascular grafts [79], substitutes for the dura mater membrane [80], facial masks for beauty [81], and in dental applications [82], among others.

Natural polymers and their derived materials are widely used in biomedical applications because of their similarities to the extracellular matrix and inherent cellular interaction. The ideal biomaterial for biomedical and cosmetics should be nontoxic, biocompatible, biodegradable, bioresorbable to support the reconstruction of new tissue without inflammation [83] and promote tissue development, besides having adequate mechanical and physical properties [84].

The biocomposite technology is used to improve the properties of natural-based materials. The selection of appropriate starting materials (biopolymers and fillers) leads to the development of new versatile biomaterials with the required properties for a wide range of specific applications. Polymeric materials used for this purpose include synthetic polymers (poly(ethylene oxide) (PEO)) that mimic natural substances, naturally occurring macromolecules, and chemically modified natural polymers (proteins and polypeptides, polysaccharides such as starch, cellulose, dextran, inulin, and chitin) [85]. Among other polysaccharides, BC has received increasing interest for cosmetic and biomedical applications [86–88].

Films, fibers, and bulk products of natural polymers such starch, chitin, chitosan, carboxymethyl cellulose (CMC), and poly(hydroxybutyrate) (PHB) can be reinforced with BC fibers, resulting in highly biocompatible materials with improved mechanical properties, compared with the polymeric matrix only. These composites have found use in wound healing, tissue engineering, food packaging, food coating, and cosmetics purposes. Among the biomedical applications, BC has been studied as a support for tissue engineering for bone and cartilage repair.

In general, some properties are desirable for scaffold materials for tissue engineering of cartilage and bone, such as appropriate porosity to support cell ingrowth, effective mass transport, and mechanical properties. BC is a material that meets the requirements for tissue engineering of cartilage, mainly due to its high strength in the wet state, as well as its *in situ* moldability and biocompatibility. BC scaffolds support chondrocyte proliferation at levels of approximately 50% of the native tissue substrate, collagen type II. Also, the chondrocytes maintain their differentiated form when growing on this material, which supports cell ingrowth [89].

Materials such as BC–hydroxyapatite nanocomposites are prepared by surface modification and biomineralization, allowing the formation of calcium-deficient hydroxyapatite crystals on the surface of the BC scaffolds, used for application in bone repair. The porous structure of the biocomposite enhances cell infiltration and leads to the integration of the scaffold into the surrounding tissue, promoting cell adherence and differentiation, an important feature for bone healing applications [90]. Nanocomposites of BC–hydroxyapatite produced by introducing the mineral phase into the culture medium during the formation of cellulose fibrils in presence of CMC showed a differentiated morphology, such as large pores, in addition to high biocompatibility and bioactivity, indicating the potential of the nanocomposite material for biomedical applications [91]. BC–hydroxyapatite membranes accelerate the bone repair when tested *in vivo* [92].

BC–starch nanocomposites can be produced through the synthesis of cellulose nanofibers in the presence of partially gelatinized starch granules, preserving the structural properties of BC, without significant changes in its mechanical behavior [93]. Also, biodegradable nanocomposites of starch reinforced with low amounts of BC shows increased modulus and tensile strength [94]. The nanocomposites have found potential application in biomedical and food packaging industries.

BC-alginate sponge composites can be used as mucosal flaps in oral tissue regeneration to cover surgical wounds. The composite is prepared blending a gel-like solution of alginate with BC slurry cross-linked by an aqueous solution of $CaCl_2$. The mixture is freeze-dried to form the BC–alginate porous sponge. The composite is a nonadherent hydrogel dressing that prevents bacterial invasion, minimizes wound dehydration, and provides drainage for wound exudates. The composite presents high mechanical strength and stability in both water and phosphate buffer solution [82].

Composite materials based on BC and chitosan prepared by *in situ* modification through the incorporation of chitosan to the culture media before the BC synthesis by microorganisms results in a composite material with glucosamine and *N*-acetylglucosamine units incorporated into the cellulose chain. The composite presents good mechanical properties, moisture retention, antimicrobial activity, and biodegradation in *in vitro* conditions [95]. Copolymers of cellulose with *N*-acetylglucosamine are obtained by direct incorporation of glucosamine and *N*-acetylglucosamine into exopolymers during the biological synthesis by *Gluconacetobacter xylinus*. These materials are susceptible to lysozyme hydrolysis, overcoming the limitations with respect to the biodegradability of cellulose-based biomaterials in the body [96]. BC–chitosan composites prepared by immersing BC membranes into chitosan solution and freeze-drying show high biocompatibility. The composites are suitable for cell attachment, suggesting its use for wound dressing and tissue engineering [97]. The incorporation of BC fibers into chitosan results in transparent films with improved mechanical properties, showing the reinforcement effect of BC on the polymeric matrix [98].

Biocomposites based on BC and collagen type I prepared *in situ* combine the positive effect of collagen in chronic wound healing with the physical properties of BC, decreasing the effect of destructive proteases and proinflammatory cytokines in wounds [99].

15.9.2.2 Application of Bacterial Cellulose to Highly Technological Products A conventional and large volume of cellulose is in demand by the papermaking and textile industries. In the papermaking industry, BC can be used as an additive to improve paper texture and resistance [100], added directly to the pulp mass as a surface coating, improving the quality of printing paper. BC fibers are used as fillers in soft wood sheets. The composite shows better tensile index, burst index, and stiffness, with a decrease in porosity and water absorption [101].

In the textile industry BC is an alternative to plant cellulose, maintaining the properties of the product. However, there is little interest in the development of this area due to the high costs of BC production and need to change the design of the production plant. Nevertheless, several processes and products for textile are protected by patents [102, 103]. For instance, high strength viscose fiber with low shrink rate and excellent elasticity used for high quality clothing fabric and industrial textiles is obtained by mixing BC with viscose fibers. A blended fiber composed of BC dissolved in ionic liquid and natural polysaccharide is used in the medical and textile fields [103].

BC is an important substrate for transparent and electrically conductive films for use in optoelectronic areas, including organic light emitting devices (LEDs) [104], thin-film transistors [105], liquid crystal displays [106], and others [107].

BC can be used as transparent conducting films to produce flexible optoelectronic devices. The flexible, transparent, and electrically conducting films are based on nanofibrous polymeric membranes and single-walled carbon nanotubes. The nanocomposites are prepared using a simple adsorption method in which the hydrated BC membrane is used as template for depositing uniformly dispersed single-walled carbon nanotubes (SWCNTs) or MWCTs [108]. The SWCNT-adsorbed BC membranes in the dry state are immersed into a transparent polymer solution of silk fibroin, generating transparent and conductive films. The polymer solution fills the spaces between the BC nanofibrils, significantly reducing the Rayleigh scattering caused by BC micropores. The electrical conductivity is conferred by carbon nanotubes. The flexibility is due to the mechanical strength of BC combined with the malleability of silk fibroin [107].

Flexible and electrically conductive polyaniline–BC nanocomposite membranes are synthesized *in situ* by oxidative polymerization of aniline using BC as matrix. The conductive particles, uniformly deposited on the surface of the BC, form a continuous nanosheath. The nanocomposites synergistically combine the electronic characteristics of polyaniline with the mechanical properties of the BC matrix, resulting in highly flexible composites with good mechanical properties and improved thermal stability. This material could be applied in sensors, flexible electrodes, and flexible displays [109].

BC membranes are useful to attain electroactive biopolymer actuators for biomedical applications. The composite is prepared by immersing dried BC membranes in an aqueous solution of lithium chloride. After surface modification by treatment with lithium chloride, a very thin gold electrode is deposited on both sides of the fabricated membrane. The electroactive actuator presents electrically driven bending deformation under fully hydrated conditions and can be electrically activated in water. These results show the potential of BC as an electroactive biopolymer in wet environments, for example, in implantable biomedical devices [110].

Regenerated BC-multiwalled carbon nanotube composite fibers can be prepared by dissolving BC in DMAc-LiCl, followed by the introduction of the MWCTs in the system. Composite fibers are prepared by wet spinning of the mixture. The spinning solutions are extruded through a spinneret into a coagulating bath. The fibers are washed with distilled water, separated by centrifugation, and dried. The composite fibers may be a precursor for the production of cellulose-based carbon fibers and have found application in medical, mechanical, and electrical materials [111].

Bionanocomposites composed of BC and small amount of poly(vinyl alcohol) (PVA) result in optical transparent nanocomposite films with excellent mechanical properties. The composite preparation can be attained *in situ*, through the direct addition of PVA into the *Acetobacterxylinum* medium or by impregnation of BC membranes with PVA solution [112].

Transparent nanocomposites based on BC and acrylic resin can be obtained by acetylation of BC. The replacement of the hydroxyl groups of cellulose by the hydrophilic acetyl group significantly reduces the hygroscopicity of BC–acrylate nanocomposites, maintaining its optical transparency and thermal stability [113, 114].

The immersion of BC membranes into acrylic or epoxy polymer matrices results in a remarkable reinforcement of the resin substrate, generating composites with high mechanical strength, low thermal expansion coefficients, and high transparency [115, 116].The reinforcement of acrylic resin is also possible through addition of BC fibers to the unmodified polymeric matrix [69].

Nanocomposites based on BC and poly(lactic acid) (PLA) can be prepared by adding the BC nanofibers in the PLA matrix [54] or immersing the BC membrane into the PLA solution [117]. The composites retain the transparency of the pure PLA film due to the nanofibrillar structure of the BC, when the cellulose content is low. The tensile strength and Young's modulus of the composites are highly increased compared with PLA. These composites can be considered for biomedical applications and food packaging, because of their transparency, mechanical properties, biodegradability, and biocompatibility.

The porous network structure of BC is used as a support structure to synthesize magnetically responsive composites. The precipitation of magnetic Ni nanoparticles within the membrane involves the immersion of BC into the solution of $NiCl_2$ followed by addition of the reductor agent solution, $NaBH_4$, in an aqueous environment, at room temperature. The Ni nanoparticles range from 10 to 60 nm and confer to the composites the coexistence of both superparamagnetic and ferromagnetic behavior. The BC does not undergo any change and retains its crystal structure, even after chemical reduction reaction. The composites have found application in magnetic printing and as magnetic tissue scaffolds [118].

15.10 CONCLUSIONS

Properties of BC such as biocompatibility make possible its application in biological and biomedical fields, where it can be used as single material or as a component of biocomposites. BC membranes in their native state have been applied, currently, in the biomedical field. A wide range of BC-based composite materials has been developed, but these are just starting to gain the attention of industries for technological applications. BC fibers have been used as a reinforcement agent for natural or synthetic polymers, generating materials with improved mechanical properties. Thanks to the natural degradation of cellulose, the association of BC with synthetic polymers helps to decrease their environmental impact.

Efforts have been made to develop new BC-based materials for sustainable and economical application in the industry. Studies about improvement of BC yield by microorganisms are of fundamental importance to reduce production costs, as well as to optimize its use and applicability for the global economy. However, its production is currently limited to the laboratory scale and a few industries in the world, with low amounts of BC. Therefore, an increase in production volume will depend on the interest of companies with the potential to optimize and develop the industrial process for BC production at competitive costs.

REFERENCES

[1] Williams, W.S., Cannon, R.E. (1989) Alternative environmental-roles for cellulose produced by acetobacter-xylinum. *Applied and Environmental Microbiology*, *55*(10), 2448–2452.

[2] Brown, J.A. (1886) On an acetic ferment which forms cellulose. *Journal of the Chemical Society, Transactions*, *49*, 432–439.

[3] Dahman, Y., Sani, A. (2010) Improvements in the production of bacterial synthesized biocellulose nanofibres using different culture methods. *Journal of Chemical Technology and Biotechnology*, *85*(2), 151–164.

[4] Deinema, M.H., Zevenhui, L.P. (1971) Formation of cellulose fibrils by Gram-negative bacteria and their role in bacterial flocculation. *Archiv Fur Mikrobiologie*, *78*(1), 42–57.

[5] Ago, M., Yamane, C., Hattoti, M., Ono, H., and Okajima, K. (2006) Characterization of morphology and physical strength for bacterial cellulose produced by an Enterobacter sp. *Sen-I Gakkaishi*, *62*(11), 258–262.

[6] Romling, U. (2002) Molecular biology of cellulose production in bacteria. *Research in Microbiology*, *153*(4), 205–212.

[7] Matthysse, A.G., Thomas, D.L., and White, A.R. (1995) Mechanism of cellulose synthesis in agrobacterium-tumefaciens. *Journal of Bacteriology*, *177*(4), 1076–1081.

[8] Ross, P., Mayer, R., and Benziman, M. (1991) Cellulose biosynthesis and function in bacteria. *Microbiological Reviews*, *55*(1), 35–58.

[9] Napoli, C., Dazzo, F., and Hubbell, D. (1975) Production of cellulose microfibrils by rhizobium. *Applied Microbiology*, *30*(1), 123–131.

[10] Iwata, T., Iwamoto, S., Kai, W.H., and Isogai, A. (2009) Elastic modulus of single cellulose microfibrils from tunicate measured by atomic force microscopy. *Biomacromolecules*, *10*(9), 2571–2576.

[11] Brown, R.M. (2004) Cellulose structure and biosynthesis: What is in store for the 21st century? *Journal of Polymer Science Part A: Polymer Chemistry*, *42*(3), 487–495.

[12] Yamada, Y., Hoshino, K., and Ishikawa, T. (1997) Taxonomic studies of acetic acid bacteria and allied organisms. 11. The phylogeny of acetic acid bacteria based on the partial sequences of 16S ribosomal RNA: The elevation of the subgenus *Gluconoacetobacter* to the generic level. *Bioscience, Biotechnology, and Biochemistry*, *61*(8), 1244–1251.

[13] Brenner, D.J., Krieg, N.R., and Staley, J.T. (2005) *Bergey's Manual of Systematic Bacteriology—The Proteobacteria*, Vol. 2, 2nd ed. New York: Springer.

[14] Lisdiyanti, P., Navarro, R.R., Uchimura, T., and Komagata, K. (2006) Reclassification of *Gluconacetobacter hansenii* strains and proposals of *Gluconacetobacter saccharivorans* sp. nov. and *Gluconacetobacter nataicola* sp. nov. (vol 56, pg 2101, 2006). *International Journal of Systematic and Evolutionary Microbiology*, *56*, 2723–2723.

[15] Trovatti, E., Serafim, L.S., Freire, C.S.R., Silvestre, A.J.D., and Neto, C.P. (2011) *Gluconacetobacter sacchari*: An efficient bacterial cellulose cell-factory. *Carbohydrate Polymers*, *86*(3), 1417–1420.

[16] Dykes, G.A., Nguyen, V.T., Flanagan, B., and Gidley, M.J. (2008) Characterization of cellulose production by a *Gluconacetobacter xylinus* strain from Kombucha. *Current Microbiology*, *57*(5), 449–453.

[17] Ryu, H.W., Kim, S.Y., Kim, J.N., Wee, Y.J., and Park, D.H. (2006) Production of bacterial cellulose by *Gluconacetobacter* sp RKY5 isolated from persimmon vinegar. *Applied Biochemistry and Biotechnology*, *131*(1–3), 705–715.

[18] Andelib, Y.A., Nuran, D.A. (2009) In Isolation of Cellulose Producing Bacteria from Wastes of Vinegar Fermentation. *Proceedings of the World Congress on Engineering and Computer Science*, San Francisco, USA, 2009, W., Ed. WCECS 2009: San Francisco, USA.

[19] Hestrin, S., Schramm, M. (1954) Synthesis of cellulose by *Acetobacter xylinum*. 2. Preparation of freeze-dried cells capable of polymerizing glucose to cellulose. *Biochemical Journal*, *58*(2), 345–352.

[20] Chao, Y.P., Ishida, T., Sugano, Y., and Shoda, M. (2000) Bacterial cellulose production by *Acetobacter xylinum* in a 50-L internal-loop airlift reactor. *Biotechnology and Bioengineering*, *68*(3), 345–352.

[21] Kouda, T., Yano, H., and Yoshinaga, F. (1997) Effect of agitator configuration on bacterial cellulose productivity in aerated and agitated culture. *Journal of Fermentation and Bioengineering*, *83*(4), 371–376.

[22] Shoda, M., Chao, Y.P., Ishida, T., and Sugano, Y. (2000) Bacterial cellulose production by *Acetobacter xylinum* in a 50-L internal-loop airlift reactor. *Biotechnology and Bioengineering*, *68*(3), 345–352.

[23] Kim, S.J., Song, H.J., Li, H.X., Seo, J.H., and Kim, M.J. (2009) Pilot-scale production of bacterial cellulose by a spherical type bubble column bioreactor using saccharified food wastes. *Korean Journal of Chemical Engineering*, *26*(1), 141–146.

[24] Masaoka, S., Ohe, T., and Sakota, N. (1993) Production of cellulose from glucose by *Acetobacter xylinum*. *Journal of Fermentation and Bioengineering*, *75*(1), 18–22.

[25] Jonas, R., Farah, L.F. (1998) Production and application of microbial cellulose. *Polymer Degradation and Stability*, *59*(1–3), 101–106.

[26] Kralisch, D., Hessler, N., Klemm, D., Erdmann, R., and Schmidt, W. (2010) White biotechnology for cellulose manufacturing-The HoLiR concept. *Biotechnology and Bioengineering*, *105*(4), 740–747.

[27] Mikkelsen, D., Flanagan, B.M., Dykes, G.A., and Gidley, M.J. (2009) Influence of different carbon sources on bacterial cellulose production by *Gluconacetobacter xylinus* strain ATCC 53524. *Journal of Applied Microbiology*, *107*(2), 576–583.

[28] Son, H.J., Jung, H.I., Jeong, J.H., Lee, O.M., Park, G.T., Kim, K.K., Park, H.C., Lee, S.M., and Kim, Y.G. (2010) Influence of glycerol on production and structural-physical properties of cellulose from *Acetobacter* sp V6 cultured in shake flasks. *Bioresource Technology*, *101*(10), 3602–3608.

[29] Jagannath, A., Kalaiselvan, A., Manjunatha, S.S., Raju, P.S., and Bawa, A.S. (2008) The effect of pH, sucrose and ammonium sulphate concentrations on the production of bacterial cellulose (Nata-de-coco) by *Acetobacter xylinum*. *World Journal of Microbiology & Biotechnology*, *24*(11), 2593–2599.

[30] Shoda, M., Bae, S., and Sugano, Y. (2004) Improvement of bacterial cellulose production by addition of agar in a jar fermentor. *Journal of Bioscience and Bioengineering*, *97*(1), 33–38.

[31] Watanabe, K., Tabuchi, M., Morinaga, Y., and Yoshinaga, F. (1998) Structural features and properties of bacterial cellulose produced in agitated culture. *Cellulose*, *5*(3), 187–200.

[32] Thompson, D.N., Hamilton, M.A. (2001) Production of bacterial cellulose from alternate feedstocks. *Applied Biochemistry and Biotechnology*, *91–3*, 503–513.

[33] Nakamura, Y., Kurosumi, A., Sasaki, C., and Yamashita, Y. (2009) Utilization of various fruit juices as carbon source for production of bacterial cellulose by *Acetobacter xylinum* NBRC 13693. *Carbohydrate Polymers*, *76*(2), 333–335.

[34] Hong, F., Qiu, K.Y. (2008) An alternative carbon source from konjac powder for enhancing production of bacterial cellulose in static cultures by a model strain *Acetobacter aceti* subsp *xylinus* ATCC 23770. *Carbohydrate Polymers*, *72*(3), 545–549.

[35] Son, H.J., Jung, H.I., Lee, O.M., Jeong, J.H., Jeon, Y.D., Park, K.H., Kim, H.S., and An, W.G. (2010) Production and characterization of cellulose by *Acetobacter* sp V6 using a cost-effective molasses-corn steep liquor medium. *Applied Biochemistry and Biotechnology*, *162*(2), 486–497.

[36] Carreira, P., Mendes, J.A.S., Trovatti, E., Serafim, L.S., Freire, C.S.R., Silvestre, A.J.D., and Neto, C.P. (2011) Utilization of residues from agro-forest industries

in the production of high value bacterial cellulose. *Bioresource Technology*, *102*, 7354–7360.

[37] OSullivan, A.C. (1997) Cellulose: The structure slowly unravels. *Cellulose*, *4*(3), 173–207.

[38] Saxena, I.M., Lin, F.C., and Brown, R.M. (1990) Cloning and sequencing of the cellulose synthase catalytic subunit gene of *Acetobacter xylinum*. *Plant Molecular Biology*, *15*(5), 673–683.

[39] Klemm, D., Heublein, B., Fink, H.P., and Bohn, A. (2005) Cellulose: Fascinating biopolymer and sustainable raw material. *Angewandte Chemie-International Edition*, *44*(22), 3358–3393.

[40] Brown, R.M. (1982) *Cellulose and Other Natural Polymer Systems: Biogenesis, Structure, and Degradation*. New York: Plenum Press.

[41] Hirai, A., Tsuji, M., and Horii, F. (2002) TEM study of band-like cellulose assemblies produced by *Acetobacter xylinum* at 4 degrees C. *Cellulose*, *9*(2), 105–113.

[42] Kondo, T., Tomita, Y. (2009) Influential factors to enhance the moving rate of *Acetobacter xylinum* due to its nanofiber secretion on oriented templates. *Carbohydrate Polymers*, *77*, 754–759.

[43] Klemm, D., Schumann, D., Kramer, F., Hessler, N., Hornung, M., Schmauder, H.P., and Marsch, S. (2006) Nanocelluloses as innovative polymers in research and application. *Polysaccharides II*, *205*, 49–96.

[44] Surma-Slusarska, B., Presler, S., and Danielewicz, D. (2008) Characteristics of bacterial cellulose obtained from *Acetobacter xylinum* culture for application in papermaking. *Fibres & Textiles in Eastern Europe*, *16*(4), 108–111.

[45] Shirk, H.G., Greathouse, G.A. (1952) Infrared spectra of bacterial cellulose. *Analytical Chemistry*, *24*(11), 1774–1775.

[46] Sugiyama, J., Persson, J., and Chanzy, H. (1991) Combined infrared and electron-diffraction study of the polymorphism of native celluloses. *Macromolecules*, *24*(9), 2461–2466.

[47] Chen, S.Y., Li, X., Hu, W.L., Shi, S.K., Shen, W., Zhang, X., and Wang, H.P. (2009) *In situ* synthesis of CdS nanoparticles on bacterial cellulose nanofibers. *Carbohydrate Polymers*, *76*(4), 509–512.

[48] El-Saied, H., El-Diwany, A.I., Basta, A.H., Atwa, N.A., and El-Ghwas, D.E. (2008) Production and characterization of economical bacterial cellulose. *Bioresources*, *3*(4), 1196–1217.

[49] Kasapis, S., Wong, S.S., and Tan, Y.M. (2009) Bacterial and plant cellulose modification using ultrasound irradiation. *Carbohydrate Polymers*, *77*(2), 280–287.

[50] Anicuta, S.-G., Dobre, L., Stroescu, M., and Jipa, I. (2010) Fourier Transform Infrared (FTIR) spectroscopy for characterization of antimicrobial films containing chitosan. *Analele UniversităŃii din Oradea Fascicula: Ecotoxicologie, Zootehnie si Tehnologii de Industrie Alimentară*, 1234–1240.

[51] Martínez-Sanz, M., Lopez-Rubio, A., and Lagaron, J.M. (2011) Optimization of the nanofabrication by acid hydrolysis of bacterial cellulose nanowhiskers. *Carbohydrate Polymers*, *85*, 228–236.

[52] Johnson, D.K., Park, S., Baker, J.O., Himmel, M.E., and Parilla, P.A. (2010) Cellulose crystallinity index: Measurement techniques and their impact on interpreting cellulase performance. *Biotechnology for Biofuels*, *3*, 1–10.

[53] Salmen, L., Akerholm, M., and Hinterstoisser, B. (2004) Characterization of the crystalline structure of cellulose using static and dynamic FT-IR spectroscopy. *Carbohydrate Research*, *339*(3), 569–578.

[54] Tome, L.C., Pinto, R.J.B., Trovatti, E., Freire, C.S.R., Silvestre, A.J.D., Neto, C.P., and Gandini, A. (2011) Transparent bionanocomposites with improved properties prepared from acetylated bacterial cellulose and poly(lactic acid) through a simple approach. *Green Chemistry*, *13*(2), 419–427.

[55] He, J.X., Cui, S.Z., and Wang, S.Y. (2008) Preparation and crystalline analysis of high-grade bamboo dissolving pulp for cellulose acetate. *Journal of Applied Polymer Science*, *107*(2), 1029–1038.

[56] Segal, L., Creely, J.J., Martin, A.E.J., and Conrad, C.M. (1959) An empirical method for estimating the degree of crystallinity of native cellulose using the X-ray diffractometer. *Textile Research Journal*, *29*(10), 786–794.

[57] Hirai, A., Tsuji, M., and Horii, F. (1998) Helical sense of ribbon assemblies and splayed microfibrils of bacterial cellulose. *Sen-I Gakkaishi*, *54*(10), 506–510.

[58] Koolman, J., Roehm, K.H. (2005) *Color Atlas of Biochemistry*, 2nd ed. Stuttgart, Germany: Thieme.

[59] Klemm, D., Philipp, B., Heinze, T., Heinze, U., and Wagenknecht, W. (1998) *Comprehensive Cellulose Chemistry. Fundamentals and Analytical Methods*, Vol. 1. Weinheim, New York, Chichester, Brisbane, Singapore, Toronto: Wiley-VCH Verlag GmbH.

[60] Guhados, G., Wan, W.K., and Hutter, J.L. (2005) Measurement of the elastic modulus of single bacterial cellulose fibers using atomic force microscopy. *Langmuir*, *21*(14), 6642–6646.

[61] Hsieh, Y.C., Yano, H., Nogi, M., and Eichhorn, S.J. (2008) An estimation of the Young's modulus of bacterial cellulose filaments. *Cellulose*, *15*(4), 507–513.

[62] Torres, F.G., Troncoso, O.P., Lopez, D., Grande, C., and Gomez, C.M. (2009) Reversible stress softening and stress recovery of cellulose networks. *Soft Matter*, *5*(21), 4185–4190.

[63] Miyamoto, T., Takahashi, S., Ito, H., Inagaki, H., and Noishiki, Y. (1989) Tissue biocompatibility of cellulose and its derivatives. *Journal of Biomedical Materials Research*, *23*(1), 125–133.

[64] Risberg, B., Helenius, G., Backdahl, H., Bodin, A., Nannmark, U., and Gatenholm, P. (2006) *In vivo* biocompatibility of bacterial cellulose. *Journal of Biomedical Materials Research. Part A*, *76A*(2), 431–438.

[65] Mazeau, K., Heux, L. (2003) Molecular dynamics simulations of bulk native crystalline and amorphous structures of cellulose. *Journal of Physical Chemistry B*, *107*(10), 2394–2403.

[66] Tischer, P.C.S.F., Sierakowski, M.R., Westfahl, H., and Tischer, C.A. (2010) Nanostructural reorganization of bacterial cellulose by ultrasonic treatment. *Biomacromolecules*, *11*(5), 1217–1224.

[67] Delmer, D.P., Amor, Y. (1995) Cellulose biosynthesis. *Plant Cell*, *7*(7), 987–1000.

[68] Delmer, D.P. (1987) Cellulose biosynthesis. *Annual Review of Plant Physiology and Plant Molecular Biology*, *38*, 259–290.

[69] Trovatti, E., Oliveira, L., Freire, C.S.R., Silvestre, A.J.D., Neto, C.P., Pinto, J.J.C.C., and Gandini, A. (2010) Novel bacterial cellulose-acrylic resin nanocomposites. *Composites Science and Technology*, *70*(7), 1148–1153.

[70] Tecson-Mendoza, E.M. (2007) Development of functional foods in the Philippines. *Food Science and Technology Research*, *13*(3), 179–186.

[71] Fontana, J.D., Desouza, A.M., Fontana, C.K., Torriani, I.L., Moreschi, J.C., Gallotti, B.J., Desouza, S.J., Narcisco, G.P., Bichara, J.A., and Farah, L.F.X. (1990) Acetobacter cellulose pellicle as a temporary skin substitute. *Applied Biochemistry and Biotechnology*, *24–5*, 253–264.

[72] Phisalaphong, M., Sanchavanakit, N., Sangrungraungroj, W., Kaomongkolgit, R., Banaprasert, T., and Pavasant, P. (2006) Growth of human keratinocytes and fibroblasts on bacterial cellulose film. *Biotechnology Progress*, *22*(4), 1194–1199.

[73] Solway, D.R. (2010) Microbial cellulose wound dressing in the treatment of skin tears in the frail elderly (vol 22, pg 17, 2010). *Wounds: A Compendium of Clinical Research and Practice*, *22*(2), A10–A10.

[74] Levinson, D.J., Solway, D.R., and Consalter, M. (2010) Microbial cellulose wound dressing in the treatment of skin tears in the frail elderly. *Wounds: A Compendium of Clinical Research and Practice*, *22*(1), 17–19.

[75] Serafica, G., Mormino, R., Oster, G.A., Lentz, K.E., and Koehler, K.P. (2008) Microbial cellulose wound dressing for treating chronic wounds.

[76] Czaja, W., Krystynowicz, A., Bielecki, S., and Brown, R.M. (2006) Microbial cellulose—The natural power to heal wounds. *Biomaterials*, *27*(2), 145–151.

[77] Brown, R.M., Czaja, W.K., Young, D.J., and Kawecki, M. (2007) The future prospects of microbial cellulose in biomedical applications. *Biomacromolecules*, *8*(1), 1–12.

[78] Klemm, D., Schumann, D., Udhardt, U., and Marsch, S. (2001) Bacterial synthesized cellulose—Artificial blood vessels for microsurgery. *Progress in Polymer Science*, *26*(9), 1561–1603.

[79] Gatenholm, P., Backdahl, H., and Risberg, B. (2011) Observations on bacterial cellulose tube formation for application as vascular graft. *Materials Science & Engineering C-Materials for Biological Applications*, *31*(1), 14–21.

[80] Mello, L.R., Feltrin, L.T., Neto, P.T.F., and Ferraz, F.A.P. (1997) Duraplasty with biosynthetic cellulose: An experimental study. *Journal of Neurosurgery*, *86*(1), 143–150.

[81] Amnuaikit, T., Chusuit, T., Raknam, P., and Boonme, P. (2011) Effects of a cellulose mask synthesized by a bacterium on facial skin characteristics and user satisfaction. *Medical Devices: Evidence and Research*, *4*, 77–81.

[82] Phisalaphong, M., Chiaoprakobkij, N., Sanchavanakit, N., Subbalekha, K., and Pavasant, P. (2011) Characterization and biocompatibility of bacterial cellulose/alginate composite sponges with human keratinocytes and gingival fibroblasts. *Carbohydrate Polymers*, *85*(3), 548–553.

[83] Risbud, M. (2001) Tissue engineering: Implications in the treatment of organ and tissue defects. *Biogerontology*, *2*(2), 117–125.

[84] Fu, X.B., Huang, S. (2010) Naturally derived materials-based cell and drug delivery systems in skin regeneration. *Journal of Controlled Release*, *142*(2), 149–159.

[85] Ottenbrite, R.M., Fadeeva, N. (1994) Polymer systems for biomedical applications. An overview. In: *Polymeric Drugs and Drug Administration*, ACS Symposium Series, ed. R.M. Ottenbrite. Washington, DC: American Chemical Society.

[86] Lina, F., Yue, Z., Jin, Z., and Guang, Y. (2011) Bacterial cellulose for skin repair materials. In: *Biomedical Engineering—Frontiers and Challenges*, ed. R. Fazel-Rezai. Rijeka: InTech.

[87] Yu, L., Dean, K., and Li, L. (2006) Polymer blends and composites from renewable resources. *Progress in Polymer Science*, *31*(6), 576–602.

[88] Belgacem, M.N., Gandini, A. (2008) *Monomers, Polymers and Composites from Renewable Resources*. Amsterdam: Elsevier.

[89] Svensson, A., Nicklasson, E., Harrah, T., Panilaitis, B., Kaplan, D.L., Brittberg, M., and Gatenholm, P. (2005) Bacterial cellulose as a potential scaffold for tissue engineering of cartilage. *Biomaterials*, *26*(4), 419–431.

[90] Zimmermann, K.A., LeBlanc, J.M., Sheets, K.T., Fox, R.W., and Gatenholm, P. (2011) Biomimetic design of a bacterial cellulose/hydroxyapatite nanocomposite for bone healing applications. *Materials Science & Engineering C-Materials for Biological Applications*, *31*(1), 43–49.

[91] Grande, C.J., Torres, F.G., Gomez, C.M., and Bano, M.C. (2009) Nanocomposites of bacterial cellulose/hydroxyapatite for biomedical applications. *Acta Biomaterialia*, *5*(5), 1605–1615.

[92] Saska, S., Barud, H.S., Gaspar, A.M.M., Marchetto, R., Ribeiro, S.J.L., and Messaddeq, Y. (2011) Bacterial cellulose hydroxyapatite nanocomposites for bone regeneration. *International Journal of Biomaterials*, Article ID 175362.

[93] Grande, C.J., Torres, F.G., Gomez, C.M., Troncoso, O.P., Canet-Ferrer, J., and Martinez-Pastor, J. (2009) Development of self-assembled bacterial cellulose-starch nanocomposites. *Materials Science & Engineering C-Biomimetic and Supramolecular Systems*, *29*(4), 1098–1104.

[94] Martins, I.M.G., Magina, S.P., Oliveira, L., Freire, C.S.R., Silvestre, A.J.D., Neto, C.P., and Gandini, A. (2009) New biocomposites based on thermoplastic starch and bacterial cellulose. *Composites Science and Technology*, *69*(13), 2163–2168.

[95] Ciechanska, D. (2004) Multifunctional bacterial cellulose/chitosan composite materials for medical applications. *Fibres & Textiles in Eastern Europe*, *12*(4), 69–72.

[96] Kaplan, D.L., Lee, J.W., Deng, F., Yeomans, W.G., Allen, A.L., and Gross, R.A. (2001) Direct incorporation of glucosamine and N-acetylglucosamine into exopolymers by *Gluconacetobacter xylinus* (=*Acetobacter xylinum*) ATCC 10245: Production of chitosan-cellulose and chitin-cellulose exopolymers. *Applied and Environmental Microbiology*, *67*(9), 3970–3975.

[97] Kim, J., Cai, Z.J., Lee, H.S., Choi, G.S., Lee, D.H., and Jo, C. (2011) Preparation and characterization of a bacterial cellulose/chitosan composite for potential biomedical application. *Journal of Polymer Research*, *18*(4), 739–744.

[98] Freire, C.S.R., Fernandes, S.C.M., Oliveira, L., Silvestre, A.J.D., Neto, C.P., Gandini, A., and Desbrieres, J. (2009) Novel transparent nanocomposite films based on chitosan and bacterial cellulose. *Green Chem.*, *11*(12), 2023–2029.

[99] Wiegand, C., Elsner, P., Hipler, U.C., and Klemm, D. (2006) Protease and ROS activities influenced by a composite of bacterial cellulose and collagen type I *in vitro*. *Cellulose*, *13*(6), 689–696.

[100] Pascoal Neto, C., Da Rocha Freire Barros, C.S., De Matos Fernandes, S.C., Domingues Silvestre, A.J., and Gandini, A. (2011) Aqueous coating compositions for use in surface treatment of cellulosic substrates.

[101] Gao, W.-H., Chen, K.-F., Yang, R.-D., Yang, F., and Han, W.-J. (2011) Properties of bacterial cellulose and its influence on the physical properties of the paper. *Bioresources*, *6*(1), 144–153.

[102] Zhu, P., Zang, Z., Wang, L., and Wei, P. (2010) High strength viscose fiber used for e.g. high quality clothes fabric, is obtained by mixture bacterial cellulose with viscose fiber spinning solution and has preset amount of bacterial cellulose.

[103] Zhu, P., Chu, Q., Zang, Z., and Wei, P. (2011) Blend fiber used in medical and textile fields, comprises bacterial cellulose and balanced natural polysaccharide.

[104] Cremona, M., Legnani, C., Vilani, C., Calil, V.L., Barud, H.S., Quirino, W.G., Achete, C.A., and Ribeiro, S.J.L. (2008) Bacterial cellulose membrane as flexible substrate for organic light emitting devices. *Thin Solid Films*, *517*(3), 1016–1020.

[105] Hwang, S.-M., Chen, H.-K., and Hwang, B.-J. (2009) Bacterial cellulose film and carbon nanotubes-like thin film structures developed from bacterial cellulose.

[106] Pei, C.H., Wang, Y., Luo, Q.P., and Peng, B.H. (2008) A novel thermotropic liquid crystalline—Benzoylated bacterial cellulose. *Carbohydrate Polymers*, *74*(4), 875–879.

[107] Jin, H.J., Kim, Y., Kim, H.S., Bak, H., Yun, Y.S., and Cho, S.Y. (2009) Transparent conducting films based on nanofibrous polymeric membranes and single-walled carbon nanotubes. *Journal of Applied Polymer Science*, *114*(5), 2864–2872.

[108] In, H.J., Jung, R., Kim, H.S., Kim, Y., Kwon, S.M., and Lee, H.S. (2008) Electrically conductive transparent papers using multiwalled carbon nanotubes. *Journal of Polymer Science Part B: Polymer Physics*, *46*(12), 1235–1242.

[109] Chen, S.Y., Hu, W.L., Yang, Z.H., Liu, L.T., and Wang, H.P. (2011) Flexible electrically conductive nanocomposite membrane based on bacterial cellulose and polyaniline. *Journal of Physical Chemistry B*, *115*(26), 8453–8457.

[110] Oh, I.K., Jeon, J.H., Kee, C.D., and Kim, S.J. (2010) Bacterial cellulose actuator with electrically driven bending deformation in hydrated condition. *Sensors and Actuators. B, Chemical*, *146*(1), 307–313.

[111] Jin, H.J., Chen, P., Kim, H.S., Kwon, S.M., and Yun, Y.S. (2009) Regenerated bacterial cellulose/multi-walled carbon nanotubes composite fibers prepared by wet-spinning. *Current Applied Physics*, *9*(2), E96–E99.

[112] Peijs, T., Gea, S., Bilotti, E., Reynolds, C.T., and Soykeabkeaw, N. (2010) Bacterial cellulose-poly(vinyl alcohol) nanocomposites prepared by an in-situ process. *Materials Letters*, *64*(8), 901–904.

[113] Nogi, M., Abe, K., Handa, K., Nakatsubo, F., Ifuku, S., and Yano, H. (2006) Property enhancement of optically transparent bionanofiber composites by acetylation. *Applied Physics Letters*, *89*(23), 233123/1–233123/3.

[114] Ifuku, S., Nogi, M., Abe, K., Handa, K., Nakatsubo, F., and Yano, H. (2007) Surface modification of bacterial cellulose nanofibers for property enhancement of optically transparent composites: Dependence on acetyl-group DS. *Biomacromolecules*, *8*(6), 1973–1978.

[115] Yano, H., Sugiyama, J., Nakagaito, A.N., Nogi, M., Matsuura, T., Hikita, M., and Handa, K. (2005) Optically transparent composites reinforced with networks of bacterial nanofibers. *Advanced Materials*, *17*(2), 153–155.

[116] Nogi, M., Yano, H. (2008) Transparent nanocomposites based on cellulose produced by bacteria offer potential innovation in the electronics device industry. *Advanced Materials*, *20*(10), 1849–1852.

[117] Jin, H.J., Kim, Y., Jung, R., and Kim, H.S. (2009) Transparent nanocomposites prepared by incorporating microbial nanofibrils into poly(L-lactic acid). *Current Applied Physics*, *9*, S69–S71.

[118] Vitta, S., Drillon, M., and Derory, A. (2010) Magnetically responsive bacterial cellulose: Synthesis and magnetic studies. *Journal of Applied Physics*, *108*(5), 539051–539057.

CHAPTER 16

Chemical Modification of Nanocelluloses

YOUSSEF HABIBI

16.1 INTRODUCTION

Cellulose constitutes the most abundant renewable polymer resource produced on earth. As fibers it has been used for thousands of years, and it is generally well known that it has been exploited as a raw chemical material to access various derivatives for nearly 150 years. What has not been known until relatively recently is that when cellulose fibers are subjected to mechanical shearing or acid hydrolysis, the fibers yield to elongated fibrillar or defect-free rod-like crystalline particles having at least one dimension in the nanoscale range. The focus on "nanotechnology" involving these cellulosic substrates has generated a tremendous level of attention that does not appear to be diminishing. These biopolymeric assemblies warrant such attention not only because of their unsurpassed quintessential physical and chemical properties, as will become evident in this chapter, but also because of their inherent renewability, sustainability and abundance, in addition to their intrinsic properties such as lightweight, nanoscale dimension, and unique morphology. Several reviews have been published recently attesting to the importance of these emerging renewable building blocks for nanomaterials [1–4].

With the presence of a large number of hydroxyl groups within their structure, these building blocks provide a unique platform for significant surface modification using various chemistries. The purpose of this chapter is to summarize, after a brief introduction and definition of nanocellulose substrates, several key advances in the chemical modification routes for the functionalization of nanocelluloses.

Biopolymer Nanocomposites: Processing, Properties, and Applications, First Edition. Edited by Alain Dufresne, Sabu Thomas, and Laly A. Pothan.

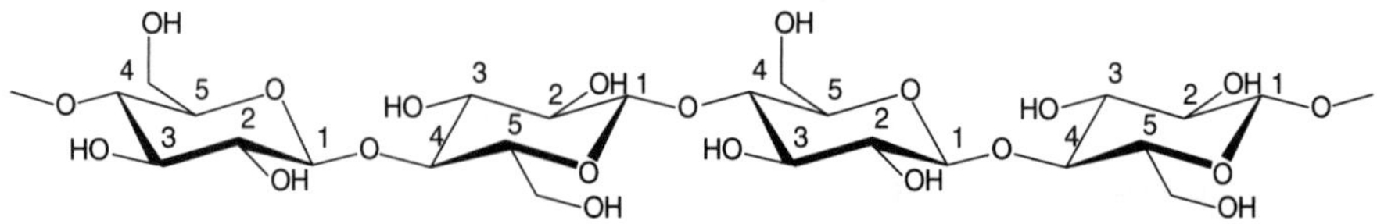

Figure 16.1 Chemical structure of cellulose.

16.2 NANOCELLULOSE SUBSTRATES

Cellulose is widely distributed in the biosphere, principally in higher plants, but also in several marine animals (e.g., tunicates) and to a lesser degree in algae, fungi, bacteria, and invertebrates. In general, cellulose is a fibrous, mechanically tough, water-insoluble biomaterial, which plays an essential role in maintaining the structure of plant cell walls while providing the key internal load resistance to facilitate uninterrupted transport mechanisms within a plant organism [5]. Regardless of its origin, cellulose may be typically characterized as a semicrystalline high molecular weight homopolymer of β-1,4 linked anhydroglucose (Fig. 16.1) [6]. In plant cell walls, the exquisite architectural arrangement of cellulose microfibrils results from the combined action of biopolymerization, spinning, and crystallization. All these events are orchestrated by specific enzymatic terminal complexes (TCs) that adopt a rosette configuration behaving as precise biological spinnerets that are credited with synthesizing up to 36 glucan chains simultaneously and in close proximity to each another. The architectural precision of the complex itself provides a template for the growing of glucan chains to co-crystallize to microfibrils that adopt a linear and rigid conformation in lieu of an amorphous array of â-glucans. On one hand, the polymer chains are assembled through van der Waals forces and both intra- and intermolecular hydrogen bonds in a hierarchical order to form elementary nanofibrils having a cross-dimensional thickness of 2–5 nm, which in turn aggregates laterally to larger macrofibers [7].

On the other hand, if the TCs are not perturbed, they can generate an interminable number of microfibrils having only a limited number of defects or amorphous regions [8, 9]. These regions are distributed on segments of the elementary fibril, which are distorted by internal strain in the fiber to undergo tilt and twist [10].

According to these morphological features, cellulose fibers can be dissociated transversally at the amorphous regions present along their axis by controlled acid hydrolysis. The acid primarily degrades the less ordered, and thus more accessible regions along the fibrils, to finally leave intact nanometric and highly crystalline defect-free rod-like fragments [3], referred hereafter as cellulose nanocrystals (CNs). The geometrical dimensions for CNs depend on the origin of cellulose fibers, but the width is generally few nanometers and the length ranges from tenths of nanometers to several micrometers.

Through a mechanical shearing process, involving sometimes pretreatment steps, aiming to beak the hydrogen bonds network, cellulose fibers can be laterally disintegrated into their substructural nanoscale units, nanofibrils called nanofibrillated cellulose (NFCs) [11]. This latter biomaterial can be also biosynthesized through microorganisms and is well known as bacterial cellulose (BC). In fact, besides being a dominant component of cell walls in plants, cellulose is also secreted extracellularly as synthesized cellulose fibers by several bacterial species [2]. BC is produced by bacteria such as *Acetobacter* species *Gluconacetobacter xylinus* by cultivation in aqueous culture media containing carbon and nitrogen sources during a period of days up to 2 weeks. The resulting cellulosic network structure is in the form of a pellicle made up of a random assembly of ribbon-shaped fibrils, less than 100 nm wide, composed of a bundle of much finer microfibrils, 2 to 4 nm in diameter. These bundles are relatively straight, continuous, and dimensionally uniform.

Nanocellulose designation refers commonly to all these types of nanometric cellulosic substrates including CNs, NFCs, and BC (Fig. 16.2).

16.3 CHEMICAL MODIFICATIONS

Taking advantage of the large number of hydroxyl groups at the surface of nanocellulose substrates, different chemical modifications have been attempted, including esterification, etherification, oxidation, silylation, amidation, and polymer grafting. Noncovalent surface modification, including the use of adsorbing surfactants and coupling agents, has been also studied. All chemical functionalization have been mainly conducted (1) to tune the surface characteristics of nanocelluloses to promote their dispersion in nonpolar organic media and/or to improve their compatibility with hydrophobic matrices in nanocomposites; (2) to introduce stable negative or positive charges on the surface of nanocellulose, mainly CNs, to obtain better electrostatic repulsion-induced dispersion, especially when exploring their self-assembly properties.

The main challenge for the chemical functionalization of the different nanocellulose substrates is to conduct it in such a way that it only changes their surface while preserving their original morphology, avoiding any polymorphic conversion, and maintaining the integrity of their native crystalline structure.

16.3.1 Noncovalent Surface Chemical Modifications

Noncovalent surface modifications of nanocelluloses are typically made via adsorption of surfactants. This approach has been introduced by Heux and coworkers [12, 13], who used surfactants consisting of the mono- and di-esters of phosphoric acid having alkylphenols tails to modify CNs. The surfactant-coated CNs dispersed very well in nonpolar solvents [13]. Detailed analyses carried out by small angle neutron scattering (SANS) revealed that the surfactant molecules formed a thin layer of about 15 Å at the surface of the CNs [14].

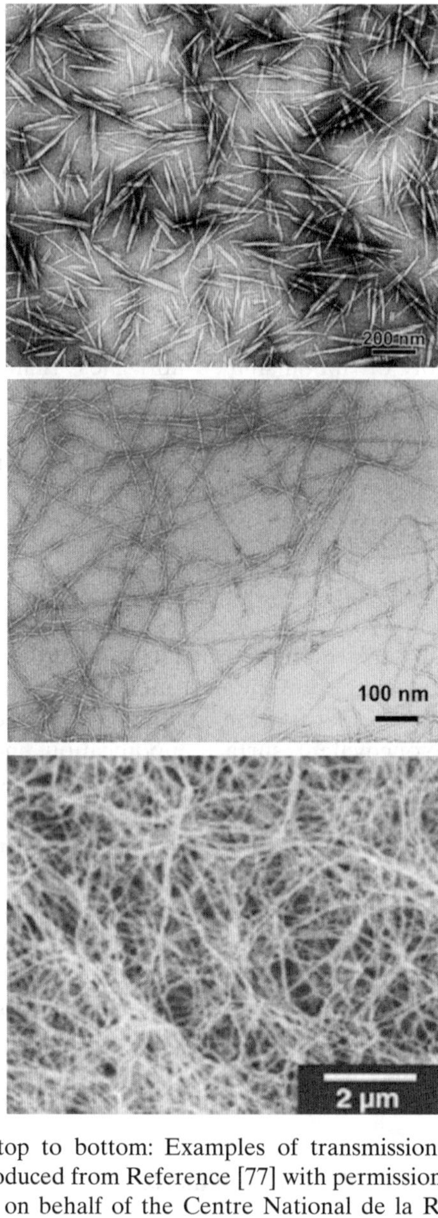

Figure 16.2 From top to bottom: Examples of transmission electron microscopy images of CNs (reproduced from Reference [77] with permission of The Royal Society of Chemistry [RSC] on behalf of the Centre National de la Recherche Scientifique [CNRS] and the RSC), NFC (reprinted with permission from Reference [28]. Copyright 2007 American Chemical Society), and BC (reused from Reference [95] with kind permission from Springer Science and Business Media).

When these surfactant-coated CNs were incorporated into isotactic polypropylene, they showed very good compatibility and acted as remarkable nucleating agents inducing the formation of the rare β-crystalline of isotactic polypropylene form in addition to the regular crystalline α-form [15]. An anionic surfactant was also used by Bondeson and Oksman [16] to enhance the dispersion of CNs in poly(lactic acid) (PLA). Other authors, Kim et al. [17] and Rojas et al. [18], reported the use of nonionic surfactants to disperse CNs in polystyrene-based composite fibers. Zhou et al. [19] recently reported a new and elegant way of noncovalent CN surface modification based on the adsorption of saccharide-based amphiphilic block copolymers. By mimicking the natural lignin–carbohydrate copolymers, they adsorbed xyloglucan oligosaccharide-poly(ethylene glycol)-polystyrene triblock copolymer onto the surface of CNs. The resulting CNs showed excellent dispersion abilities in nonpolar solvents. Another biomimetic approach involving an environmentally friendly synthetic pathway was developed for the production of a high-strength composite material consisting of self-assembled and nanostructured BC coated with hydroxyethylcellulose (HEC). This composite, having analogous hierarchical organization of wood or tendon, was simply generated by growing BC nanofibers in the presence of HEC [20]. Prior to this study, Brown and Laborie [21] used the same bioengineering strategy to elaborate polyethylene oxide (PEO)/BC composites. Films made of oxidized NFCs were modified by adsorption of cationic surfactant cetyltrimethylammonium bromide (CTAB) [22]. The interaction between the NCFs and surfactant were strong enough that the surfactant cannot be removed after thorough rinsing with water.

16.3.2 TEMPO-Mediated Oxidation

(2,2,6,6-tetramethylpiperidine-1-oxyl)-mediated (or TEMPO-mediated) oxidation of CNs has been used to convert the hydroxymethyl groups present on their surface to their carboxylic form. This oxidation reaction, which is highly discriminative of primary hydroxyl groups, is also "green" and simple to implement. It involves the application of a stable nitroxyl radical, the 2,2,6,6-tetramethylpiperidine-1-oxyl (TEMPO), in the presence of NaBr and NaOCl (see Fig. 16.3, top). The use of this technique has been the subject of a number of reports since it was first introduced by de Nooy et al. [23], who were the first to show that only the hydroxymethyl groups of polysaccharides were oxidized, while the secondary hydroxyls remained unaffected. In fact, TEMPO-mediated oxidation of CNs involves a topologically confined reaction sequence, and as a consequence of the twofold screw axis of the cellulose chain, only half of the accessible hydroxymethyl groups is available to react, whereas the other half is buried within the crystalline particle (Fig. 16.3, bottom).

TEMPO-mediated oxidation of CNs, obtained from HCl acid hydrolysis of cellulose fibers, was first reported by Araki et al. [24] as an intermediate step to promote the grafting of polymeric chains. These authors demonstrated that after TEMPO-mediated oxidation, the CNs maintained their initial morphological integrity and formed a homogenous suspension when dispersed in

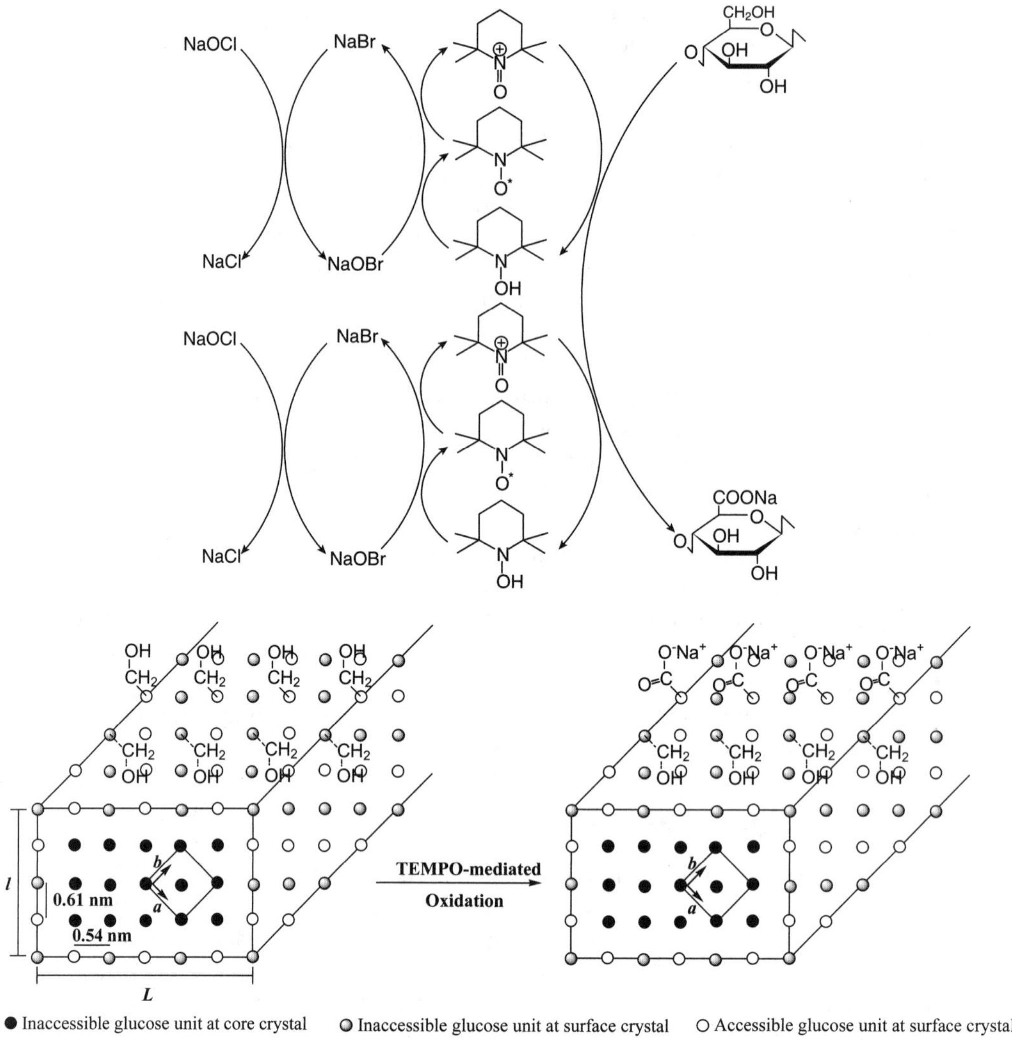

Figure 16.3 Scheme of TEMPO-mediated oxidation mechanism of the hydroxymethyl groups of cellulose (top reaction scheme) and cross-sectional representation of cellulose crystal indicating the occurrence of surface TEMPO-mediated oxidation of available hydroxyl groups (bottom surface crystal figures).

water. The basis for these latter observations was the presence of the newly installed carboxyl groups that imparted negative charges at the CN surface and thus induced electrostatic stabilization. Similar observations were reported by Montanari et al. [25], who also showed that during excessive TEMPO-mediated oxidation a decrease of the crystal size occurred resulting from the partial delamination of cellulose chains that are extant on the surface.

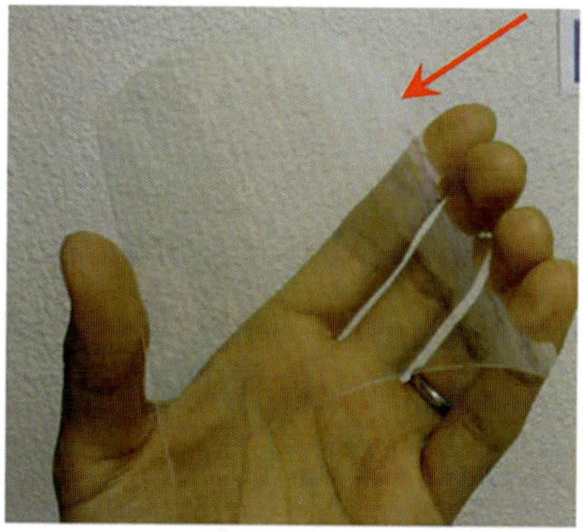

Figure 2.11 Transparent thin (60 μm) DCP film reinforced with CNFs.

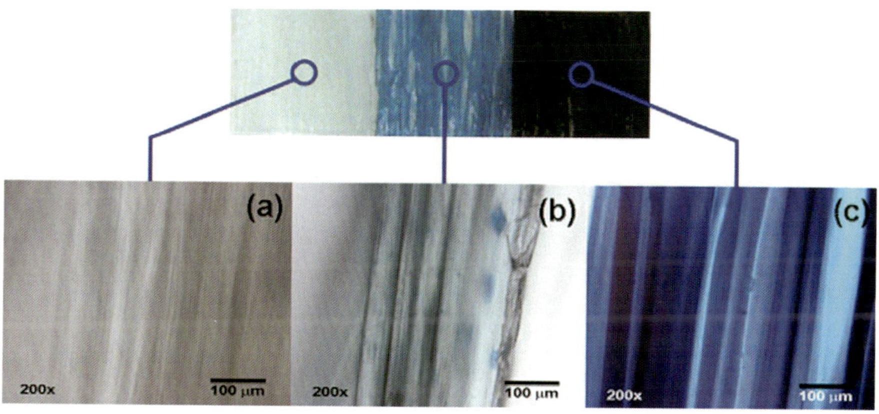

Figure 5.1 Representative photographic (top panel) and corresponding optical microscopic (bottom panel; magnification 200×) images of (a) neat alginate yarns, (b) alginate nanocomposite yarns containing 0.6% w/w of chitosan whiskers, and (c) neat chitosan yarns, after they had been stained with 0.01% w/v Amido Black 10B aqueous solution for 12 hours [4].

Biopolymer Nanocomposites: Processing, Properties, and Applications, First Edition. Edited by Alain Dufresne, Sabu Thomas, and Laly A. Pothan.

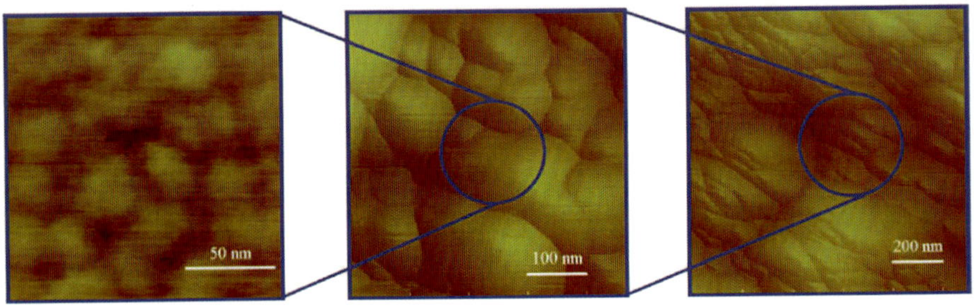

Figure 5.12 Atomic force microscope images in tapping mode illustrating multilevel structure in ChiHAP50, PgAHAP50, and ChiPgAHAP50 nanocomposites. Images are illustrative of the multilevel structure and are not to scale [77].

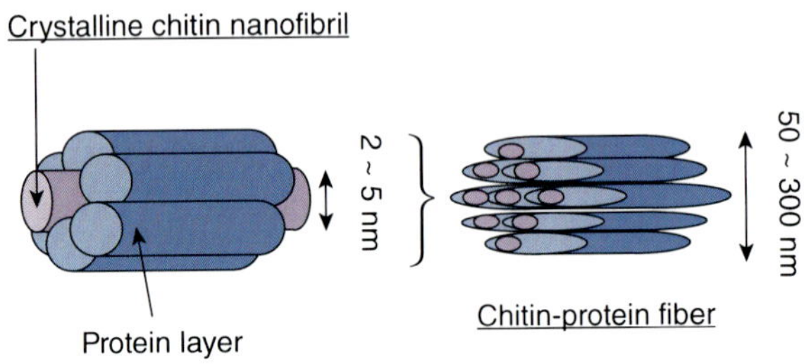

Figure 7.2 Presence of native chitin in crustacean shells.

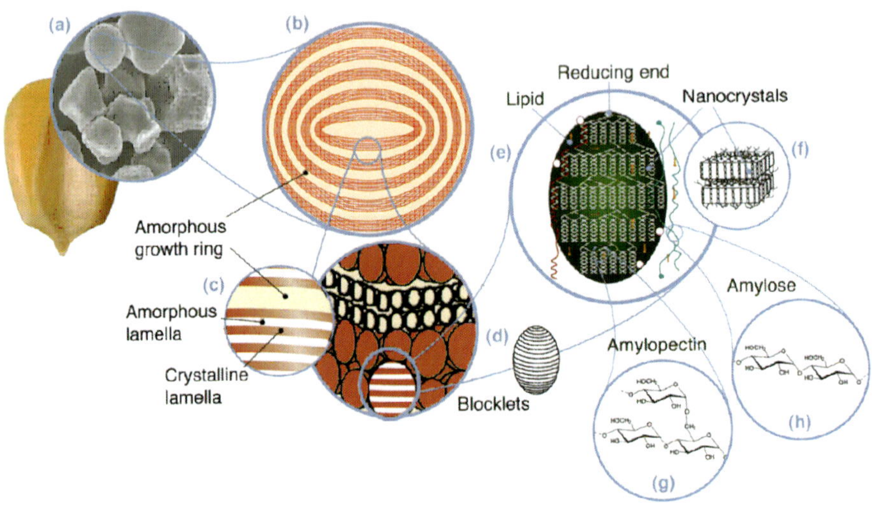

Figure 10.1 Starch multiscale structure: (a) starch granules from normal maize (30 μm), (b) amorphous and semicrystalline growth rings (120–500 nm), (c) amorphous and crystalline lamellae (9 nm), magnified details of the semicrystalline growth ring, (d) blocklets (20–50 nm) constituting a unit of the growth rings, (e) amylopectin double helices forming the crystalline lamellae of the blocklets, (f) nanocrystals: other representation of the crystalline lamellae called starch nanocrystals when separated by acid hydrolysis, (g) amylopectin's molecular structure, and (h) amylose's molecular structure (0.1–1 nm). Reprinted with permission from Reference [29]. Copyright 2010 American Chemical Society.

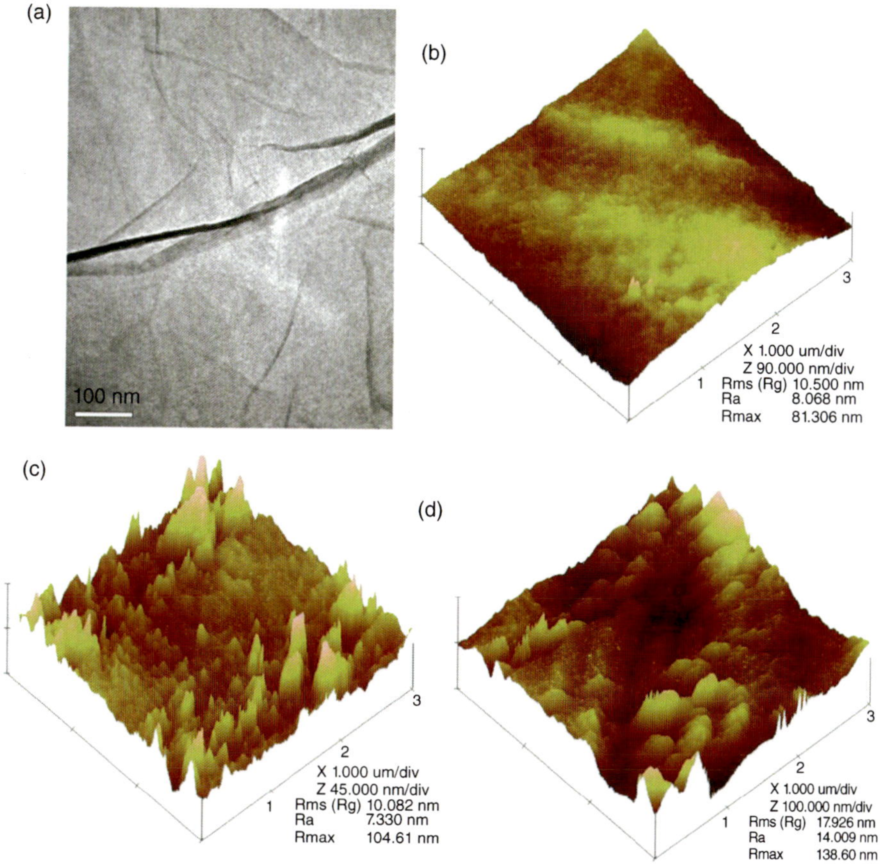

Figure 11.2 TEM image of graphene oxide (GO) powder (a); AFM images of glycerol plasticized pea starch (GPS) (b), and the GPS/GO nanocomposite films containing 3.0 wt% (c), and 5.0 wt% (d) GO. Reprinted with permission from Reference [27].

Figure 13.4 Toast packaged with a biodegradable film based on cassava starch formulated with glycerol, sucrose, and inverted sugar as plasticizers. Reprinted with permission from Reference [27].

Figure 14.4 NFC film and 0.4% consistency NFC gel produced at the Paper and Fiber Research Institute (PFI), Trondheim, Norway, being handled by Kristin Syverud (left) and Ingebjorg Leirset (right), respectively. The photograph was taken by Oddbjørn Svarlien and kindly supplied by Kristin Syverud.

Figure 16.4 Aqueous 0.53% (w/v) suspensions of cellulose nanocrystals observed between crossed polarizers after production by HCl-catalyzed hydrolysis (left), and after their oxidation via TEMPO-mediated reactions (right) (taken from Reference [26]).

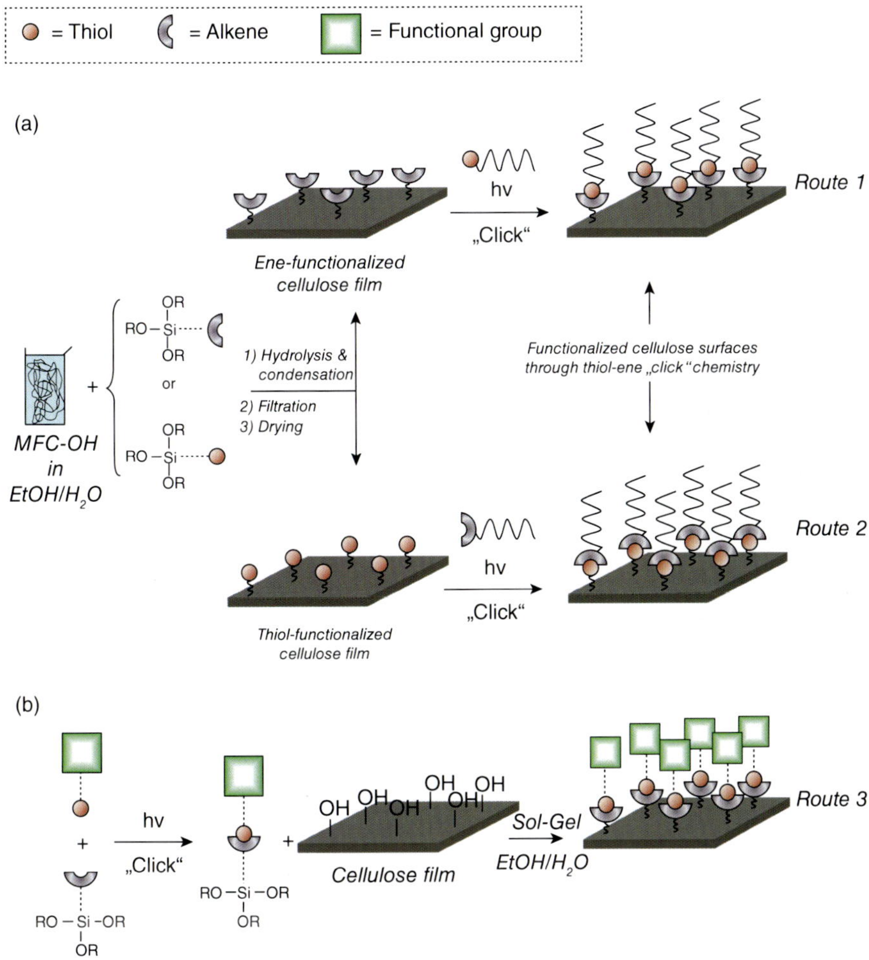

Figure 16.7 Schematic illustration of NFC-based films surface functionalization using click chemistry. (a) Synthesis of ene- and thiol-functionalized films with alkoxysilane molecules, and their subsequent "click" coupling reactions with thiol- (Route 1) and ene- (Route 2) molecules, respectively. (b) Synthesis of a functional alkoxysilane molecule using thiol-ene "click" chemistry, and its subsequent coupling reaction with a cellulose film through a sol–gel process (Route 3) (reproduced from Reference [56] with permission of The Royal Society of Chemistry).

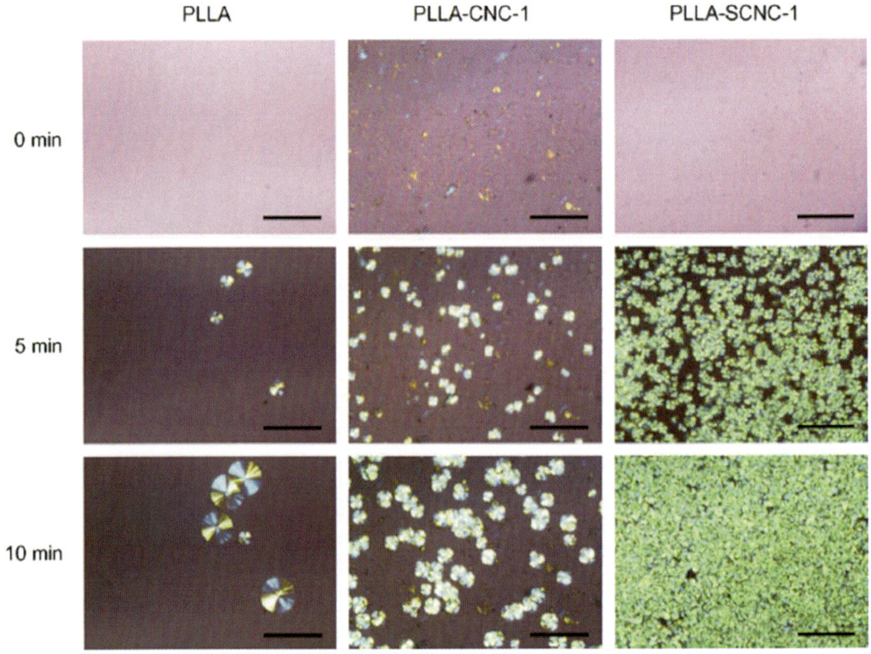

Figure 18.6 Polarized optical microscopy images of PLLA, 1 wt% CNW/PLLA (PLLA-CNC-1 in this figure), and 1 wt% SCNW/PLLA (PLLA-SCNC-1 in this figure) acquired on the 0, 5th, and 10th minute at 125°C after quenched from melt at 210°C. Scale bar: 200 μm (reprinted with permission from Pei et al. [72]. Copyright 2010 Elsevier Ltd.)

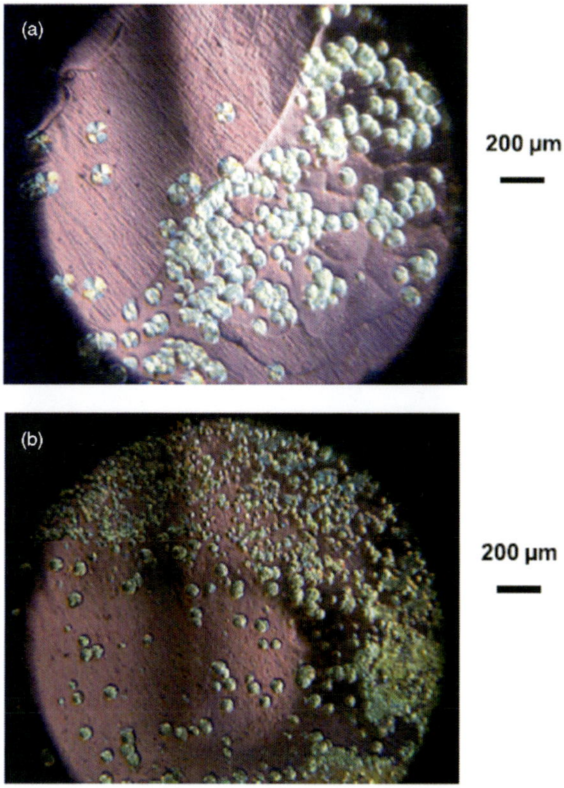

Figure 18.10 POM micrographs obtained after isothermal crystallization for 100 seconds at 53.6°C for a POE-based film with (a) 0 and (b) 10 wt% cellulose nanowhiskers (reprinted with permission from Azizi Samir et al. [47]. Copyright 2004 Elsevier Ltd.).

Figure 18.11 Crystallization of polypropylene against cellulose nanocrystal film. i-PP initially melt at 220°C. PLM images after 5 minutes (left) and after 20 minutes (right) at 136°C. The right image at higher magnification shows in detail the edge of the CNWs film in contact with i-PP melt. Scale bars: 200 µm (reprinted with permission from Gray [92]. Copyright 2007 Springer Science+Business Media B.V.)

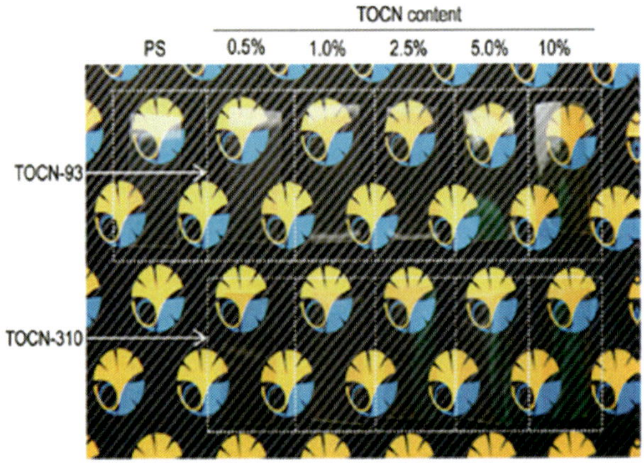

Figure 19.4 Transparent TOCN-93/PS and TOCN-310/PS nanocomposite films. Reprinted with permission from Reference [60].

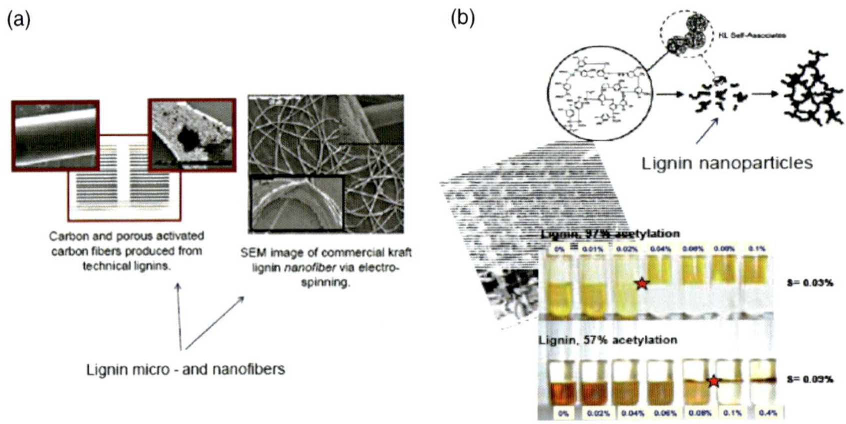

Figure 23.1 Possibilities of obtaining lignin-based nanofibers (adapted from Reference [16]).

Habibi et al. [26] performed TEMPO-mediated oxidation of CNs obtained from HCl hydrolysis of cellulose fibers from tunicate and showed that TEMPO-mediated oxidation did not compromise the morphological integrity of CNs or their native crystallinity. Based on the supramolecular structure, morphology, and crystallographic parameters of the CN, these authors demonstrated that various degrees of oxidation can be predicted and achieved by using specific amounts of the primary oxidizing agent, that is, NaOCl (see Fig. 16.3 bottom).

When dispersed in water, TEMPO-oxidized or carboxylated CNs suspensions display birefringence patterns and do not flocculate or sediment owing to the polyanionic character imparted by the negative charges on the CN surfaces (see Fig. 16.4).

TEMPO-mediated oxidation represents a cost-effective chemical pretreatment facilitating the production of NFCs before the mechanical shearing. The oxidation of cellulosic fibers creates carboxyl groups on their surface and within the individual microfibril. The created negative charges promote the individualization of the fibers. Therefore, the TEMPO-oxidized cellulose fibers can be easily converted, utilizing mechanical shearing, to transparent and

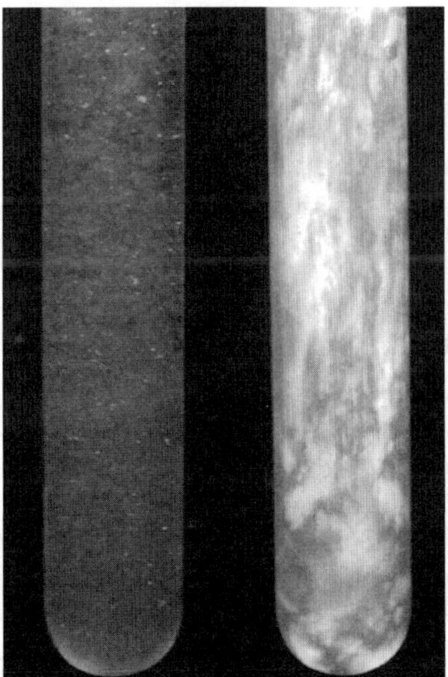

Figure 16.4 Aqueous 0.53% (w/v) suspensions of cellulose nanocrystals observed between crossed polarizers after production by HCl-catalyzed hydrolysis (left), and after their oxidation via TEMPO-mediated reactions (right) (taken from Reference [26]). See color insert.

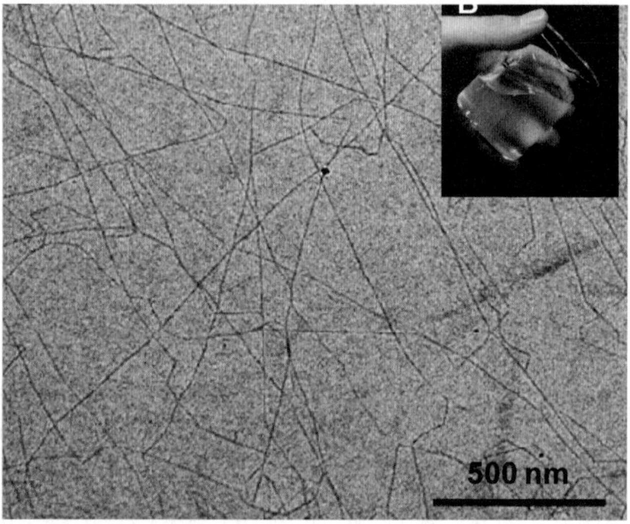

Figure 16.5 Transmission electron micrograph of the TEMPO-mediated oxidized cellulose nanofibrils, and the corresponding highly viscous and transparent dispersion at 0.3% (reprinted with permission from Reference [27]. Copyright 2009 American Chemical Society).

highly viscous dispersions in water, consisting of highly crystalline individual nanofibers [27–30]. It was shown that the optimal conditions for the processing were attained at pH 10, generating cellulose nanofibers 3–4 nm in width and a few microns in length (Fig. 16.5).

16.3.3 Cationization

Positive charges can also be easily introduced on the surface of CNs; for example, weak or strong ammonium-containing groups, such as epoxypropyltrimethylammonium chloride (EPTMAC) can be grafted onto the CN surfaces [31]. Such surface cationization proceeds via a nucleophilic addition of the alkali-activated cellulose hydroxyl groups to the epoxy moiety of EPTMAC and leads to stable aqueous suspensions of CNs with unexpected thixotropic gelling properties. Shear birefringence was observed, but no liquid crystalline chiral nematic phase separation was detected for these cationic CNs, most likely owing to the high viscosity of the suspension. Similar reaction was performed recently on cellulose fibers before their mechanical shearing in order to produce at low energy consumption NFCs [32, 33]. Indeed, cellulose pulp was cationized in an etherification reaction with EPTMAC or chlorocholine chloride and consequently mechanically disintegrated to yield trimethylammonium-modified NFCs.

16.3.4 Esterification

Homogeneous and heterogeneous acetylation of model CNs extracted from *Valonia* and tunicate has been studied by Sassi and Chanzy [34] by using acetic anhydride in acetic acid. Their ultrastructural study, carried out by transmission electron microscopy (TEM) imaging and X-ray diffraction, showed that the reaction proceedes by a reduction of the diameters of the crystals, while only a limited reduction in CN lengths was observed. It has been suggested that the reaction involved a nonswelling mechanism that affected only the cellulose chains localized at the crystal surface. In the case of homogeneous acetylation, the partially acetylated molecules immediately partitioned into the acetylating medium as soon as they were sufficiently soluble; while in heterogeneous conditions the cellulose acetate remained insoluble and surrounded the crystalline core of unreacted cellulose chains. BC nanofibers were surface acetylated using acetic anhydride and sulfuric acid as catalyst, with the aim of modifying their surface properties in order to enhance their compatibility with polylactide (PLA) [35]. The nanofibers underwent no structural and/or morphological changes under the applied reaction conditions. At only 6% acetylated BC loading, the obtained bionanocomposites showed considerable improvement in thermal and mechanical properties, as evidenced by the significant increase in both elastic and Young's moduli, and in the tensile strength. These nanocomposites also showed low hygroscopicity and considerable transparency features reported.

The simultaneous occurrence of cellulose hydrolysis and acetylation of hydroxyl groups has been also reported. Fischer esterification of hydroxyl groups simultaneously with the hydrolysis of amorphous cellulose chains has been introduced as a viable one-pot reaction methodology that allows isolation of acetylated CNs in a single-step process (Fig. 16.6) [36, 37].

An environmentally friendly CN surface acetylation route was recently developed by Yuan et al. [38] involving low reagent consumption and a simple-to-apply procedure. The method used alkyenyl succinic anhydride aqueous emulsions as a template. The emulsions were simply mixed with CN suspensions and freeze-dried, and the resulting solid was heated to 105°C. The obtained derivative conferred to the acylated CNs a highly hydrophobic character because they were easily dispersible in solvents with widely different polarities as measured by the respective dielectric constant, ε; for example, they were not only dispersible in dimethyl sulfoxide (DMSO) having a very high ε of 46.45, but also in 1,4-dioxane, which has a quite low ε of 2.21. Berlioz et al. [39] have reported recently a new and highly efficient solvent-free synthetic method for almost complete surface esterification of BC and CNs extracted from tunicate, leading to highly substituted nanocellulose esters. The reaction of fatty acids such as palmitoyl chloride was carried out on nanocellulose substrates freeze-dried or dried via a gas-phase process. It has been shown by SEM and X-ray diffraction analyses that the esterification proceeded from the surface of the substrate to the crystal core. Under moderate

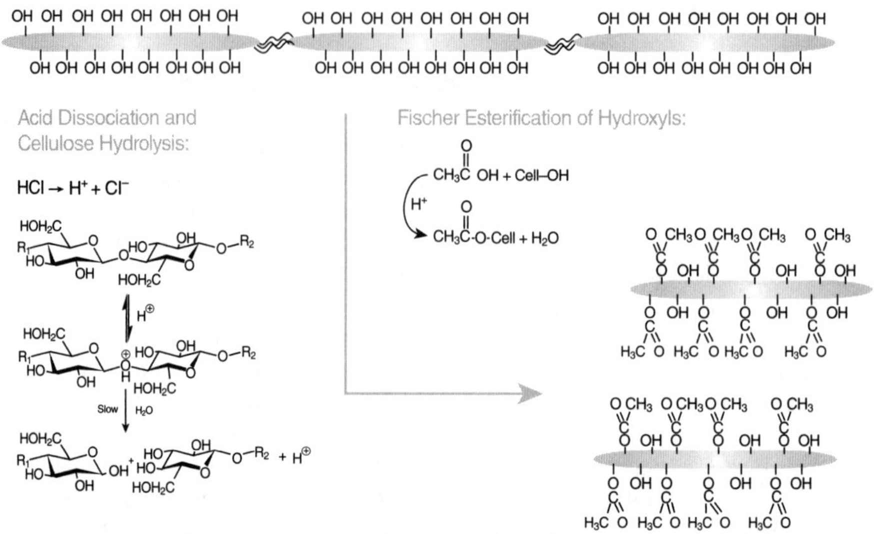

Figure 16.6 Reaction scheme illustrating the one-pot cellulose hydrolysis and esterification of hydroxyl groups. (reprinted with permission from Reference [36]. Copyright 2008 American Chemical Society).

conditions, the surface was fully reacted, whereas the original morphology was maintained, and the core of the crystal remained unmodified.

Esterification of CNs by reacting organic fatty acid chlorides, having different lengths of the aliphatic chain (C_{12} up to C_{18}), has also been reported with a grafting density high enough such that the fatty acids with a backbone of 18 carbons were able to crystallize on the surface CNs [40]. A straightforward method for the surface acetylation of cellulose nanowhiskers by transesterification of vinyl acetate and using potassium carbonate as catalyst was proposed by Çetin et al. [41]. Under moderate conditions, only the surface of the nanowhiskers was modified, while their dimensions and crystallinity remained unchanged. With increasing reaction time, diffusion mechanisms controlled the rate, leading to nanowhiskers with higher levels of acetylation, smaller dimensions, and lower crystallinity.

BC nanofibrils were also subjected to acetylation with organic acids having different chain lengths such as acetic, hexanoic, and dodecanoic acids [42, 43]. The hydrophobic nature of these modified BC was enhanced and showed a good compatibility when incorporated in polylactide nanocomposites.

16.3.5 Silylation

Cellulose whiskers resulting from the acid hydrolysis of tunicate have been partially silylated by a series of alkyldimethylchlorosilanes, with the carbon backbone of the alkyl moieties ranging from a short carbon length of isopropyl to longer lengths represented by n-butyl, n-octyl, and n-dodecyl [44]. It has been demonstrated that with a degree of substitution (DS) between 0.6–1, the whiskers became readily dispersible in solvents of low polarity (such as tetrahydrofuran [THF]), leading to stable suspensions with birefringent behavior, while their morphological integrity was preserved. However, at high silylation (DS greater than 1), the chains in the core of the crystals became silylated, resulting in the disintegration of the crystal and consequently the loss of original morphology.

A mild silylation protocol was applied to surface silylate dispersed NFCs extracted from parenchymal cell cellulose. These silylated NFCs, with isopropyl dimethylchlorosilane, maintained the same morphological features as those of the underivatized samples and were dispersible into a nonpolar solvent to yield stable suspensions that did not flocculate. The rheological properties of their suspension in methyl oleate showed a marked shear-thinning effect. Thus, by silylation, the NFCs have acquired an inherent flexibility with the result that their suspensions present the rheological behavior of polymer solutions [45].

Andresen et al. [46] have reported the successful surface silylation of NFCs with chlorodimethyl isopropylsilane. At moderated DS (between 0.6 and 1), hydrophibized silylated NCFs kept their morphological integrity and were able to stabilize water-in-oil emulsions [47]. NFCs were also grafted with a quaternary ammonium silane derivative (e.g., octadecyldimethyl(3-trimethoxysilylpropyl)ammonium chloride) through a simple adsorption-curing process. Films prepared from these hydrophibized NFCs showed substantial nonleaching-type antibacterial capacity [48].

Surface-trimethyl silylation of CNs extracted from bacterial cellulose and their resulting cellulose acetate butyrate [49] or polysiloxane [50] based nanocomposites were also investigated. CNs functionalized by partial silylation through coupling with n-dodecyldimethylchlorosilane showed a very good dispersion in poly(L-lactide) and acted as nucleating agents and hence accelerate the crystallization rate of PLLA [51]. Similarly, the coupling of either CNs or NFCs with the 3-aminopropyltriethoxysilane enhanced their compatibility with PLA, improving significantly the mechanical performances of the final nanocomposites [52]. CNs were straightforwardly, after their extraction, coupled with functional trialkoxysilanes bearing various organic moieties (alkyl, amino and methacryloxy) [53]. The reaction was conducted efficiently in aqueous medium and at room temperature. After freeze-drying, the resulting silylated CNs were successfully incorporated into PLA by melt extrusion without any alterations of their nanostructure after melt processing. The thermomechanical properties of the resulting bionanocomposites were shown to be enhanced when silylated CNs were used as nanofillers.

Silylation was also attempted in order to tune the wetting properties towards nonpolar liquids of preformed NFC-based aerogels. Indeed, perfluorodecyltrichlorosilane was grafted by chemical vapor deposition method [54].

Silylation was also exploited as intermediate step for further functionalization. Aminosilanes were first grafted onto CNs and serve as reactive sites to covalently attach fluorescent moieties [55]. More recently, ene- or thiol-functionalized silanes were also covalently grafted on NFC-based films and consequently clicked with appropriate moieties through thiol-ene click chemistry [56].

16.3.6 Urethanization

Urethanization proceeds by reacting an isocayante with hydroxyl groups available at the surface of nanocelluloses to create a urethane linkage. Coupling CNs and NFCs with N-octadecyl isocyanate, via a bulk reaction in toluene, has also been reported to enhance their dispersion in organic medium and compatibility with polycaprolactone, which significantly improved the stiffness and ductility of the nanocomposites [57, 58]. Isocyanate-rich CNs were prepared by coupling CNs with 1,6-hexamethylene diisocyanate (HDI) by means of *in situ* polymerization. These isocyanate-rich CNs were incorporated onto segmented thermoplastic polyurethane elastomers by casting evaporation technique. Modified CNs were effectively dispersed in the polyurethane matrix and interacted with matrix hard phase, promoting physical association with hard segments, enhancing stiffness, and dimensional stability [59]. Isocyanate-mediated coupling was also utilized to promote the grafting of isocyanate-functionalized polymers. Likewise, thiol-based isocyanate (isothiocyanate) were grafted on the surface of CNs to attach pH-sensitive dyes [60].

16.3.7 Amidation

Most of amidation-based couplings were realized on carboxylic groups of pre-oxidized nanocellulose substrates. The covalent attachment of amine derivatives on the surface of nanocelluloses was achieved via a carbodiimide-mediated amidation reaction without the alteration of their native morphological or crystalline properties. Various amine derivatives such as 4-amino-TEMPO [61], benzylamine, hexylamine, dodecylamine, and Jeffamine [62] were grafted on CNs or NFCs, respectively. All these amidated nanocelluloses provided a very good dispersion in nonpolar solvents, attesting to the increase of their hydrophobic character after the modification. This reaction was also exploited to graft amine-terminated polymers as will be described later.

16.3.8 Clicking

"Clicking" approach on CNs was first introduced by Filpponen and Argyropoulos [63], who attempted to link together CNs. Two sets of CNs bearing an azide derivative for one set and an alkyne derivative for the other set were prepared through amidation of TEMPO-mediated oxidized CNs. Subsequently, the Cu(I)-catalyzed Huisgen 1,3-dipolar cycloaddition-mediated click reaction, between the azide and the alkyne surface-activated CNs, was employed, bringing together the nanocrystalline materials in a unique regularly packed arrangement. A similar procedure was followed by Sadeghifar et al. [64]. Recently, Feese et al. [65] also employed Cu(I)-catalyzed Huisgen–Meldal–Sharpless 1,3-dipolar cycloaddition reaction to surface immobilize cationic porphyrin. CNs–porphyrin showed excellent efficiency toward the photodynamic inactivation of *Mycobacterium smegmatis* and *Staphylococcus aureus*, albeit only slight activity against *Escherichia coli*. Similarly, an imidazolium salt was grafted to CNs using Cu(I)-catalysed azide-alkyne cycloaddition [66]. The authors demonstrated the feasibility of exchanging the bromide anion to bistriflimide and an anionic dye. This approach may provide the opportunity to easily access a wide variety of ion exchange systems or catalysts using CNs as a support medium.

CNs were converted into ratiometric pH-sensing nanosized particles by dual fluorescent labeling in a three-step chemical procedure involving thiol-ene-mediated click reaction [60]. In this procedure, an amine group was introduced via esterification followed by a thiol-ene click reaction. This amine group consequently served as a reactive site to attach succinimidyl ester dyes that are pH-sensitive. Tingaut et al. [56] have developed a chemical strategy following different routes implicating thiol-ene click reaction to chemically modify NFC-based films (Fig. 16.7). A thiol-functionalized silane was first grafted on the surface of NFC films and then ene-terminated molecules were photochemically clicked or vice versa. The route comprising the clicking of ene-activated NFC films with the silane–thiol was greatly faster. In a different route, the silane–thiol and ene-terminated molecules were first clicked together and then grafted through silylation usng a sol–gel process to the surface of NFC films.

16.3.9 Polymer Grafting

Polymer grafting, on the surface of nanocelluloses, has been carried out by using two main strategies, namely "grafting onto" or "grafting from." The grafting onto approach involves attachment onto hydroxyl groups at the nanocellulose surface of presynthesized and eventually functionalized polymer chains by using a coupling agent. In the "grafting from" approach, the polymer chains are formed by *in situ* surface-initiated polymerization eventually with immobilized initiators on the substrate.

16.3.9.1 Grafting Onto The "grafting onto" approach was used by Ljungberg et al. [67] to graft maleated polypropylene (PPgMA) onto the surface of tunicate extracted CNs. The resulting grafted nanocrystals showed very good compatibility and high adhesion when dispersed in atactic polypropylene. Araki et al. [24] and Vignon et al. [68] studied the grafting of amine-terminated polymers on the surface of TEMPO-mediated oxidized CNs by using a peptide

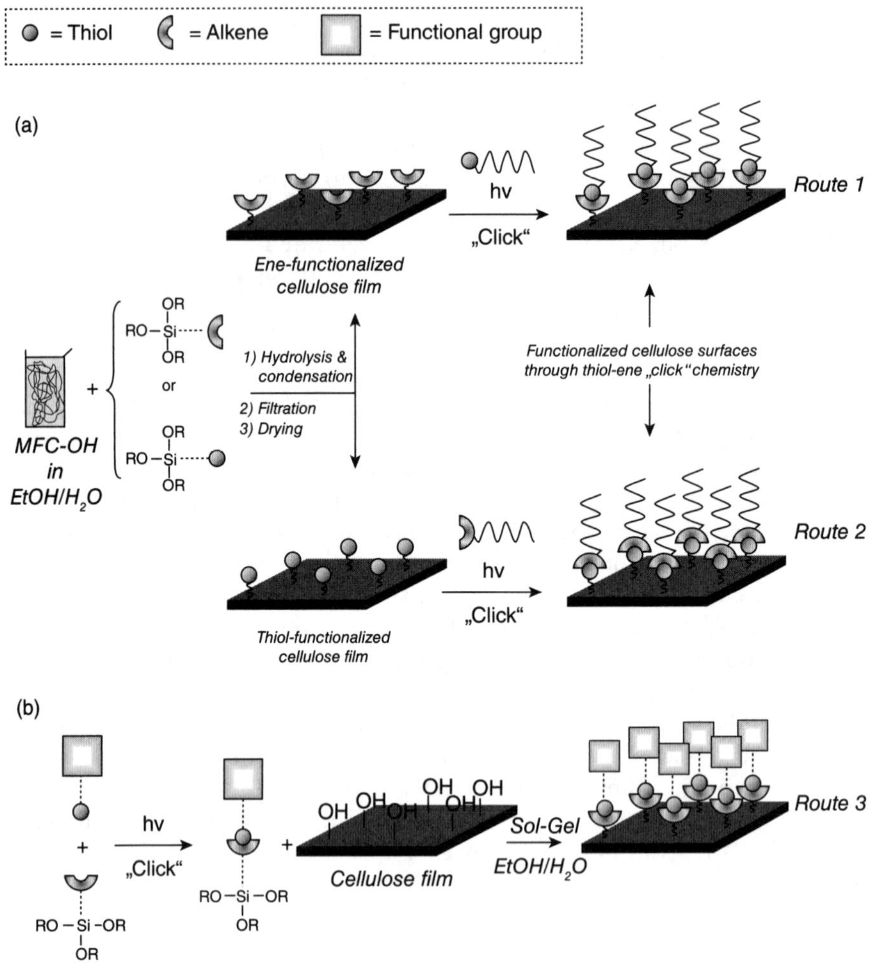

Figure 16.7 Schematic illustration of NFC-based films surface functionalization using click chemistry. (a) Synthesis of ene- and thiol-functionalized films with alkoxysilane molecules, and their subsequent "click" coupling reactions with thiol- (Route 1) and ene- (Route 2) molecules, respectively. (b) Synthesis of a functional alkoxysilane molecule using thiol-ene "click" chemistry, and its subsequent coupling reaction with a cellulose film through a sol–gel process (Route 3) (reproduced from Reference [56] with permission of The Royal Society of Chemistry). See color insert.

coupling process catalyzed by carbodiimide derivatives in water. The same approach was implemented by Mangalam et al. [69], who grafted DNA oligomers and on the surface of CNs. Harrisson et al. [70] and Azzam et al. [71] used dimethylformamide (DMF) rather than water as reaction medium to graft amine-terminated poly(styrene) or poly(tert-butyl acrylate) and Jeffamine copolymers, respectively. Epoxy-terminated poly(ethylene oxide) was successfully grafted onto CNs through a nucleophilic addition of the alkali-activated cellulose hydroxyl groups to the epoxy moieties resulting in ether linkage. PEO-grafted CNs formed a stable colloidal suspension in water that remained well dispersed, and a chiral nematic phase was observed upon concentration of these aqueous suspension [72]. Similarly, presynthesized epoxy-terminated copolymer of PLA and glycidyl methacrylate was grafted on the surface of BC nanofibers in an attempt to reduce their hydrophilicity before their incorporation in PLA [73].

The grafting of poly(caprolactone) (PCL) having different molecular weights on the surface of CNs has been achieved by using isocyanate-mediated coupling [74]. These authors reported reaching a grafting density that was high enough that the grafted PCL-chains were able to crystallize at the surface of CNs. Similar efforts were made by Cao et al. [75], who reported on the isocyanate-catalyzed grafting of presynthesized waterborne polyurethane polymers via a one-pot process. Such crystallization provoked co-crystallizations of the free chains of the respective polymer matrices during CN-based nanocomposite processing. Furthermore, this co-crystallization phenomenon induced the formation of a co-continuous phase between the matrix and filler, which significantly enhanced the interfacial adhesion and consequently contributed to a highly improved mechanical strength of the resulting nanocomposites. A similar approach was followed by Pei et al. [76] to prepare polyurethane/CNs nanocomposites with ultrahigh tensile strength and stain-to-failure with strongly improved modulus.

16.3.9.2 Grafting From The grafting "from approach" applied to CNs was first reported by Habibi et al. [77], who grafted PCL on the surface of CNs via ring-opening polymerization (ROP) by using stannous octoate ($Sn(Oct)_2$) as grafting and polymerization agent. Also, the interfacial grafting of PCL as well as PLA from the surface of CNs was studied and the resulting hydrids were incorporated in the respective matrix [78, 79]. Enhanced mechanical properties and better thermal stability were obtained originating from co-crystallization and entanglement between bounded chains and those from the matrix. Lönnberg et al. [80, 81] used the same approach to graft PCL on the surface of NFCs, and they were able to process films made with PCL-grafted NFC together with PCL films in bilayer laminates using hot press. Through peeling tests, they evidenced the occurrence of physical entanglement between PCL chains grafted at the NFC surface and the unbounded ones from the matrix. This entanglement depends strongly on the length of grafted PCL chains [82]. Likewise, Chen et al. [83] and Lin et al. [84] conducted similar grafting reactions under microwave irradiation to enhance the grafting efficiency. *In situ* polymerization of furfuryl alcohol from the surface of cellulose whiskers was studied by Pranger and Tannenbaum [85]. In this case, the polymerization was catalyzed by sulfonic acid residues present at the surface of CNs. At elevated temperatures, the sulfonic acid groups were de-esterified and consequently released into the medium to catalyze *in situ* the polymerization. Yi et al. [86] and Morandi et al. [87] propagated polystyrene brushes via atom transfer radical polymerization (ATRP) on the surface of CNs with ethyl 2-bromoisobutyrate as the initiator agent. Similarly, other vinyl monomers, mainly acrylic monomers such as N-isopropylacrylamide [88] and acrylic acid [89], were also polymerized from the surface of CNs to produce thermo-responsive and pH-sensitive substrates, respectively. An amphotropic polymer bearing azo moieties, poly(6-[4-(4-methoxyphenylazo)phenoxy] hexyl methacrylate) (PMMAZO), was successfully attached and grown from the surface of CNs following the same strategy [90]. PMMAZO-grafted CNs exhibited both types of liquid crystal formation, thermotropic, and lyotropic as they showed smectic-to-nematic transition at 95°C and nematic-to-isotropic transition at 135°C, and exhibited analogous lyotropic liquid-crystalline phase behavior above 135°C. It also showed a lyotropic nematic phase in chlorobenzene above a concentration of 5.1 wt%. Grafting of polyaniline from CNs was achieved by *in situ* polymerization of aniline onto CNs, in hydrochloric acid aqueous solution, via an oxidative polymerization by using ammonium peroxydisulfate as the initiator [91]. Cast films of these polyaniline-grafted CNs showed conductivity close to the conducting polymer. They have been also incorporated in memory-shaped polyurethane showing a stepwise change in thermal, rheological, and electrical properties in the range of 4–10 wt% CNs, while maintaining the shape-responsiveness of the nanocomposites [92]. Likewise NFCs were grafted in aqueous solution using a redox initiated free radical polymerization with different (meth)acrylate monomers but using Cerium ammonium nitrate as initiator. The reaction proceeds first by the oxidation of 1,2-glycol groups generating free radicals on the cellulose backbone followed by the grafting of vinylic monomers. The nanostructure of the NFCs was preserved during the synthesis, whereas the density of the grafted polymers varied from a thin coating to a continuous matrix completely enveloping the fibrils enhancing their hydrophobic character [93]. This strategy was first used by Stenstad et al. [94] to graft glycidyl methacrylate on the surface of NFCs.

16.4 CONCLUSIONS

Over the last decade, nanocelluloses, including cellulose NCs or nanowhiskers, NFCs, and BC, have become fascinating building blocks for the design of new biomaterials. They can be used as they are in different applications involving hydrophilic media, which restrict their uses. However, chemical modifications are essential, sometime unavoidable, to adapt their interfacial properties in various applications or to adjust their hydrophilic–hydrophobic balance. As nanocelluloses offer an excellent platform for chemical modification regarding their regularity and the presence of a large number of hydroxyl groups at their surface, various chemistries have been developed, aiming to surface-modify them to confer to them the desired property, extending therefore their uses to highly sophisticated applications.

REFERENCES

[1] Eichhorn, S., Dufresne, A., Aranguren, M., Marcovich, N., Capadona, J., Rowan, S., Weder, C., Thielemans, W., Roman, M., Renneckar, S., Gindl, W., Veigel, S., Keckes, J., Yano, H., Abe, K., Nogi, M., Nakagaito, A., Mangalam, A., Simonsen, J., Benight, A., Bismarck, A., Berglund, L., and Peijs, T. (2010) Review: Current international research into cellulose nanofibres and nanocomposites. *Journal of Materials Science*, *45*, 1–33.

[2] Gatenholm, P., Klemm, D. (2010) Bacterial nanocellulose as a renewable material for biomedical applications. *MRS Bulletin*, *35*, 208–213.

[3] Habibi, Y., Lucia, L.A., and Rojas, O.J. (2010) Cellulose nanocrystals: Chemistry, self-assembly, and applications. *Chemical Reviews*, *110*, 3479–3500.

[4] Siró, I., Plackett, D. (2010) Microfibrillated cellulose and new nanocomposite materials: A review. *Cellulose*, *17*, 459–494.

[5] Fengel, D., Wegener, G. (1983) *Wood, Chemistry, Ultrastucture, Reactions*. New York: Walter de Gruyter.

[6] French, A.D., Bertoniere, N.R., Brown, R.M., Chanzy, H., Gray, D., Hattori, K., and Glasser, W. (2004) Cellulose. In: *Kirk-Othmer Encyclopedia of Chemical Technology*, 5th ed, ed. A. Seidel. Hoboken, NJ: John Wiley & Sons, pp. 360–394.

[7] Somerville, C. (2006) Cellulose synthesis in higher plants. *Annual Review of Cell and Developmental Biology*, *22*, 53–78.

[8] Brown, R.M. (1996) The biosynthesis of cellulose. *Journal of Macromolecular Science, Part A*, *33*, 1345–1373.

[9] Brown, R.M. (2004) Cellulose structure and biosynthesis: What is in store for the 21st century? *Journal of Polymer Science Part A: Polymer Chemistry*, *42*, 487–495.

[10] Rowland, S.P., Roberts, E.J. (1972) The nature of accessible surfaces in the microstructure of cotton cellulose. *Journal of Polymer Science Part A-1: Polymer Chemistry*, *10*, 2447–2461.

[11] Spence, K., Habibi, Y., and Dufresne, A. (2011) Nanocellulose-based composites. In: *Cellulose Fibers: Bio- and Nano-Polymer Composites*, eds. S. Kalia, B.S. Kaith, and I. Kaur. Berlin: Springer, pp. 179–213.

[12] Heux, L., Bonini, C. (2000) Microfibrillated and/or microcrystalline dispersion, in particular of cellulose, in an organic solvent.

[13] Heux, L., Chauve, G., and Bonini, C. (2000) Nonflocculating and chiral-nematic self-ordering of cellulose microcrystals suspensions in nonpolar solvents. *Langmuir*, *16*, 8210–8212.

[14] Bonini, C., Heux, L., Cavaillé, J.-Y., Lindner, P., Dewhurst, C., and Terech, P. (2002) Rodlike cellulose whiskers coated with surfactant: A small-angle neutron scattering characterization. *Langmuir*, *18*, 3311–3314.

[15] Ljungberg, N., Cavaillé, J.Y., and Heux, L. (2006) Nanocomposites of isotactic polypropylene reinforced with rod-like cellulose whiskers. *Polymer*, *47*, 6285–6292.

[16] Bondeson, D., Oksman, K. (2007) Dispersion and characteristics of surfactant modified cellulose whiskers nanocomposites. *Composite Interfaces*, *14*, 617–630.

[17] Kim, J., Montero, G., Habibi, Y., Hinestroza, J.P., Genzer, J., Argyropoulos, D.S., and Rojas, O.J. (2009) Dispersion of cellulose crystallites by nonionic surfactants in a hydrophobic polymer matrix. *Polymer Engineering and Science*, *49*, 2054–2061.

[18] Rojas, O.J., Montero, G.A., and Habibi, Y. (2009) Electrospun nanocomposites from polystyrene loaded with cellulose nanowhiskers. *Journal of Applied Polymer Science*, *113*, 927–935.

[19] Zhou, Q., Brumer, H., and Teeri, T.T. (2009) Self-organization of cellulose nanocrystals adsorbed with xyloglucan oligosaccharide-poly(ethylene glycol)-polystyrene triblock copolymer. *Macromolecules*, *42*, 5430–5432.

[20] Zhou, Q., Malm, E., Nilsson, H., Larsson, P.T., Iversen, T., Berglund, L.A., and Bulone, V. (2009) Nanostructured biocomposites based on bacterial cellulosic nanofibers compartmentalized by a soft hydroxyethylcellulose matrix coating. *Soft Matter*, *5*, 4124–4130.

[21] Brown, E.E., Laborie, M.-P.G. (2007) Bioengineering bacterial cellulose/poly(ethylene oxide) nanocomposites. *Biomacromolecules*, *8*, 3074–3081.

[22] Syverud, K., Xhanari, K., Chinga-Carrasco, G., Yu, Y., and Stenius, P. (2011) Films made of cellulose nanofibrils: Surface modification by adsorption of a cationic surfactant and characterization by computer-assisted electron microscopy. *Journal of Nanoparticle Research*, *13*, 773–782.

[23] de Nooy, A.E.J., Besemer, A.C., and van Bekkum, H. (1994) Highly selective tempo mediated oxidation of primary alcohol groups in polysaccharides. *Recueil des Travaux Chimiques des Pays-Bas*, *113*, 165–166.

[24] Araki, J., Wada, M., and Kuga, S. (2000) Steric stabilization of a cellulose microcrystal suspension by poly(ethylene glycol) grafting. *Langmuir*, *17*, 21–27.

[25] Montanari, S., Rountani, M., Heux, L., and Vignon, M.R. (2005) Topochemistry of carboxylated cellulose nanocrystals resulting from TEMPO-mediated oxidation. *Macromolecules*, *38*, 1665–1671.

[26] Habibi, Y., Chanzy, H., and Vignon, M. (2006) TEMPO-mediated surface oxidation of cellulose whiskers. *Cellulose*, *13*, 679–687.

[27] Saito, T., Hirota, M., Tamura, N., Kimura, S., Fukuzumi, H., Heux, L., and Isogai, A. (2009) Individualization of nano-sized plant cellulose fibrils by direct surface carboxylation using TEMPO catalyst under neutral conditions. *Biomacromolecules*, *10*, 1992–1996.

[28] Saito, T., Kimura, S., Nishiyama, Y., and Isogai, A. (2007) Cellulose nanofibers prepared by TEMPO-mediated oxidation of native cellulose. *Biomacromolecules*, *8*, 2485–2491.

[29] Saito, T., Nishiyama, Y., Putaux, J.-L., Vignon, M., and Isogai, A. (2006) Homogeneous suspensions of individualized microfibrils from TEMPO-catalyzed oxidation of native cellulose. *Biomacromolecules*, *7*, 1687–1691.

[30] Saito, T., Okita, Y., Nge, T.T., Sugiyama, J., and Isogai, A. (2006) TEMPO-mediated oxidation of native cellulose: Microscopic analysis of fibrous fractions in the oxidized products. *Carbohydrate Polymers*, *65*, 435–440.

[31] Hasani, M., Cranston, E.D., Westman, G., and Gray, D.G. (2008) Cationic surface functionalization of cellulose nanocrystals. *Soft Matter*, *4*, 2238–2244.

[32] Ho, T., Zimmermann, T., Hauert, R., and Caseri, W. (2011) Preparation and characterization of cationic nanofibrillated cellulose from etherification and high-shear disintegration processes. *Cellulose*, *18*, 1391–1406.

[33] Olszewska, A., Eronen, P., Johansson, L.-S., Malho, J.-M., Ankerfors, M., Lindström, T., Ruokolainen, J., Laine, J., and Österberg, M. (2011) The behaviour of cationic nanofibrillar cellulose in aqueous media. *Cellulose*, *18*, 1213–1226.

[34] Sassi, J.-F., Chanzy, H. (1995) Ultrastructural aspects of the acetylation of cellulose. *Cellulose*, *2*, 111–127.

[35] Tome, L.C., Pinto, R.J.B., Trovatti, E., Freire, C.S.R., Silvestre, A.J.D., Neto, C.P., and Gandini, A. (2011) Transparent bionanocomposites with improved properties prepared from acetylated bacterial cellulose and poly(lactic acid) through a simple approach. *Green Chemistry*, *13*, 419–427.

[36] Braun, B., Dorgan, J.R. (2008) Single-step method for the isolation and surface functionalization of cellulosic nanowhiskers. *Biomacromolecules*, *10*, 334–341.

[37] Sobkowicz, M.J., Braun, B., and Dorgan, J.R. (2009) Decorating in green: Surface esterification of carbon and cellulosic nanoparticles. *Green Chemistry*, *11*, 680–682.

[38] Yuan, H., Nishiyama, Y., Wada, M., and Kuga, S. (2006) Surface acylation of cellulose whiskers by drying aqueous emulsion. *Biomacromolecules*, *7*, 696–700.

[39] Berlioz, S., Molina-Boisseau, S., Nishiyama, Y., and Heux, L. (2009) Gas-phase surface esterification of cellulose microfibrils and whiskers. *Biomacromolecules*, *10*, 2144–2151.

[40] de Menezes, A.J., Siqueira, G., Curvelo, A.A.S., and Dufresne, A. (2009) Extrusion and characterization of functionalized cellulose whiskers reinforced polyethylene nanocomposites. *Polymer*, *50*, 4552–4563.

[41] Çetin, N.S., Tingaut, P., Oezmen, N., Henry, N., Harper, D., Dadmun, M., and Sebe, G. (2009) Acetylation of cellulose nanowhiskers with vinyl acetate under moderate conditions. *Macromolecular Bioscience*, *9*, 997–1003.

[42] Lee, K.-Y., Blaker, J.J., and Bismarck, A. (2009) Surface functionalisation of bacterial cellulose as the route to produce green polylactide nanocomposites with improved properties. *Composites Science and Technology*, *69*, 2724–2733.

[43] Lee, K.-Y., Quero, F., Blaker, J., Hill, C., Eichhorn, S., and Bismarck, A. (2011) Surface only modification of bacterial cellulose nanofibres with organic acids. *Cellulose*, *18*, 595–605.

[44] Goussé, C., Chanzy, H., Excoffier, G., Soubeyrand, L., and Fleury, E. (2002) Stable suspensions of partially silylated cellulose whiskers dispersed in organic solvents. *Polymer*, *43*, 2645–2651.

[45] Goussé, C., Chanzy, H., Cerrada, M.L., and Fleury, E. (2004) Surface silylation of cellulose microfibrils: Preparation and rheological properties. *Polymer*, *45*, 1569–1575.

[46] Andresen, M., Johansson, L.-S., Tanem, B.S., and Stenius, P. (2006) Properties and characterization of hydrophobized microfibrillated cellulose. *Cellulose*, *13*, 665–677.

[47] Andresen, M., Stenius, P. (2007) Water-in-oil emulsions stabilized by hydrophobized microfibrillated cellulose. *Journal of Dispersion Science and Technology*, *28*, 837–844.

[48] Andresen, M., Stenstad, P., Møretrø, T., Langsrud, S., Syverud, K., Johansson, L.-S., and Stenius, P. (2007) Nonleaching antimicrobial films prepared from surface-modified microfibrillated cellulose. *Biomacromolecules*, *8*, 2149–2155.

[49] Grunert, M., Winter, W.T. (2002) Nanocomposites of cellulose acetate butyrate reinforced with cellulose nanocrystals. *Journal of Polymers and the Environment*, *10*, 27–30.

[50] Grunert, M., Winter, W.T. (2000) Progress in the development of cellulose reinforced nanocomposites. *Polymeric Materials Science and Engineering*, *82*, 232–233.

[51] Pei, A., Zhou, Q., and Berglund, L.A. (2010) Functionalized cellulose nanocrystals as biobased nucleation agents in poly(L-lactide) (PLLA)—Crystallization and mechanical property effects. *Composites Science and Technology*, *70*, 815–821.

[52] Frone, A.N., Berlioz, S., Chailan, J.F., Panaitescu, D.M., and Donescu, D. (2011) Cellulose fiber-reinforced polylactic acid. *Polymer Composites*, *32*, 976–985.

[53] Raquez, J.M., Murena, Y., Goffin, A.L., Habibi, Y., Ruelle, B., DeBuyl, F., and Dubois, P. (2012) Surface-modification of cellulose nanowhiskers and their use as nanoreinforcers into polylactide: A sustainably-integrated approach. *Composites Science and Technology.*

[54] Aulin, C., Netrval, J., Wagberg, L., and Lindstrom, T. (2010) Aerogels from nanofibrillated cellulose with tunable oleophobicity. *Soft Matter*, *6*, 3298–3305.

[55] Yang, Q., Pan, X. (2010) A facile approach for fabricating fluorescent cellulose. *Journal of Applied Polymer Science*, *117*, 3639–3644.

[56] Tingaut, P., Hauert, R., and Zimmermann, T. (2011) Highly efficient and straightforward functionalization of cellulose films with thiol-ene click chemistry. *Journal of Materials Chemistry*, *21*, 16066–16076.

[57] Siqueira, G., Bras, J., and Dufresne, A. (2009) Cellulose whiskers versus microfibrils: Influence of the nature of the nanoparticle and its surface functionalization on the thermal and mechanical properties of nanocomposites. *Biomacromolecules*, *10*, 425–432.

[58] Siqueira, G., Bras, J., and Dufresne, A. (2010) New process of chemical grafting of cellulose nanoparticles with a long chain isocyanate. *Langmuir*, *26*, 402–411.

[59] Rueda, L., Fernández d'Arlas, B., Zhou, Q., Berglund, L.A., Corcuera, M.A., Mondragon, I., and Eceiza, A. (2011) Isocyanate-rich cellulose nanocrystals and their selective insertion in elastomeric polyurethane. *Composites Science and Technology*, *71*, 1953–1960.

[60] Nielsen, L.J., Eyley, S., Thielemans, W., and Aylott, J.W. (2010) Dual fluorescent labelling of cellulose nanocrystals for pH sensing. *Chemical Communications*, *46*, 8929–8931.

[61] Follain, N., Marais, M.-F., Montanari, S., and Vignon, M.R. (2010) Coupling onto surface carboxylated cellulose nanocrystals. *Polymer*, *51*, 5332–5344.

[62] Lasseuguette, E. (2008) Grafting onto microfibrils of native cellulose. *Cellulose*, *15*, 571–580.

[63] Filpponen, I., Argyropoulos, D.S. (2010) Regular linking of cellulose nanocrystals via click chemistry: Synthesis and formation of cellulose nanoplatelet gels. *Biomacromolecules*, *11*, 1060–1066.

[64] Sadeghifar, H., Filpponen, I., Clarke, S., Brougham, D., and Argyropoulos, D. (2011) Production of cellulose nanocrystals using hydrobromic acid and click reactions on their surface. *Journal of Materials Science*, *46*, 7344–7355.

[65] Feese, E., Sadeghifar, H., Gracz, H.S., Argyropoulos, D.S., and Ghiladi, R.A. (2011) Photobactericidal porphyrin-cellulose nanocrystals: Synthesis, characterization, and antimicrobial properties. *Biomacromolecules*, *12*, 3528–3539.

[66] Eyley, S., Thielemans, W. (2011) Imidazolium grafted cellulose nanocrystals for ion exchange applications. *Chemical Communications*, *47*, 4177–4179.

[67] Ljungberg, N., Bonini, C., Bortolussi, F., Boisson, C., Heux, L., and Cavaillé, J.-Y. (2005) New nanocomposite materials reinforced with cellulose whiskers in atactic polypropylene: effect of surface and dispersion characteristics. *Biomacromolecules*, *6*, 2732–2739.

[68] Vignon, M., Montanari, S., and Habibi, Y. (2004) Crystalline polysaccharide derivatives, their production and their applications.

[69] Mangalam, A.P., Simonsen, J., and Benight, A.S. (2009) Cellulose/DNA hybrid nanomaterials. *Biomacromolecules*, *10*, 497–504.

[70] Harrisson, S., Drisko, G.L., Malmstrom, E., Hult, A., and Wooley, K.L. (2011) Hybrid rigid/soft and biologic/synthetic materials: Polymers grafted onto cellulose microcrystals. *Biomacromolecules*, *12*, 1214–1223.

[71] Azzam, F., Heux, L., Putaux, J.-L., and Jean, B. (2010) Preparation by grafting onto, characterization, and properties of thermally responsive polymer-decorated cellulose nanocrystals. *Biomacromolecules*, *11*, 3652–3659.

[72] Kloser, E., Gray, D.G. (2010) Surface grafting of cellulose nanocrystals with poly(ethylene oxide) in aqueous media. *Langmuir*, *26*, 13450–13456.

[73] Li, Z.Q., Zhou, X.D., and Pei, C.H. (2010) Synthesis of PLA-co-PGMA copolymer and its application in the surface modification of bacterial cellulose. *International Journal of Polymeric Materials*, *59*, 725–737.

[74] Habibi, Y., Dufresne, A. (2008) Highly filled bionanocomposites from functionalized polysaccharide nanocrystals. *Biomacromolecules*, *9*, 1974–1980.

[75] Cao, X., Habibi, Y., and Lucia, L.A. (2009) One-pot polymerization, surface grafting, and processing of waterborne polyurethane-cellulose nanocrystal nanocomposites. *Journal of Materials Chemistry*, *19*, 7137–7145.

[76] Pei, A., Malho, J.-M., Ruokolainen, J., Zhou, Q., and Berglund, L.A. (2011) Strong nanocomposite reinforcement effects in polyurethane elastomer with low volume fraction of cellulose nanocrystals. *Macromolecules*, *44*, 4422–4427.

[77] Habibi, Y., Goffin, A.-L., Schiltz, N., Duquesne, E., Dubois, P., and Dufresne, A. (2008) Bionanocomposites based on poly((-caprolactone)-grafted cellulose nanocrystals by ring-opening polymerization. *Journal of Materials Chemistry*, *18*, 5002–5010.

[78] Goffin, A.-L., Raquez, J.-M., Duquesne, E., Siqueira, G., Habibi, Y., Dufresne, A., and Dubois, P. (2011) From interfacial ring-opening polymerization to melt processing of cellulose nanowhisker-filled polylactide-based nanocomposites. *Biomacromolecules*, *12*, 2456–2465.

[79] Goffin, A.L., Raquez, J.M., Duquesne, E., Siqueira, G., Habibi, Y., Dufresne, A., and Dubois, P. (2011) Poly((-caprolactone) based nanocomposites reinforced by surface-grafted cellulose nanowhiskers via extrusion processing: Morphology, rheology, and thermo-mechanical properties. *Polymer*, *52*, 1532–1538.

[80] Lönnberg, H., Fogelström, L., Azizi Samir, M.A.S., Berglund, L., Malmström, E., and Hult, A. (2008) Surface grafting of microfibrillated cellulose with poly(ε-caprolactone)—Synthesis and characterization. *European Polymer Journal*, *44*, 2991–2997.

[81] Lönnberg, H., Larsson, K., Lindström, T., Hult, A., and Malmström, E. (2011) Synthesis of polycaprolactone-grafted microfibrillated cellulose for use in novel bionanocomposites-influence of the graft length on the mechanical properties. *ACS Applied Materials & Interfaces*, *3*, 1426–1433.

[82] Lönnberg, H., Fogelström, L., Zhou, Q., Hult, A., Berglund, L., and Malmström, E. (2011) Investigation of the graft length impact on the interfacial toughness in a cellulose/poly(ε-caprolactone) bilayer laminate. *Composites Science and Technology*, *71*, 9–12.

[83] Chen, G., Dufresne, A., Huang, J., and Chang, P.R. (2009) A novel thermoformable bionanocomposite based on cellulose nanocrystal-graft-poly(epsilon-caprolactone). *Macromolecular Materials and Engineering*, *294*, 59–67.

[84] Lin, N., Chen, G., Huang, J., Dufresne, A., and Chang, P.R. (2009) Effects of polymer-grafted natural nanocrystals on the structure and mechanical properties of poly(lactic acid): A case of cellulose whisker-graft-polycaprolactone. *Journal of Applied Polymer Science*, *113*, 3417–3425.

[85] Pranger, L., Tannenbaum, R. (2008) Biobased nanocomposites prepared by *in situ* polymerization of furfuryl alcohol with cellulose whiskers or montmorillonite clay. *Macromolecules*, *41*, 8682–8687.

[86] Yi, J., Xu, Q., Zhang, X., and Zhang, H. (2008) Chiral-nematic self-ordering of rodlike cellulose nanocrystals grafted with poly(styrene) in both thermotropic and lyotropic states. *Polymer*, *49*, 4406–4412.

[87] Morandi, G., Heath, L., and Thielemans, W. (2009) Cellulose nanocrystals grafted with polystyrene chains through surface-initiated atom transfer radical polymerization (SI-ATRP). *Langmuir*, *25*, 8280–8286.

[88] Zoppe, J.O., Habibi, Y., Rojas, O.J., Venditti, R.A., Johansson, L.-S., Efimenko, K., Osterberg, M., and Laine, J. (2010) Poly(N-isopropylacrylamide) brushes grafted from cellulose nanocrystals via surface-initiated single-electron transfer living radical polymerization. *Biomacromolecules*, *11*, 2683–2691.

[89] Majoinen, J., Walther, A., McKee, J.R., Kontturi, E., Aseyev, V., Malho, J.M., Ruokolainen, J., and Ikkala, O. (2011) Polyelectrolyte brushes grafted from cellulose nanocrystals using Cu-mediated surface-initiated controlled radical polymerization. *Biomacromolecules*, *12*, 2997–3006.

[90] Xu, Q., Yi, J., Zhang, X., and Zhang, H. (2008) A novel amphotropic polymer based on cellulose nanocrystals grafted with azo polymers. *European Polymer Journal*, *44*, 2830–2837.

[91] Mattoso, L.H.C., Medeiros, E.S., Baker, D.A., Avloni, J., Wood, D.F., and Orts, W.J. (2009) Electrically conductive nanocomposites made from cellulose nanofibrils and polyaniline. *Journal of Nanoscience and Nanotechnology*, *9*, 2917–2922.

[92] Auad, M.L., Richardson, T., Orts, W.J., Medeiros, E.S., Mattoso, L.H.C., Mosiewicki, M.A., Marcovich, N.E., and Aranguren, M.I. (2011) Polyaniline-modified cellulose nanofibrils as reinforcement of a smart polyurethane. *Polymer International*, *60*, 743–750.

[93] Littunen, K., Hippi, U., Johansson, L.-S., Österberg, M., Tammelin, T., Laine, J., and Seppälä, J. (2011) Free radical graft copolymerization of nanofibrillated cellulose with acrylic monomers. *Carbohydrate Polymers*, *84*, 1039–1047.

[94] Stenstad, P., Andresen, M., Tanem, B.R., and Stenius, P. (2008) Chemical surface modifications of microfibrillated cellulose. *Cellulose*, *15*, 35–45.

[95] Nakagaito, A.N., Iwamoto, S., and Yano, H. (2005) Bacterial cellulose: The ultimate nano-scalar cellulose morphology for the production of high-strength composites. *Applied Physics A: Materials Science & Processing*, *80*, 93–97.

CHAPTER 17

Cellulose-Based Nanocomposites: Processing Techniques

ROBERT A. SHANKS

17.1 INTRODUCTION

The purpose of this chapter is to review cellulose-based nanocomposites where the nano-phase component can be cellulose nanofibers or cellulose nanocrystals. The matrix phase can be comprised of cellulose, a cellulose derivative, or more typically a synthetic or bioderived polymer. The matrix phase can be thermoplastic or thermoset. An alternative is where the nanoparticles are not derived from cellulose, while a second dispersed phase or a matrix phase is constituted of cellulose. The cellulose can be derived from many sources with the choice being related to the locality where the composites will be made. This chapter concentrates on processing techniques for the cellulose. Details of specific polymers and composites are discussed in other chapters.

Cellulose is an abundant natural and renewable resource with high performance structural properties that is available in many forms, often in the presence of other natural materials of varying composition. Cellulose is the structural material of all plants. Commercial grades of cellulose are derived from trees, bast fibers, bamboo, and agricultural by-products. Cellulose has been consolidated into wood, including lignin flours to provide new materials from renewable resources with the same chemical structures within [1]. Wood, bamboo, and some other products are used directly in buildings, furniture, and transport, with little treatment other than heating and drying to increase dimensional stability. Other cellulose products, such as those from flax, hemp,

Biopolymer Nanocomposites: Processing, Properties, and Applications, First Edition. Edited by Alain Dufresne, Sabu Thomas, and Laly A. Pothan.

sisal, kenaf, jute, and waste crop fibers, require separation and purification of fibers to form paper, textiles, and many other cellulose and derivative products. Natural fibers have received much attention because of their relatively low density, and nonabrasive, nontoxic, and biodegradable nature [2]. Natural fibers are available in many forms with comparable material properties and high molding flexibility. They are environmentally friendly, making them substitute for reinforcing fillers such as mica, calcium carbonate, and glass. Biocomposites can be designed to make high performance commodity applications. Some examples are electroactive papers, fuel cell membranes, controlled drug release mechanisms, and biosensors [3].

Cellulose fibers can also be employed for the preparation of hybrid composites along with another polymer and a second complementary mineral filler reinforcement. A hybrid composite of polyurethane, cellulose, or synthetic polymer fibers and fly-ash was prepared as a sandwich of interconnecting ribs between two planar surfaces. The composite panels were of relatively low density suitable for walls, floors, ceilings, and roofs [4].

Separation and purification of cellulose require removal of unwanted materials from cellulose. An initial step may be retting where other materials are degraded by microorganisms faster than cellulose fibres. Plant structures can be disrupted by physical threshing, decortication, or steam explosion to release cellulose fibers from accompanying materials. Solvent extraction is a further step to remove waxes with acetone or related organic solvent. Lignin is a phenolic polymer that can be regarded as an adhesive matrix for cellulose fibers. In other plant materials, pectins and hemicellulose are found. These latter polysaccharides contain glucuronic acid and related carboxylic acids from other monosaccharides. Lignin, pectins, and hemicellulose are extracted with varying concentrations of (10 to 20%·w/v) sodium hydroxide solution.

Pectins can be removed from cellulose fibers by enzyme bioscouring with pectate lyase. Enzyme treatment occurs under mild conditions, 55°C and pH = 8.5 in a pH-buffered water dispersion. A longer treatment time improved the removal of low methoxy pectin. No crystalline transformation of the cellulose was observed. The surface was observed to become smoother, with increased BET surface area and increased pore size [5].

During extraction with sodium hydroxide, cellulose is not dissolved; however it is swollen, allowing the crystal structure of cellulose to change from type I to type II. This means that crystal type I or native cellulose, where the cellulose chains are parallel and aligned in the state that they were biosynthesized, is a metastable form. Cellulose type II or textile cellulose, where the cellulose chains are antiparallel with tighter packing, is a thermodynamically more stable state. The process of conversion of native to textile cellulose using sodium hydroxide solution is called mercerization.

Cellulose cannot be melted for processing. Heat will volatilize the moisture content and cause the structure to become brittle or friable. The equilibrium moisture content depends upon humidity, typically in a range between 8% w/w and 12%·w/w. Water molecules are part of the crystal structure. Processing of

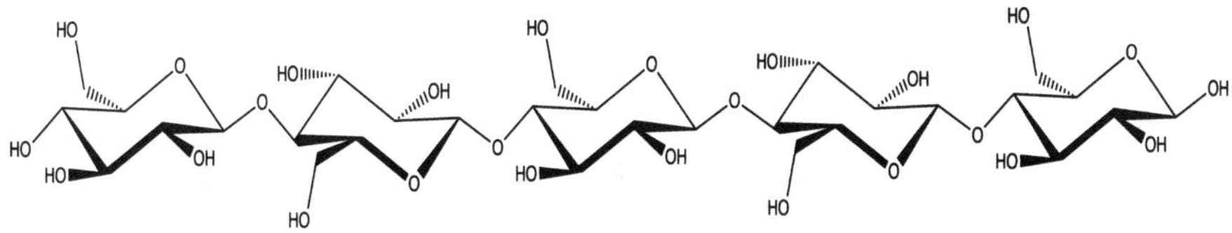

Figure 17.1 A representative segment of cellulose structure.

cellulose requires dissolution, and there are few suitable solvents. Traditionally dissolving cellulose requires complexation or derivatization.

The first method for dissolving cellulose was via tetra-ammonia copper(II) ($(Cu(NH_3)_4)^{2+}$) complexation in water to form a deep blue viscous solution. The complex is not stoichiometric, although there are optimum ratios of copper(II), ammonium, cellulose, and pH. The complex is dissociated in acid solution to regenerate the cellulose. This regeneration can be performed via extruding a thin stream of complex into dilute sulfuric acid to form regenerated cellulose fibers, viscose, or rayon. The second method is to form a sodium xanthate derivative by reaction with sodium hydroxide and carbon disulfide. The derivative is water-soluble and the cellulose solution can be purified by filtration before extruding into a dilute sulfuric acid regeneration bath to form regenerated cellulose fiber, the same as with the cuprammonium method. The xanthate method is the preferred method, although there are now more effective solvents for cellulose and production of regenerated cellulose fibers.

17.1.1 Cellulose Structure and Properties

Cellulose consists of 1,4-anhydroglucopyranose repeat units (Fig. 17.1). The sequence of cyclic polyhydroxy repeat units is remarkable in that all of the hydroxyl groups are equatorial with respect to the six-membered ring. The link from carbons 1 to 4 via an acetal group is also equatorial; that is, the acetal group is in the beta configuration. These structural features confer regularity on the cellulose chain, allowing close packing between chains in the form of crystallinity (Fig. 17.2). Regularity, close packing, and crystallinity confer high modulus and strength.

The pendant equatorial hydroxyl groups are important for further stiffening and strengthening of the cellulose chains internally by forming hydrogen bonds along the axis of the chain between adjacent glucose units. These intramolecular hydrogen bonds prevent rotation of the chain about the 1–4 acetal links. Rotation is prevented by the pyranose ring structures. The combination of ring structures and intramolecular hydrogen bonding makes cellulose molecules completely rigid. Other hydroxyl groups form intermolecular hydrogen bonds between adjacent chains, ensuring the integrity of the crystal structure. The intermolecular hydrogen bonds assist and stabilize the crystalline structure.

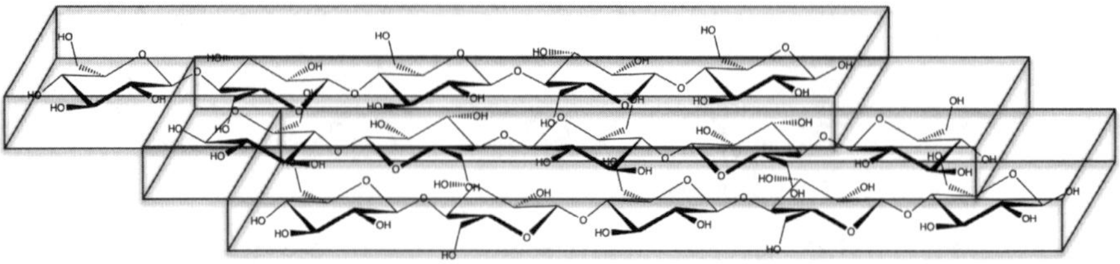

Figure 17.2 Rigid cellulose chain segments stack together in crystals.

Chemical properties of cellulose are defined by the hydroxyl and acetal groups. Acetal groups are resistant to alkali though hydrolyzed by acid. Thus, cellulose is hydrolyzed by acid and hydrolysis occurs faster in amorphous or less ordered crystalline regions. In the more regular crystalline region, hydrolysis is slow due to inhibited permeation of acid. The three hydroxyl groups on each repeat unit are more acidic than typical alcohol groups, though less acidic than phenolic hydroxyl groups. If cellulose is treated with an alkali such as sodium hydroxide or a base such as sodium carbonate, then some of the hydroxyl groups are dissociated into alkoxide anions within the examples given a sodium cation. Formation of alkoxide anions is important for many reactions of cellulose, such as where the alkoxides act as nucleophiles for acetylation, carboxymethylation, hydroxylpropylation, reaction with caprolactone, and coupling with reactive dyes via chlorotriazine or vinylsulfone coupling groups on the dye.

Alkoxide formation with sodium hydroxide disrupts hydrogen bonding and hence the crystal structure. This is the basis of mercerization where hydrogen bond disruption creates swelling and the crystal structure transforms from type I, native cellulose, to type II, textile cellulose. After the crystal transformation, the swelling is removed by washing and neutralization and the fibers remain in the type II crystalline form. Mercerization occurs because type I crystals are less thermodynamically stable than type II crystals. Initial formation of type I crystals is due to the mechanism of formation of cellulose and immediate clustering into crystals. Swelling can be performed using acid, although there is competition from random scission due to hydrolysis of acetal links.

17.1.2 Cellulose Solubility

Cellulose molecules are bound together by hydrogen bonds and molecular geometry gives tight packing and rigid molecules. All of these factors are responsible for cellulose being mainly insoluble. Cellulose can dissolve when it can form a complex that disrupts internal hydrogen bonding. A soluble complex is formed with tetra-ammonia copper(II) hydroxide as discussed previously. Highly polar liquids can dissolve cellulose since they compete

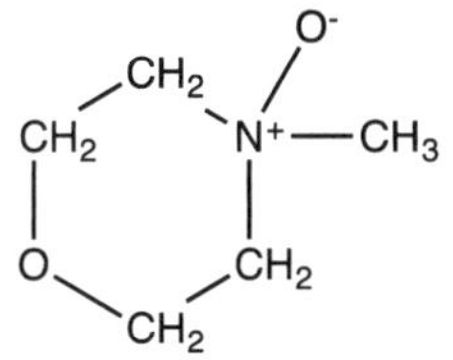

Figure 17.3 N-methylmorpholine-N-oxide structure.

favorably for the hydrogen bonds. Such a solvent is N-methylmorpholine-N-oxide (NMMO) (Fig. 17.3). NMMO solutions of cellulose are used to prepare regenerated microfibers such as designated by the trade names Lyocell and Tencel. The fibers are regenerated by extruding cellulose–NMMO solution into water where the cellulose is precipitated. NMMO has a high boiling temperature, and it can be recovered from the regeneration water bath. Lyocell fiber is more uniform and have a smaller diameter than the traditional Rayon.

Lyocell staple fibers and filaments are suitable for composite reinforcement because they are superior to other cellulose fibers, such as compared with flax, due to purity and perfection [6]. Further enhancement of Lyocell type fibers can be made by inclusion of multiwalled carbon nanotubes (MWCNT) into the fibers. MWCNT can contribute higher mechanical performance for application of the fibers and composites, and electrical conductivity, with draw ratio being a determinant of properties [7].

NMMO has been used to prepare cellulose nanoparticles from a solution of cellulose via addition of a nonsolvent (water) under rapid mixing so that the cellulose was coagulated into nanoparticles. Hollow cellulose nanoparticles were produced by the same method by self-encapsulation in a poly(acrylic acid) hydrogel layer. Inorganic particles were included, and they functioned as templates for core-shell inorganic–cellulose nanoparticles [8].

NMMO contains an internal salt structure on the N-oxide function. By analogy with NMMO, cellulose is soluble in other strongly complexing liquids such as ionic liquids, an example being imidazolium chloride. Ionic liquids are strongly interacting and compete well with intermolecular hydrogen bonding to separate and dissolve cellulose. Dilution with water decreases solvation of cellulose that precipitates in the more stable type I crystalline polymorph. Cellulose solutions are used to determine cellulose molar mass by dilute solution viscometry using a Uhbelodhe suspended-level viscometer.

Cellulose solutions in ionic liquids are used for characterization using typical solution-based physical methods, which help determine their interactions [9].

1-butyl-3-Methylimidazolium chloride (Fig. 17.4) was used to prepare concentrated solutions of cellulose and amylose, and dynamic rheological measurements were used to detect molar mass between entanglements.

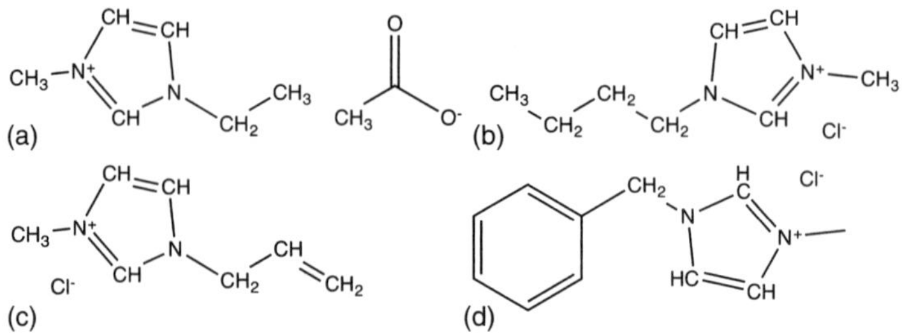

Figure 17.4 Structure of ionic liquids (a) 1-ethyl-3-methylimidazolium acetate, (b) 1-butyl-3-methylimidazolium chloride, (c) 1-allyl-3-methylimidazolium chloride, and (d) 1-benzyl-3-methylimidazolium chloride.

Extrapolation to zero solvent was used to calculate molar mass between entanglements in the quasi-melt state, and the results were found to be consistent with values of polysaccharides in a random coil conformation such as agarose and gellan. The molar mass between entanglements for amylose was found to be larger due to the helical conformation of its chains [10].

The solution state of cellulose in various solvents including NMMO and ionic liquids with varying anions was characterized using rheology. Zero-shear viscosities, storage and loss moduli, time–temperature master curves, and complex viscosities were measured using comparable molar ratios between cellulose and solvent. Anisotropic properties of concentrated cellulose solutions were characterized using polarized light microscopy and rheology with variation of solvent, concentration, and temperature [11].

Ionic liquid (1-ethyl-3-methylimidazolium acetate, Fig. 17.4) was used to swell poplar wood to permeate enzymes and catalysts deep into the wood structure for improved pretreatments and accelerated cellulose hydrolysis. Silver and gold nanoparticles were added as probes to detect their depth penetration into the wood, and they remained in the wood after contracting the ionic liquid-swollen wood with water [12].

Corn husk cellulose was dissolved in 1-allyl-3-methylimidazolium chloride and 1-ethyl-3-methylimidazolium acetate (Fig. 17.4) to form solutions that were used to cast cellulose films that were found to contain cellulose type II crystals. The films exhibited high tensile modulus and strength, and the ionic liquids were able to be recycled [13].

Cellulose has been described previously in this chapter as being swollen by sodium hydroxide solution. Due to internal hydrogen bonding being strong, cellulose can exhibit an upper critical solution temperature. This means that cellulose is likely to be more soluble in sodium hydroxide solutions at low temperatures. Upon heating, the internal hydrogen bonds are more thermally stable than complexes or salt formation and swelling with sodium hydroxide solution. Dilute solutions of cellulose in sodium hydroxide solution are formed by cyclic freeze–thawing of the mixture where the thaw temperature is kept below about 10°C. Cellulose can be precipitated from the solution by heating or neutralizing.

The proposition that cellulose in sodium hydroxide has a lower critical solution temperature comes from the extensive internal hydrogen bonding supported by regularity of cellulose. Heating disrupts these cellulose interactions less than random interactions, hydrogen bonding, and salt formation with sodium hydroxide. At low temperatures these sodium hydroxide-initiated interactions can predominate and affect a low level of solubility. Such observations are consistent with lower critical solution temperatures of other hydrogen-bonded polymers in polar media. Cellulose solubility is enhanced through more strongly interacting solvents such as ionic liquids. NMMO is essentially an ionic liquid, though a zwitterionic liquid due to its N-oxide functionality compared with those having a separate complex organic cation and an anion such as halogen.

A problem with cellulose fibers is that they contain moisture that varies with humidity, temperature, and equilibration time in the presence or absence of water. The water content and its variation is a limitation for the application of cellulose composites. Completely dry cellulose fibers are brittle, and they will regain moisture. A solution is to replace the moisture with other hydrophilic molecules or grafted polymer. Dry cellulose was saturated with acrylonitrile with a thermal initiator, and then the acrylonitrile was polymerized. The resulting polyacrylonitrile grafted cellulose was extracted to remove surface-adsorbed polyacrylonitrile. The grafted fibers were then used to prepare composites. The mechanical properties performed by a single fiber test method were strongly influenced by the cellulose structure, lateral index of crystallinity, and fraction of grafting [14]. Similar interstitial polymerization was performed using butyl acrylate, 2-ethylhexyl acrylate and methyl methacrylate to replace the water in flax cellulose fibers. Properties of composites formed from the treated fibers confirmed that adhesion between matrix and fibers was enhanced by the interstitial polymerization. The composites were subjected to moist environmental conditions to test for development of mold and fungi, and the acetone-washed polymerization-treated flax composites were resistant to these growths [15].

Cellulose was used as a template for layer-by-layer deposition of titania–poly(vinyl alcohol) (PVA) with nanometer precision. The cellulose was extracted with sodium hydroxide–urea, resulting in hierarchical porous titania–PVA composite sheets that replicated the morphology of the template cellulose, flexibility, and solvent swelling characteristics. Removal of the titania with acid left a hierarchical PVA sheet. The method allows the design and production of biomimetic hybrid materials [16].

Natural polymers such as cellulose, starch, and proteins are not a special category of polymers; they have the same structure–property relationships as their synthetic polymer relatives after their functionality and stereochemistry have been considered. While no new theories are necessary, cellulose is unique

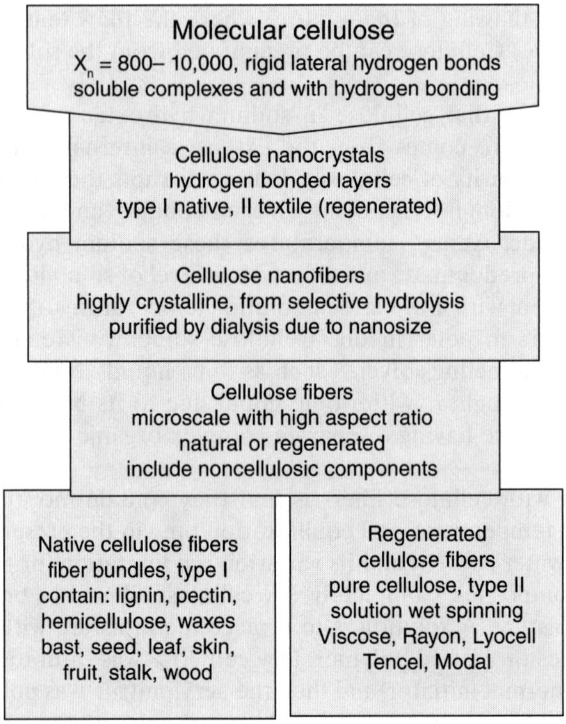

Figure 17.5 Hierarchical structures of cellulose from molecule to superstructures.

in its structural complexity, regularity, intra-, and intermolecular hydrogen bonding. The cellulose structure resists solution formation in all but the most strongly interacting solvents. Powerful, crude tools are needed to overcome the delicate intricacies of nature. Figure 17.5 shows a hierarchical view of cellulose structures from the molecule to complex multicomponent crystalline fibrous compositions.

17.1.3 Electrospun Cellulose Fibers

Cellulose fibers with diameters in the nanometer range are prepared by electrospinning. The solvents that are suitable for cellulose solutions are ionic and can be used to form filaments under a voltage differential between capillary and target screen or roller. About 300 V DC is typical to accelerate a cellulose solution and cause a narrow diameter on the target. The mesh of nanofibers formed on the target can be used for composite formation by impregnating with a compatible matrix polymer preferably as a solution, chain extendable, or cross-linkable prepolymer.

The small diameter and orienting directional flow of the solution as it accelerates to the target produce highly regular fibers with high mechanical performance. The electrospinning process produces a dense packed nonwoven mat that creates high in-plane modulus and strength in composites produced from it. Cellulose can be dissolved and electrospun from two solvent systems: lithium chloride (LiCl)–N,N-dimethylacetamide (DMA) and N-methylmorpholine-N-oxide (NMMO)–water. Nonwoven mats of nanodiameter cellulose fibers (250–750 nm) have been obtained. The cellulose fibers vary from amorphous to semicrystalline, depending on solvent and conditions [17]. Electrospun cellulose fiber mats have sufficiently dense packing to be considered for direct application as filters and membranes without a matrix phase. A matrix phase is preferred if the intended application requires mechanical properties.

Cellulose nonwoven fibers were electrospun from cellulose in an ionic liquid, 1-butyl-3-methylinmidazolium chloride (Fig. 17.4) in a constant temperature chamber to prevent solidification of the ionic liquid. The fibers had diameters in the range of 500–800 nm [18]. Cellulose nanofibers have been electrospun from solution in pure ionic liquid or a binary mixture of ionic liquids at ambient temperatures. Electrospinning of 8%·w/w cellulose in 1-ethyl-3-methylimidazolium acetate (Fig. 17.4) produced electrospun fibers with average diameters of 470 nm while binary mixtures of 1-ethyl-3-methylimidazolium acetate and 1-decyl-3-methylimidazolium chloride resulted in average diameters of 120 nm. The electrospinning was carried out with direct coagulation of the fibers in a water bath from which the ionic solvent could be recovered [19].

Oriented cellulose nanofibers were prepared by electrospinning on a rotating cylinder collector. An optically transparent (88–92% transmission) epoxy resin composite reinforced with the electrospun cellulose nanofiber mats was prepared by solution impregnation with 5–30%·w/w fibers. Wetting, impregnation, and strong interfacial adhesion were obtained. Modulus and strength increased significantly, and storage modulus increased at temperatures above the glass transition temperature [20].

17.1.4 Microcrystalline Cellulose

Microcrystalline cellulose (MCC) is formed by partial hydrolysis of native cellulose materials such as cotton. Cellulose is suspended in sulfuric acid for about an hour for hydrolysis to occur, then the remaining highly crystalline cellulose is separated by filtration or centrifugation. MCC dispersion requires high shear because of its tendency to flocculate, which is the basis of its thixotropic behavior. Suspensions for MCC are thixotropic and useful in many food- and solution-based products. The ability of MCC to flocculate and absorb or complex with other polar components within a composite suggests its use as a third phase or interphase in composites containing other cellulose fillers or matrix. MCC is not expected to be a reinforcing phase in the same way as are cellulose fibers; however, it can contribute via its binding characteristics.

Ball milling has been used to decrease the size of cellulose fibers and increase surface area as measured by a BET gas absorption technique. During ball milling, fiber bundles were disrupted and the crystallinity of the cellulose was reduced. An increase in solvent polarity, solvent–fiber ratio, agitation speed, and drying rate resulted in the rearrangement of ball-milled cellulose crystalline structures to a greater extent with increase in BET surface area [21].

Cellulose fibers were obtained by partial acid hydrolysis of MCC or mechanical disintegration of regenerated wood fibers. The fibers were surface treated with 3-aminopropyltriethoxysilane (APS) for compatibilization with poly(lactic acid) (PLA). Cellulose–PLA composites were prepared by melt compounding, and enhanced modulus and relaxation phenomena were found, particularly with the cellulose derived from MCC due to the higher aspect ratio of these fibers and crystallization changes in the PLA [22].

Composites of MCC and cellulose nanofiber were prepared in polypropylene (PP). Single batch mixing was compared with multiple batch mixing to evaluate processing contribution to dispersion of the cellulose. Thermogravimetry (TGA) was used to measure dispersion according to the distribution of results for PP and cellulose contents, which also translated into variability of mechanical properties [23].

17.1.5 Bacterial Cellulose

Ultrathin cellulose fibers are produced by bacteria. These natural nanofibers have highly ordered crystalline structures that are suitable for composites and water suspension gelation. Bacterial cellulose application is limited by it being only available in small quantities. A high structural perfection contributes enhanced modulus and strength compared with typical micrometer diameter cellulose fibers. Bacterial cellulose has unique properties due to its highly ordered structure, and it can be used as a substrate for composites or can be chemically functionalized [24].

Magnetic membranes were formed from bacterial cellulose using NMMO as solvent and combining the solution with ferrites. Thin sections of bacterial cellulose showed the magnetic particles located along cellulose microfibrils that were assumed to be sites for their nucleation [25]. Cellulose nanocrystals from bacterial cellulose were prepared and surface-modified using trimethylsilylation, then used to reinforce cross-linked poly(dimethylsiloxane) [26].

Bacterial cellulose sheets impregnated with phenol–formaldehyde resin were used to produce high modulus and strength composites by compression molding. The high modulus was a result of uniform, continuous, and aligned nanoscalar network of cellulosic fibers oriented in-plane by compression of bacterial cellulose pellicles [27].

Bacterial cellulose–silica hybrids were prepared from bacterial cellulose membranes and tetraethoxysilane (TEOS) under neutral pH conditions at ambient temperature. Macroscopically homogeneous membranes were obtained containing silica spheres. The silica spheres were readily recovered after removing the cellulose. The new hybrids were thermally stable and displayed emission under ultraviolet (UV) excitation due to oxygen-related defects at the silica surface and the emission color was dependent on excitation wavelength [28]. PVA was reinforced with bacterial cellulose by forming composites in a coagulating bath and cross-linking the material with formaldehyde. The resulting nanocomposite hydrogels exhibited enhanced mechanical strength, thermal resistance, and limiting swelling ratio [29].

17.1.6 Nanocellulose Formation

Cellulose nanofibers and nanocrystals are formed by selective degradation. Acid hydrolysis continued to about 50% mass loss leaving nanoparticles at first nanofibers, then nanocrystals. Suspensions of nanocellulose are transparent because the nanoparticles are too small to scatter light. The particles are too small to be filtered; however, they can be separated from the soluble hydrolysis products by dialysis to form a suspension of nanocellulose in water. Gels can be formed by freeze-drying nanocellulose suspensions.

Dissolved, emulsified, or dispersed polymer can be combined with nanocellulose to form composites after removal of the water. The nanocellulose particles are stabilized by polymer adsorption. Due to the large surface area to volume ratio of the cellulose, considerable reinforcement is found at relatively low concentrations, such as about 1–5%·v/v.

Cellulose nanofibers or whiskers, prepared from tunicate fibers, have been used to form aerogels formed through freeze-drying suspensions of the fibers in water. An ultralow density layered superstructure was formed from templating the ice crystal morphology, although the neat aerogels were fragile. Inorganic–organic hybrid aerogels were formed comprising montmorillonite and cellulose nanofibers. These aerogels exhibited compressive strengths that were higher than predicted by additivity calculation. The cause of this synergy was enhanced formation of three-dimensional network structures during freeze-drying [30].

Microfibrillated cellulose (MFC) is a web-like cluster of cellulose fibrils with diameters in the range of 10–100 nm. MFC was dispersed in an acetone solution of CA to give an all-cellulose nanocomposite that was improved if the MFC were treated with APS. MFC doubled the tensile modulus of CA and tensile strength was similarly enhanced [31].

Acid hydrolyzed cellulose suspensions in water were found to form biphasic structure within a narrow concentration range where the upper and lower phases were isotropic and anisotropic (chiral nematic). Polarizing optical microscopy revealed that the chiral nematic pitch of the anisotropic phase decreased with increasing concentration, which was caused by a decrease in the electrostatic double layer thickness of the individual nanofibers, due to increased intermolecular interaction. Chiral nematic textures were observed using scanning electron microscopy (SEM). SEM was suitable for study of nanofiber orientation within the liquid crystalline phase [32].

17.2 CELLULOSE NANOCOMPOSITES

Cellulose derivatives are more readily dissolved or processed than cellulose and therefore suited for nanocomposite preparation with available nanoparticles, platelets, and fibers. Cellulose acetate (CA), cellulose acetate–butyrate (CAB), plasticized with triethyl citrate have been extruded into nanocomposites with modified layered clay. The cellulose–clay interface was enhanced when maleic anhydride-grafted CAB was used, and CA or CAB was preplasticized before clay addition [33]. Exfoliation of the clay was confirmed and improved when maleic anhydride-grafted CAB was used as compatibilizer and mechanical properties were enhanced in the clay nanocomposites [34].

17.2.1 Thermoplastic Composites

The main thermoplastic used with cellulose fibers is PP. Several methods are used to disperse cellulose in PP, depending on the form of the cellulose. Particulate cellulose from wood flour to nanocellulose can be dispersed in melted PP by extrusion or batch mixing. The cellulose–PP dispersions are then formed by injection molding, thermoforming, or further extrusion.

Long fiber forms of cellulose require different techniques to maintain the cellulose structure. Long cellulose fibers such as those from cotton, flax, hemp, kenaf, and other plant materials can be first formed into felted (nonwoven or needle punched) or woven mats. The mats must be impregnated with PP prior to formation of a composite under heat and pressure. One method is to scatter PP powder into a cellulose mat under vibration, so as to mix the PP powder with the fiber mesh. The PP is melted and a composite formed by passing the material through heated rollers. Another method is to intersperse or weave PP fibers into the cellulose fibers. The combined fiber mat is heated under pressure to melt and diffuse the PP while consolidating the composite. Melting and consolidation is a maximum at the surfaces, which are critical to mechanical performance. At temperatures above 200°C cellulose degrades, so at this temperature, a time of about 1–4 minutes is suitable at moderate pressure to enable flow of the PP between fibers.

MAPP has been used to prepare composites with MCC where the MCC was pretreated with stearic acid, silicone oil, or alkyltitanate coupling agent to decrease the hydrophobicity of the MCC surface and facilitate processing, allowing MCC to show improved dispersion. Tensile mechanical properties increased with MCC content, PP creep was reduced, and storage and loss moduli as well as glass transition temperature were increased in the composites [35].

The methods described are suitable for other thermoplastics for which the melting and molding temperatures of the thermoplastic are preferably less than 200°C. The limiting conditions for composite formation depend on the time, temperature, and pressure combination chosen and usually an optimum will be found. At higher temperatures cellulose dehydrates, with loss of physically and chemically bonded water. Chemical dehydration products give yellowing and further darkening of the cellulose fibers.

Cellulose microfibrils possess unique properties compared with existing fibers: high strength, formability, and geometrical complexity at very small scale, low density, and abrasive resistance, biomedical compatibility, and possibility of recycling. Cellulose microfibrils has been combined with PP in conjunction with a silane coupling agent to form composites with enhanced compatibility and mechanical properties [36].

Bacterial poly(hydroxybutyrate-co-hydroxyvalerate) (PHBV) was reinforced with cellulose nanowhiskers (CNW) using solvent-casting. CNW were prepared from microcrystalline cellulose using sulfuric acid hydrolysis. Tensile modulus and strength of PHBV were enhanced and molecular mobility was constrained by the CNW in the vicinity of the CNW surfaces due to strong interactions between the components [37].

17.2.2 Thermosetting Composites

Cellulose nanoparticles can be dispersed in epoxy resins, unsaturated polyesters, and vinyl ester resins. Laboratory-scale dispersions can be made using ultrasonication, while high shear impellor dispersers are suitable, though less efficient, at various batch sizes. A persistent problem is entrapment of air bubbles, so some form of degassing is required before the gel time is reached. Cyclic application of vacuum is suitable for degassing. Alternatively centrifuging separates bubbles on a laboratory scale. Generally, thermal cure during dispersion must be prevented by dispersion in the main phase only such as an epoxy resin without the hardener. This is readily achieved if only a small proportion of curing agent is to be added so the dispersion is uniform and not too viscous for further additions. Unsaturated polyester and vinyl ester resins require a small addition of initiator so the volume of resin available for cellulose dispersion is large. If the initiator has a high temperature for radical formation, then it could be added in one step along with the filler.

There is potential for epoxy to react with the slightly acidic hydroxyl groups on cellulose and form a strong interface. Some hydroxyl sited radicals may form and create bonds with the radical initiated resins, such as occurs more specifically with ceric ion-initiated grafting of vinyl monomers onto cellulose. MFC as a reinforcement for phenol–formaldehyde resin provided high modulus and strength due in part to the formation of nano-order scale interconnected fibrils with high surface area in a web-like network structure. MFC with its extremely high strength formed composites with outstanding mechanical properties [38].

17.2.3 Cellulose Derviative–Cellulose Composites

The matrix can be CA, CAB, methylcellulose, or other soluble cellulose derivative so that a solution can be formed in which to disperse cellulose nanoparticles. After evaporation of solvent, the composites were consolidated using a heat press to remove voids and relax the structure. CAB was used to form hemp fiber composites plasticized with tributyl citrate, where the volume fraction of hemp was 0.4. Prior to composite formation, the hemp fibers were purified with pectate lyase enzyme and the composites after this treatment were superior to those from untreated or alkali treated hemp. Shorter fibers gave the most enhanced properties unexpectedly due to more efficient wetting and better consolidation [39].

17.2.4 All-Cellulose Composites

All-cellulose composites are those in which the matrix is cellulose and the dispersed reinforcing phase is cellulose fibers. The matrix cellulose must be dissolved to enable dispersal of the nanoparticles of cellulose since cellulose cannot be melted. All-cellulose composites were prepared from hemp fibers purified by acetone extraction and sodium hydroxide treatment that were impregnated with a solution of hemp cellulose in NMMO to which sufficient water had been added to prevent further cellulose dissolution. The composite was formed by coagulation of the cellulose solution in ethanol–water, removal of NMMO by washing with ethano–water, followed by drying and hot pressing to consolidate the structure. Morphology of the fibers was modified during the processing. The mechanical properties of the composites depended on size, surface area, crystallinity, and the structural swelling of the fibers [40]. In these composites the cellulose matrix provided adhesion, as do lignin, pectin, and hemicelluloses in plant fibers.

All-cellulose composites have been prepared based on ramie fibers impregnated with a cellulose solution in sodium hydroxide and urea in water. Good compatibility was observed between the fibers and matrix as expected from their same chemical structures. The properties of the cellulose matrix were enhanced by the presence of the independently crystallized cellulose in the fibers [41]. DMAC has been used as a solvent for partially dissolving MCC that formed gels after precipitation. During dissolution the MCC fibers were progressively separated into thinner crystals that finally remained in the composite with the dissolved cellulose forming the matrix. Mechanical properties and morphology of the all-cellulose composites were controlled by rate of precipitation, initial cellulose concentration, dissolution time, and precipitation conditions. Preparation conditions limiting defects produced the highest tensile modulus and strength [42].

Lyocell and high modulus and strength cellulose fibers were used to prepare all-cellulose composites by a selective dissolution method that depended on immersion time in the solvent. The all-cellulose composites demonstrated a suitable route to form biocomposites with mechanical and thermal properties that can be varied depending on selection of fibers and preparation parameters [43].

17.3 FUTURE DIRECTIONS FOR CELLULOSE-BASED MATERIALS

Cellulose has been used as reinforcement and matrix in many materials due to its broad range of properties that provide strength, rigidity, and toughness; in addition it can be shaped, cut, fastened, and it is durable. Cellulose has a multiscale morphology from large timber structures, fiber bundles, fibers, nanofibers, nanocrystals to individual molecules. Many composites have incorporated wood flour and cellulose fibers from wood, bast fibers, and waste agricultural materials. Recent developments are moving to smaller scales because increased knowledge of nanomaterials has shown the high performance and efficiency that can be gained from materials with small size, huge surface area, and defect-free morphology. There is a distinction between crystals growing within a matrix material and a separate nanocrystalline phase added as a filler. Cellulose nanocrystals added and dispersed in a matrix have more perfect structure and molecular orientation. This is because they have survived preparation by selective acid hydrolysis of larger cellulose structures. This preparation from cellulose fibers ensures orientation. Similarly, cellulose nanocrystals are from the most perfect structures within plant fibers and wood.

The current trend towards materials from renewable resources demonstrates the high performance of cellulose fibers. Purification of native fibers has been practiced in the textile industry. Enzymic methods begin with retting and move to specific enzymes such as pectinase. Newer methods of purification and regeneration are based on recyclable efficient solvents such as NMMO and ionic liquids. The methods are now applied to preparation of all-cellulose composites. Hybrid composites are formed from cellulose fibers and inorganic-based nanoparticles, where long fibers can be used in high volume fraction with the nanoparticles, providing another level of reinforcement with their high surface area and physical network formation. This trend culminates in the nano-phase consisting of cellulose nanostructures or the nano-phase providing the total reinforcement. The low percolation threshold of nanoparticles, particularly nanofibers, enhances the properties compared with individually dispersed particles.

17.4 CONCLUSION

Cellulose is the most abundant polymer evolved for structural plant components in semicrystalline morphologies often present in biocomposites and fibers. Application of cellulose in modified and semisynthetic materials has been commercialized and new materials are continuously being developed because of high performance and availability from renewable natural resources. Extension to nanocomposites is biomimetic because natural cellulose materials are constructed from the nanoscale. At the nanoscale, morphologies approach perfection and the interface is much larger for stress transfer. Cellulose nanofibers and nanocrystals have been formed by partial degradation of larger cellulose entities. Combining cellulose fibers with other nanofillers has been used to form hybrid nanocomposites. Cellulose has been included in the disperse phase, matrix phase, or both phases. Cellulose nanocomposites exhibit exceptional mechanical properties with relatively low density; they are biodegradable and can replace many synthetic polymer composites and natural materials that have not been designed for specific purposes.

ABBREVIATIONS

APS	3-aminopropyltriethoxysilane
BET	Brunauer–Emmett–Teller (gas absorption surface area measurement)
CA	cellulose acetate
CAB	cellulose acetate–butyrate
CNT	carbon nanotubes
CNW	cellulose nanowhiskers
DMAC	N,N-dimethylacetamide
MAPP	maleic anhydride-grafted polypropylene
MCC	microcrystalline cellulose
MFC	microfibrillated cellulose
MWCNT	multiwalled carbon nanotubes
NMMO	N-methylmorpholine-N-oxide
PHBV	poly(hydroxybutyrate-co-hydroxyvalerate)
PLA	poly(lactic acid)
PP	polypropylene
PVA	poly(vinyl alcohol)
TEOS	tetraethoxysilane
TGA	thermogravimetry

REFERENCES

[1] Cipriani, G., Salvini, A., Baglioni, P., and Bucciarelli, E. (2010) Cellulose as a renewable resource for the synthesis of wood consolidants. *Journal of Applied Polymer Science*, *118*, 2939–2950.

[2] Kalia, S., Kaith, B.S., and Kaur, I. (2009) Pretreatments of natural fibers and their application as reinforcing material in polymer composites: A review. *Polymer Engineering and Science*, *49*, 1253–1272.

[3] Pandey, J.K., Ahn, S.H., Lee, C.S., Mohanty, A.K., and Misra, M. (2010) Recent advances in the application of natural fiber based composites. *Macromolecular Materials and Engineering*, *295*, 975–989.

[4] Fraval, H.R., Emens, G.W. (2001) Manufacture of prefabricated structural building panels using high strength lightweight fiber ash composite. *PCT Int. Appl.* (50 pp.). WO: 0181267 (Aerock, Inc., USA).

[5] Ouajai, S., Shanks, R.A. (2005) Morphology and structure of hemp fibre after bioscouring. *Macromolecular Bioscience*, *5*, 124–134.

[6] Mieck, K.P., Lutzkendorf, R., Reussmann, T., Nechwatal, A., and Eilers, M. (2001) Lyocell fibers—State of development in view of their application in composites. *Gummi, Fasern, Kunststoffe*, *54*, 800–805.

[7] Lu, J., Zhang, H., Jian, Y., Shao, H., and Hu, X. (2012) Properties and structure of MWNTs/cellulose composite fibers prepared by Lyocell process. *Journal of Applied Polymer Science*, *123*, 956–961.

[8] Nelson, K., Deng, Y. (2007) Encapsulation of inorganic particles with nanostructured cellulose. *Macromolecular Materials and Engineering*, *292*, 1158–1163.

[9] Zakrzewska, M.E., Bogelaukasik, E., and Bogelaukasik, R. (2010) Solubility of carbohydrates in ionic liquids. *Energy & Fuels*, *24*, 737–745.

[10] Horinaka, J.-I., Yasuda, R., and Takigawa, T. (2011) Entanglement properties of cellulose and amylose in an ionic liquid. *Journal of Polymer Science Part B: Polymer Physics*, *49*, 961–965.

[11] Kosan, B., Schwikal, K., and Meister, F. (2010) Solution states of cellulose in selected direct dissolution agents. *Cellulose*, *17*, 495–506.

[12] Lucas, M., Macdonald, B.A., Wagner, G.L., Joyce, S.A., and Rector, K.D. (2010) Ionic liquid pretreatment of poplar wood at room temperature: Swelling and incorporation of nanoparticles. *ACS Applied Materials & Interfaces*, *2*, 2198–2205.

[13] Cao, Y., Li, H., Zhang, Y., Zhang, J., and He, J. (2009) Structure and properties of novel regenerated cellulose films prepared from cornhusk cellulose in room temperature ionic liquids. *Journal of Applied Polymer Science*, *116*, 547–554.

[14] Ouajai, S., Hodzic, A., and Shanks, R.A. (2004) Morphological and grafting modification of natural cellulose fibers. *Journal of Applied Polymer Science*, *94*, 2456–2465.

[15] Shanks, R.A., Hodzic, A., and Ridderhof, D. (2006) Composites of poly(lactic acid) with flax fibers modified by interstitial polymerization. *Journal of Applied Polymer Science*, *99*, 2305–2313.

[16] Gu, Y., Huang, J. (2009) Fabrication of natural cellulose substance derived hierarchical polymeric materials. *Journal of Materials Chemistry*, *19*, 3764–3770.

[17] Kima, C.W., Kima, D.S., Kanga, S.Y., Marquezb, M., and Joo, Y.L. (2006) Structural studies of electrospun cellulose nanofibers. *Polymer*, *47*, 5097–5107.

[18] Quan, S.L., Kang, S.G., and Chin, I.J. (2010) Characterization of cellulose fibers electrospun using ionic liquid. *Cellulose*, *17*, 223–230.

[19] Freire, M.G., Teles, A.R.R., Ferreira, R.A.S., Carlos, L.D., Lopes-da-Silva, J.A., João, A.P., and Coutinho, J.A.P. (2011) Electrospun nanosized cellulose fibers using ionic liquids at room temperature. *Green Chemistry*, *13*, 3173–3183.

[20] Liao, H., Wu, Y., Wu, M., Zhan, X., and Liu, H. (2010) Aligned electrospun cellulose fibers reinforced epoxy resin composite films with high visible light transmittance. *Cellulose*, *19*, 111–119.

[21] Ouajai, S., Shanks, R.A. (2006) Solvent and enzyme induced recrystallization of mechanically degraded hemp cellulose. *Cellulose*, *13*, 31–44.

[22] Frone, A.N., Berlioz, S., Chailan, J.F., Panaitescu, D.M., and Donescu, D. (2011) Cellulose fiber-reinforced polylactic acid. *Polymer Composites*, *32*, 976–985.

[23] Yang, H.-S., Gardner, D., and Nader, J. (2011) Dispersion evaluation of microcrystalline cellulose/cellulose nanofibril-filled polypropylene composites using thermogravimetric analysis. *Journal of Thermal Analysis and Calorimetry*, *103*, 1007–1015.

[24] Chen, P., Cho, S.Y., and Jin, H.J. (2010) Modification and applications of bacterial celluloses in polymer science. *Macromolecular Research*, *18*, 309–320.

[25] Sourty, E., Ryan, D.H., and Marchessault, R.H. (1998) Characterization of magnetic membranes based on bacterial and man-made cellulose. *Cellulose*, *5*, 5–17.

[26] Grunert, M., Winter, W.T. (2000) Progress in the development of cellulose reinforced nanocomposites. *Polymeric Materials Science and Engineering*, *82*, 232.

[27] Nakagaito, A.N., Iwamoto, S., and Yano, H. (2005) Bacterial cellulose: The ultimate nano-scalar cellulose morphology for the production of high-strength composites. *Applied Physics A: Materials Science and Processing*, *80*, 93–97.

[28] Barud, H.S., Assunção, R.M.N., Martines, M.A.U., Dexpert-Ghys, J., Marques, R.F.C., Messaddeq, Y., and Ribeiro, S.J.L. (2008) Bacterial cellulose–silica organic–inorganic hybrids. *Journal of Sol-Gel Science and Technology*, *46*, 363–367.

[29] Yang, L., Zhang, H.Y., Yang, O., and Lu, D.N. (2012) Bacterial cellulose–poly(vinyl alcohol) nanocomposite hydrogels prepared by chemical crosslinking. *Journal of Applied Polymer Science*, *126*, E245–E251.

[30] Gawryla, M.D., Berg, O.V.D., Weder, C., and Schiraldi, D.A. (2009) Clay aerogel/cellulose whisker nanocomposites: A nanoscale wattle and daub. *Journal of Materials Chemistry*, *19*, 2118–2124.

[31] Lu, J., Drzal, L.T. (2010) Microfibrillated cellulose/cellulose acetate composites: Effect of surface treatment. *Journal of Polymer Science Part B: Polymer Physics*, *48*, 153–161.

[32] Miller, A.F., Donald, A.M. (2003) Imaging of anisotropic cellulose suspensions using environmental scanning electron microscopy. *Biomacromolecules*, *4*, 510–517.

[33] Park, H.M., Mohanty, A.K., Drzal, L.T., Lee, E., Mielewski, D.F., and Misra, M. (2006) Effect of sequential mixing and compounding conditions on cellulose acetate/layered silicate nanocomposites. *Journal of Polymers and the Environment*, *14*, 27–35.

[34] Park, H.M., Liang, X.M., Mohanty, A.K., Misra, M., and Drzal, L.T. (2004) Effect of compatibilizer on nanostructure of the biodegradable cellulose acetate/organoclay nanocomposites. *Macromolecules*, *37*, 9076–9082.

[35] Spoljaric, S., Genovese, A., and Shanks, R.A. (2009) Polypropylene-microcrystalline cellulose composites with enhanced compatibility and properties. *Composites Part A: Applied Science and Manufacturing*, *40*, 791–799.

[36] Panaitescu, D.M., Donescu, D., Bercu, C., Vuluga, D.M., Iorga, M., and Ghiurea, M. (2007) Polymer composites with cellulose microfibrils. *Polymer Engineering & Science*, *47*, 1228–1234.

[37] Ten, E., Turtle, J., Bahr, D., Jiang, L., and Wolcott, M. (2010) Thermal and mechanical properties of poly(3-hydroxybutyrate-co-3-hydroxyvalerate)/cellulose nanowhiskers composites. *Polymer*, *51*, 2652–2660.

[38] Nakagaito, A.N., Yano, H. (2005) Novel high-strength biocomposites based on microfibrillated cellulose having nano-order-unit web-like network structure. *Applied Physics A: Materials Science and Processing*, *80*, 155–159.

[39] Ouajai, S., Shanks, R.A. (2009) Biocomposites of cellulose acetate butyrate with modified hemp cellulose fibres. *Macromolecular Materials and Engineering*, *294*, 213–221.

[40] Ouajai, S., Shanks, R.A. (2009) Preparation, structure and mechanical properties of all-hemp cellulose biocomposites. *Composites Science and Technology*, *69*, 2119–2126.

[41] Yang, Q., Lue, A., and Zhang, L. (2010) Reinforcement of ramie fibers on regenerated cellulose films. *Composites Science and Technology*, *70*, 2319–2324.

[42] Duchemin, B.J.C., Newman, R.H., and Staiger, M.P. (2009) Structure-property relationship of all-cellulose composites. *Composites Science and Technology*, *69*, 1225–1230.

[43] Soykeabkaew, N., Nishino, T., and Peijs, T. (2009) All-cellulose composites of regenerated cellulose fibres by surface selective dissolution. *Composites Part A: Applied Science and Manufacturing*, *40*, 321–328.

CHAPTER 18

Morphological and Thermal Investigations of Cellulosic Bionanocomposites

ANAYANCY OSORIO-MADRAZO and MARIE-PIERRE LABORIE

18.1 INTRODUCTION

The natural renewability, material properties, and biodegradability of cellulose nanofibers (CNFs) have made these materials an alternative of choice for manufacturing environmentally friendly composites. Following the reports of Favier et al. [1, 2] on the reinforcement potential of CNFs almost 20 years ago, research activities and discoveries in this field of research have intensified, shown by the number of recent review articles [3–16].

CNFs of various dimensions and morphologies can be obtained from plant [17], algal [18–20], bacterial [21, 22], and animal sources [23]. During cellulose biosynthesis, glucan chains arrange in a parallel fashion and their interaction leads to the formation of a semi-crystalline structure comprising cellulose I allomorph, which assembles into cellulose microfibrils [24–30]. Microfibril lengths can reach tens of microns. Beside the crystalline regions, native cellulose contains amorphous domains, which give the opportunity to process native cellulose materials into CNFs. While the terminology on CNFs is still in the course of being normalized, one generally defines two types of families as cellulose nanowhiskers (CNWs) or microfibrillated cellulose (MFC). CNWs or cellulose nanocrystals (CNC) are obtained by acid hydrolysis of the amorphous fraction of cellulose [31]. The liberated rod-like monocrystals present high stiffness (114–140 GPa) and high strength [32–36]. Both origin and hydrolysis conditions govern the final dimensions of CNWs [17]. Alternatively, MFC or nanofibrillated cellulose (NFC) results from "peeling" off native cellulose

Biopolymer Nanocomposites: Processing, Properties, and Applications, First Edition. Edited by Alain Dufresne, Sabu Thomas, and Laly A. Pothan.

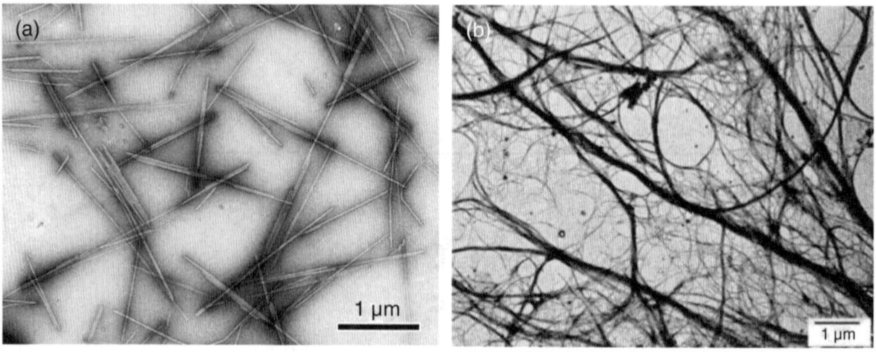

Figure 18.1 TEM images of (a) negatively stained *Cladophora* sp. cellulose whiskers (reprinted with permission from Osorio-Madrazo et al. [20]. Copyright 2012 American Chemical Society); (b) microfibrillated cellulose (MFC) from *Opuntia-ficus indica* (reprinted with permission from Malainine et al. [105]. Copyright 2005 Elsevier Ltd.).

microfibrils into an entangled and hairy network of nanofibers presenting a broad range of sizes (Fig. 18.1b) [5, 37, 38]. The delamination of the microfibril bundles results from the action of mechanical energy by means of high pressure homogenizer, microfluidizer, cryo-crushing, refining, or ultrasonication, often in combination with a chemical or biological modification [5, 37, 38]. CNFs can also be biosynthesized in the form of continuous ribbons by microorganisms and are generally termed bacterial or microbial cellulose [14, 21]. Spinning cellulose solutions is another synthetic method to produce continuous single nanofibers. These latter two types of CNFs are not considered in this review, which focuses on bionanocomposites reinforced by MFC and CNWs.

In relation to their high surface area, high aspect ratio, and high crystallinity, MFC and CNWs improve the mechanical and barrier properties of polymer matrices [3, 4], thanks to the formation of a percolating network (for CNWs) and to an inherent entangled network morphology (for MFC) [39]. In other words, CNFs are very efficient fillers to control the properties of polymer matrices. As for filled polymers, it is well established that the thermal properties of polymers, namely glass rubber transition temperature (T_g), melting temperature (T_m), equilibrium viscoelastic properties, and thermal stability, are sensitive to the addition of fillers [40]. The magnitude of filler effect is a function of the interfacial interactions and compatibility between polymer matrix and filler [41]. CNW and MFC fillers are no exception and the literature abounds on this topic, although at a first glance general trends are not evident. This is no surprise if one considers the expected large specific surface area of CNFs, estimated as 10–100 times larger than that of native wood fiber [42], and the large resulting interfacial contact with a polymer matrix. Besides, these thermal properties have a strong practical significance for the macroscopic performance of nanocomposites, in particular for the mechanical, barrier, and biodegradation properties [43–47]. For example, in a polymer matrix, a higher crystallinity index, which might be brought about by the presence of fibers, will influence mechanical and barrier properties [43, 48, 49]. Additionally, in composites the main thermal transitions of a polymer matrix reflect the morphologies of both the matrix and the matrix/filler interphase [40, 50]. Indeed, the relaxation behavior of a polymer matrix as well as the kinetics and thermodynamics of phase formation and transitions are all influenced by the scale of miscibility between the polymer matrix and the reinforcement. In the case of nanosized reinforcements such as CNWs and MFC, significant effects might be expected, providing a molecular lens to the interfacial adhesion and to the phase miscibility in such nanocomposites [50–54]. Therefore, aside from the practical significance of thermal properties, their diagnostic capability for assessing filler/matrix miscibility, scale of interactions, and microstructure at different length scales is particularly interesting.

This chapter highlights the impact of CNFs on the thermal properties of polymers and provides insight into the composite morphologies down to the molecular scale. With a particular focus on semicrystalline biodegradable polymers, this review attempts to decipher the role of CNF surface attributes, shape, and loading amounts on morphological and thermal properties. Thus, this chapter first briefly reviews the theoretical framework of polymer transitions and crystallization kinetics for semicrystalline polymers. Then, morphology and thermal behavior (other than thermal stability) of specific cellulose/biopolymer matrix materials are discussed.

18.2 THEORETICAL CONSIDERATIONS ON THE TRANSITIONS AND THE CRYSTALLIZATION KINETICS OF THERMOPLASTIC POLYMERS

Semicrystalline polymers, with their amorphous and crystallizable fractions, experience several thermal transitions; some of these are characteristic of only one fraction and others are influenced by both. The amorphous polymer fraction is characterized by a main chain segmental motion, denoted β (or α) relaxation with its characteristic glass-rubber transition temperature, T_g, which appears much broader in semicrystalline polymers than in amorphous polymers, due to the constraining effects of the crystals. On the other hand, subglass processes, which involve small motional volumes, sense the structure within only 1–2 nm and are generally not affected by the crystals. In addition to these relaxations, melting and crystallization are associated with the crystallizable fraction and the characteristic temperatures, T_m and T_c, respectively. In polymer blends these thermal transitions are of prime importance as they dictate the physical state and thermomechanical behavior of the blend for a given thermal energy. In addition, they directly relate to the scale of miscibility [40, 41, 51–53]. Blends that are miscible at the 10–20 nm scale exhibit one single T_g, intermediate to those of the parent polymers [50]. In these systems, the T_g dispersion is also often broadened toward low frequency. Alternatively, phase separation with domain sizes greater than ca. 20 nm is reflected by the detection of the individual T_g of the parent polymers. Similar morphological manifestations are expected in composites reinforced with rigid fillers. That is, when a rigid filler is intimately dispersed within a polymer matrix, T_g generally increases and broadens in relation to the intensity of filler/matrix interfacial interactions [40]. Thus, good mixing of stiff fillers into a polymer matrix constrains the main chain motions of the polymer matrix and increases motional heterogeneity due to concentration fluctuations and strength of attractive interactions. Alternatively, composites where fillers agglomerate or develop little interfacial adhesion with the polymer matrix leave segmental relaxations and local motions unchanged [40].

In the crystallizable fraction, polymer blend miscibility will also affect the resulting crystalline morphology and the kinetics of the formation of this phase. Indeed, for a semicrystalline polymer, the interactions with another miscible polymer can significantly disturb the crystallization process and the resulting morphology [41]. For example, blends, which are miscible in the melt, experience a reduction in the equilibrium melting temperature, T_m°, indicative of lower crystal perfection and favorable energy of mixing. Indeed, equilibrium melting point directly relates to the Flory–Huggins interaction parameter χ_{12} [54]. When specific intermolecular interactions between two polymers are in action, they hinder chain mobility in the melt and T_c decreases as a result [55]. For semicrystalline polymer composites, the situation is that of a crystallization process taking place in the presence of a solidified dispersed phase; that is, heterogeneous nucleation dominates over homogeneous nucleation and the matrix consequently crystallizes more rapidly [56]. Such behavior has indeed often been observed in polymers filled with natural fibers, where transcrystalline layers (TCL) and higher crystallinity indices reflect the fiber nucleating effect [57, 58]. Common empirical models to describe crystallization comprise the Avrami bulk crystallization kinetics [59–61] and the Lauritzen–Hoffman secondary nucleation kinetics [62]. The former is based on the time (t) evolution of volumetric crystallinity X_c:

$$\ln[-\ln(-X_c(t))] = \ln k + n\ln[t] \tag{18.1}$$

where $X_c(t)$ ranges from 0 to 1, n is the Avrami exponent, which depends on both the nucleation mechanism and density and therefore portrays the crystal shape, and k is the crystallization rate constant. In practice, the Avrami model of bulk crystallization, which only applies up to the point of crystallite impingement and thickening (generally, for $X_c(t) < 0.5$), does not differentiate between nucleation and lamellar growth rates, although these independently contribute to bulk crystallization rate. On the other hand, the Lauritzen–Hoffman secondary nucleation theory considers two contributions to the linear lamellar growth rate (G) with a mobility component related to the activation energy for chain reptation U^* and the temperature at which all flow and reptation ceases T_∞, and with a nucleation term where K_g stands for the nucleation constant growth rate and f is a constant [62]:

$$\ln G = \ln G_0 - \frac{U^*}{R(T_C - T_\infty)} - \frac{K_g}{T_C \Delta T f} \tag{18.2}$$

To evaluate both morphology and scale of miscibility in polymer blends and composites, thermal analysis and microscopy are well suited. The former helps monitor thermal transitions and thus the dimension of motional domains, which hinges on the scale of miscibility. The latter allows direct visualization of phase morphology and crystallites. These tools have therefore been most widely used for the thermal and morphological characterization of CNF-reinforced composites.

18.3 THERMAL BEHAVIOR AND MORPHOLOGY OF CRYSTALLIZABLE BIOPOLYMERS MODIFIED WITH CELLULOSE NANOFIBERS

18.3.1 Polyhydroxyalkanoate/Cellulose Nanocomposites

Polyhydroxyalkanoates (PHA) are biodegradable polyesters produced by bacteria as intracellular storage material. Their high mechanical properties explain the large interest in this family of polymers as a possible biodegradable alternative to polyolefins. Some of the first attempts at reinforcing polymers with CNFs have therefore focused on PHA/CNF nanocomposite systems. Dufresne et al. [46] solvent cast elastomeric medium-chain-length polyhydroxyalkanoates (Mcl-PHA) latex with CNWs with a view to assessing crystalline morphology and the occurrence of transcrystallization [46]. In this work, the Mcl-PHA and the corresponding composites were prepared in both amorphous and semicrystalline morphologies. As expected, for the neat polymers (i.e., without CNW), the storage modulus (E′) drop at T_g was significantly higher in the amorphous PHA than in the semicrystalline system, where crystalline regions acted as physical cross-links (Fig. 18.2). Whiskers increased the storage modulus of the PHA significantly, especially in the rubbery region and more so for the amorphous system than for the semicrystalline analogue. While the matrix T_g was not affected by CNW addition, damping was overall reduced, reflecting the reduced polymer mobility in the presence of whiskers. Interestingly, in the semicrystalline nanocomposites, E′ dropped irremediably with the breakup of the crystalline regions at T_m, whereas the CNW-induced-stabilization was still apparent for the amorphous polymers [46]. It is nowadays well established that thermal stabilization arises from the formation of a percolation network of CNWs [2, 63], and this is especially noticeable in low modulus rubbery polymers. In glassy or rigid polymers, a simple filler additive effect is expected

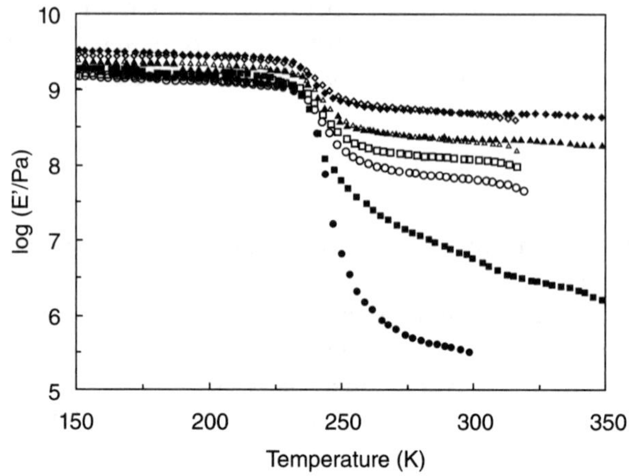

Figure 18.2 Dynamic thermomechanical analysis of amorphous (filled symbols) and semicrystalline (open symbols) CNW/PHA composites filled with 0 (○●), 1 wt% (□■), 3 wt% (△▲), and 6 wt% (◇◆) of cellulose nanowhiskers (reprinted with permission from Dufresne et al. [46]. Copyright 1999 American Chemical Society).

based on the Halpin and Kardos model [64]. In this work, the authors made the proposition that the large modulus drop at T_m reflected the close association between whiskers and crystalline polymer regions [46]; that is, it was proposed that the cellulose percolation network formed within a transcrystalline layer.

Recently Wolcott et al. revisited the morphology of solvent cast PHA/CNW nanocomposites using the copolymer poly(3-hydroxybutyrate-co-3-hydroxyvalerate) (PHBV) [65–68]. Differential scanning calorimetry (DSC) heating scans revealed that freshly casted PHBV matrix and nanocomposites (1–5 wt% CNW) were fully crystallized, exhibiting only a double melting point ascribed to melting and recrystallization (Fig. 18.3b) [65]. After erasing the thermal history of the samples, T_g and T_c were additionally detected (Fig. 18.3a). In this case, T_g was not affected by CNW addition. When CNWs were modified with polyethylene glycol for compatibilization, the T_g of the matrix was also unchanged, thus evidencing phase-separated morphologies [66]. In contrast, for the solvent cast composites, T_c was found to decrease with increasing CNWs, suggesting that CNWs have a nucleating effect on PHBV.

Polarized optical microscopy (POM) of neat matrix and nanocomposites confirmed CNW nucleating effect with the presence of smaller and more numerous spherulites (Fig. 18.4). Others confirmed the nucleating effect of CNWs on PHBV and showed expedited crystallization kinetics with up to 10 wt% loading with Avrami kinetics [69]. A subsequent study of the same system clearly identified the dual effect of CNWs for PHBV, as heterogeneous nucle-

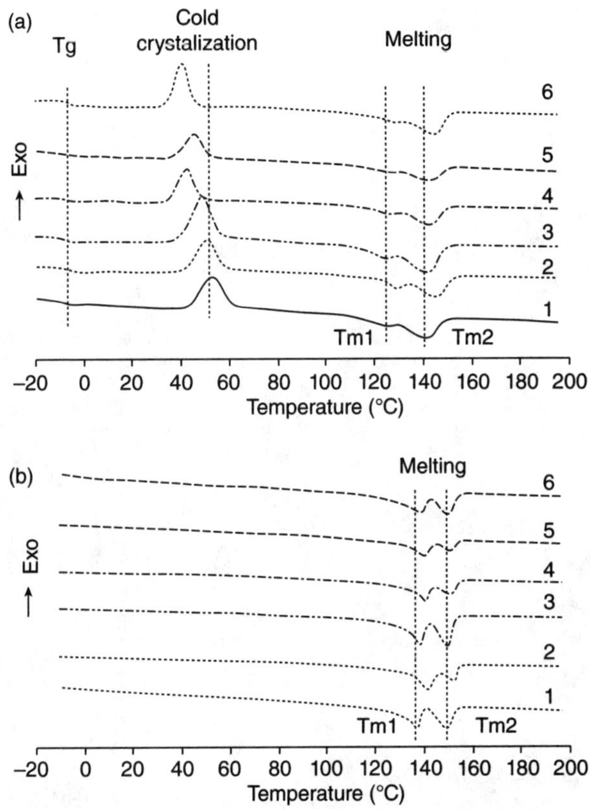

Figure 18.3 DSC thermograms of neat PHBV and CNW/PHBV nanocomposites from the second (a) and first (b) heating scans. 1) corresponds to neat PHBV, 2)–6) corresponds to 1 wt%, 2 wt%, 3 wt%, 4 wt%, and 5 wt% CNW/PHBV nanocomposites, respectively (reprinted with permission from Ten et al. [65]. Copyright 2010 Elsevier Ltd.)

ating agents on the one hand and as confinement agents on the other hand [68]. As a result, an increase or a decrease of crystallization rate could be observed, depending on loading and crystallization temperature. In these recent studies [65, 68, 69], the original proposal for epitaxial growth of Dufresne et al. [46] appears to be corroborated.

18.3.2 Poly (Lactic Acid)/Cellulose Nanocomposites

Because it can be prepared from agricultural products, PLA has become very popular in the manufacture of cellulose bionanocomposites [70–87]. PLA is a biodegradable and compostable polymer, produced from ring-opening polymerization of lactide, which due to its chirality, can result in a semicrystalline form, poly(L-lactide) (PLLA), or a mostly amorphous form, poly(D-lactide)

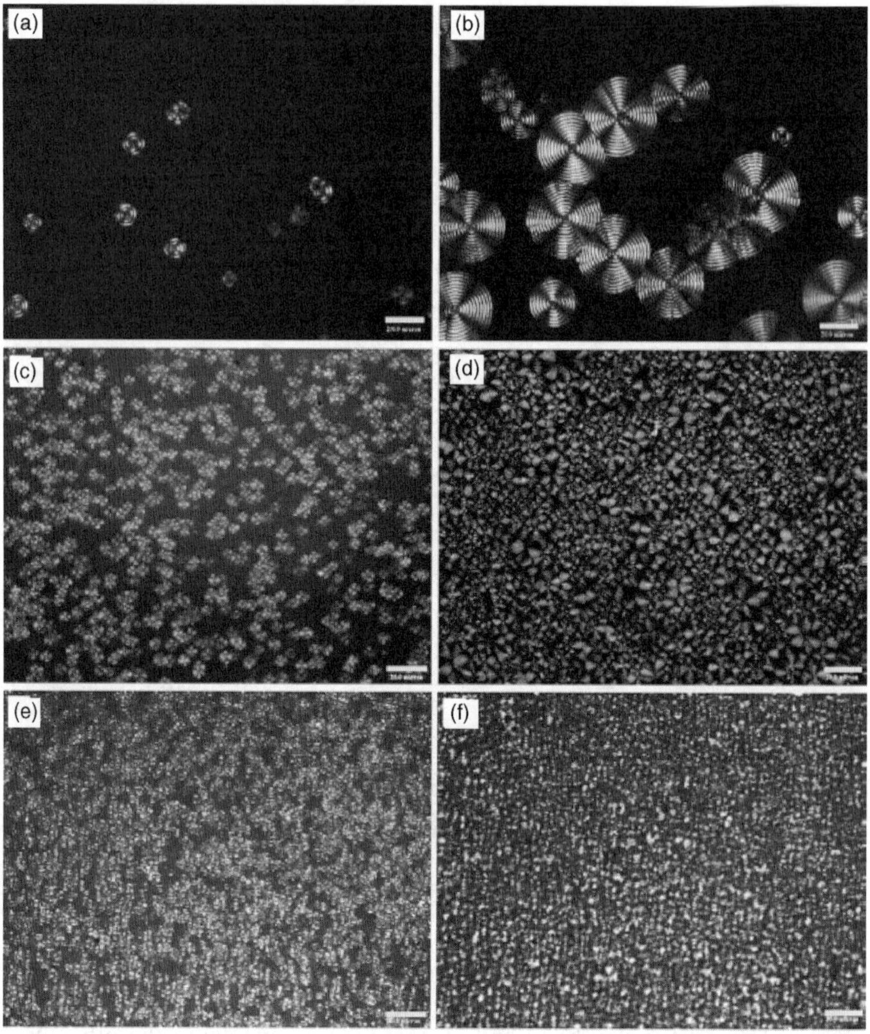

Figure 18.4 POM micrographs of isothermal crystallization of neat PHBV and CNW/PHBV nanocomposites at 30°C for different crystallization times (a, c, e: 5 minutes; b, d, f: 20 minutes). (a, b) Neat PHBV; (c, d) 1 wt% CNW/PHBV nanocomposite; (e, f) 5 wt% CNW/PHBV nanocomposite. Scale bar: 20 μm (reprinted with permission from Ten et al. [65]. Copyright 2010 Elsevier Ltd.).

(PDLA). One of the inherent limitations of these polymers, aside from relatively low impact properties, is the slow crystallization rate and resulting low degree of crystallinity, especially when the D–L lactide is incorporated in the polymer chain. Thus, many studies have attempted to improve PLA properties with addition of fillers including CNFs [80–87]. Okubo et al. [75] prepared PLA/MFC nanocomposites in low filler content (1–2%) by combining three-

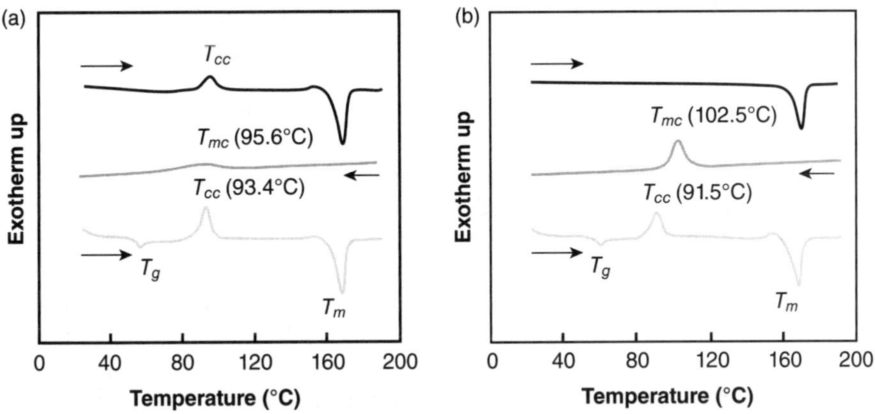

Figure 18.5 DSC thermogram of (a) neat PLA and (b) 10 wt% MFC/PLA nanocomposite (reprinted with permission from Suryanegara et al. [70]. Copyright 2009 Elsevier Ltd.).

roll mill and solution casting processing. DSC analyses of neat PLA and nanocomposites indicated that for solvent cast composites with such low MFC content, the fillers had no discernible effect on polymer crystallization and morphology. With higher cellulose loading, however, a prominent effect on PLA crystallization and morphology has been repeatedly observed [70]. Indeed Suryanegara et al. [70] prepared PLA/MFC nanocomposites with nanofiber loadings ranging from 3 wt% to 20 wt% using a two-step procedure based on solution-casting, MFC kneading by a twin rotary roller mixer, and hot-pressing. The freshly hot-pressed samples were either quenched in liquid nitrogen into fully amorphous materials or additionally annealed at 100°C into highly crystalline composites. For the amorphous PLA and the corresponding composites, T_g, cold crystallization temperature (T_{cc}) and T_m were observed in the first DSC heating scan, as well as a melt crystallization point (T_{mc}) upon cooling (Fig. 18.5). The 10 wt% MFC composite showed a higher melt temperature (102.5°C) than the neat PLA (95.6°C) and much higher crystallinity after cooling (39.4%) compared with 17.3% for the neat PLA. This indicated that PLA crystallization started sooner in the presence of MFC and developed to a greater extent. While T_g was not detected with DSC in the most crystalline composites, the dynamic mechanical analysis (DMA) traces suggested a slightly increased T_g in relation to the MFC loading, in agreement with other reports for PDLA. Thus, in highly filled nanocomposites MFC can facilitate and augment PLA crystallization, that is, serve as a nucleating agent in addition to rigidifying the amorphous phase, a possible consequence of the higher crystallinity index of the composites. Better thermal and mechanical properties result from these morphologies, in particular above the matrix T_g, where a significant thermal stabilization effect of MFC occurs for well-dispersed composites [81–83].

Partially annealed and crystallized PLA/MFC composites with 10 wt% loadings resulted in higher crystallinity (43%) and better mechanical properties at high temperatures when compared to the amorphous analogues [71]. The apparently contrasting effect of CNF suggests that as for PHA, fiber loading and possibly other factors play a large role in the composite morphology and thermal properties. In fact, surface chemistry and dispersion of CNFs should also impact the thermal and morphological properties of nanocomposites. With this hypothesis, several authors have attempted to compatibilize CNFs with PLA through surface grafting or adsorption [72, 85, 86]. For example, Pei et al. [72] solution cast a series of PLLA composites filled with 1 wt% and 2 wt% neat or silylated cellulose CNWs. Improved whiskers dispersion in PLLA was observed for the silylated whiskers. For all samples, T_m was unchanged by whisker additions, a result consistent with previous reports with similarly low CNF loadings. The degree of crystallinity increased marginally with the addition of neat CNWs (up to 16.4%) compared with neat PLLA (at 14.3%), but very significantly (up to 30%) with that of silylated CNWs [72]. Enhanced crystallization for silylated CNW/PLLA nanocomposites was also evident from cooling scans. Thus, even at low loading, well-dispersed and compatibilized whiskers caused a strong nucleating effect. The DSC traces presented in this work also suggest that silylated CNWs significantly increased PLLA T_g. Isothermal crystallization studies with DSC and POM confirmed the better dispersion and more dramatic impact on crystallization of the silylated CNWs (Fig. 18.6). Nucleus density and nucleation rate were generally enhanced by whisker addition but more so with the silylated whiskers, presumably as a result of better dispersion and higher surface area available for heterogeneous nucleation.

Isothermal crystallization and Avrami parameters, namely bulk crystallization rate constant, k, and exponent, n, corroborated these observations. K increased most significantly with the addition of silylated whiskers [72]. Also, n value increased from ca. 3 for neat PLLA to 3–3.5 with neat CNW addition and 3.8–4.9 with silylated-CNW addition. The shift from homogenous nucleation associated with three-dimensional crystal growth in neat PLA to heterogeneous nucleation and epitaxial crystallization was thus more marked with the well-dispersed silylated CNWs. In another recent study, acetylating CNWs allowed uniform CNW dispersion in PLA and strong interfacial adhesion [88]. The major role of the interfacial area and filler/matrix compatibility in thermal and mechanical properties was further confirmed in this study [88]. The positive impact of compatibilization and interfacial adhesion between CNF and PLA for the thermal properties is corroborated by the work of Tingaut et al. on acetylated MFC [86]; that is, acetylation induced better MFC dispersion in the matrix, causing an increase in T_g in some cases, while a slight T_g depression was observed for neat MFC. Other attempts to disperse CNF in the PLA matrix have consisted of using polymeric compatibilizer, such as poly(vinyl alcohol) (PVA) [80]. However, CNWs apparently concentrated in the PVA phase, which separated from the PLA phase. This micrometer-scale phase-separated mor-

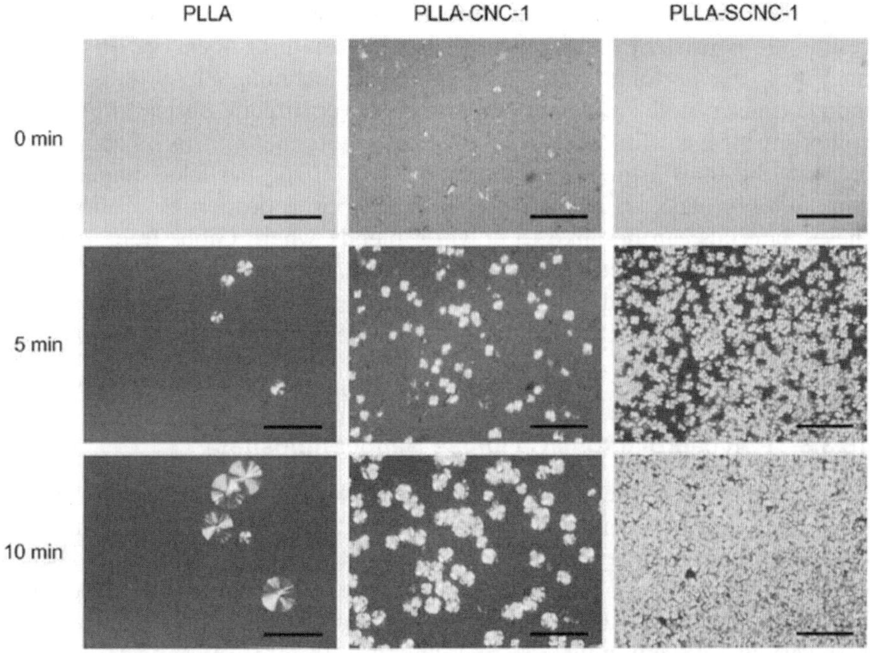

Figure 18.6 Polarized optical microscopy images of PLLA, 1 wt% CNW/PLLA (PLLA-CNC-1 in this figure), and 1 wt% SCNW/PLLA (PLLA-SCNC-1 in this figure) acquired on the 0, 5th, and 10th minute at 125°C after quenched from melt at 210°C. Scale bar: 200 μm (reprinted with permission from Pei et al. [72]. Copyright 2010 Elsevier Ltd.) See color insert.

phology resulted in little improvement of the thermal properties. Low molecular weight compounds, such as tert-butanol, have also been used as surfactant to compatibilize matrix and fiber phases [85]. Tert-butanol adsorption on CNWs was very effective at compatibilizing the nanofillers with PLA. A spectacular increase in T_g (+20°C) for the compatibilized component was observed from DMTA, although the thermal stabilization of the modulus was rather moderate. Perhaps the chemical treatment of the whiskers or the low whisker loading (5 wt. %) did not allow creating a percolating network. Recently, lactide ring-opening polymerization was initiated from the surface of whiskers and allowed preparing PLA nanocomposites with excellent properties via good interfacial adhesion and formation of a percolation network [78].

To summarize, the thermal properties of CNF-filled PLA depend on composition and on microstructure, as well as on filler/matrix compatibility. To be sure, CNF loading, and surface and interfacial chemistries play an important role in the morphology and resulting properties of the nanocomposites. CNFs are, in general, nucleating agents for PLAs, and this effect is enhanced by a good dispersion, that is, a large surface area available for nucleation. These results are consistent with the knowledge that natural fibers are, in general, nucleating agents for thermoplastic polymers, including PLAs [89], and can induce transcrystallization, although it was never directly observed for nanocomposites using cellulose. It is interesting to note that interfacial compatibility with the matrix fine-tunes the magnitude of the particles' impact. That compatibility plays an important role in composite properties confirms that these materials can be alike to polymer blends, where Gibbs free energy of mixing is a driving factor of morphology and thermal properties. From the practical sense, the addition of CNFs in slow crystallizing polymers such as PLA is a good avenue for tailoring both processing and thermomechanical properties.

18.3.3 Polycaprolactone/Cellulose Nanocomposites

Polycaprolactone is a semicrystalline polymer with T_g and T_m around –50° and –60°C, respectively. It is also of significant interest for the development of bionanocomposites due to its biodegradability and biocompatibility. In a series of papers on such nanocomposites, Siqueira et al. highlighted the differences between CNW- and MFC-based nanocomposites [42, 90, 91]. PCL was filled with various CNF amounts (3%, 6%, 9%, and 12% of CNWs and MFC), both unmodified and grafted with C18 chains in degree of substitution ranging from 0.07 to 0.09. The tensile properties of the bionanocomposites showed different trends with the different fillers. Addition of the modified CNWs and to a greater extent the modified MFC in PCL enhanced Young's modulus, while strength and strain at break were reduced [91]. For all nanocomposites, that is, based on modified and unmodified CNWs as well as modified MFC, T_g significantly increased and surprisingly more so with the unmodified whiskers. The crystallinity index and T_c also increased, indicating facilitated crystallization and resulting in stiffer materials. A thorough examination of nanocomposites with 9 wt% of C18 modified nanofibers helped shed further light on the competing effects of surface chemistry, surface area, and chain mobility on morphology and thermal properties. For isothermally crystallized composites, broadening and splitting of the melt endotherm was observed in a DSC heating scan (Fig. 18.7) and tentatively assigned to the coexistence of bulk spherulites and of a TCL with lower crystal perfection at the filler surface as suggested in other systems [92]. Together with the higher crystallinity, lower crystallization activation energy especially with the modified CNWs, a depression equilibrium melting point of ca. 10°C was measured; this was again ascribed to the nucleating effect of the CNF and to a favorable interaction between the two components.

As for PLA composites, PCL composites with high filler loadings experienced faster crystallization (Fig. 18.8); Avrami rate constant increased by two orders of magnitude, resembling the effect previously observed in PCL filled with carbon nanotubes [93] and nanoclay [94].

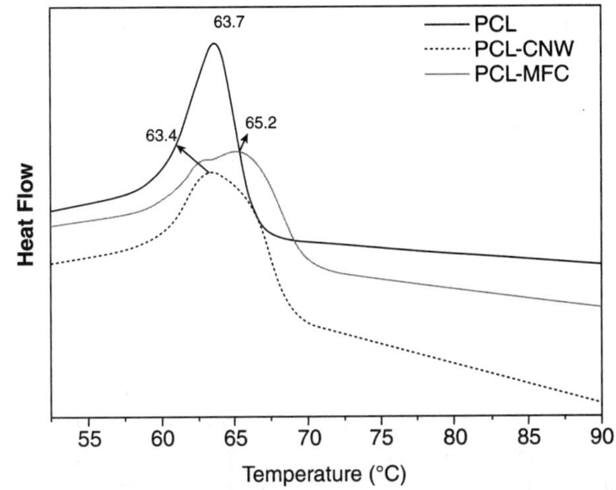

Figure 18.7 Representative melting endotherms for PCL and PCL-based nanocomposites reinforced with 12 wt% modified CNWs and MFC crystallized at T_c = 46°C (reprinted with permission from Siqueira et al. [90]. Copyright 2011 Elsevier Ltd.)

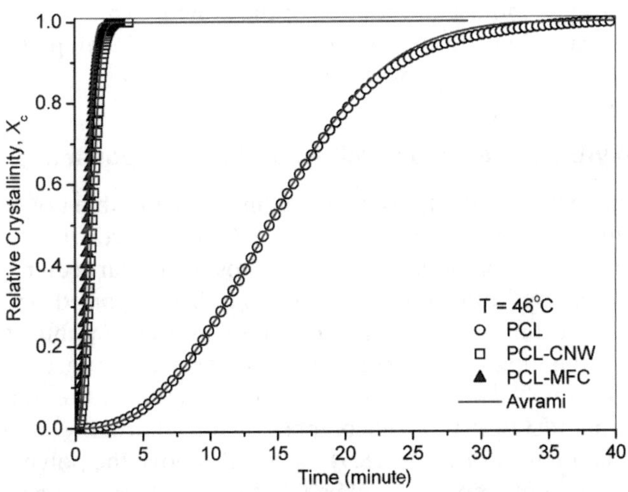

Figure 18.8 Relative crystallinity as a function of time for pure PCL and PCL-nanocomposites reinforced with 12 wt% of modified sisal CNWs or modified sisal MFC crystallized at T_c = 46°C. Experimental data points are compared with the Avrami model (reprinted with permission from Siqueira et al. [90]. Copyright 2011 Elsevier Ltd.).

Based on the Lauritzen–Hoffman secondary nucleation, nucleation parameter and lamellar growth rate were found to decrease with the incorporation of modified CNFs, confirming facilitated nucleation and lower chain mobility. It was proposed that the confinement effect of CNF decreases the overall mobility of the PCL chains in nanocomposites, hindering chain diffusion to the crystallization front [95]. In fact, earlier studies by Lönnberg et al. demonstrated that when PCL is grafted onto MFC, the anchored PCL chains take longer to crystallize due to hindered mobility [96, 97]. Along these lines, earlier literature suggested that entanglements in MFC could help maintain the cohesion of the cellulose network, offering a more rigid local environment than CNWs [39].

In another study, Lee et al. [98] mixed MFC with PCL and/or polypropylene (PP) to yield hybrid composites with varied compositions. Good dispersion of the MFC in the polymers was achieved by first kneading MFC with PCL at a relatively low melting temperature (90°C) and further kneading the freeze-dried PCL/MFC composite with PP at 180°C. Addition of maleic anhydride-grafted PP (MAPP) contributed to improving the distribution of MFC/PCL in PP. As before, the addition of MFC to the PCL matrix induced a slight increase in T_c, that is, the crystallization started earlier. On the other hand, MFC reduced crystallinity, leaving the T_m rather unchanged. The authors concluded that the reduced crystallinity in both PP and PCL phases might arise from the restricted molecular mobility of the polymer chains again. It is to be noted in this case that a complex morphology likely developed with ternary polymer blends compounded with a filler.

18.3.4 Poly(Ethylene Oxide)/Cellulose Nanocomposites

Poly(ethylene oxide) (PEO), a semicrystalline hydrosoluble polyether with T_g and T_m found approximately at −60°C and 70°C, respectively, has been assessed for the development of cellulose nanocomposites in particular for batteries and biomedical applications. Dufresne's group first reported on the thermal and morphological properties of PEO/CNF in a series of publications [39, 47, 99, 100]. Confounding effects were reported for such nanocomposites filled with up to 0–30 wt% [47]. As observed in many systems, the modulus of the semicrystalline PEO/CNW nanocomposites was thermally stabilized by a CNW percolation network (Fig. 18.9), especially above the polymer T_g and T_m, which were apparently not affected by CNWs, except for loadings higher than 10%, where T_m started to decrease. In the dynamic mechanical analysis of the composites, the main effect of the whiskers was observed at higher temperatures, that is, above the T_m. While the storage modulus E' of the neat PEO dropped irremediably during the melting, the addition of whiskers thermally stabilized the material above the T_m in an appreciable manner and already for CNW concentrations as low as 3 wt%.

In contrast to other polymer systems, however, T_c and degree of crystallinity were found to decrease. This behavior was ascribed to an "antinucleating"

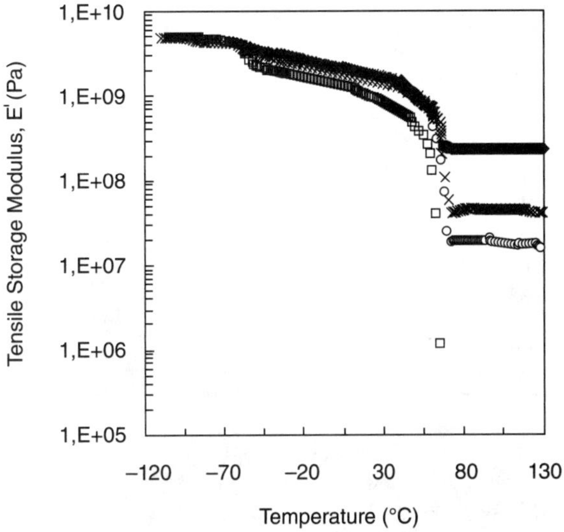

Figure 18.9 Dynamic thermomechanical analysis of POE-based composites filled with 0 (□), 3 (○), 6 (x), and 10 wt% (◇) of CNWs (reprinted with permission from Azizi Samir et al. [47]. Copyright 2004 Elsevier Ltd.).

effect, whereby the well-known miscibility between cellulose and PEO was proposed to restrict polymer chain mobility [101, 102]. On the other hand, POM supported, as for many other polymer systems, a nucleating effect with the occurrence of smaller and more numerous spherulites in composites compared with the neat PEO (Fig. 18.10). CNWs were proposed to interfere with spherulite growth, thus being ejected into the interspherulitic space. However, the spherulitic growth rate, as measured from the POM images, revealed no effect of CNWs on growth rate. The apparently diverging results on CNF effect on semicrystalline polymers highlight the complexity of the morphology of these multiphase systems.

When the same system was supplemented with lithium imide salt, no particular effect on the thermal properties was observed [99]. On the other hand, tetra(ethylene) glycol dimethyl ether (TEGDME) demonstrated a strong plasticizing effect, significantly depressing T_g [100]. The thermomechanical behavior of the salt-free composites was similar to the composite polyelectrolyte containing lithium salt. As for the unplasticized PEO, the addition of whiskers to the TEGDME/PEO material induced an outstanding increase in the storage modulus in the rubbery state [47]. Here again, the mechanical stability of the composites above the melting point of the matrix (50°C) indicated the establishment of a percolating whiskers network within the matrix. A complex phase-separated nanocomposite with a plasticizer rich and a plasticizer poor domain was proposed to explain the thermomechanical data.

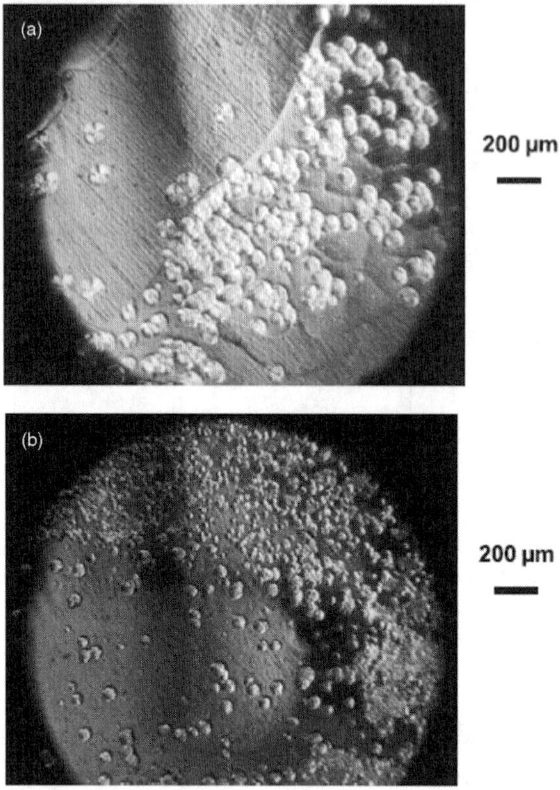

Figure 18.10 POM micrographs obtained after isothermal crystallization for 100 seconds at 53.6°C for a POE-based film with (a) 0 and (b) 10 wt% cellulose nanowhiskers (reprinted with permission from Azizi Samir et al. [47]. Copyright 2004 Elsevier Ltd.). See color insert.

18.4 CONCLUSIONS AND DISCUSSION

The survey of a few biopolymer/CNF systems illustrates the diversity and complexity of morphologies and thermal properties in these composite systems. In general, a few trends can be delineated:

- Thermal stabilization occurs especially in low modulus matrices as, for example, in the case of amorphous polymers in their rubbery state and above T_m for semicrystalline polymers. This is due to the rigid nature of a hydrogen-bonded network of CNFs, whether it arises from CNWs or MFC. Higher thermal stabilization of flexible and hairy microfibrils is apparently observed and ascribed to a tangling effect in such nanofiber network morphologies [39]. In some semicrystalline matrices, the thermal

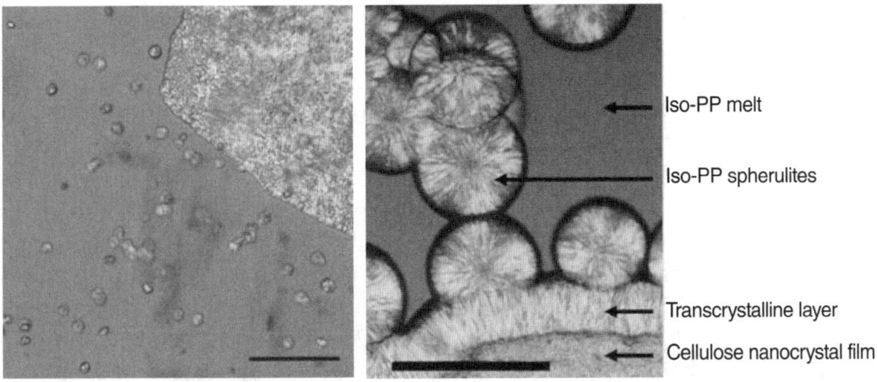

Figure 18.11 Crystallization of polypropylene against cellulose nanocrystal film. i-PP initially melt at 220°C. PLM images after 5 minutes (left) and after 20 minutes (right) at 136°C. The right image at higher magnification shows in detail the edge of the CNWs film in contact with i-PP melt. Scale bars: 200 μm (reprinted with permission from Gray [92]. Copyright 2007 Springer Science+Business Media B.V.) See color insert.

stabilization disappears suddenly at the matrix T_m, suggesting that the CNF network is imbedded within the matrix crystalline regions through transcrystallization. Depending on the state of the polymer and on the competition between interfacial and interfiber interactions, thermal stabilization might be described either with a simple mean field theory as the Halpin–Kardos model [64] or with the percolation network model [2, 63].

- Addition of CNFs dramatically alters the formation of crystalline structures in polymers. While many researchers refer to the heterogeneous nucleating effect of CNFs with the development of a TCL, there is only one demonstration of the ability of cellulose I CNWs to generate such TCL. Gray [92] imaged a TCL of isotactic PP on a cellulose I CNW film and revealed that the TCL developed much sooner than bulk spherulites resulting in true interfacial crystallization (Fig. 18.11).

However, such crystalline superstructures arising from interfacial crystallization have only been inferred in CNF nanocomposites based on thermal properties. When one compares with the general morphology of nanocomposites, for example, prepared with carbon nanotubes or clay, there is an abundance of literature that demonstrates that such nanofillers give rise to unique morphologies [103]. Aside from TCL, shish–kebab, shish–calabash are common superstructures resulting from interfacial crystallization in nanocomposites. These have not been observed to date with CNF-reinforced polymers and might also be of particular significance for the properties of cellulose nanocomposites. Indeed, interfacial crystallization is now established as a major contribution to nanocomposite performance [103]. Further studies on the nanoscale morphology of such nanocomposites are therefore necessary to better understand the fundamental role of CNFs on interfacial crystallization.

- In general, CNFs are reported to expedite crystallization and foster crystallinity, but the opposite situation is also sometimes observed. A trend seems to emerge from the body of literature that low loading of CNFs and good interfacial compatibility enhance crystallization while high loadings of CNFs might hinder crystallization [68, 90]. It is interesting to note that while comparable results in composites with microsized cellulose fibers are observed [89], the magnitude of the effects are different since very large proportions of microsized fillers (25 wt%) have been required to observe such nucleating effects; in contrast in well-dispersed nanocomposites, 1–2% CNW content suffices to dramatically enhance crystallization. This difference highlights the impact of surface area on polymer crystallization. Whiskers have been proposed to have approximately a 10-fold larger surface area (533 m^2/g) than MFC (51 m^2/g), which in turn is assessed to have a much larger surface area than the native wood fiber (2–10 m^2/g) [42]. It is therefore no surprise when one considers the whole body of literature that whiskers, which have a higher expected surface area than MFC and wood fibers, have a more significant effect at the smallest loadings, followed by MFC and then by larger cellulose fibers. Such behavior would be consistent with general observations in other nanocomposites. For example, in montmorillonite nanocomposites, the nucleating power of nanoparticle dominates at low loadings and facilitate crystallization, while confinement and hindrance in polymer chain mobility appear to dominate at high loadings, thus hampering crystallization [104].

The studies reviewed above further highlight the fact that surface area alone does not dictate the thermal properties and impact on matrix morphology. Interfacial adhesion and dispersion on the nanometer scale are most critical. In fact, the system can be similar to polymer blend systems whereby the free energy of mixing dictates the scale of miscibility, in essence governing both morphology and the resulting material performance. For nanocomposites it is also debated whether the nanoparticles tailor the formation of crystalline morphology or whether it is the polymer that segregates the nanoparticles in order to pursue crystallization in a minimally perturbed form [104]. To be sure, the field of cellulosic bionanocomposites will keep benefiting from studies and investigations relying on the principles of polymer physics, as there still is much to be revealed on the morphology development of these new material systems.

ACKNOWLEDGMENTS

The authors would like to thank Prof. Wolfgang Glasser for helpful suggestions on the manuscript as well as Mrs. Jia Mao and Mr. Michel Brioude for support.

REFERENCES

[1] Favier, V., Canova, G.R., Cavaillé, J.Y., Chanzy, H., Dufresne, A., and Gauthier, C. (1995) Nanocomposite materials from latex and cellulose whiskers. *Polymers for Advanced Technologies*, *6*, 351–355.

[2] Favier, V., Chanzy, H., and Cavaillé, J.Y. (1995) Polymer nanocomposites reinforced by cellulose whiskers. *Macromolecules*, *28*, 6365–6367.

[3] Cao, X., Habibi, Y., Magalhães, W., Rojas, O.J., and Lucia, L.A. (2011) Cellulose nanocrystals-based nanocomposites: Fruits of a novel biomass research and teaching platform. *Current Science*, *100*, 11762–1176.

[4] Azizi Samir, M.A.S., Alloin, F., and Dufresne, A. (2005) Review of recent research into cellulosic whiskers, their properties and their application in nanocomposite field. *Biomacromolecules*, *6*, 612–626.

[5] Siró, I., Plackett, D. (2010) Microfibrillated cellulose and new nanocomposite materials: A review. *Cellulose*, *17*, 459–494.

[6] Eichhorn, S., Dufresne, A., Aranguren, M., Marcovich, N., Capadona, J., Rowan, S., Weder, C., Thielemans, W., Roman, M., Renneckar, S., Gindl, W., Veigel, S., Keckes, J., Yano, H., Abe, K., Nogi, M., Nakagaito, A., Mangalam, A., Simonsen, J., Benight, A., Bismarck, A., Berglund, L., and Peijs, T. (2010) Review: Current international research into cellulose nanofibres and nanocomposites. *Journal of Materials Science*, *45*, 1–33.

[7] Dufresne, A. (2008) Polysaccharide nano crystal reinforced nanocomposites. *Canadian Journal of Chemistry*, *86*, 484–494.

[8] Abdul Khalil, H.P.S., Bhat, A.H., and Ireana Yusra, A.F. (2012) Green composites from sustainable cellulose nanofibrils: A review. *Carbohydrate Polymers*, *87*, 963–979.

[9] Hubbe, M.A., Rojas, O.J., Lucia, L.A., and Sain, M. (2008) Cellulosic nanocomposites: A review. *Bioresources*, *3*, 929–980.

[10] Habibi, Y., Lucia, L.A. (2012) Nanocelluloses: Emerging building blocks for renewable materials. In: *Polysaccharide Building Blocks*. Hoboken, NJ: John Wiley & Sons, Inc., pp. 105–125.

[11] Pandey, J.K., Kumar, A.P., Misra, M., Mohanty, A.K., Drzal, L.T., and Palsingh, R. (2005) Recent advances in biodegradable nanocomposites. *Journal of Nanoscience and Nanotechnology*, *5*, 497–526.

[12] de Souza Lima, M.M., Borsali, R. (2004) Rodlike cellulose microcrystals: Structure, properties, and applications. *Macromolecular Rapid Communications*, *25*, 771–787.

[13] Siqueira, G., Bras, J., and Dufresne, A. (2010) Cellulosic bionanocomposites: A review of preparation, properties and applications. *Polymers*, *2*, 728–765.

[14] Klemm, D., Kramer, F., Moritz, S., Lindström, T., Ankerfors, M., Gray, D., and Dorris, A. (2011) Nanocelluloses: A new family of nature-based materials. *Angewandte Chemie International Edition*, *50*, 5438–5466.

[15] Moon, R.J., Martini, A., Nairn, J., Simonsen, J., and Youngblood, J. (2011) Cellulose nanomaterials review: Structure, properties and nanocomposites. *Chemical Society Reviews*, *40*, 3941–3994.

[16] Lin, N., Huang, J., and Dufresne, A. (2012) Preparation, properties and applications of polysaccharide nanocrystals in advanced functional nanomaterials: A review. *Nanoscale*, *4*, 3274–3294.

[17] Elazzouzi-Hafraoui, S., Nishiyama, Y., Putaux, J.-L., Heux, L., Dubreuil, F., and Rochas, C. (2007) The shape and size distribution of crystalline nanoparticles prepared by acid hydrolysis of native cellulose. *Biomacromolecules*, *9*, 57–65.

[18] Sugiyama, J., Vuong, R., and Chanzy, H. (1991) Electron diffraction study on the two crystalline phases occurring in native cellulose from an algal cell wall. *Macromolecules*, *24*, 4168–4175.

[19] Nishiyama, Y., Shigenori, K., Masahisa, W., and Takeshi, O. (1997) Cellulose microcrystal film of high uniaxial orientation. *Macromolecules*, *30*, 6395–6397.

[20] Osorio-Madrazo, A., Eder, M., Rueggeberg, M., Pandey, J.K., Harrington, M.J., Nishiyama, Y., Putaux, J.-L., Rochas, C., and Burgert, I. (2012) Reorientation of cellulose nanowhiskers in agarose hydrogels under tensile loading. *Biomacromolecules*, *13*, 850–856.

[21] White, A.R., Brown, R.M. (1981) Enzymatic hydrolysis of cellulose: Visual characterization of the process. *Proceedings of the National Academy of Sciences of the United States of America*, *78*, 1047–1051.

[22] Brown, E.E., Laborie, M.-P.G. (2007) Bioengineering bacterial cellulose/poly(ethylene oxide) nanocomposites. *Biomacromolecules*, *8*, 3074–3081.

[23] Angles, M.N., Dufresne, A. (2000) Plasticized starch/tunicin whiskers nanocomposites. 1. Structural analysis. *Macromolecules*, *33*, 8344–8353.

[24] Delmer, D.P. (1999) Cellulose biosynthesis: Exciting times for a difficult field of study. *Annual Review of Plant physiology and Plant Molecular Biology*, *50*, 245–276.

[25] Reiter, W.-D. (2002) Biosynthesis and properties of the plant cell wall. *Current Opinion in Plant Biology*, *5*, 536–542.

[26] Chanzy, H. (1990) Aspects of cellulose structure. In: *Cellulose Sources and Exploitation*, eds. J.F. Kennedy, G.O. Philips, and P.A. Williams. New York: Ellis Horwoood Ltd.

[27] Marchessault, R.H. (1989) Cellulose and wood-chemistry and technology. In: *Proceedings of the Tenth Cellulose Conference*, ed. C. Schuerch. New York: Wiley.

[28] Sarko, A., Muggli, R. (1974) Packing analysis of carbohydrates and polysaccharides. III. *Valonia* cellulose and cellulose II. *Macromoloecules*, *7*, 486–494.

[29] Gardner, K.H., Blackwell, J. (1974) The structure of native cellulose. *Biopolymers*, *13*, 1975–2001.

[30] Nishiyama, Y., Langan, P., and Chanzy, H. (2002) Crystal structure and hydrogen-bonding system in cellulose Iβ from synchrotron X-ray and neutron fiber diffraction. *Journal of the American Chemical Society*, *124*, 9074–9082.

[31] Rånby, B.G. (1952) Physicochemical investigations on animal cellulose (tunicin). *Arkiv for Kemi*, *4*, 241–248.

[32] Hepworth, D.G., Bruce, D.M. (2000) A method of calculating the mechanical properties of nanoscopic plant cell wall components from tissue properties. *Journal of Materials Science*, *35*, 5861–5865.

[33] Hsieh, Y.C., Yano, H., Nogi, M., and Eichhorn, S. (2008) An estimation of the Young's modulus of bacterial cellulose filaments. *Cellulose*, *15*, 507–513.

[34] Sakurada, I., Nukushina, Y., and Ito, T. (1962) Experimental determination of the elastic modulus of crystalline regions in oriented polymers. *Journal of Polymer Science*, *57*, 651–660.

[35] Šturcová, A., Davies, G.R., and Eichhorn, S.J. (2005) Elastic modulus and stress-transfer properties of tunicate cellulose whiskers. *Biomacromolecules*, *6*, 1055–1061.

[36] Nishino, T., Takano, K., and Nakamae, K. (1995) Elastic modulus of the crystalline regions of cellulose polymorphs. *Journal of Polymer Science Part B: Polymer Physics*, *33*, 1647–1651.

[37] Spence, K., Venditti, R., Rojas, O., Habibi, Y., and Pawlak, J. (2011) A comparative study of energy consumption and physical properties of microfibrillated cellulose produced by different processing methods. *Cellulose*, *18*, 1097–1111.

[38] Pääkkö, M., Ankerfors, M., Kosonen, H., Nykänen, A., Ahola, S., Österberg, M., Ruokolainen, J., Laine, J., Larsson, P.T., Ikkala, O., and Lindström, T. (2007) Enzymatic hydrolysis combined with mechanical shearing and high-pressure homogenization for nanoscale cellulose fibrils and strong gels. *Biomacromolecules*, *8*, 1934–1941.

[39] Azizi Samir, M.A.S., Alloin, F., Paillet, M., and Dufresne, A. (2004) Tangling effect in fibrillated cellulose reinforced nanocomposites. *Macromolecules*, *37*, 4313–4316.

[40] Ferry, J.D. (1980) *Viscoelastic Properties of Polymers*. New York: John Wiley & Sons.

[41] Utracki, L.A. (2003) *Polymer Blends Handbook*. Dordrecht: Kluwer Academic.

[42] Siqueira, G., Bras, J., and Dufresne, A. (2009) New process of chemical grafting of cellulose nanoparticles with a long chain isocyanate. *Langmuir*, *26*, 402–411.

[43] Ludueña, L., Vázquez, A., and Alvarez, V. (2012) Effect of lignocellulosic filler type and content on the behavior of polycaprolactone based eco-composites for packaging applications. *Carbohydrate Polymers*, *87*, 411–421.

[44] Alvarez, V.A., Ruseckaite, R.A., and Vázquez, A. (2006) Degradation of sisal fibre/Mater Bi-Y biocomposites buried in soil. *Polymer Degradation and Stability*, *91*, 3156–3162.

[45] Kowalczyk, M., Piorkowska, E., Kulpinski, P., and Pracella, M. (2011) Mechanical and thermal properties of PLA composites with cellulose nanofibers and standard size fibers. *Composites Part A: Applied Science and Manufacturing*, *42*, 1509–1514.

[46] Dufresne, A., Kellerhals, M.B., and Witholt, B. (1999) Transcrystallization in Mcl-PHAs/cellulose whiskers composites. *Macromolecules*, *32*, 7396–7401.

[47] Azizi Samir, M.A.S., Alloin, F., Sanchez, J.-Y., and Dufresne, A. (2004) Cellulose nanocrystals reinforced poly(oxyethylene). *Polymer*, *45*, 4149–4157.

[48] Sanchez-Garcia, M.D., Gimenez, E., and Lagaron, J.M. (2008) Morphology and barrier properties of solvent cast composites of thermoplastic biopolymers and purified cellulose fibers. *Carbohydrate Polymers*, *71*, 235–244.

[49] Lavoine, N., Desloges, I., Dufresne, A., and Bras, J. (2012) Microfibrillated cellulose—Its barrier properties and applications in cellulosic materials: A review. *Carbohydrate Polymers*, *90*, 735–764.

[50] Gedde, U.W. (1995) *Polymer Physics*. London, UK: Chapman & Hall.

[51] Nielsen, L.E. (1967) Mechanical properties of particulate-filled systems. *Journal of Composite Materials*, *1*, 100–119.

[52] Nielsen, L.E. (1969) Dynamic mechanical properties of filled polymers. *Applied Polymer Symposium*, *12*, 249–265.

[53] Nielsen, L.E., Landel, R.F. (1994) *Mechanical Properties of Polymers and Composites*, 2nd ed. New York: Marcel Dekker, p. 50.

[54] Flory, P.J. (1953) *Principles of Polymer Chemistry*. New York: Cornell University Press.

[55] Lo, C.T., Seifert, S., Thiyagarajan, P., and Narasimhan, B. (2004) Phase behavior of semicrystalline polymer blends. *Polymer*, *45*, 3671–3679.

[56] Nadkarni, V.M., Jog, J.P. (1991) Crystallization behavior in polymer blends. In: *Two-Phase Polymer Systems*, ed. L.A. Utracki. Munich–New York: Hanser Publishers.

[57] Quan, H., Li, Z.-M., Yang, M.-B., and Huang, R. (2005) On transcrystallinity in semi-crystalline polymer composites. *Composites Science and Technology*, *65*, 999–1021.

[58] Jenckel, K., Teege, E., and Hinrichs, W. (1952). *Kolloid-Z*, *129*, 19–24.

[59] Avrami, M. (1939) Kinetics of phase change. I general theory. *The Journal of Chemical Physics*, *7*, 1103–1112.

[60] Avrami, M. (1940) Kinetics of phase change. II transformation-time relations for random distribution of nuclei. *Journal of Chemical Physics*, *8*, 212–224.

[61] Avrami, M. (1941) Granulation, phase change, and microstructure kinetics of phase change. III. *Journal of Chemical Physics*, *9*, 177–184.

[62] Hoffman, J.D., Davis, G.T., and Lauritzen, J.I. (1976) The rate of crystallization of linear polymers with chain folding. In: *Treatise on Solid State Chemistry*, ed. N.B. Hannay. New York: Plenum Press, pp. 497–614.

[63] Takayanagi, M., Uemura, S., and Minami, S. (1964) Application of equivalent model method to dynamic rheo-optical properties of crystalline polymer. *Journal of Polymer Science Part C: Polymer Symposia*, *5*, 113–122.

[64] Halpin, J.C., Kardos, J.L. (1972) Moduli of crystalline polymers employing composite theory. *Journal of Applied Physics*, *43*, 2235–2241.

[65] Ten, E., Turtle, J., Bahr, D., Jiang, L., and Wolcott, M. (2010) Thermal and mechanical properties of poly(3-hydroxybutyrate-co-3-hydroxyvalerate)/cellulose nanowhiskers composites. *Polymer*, *51*, 2652–2660.

[66] Long, J., Morelius, E., Jinwen, Z., Wolcott, M., and Holbery, J. (2008) Study of the poly(3-hydroxybutyrate-co-3-hydroxyvalerate)/cellulose nanowhisker composites prepared by solution casting and melt processing. *Journal of Composite Materials*, *42*, 2629–2645.

[67] Ten, E., Bahr, D.F., Li, B., Jiang, L., and Wolcott, M.P. (2012) Effects of cellulose nanowhiskers on mechanical, dielectric, and rheological properties of poly(3-hydroxybutyrate-co-3-hydroxyvalerate)/cellulose nanowhisker composites. *Industrial & Engineering Chemistry Research*, *51*, 2941–2951.

[68] Ten, E., Bahr, D.F., Jiang, L., and Wolcott, M.P. (2012) Crystallization kinetics of poly(3-hydroxybutyrate-co-3-hydroxyvalerate)/cellulose nanowhiskers composites. *Carbohydrate Polymers*, *90*, 541–550.

[69] Yu, H.-Y., Qin, Z.-Y., and Zhou, Z. (2011) Cellulose nanocrystals as green fillers to improve crystallization and hydrophilic property of poly(3-hydroxybutyrate-

co-3-hydroxyvalerate). *Progress in Natural Science: Materials International*, *21*, 478–484.

[70] Suryanegara, L., Nakagaito, A.N., and Yano, H. (2009) The effect of crystallization of PLA on the thermal and mechanical properties of microfibrillated cellulose-reinforced PLA composites. *Composites Science and Technology*, *69*, 1187–1192.

[71] Suryanegara, L., Nakagaito, A., and Yano, H. (2010) Thermo-mechanical properties of microfibrillated cellulose-reinforced partially crystallized PLA composites. *Cellulose*, *17*, 771–778.

[72] Pei, A., Zhou, Q., and Berglund, L.A. (2010) Functionalized cellulose nanocrystals as biobased nucleation agents in poly(l-lactide) (PLLA)—Crystallization and mechanical property effects. *Composites Science and Technology*, *70*, 815–821.

[73] Fukuzumi, H., Saito, T., Iwata, T., Kumamoto, Y., and Isogai, A. (2008) Transparent and high gas barrier films of cellulose nanofibers prepared by TEMPO-mediated oxidation. *Biomacromolecules*, *10*, 162–165.

[74] Okubo, K., Fujii, T., and Yamashita, N. (2005) Improvement of interfacial adhesion in bamboo polymer composite enhanced with micro-fibrillated cellulose. *JSME International Journal Series A Solid Mechanics and Material Engineering*, *48*, 199–204.

[75] Okubo, K., Fujii, T., and Thostenson, E.T. (2009) Multi-scale hybrid biocomposite: Processing and mechanical characterization of bamboo fiber reinforced PLA with microfibrillated cellulose. *Composites Part A: Applied Science and Manufacturing*, *40*, 469–475.

[76] Chakraborty, A., Sain, M., Kortschot, M., and Cutler, S. (2007) Dispersion of wood microfibers in a matrix of thermoplastic starch and starchpolylactic acid blend. *Journal of Biobased Materials and Bioenergy*, *1*, 71–77.

[77] Wang, B., Sain, M. (2007) The effect of chemically coated nanofiber reinforcement on biopolymer based nanocomposites. *Bioresources*, *2*, 371–388.

[78] Braun, B., Dorgan, J.R., and Hollingsworth, L.O. (2012) Supra-molecular EcoBioNanocomposites based on polylactide and cellulosic nanowhiskers: Synthesis and properties. *Biomacromolecules*, *13*, 1991–2218.

[79] Goffin, A.-L., Raquez, J.-M., Duquesne, E., Siqueira, G., Habibi, Y., Dufresne, A., and Dubois, P. (2011) From interfacial ring-opening polymerization to melt processing of cellulose nanowhisker-filled polylactide-based nanocomposites. *Biomacromolecules*, *12*, 2456–2465.

[80] Bondeson, D., Oksman, K. (2007) Polylactic acid/cellulose whisker nanocomposites modified by polyvinyl alcohol. *Composites Part A: Applied Science and Manufacturing*, *38*, 2486–2492.

[81] Iwatake, A., Nogi, M., and Yano, H. (2008) Cellulose nanofiber-reinforced polylactic acid. *Composites Science and Technology*, *68*, 2103–2106.

[82] Jonoobi, M., Harun, J., Mathew, A.P., and Oksman, K. (2010) Mechanical properties of cellulose nanofiber (CNF) reinforced polylactic acid (PLA) prepared by twin screw extrusion. *Composites Science and Technology*, *70*, 1742–1747.

[83] Nakagaito, A.N., Fujimura, A., Sakai, T., Hama, Y., and Yano, H. (2009) Production of microfibrillated cellulose (MFC)-reinforced polylactic acid (PLA)

nanocomposites from sheets obtained by a papermaking-like process. *Composites Science and Technology*, *69*, 1293–1297.

[84] Oksman, K., Mathew, A.P., Bondeson, D., and Kvien, I. (2006) Manufacturing process of cellulose whiskers/polylactic acid nanocomposites. *Composites Science and Technology*, *66*, 2776–2784.

[85] Petersson, L., Kvien, I., and Oksman, K. (2007) Structure and thermal properties of poly(lactic acid)/cellulose whiskers nanocomposite materials. *Composites Science and Technology*, *67*, 2535–2544.

[86] Tingaut, P., Zimmermann, T., and Lopez-Suevos, F. (2009) Synthesis and characterization of bionanocomposites with tunable properties from poly(lactic acid) and acetylated microfibrillated cellulose. *Biomacromolecules*, *11*, 454–464.

[87] Luiz de Paula, E., Mano, V., and Pereira, F.V. (2011) Influence of cellulose nanowhiskers on the hydrolytic degradation behavior of poly(d,l-lactide). *Polymer Degradation and Stability*, *96*, 1631–1638.

[88] Lin, N., Huang, J., Chang, P.R., Feng, J., and Yu, J. (2011) Surface acetylation of cellulose nanocrystal and its reinforcing function in poly(lactic acid). *Carbohydrate Polymers*, *83*, 1834–1842.

[89] Mathew, A.P., Oksman, K., and Sain, M. (2006) The effect of morphology and chemical characteristics of cellulose reinforcements on the crystallinity of polylactic acid. *Journal of Applied Polymer Science*, *101*, 300–310.

[90] Siqueira, G., Fraschini, C., Bras, J., Dufresne, A., Prud'homme, R., and Laborie, M.-P. (2011) Impact of the nature and shape of cellulosic nanoparticles on the isothermal crystallization kinetics of poly(ε-caprolactone). *European Polymer Journal*, *47*, 2216–2227.

[91] Siqueira, G., Bras, J., and Dufresne, A. (2008) Cellulose whiskers versus microfibrils: Influence of the nature of the nanoparticle and its surface functionalization on the thermal and mechanical properties of nanocomposites. *Biomacromolecules*, *10*, 425–432.

[92] Gray, D. (2008) Transcrystallization of polypropylene at cellulose nanocrystal surfaces. *Cellulose*, *15*, 297–301.

[93] Chen, E.-C., Wu, T.-M. (2007) Isothermal crystallization kinetics and thermal behavior of poly(ε-caprolactone)/multi-walled carbon nanotube composites. *Polymer Degradation and Stability*, *92*, 1009–1015.

[94] Ludueña, L.N., Vazquez, A., and Alvarez, V.A. (2008) Crystallization of polycaprolactone–clay nanocomposites. *Journal of Applied Polymer Science*, *109*, 3148–3156.

[95] Ho, R.-M., Chiang, Y.-W., Lin, C.-C., and Huang, B.-H. (2005) Crystallization and melting behavior of poly(ε-caprolactone) under physical confinement. *Macromolecules*, *38*, 4769–4779.

[96] Lönnberg, H., Zhou, Q., Brumer, H., Teeri, T.T., Malmström, E., and Hult, A. (2006) Grafting of cellulose fibers with poly(ε-caprolactone) and poly(l-lactic acid) via ring-opening polymerization. *Biomacromolecules*, *7*, 2178–2185.

[97] Lönnberg, H., Fogelström, L., Berglund, L., Malmström, E., and Hult, A. (2008) Surface grafting of microfibrillated cellulose with poly(ε-caprolactone)—Synthesis and characterization. *European Polymer Journal*, *44*, 2991–2997.

[98] Lee, S.-H., Teramoto, Y., and Endo, T. (2011) Cellulose nanofiber-reinforced polycaprolactone/polypropylene hybrid nanocomposite. *Composites Part A: Applied Science and Manufacturing*, *42*, 151–156.

[99] Azizi Samir, M.A.S., Alloin, F., Gorecki, W., Sanchez, J.-Y., and Dufresne, A. (2004) Nanocomposite polymer electrolytes based on poly(oxyethylene) and cellulose nanocrystals. *The Journal of Physical Chemistry B*, *108*, 10845–10852.

[100] Azizi Samir, M.A.S., Mateos, A.M., Alloin, F., Sanchez, J.-Y., and Dufresne, A. (2004) Plasticized nanocomposite polymer electrolytes based on poly(oxyethylene) and cellulose whiskers. *Electrochimica Acta*, *49*, 4667–4677.

[101] Guo, Y.Q., Liang, X.H. (1999) The miscibility of cellulose-polyethylene glycol blends. *Journal of Macromolecular Science, Part B. Physics*, *38*, 439.

[102] Nishio, Y., Hirose, N., and Takahashi, T. (1989) Thermal analysis of cellulose/poly(ethylene oxide) blends. *Polymer Journal*, *21*, 347–351.

[103] Ning, N., Fu, S., Zhang, W., Chen, F., Wang, K., Deng, H., Zhang, Q., and Fu, Q. (2011) Realizing the enhancement of interfacial interaction in semicrystalline polymer/filler composites via interfacial crystallization. *Progress in Polymer Science.*

[104] Khan, J., Harton, S.E., Akcora, P., Benicewicz, B.C., and Kumar, S.K. (2009) Polymer crystallization in nanocomposites: Spatial reorganization of nanoparticles. *Macromolecules*, *42*, 5741–5744.

[105] Malainine, M.E., Mahrouz, M., and Dufresne, A. (2005) Thermoplastic nanocomposites based on cellulose microfibrils from *Opuntia ficus-indica* parenchyma cell. *Composites Science and Technology*, *65*, 1520–1526.

CHAPTER 19

Mechanical Properties of Cellulose-Based Bionanocomposites

B. DEEPA, SAUMYA S. PILLAI, LALY A. POTHAN, and SABU THOMAS

19.1 INTRODUCTION

The most recent trends in natural fiber composites comprise the introduction of cellulose nanofibers (CNFs) for biopolymeric composite materials. Biocomposites based on cellulose nanoparticles are interesting for many branches of industry. The increasing number of publications during the recent years including reviews [1–14] and books [15–19] reflect the growing importance of these new biocomposites. Cellulose nanoparticles (CNs), generally known as nanocellulose, have garnered interest from the scientific community because of their biodegradability, sustainability, easy availability, and related characteristics such as a very large surface-to-volume ratio and outstanding mechanical, electrical, and thermal properties. CNs also have a high aspect ratio, low density (1.6 g/cm^3), high elastic modulus (~140 GPa), and a reactive surface of –OH side groups that facilitates grafting chemical species to achieve different surface properties (surface functionalization). Surface functionalization allows the tailoring of particle surface chemistry to facilitate self-assembly, controlled dispersion within a wide range of matrix polymers, and control of both the particle–particle and particle–matrix bond strength. These unique characteristics, along with the remarkable suitability for surface functionalization, make nanocellulose an ideal candidate for improving the mechanical properties of the host material.

The incorporation of only a small amount of nanofibers to base polymers confers improved properties that make them usable in automotive, construction, packaging, display, water treatment, and medical applications [20–25]. Properties that have been shown to improve substantially are mechanical

Biopolymer Nanocomposites: Processing, Properties, and Applications, First Edition. Edited by Alain Dufresne, Sabu Thomas, and Laly A. Pothan.

properties (e.g., tensile strength, elastic modulus, and dimensional stability), thermomechanical properties, and permeability (e.g., gases, water, and hydrocarbons). In addition, the thermal stability and heat distortion temperature, flame retardancy and smoke emissions, chemical resistance, surface appearance, physical weight, and electrical conductivity of the composites are found to improve.

Materials based on CNFs show interesting characteristics. For instance, nanopaper made from wood CNFs shows high resistance to fracture and can have a modulus of 13.2 GPa, a tensile strength above 200 MPa, and a strain to failure of 10% [26], much superior to microstructured paper and paperboard materials. Tailored cellulose nanocomposites with nanostructured matrix distribution can even show a tensile strength above 300 MPa [27] for the in-plane random fibril orientation case. Nanofibers can also be used in high porosity cellulose aerogels and foams, in optically transparent biocomposites, as oxygen barriers, to reinforce the cell walls of polymer foams, and as templates for magnetic nanoparticle hybrids [28–34].

Some CN composites produced to date are transparent, have tensile strengths greater than cast iron, and have very low coefficients of thermal expansion (CTE). For example, Okahisa et al. [35] fabricated an organic light-emitting diode (OLED) on flexible, low CTE, and optically transparent wood–cellulose nanocomposites. Nanocomposites were prepared using matrices with different Young's moduli. With the same fiber content, the nanocomposites using lower Young's modulus matrix resin exhibited lower CTE values than those using higher Young's modulus matrix resins. This led to the development of nanocomposites with a very low CTE while having high flexibility and ductile properties, which opens up many possibilities for the application of OLEDs in flexible, transparent displays. Recently, a cellulose-based nanocomposite material has been investigated as a flexible humidity and temperature sensor [36].

Nanocellulose-reinforced biocomposites could soon provide advanced performance, durability, value, service life, and utility while at the same time being a fully sustainable technology. Potential applications include, but are not limited to, barrier films, antimicrobial films, transparent films, flexible displays, reinforcing fillers for polymers, biomedical implants, pharmaceuticals, drug delivery, fibers and textiles, templates for electronic components, separation membranes, batteries, supercapacitors, and electroactive polymerss. This chapter reviews some recent studies devoted to the mechanical properties of CNF-reinforced biopolymer composites.

19.2 MECHANICAL PERFORMANCE OF BIONANOCOMPOSITES

It is important to be knowledgeable of certain mechanical properties of nanocomposites to be able to exploit their potential. Among these properties are the tensile, flexural, impact, dynamic mechanical, and creep properties.

Three main parameters were reported to affect the mechanical properties of materials: (1) the morphology and dimensions of the nanoparticles, (2) the processing method, and (3) the microstructure of the matrix and matrix/filler interactions.

For cellulose nanocrystals occurring as rod-like nanoparticles, the geometrical aspect ratio, defined as the length-to-diameter (L/d), is a major factor that controls the mechanical properties of nanocomposites and determines the percolation threshold value.This factor is linked to the source of cellulose and whiskers and also its preparation conditions. Fillers with a high aspect ratio give the best reinforcing effect.It was reported that the highest modulus increase in the rubbery state of the poly(Sco-BuA) matrix and thermal stability were obtained with tunicin whiskers (L/d ~67) in comparison with bacterial (L/d ~60) orAvicel whiskers (L/d ~10) [37, 38]. de Rodriguez and coworkers [39] studied sisal nanowhiskers with a high aspect ratio as filler in the nanocomposites with polyvinyl acetate (PVAc) as the matrix. They found that the high aspect ratio could ensure percolation, resulting in mechanical improvements and thermal stability at lower fiber loadings. The flexibility and tangling possibility of the nanofibers also plays an important role [40–42]. It was reported that entangled microfibrillated cellulose (MFC) induces a higher reinforcing effect than straight whiskers, whereas the elongation at break was lower.

The processing method conditions present the possibility of the formation of a continuous whiskers network and determine the final properties of the nanocomposite material. Slow processes such as casting/evaporation were reported to give the highest mechanical performance materials compared with freeze-drying/molding and freeze-drying/extruding/molding techniques. During slow water evaporation, because of Brownian motions in the suspension or solution (whose viscosity remains low, up to the end of the process when the latex particle or polymer concentration becomes very high), the rearrangement of the nanoparticles is possible. They have significant time to interact and connect to form a continuous network, which is the basis of their reinforcing effect. The resulting structure is completely relaxed and direct contacts between nanoparticles are then created. Conversely, during the freeze-drying/hot-pressing process, the nanoparticle arrangement in the suspension is first frozen, and then, during the hot-pressing stage, because of the polymer melt viscosity, the particle rearrangements are strongly limited.

The microstructure of the matrix and the resulting competition between matrix/filler and filler/filler interactions also affect the mechanical behavior of cellulose nanocrystals reinforced nanocomposites. Classical composite science tends to favor matrix/filler interactions as a fundamental condition for optimal performance. But, for cellulose whisker-based composite materials, the opposite trend is generally observed when the material is processed via casting/evaporation method. This unusual behavior is ascribed to the originality of the reinforcing phenomenon of cellulosic nanoparticles resulting from the formation of a hydrogen-bonded percolating network. However, when using another processing route than casting/evaporation in aqueous medium, the dispersion of the hydrophilic filler in the polymeric matrix is also involved and improved filler/matrix interactions generally lead to higher mechanical properties [43]. In nonpercolating systems, for instance for materials processed from freeze-dried cellulose nanocrystals, strong matrix/filler interactions enhance the reinforcing effect of the filler [44].

19.2.1 Tensile Properties

The tensile properties are among the most widely tested properties of natural fiber reinforced composites. The fiber strength can be an important factor regarding the selection of a specific natural fiber for a specific application. A tensile test reflects the average property through the thickness, whereas a flexural test is strongly influenced by the properties of the specimen closest to the top and bottom surfaces. The stresses in a tensile test are uniform throughout the specimen cross section, whereas the stresses in flexure vary from zero in the middle to maximum in the top and bottom surfaces. Recent studies that focus on the influence of cellulose nanofibres on the mechanical properties of biopolymer composites are discussed below.

Sehaqui et al. [34] recentlyinvestigated the strength-enhancing potential of CNFs in cellulose fiberboard type of biocomposites. Wood cellulose biocomposites with fiber network structures at both micro- and nanoscale were successfully processed by filtering and drying of a fibrous water suspension. The presence of nanofibers increased the tensile strength and work to fracture of the composite considerably. The strength increased from 98 MPa (wood fiber reference) to 160 MPa (10% NFC, 90% wood fibers by weight), and the work to fracture was more than doubled (from 1.7 to 4.4 MJ/m^3). The reason for this strong improvement is the presence of fiber network structures at two length scales (micro and nano). The uniaxial stress–strain curves in tension are presented in Figure 19.1.

Kaushik et al. [45] examined the mechanical properties of bionanocomposites using CNFs from wheat straw and plasticized maize starch. The resulting nanocomposite showed the highest tensile modulus (about 220 MPa) at 15% fiber content and the highest yield strength (about 6.5 MPa) at that same fiber content.

Lu et al. [46] investigated the mechanical properties of the ramie cellulose nanocrystallites (RN)-reinforced plasticized starch biocomposites. Different fractions of RN (0–40 wt%) were used to form composites with the glycerol plasticized starch. Tensile strength and Young's modulus of the plasticized starch/RN composites,conditioned at 50% relative humidity, increased from 2.8 to 6.9 MPa and from 56 to 480 MPa, respectively, as RN content increased from 0 % to 40 wt%, but the elongation at break decreased from 94.2% to 13.6%. Results revealed that there are synergistic interactions between fillers and between filler and plasticized starch matrix, and these play a key role in reinforcing the composites.

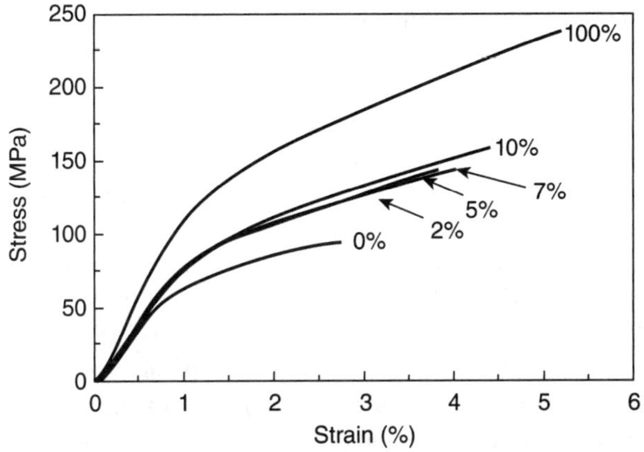

Figure 19.1 Tensile stress–strain curves of the wood fiber/NFC composites. 100% NFC refers to the nanopaper sample. Reprinted with permission from Reference [34].

Alemdar et al. [47] prepared nanocomposites from wheat straw nanofibers and thermoplastic starch (TPS) from modified potato starch by the solution-casting method. The tensile strength and modulus were significantly enhanced in the nanocomposite films, which could be explained by the uniform dispersion of nanofibers in the polymer matrix. The modulus of the TPS increased from 111 to 271 MPa with maximum (10 wt%) nanofiber filling.

In another interesting study, novel biomimetic nanocomposite foams were prepared by Svagan et al. [32] based on cellulose nanofibrils (10–70 wt%) in an amylopectin matrix and the use of a lyophilization process. Compared with unreinforced amylopectin foam, a significant improvement in modulus and yield strength was observed. This improvement is due to improved mechanical properties of the nanofibril reinforced foam cell wall and the microcellular structure.

Azeredo et al. [48] studied the tensile properties of novel nanocomposites using CNFs in different concentrations (up to 36 g/100 g) as nano-reinforcement in mango puree-based edible films for packaging applications. The study revealed that CNF increased tensile strength from 4.06 to 8.09 MPa, and Young's modulus from 19.85 to 322.05 MPa, especially at higher concentrations, which suggest the formation of a fibrillar network within the matrix.

In a study by Iwatake et al. [49] cellulose nanofibres was premixed with poly(lactic acid) (PLA) using acetone and the mixture was kneaded after the removal of the solvent to attain dispersion. As a consequence of uniform distribution the Young's modulus and tensile strength of PLA increased by 40% and 25%, respectively, without a reduction of yield strain at a fiber content of 10 wt%. Suryanegara et al. [50] applied the same method but then exchanged acetone for dichloromethane and showed that crystallization of PLA increases the strength and Young's modulus of the developed nanocomposites without significant reduction in the strain at break. These authors also noticed that CNFs can act as a nucleating agent for the crystallization of PLA.

PLA and cellulose fibrils nanocomposites have been developed by Qu et al. [51]. Bleached wood pulp was used as the fiber and commercial-grade PLA as the matrix. Polyethylene glycol (PEG) was added to the matrix as a compatibilizer to improve the interfacial bonding between the matrix and the fiber. The composites were obtained by solvent-casting methods from N, N, Dimethylacetamide (DMAC). After addition of cellulose nanofibrils to the PLA matrix, the composite has tensile strength of 30 MPa with 2.5% elongation, which is lower than that of pure PLA. Addition of PEG to the blend of PLA and cellulose nanofibrils resulted in a significant improvement in tensile strength and percent elongation of the composites. The tensile strength and the elongation (of PLA + PEG + nanofibrils) increased by 28.2% and 25%, respectively, compared with pure PLA, and increased by 56.7% and 60% compared with the PLA/cellulose nanofibrils.

Bondeson and Oksman [52] used commercial PLA as the matrix with cellulose whiskers treated with anionic surfactant (5 wt%), 10 wt%), and 20 wt%) as reinforcement. The compounded materials were extruded in three steps with pelletizing between the first and the second steps. Extruded nanocomposites were compression molded and characterized. The tensile strength and elongation at break decreased for this composite compared with PLA. This lack of improvement in the mechanical properties of the composite was attributed to poor adhesion between the fiber and the matrix, which is directly related to the nature of the fiber and the type of mechanical treatment given.

Cao et al. [53] have reported nanocomposite materials reinforced with flax cellulose nanocrystals in waterborne polyurethane. The films showed a significant increase in Young's modulus and tensile strength from 0.51 to 344 MPa and 4.27 to 14.96 MPa, respectively, with increasing filler content from 0 wt% to 30 wt%, with better dispersion.

A new route to enhance the wet properties of chitosan-acetic-acid-salt films by adding cellulose nanofibers at 5 wt% loading has been reported by Nordqvist et al. [54]. Results revealed that the addition of cellulose nanofibres to weak wet chitosan matrix increased the wet strength from 12% to 25% of the dry strength and the wet modulus increased from 0.4% to 3% of the dry modulus.

Tensile properties of nanocellulose-reinforced poly(vinyl alcohol) (PVA) nanocomposites has been reported by Cho et al. [55]. The tensile modulus and strength of the nanocomposites were improved with an increase in the nanocellulose content, but decreased at the nanocellulose content of 7 wt%. The highest modulus and strength occurred at the 5 wt% nanocellulose content, which showed 60% and 28% increase in its reinforcement, respectively. Figure 19.2 and Figure 19.3 show the tensile strength and tensile modulus of PVA nanocomposite as the function of the nanocellulose content, respectively.

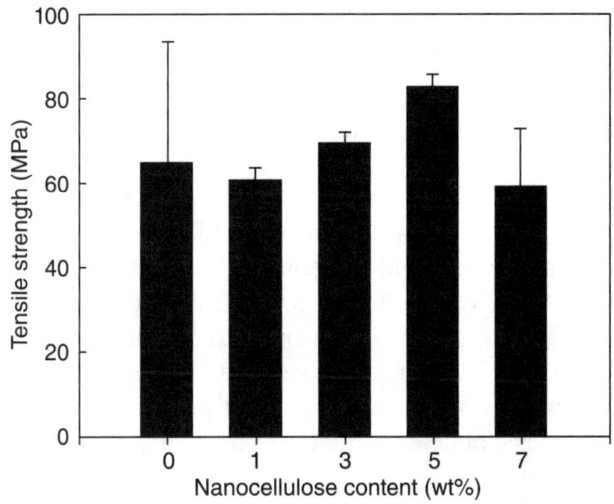

Figure 19.2 Tensile strength of PVA nanocomposites as the function of the nanocellulose content. Reprinted with permission from Reference [55].

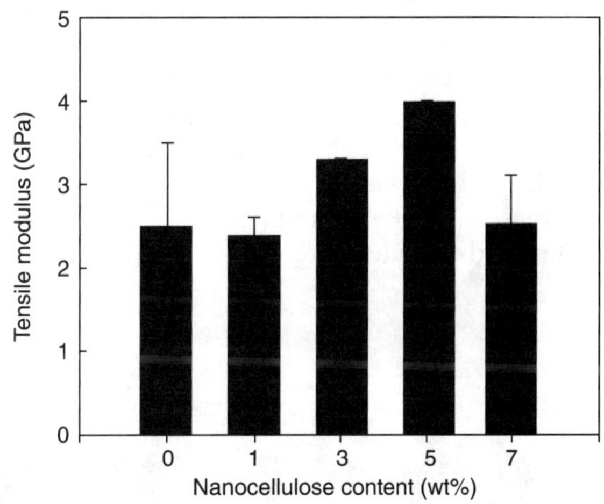

Figure 19.3 Tensile modulus of PVA nanocomposite as the function of the nanocellulose content. Reprinted with permission from Reference [55].

Leitner et al. [56] prepared PVA nanocomposites with a range of nanocellulose contents (0–90 wt%). At a cellulose content of 50 wt%, the modulus of elasticity of PVOH increased by a factor of 20 and tensile strength increased by a factor of 3.5. Both parameters increased further at cellulose contents of 70 wt% and 90 wt%, respectively. It was also reported that tensile strength increased linearly as a function of filler content, which suggests that cellulose content was the major determinant of strength in these composites.

A blend containing 10% CNFs obtained from various sources, such as flax bast fibers, hemp fibers, kraft pulp, or rutabaga and 90% PVA was used for making nanofiber-reinforced composite material by a solution-casting procedure [57].Both tensile strength and Young's modulus were improved compared to neat PVOH film, with a pronounced four- to fivefold increase in Young's modulus observed. Similarly, Wang and Sain [58] reported that soybean stock-based nanofiber-reinforced PVOH films (up to 10% nanofiber content) demonstrated a twofold increase in tensile strength when compared with films without filler. With a higher loading of MFC, the relative enhancement of mechanical properties was even more remarkable. Bruce et al. [59] reported approximately five times higher tensile strength for PVOH containing 50 wt% MFC when compared to the neat polymer. Such improvement in mechanical properties can be explained by strong interfacial bonding between the CNFs and PVOH.

Fujisawa et al. [60] investigated the outstanding nanoreinforcement effect of 2,2,6,6-tetramethylpiperidine-1-oxyl (TEMPO)-oxidized cellulose nanofibrils (TOCNs) in a polystyrene (PS) matrix. TOCN/PS nanocomposite films containing 0–10 wt% TOCN with free carboxyl groups (TOCN-COOH) were prepared successfully using TOCN-93 and TOCN-310 of different aspect ratios. Figure 19.4 shows the transparent TOCN-93/PS and TOCN-310/PS nanocomposite films. Results revealed that TOCNs provide a superior reinforcement effect even at low addition levels as a result of the unique characteristics of TOCNs such as high aspect ratio, high crystallinity, and nanodispersibility in the PS matrix. Both the tensile strength and elastic modulus were increased by TOCN addition (Fig. 19.5). The TOCN-310/PS nanocomposite films had a tendency to display a higher tensile strength and

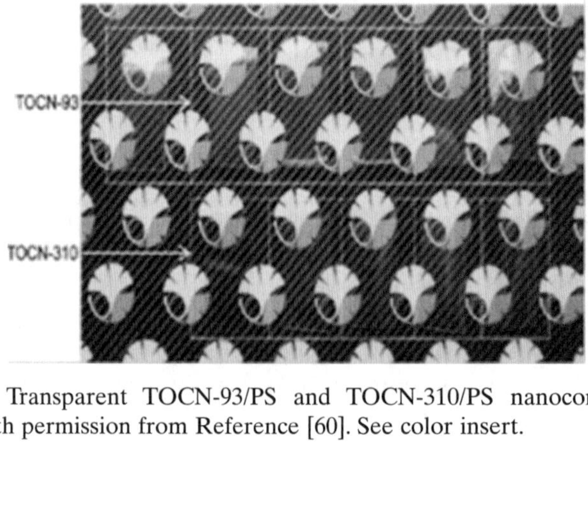

Figure 19.4 Transparent TOCN-93/PS and TOCN-310/PS nanocomposite films. Reprinted with permission from Reference [60]. See color insert.

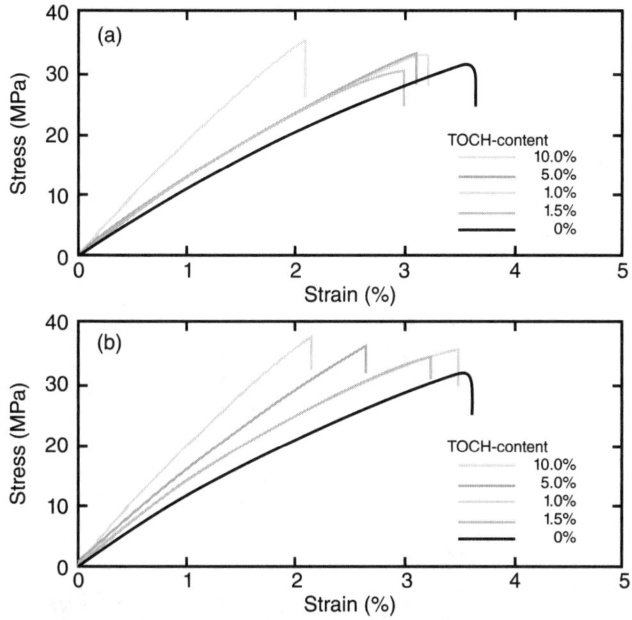

Figure 19.5 Stress–strain curves of (a) TOCN-93/PS and (b) TOCN-310/PS nanocomposite. Reprinted with permission from Reference [60].

elastic modulus than those of the TOCN-93/PS films, and the nanocomposite film containing 10 wt% TOCN-310 exhibited a 91% higher elastic modulus and 18% higher tensile strength than the neat PS, whereas the elongation at break decreased to 31% lower than that of the PS.

Self-standing films with sufficient light transparency and flexibility were prepared from softwood and hardwood TOCN/water dispersions have been reported by Fukuzumi et al. [61]. The tensile strengths and elastic moduli of the films were 200–300 MPa and 6–7 GPa, respectively, which were higher than those of cellophane films (Fig. 19.6). Both the tensile strength and elastic modulus of a PVA film were remarkably improved by 20% TOCN addition. Atomic force microscopy (AFM) observation of TOCN films showed that the film surface consists of randomly assembled nanofibers (Fig. 19.7). Results confirmed that distribution control of individualized TOCN to obtain more homogeneous distribution in a composite is important to achieve the effect of high aspect ratio TOCNs on the mechanical properties.

Trovatti et al. [62] prepared novel bionanocomposite films with improved thermal and mechanical properties by casting water-based suspensions of pullulan and nanofibrillated cellulose. The effect of the addition of glycerol, as a plasticizer, on the properties of the materials was also evaluated. Results revealed that the developed bionanocomposites exhibit considerable

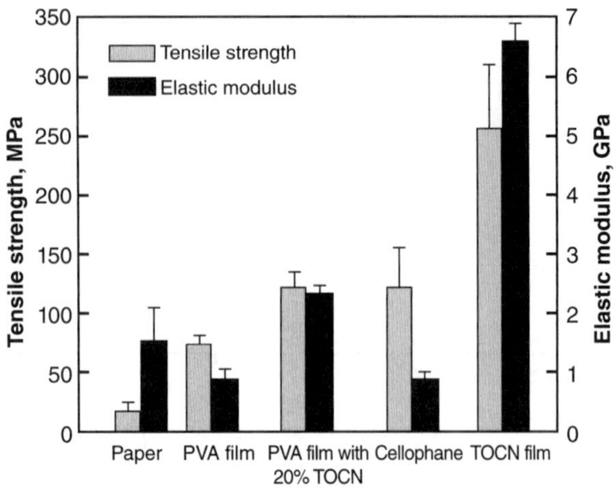

Figure 19.6 Tensile strength and elastic modulus of TOCN film, 20% TOCN/PVA composite film, and others. Reprinted with permission from Reference [61].

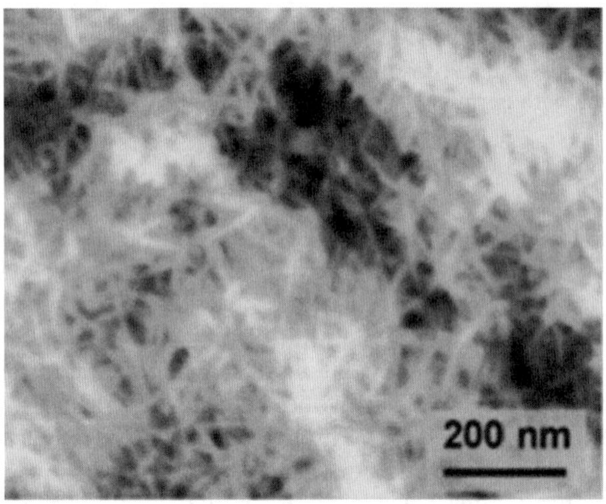

Figure 19.7 AFM image of the surface of a nanofiber film prepared from TEMPO-oxidizedsoftwood cellulose. Reprinted with permission from Reference [61].

improvements in mechanical properties (increments of up to 5500% and 8000% in Young's modulus and tensile strength, respectively, for films plasticized with glycerol)when compared to the unfilled pullulan films.

Biobased green composites with high performance were prepared from PLA and microcrystalline cellulose (MCC) fibers grafted with L-lactic acid oligomers (g-MC) have been reported by Xiao et al. [63]. The improved compatibility between g-MC and PLA caused by grafting results in more enhanced

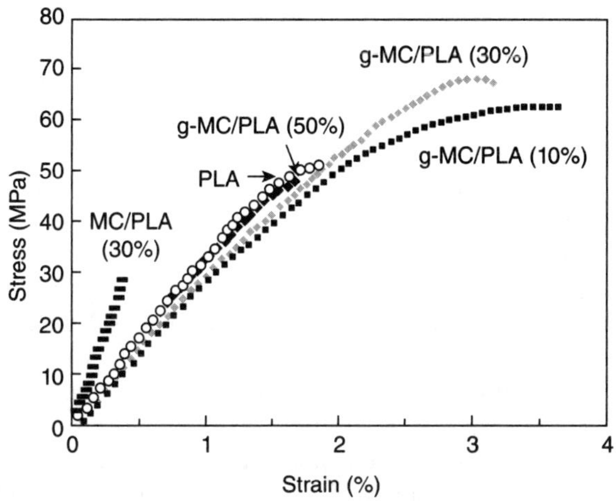

Figure 19.8 Typical stress–train curves of the pure PLA, the MC/PLA composite, and the g-MC/PLA composites. Reprinted with permission from Reference [63].

mechanical properties than pure PLA, with a high tensile strength of 70 MPa and a higher elongation at breakage. The stress/strain properties of the composites are shown in Figure 19.8. The examples cited above revealed that the application of nanocellulosic fillers improves polymer mechanical properties such as tensile strength and modulus in a more efficient manner than is achieved in conventional micro- or macrocomposite materials.

19.2.2 Flexural and Impact Properties

The flexural stiffness is a criterion of measuring deformability. The flexural stiffness of a structure is based on two essential properties: the first is the elastic modulus (stress per unit strain) of the material that composes it, and the second is the moment of inertia, a function of the cross-sectional geometry.

Impact strength is the ability of a material to resist fracture under stress applied at high speed. The impact strength of the composite depends on different grades of the fibers in use and is inversely proportional to its adhesion mechanism to the matrix. At impact the fiber is accompanied by three primary mechanisms, namely debonding, pull out, and fracture. It may be assumed that the strain energy that is released by fiber debonding and fracture is proportional to the fiber length and diameter. It may thus be concluded better fiber/matrix adhesion results in its worst impact strength. The following section highlights some recent studies on the flexural and impact properties of nanocellulosic fillers as a reinforcing material in bionanocomposite production.

Takagi and Asano [64] investigated the effect of processing conditions on the mechanical properties and internal microstructures of composites consisting of a dispersion-type biodegradable resin made from esterified starch and CNFs. All samples with nanofiber loading of 70 wt% were prepared by hot-pressing at 140°C and pressures of 10–50 MPa. It was found that the density of the composites increased with increased molding pressure. Density was used as an indicator for the mechanical strength of the composites. Similar densities were measured for vacuum-treated and extra-stirred samples, the latter showing significantly higher flexural strength, which was explained by difference between their internal microstructure and fiber dispersion.

A high-strength elastomeric nanocomposite has successfully been prepared by dispersing MC in a polyurethane matrix [65]. The resulting nanocomposites show increased strain-to-failure in addition to increased stiffness and strength compared with the unfilled polyurethane. The optimal composite contained 5 wt% cellulose. The average true strength for this composition was 257 MPa, compared with 39 MPa for the neat polyurethane, which showed the highest strain-to-failure. The improvements of stiffness, strength, as well as strain-to-failure are believed to be due to good interaction, by both covalent and hydrogen bonds, between the polyurethane and the cellulose nanofibrils.

Huang and Netravali investigated the flexural properties of biodegradable and environment friendly green composites based on micro/nanosized bamboo fibrils. Results revealed that these composites possessing moderate strength and stiffness [66]. Recently novel aerogels (or aerocellulose) based on all-cellulose composites were prepared by partially dissolving MCC in an 8 wt% LiCl/DMAc solution [67]. Cellulose gels were precipitated and then processed by freeze-drying to maintain the openness of the structure. The density of aerocellulose increased with the initial cellulose concentration and ranged from 116 to 350 kg/m^3. Aerocellulose with relatively high mechanical properties were successfully produced. The flexural strength and modulus of the aerocellulose was measured up to 8.1 and 280 MPa, respectively. Generally, the specific flexural strength and stiffness increase with the content of highly crystalline cellulose.

In order to achieve improved mechanical properties in polymer nanocomposites, good filler–matrix interaction is essential. A challenge associated with using nanocellulose in composites is the lack of compatibility with hydrophobic polymers and various chemical modification methods such as esterification, cationization, carboxylation, silylation, and polymer grafting have been explored in order to address this hurdle (Fig. 19.9) [68–73]. Most of these focused on the improvement of its dispersability and compatibility in different solvents or matrices that are suitable in the production of nanocomposites. For instance, Nakahara et al. [74] suggested an interesting method to overcome the problem of nanocellulose–hydrophobic polymer incompatibility. These authors simultaneously cut conifer kraft pulp fibers and PLA fibers into 1-mm length and dispersed them in water. The mixture was then treated in a refiner and a homogenizer to achieve the desired nanoscale structure, followed by filtering, pulverizing, and finally injection molding. The resulting test piece showed a flexural modulus of 5.5 GPa and a bending strength of 60 MPa.

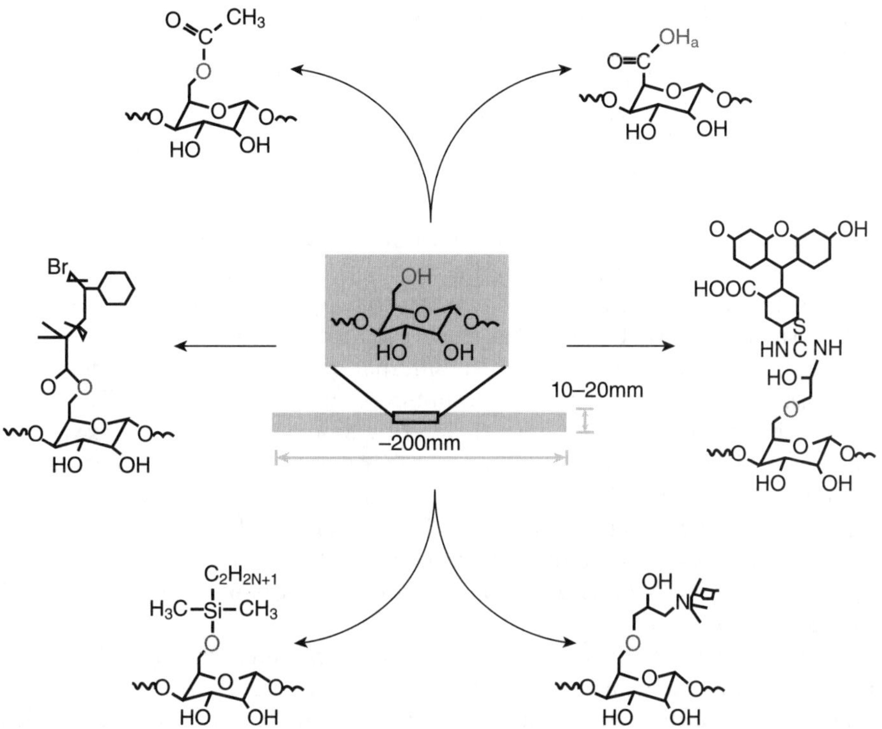

Figure 19.9 Schematic diagram illustrating the various types of chemical modifications on NCC surface. Reprinted with permission from Reference [73].

In another study, Nakahara [75] grafted nanocellulose with either PVOH or PLA, followed by dewatering and kneading with PLA. The resulting materials were then compression molded. The product possessed flexural modulus of 5.1–5.7 GPa, flexural strength of 100–105 MPa, and an impact strength of ~26 J/m. Hashiba [76] succeeded in fabricating a thermoplastic resin composite by mixing an aqueous dispersion of nanocellulose with PLA resin in an agitator. The mixture was then pulverized and injection molded to give a test piece with high flexural strength and modulus (110 MPa and 4 GPa, respectively).

19.2.3 Thermomechanical Properties

Thermal degradation of cellulose materials or the reduction in mechanical properties at elevated temperatures is one of the major issues that limit nanocellulose applications. The onset of thermal degradation of CNs typically occurs at ~200–300°C and provides an upper limit to the application and processing temperatures appropriate for nanocellulose-based products. However, if CNs have been chemically modified (e.g., sulfate esters introduced during hydrolysis by H_2SO_4 or carboxylation via TEMPO-based oxidation) the onset temperature of thermal degradation will likely change. For example, for NFC neat and modified films, Fukuzumi et al. [61] showed that the onset of thermal degradation, as measured by thermal gravimetric analysis (TGA) of modified films made from TEMPO-oxidized NFCs were ~100°C lower than those of neat NFC films (~300°C).

19.2.3.1 Dynamic Mechanical Thermal Analysis It has been the most reported technique for evaluating the response of the mechanical properties of cellulose nanostructures and composites to temperature. Dynamic mechanical thermal analysis (DMTA) has been used to characterize changes in the loss and storage moduli (E′), and the internal friction factor as a function of increasing temperature, typically within the range of 100 to ~200°C. The internal friction factor (loss modulus divided by storage modulus), assesses the damping of the material, while the storage moduli assesses the stiffness of the material. DMTA has been used to test nanocellulose-reinforced polymer matrix composites neat CN films [50, 77] and modified/composite CN films [78]. Recent studies on the thermal stability of CN-reinforced polymer matrix composites, neat CN films, and modified/composite CN films are briefly summarized below.

Fujisawa et al. [60] investigated the storage modulus of TOCN/PS nanocomposite films was characterized by DMA analysis (Fig. 19.10). DMA showed that the storage modulus of the TOCN/PS films increased significantly with TOCN content above the glass transition temperature of PS by the formation of an interfibrillar network structure of TOCNs in the PS matrix, based on percolation theory. The modulus of nanocomposite films maintained high values of up to ~230°C, whereas the neat PS fractured at ~150°C. The dimensional stability of PS subjected to temperature was enhanced significantly owing to the TOCN–matrix or TOCN–TOCN interactions. Compared with other fillers such as carbon nanotubes, TOCNs provide a superior reinforcement effect even at low addition levels.

The DMTA analysis of novel bionanocomposites using CNFs as reinforcement in PLA using acetone have been reported [49]. Results showed that the storage modulus of the composites was kept constant above the glass transition temperature of the matrix polymer. In another study, Suryanegara et al. [50] applied the same method but then exchanged acetone for dichloromethane and showed that the resulting nanocomposites had improved storage modulus when compared to neat PLA.

Sehaqui et al. [34] studied the dynamic thermal mechanical properties of the wood fiber/NFC composites. Three materials were studied, namely the reference biocomposite (0% NFC), the 5% NFC composite, and the 100% NFC nanopaper. The level of the E′ data at room temperature is similar to the static tensile moduli. Results revealed that cellulose nanocomposites can be very stiff under conditions of low mobility and exhibited high modulus values at low temperatures. Moduli are also well preserved until temperatures as high

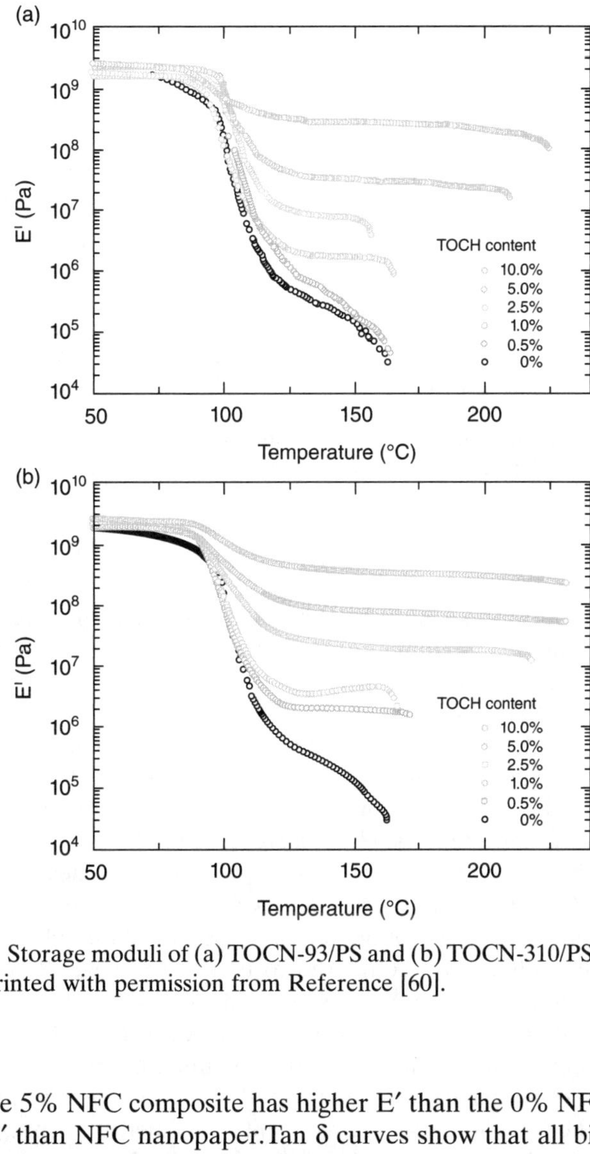

Figure 19.10 Storage moduli of (a) TOCN-93/PS and (b) TOCN-310/PS nanocomposite films. Reprinted with permission from Reference [60].

as 200°C. The 5% NFC composite has higher E′ than the 0% NFC composite and lower E′ than NFC nanopaper.Tan δ curves show that all biocomposites have three different relaxations. Data for storage modulus E′ and tan δ are presented in Figure 19.11.

Xiao et al. [63] investigated the temperature dependence of the dynamic mechanical properties of the composite films with 10 wt% and 30 wt% MC fibers grafted with L-lactic acid oligomers (g-MC). It has been found that the storage modulus (E′) decreases sharply at glass transition temperature (Tg). Both PLA and the g-MC/PLA composites show an increase in E′ at ~100°C. This is probably due to the crystallization of PLA that occurs in that temperature range. The presence of PLA crystals in g-MC/PLA composites and PLA

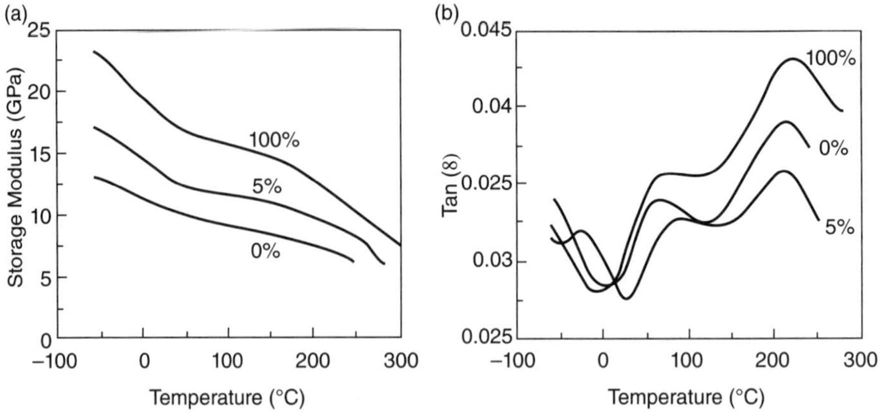

Figure 19.11 Storage modulus E′ (a) and tan δ (b) as a function of temperature for the 0% NFC, 5% NFC, and 100% NFC materials. Reprinted with permission from Reference [34].

raises their moduli. Figure 19.12a shows the storage modulus (E′) of the samples as a function of temperature. From Figure 19.12a, the crystallization temperature (Tc) of the g-MC/PLA composites are in the range of 90–100°C, while the Tc of PLA is in the range of 105–110°C. The loss factor (tan δ), which characterizes the length of mobile chain segments, changes inversely with E′ as temperature changes (Fig. 19.12b). As the Tg occurs, the mobility of the PLA chains in the composites increases, as reflected by the rapid increase of tan δ at 50–60°C and then it slightly decreases at ~100°C.

Cho et al. [55] investigated the thermomechanical properties of nanocellulose-reinforced PVA nanocomposites by DMA analysis. The DMA result shows a significant increase of the storage modulus (E′) of about 74% at the 3 wt% nanocellulose loading level compared to that of the neat PVA film at 25°C (Fig. 19.13). Figure 19.14 shows the tan δ of the nanocomposites reinforced with various amounts of the nanocellulose as a function of temperature. As the nanocellulose content increased, the maximum tan δ slightly increased at 1 wt% level and then decreased up to the 5 wt% content followed by a decrease at the 7 wt% content. This result confirmed that that a low tensile strength results in a decreased E′, which subsequently increases the tan δ.

Dynamic mechanical analysis (DMA) of nanocomposites from wheat straw nanofibers and TPS from modified potato starch has been reported. Results revealed that the Tg of the nanocomposites was shifted to higher temperatures with respect to the pure TPS [47]. Lu et al. [79] observed no further improvement of mechanical properties when MFC was applied above 10 wt%. In their study, MFC suspension was added at 1 wt%, 5 wt%, 10 wt%, and 15 wt% loadings to the PVOH matrix. A steady increase in film modulus and strength was observed until a plateau was reached at 10 wt% MFC. DMA showed an

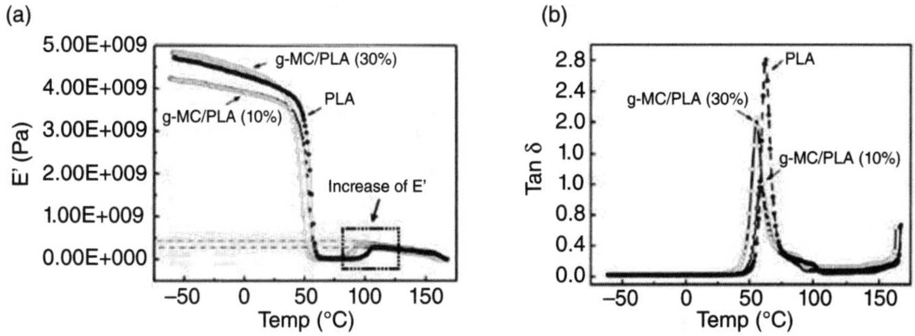

Figure 19.12 Dynamic mechanical properties of the pure PLA and the g-MC/PLA composites with different g-MC contents. (a)Temperature dependence of E′; (b) Temperature dependence of tan δ. Reprinted with permission from Reference [63].

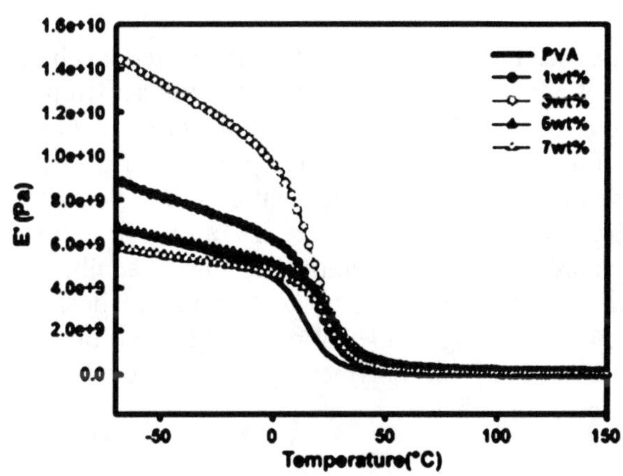

Figure 19.13 Storage tensile modulus E′ versus the temperature (°C) for nanocellulose-reinforced PVA nanocomposites. Reprinted with permission from Reference [55].

increase of storage tensile modulus in the glassy state with increasing MFC content. The thermal stability of the PVOH composite films was slightly increased with the addition of MFC.

Many studies on tensile, flexural, impact, and thermomechanical properties highlighted in this chapter gave an insight into the effect of processing conditions, chemical modification, and types of matrices used for composites on the mechanical performance of cellulose bionanocomposites. Tensile, flexural, and

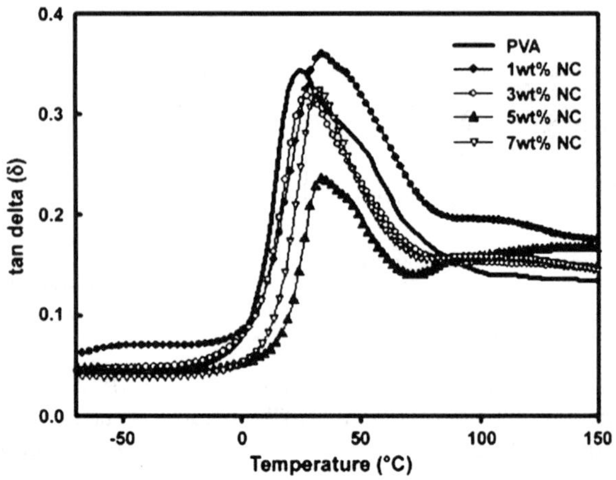

Figure 19.14 Tan δ as the function of the temperature (°C) for nanocellulose-reinforced PVA nanocomposites. Reprinted with permission from Reference [55].

impact properties are the most commonly investigated mechanical properties of cellulose nanofiber-reinforced biocomposites. Impact strength is one of the undesirable weak points of these materials in terms of mechanical performance. Besides these tensile, flexural and impact properties, dynamic mechanical behavior is also investigated. Interfacial adhesion between natural fibers and matrix will remain the key issue in terms of overall performance, since it dictates the final properties of the composites. Improved fiber matrix adhesion and thereby improved mechanical properties to the desired level can be accomplished by engineering a superior processing condition, in preparing the biocomposites, by altering the polymer architecture of the matrix or by the surface treatment of the fiber.

19.3 CONCLUSION

Nanocellulose is a promising renewable biomaterial that can be used as a reinforcing component in high performance bionanocomposites. The combination of biodegradable nanocellulose and biodegradable, renewable polymers is particularly attractive from an environmental point of view. It is shown that nanocellulose has a distinct advantage for preparing nanocomposites with outstanding thermal and mechanical properties. Recent studies on tensile, flexural, impact, and thermomechanical properties highlighted in this chapter gave an insight into the many potential applications of nanocellulosic fillers as a reinforcing material for bionanocomposite production. In order to achieve improved mechanical properties in polymer nanocomposites, good filler–

matrix interaction is essential because it dictates the overall performance of the composites. Consequently, more research targeting novel, environmentally friendly methods of modification, as well as an understanding of the mechanism of reactions occurring at the CNF–polymer matrix interface, is now required.

REFERENCES

[1] Eichhorn, S.J. (2011) Cellulose nanowhiskers: Promising materials for advanced applications. *Soft Matter*, *7*(2), 303–315.

[2] Eichhorn, S.J., Dufresne, A., Aranguren, M., Marcovich, N.E., Capadona, J.R., Rowan, S.J., Weder, C., Thielemans, W., Roman, M., Renneckar, S., Gindl, W., Veigel, S., Keckes, J., Yano, H., Abe, K., Nogi, M., Nakagaito, A.N., Mangalam, A., Simonsen, J., Benight, A.S., Bismarck, A., Berglund, L.A., and Peijs, T. (2010) Review: Current international research into cellulose nanofibers and nanocomposites. *Journal of Material Science*, *45*(1), 1–33.

[3] Habibi, Y., Lucia, L.A., and Rojas, O.J. (2010) Cellulose nanocrystals: Chemistry, self-assembly, and applications. *Chemical Reviews*, *110*(6), 3479–3500.

[4] Sunday, A.O., Lawal, G.I., Sambo, B.A., and Emmanuel, A.I. (2012) Review of green polymer nanocomposites. *Journal of Minerals and Materials Characterization and Engineering*, *11*(4), 385–416.

[5] Abdul Khalil, H.P.S., Bhat, A.H., and IreanaYusra, A.F. (2012) Green composites from sustainable cellulose nanofibrils: A review. *Carbohydrate Polymers*, *87*(2), 963–979.

[6] Siro, I., Plackett, D. (2010) Microfibrillated cellulose and new nanocomposite materials: A review. *Cellulose*, *17*(3), 459–494.

[7] Faruk, O., Bledzki, A.K., Fink, H.P., and Sain, M. (2012) Biocomposites reinforced with natural fibers: 2000-2010. *Progress in Polymer Science*, *37*(11), 1552–1596.

[8] Satyanarayana, K.G., Arizaga, G.G.C., and Wypych, F. (2009) Biodegradable composites based on lignocellulosic fibers—An overview. *Progress in Polymer Science*, *34*(9), 982–1021.

[9] Venkateshwaran, N., Elayaperumal, A. (2010) Banana fiber reinforced poly-mer composites—A review. *Journal of Reinforced Plastics and Composites*, *29*(15), 2387–2396.

[10] John, M.J., Thomas, S. (2008) Review-biofibres and biocomposites. *Carbohydrate Polymers*, *71*(3), 343–364.

[11] Moon, R.J., Martini, A., Nairn, J., Simonsen, J., and Youngblood, J. (2011) Cellulose nanomaterials review: Structure, properties and nanocomposites. *Chemical Society Review*, *40*(7), 3941–3994.

[12] Ioelowich, M. (2008) Cellulose as a nanostructured polymer: A short review. *Bioresources*, *3*(4), 1403–1418.

[13] Siqueira, G., Bras, J., and Dufresne, A. (2010) Cellulosic bionanocomposites: A review of preparation, properties and applications. *Polymer*, *2*(4), 728–765.

[14] Isogai, A., Saito, T., and Fukuzumi, H. (2011) TEMPO-oxidized cellulose nanofibers. *Nanoscale*, *3*, 71–85.

[15] Pickering, K. (2008) *Properties and Performance of Natural-Fibre Composites*. Cambridge, UK: Woodhead Publishing.

[16] Thomas, S., Pothan, L.A., eds. (2009) *Natural Fibre Reinforced Polymer Composites: From Macro to Nanoscale*. Philadelphia: Old City Publishing.

[17] Staiger, M.P., Tucker, N. (2008) Natural-fibre composites in structural applications. In: *Properties and Performance of Natural-Fibre Composites*, ed. K. Pickering. Cambridge, UK: Woodhead Publishing, pp. 269–300.

[18] Mohanty, A.K., Misra, M., and Drzal, L.T., eds. (2005) *Natural Fibers, Biopolymers, and Biocomposites*. Boca Raton, FL: Taylor & Francis Group.

[19] Kalia, S., Kaith, B.S., and Kaur, I., eds. (2011) *Cellulose Fibers: Bio- and Nano-Polymer Composites. Green Chemistry and Technology*. Heidelberg Dordrecht, London, New York: Springer.

[20] Ummartyotin, S., Juntaro, J., Sain, M., and Manuspiya, H. (2012) Development of transparent bacterial cellulose nanocomposite film as substrate for flexible organic light emitting diode (OLED) display. *Industrial Crops and Products*, *35*(1), 92–97.

[21] Berglund, L.A., Peijs, T. (2010) Cellulose biocomposites–From bulk moldings to nanostuctured systems. *MRS Bulletin*, *35*, 201–207.

[22] Ly, B., Thielemans, W., Chaussy, D., and Belgacem, M.N. (2008) Surface functionalization of cellulose fibers and their incorporation in renewable polymeric matrices. *Composite Science and Technology*, *68*, 3193–3201.

[23] Mangalam, A.P., Simonsen, J., and Benight, A.S. (2009) Cellulose/DNA hybrid nanomaterials. *Biomacromolecules*, *10*, 497–504.

[24] Mattoso, L.H.C., Medeiros, E.S., Baker, D.A., Avloni, J., Wood, D.F., and Orts, W.J. (2009) Electrically conductive nanocomposites made from cellulose nanofibrils and polyaniline. *Journal of Nanoscience and Nanotechnology*, *9*, 2917–2922.

[25] Hongyang, M., Hsiao, B.S., and Chu, B. (2012) Ultrafine cellulose nanofibers as efficient adsorbents for removal of UO2 $^{2+}$ in water. *ACS Macro Letters*, *1*, 213–216.

[26] Henriksson, M., Berglund, L.A., Isaksson, P., Lindstrom, T., and Nishino, T. (2008) Cellulose nanopaper structures of high toughness. *Biomacromolecules*, *9*, 1579–1585.

[27] Zhou, Q., Malm, E., Nilsson, H., Larsson, P.T., Iversen, T., Berglund, L.A., and Bulone, V. (2009) Nanostructured biocomposites based on bacterial cellulosic nanofibers compartmentalized by a soft hydroxyethylcellulose matrix coating. *Soft Matter*, *5*, 4124–4130.

[28] Paakko, M., Vapaavuori, J., Silvennoinen, R., Kosonen, H., Ankerfors, M., Lindstrom, T., Berglund, L.A., and Ikkala, O. (2008) Long and entangled native cellulose I nanofibers allow flexible aerogels and hierarchically porous templates for functionalities. *Soft Matter*, *4*, 2492–2499.

[29] Sehaqui, H., Salajkova, M., Zhou, Q., and Berglund, L.A. (2010) Mechanical performance tailoring of tough ultra-high porosity foams prepared from cellulose I nanofiber suspensions. *Soft Matter*, *6*, 1824–1832.

[30] Yano, H., Sugiyama, J., Nakagaito, A.N., Nogi, M., Matsuura, T., and Hikita, M. (2005) Optically transparent composites reinforced with networks of bacterial nanofibers. *Advanced Materials*, *17*, 153–155.

[31] Nogi, M., Iwamoto, S., Nakagaito, A.N., and Yano, H. (2009) Optically transparent nanofibre paper. *Advanced Materials*, *21*, 1595–1598.

[32] Svagan, A.J., Samir, M.A.S.A., and Berglund, L.A. (2008) Biomimetic foams of high mechanical performance based on nanostructured cell walls reinforced by native cellulose nanofibers. *Advanced Materials*, *20*, 1263–1269.

[33] Olsson, R.T., Samir, M.A.S.A., Salazar-Alvarez, G., Belova, L., Ström, V., Berglund, L.A., Ikkala, O., Nogués, J., and Gedde, U.W. (2010) Making flexible magnetic aerogels and stiff magnetic nanopaper using cellulose nanofibers as templates. *Nature Nanotechnology*, *5*, 584–588.

[34] Sehaqui, H., Allais, M., Zhou, Q., and Berglund, L.A. (2011) Wood cellulose biocomposites with fibrous structures at micro- and nanoscale. *Composites Science and Technology*, *71*, 382–387.

[35] Okahisa, Y., Yoshida, A., Miyaguchi, S., and Yano, H. (2009) Optically transparent wood–cellulose nanocomposite as a base substrate for flexible organic light-emitting diode displays. *Composit Science and Technology*, *69*, 1958–1961.

[36] Mahadeva, S.K., Yun, S., and Kim, J. (2011) Flexible humidity and temperature sensor based cellulose-polypyrrolenanocomposite. *Sensors and Actuators. A Physical*, *165*, 194–199.

[37] Samir, M.A.S.A., Alloin, F., and Dufresne, A. (2005) Review of recent research into cellulosic whiskers, their properties and their application in nanocomposite field. *Biomacromolecules*, *6*, 612–626.

[38] Dufresne, A. (2006) Comparing the mechanical properties of high performance polymer nanocomposites from biological sources. *Journal of Nanoscienceand Nanotechnology*, *6*, 322–330.

[39] de Rodriguez, N.L.G., Thielemans, W., and Dufresne, A. (2006) Sisal cellulose whiskers reinforced polyvinyl acetate nanocomposites. *Cellulose*, *13*, 261–270.

[40] Samir, M.A.S.A., Alloin, F., Paillet, M., and Dufresne, A. (2004) Tangling effect in fibrillated cellulose reinforced nanocomposites. *Macromolecules*, *37*, 4313–4316.

[41] Bendahou, A., Habibi, Y., Kaddami, H., and Dufresne, A. (2009) Physico-chemical characterization of palm from Phoenix dactylifera-L, preparation of cellulose whiskers and natural rubber-based nanocomposites. *Journal of Biobased Materials and Bioenergy*, 81–90.

[42] Siqueira, G., Bras, J., and Dufresne, A. (2009) Cellulose whiskers versus microfibrils: Influence of the nature of the nanoparticle and its surface functionalization on the thermal and mechanical properties of nanocomposites. *Biomacromolecules*, *10*, 425–432.

[43] deMenezes, J.A., Siqueira, G., Curvelo, A.A.S., and Dufresne, A. (2009) Extrusion and characterization of functionalized cellulose whiskers reinforced polyethylene nanocomposites. *Polymer*, *50*, 4552–4563.

[44] Ljungberg, N., Bonini, C., Bortolussi, F., Boisson, C., and Heux, L. (2005) New nanocomposite materials reinforced with cellulose whiskers in atactic polypropylene: Effect of surface and dispersion characteristics. *Biomacromolecules*, *6*, 2732–2739.

[45] Kaushik, A., Singh, M., and Verma, G. (2010) Green nanocomposites based on thermoplastic starch and steam exploded cellulose nanofibrils from wheat straw. *Carbohydrate Polymers*, *82*, 337–345.

[46] Lu, Y., Weng, L., and Cao, X. (2006) Morphological, thermal and mechanical properties of ramie crystallites–reinforced plasticized starch biocomposites. *Carbohydrate Polymers*, *63*, 198–204.

[47] Alemdar, A., Sain, M. (2008) Biocomposites from wheat straw nanofibers: Morphology, thermal and mechanical properties. *Composite Science and Technology*, *68*, 557–565.

[48] Azeredo, H.M.C., Mattoso, L.H.C., Wood, D., Williams, T.G., Avena-Bustillos, R.J., and Mchugh, T.H. (2009) Nanocomposite edible films from mango puree reinforced with cellulose nanofibers. *Journal of Food Science*, *74*, 31–35.

[49] Iwatake, A., Nogi, M., and Yano, H. (2008) Cellulose nanofiber-reinforced polylactic acid. *Composite Science and Technology*, *68*, 2103–2106.

[50] Suryanegara, L., Nakagaito, A.N., and Yano, H. (2009) The effect of crystallization of PLA on the thermal and mechanical properties of microfibrillated cellulose-reinforced PLA composites. *Composite Science and Technology*, *69*, 1187–1192.

[51] Qu, P., Gao, Y., Wu, G., and Zhang, L. (2010) Nanocomposites of poly (lactic acid) reinforced with cellulose nanofibrils. *BioResources*, *5*, 1811–1823.

[52] Bondeson, D., Oksman, K. (2007) Dispersion and characteristics of surfactant modified cellulose whiskers nanocomposites. *Composite Interface*, *14*, 617–630.

[53] Cao, X., Dong, H., and Li, C.M. (2007) New nanocomposite materials reinforced with flax cellulose nanocrystals in waterborne polyurethane. *Biomacromolecules*, *8*, 899–904.

[54] Nordqvist, D., Idermark, J., and Hedenqvist, M.S. (2007) Enhancement of the wet properties of transparent chitosan-acetic-acidsalt films using microfibrillated cellulose. *Biomacromolecules*, *8*, 2398–2403.

[55] Cho, M.J., Park, B.D. (2011) Tensile and thermal properties of nanocellulose-reinforced poly (vinyl alcohol) nanocomposites. *Journal of Industrial and Engineering Chemistry*, *17*, 36–40.

[56] Leitner, J., Hinterstoisser, B., Wastyn, M., Keckes, O., and Gindl, W. (2007) Sugar beet cellulose nanofibril-reinforced composites. *Cellulose*, *14*, 419–425.

[57] Bhatnagar, A., Sain, M. (2005) Processing of cellulose nanofiber-reinforced composites. *Journal of Reinforced Plastics and Composites*, *24*, 1259–1268.

[58] Wang, B., Sain, M. (2007) Dispersion of soybean stock-based nanofiber in a plastic matrix. *Polymer International*, *56*, 538–546.

[59] Bruce, D.M., Hobson, R.N., Farrent, J.W., and Hepworth, D.G. (2005) High-performance composites from low-cost plant primary cell walls. *Composite Part A-Applied Science and Manufacturing*, *36*, 1486–1493.

[60] Fujisawa, S., Ikeuchi, T., Takeuchi, M., Saito, T., and Isogai, A. (2012) Superior reinforcement effect of TEMPO-oxidized cellulose nanofibrils in polystyrene matrix: Optical, thermal, and mechanical studies. *Biomacromolecules*, *13*(7), 2188–2194.

[61] Fukuzumi, H., Saito, T., Wata, T., Kumamoto, Y., and Isogai, A. (2009) Transparent and high gas barrier films of cellulose nanofibers prepared by TEMPO-mediated oxidation. *Biomacromolecules*, *10*, 162–165.

[62] Trovatti, E., Fernandes, S.C.M., Rubatat, L., da Silva, D.P., Freire, C.S.R., Silvestre, A.J.D., and Neto, C.S. (2012) Pullulan-nanofibrillated cellulose composite films with improved thermal and mechanical properties. *Composite Science and Technology*, *72*(13), 1556–1561.

[63] Xiao, L., Mai, Y., He, F., Yu, L., Zhang, L., Tang, H., and Yang, G. (2012) Bio-based green composites with high performance from poly(lactic acid) and surface-modified microcrystalline cellulose. *Journal of Material Chemistry*, *22*, 15732–15739.

[64] Takagi, H., Asano, A. (2008) Effects of processing conditions on flexural properties of cellulose nanofiber reinforced "green" composites. *Composite Part A-Applied Science and Manufacturing*, *39*, 685–689.

[65] Wu, Q., Henriksson, M., Liu, X., and Berglund, L.A. (2007) A high strength nanocomposite based on microcrystalline cellulose and polyurethane. *Biomacromolecules*, *8*(12), 3687–3692.

[66] Huang, X., Netravali, A. (2009) Biodegradable green composites made using bamboo micro/nanofibrils and chemically modified soy protein resin. *Composie Science and Technology*, *69*, 1009–1015.

[67] Duchemin, B. (2008) Structure, property and processing relationships of all-cellulose composites. PhD thesis: Mechanical Engineering University of Canterbury Christchurch, New Zealand.

[68] Gouss'e, C., Chanzy, H., Excoffier, G., Soubeyrand, L., and Fleury, E. (2002) Stable suspensions of partially silylated cellulose whiskers dispersed in organic solvents. *Polymer*, *43*, 2645–2651.

[69] Habibi, Y., Chanzy, H., and Vignon, M. (2006) TEMPO-mediated surface oxidation of cellulose whiskers. *Cellulose*, *13*, 679–687.

[70] Hasani, M., Cranston, E.D., Westman, G., and Gray, D.G. (2008) Cationic surface functionalisation of cellulose nanocrystals. *Soft Matter*, *4*, 2238–2244.

[71] Morandi, G., Heath, L., and Thielemans, W. (2009) Cellulose nanocrystals grafted with polystyrene chains through surface-initiated atom transfer radical polymerisation (SI-ATRP). *Langmuir*, *25*, 8280–8286.

[72] Braun, B., Dorgan, J.R. (2009) Single-step method for the isolation and surface functionalisation of cellulosic nanowhiskers. *Biomacromolecules*, *10*, 334–341.

[73] Peng, B.L., Dhar, N., Liu, H.L., and Tam, K.C. (2011) Chemistry and applications of nanocrystallinecellulose and its derivatives: A nanotechnology perspective. *Canadian Journal of Chemical Engineering*, *89*(5), 1191–1206.

[74] Nakahara, S., Saito, E., and Asada, T. (2008) Manufacture of microfibrillated cellulose-polymer composites with high mechanical strength, aqueous dispersions therefore, and manufacture of moldings comprising them. Jpn Kokai Tokkyo Koho; 2007-82938:10.

[75] Nakahara, S. (2008) Resin composite materials containing surface- treated microfibrillated cellulose (MFC) reinforcement, their manufacture, and their articles. Jpn Kokai Tokkyo Koho; 2007-17153:10.

[76] Hashiba, M. (2009) Thermoplastic resin compositions containing cellulose nanofibers with good bending properties. PCT Int Appl.; 2008-JP58502; 2007-195163:24.

[77] Svagan, A.J., Samir, M.A.S.A., and Berglund, L.A. (2007) Biomimetic polysaccharide nanocomposites of high cellulose content and high toughness. *Biomacromolecules*, *8*, 2556–2563.

[78] Henriksson, M., Berglund, L.A. (2007) Structure and properties of cellulose nanocomposite films containing melamine formaldehyde. *Journal of Applied Polymer Science*, *106*, 2817–2824.

[79] Lu, J., Wang, T., and Drzal, L.T. (2008) Preparation and properties of microfibrillated cellulose polyvinyl alcohol composite materials. *Compos Part A-Appled Science and Manufacturing*, *39*, 738–746.

CHAPTER 20

Review of Nanocellulosic Products and Their Applications

JOE ASPLER, JEAN BOUCHARD, WADOOD HAMAD, RICHARD BERRY, STEPHANIE BECK, FRANÇOIS DROLET, and XUEJUN ZOU

20.1 INTRODUCTION

20.1.1 Background and Scope of This Review

The growing demand for green and sustainable products made from natural and renewable raw materials has created a worldwide interest in the development of high performance, high value nanocellulosics. Like many other nanomaterials, nanocellulosic products have unique properties, with the potential to develop new products or significantly enhance the performance of existing products. One important difference from many other nanomaterials is that nanocellulosic products are made from natural raw materials that have the potential to be harvested in a sustainable manner.

Hundreds of papers and patents on micro- and nanocellulosic products have been published in the last 10 years alone. Table 20.1 defines the basic terms and products discussed in this review. It must be noted that there is considerable overlap in technical and commercial definitions. Many of the names are arbitrary distinctions. Discussions (and sales literature) for many products—even those considered to be proprietary—overlap with each other in terms of their properties and projected end-uses.

As noted in Table 20.1, "micro" cellulosic products are distinguished by particle sizes in the scale of many microns, and "nano" cellulosic products are distinguished by particle sizes in the scale of nanometers. Some products may be nanoscale in one or two dimensions only, so their classification is still

Biopolymer Nanocomposites: Processing, Properties, and Applications, First Edition. Edited by Alain Dufresne, Sabu Thomas, and Laly A. Pothan.

TABLE 20.1 Competing Micro- and Nanocellulosic Products. See Also Appendix A

Product	Preparation	Typical Size Range	Typical Aspect Ratio	Degree of Polymerization	Commercial Status and Market Size
Powdered cellulose	Mechanical milling of high purity cellulose fibers, with or without sodium hypochlorite and caustic soda	20–50 μm in diameter Up to 400 μm in length	Up to 20	Up to 1400	Long established
Microcrystalline cellulose (MCC)	Mild acid hydrolysis of cellulose fibers	50–300 μm	~1–3	150–350	Long established 50,000 tpy
Microfibrillated cellulose (MFC); cellulose fibers with nanofibrillation (CFN)	MCC or fibers with high-pressure homogenization, possibly aided by chemical or enzymatic treatment	10–100 nm wide, 1–100 μm long	Up to 100	Up to 1000	Recently commercialized
Nanofibrillated cellulose (NFC)	Enzymatic or chemical treatment of MCC or fibers; with homogenization	4–30 nm wide; 70 nm–5 μm long	Up to 100	Up to 900	Innventia (Sweden): Pilot plant inaugurated February 2011 ~100 kg/day [51] Tokyo University with Kao Corporation and Nippon Paper (scale unknown)
Nanocrystalline cellulose (NCC) (cellulose whiskers)	Strong acid hydrolysis with homogenization	3–5 nm wide elementary fibril level up to 200 nm long	~20	Up to 300	CelluForce (FPInnovations—Domtar) 3 kg/day—operational 1 tonne/day demonstration plant: in operation January 2012

somewhat open to debate. Furthermore, the term fibrillation itself causes problems of terminology. The entire process of pulp development, from the original native wood fibers, is one of progressive fibrillation, as shown in the illustrations to this chapter. Perhaps only the cellulose whiskers or nanocrystalline cellulose (NCC) particles are the only nonfibrillated materials—but only because these elementary rods have had all visibly fibrillar material removed by a combination of intensive chemical and physical action.

Several important technical reviews have also recently been published. A key review by Eichhorn et al. summarized the technical contributions from many different groups working in the field [1]. Hubbe et al. [2] covered nanocomposite systems, with particular emphasis on surface derivatization of cellulosic material to make it more compatible with dispersing media such as polymers. The most recent reviews cover the production and structure of the different forms of micro- and nanofibrillated cellulose (NFC), as well as their abilities to act as reinforcing agents [3–5]. In the most recent review, Peng et al. [6] described a comprehensive range of surface-modified NCC products, along with the potential of NCC to be used as a reinforcing filler or support for different products.

Yano [7] has summarized efforts in Japan. This review covered the work of several research groups, including universities (Kyoto, Tokyo), paper manufacturers (Oji, Nippon Paper), chemical companies, and others. The areas of research in Japan include microfibrillated cellulose (MFC), NFC, and NCC. Products under investigation include nanocellulose/polymer composites, biofilms, and packaging barrier layers. Another review co-authored by three of the key institutions carrying out research in the field (McGill University, Innventia, and the University of Jena) [8] summarized three basic forms of micro- and nanocellulose: MFC, NCC, and bacterial nanocellulose.

These reviews certainly provide a comprehensive summary of recent research findings in micro- and nanocellulosic materials. However, the focus has been mainly on producing various new materials and exploring their potential applications, and less on highlighting their differences in properties and assessing their real potential for commercial applications. The purpose of this review is not to repeat the wealth of recent technical material, but rather to compare and contrast the range of micro- and nanocellulosic products in terms of the broad continuum of properties, their current and projected markets, the extent to which new products may compete with each other, the extent to which a new product may replace an existing product, and the extent to which new cellulosic derivatives will create new product lines. This review is intended to provide background information for industry and governments to develop R&D strategies and make business decisions and as guidance for researchers in focusing their research efforts.

The future in expanding and maximizing forest resource value is illustrated schematically in Figure 20.1 [9]. The left side of the image shows the traditional "re-engineering for value" of the forest industry. By further disassembly of the wood into micro, nano, and fundamental chemical components as we move to

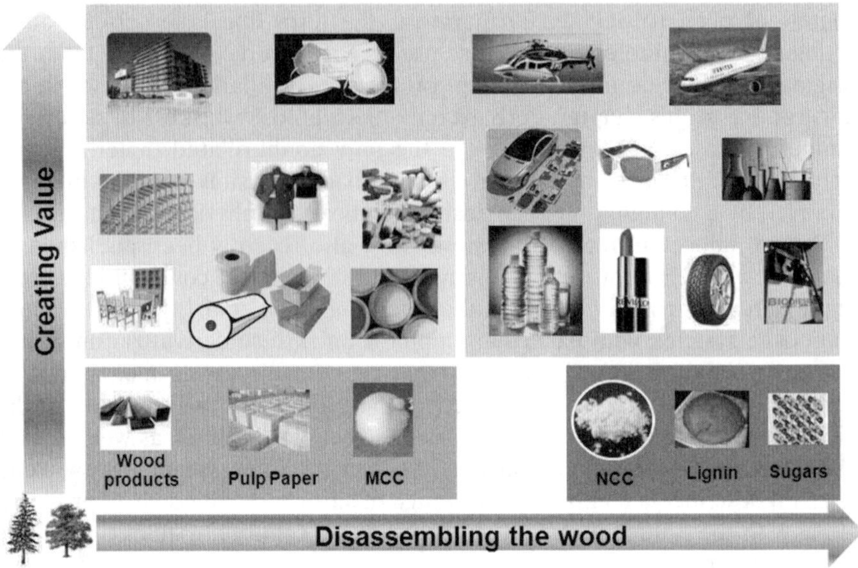

Figure 20.1 Expanding value by expanding the range of products from the forest industry.

the right of the image, we expand the applications and increase the potential value creation for the forest industry.

20.1.2 Brief Description of Nanocellulosic Products and How They Compare with Microcellulosic Products

Attempts are being made through the International Organization for Standardization (ISO) to regularize the situation [10], but these discussions are still in a very early stage. A recent article co-authored by the National Institute of Standards and Technology in the United States was also devoted to the techniques required for the metrology of nanoparticles [11]. This article noted the potential health hazards of carbon nanotubes (CNT) compared with the relatively benign nature of NCC.

In addition to significant differences in physical properties, the values for the degree of polymerization (DP) of the different forms of cellulose cover a very wide range (Table 20.1). DP depends on the source material (hardwood, softwood, cotton, etc.) and the degree or intensity of chemical and mechanical treatment. In general, a more severe chemical treatment yields a lower DP. A product with a higher aspect ratio will also tend to have a higher DP.

It is also important to note that production of various micro- and nanocellulosic products on the gram scale in a research environment is straightforward, using conventional wet chemistry combined with readily available dispersion and homogenization equipment. Many of these research products are far from commercial production. Appendix A gives a more detailed summary of worldwide research, development, and production of nanocellulosics.

Micro- and nanocellulosic products are distinguished by the following:

1. They are nearly pure cellulose, from tree fiber, plant, or bacterial sources, sometimes with surface derivatization.
2. They are prepared by a variety of treatments, from purely mechanical breakdown of high purity cellulose fibers to a combination of mechanical treatment with either chemical or enzymatic assistance.
3. They are distinguished in terms of particle size, shape, structure, and surface properties.
4. There exists a continuum of structure and size, from the still-recognizable fibrous structure of powdered and microcrystalline cellulose (MCC) to the fibrillar structure of MFC and NFC to the nanoscale, rod-like, elementary cellulose particles of NCC

Powdered cellulose and MCC have been in commercial use for many decades. Other products, such as MFC, are newly arrived commercial products still looking for markets. These markets might be as new products, or as substitutes for existing products. The newest products, such as NFC and NCC, are still under intensive development and are only beginning to look for commercial outlets. The increasing interest in nanocellulosic products and the large amount of funding provided by governments around the world are accelerating the development of these products.

The continuum of particle size and shape also extends to product properties. In some cases, the newer product may be in a position to replace an older product, but only if there is an advantage in price and/or performance. At the same time, newer products possess properties that can be used to create novel end-uses.

20.1.3 Summary of Patents in the Field

Figure 20.2a shows the results of a patent search using terms including nanofibrillated cellulose, nanocrystalline cellulose, and related terms, from 1997 to early December 2011. There has been a dramatic increase in the number of patents since 2008, reflecting the rapidly growing interest in nanocellulosic materials.

The numbers are somewhat different from those shown in an earlier review covering the period to mid-2009 [4]. The difference may have been due to different search criteria or due to the evolution of, for example, patent applications to final patents. No attempt was made to find patents on MCC, since this

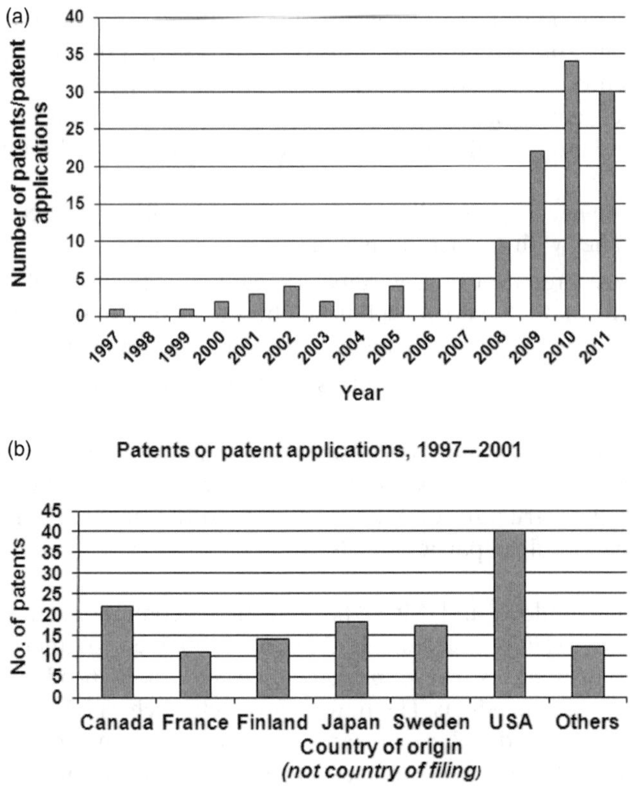

Figure 20.2 (a) Number of patents/patent applications, 1997–December 2011. (b) Patents by country of origin 1997–December 2011.

is a much older field, with hundreds of patents. Figure 20.2b breaks the patents down by country of origin (not country of filing).

Many patents are deliberately vague as to whether the product in question is a true nanomaterial. Given the nature of patent writing, it is typical for the patents to give the broadest possible size range of the product, from the true nanoparticle scale of 4–5 nm in cross section, up to hundreds of nanometers. As an example, US Patent Application 2010/0162926 [12] describes a cement board produced using NCC. Yet the text of the patent describes their so-called NCC as a product with a fiber width from 50 nm to 5 μm. These product dimensions are more in line with those of MFC.

Figure 20.3 shows a breakdown of the patent list by approximate end-use. Some patents are very clear as to their intended end-use. Other patents are deliberately vague, while patents without specific end-uses are listed as "preparation" patents. Two key areas targeted for applications are polymer composites and medical and cosmetic material, representing two extremes: high

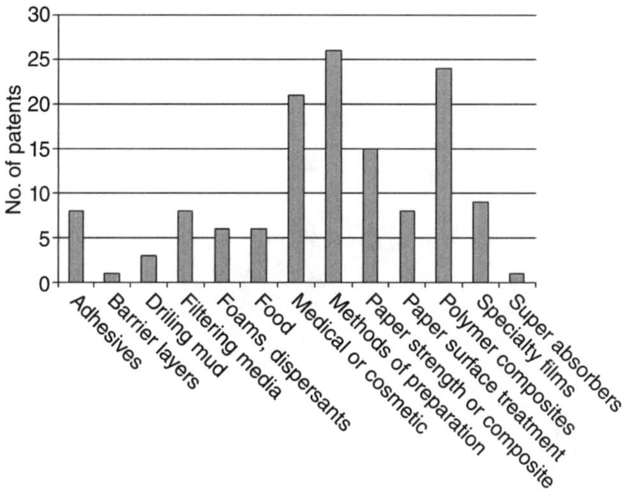

Figure 20.3 Number of patents by end-use, 1997–December 2011.

volume but low value products, and low volume but high value products. In the former case, small amounts of nanocellulosic products may be added as a reinforcement additive and in the latter case, nanocellulosic products are used as a major component.

20.2 MICRO- AND NANOCELLULOSIC PRODUCTS TODAY

Production of micro- and nanocellulosic products begins with high purity cellulose fibers. These include cotton and high purity bleached kraft and sulfite pulps.

Figure 20.4 [13] summarizes the aspect ratio versus the fiber diameter for key micro- and nanocellulosic products. The implications for this will be discussed in detail throughout this review. Of particularly note are the position of NCC and the related TEMPO-oxidized cellulose nanocrystals (TOCN) as very small cross-section, high aspect ratio materials.

Figure 20.5 shows relative scales of the products from the tree to the single fiber level to the micro level to the nano level, ending with the glucose polymer chains. Depending on the source, cellulose fibers typically have dimensions of approximately 50–100 μm in width and 2 mm in length. The cellulose fiber is composed of amorphous (disordered) zones and crystalline zones.

Figure 20.6 shows schematically how different nanocellulosic products can be produced from wood pulp. In particular, we note that a range of chemical treatments combined with mechanical treatments can be used to produce MFC, NFC, and NCC. Figure 20.7 shows the "ultimate" in cellulose nanostructure [14].

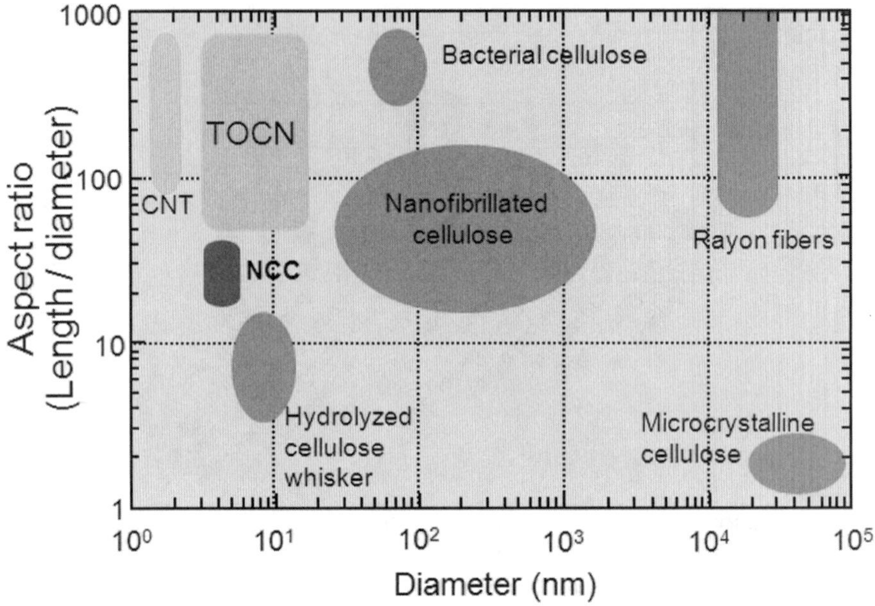

Figure 20.4 Aspect ratio versus diameter for key "micro" and "nano" products. TOCN, TEMPO-oxidized nanofibers; NCC; nanocrystalline cellulose; CNT, carbon nanotubes [9] (modified from Reference [13]). Reprinted with permission of the author.

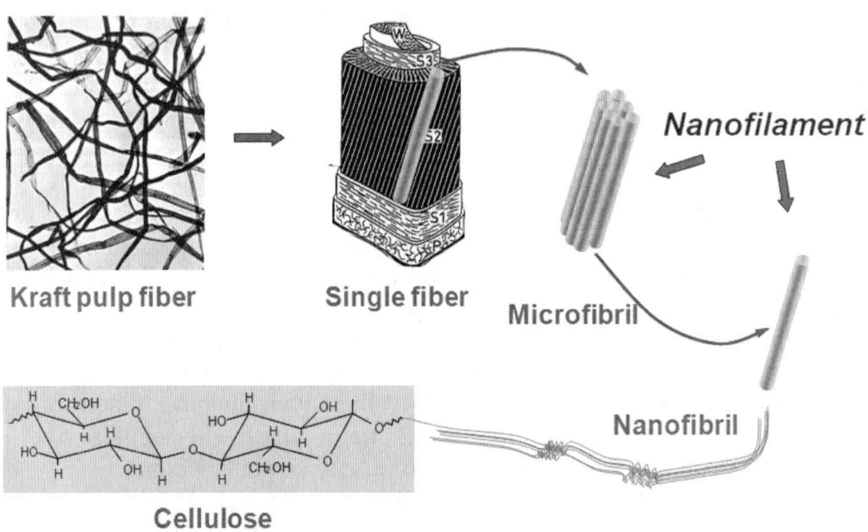

Figure 20.5 Cellulose structure, from the single fiber level to the fibril, microfibril, nanofibril, and cellulose polymer chains.

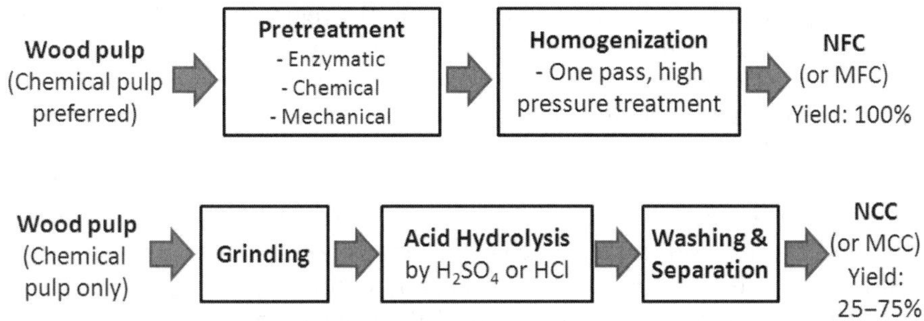

Figure 20.6 Manufacturing different nanocellulosic products from wood pulps. The yield is based on the dried pulp weight.

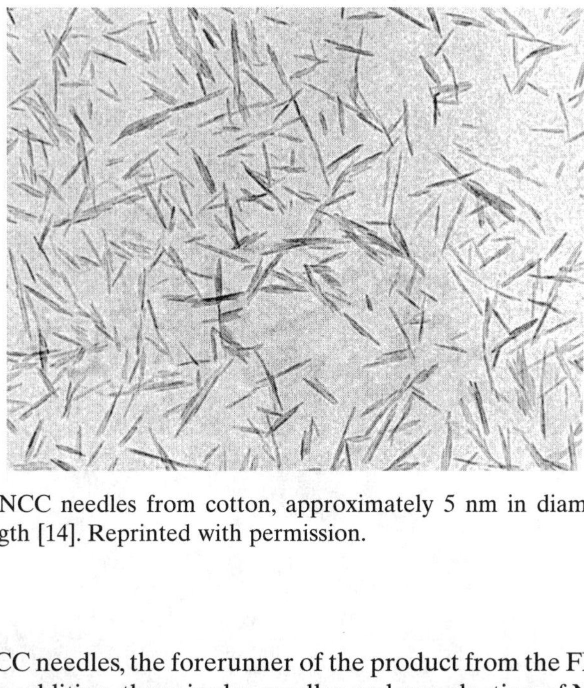

Figure 20.7 NCC needles from cotton, approximately 5 nm in diameter and 150–200 nm in length [14]. Reprinted with permission.

These are NCC needles, the forerunner of the product from the FPInnovations pilot plant. In addition, there is also smaller scale production of NCC (in gram quantities) at different universities, as shown in Appendix A.

20.2.1 Powdered Cellulose and Microcrystalline Cellulose

Powdered cellulose was the first "micro" cellulose product. As the name implies, the material is produced by mechanically grinding high purity cellulose fibers. The fibers may also be chemically treated as well (sodium hypochlorite plus sodium hydroxide), to assist in depolymerization of the cellulose.

Typically, powdered cellulose particles are in the region from 60 to 400 μm in length (depending on the grade), while preserving the typical fiber diameter of 30–50 μm. Powdered cellulose is still used as a binder and texturizer for food products and cosmetics, but has been significantly supplanted by MCC and other cellulose derivatives.

MCC has been known since the late 1800s. Also called hydrocellulose, MCC is produced by the controlled acid hydrolysis of high purity cellulose fibers. The acid hydrolysis dissolves much of the amorphous portion of the cellulose, but maintains the crystalline areas relatively untouched (e.g., [15]). This provides one distinguishing feature of MCC: the product retains the approximate appearance of the original cellulose fibers, if partially disintegrated, as seen in Figure 20.8 [16]. The reason is that the relatively low acid concentration and relatively low duration of the hydrolysis reaction (2.5N HCl, 15 minutes, at 105°C) only partially degrades the amorphous zones of the cellulose fibers.

MCC has been used in commercial and consumer products for more than 50 years. MCC has been specifically approved for use in food and other products by regulatory bodies such as the Food and Drug Administration (United States) and the European Community, and is subject to health standards by the World Health Organization (WHO) and United Nations Food and Agricultural Organization (FAO; e.g., [17]).

Commercial prices of these materials depend very much on factors such as quantity, degree of development, purity, source country, and particular state of the economy. As one example, the price of pharmaceutical-grade MCC is much higher than the price of MCC used as an industrial thickener or viscosity modifier.

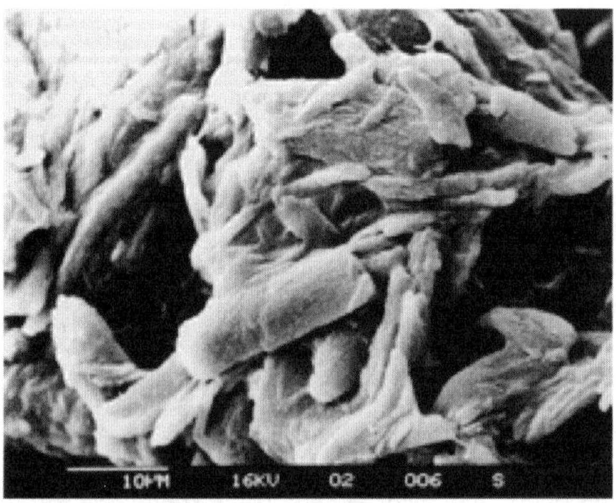

Figure 20.8 Commercial microcrystalline cellulose [16]. Reprinted with permission.

The average particle size of commercial MCC is between 50 and 350 µm. Some manufacturers offer a range of particle sizes, depending on the end-use (e.g. [18]). Since MCC particles are (very approximately) spherical, the aspect ratio of MCC is very low, typically 1–3. Common uses for MCC include:

- Pharmaceutical excipient—the term for nonmedicinal ingredients such as binders and bulking agents. In particular, MCC is considered to be far superior to powdered cellulose for this end-use [19].
- Cosmetics and other skin care products: binder and texturizer.
- Food additive: emulsifying agent, thickener/rheology modifier, and fat replacement agent.
- Edible coatings (e.g., sausage casings).

MCC provides an excellent example of how one cellulose derivative with improved properties replaced another cellulose derivative. Powdered cellulose was initially used as a thickener and viscosity modifier in food products. Due to poor "mouthfeel" of powdered cellulose (a gritty or other undesirable texture, in the terminology of the food industry), the finer MCC is considered to be a superior food additive.

Although a mature commercial product, the development and improvement of MCC is still ongoing. Changes that have been made in recent years, or that are under development and consideration, include:

- A shift from the original use of high purity cotton linters as the raw material for MCC to fully bleached kraft pulp [20].
- The use of other materials (including agricultural waste) as the raw material, including bagasse, corn [21], cotton [21], sisal [22], and bacterial cellulose.
- The use of MCC as the raw material for the preparation of micro- and nanocellulose.
- The use of enzymes or conditions of milder acidity to produce MCC [21].

20.2.2 Microfibrillated Cellulose and Related Products

The next step downward in size was the development of a material known by a variety of names, particularly MFC [23, 24]. This material was originally produced by additional mechanical action (especially high pressure homogenization) on MCC. The high shear of mechanical homogenization (helped by pre-cutting pulp sheets to lower the average fiber length) separates the cellulose fiber wall into its constituent microfibrils. Chemical treatment reduces the energy costs, while lowering the average DP. An example of MFC is given in Figure 20.9 [25], showing the micron-scale length, the sub-micron fiber width, and the fibrillar substructure.

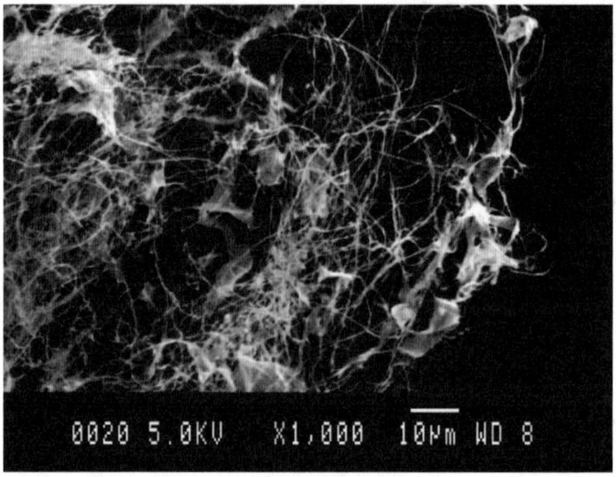

Figure 20.9 Freeze-dried microfibrillated cellulose. From reference [25]. Reprinted with permission of John Wiley & Sons, Inc.

In an early, key paper [26], a variety of cellulosic and noncellulosic fibers were subjected to a "super-grinding" technique, designed to break the fiber into longitudinal nanofibers. Cellulose source materials included wood pulp, cotton, and cellulosic material from the exterior mantle of the tunicate marine organism or "sea squirt." The material was on the scale of 20–90 nm in diameter, and formed strong, translucent films. The authors described these various products as "microfibrillated material," although this definition is open to question. In particular, other workers with tunicate-derived nanocellulose have pointed out that the tunicate-derived material comprises true cellulose whiskers or NCC [27].

Just as MCC partially replaced the coarser powdered cellulose as a food additive, MCC in its turn is considered to be too coarse for some purposes. A patent issued in 2000 discussed an "ultrafine microcrystalline cellulose" [28] with sub-micron particle size, which was considered to be an advantage in food products. This is effectively an MFC product. The degree of microfibrillation and corresponding increase in specific surface area means that MFC dispersions form stable gels at low concentrations [24].

In a patent issued in 2001, Matsuda et al. [29] described a "supermicrofibrillated cellulose" (SMFC). Their SMFC was defined as having an average fiber length of between 50 and 100 μm. The production technique appeared to involve further mechanical treatment of the original MCC. Claims included the production of dyed cellulosic particles, for the production either of colored paper or of colored coating pigments.

Even more recent work has suggested that MFC (in the 30–150 μm range) can be used as a strengthening agent in paper [30]. One can assume that the high specific surface area of the MFC allows it to play the same role as well-fibrillated pulp fines, only much more efficiently. It is likely that the very high specific surface area of MFC combined with its hydroxyl-rich surface gives it a much higher bonding potential than conventional pulp fibers or fines. NCC is not an effective paper strength additive due to its small particle size and resulting difficulty in retention within the fiber network structure. As noted below, there is also interest in using NFC for this purpose.

20.2.2.1 Commercial Microfibrillated Cellulose Products J. Rettenmaier & Söhne, a long-established supplier of MCC and other high purity cellulosic products, has introduced "cellulose fibers with nanofibrillation" (CFNF). In reality, this is another MFC product [31]. This is sold under the trade name Arbocel NFC B 1011. Fiber thickness is estimated to be 1–5 μm, and fiber length is between 200 and 300 μm. The product is supplied as a 23% solids gel/dispersion in water. It has been noted that the dried material is difficult to redisperse in water. The preparation technique is proprietary, but appears to be a further mechanical treatment of MCC. So far the recommended end-uses of this material are extensions of existing MCC end-uses. However, as noted in Appendix A, Rettenmaier is also part of the European Community program, SUNPAP (Scaling Up Nano Particles in Modern Papermaking).

Borregaard ChemCell of Norway introduced "Power Fiber." From their published literature and illustrations, this is another MFC product [32]. Initial marketing information suggests that this manufacturer is targeting the food additive sector, likely as a replacement for MCC. As shown in Appendix A, through industrial partners, Borregaard may also expand its horizons to include areas such as polymer reinforcement, adhesives, and paper strength development. Borregaard is collaborating with SCA of Sweden on the commercialization of MFC products. In this light, a related patent from SCA Hygiene Products AB describes the use of MFC as an absorbent material in personal hygiene products [33].

Daicel, a long-established supplier of MCC and other cellulosic products in Japan, has introduced a version of NFC, under the trade name Celish [34]. This product is also provided in a range of average length, diameter, and aspect ratio. This appears to be a true nanofibrillated product. The average fiber thickness is between 0.01 and 0.10 μm, and the average length is about 2 μm or more. The product is supplied as a paste of up to 35% solids. As with the Arbocel nanofibrillated product, there can be difficulties in redispersing the dried product, although this problem is surmountable [35, 36]. As shown in Appendix A, Daicel is also cooperating with the Kyoto University in developing other nanocellulosic products.

In Daicel's process, fibers are separated by a refiner, followed by a high pressure homogenizer. Chemical or enzymatic assistance in the process is likely, but not stated. The manufacturer's schematic diagram shows a typical production path, from fiber to microfibrillation to nanofibrillation [34].

The viscosity of Daicel's Celish is similar to that of other cellulosic gums used as viscosity modifiers, with greater viscosity stability as a function of temperature. This viscosity stability is used as a marketing point. The producer's marketing also advertises the material as a substitute for MCC (rheology control and humectant). A recent report [37] described the ability of one form of the Celish product to act as a reinforcing agent in an esterified starch-based plastic film.

In May 2011, StoraEnso announced the start of production of a "pre-commercial" MFC plant, scheduled to begin operation at the end of 2011 [38]. UPM of Finland recently announced the start of pilot-scale production of "fibril cellulose" [39]. According to information supplied by UPM, this is a form of MFC, and is intended for use as a reinforcing material, for paper, coatings, varnishes, and other products [40, 41].

20.2.2.2 Microfibrillated Cellulose Derivatization With its hydroxyl-rich surface of pure cellulose, MFC is also amenable to surface derivatization and graft copolymerization (e.g., [42]). Examples include reaction with hydrophobic AKD, extreme hydrophobic silylation with reactive alkyl chlorosilanes [43], and reaction with various cellulose ethers, and vinyl and acrylic monomers (e.g. [44],). In general, these various surface treatments and polymer additions have the goal of improving the dispersibility of the nanofibers in different polar and nonpolar media (e.g., [45]).

Zhang et al. described [46] what they called "spherical cellulosic nanoparticles," "nanospheres," or "nanoballs." From their synthetic procedure, from the roughly spherical shape of their product, and particle size (diameter ranging from 60 nm to 570 nm), it appears that this is another form of MFC. In a companion work [47], the same group showed that cellulose whiskers (i.e., NCC) were effective as reinforcement of latex films, while nanoballs and hardwood kraft fibers were not. The most likely explanation is the higher aspect ratio and higher specific surface area of the true cellulose whiskers, compared with the larger and coarser nanoballs or kraft fibers.

Zimmermann et al. [48] compared MFC prepared by mechanical disintegration with MFC prepared by chemically assisted separation. Their mechanically derived MFC had diameters in the range of 20–100 nm and lengths of up to several tens of microns. Using pretreatment with sulfuric acid before mechanical separation, they prepared microfibrils that were much reduced in size, with average diameters of 50 nm and lengths still in the order of a few microns.

They noted a critical distinction: *agglomerated* microfibrils are detrimental to reinforcement performance, while *networked* microfibrils are very desirable. With this in mind, they claimed that their microfibrils prepared *without* acid pretreatment are better reinforcing agents. It should be noted that the scale of their product is still much larger than that of NCC. The modulus of elasticity of their pure MFC films was about 5–6 GPa, which is not far from that of wood itself (typically 8–10 GPa). They also prepared films of hydroxypropylcellulose

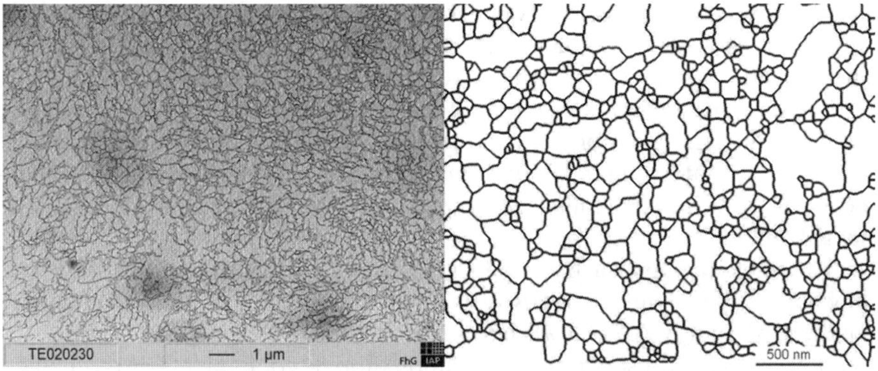

Figure 20.10 Left: TEM micrograph—Transverse section of HPC film with a fibril content of 20%, showing network of microfibrillated material. Right: Model of microfibrillar network. From Reference [49], reproduced with permission. Copyright Wiley-VCH Verlag GmbH & Co. KGaA.

(HPC) and of poly(vinyl alcohol) (PVA) with their MFC, and found improved mechanical properties.

As further evidence of their hypothesis, in an accompanying paper [49], the same group used transmission electron microscopy (TEM) images to show that at a high fibril content, although their fibrils are well dispersed, they form a mesh-like network, rather than the conventional thinking of aligned rods (Fig. 20.10).

The group of López-Suevos et al. [50] further explored different forms of MFC as reinforcement aids in poly(vinyl acetate) adhesives. They used mechanically treated MFC, chemically treated MFC, and MFC that was subjected to chemical treatment followed by mechanical treatment. Their reference material was their original hardwood kraft pulp. In most cases, the MFC provided greatly improved physical properties, in the order: combined chemical/mechanical MFC > mechanical treatment only > chemical treatment only > original kraft fibers. The advantage of the chemical/mechanical MFC was ascribed to the formation of the most extensive network. However, the authors noted that their work is not complete and that excessively hydrophilic MFC in an adhesive may cause problems in the presence of high moisture or humidity.

20.2.3 Nanofibrillated Cellulose

In the downward progression in size starting with MCC and then MFC, NFC is next. This material has been of particular interest to many research groups, particularly Innventia, Sweden. Innventia [51] recently announced the inauguration of a pilot plant for producing NFC and MFC at a scale of about

100 kg/day. As shown below, while the FPInnovations NCC product and the Innventia NFC/MFC products each possess unique areas of potential end-use, there is also an area of overlap where competition may be inevitable.

Innventia has also taken the approach of providing a range of MFC and NFC micro- and nanofibrillated products, with particle sizes that are smaller than the present commercial MCC products, yet much larger than NCC. Publications from Innventia have shown that their MFC may be as important as their NFC.

The Innventia approach started with purely mechanical homogenization, as developed at ITT in the 1980s (e.g., [24]). This approach used an excessive amount of energy (27,000 kWh/tonne), later [52] shown to be close to the theoretical value for separation of cellulose into its components by mechanical means alone. Further development reduced the energy requirement to 2000 kWh/tonne, by means of enzymatic pretreatment.

The Swiss research organization EMPA (a branch of ETH, the Federal Institute of Technology in Switzerland) has taken a similar approach, promoting research and development of MFC/NFC products [53]. Some of their work has already been referred to [48, 49]. They have affiliations with universities in Germany and in Sweden. Their stated areas of application are quite broad, including superabsorbers, biomedical applications, polymer reinforcement agents, and adhesive components. According to their web site, they are not yet at the scale of kg/day production.

Researchers at PFI, Norway, produce NFC and MFC on the laboratory scale. They showed that small amounts of NFC can counter the strength losses encountered in highly filled supercalendered mechanical printing grades, particularly with the addition of both clay filler and NFC reinforcement additive to the surface layers [54]. They are also interested in the use of MFC as a barrier layer [55]; an area of active research for NFC and NCC as well.

As noted above, there is interest in using micro- and nanocellulosic products to enhance paper strength, especially for paper with very a high mineral filler content (e.g., [30]). Several patents have been issued in this area (e.g., [56]). TEMPO-oxidized NFC is particularly effective [57], presumably due to the improved hydrogen bonding imparted by the oxidation process.

20.2.4 Nanocrystalline Cellulose—FPInnovations Approach

As discussed in the Introduction and as shown in Figure 20.5 and Figure 20.7, NCC makes up the elementary structural unit of cellulose fibers. Unlike the larger MCC, MFC, and NFC particles, NCC particles can no longer be referred to as "fibrillated."

Although first described more than 50 years ago [58], the properties of NCC were only seriously examined by Paprican (now FPInnovations) researchers at the McGill University Pulp and Paper Centre in the mid-1990s (e.g., [14]). In particular, they explored the ability of NCC to form ordered dispersions, similar in form to cholesteric liquid crystals [59]. The iridescent colors of the liquid crystalline state are maintained in the dried NCC films. These colors are controllable through, for example, ionic strength or intensity of the sonication pretreatment of the casting solution [60].

From a laboratory curiosity on the gram level, FPInnovations developed an in-house pilot plant capable of producing NCC on the kilogram scale, a requirement for serious industrial development. In July 2010, FPInnovations and Domtar Inc. announced [61] the construction of a 1 tonne per day demonstration plant to manufacture different forms of NCC, in Windsor, Quebec, operational in January 2012 under the name CelluForce [62].

NCC particles are about 4–7 nm in cross section, while typical mean particle length is about 110 nm from bleached kraft pulp and 150 nm from cotton fibers. The yield, average particle length, and the particle length distribution also depend on the hydrolysis conditions. Typically, NCC is produced by treating cellulose fibers with concentrated (64%) sulfuric acid, at 45°C for about 1 hour, followed by ultrasonication or mechanical homogenization. Concentrated hydrochloric acid can also be used but produces uncharged and unstable dispersions that do not lead to the organization of a chiral nematic structure in suspension, and does not produce the interesting color effects for which NCC is now famous. As noted previously, NCC consists of true crystalline "whiskers," with minimal amounts of residual amorphous material, which is removed by the severe acid hydrolysis.

Yet another possible end-use for NCC was shown in a recent patent from FPInnovations [63]. In this patent, NCC was used as the cross-linking/reinforcing agent in hydrogels, produced by copolymerization with hydrophilic monomers.

In common with MFC, NCC, and other forms of cellulose, NCC is also amenable to modification of its surface properties by reaction with other materials. For example, due to the hydrophilic surface of NCC, it will disperse with difficulty in both hydrophobic liquids and hydrophobic solid matrices. However, NCC can be reacted with hydrophobic monomers, yielding a surface that is hydrophobic. A hydrophobic NCC surface maintains the reinforcing ability of the NCC whiskers, yet is compatible with hydrophobic/nonpolar liquids and solids (e.g., [64]).

As noted in Table 20.2, NCC may also be a useful additive to reinforce textile fibers. This was shown in recent work co-authored by North Carolina State University and the Helsinki University of Technology [65] in which NCC was used as an effective additive to strengthen fibers of PVA.

Just as other forms of nanocellulose can act as adhesives (e.g., [50]), NCC has also been shown [66] to be an adhesive for paper and wood. Both the acid form and the monovalent cationic form (e.g., sodium NCC) have been shown to be effective.

20.2.4.1 Nanocrystalline Cellulose as a Pharmaceutical Excipient

As noted in the Introduction, cellulose products such as MCC have a long history as pharmaceutical excipients (i.e., the nonmedicinal ingredients of a

TABLE 20.2 NCC versus MFC/NFC

Performance/Applications	MFC/NFC	NCC	Notes
Strength properties (film)	Tough	Strong, but brittle (flexibility can be enhanced with plasticizing polymer)	Pure NFC film may have a higher flexibility
Optical effects/UV protection (film)	No	Yes	NCC is unique. Iridescence can be controlled [60]
Transparency (film)	Yes	Yes	
Redispersibility (powder or film)	Yes (with chemical modification)	Yes	As described in Reference [35].
Water resistance (film)	Probable	Yes (by controlling ions or by surface derivatization)	As described in Reference [35].
Rheology (suspension)	Strongly pseudo-plastic (shear thinning); function of manufacturing conditions	Can be prepared as a strongly shear thinning dispersion	NCC rheology can be easily controlled
Formation of gel	Yes	Yes (usually around 10% solids, but adjustable)	NCC, with a higher surface area, may have an advantage
Use as pharmaceutical excipient	Probable	Probable	Report on NCC as an excipient [67]
Use in skin care/cosmetics	Probable	Probable	No research has been reported
Use as absorbent/super absorbent products	Probable	Probable	No research has been reported
Use in aerogels/nanofoams	Yes	Yes	NCC, with a higher surface area and smaller particle size in all dimensions, may have an advantage

Use for polymer reinforcement (synthetics)	Yes	Yes	Higher aspect ratio of NFC may be advantageous, but NCC has a high inherent strength; dispersibility and compatibility are common issues to be addressed
Use for polymer reinforcement (biopolymers such as PLA)	Yes	Yes	Same as above
Rubber reinforcement	Yes	Uncertain—will depend on dispersion conditions	Low-end applications (if the amount is large)
Use as catalyst carrier or carrier for producing other nanomaterials	Unknown	Yes	Higher specific surface area of NCC is a clear advantage.
Use for reinforcing individual textile fibers	Probably not due to greater particle length	Yes	Small size and high specific surface area of NCC should help
Paper strength additive	Moderate (MFC product)	No, as it is difficult to retain	Low-end application
Lint reduction/surface strength agent	Yes	Yes	Low-end application
Paper coating additive/ binder replacement	Yes	Yes	NCC has an advantage with its smaller size
Paper coating pigment replacement	Yes	Probably not	Low-end application as most coating pigments are inexpensive
Support for dye or colorant	Yes (scale appropriate to papermaking)	Yes (scale appropriate to ink pigments and other coatings)	No research has been reported

(*Continued*)

TABLE 20.2 (*Continued*)

Performance/Applications	MFC/NFC	NCC	Notes
Improve water absorption and ink holdout during printing	Yes	Yes	Demonstrated in lab studies
Food packaging/barrier layers	Probable	Probable	
Decorative films/consumer products/security sign	No	Yes	Iridescence of NCC film is a clear advantage
Printed sensors	Unknown	Yes	Research projects have been proposed by universities
Compatible with varnishes	Yes	Yes	NCC has been shown to improve properties of wood varnishes [78]
Compatible (blending) with water-based printing inks and printing varnishes	Yes	Yes	
Compatible as printing substrate with water-based printing inks and printing varnishes	Yes	Yes	Improved print quality with both aqueous ink jet and water-based flexo
Compatible with solvent-based printing inks	Probable (with surface derivatization)	Probable (with surface derivatization)	

pharmaceutical product). In general, MCC serves as no more than a useful but inert filler and binder. Work co-authored by FPInnovations [67] showed that NCC has important additional advantages over MCC. Its highly charged surface and much greater specific surface area make it capable of absorbing drugs and then releasing them in a controlled manner over a period of time.

20.2.4.2 TEMPO Oxidation Several research groups have published work on the use of the TEMPO (2,2,6,6-tetramethylpiperidine-1-oxyl) radical as a selective catalytic oxidizing agent in the presence of sodium chlorite. This provides an alternative route to forming NCC by decreasing the amount of mechanical energy needed. TEMPO oxidation converts the primary alcohol groups in cellulose (at the C6 position) to aldehydes and then to carboxylic acid groups. Combined with mechanical action, this produces what has been termed TOCN.

Surface derivatization (e.g., hydrophobization by alkyl ketene dimer) can readily be done on the TEMPO-oxidized surfaces with their high carboxyl content. Even if TOCNs possess negative charges like sulfated NCC, they do not form highly ordered chiral nematic structures in suspension because of their excessive length. On the other hand, the surface of TEMPO-oxidized nanofibrils has a very high concentration of hydroxyl (-OH) and carboxyl (-COOH) groups, which provide an advantage in terms of forming hydrogen bonds with polar polymers [68].

TEMPO-oxidized fibers also require less mechanical energy than other forms of chemical pretreatment [69]. Films from TEMPO-oxidized material combine high transparency with high tensile strength, flexibility, toughness [70], and low gas transmission rates [71].

One possible benefit of the (relatively) mild treatment by TEMPO was shown in a recent paper in which the authors produced what we may call "sub" nanoscale cellulose particles [72]. By a combination of TEMPO oxidation and ultrasonication, the authors were able to "peel back" cellulose whiskers (about 4 nm cross section) into sub-fibrils or "molecular sheets" of about 1 nm average thickness and a length of up to 500 nm.

20.2.5 Other Canadian Sources of Nanocrystalline Cellulose

As noted in Appendix A, in addition to FPInnovations, at least two other groups in Canada are interested in producing NCC on a large scale. These include a joint venture including the Alberta Government, the University of Alberta, and Alberta-Pacific Forest Industries [73, 74]. In the latest development of the Alberta initiative, a pilot plant to produce up to 100 kg/week from wood pulp as well as from agricultural residues such as wheat straw has recently been announced [75].

Bio Vision Technology of Nova Scotia [76] works in cooperation with the National Research Council of Canada—Biotechnology Research Institute [77]. Although their major focus is on biorefinery, they are also working on the production of NCC and can supply an NRC-developed "carboxylated NCC"—without providing details of the production path or of the final product properties. Although it is the stated goal of Bio Vision to go into full scale production of NCC (as part of an overall biorefinery plan), no production facilities have been announced as yet.

20.3 THE BUSINESS CASE FOR NANOCRYSTALLINE CELLULOSE AND OTHER NANOCELLULOSIC MATERIALS

The market value for a potential end-use must always be considered. As shown in this report and elsewhere, NCC and other nanocellulosic products can successfully provide many technical benefits. The challenge is to assess the potential for value creation and to develop business cases and markets. For nanocellulosics to succeed in a given product, one or more of the following criteria must be met:

1. To simply replace another additive, with no improvement in product performance, a nanocellulosic material must provide a cost advantage: for example, replacing a larger amount of an expensive additive with a smaller amount of nanocellulosic material.
2. A nanocellulosic material may provide a clear environmental advantage, for example, replacing an additive being phased out for health or environmental issues. A simple statement that NCC is "green" or "a renewable product" may not suffice if this also results in a higher cost.
3. Improved product performance must be quantifiable. A significant improvement in performance may well justify a higher cost. This is not an absolute, if the existing product is in a mature market and current performance is "good enough."
4. An entirely new product, either something unavailable in the past or something that could only be produced with much more complex technology, is also an advantage, again if it meets the needs of a potential customer. The special optical properties of NCC provide a good example here.

As also shown in this review and elsewhere, NCC can be produced in different forms, including:

- High versus low yield.
- Production via sulfuric acid, hydrochloric acid, or TEMPO oxidation, each with its own particular surface properties.
- Post-production derivatization to provide properties such as compatibility with hydrophobic materials and nonpolar media.

The guidelines regarding market value also apply to the different forms of NCC and other nanocellulosics. Since any modification of nanocellulosic materials will increase the cost, we may expect that the first commercial application of the NCC products will not involve special treatment or derivatization. As shown below, even nonderivatized NCC products are not all equal in their ability to disperse in pure water.

20.4 COMPARISON: NANOCRYSTALLINE CELLULOSE VERSUS MICROFIBRILLATED CELLULOSE AND NANOFIBRILLATED CELLULOSE

It is useful to compare NCC products with MFC/NFC products. Many of these product areas are summarized in Table 20.2, which includes technical advantages/disadvantages as they are currently understood.

The ability of NCC to improve the properties of industrial coatings such as wood varnishes has already been shown by the Wood Products Division of FPInnovations [78]. In this work, adhesion, scratch resistance, and impact resistance were all improved by the addition of between 1% and 3% of NCC. A recent patent from Finland made similar claims for the use of NFC in paints and varnishes [79].

A number of publications from Innventia and elsewhere have shown some applications of MFC/NCC. The next sections summarize these published results. It should be noted that most of these studies were driven by technical potential and possibilities created by nanocellulosic materials, not by market demand and value creation.

20.4.1 Tough Nanostructure Film

Henriksson et al. [80] produced a "nanopaper" from a microfibrillated cellulose. Their product showed a very high tensile strength and deformation to failure. A relatively large particle was preferred, due to the greater probability of entanglements. These films are virtually transparent. On the one hand, the films of the Innventia product lack the unique color and iridescent properties of NCC films. On the other hand, with suitable additives, NCC can also produce transparent films, without color effects [81].

VTT, Finland, also recently reported on the preparation of films of NFC on a pilot plant/semicommercial scale [82].

20.4.2 Lint Reduction Additive

Ankerfors et al. [83, 84] demonstrated that MFC is an effective surface additive to reduce offset linting. A surface treatment of 1 g/m^2 of starch was effective, while a 1 g/m^2 coating of MFC was somewhat more effective. However, a mixture containing 0.5 g/m^2 of each was most effective of all. The authors also noted that a thin film of MFC at the sheet surface created a weak boundary layer *within* the sheet, which could cause cohesive failure beneath the sheet surface.

Surface treatment provides a good example of the limits of nanocellulosics. In general, nanocellulosic products may serve as an effective lint reduction aid in publication papers. However, for many producers of publication grades of paper, the capital cost of any kind of surface addition is prohibitive. For those who do have the necessary equipment, other much less expensive materials such as starch are already in use, and provide satisfactory performance.

20.4.3 Nanofoams

There has been a considerable amount of interest in using MFC/NFC in the production of foam products based on renewable resources. Much of this work has been done at Innventia, in association with KTH (Royal Institute of Technology, Stockholm). Svagan et al. [85] showed that particles of the size of native cellulose fibers are foam breakers, not foam stabilizers, in particular if the particles are thicker than the desired foam wall. This group produced starch foams from amylopectin, with cellulose microfibrils (4–10 nm in cross section) as a reinforcing agent (Fig. 20.11). Another critical point is a homogenous dispersion of the reinforcing nanofibers. In related work, they made foams from starch and MFC [86], with a typical diameter of 30 nm and several micrometers in length.

Other research at KTH and Innventia has also addressed the issue of MFC and NFC as foam stabilizers (e.g., [87, 88]). A recent report [89] showed aerogel foams with densities as low as 0.078 g/cm^3 and specific surface areas

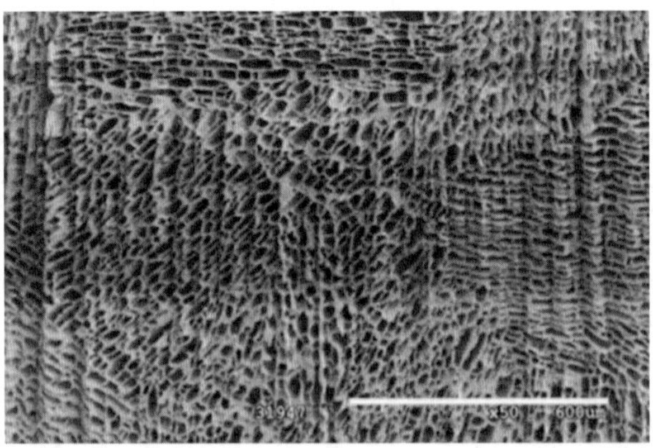

Figure 20.11 Cross section of reinforced nanofibril–starch foam. Scale bar ~600 μm. From Reference [85], reproduced with permission. Copyright Wiley-VCH Verlag GmbH & Co. KGaA.

as high as 605 m^2/g, using an NCC material. The production path for their nanofoam product used only water, ethanol, and supercritical CO_2. Given its high surface area and much smaller particle size, NCC could likely have the advantage over NFC in the production of nanofoams. The Oregon State University College of Forestry similarly prepared aerogels using TEMPO-oxidized NCC [90]. Lapidot of the Hebrew University and Melodea, Israel [91, 92], achieved achieved NCC aerogel foam densities as low as 0.025 g/cm^3, using materials that included paper machine scrap.

Eriksen [93] showed a simple and useful schematic of the size range of MCF/NCF material. He rated nanomaterials in size order from the coarsest (up to 150 μm)—paper strength enhancement; food additives; paper surface modification; polymer reinforcement; film formation; barrier formation—to the finest (true nanoscale particles), for emulsion stabilization.

Researchers at the University of Maine have used multiple approaches to prepare NFC, followed by examination of its potential for a variety of end-uses. This work is sponsored (at least in part) by the US federal and the Maine state governments [94].

As described in Reference [95], Saddiqui started with bleached softwood pulp, and used cellulase enzymatic treatment combined with homogenization in a commercial disintegrator. High consistency homogenization was a critical factor in reducing energy consumption (Fig. 20.12). Enzymatic pretreatment was particularly seen as an aid to processing. The work in Maine showed that enzymatic pretreatment gave a more uniform product, with lower energy consumption. $MgCl_2$ pretreatment was much less expensive in chemical costs, but at a higher energy cost.

Hamada and Bousfield [96] used both enzymatic pretreatment and pretreatment with $MgCl_2$ to prepare different types of NFC. After coating papers with NFC, they printed with water-based flexographic inks, and with water-based ink

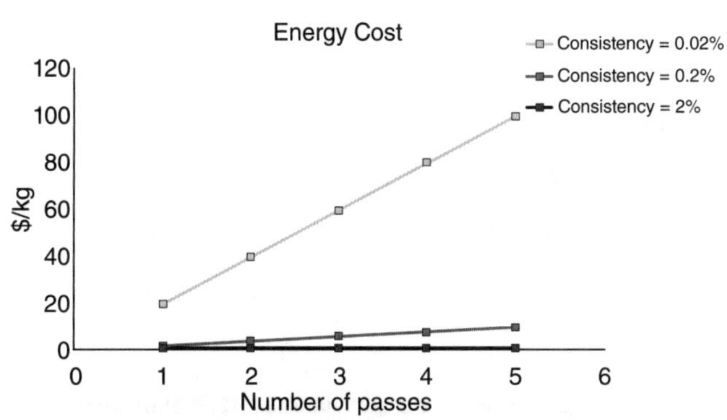

Figure 20.12 Energy cost for laboratory-scale production of NFC. The energy costs are particularly dependent on the consistency [95]. Reprinted with permission.

jet inks (both dye-based and pigmented). The visual quality of pigmented inks was greatly improved, perhaps by the formation of a hydrophilic, physical barrier layer. However, it is important to note that a coating layer of as little as 1 g/m^2 means the use of about 20 kg of NFC per tonne based on a 100 g/m^2 paper is clearly not cost-effective at current prices.

In a related study, Hamada et al. [97] combined prepared heavier coatings in which NFC acted either as coating pigment (preponderance of NFC in the coating), or as coating binder (preponderance of clay blended with a small amount of NFC). NFC as binder in a clay coating did greatly improve the ink holdout, but there was no advantage of the NFC over much less expensive PVA binders.

In work related to the University of Maine approach, StoraEnso recently reported a paper coating based on traditional coating pigments such as clay, with MFC as the binder [98]. The patent also allowed for the possibility of using other cellulose-based materials such as carboxymethylcellulose as additional binders.

20.4.4 Nanocellulose Barrier Layers

The ability of a film to form a barrier is critical in the packaging industry. The most important properties are the oxygen transmission rate (OTR) and the water vapor transmission rate (WVTR). Common, hydrophobic polyolefin

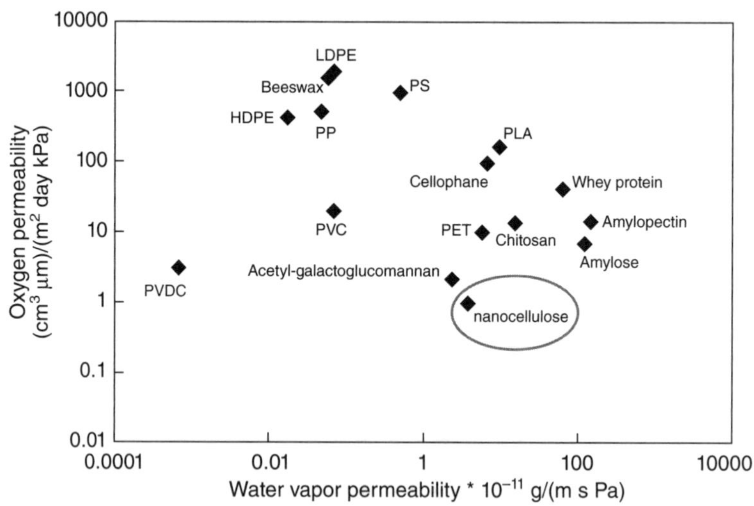

Figure 20.13 Oxygen and water vapor permeability of various materials [99]. Reprinted with permission. Copyright J. Wiley & Sons.

films such as polypropylene and polyethylene have a low WVTR but a high OTR. Conversely, many cellulosic films, including cellophane (regenerated cellulose) and NCC, have a low OTR but have a very high WVTR.

Figure 20.13 shows the barrier properties (OTR vs. WVTR) for many materials including cellophane and nanocellulose [99]. Many researchers have carried out equivalent measurements on many films, including different forms of nanocellulose. Such data may not be interchangeable, due to different measuring conditions including temperature, relative humidity, and film thickness (e.g., [100]).

The low OTR of nanocellulose films is due to the low porosity of the well-packed and ordered material. However, the inherent sensitivity of cellulosic materials to water vapor yields a high WVTR (via moisture diffusion through cellulose). The technical challenge is to address this fundamental problem, possibly by making the NCC particles hydrophobic, or by combining NCC with other polymer films that have a low WVTR, but a high OTR. This could yield a product with the desired properties of both low OTR and low WVTR.

Aulin et al. [101] prepared pure MFC films, and found OTR values similar to those measured for NCC at FPInnovations. Various workers (e.g., [3]) have measured barrier properties of various polymers with and without NCC, MFC, or NFC as reinforcement additive. In addition to the improved strength properties, the reinforcing material generally provided significant improvements in barrier properties.

In the work from the University of Tokyo [71], poly(lactic acid) (PLA) films showed very large improvements in oxygen barrier properties when coated with TEMPO-oxidized material, with dimensions of approximately 3–4 nm cross section and lengths in the microns range. The same authors suggested that hydrophobized NCC could reduce the WVTR, but did not actually publish any data. Similarly, Sanchez-Garcia and Lagaron [102] showed that as little as 1% of "cellulose nanowhiskers" (NCC) blended into PLA not only improved strength properties, but also reduced both oxygen and water vapor transmission. The group had previously shown [103] that 5% of a nanoclay product blended into PLA showed a much smaller reduction in water vapor and oxygen transmission through PLA.

Modified atmosphere packaging (MAP) refers to the practice of storing produce under conditions of reduced oxygen and/or elevated carbon dioxide levels as a means of prolonging shelf life during storage and transportation (e.g., [104]). Syverud and Stenius [105] showed that pure films of NFC of at least 35 g/m^2 basis weight "fulfilled the requirements for oxygen transmission rate in modified atmosphere packaging."

To conclude, NCC and other forms of nanocellulose show promise as barrier-forming materials, particularly for oxygen. However, using cellulosics to form a barrier to water vapor is a challenge. Among various nanocellulosic materials, the published literature is too scanty to prove the advantage of one product over another. The following work would be required to show which nanocellulosic product (if any) has the advantage:

1. Measurement of oxygen and water vapor transmission through MFC, NFC, and NCC films under identical conditions of film thickness, temperature, and relative humidity.
2. Since cellulosic films have a good record in forming oxygen barriers, and synthetic polymer films have a good record in forming barriers against water vapor transmission, it is possible that the ideal barrier film may be a composite, a laminate, or a coating of nanocellulose onto another polymer. It would then be highly desirable to measure oxygen and water vapor transmission through thin layers of MFC, NFC, and NCC coated onto representative commercial polymers, again under identical conditions of measurement.
3. Measurement of OTR and WVTR through films of hydrophobized NCC. This would define whether a hydrophobized NCC alone could serve as a barrier layer for both water vapor and oxygen.

20.5 OTHER COMMERCIAL APPLICATIONS FOR NANOCRYSTALLINE CELLULOSE AND NANOFRIBRILLATED CELLULOSE

20.5.1 Problems in Redispersing Nanocellulose

Any additive meant to enhance performance or add functionality must distribute itself uniformly, without coagulating or forming agglomerations. That is, a uniform distribution is essential, either in the aqueous media that may be the starting point in the manufacturing process, or in a final solid matrix. It has repeatedly been reported that it can be difficult to redisperse dry micro- and nanocellulosic products in water [106]. This had led to problems such as undesired agglomeration of suspensions in aqueous media or solid products.

As one example, so-called water-based commercial products such as ink, paints, and varnishes are a complex mixture of water, co-solvents, and other water-soluble or water-dispersible components. Experience has shown that while NCC may disperse in pure water, the same material may have problems in dispersing in the presence of all of the ingredients within a "water-based" commercial product.

Agglomeration in solid media such as polymers is likely the cause of reinforcement failure here as well. Presumably, the poor redispersibility is due to the very strong hydrogen bonding between particles, which is maximized in the bone-dry product. This is a form of the well-known "hornification" process whereby aged and dried cellulose fibers and papers rewet only with difficulty [107]. Several possible solutions have been proposed.

Eyholzer et al. [108] proposed carboxymethylation of NFC to low levels of substitution–not sufficient to make the NFC lose its structure and become water-soluble, but enough to disrupt the hydrogen bonding so that the dry NFC will redisperse. However, this approach would increase the NFC produc-

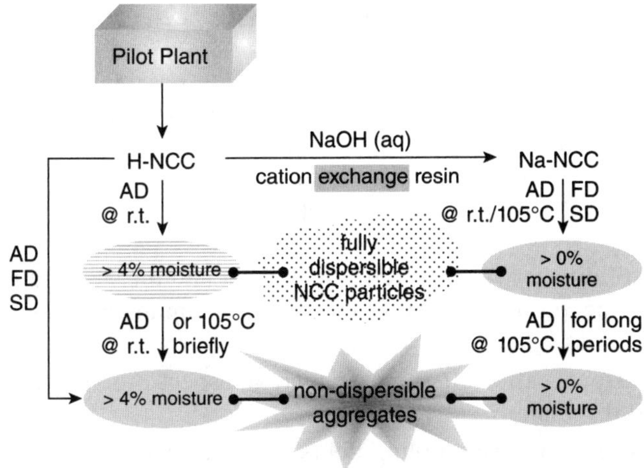

Figure 20.14 Dispersibility of NCC under different drying conditions and with different counterions [35].

tion cost. Nonuniform carboxymethylation may also produce zones that are still poorly dispersible.

Beck et al. [35, 36] showed the acid-form NCC can be made spontaneously water-dispersible in two ways, as shown in Figure 20.14. One procedure is to prevent excessive hydrogen bonding by simply not letting the material become extremely dry (keeping moisture content >4%). Another procedure is to convert the acid form $–HSO_4$ to a form with a large counterion (e.g., $-SO_4^-$ Na^+), which presumably also prevents excessive hydrogen bonding and agglomeration. The advantage of these procedures is that they are relatively inexpensive and simple to implement in an online process.

20.5.2 Nanocrystalline Cellulose as a Rheology-Modifying Agent

Just as the longer established MCC and its derivatives have been used for many years as rheology modifiers in, for example, food products and cosmetics, NCC may be able to serve the same purpose. Boluk and Zhao [109] described the use of NCC as a thickening or rheology-modifying agent in aircraft de-icing fluids.

20.5.3 Polymer Reinforcement Additives

As shown in Figure 20.3 (for patents) and elsewhere in this review (for research papers), the single largest body of work on cellulosic fibers, microfibrils, and nanofibrils has been as polymer reinforcement. We have already discussed several instances in which micro- and NFC have been used as a polymer reinforcement material.

In this section, we will address the competing pros and cons for different types of micro- and nanocellulosic material as reinforcement additives. Work has been done in many potential products, including films and structures, rubber, and even dried food. The requirements for reinforcing agents (summarizing a considerable amount of work) include:

- Compatibility between the reinforcing fiber and the suspending polymeric medium, necessary for chemical wetting (maximum adhesion between the components).
- A high aspect ratio, although too high an aspect ratio can be detrimental to good dispersion.
- Dispersion of the reinforcing fibers (related to the chemical wetting, to avoid agglomeration).
- The ability to form a three-dimensional lattice or percolation network, or the ability to form large-scale entanglements.
- In many cases, reinforcing fillers already exist, with satisfactory performance at an acceptable price. For nanocellulosics to succeed under such conditions, they must provide either a distinct technical advantage or a distinct environmental advantage.

There has also been some debate in the literature about the appropriate particle size. There are instances where a particle is physically too large, for example, for thin film or foam reinforcement. At the same time, given the ability of MFC, NCC, and related material to form extended three-dimensional networks, cases where one product has an advantage over another may really be questions of compatibility, especially dispersion versus agglomeration.

In relatively early work, the original commercial powdered cellulose was shown to be an effective reinforcement for extrusion- and injection-molded starch derivatives [110]. The reinforcing material was powdered cellulose, with dimensions of 20×60 μm (Arbocel B600), 20×300 μm (Arbocel BC 200), and 20×900 μm (Arbocel B400). The longer fiber material was considered to be better for this purpose.

The group of Nakagaito and Yano has published several articles on the subject (e.g., [111]). A critical conclusion by this group includes the fact that the degree of fibrillation that we normally associate with well-refined pulp fibers is not sufficient to provide adequate polymer reinforcement. It requires fibrillation at the micro or nano scale to achieve this desired effect.

Another critical paper used cellulose whiskers (effectively NCC) [112] with an aspect ratio of about 100, derived from the tunicate organism. These were blended with a styrene–butyl acrylate resin under different conditions, providing significant improvements in modulus and tensile strength.

The current literature demonstrates that the surface chemistry of the nanofibers is critical to dispersibility (e.g., polar vs. nonpolar polymer). This affects the wetting between the reinforcing fibers and the surrounding polymeric matrix. Another critical factor is the tendency of the reinforcing fibers to self-agglomerate rather than disperse in the polymeric matrix. This is also a function partially of the suspending medium and partially of the surface chemical treatment of the reinforcing fibers. The degree of fibrillation may also be critical, depending on the optimum amount of three-dimensional entanglement that may be required.

Kim et al. [113] showed at least one possible solution to the problem of compatibility between polar nanocellulose and nonpolar supporting media (including the problem of nanocellulose aggregation). They used a commercial surfactant to disperse NFC (particle size 10–20 nm × 200–500 nm) in a polystyrene film. They also noted that their improvement in mechanical properties were relatively modest, given the inherent strength of pure polystyrene films.

As part of a detailed patent on the fabrication of polymer composites for use in, for example, dialysis membranes, Jovanovic et al. [114] presented their solution to the problem of dispersing NCC in a polymer matrix. Starting with NCC in aqueous dispersion, they used the well-known procedure of solvent exchange to prepare NCC dispersions in organic solvents. The solvent-dispersed NCC was then polymerized *in situ* with the appropriate monomeric material, to produce the filled composite polymer. A similar product was also reported by Li et al. [115], without the solvent exchange of the NCC out of the original aqueous dispersion.

20.5.3.1 Reinforcement of Dried Food One interesting application of nanocellulose is in the strength reinforcement of dried food, a form of polymer reinforcement [116]. In this study, the authors used a commercial "cellulose with nanofibrillation" to improve the physical properties of a dried fruit product. Their product was approximately 7 nm cross section × 83 nm length. This is similar to a true NCC. Addition of 10% by weight of this product to dried fruit provided a significant strength increase and also a modest water vapor barrier improvement.

This particular end-use raises a critical point, since the regulations of the European Parliament governing MCC in food products state that the particle size cannot be less than 5 μm [117]. Whether modifying this regulation would be a minor technical detail or a major political issue is, of course, a question for those dealing with regulators. In general, there is considerable concern about the use of nanoscale material in consumer products, particularly food products [118, 119].

20.5.3.2 Rubber Reinforcement Additives Bendahou et al. [120] evaluated "cellulose whiskers" (effectively NCC) and MFC as reinforcing agents in natural rubber. The authors claimed that the MFC was a superior reinforcing agent to the NCC whiskers. They suggested that this was due to the greater ease of forming entanglements for the MFC. This conclusion may not be correct, considering the following discussion on the importance of particle dispersion.

In recent work at the University of Oregon, Bai [121] showed that MCC, especially MCC that has been surface modified to be more hydrophobic, can replace up to 25% of the silica used as filler for rubber tires with no serious effects on mechanical properties. The same study claimed that NCC is not an effective reinforcing additive. However, it was acknowledged that the problem may have been due to the poor dispersibility of freeze-dried NCC and that dispersing NCC in the aqueous latex mixture before cross-linking and hardening might have been more effective.

The potential for cellulosic materials to substitute for even a fraction of the silica used in rubber reinforcement requires a reality check. At present, products such as precipitated silica and fumed silica already provide adequate performance as rubber reinforcing fillers, at a lower price. Clearly, MCC, NFC, NCC, and related cellulose derivatives costing several times more will not substitute for the currently used silica fillers in rubber. However, NCC and NFC may be competitive with the higher priced nanosilica products. Carbon black, also widely used to reinforce rubber, is even lower in price than silica.

20.5.3.3 Polyurethane Reinforcement Additive Polyurethane is an important industrial polymer, particularly as an elastomer (synthetic rubber). Polyurethane has an advantage in dispersing cellulosic materials, due to its polar surface. Wu et al. [122] polymerized polyurethane *in situ* with MFC (prepared by chemical pretreatment of commercial MCC). They noted that about 5% of cellulosic material as reinforcement additive was optimal for improving tensile properties (high tensile strength and high strain to failure). They also noted that untreated MCC without microfibrillation is not suitable as polymer reinforcement. This provides evidence that the surface chemistry of the reinforcing material may be as important as its structure. Both the nanofibrils and the polymer molecules become strongly oriented in the loading direction, providing extra stiffening. Finally, they noted that the strength of the network is due not only to strong hydrogen bonds, but due to the creation of even stronger chemical bonds between the cellulose microfibrils and the polyurethane molecules. This may be analogous to the formation of chemical bonds between carboxyl groups on fatty acids (found for example in printing ink vehicles) and the hydroxyl groups on the cellulose backbone.

A recent thesis took polyurethane reinforcement from the MFC scale to the NCC scale [123]. This work also showed that NCC is an effective reinforcing agent for polyurethane, either by polymerization with the pre-polymers *in situ*, or by blending with the finished polymer.

It is difficult to compare the properties of the MFC-reinforced polyurethane in reference [122] to those of the NCC-reinforced polyurethane in reference [123]. There are uncontrolled differences such as variations in the source, molecular weight, and structure of the oligomers used in the initial polyurethane preparation. The MFC work used much higher concentrations of MFC—up to 10% by weight. The NCC work only went as high as 2% by weight of NCC. With that in mind, comparing the two sets of data showed that the Young's modulus of the two filled products was similar, but the NCC-filled product had both a lower tensile strength and a lower % stretch to failure.

20.5.3.4 Poly(Lactic) Acid PLA is of interest to major suppliers of polymer films and polymer products as a green product. However, PLA is too brittle and possesses insufficient heat resistance to substitute for more common synthetic polymers. Different groups have used both MFC and NCC to reinforce PLA films.

The group at Kyoto University has shown that MFCcan be used to reinforce PLA [124, 125]. Their MFC is typically in the range of 20–50 nm in diameter by several micrometers in length. At an MFC content of 10%, Young's modulus and tensile strength of PLA samples were significantly increased, without reducing the yield strength. By comparison, unrefined kraft pulp had a detrimental effect on mechanical properties while refined kraft pulp resulted in only marginal improvements. In another work [126] the same group blended PLA and NFC fibers in a handsheet machine. The resulting "sheets" were then compression moulded at high temperature and pressure into the final composite material.

Comparing the work of filled PLA done by the Kyoto group using MFC and by FPInnovations using NCC provides an important conclusion. Both MFC and NCC are effective reinforcing agents for PLA. However, the data show that using MFC to improve the strength of PLA requires ten times more material compared with the 2–3% of NCC needed to gain the same benefit.

20.5.4 Nanocrystalline Cellulose versus Inorganic Fillers

NCC has to compete with nanoinorganic fillers such as nanoclay, nanosilica, and nanoalumina. Nanoinorganics have been the subject of ongoing research for at least 20 years [127]. However, NCC has the advantages (at least) of being a nontoxic and more green material.

The ultimate goal of all nanoscale fillers is the same: to improve mechanical properties by reinforcing a polymer network (including coatings). Nanoclays are less costly than NCC. On the other hand, nanosilicas and nanoaluminas range in price from roughly equal to the price of NCC to several times higher, depending on the grade.

So far, little work direct comparison has been done between the properties of polymers filled with NCC and polymers filled with nanoinorganic materials. As already shown [102, 103] the group of Sanchez-Garcia demonstrated that NCC is superior at reducing the rate of water vapor and oxygen transmission through films of PLA compared with a nanoclay product.

As already noted [78], the Wood Products Division of FPInnovations has shown that NCC can improve the properties of industrial varnishes for wood products. The same group had previously [128, 129] examined nanosilica and nanoalumina as varnish additives. In all cases, the degree of dispersion of the nanomaterial is critical to the end-use, giving NCC an advantage at least in aqueous varnishes. In general we may say that NCC is at least competitive with nanosilica and nanoalumina, and for at least some measured properties, NCC is superior.

Several groups have looked at the combination of NCC with nanoinorganic fillers. Yano et al. [130] prepared a composite material into which silica nanoparticles were "mingled" with bacterial cellulose microfibrils. The modulus and strength of this composite were increased significantly compared with the cellulose film itself.

Shopsowitz et al. [131] suspended organosilicon compounds such as tetramethoxy silane $Si(OCH_3)_4$ in aqueous NCC, generating the chiral nematic organization that leads to iridescent colors that we now associate with NCC films. After high temperature pyrolysis of the organic material, the silica films have retained the chiral nematic footprint and the associated specific wavelength light reflection of the original NCC. This new silica structure is mesoporous because of the empty spaces left from NCC calcination. This suggests a means by which NCC can help create new products in combination with silica and other inorganic compounds. In a more recent paper, they have been able to load these open structures with silver nanoparticles [132].

Although not at the moment an end-use issue, the yield of NCC from the cellulose fibers can be as low as 50%. Therefore, there is also considerable opportunity to recover valuable chemical by-products. Jemaa et al. [133] demonstrated that monomeric and oligomeric sugars can be isolated from the NCC waste stream. These have the potential to add considerable value to any nanocellulose production process. In the same way, it is a matter of both economic and environmental concern that the sulphuric acid used in the preparation of NCC be recycled for reuse [133], or for the production of additional by-products [134].

20.5.5 Printed Electronics

For the last 10 years, there has been considerable interest in the use of conventional printing (offset, flexography, ink jet, and rotogravure) to fully print electronic products (e.g., organic light emitting diodes, solar power panels, RFID circuits, and more), at a very low cost. Paper and other cellulose derivatives have themselves been considered as substrates [135]. Because of problems in obtaining absolute fidelity of printed features—even more stringent than the requirements for very high quality color imaging—most researchers have been concentrating on plastic films such as poly(ethylene terephalate) (PET).

Most high quality plastic films have the smoothness and low porosity required to print very finely detailed electronic circuit components. However, these films have high coefficients of thermal expansion. This is absolutely unacceptable for feature placement in printed electronics. Recent work [136, 137] has shown that conventional polymer films either laminated or filled with NFC have an extremely low coefficient of thermal expansion. Films made from TEMPO-prepared nanofiber also showed very low coefficients of thermal expansion [71]. Composite sheets made from NFC and mineral filler such as kaolin clay or precipitated calcium carbonate gave very smooth surfaces, with properties (including dimensional changes with relative humidity) similar to those of a reference PET sheet [138]. New opportunities for plastic films designed for printed electronics may result from the combination of nanocellulose with synthetic plastic films.

Richardson [123] prepared conductive NCC by polymerizing aniline in the presence of NCC, creating a graft copolymer or at least a uniform coating of polyaniline on the NCC surface. This may have applications in printed electronics.

In a similar vein, Lee et al. showed that a "cellulose nanofiber sheet," likely a sheet of NFC, could be used as a separator material in a lithium ion battery, particularly due to its superior thermal stability [139].

Another potential use of NCC and related materials could be in providing a "template" for semiconducting materials. Padalkar et al. [140] precipitated materials such as cadmium sulphide onto "cellulose nanocrystals," particles either of NCC or of somewhat larger cellulose nanofibrils. However, no end-use properties were reported.

In summary, the recent research activities have identified many interesting and even exciting potential applications for nanocellulose products. This has helped attract significant interests from various end-users and funding from governments. However, a strong value proposition and business case must be defined, market size estimated, and partners/end-used recruited to develop real application opportunities and accelerate the commercialization.

20.6 CONCLUDING REMARKS

There has been an increasing interest in developing micro- and nanocellulosic products, leading to very extensive R&D efforts around the globe to develop and demonstrate manufacturing technologies for these products. At the same time, a large and growing body of information on potential end-uses of nanocellulose has been accumulated. This includes the NCC products now under advanced development at FPInnovations and CelluForce, Canada, as well as the MFC/NFC products of other organizations, such as Innventia, Sweden, and UPM and StoraEnso in Finland.

Although the field is still in its early development stages, it is clear that there is some overlap in properties from MCC (largest) to MFC, NFC, and NCC. There is also a certain continuum in size, at least in the fibrillated material such as MCC, MFC, and NFC. In some potential product areas, NCC has a clear technical advantage. In other areas, the data are not as clear, and further work is required.

Many potential applications for nanocellulosic materials have been identified and demonstrated in the lab. However, for each case, the challenge is to demonstrate a value proposition and to develop a strong business case. The involvement and partnership with the end-users are critical to bringing these new nanomaterials into the market.

APPENDIX A

This Appendix summarizes worldwide research, development, and production in the area of nanocellulose. It is worth noting that some players are involved in more than one research consortium. For example, Innventia (Sweden) has its own NFC program, and also coordinates the European Community's SustainComp program. VTT (Finland) is involved in the Finnish Centre for Nano Cellulosics, and also coordinates the European Community's SUNPAP (Scaling Up Nano Particles in Modern Papermaking) program. Similarly, FPInnovations (Canada) has its own NCC program in addition to a strong involvement in the Canadian ArboraNano program sponsored by the Government of Canada and the Government of Quebec and the FPInnovations/Domtar CelluForce joint venture .

In addition to systematic (often government funded) programs, there is also a considerable amount of small-scale, university-level research. Many such small-scale cases are presented within the body of this review. Commercial products and efforts are shown in Table 20.A.1. Governmental and other research groups are shown in Table 20.A.2.

ACKNOWLEDGMENTS

The authors gratefully acknowledge financial support from Natural Resources Canada under the Transformative Technology Program. We would like to thank Drs. Raman Nayar and Pierre LePoutre for valuable comments made during the review of this report.

TABLE 20.A.1 Commercial Products

Organization	Product	Commercial
Daicel	MFC	Yes
J. Rettenmaier & Söhne Inc.	MFC	Yes
Borregaard Biorefinery	MFC	Yes

TABLE 20.A.2 Industrial, University, and Government Research and Development

Organization	Products	Production Capabilities	Application Being Developed
FPInnovations, Canada, in partnership with Domtar (CelluForce) Also in partnership with ArboraNano (Governments of Canada and of Quebec)	Nanocrystalline cellulose (NCC)	1 kg/day pilot plant (batch) 3 kg/day (continuous) pilot plant 1 tonne per day demonstration plant Official opening in January 2012 (CelluForce)	Paints and coatings Films and barriers Textiles Polymer composites (including biopolymers) Excipients Cosmetics
Innventia, Sweden, in partnership withKTH, Stockholm and Aalto University, Helsinki Many Scandinavian and European producers	MFC and NFC	100 kg/day pilot plant in operation, February 2011 [51]	Polymer reinforcement Paper bulk and surface strength Coatings Superhydrophobic materials Nanofoams Superabsorbers
Borregaard Biorefinery, Norway, partnered with SCA, Sweden Reichhold Chemicals	MFC and NFC	100–200 kg/day pilot plant under construction	Paper strength and printability improvement Adhesives and polymer composites
Finnish Centre for Nano Cellulosics, Finland, in partnership with Aalto University, VTT, UPM	MFC and NFC	Details unknown	Paper and board strength and reinforcement Industrial coatings Foams, composites Adhesives Superabsorbers Films
PFI, Norway	NFC and MFC	Details unknown	Composite materials, paper strength, barrier properties

(*Continued*)

TABLE 20.2 ***(Continued)***

Organization	Products	Production Capabilities	Application Being Developed
EMPA, Switzerland	MFC and NFC	Small	Superabsorbers, biomedical, polymer reinforcement and composites, adhesives
US Dept of Agriculture and US Forest Products Laboratory	NFC and NCC (including TEMPO oxidation)	Support to US universities	Preparation and characterization of materials
Virginia Tech; Purdue University; University of Tennessee; Georgia Tech; Cornell University; Auburn University			Composites
Alberta Innovates and NINT, Canada, in partnership with Alpac	NCC	Plan for 100 kg/week pilot plant announced in July 2011	
INP-Pagora (Grenoble), France	NCC	Lab scale (grams)	Composites
CERMAV, (Grenoble) France	NCC	Lab scale (grams)	Fundamental Electromagnetic alignment
SustainComp, European Community Project, Coordinated by Innventia, Sweden	MFC/NFC	Unknown	
SunPap, European Community effort, coordinated by VTT, Finland Partnered with Stora Enso, Ahsltrom, UPM, Rettenmaier	NFC	Commercial material supplied by Rettenmaier	Papermaking implications

Kyoto University, University of Tokyo, Oji Paper, Nippon Paper, and others—Japan	Bacterial nanocellulose, MFC TEMPO-oxidized NCC in a variety of end-uses including biocomposites	Lab scale (grams) Commercial MFC supplied by Daicel and others	Composite materials (including biopolymer blends) Optical/thermal properties of clear films
University of Maine, with Maine state government and US federal governments	NFC (enzymatic, chemical, or mechanical)	Laboratory—pilot plant scale	Paper surface treatment Polymer reinforcement Paper coating additive
University of Oregon	MCC and NCC	Laboratory scale production	Polymer reinforcement, including rubber NCC aerogels Barrier layers
Bio Vision Technology, Canada; in partnership with the National Research Council of Canada—Biotechnology Research Institute	NCC; part of the overall, stated strategy of Bio Vision for a cellulose-based biorefinery to produce other chemicals plus energy from cellulose	Pilot production is claimed to start soon, but no details have been provided	Focus on production; application to be explored by partners
UPM, Finland	"Fibril Cellulose" described as a range of products from MFC to NFC	Pilot production started in November 2011; scale not given	Focus on papermaking, concrete additive, and paints/varnishes.
StoraEnso, Finland	MFC "pre-commercial plant"	Production claimed to begin at the end of 2011; scale not given	Focus on papermaking, reinforcing material, and barrier layers
Melodea, Israel	NCC derived from waste cellulosic material	Scale not given	NCC-based nanofoams (e.g., [91])

REFERENCES

[1] Eichhorn, S., et al. (2010) Review: Current international research into cellulose nanofibres and nanocomposites. *J. Mater. Sci.*, *45*, 1.

[2] Hubbe, M.A., Rojas, O.J., Lucia, L.A., and Sain, M. (2008) Cellulosic nanocomposites: A review. *BioResources*, *4*, 929.

[3] Dufresne, A. (2008) Polysaccharide nanocrystal reinforced nanocomposites. *Can. J. Chem.*, *86*, 484.

[4] Siró, I., Plackett, D. (2010) Microfibrillated cellulose and new nanocomposite materials: A review. *Cellulose*, *17*, 459.

[5] Habibi, Y., Lucia, L.A., and Rojas, O.J. (2010) Cellulose nanocrystals: Chemistry, self-assembly, and applications. *Chem. Rev.*, *110*, 3479.

[6] Peng, B.L., Dhar, N., Liu, H.L., and Tam, K.C. (2011) Chemistry and applications of nanocrystalline cellulose and its derivatives: A nanotechnology perspective. *Can. J. Chem. Eng.*, *89*, 1191.

[7] Yano, H. (2011) Current situation of nanocellulose in Japan, The 2011 Tappi International Conference on Nanotechnology for Renewable Resources, Washington, DC, June 2011, Tappi Press, Atlanta.

[8] Klemm, D., Kramer, F., Moritz, S., Lindström, T., Ankerfors, M., Gray, D.G., and Dorris, A. (2011) Nanocelluloses: A new family of nature-based materials. *Angew. Chem. Int. Ed.*, *50*, 5438.

[9] Hamel, J. (2011) Presented at the 97th Annual Meeting of PAPTAC, Montreal.

[10] Sneck, A. (2011) New approach to classification of cellulosic fibrils and suitable methods for their characterization, The 2011 Tappi International Conference on Nanotechnology for Renewable Resources, Washington, DC, June 2011, Tappi Press, Atlanta.

[11] Postek, M.T., et al. (2011) Development of the metrology and imaging of cellulose nanocrystals. *Meas. Sci. Technol.*, *22*, 024005.

[12] Thomson, S.L., O'Callaghan, D.J., Westland, J.A., and Su, B. (2010) Method of making a fiber cement board with improved properties and the product, US Patent Application 2010/0162926, July 1, 2010, assigned to Weyerhaeuser Co.

[13] Isogai, A. (2009) Individualization of nano-sized plant cellulose fibrils achieved by direct surface carboxylation using TEMPO catalyst, The 2009 TAPPI Nanotechnology Conference, Edmonton, Alberta, June 24–26, 2009.

[14] Dong, X., Revol, J.-F., and Gray, D.G. (1998) Effect of microcrystallite preparation on the formation of colloid crystals of cellulose. *Cellulose*, *5*, 19.

[15] Battista, O.A. (1966) Method of preparing stable aqueous dispersion-forming cellulosic aggregates, Assigned to FMC Corp., U.S. Patent 3,251,824, May 17, 1966.

[16] http://www.nbent.com/products.htm (accessed April 2, 2013).

[17] FAO/WHO Food Standards (2012) Microcrystalline cellulose (Cellulose Gel), http://www.codexalimentarius.net/ (accessed April 1, 2013).

[18] Pharmacopeial specifications, http://www.nbent.com/pharmacopial.htm (accessed April 2, 2013).

[19] Dukic-Ott, A., Thommes, M., Remon, J.-P., Kleinebudde, P., and Vervaet, C. (2009) Production of pellets via extrusion–spheronisation without the incorporation of microcrystalline cellulose: A critical review. *Eur. J. Pharmaceut. Biopharmaceut.*, *71*, 38.

[20] Laka, M., Chernyavskaya, S. (2007) Obtaining microcrystalline cellulose from softwood and hardwood pulp. *Bioresources*, *2*(3), 583.

[21] Agblevor, F., Ibrahim, M., and El-Zawawy, W. (2007) Coupled acid and enzyme mediated production of microcrystalline cellulose from corn cob and cotton gin waster. *Cellulose*, *14*, 247.

[22] Moran, J., Alvarez, V., Cyras, V., and Vazquez, A. (2008) Extraction of cellulose and preparation of nanocellulose from sisal fibres. *Cellulose*, *15*, 149.

[23] Turbak, A., Snyder, F., and Sandberg, K. (1982) Food products containing microfibrillated cellulose, Assigned to ITT, US Patent 4,341,807, July 27, 1982.

[24] Turbak, A., Snyder, F., and Sandbertg, K. (1983) Microfibrillated cellulose, a new cellulose product: Properties, uses, and commercial potential. *J. Appl. Polym. Sci.*, *37*, 815.

[25] Henriksson, M., Berglund, L.A. (2007) Structure and properties of cellulose nanocomposite films containing melamine formaldehyde. *J. Appl. Polym. Sci.*, *106*, 2817.

[26] Taniguchi, T., Okamura, J. (1998) New films produced by microfibrillated natural fibers. *Polym. Int.*, *47*, 291.

[27] Said, M.A.S.A., Alloin, F., Sanchez, J.-Y., El Kissi, N., and Dufresne, A. (2004) Preparation of cellulose whiskers reinforced nanocomposites from an organic medium suspension. *Macromolecules*, *37*, 1386.

[28] Venables, A., Buliga, G., Dell, S., and Colliopoulos, J. (2000) Ultra-fine microcrystalline cellulose compositions and processes, Assigned to FMC Corp., US Patent 6,037,380, March 14, 2000.

[29] Matsuda, Y., Hirose, M., and Ueno, K. (2001) Super microfibrillated cellulose, process for producing the same, and coated paper and tinted paper using the same, Assigned to Tokoshu Paper Co., US Patent 6,183,596, February 6, 2001.

[30] Eriksen, Ø., Syverud, K., and Gregersen, Ø. (2008) The use of microfibrillated cellulose produced from kraft pulp as strength enhancer in TMP paper. *Nord. Pulp. Paper Res. J.*, *23*, 299.

[31] J. Rettenmaier & Söhne GmbH, http://www.jrs.de (accessed July 15, 2010).

[32] New fibre product from borregaard, http://www.orkla.com (accessed August 30, 2010).

[33] Theliander, H., Wernersson, F., Hanson, C., Gustafson, I., and Falk, T. (2011) Absorbent article comprising a composite material, WIPO Patent Application WO/2011/090425, Assigned to SCA Hygiene Products AB, July 28, 2011.

[34] Daicel Chemical Industries, Ltd. http://www.daicel.com/en/research/activity.html (accessed April 1, 2013).

[35] Beck, S., Bouchard, J., and Berry, R. (2012) Dispersibility in water of dried nanocrystalline cellulose, *Biomacromolecules*, *13*, 1486.

[36] Beck, S., Bouchard, J., and Berry, R. (2010) A method to control the dispersibility and barrier properties of dried nanocrystalline cellulose in solutions of different

pH and ionic strength, US Patent Application 2010/0148118 Assigned to FPInnovations, June 17, 2010.

[37] Takagi, H., Asano, A. (2007) Characterization of "Green" composites reinforced by cellulose nanofibers. *Key Eng. Mater.*, *334–335*, 389.

[38] http://www.storaenso.com/media-centre/press-releases/2011/05/Pages/stora-enso-makes-a-groundbreaking.aspx (accessed December 29, 2011).

[39] UPM starts pre-commercial production of fibril cellulose, http://www.paperage.com/2011news/11_15_2011upm_fibril_cellulose.html (accessed December 20, 2011).

[40] Kapanen, A., Vartiainen, J., Lappalainen, T., Korhonen, P., and Vikman, M. (2011) Toxicity and characteristics of microfibrillated cellulose in kinetic luminescent bacteria test environment, VTT Technical Research Centre of Finland.

[41] UPM Viewpoint: UPM has started pre-commercial production of fibril cellulose, http://www.youtube.com/watch?v=UyOO4J7eQao (accessed December 21, 2011).

[42] Stenstad, P., Andresen, M., Tanem, B., and Stenius, P. (2008) Chemical surface modifications of microfibrillated cellulose. *Cellulose*, *15*, 35.

[43] Andresen, M., Johansson, L.S., Tanem, B.J., and Stenius, P. (2006) Properties and characterization of hydrophobized microfibrillated cellulose. *Cellulose*, *13*, 665.

[44] Lönnberg, H., Fogelström, L., Said, M.A., Berglund, L., Malström, E., and Hult, A. (2008) Surface grafting of microfibrillated cellulose with poly(ε-caprolactone)—Synthesis and characterization. *Eur. Polym. J.*, *44*, 2991.

[45] Bergenstråhle, M., Mazeau, K., and Berglund, L.A. (2008) Molecular modeling of interfaces between cellulose crystals and surrounding molecules: Effects of caprolactone surface grafting. *Eur. Polym. J.*, *44*, 3662.

[46] Zhang, J., Elder, T.J., Pu, Y., and Ragauskas, A.J. (2007) Facile synthesis of spherical cellulose nanoparticles. *Carbohydr. Polym.*, *69*, 607.

[47] Pu, Y., Zhang, J., Elder, T., Deng, Y., Gtenholm, P., and Ragauskas, A.J. (2007) Investigation into nanocellulosics vs. acacia reinforced acrylic films. *Compos. Part B: Eng.*, *38*, 360.

[48] Zimmermann, T., Pöhler, E., and Geiger, T. (2004) Cellulose fibrils for polymer reinforcement. *Adv. Eng. Mater.*, *6*, 754.

[49] Zimmermann, T., Pöhler, E., and Schwaller, P. (2005) Mechanical and morphological properties of cellulose film reinforced nanocomposites. *Adv. Eng. Mater.*, *7*, 1156.

[50] López-Suevos, F., Eyholzer, E., Bordeanu, N., and Richter, K. (2010) DMA analysis and wood bonding of PVAc latex reinforced with cellulose nanofibrils. *Cellulose*, *17*, 387.

[51] Nanocellulose—for the first time on a large scale, http://www.innventia.com/en/About-us/News1/Press-Release-Nanocellulose—for-the-first-time-on-a-large-scale/ (accessed April 1, 2013).

[52] Chakraborty, A., Sain, M., Kortschot, M., and Ghosh, S. (2007) Modelling energy consumption for the generation of microfibres from bleached kraft pulp fibres in a PFI mill. *Bioresources*, *2*, 210.

[53] Cellulose nanocomposites, http://www.empa.ch/plugin/template/empa/*/39597 (accessed November 18, 2010).

[54] Mörseburg, K., Chinga-Carrasco, G. (2009) Assessing the combined benefits of clay and nanofibrillated cellulose in layered TMP-based sheets. *Cellulose*, *16*, 795.

[55] Hult, E.L., Iotti, M., and Lenes, M. (2010) Efficient approach to high barrier packaging using microfibrillar cellulose and shellac. *Cellulose*, *17*, 575.

[56] Hentze, H.-P., Sievanen, J., Kettle, J., Kulachenko, A., Korpela, A., Ketoja, J., Hellen, E., Hjelt, T., Hiltunen, J., Turunen, E., and Sneck, A. (2010) Method of manufacturing paper and products obtained by the method. Assigned to UPM-Kymmene Corp and Stora Emnso OYJ, World Intellectual Property Organization, WO 2010/142846, December 16, 2010.

[57] Henriksson, M., Berglund, L., Bulone, V., and Zhou, Q. (2010) Method of producing and the use of microfibrillated paper, World Intellectual Property Organization WO 2010/134868, Assigned to Swetree Technologies AB, November 25, 2010.

[58] Marchessault, R.H., Morehead, F.F., and Walter, N.M. (1959) Liquid crystalline systems from fibrillar polysaccharides. *Nature*, *184*, 632.

[59] Revol, J.-F., Godbout, D.L., and Gray, D.G. (1997) Solidified liquid crystal of cellulose with optically variable properties, US Patent 5629055, Assigned to Pulp and Paper Research Institute of Canada, May 13, 1997.

[60] Beck, S., Bouchard, J., and Berry, R. (2011) A new method to control iridescence colour in solid films of nanocrystalline cellulose. *Biomacromolecules*, *12*, 167–172.

[61] CelluForce starts operations in Windsor, Quebec, http://foresttalk.com/index.php/2011/12/12/celluforce-starts-operations-in-windsor-quebec/ (accessed April 1, 2013).

[62] http://www.celluforce.com (accessed June 6, 2011).

[63] Su, S., Hamad, W.Y. (2011) Nanocomposite hydrogel and method for preparing it, for industrial and medical applications, US Patent Application 2011/0182990, Assigned to FPInnovations, July 28, 2011.

[64] Hamad, W.Y., Su, S. (2011) Thermoplastic nanocomposite material based on nanocrystalline cellulose (NCC), US Patent Application 20110201755, Assigned to FPInnovations, August 18, 2011.

[65] Peresin, M.S., Habibi, Y., Zoppe, J.O., Pawlak, J.J., and Rojas, O.J. (2010) Nanofiber composites of poly vinyl alcohol and cellulose nanocrystals: Manufacture and characterization. *Biomacromolecules*, *11*, 674.

[66] Jiang, Z.-H., Berry, R., Bouchard, J., and Audet, A. (2011) Adhesion with nanocrystalline cellulose, US Patent Application 20110293932, assigned to FPInnovations, December 1, 2011.

[67] Jackson, J.K., Letchford, K.L., Wasserman, B.J., Ye, L., Hamad, W.Y., and Burt, H.M. (2011) The use of nanocrystalline cellulose for the binding and controlled release of drugs. *Int. J. Nanomedicine*, *6*, 321.

[68] Li, Z., Renneckar, S., and Barone, J.R. (2010) Nanocomposites prepared by *in situ* enzymatic polymerization of phenol with TEMPO-oxidized nanocellulose. *Cellulose*, *17*, 57.

[69] Ahola, S. (2008) Properties and interfacial behaviour of cellulose nanofibrils, PhD Thesis, Helsinki University of Technology.

[70] Saito, T., Hirota, M., Tamura, N., Kimura, S., Fukuzumi, H., Heux, L., and Isogai, A. (1992) Individualization of nano-sized plant cellulose fibrils by direct surface

carboxylation using TEMPO catalyst under neutral conditions. *Biomacromolecules*, *10*, 2009.

[71] Fukuzumi, H., Saito, T., Iwata, T., Kumamoto, Y., and Isogai, A. (2009) Transparent and high gas barrier films of cellulose nanofibers prepared by TEMPO-mediated oxidation. *Biomacromolecules*, *10*, 162.

[72] Li, Q., Rennecakr, S. (2009) Molecularly thin nanoparticles from cellulose—Isolation of sub-microfibrillar structures. *Cellulose*, *16*, 1025.

[73] Alberta nanotechnology strategy, http://www.albertatechfutures.ca/LinkClick.aspx?fileticket=m3PeyCuAOMY%3D&tabid=150 (accessed March 15, 2011).

[74] Tetreau, J.P. (2010) Impact of nanotechnology in Alberta: NCC, October 30, 2010, http://www.thecis.ca/cms3/userfiles/Image/NCC%20CaseStudy%20V11%2030Oct2010.pdf (accessed March 15, 2011).

[75] New nano material developed from wood and straw pulp is stronger than steel, http://foresttalk.com/index.php/2011/07/06/new-nano-material-developed-from-wood-and-straw-pulp-is-stronger-than-steel/ (accessed August 12, 2011).

[76] BioVision Technology Inc. http://www.biovisiontechnology.com/ (accessed April 2, 2013).

[77] Nano-polymer could reduce carbon footprint, National Research Council of Canada, http://www.nrc-cnrc.gc.ca/eng/news/nrc/2011/02/01/nano-biovision.html (accessed March 1, 2011).

[78] Landry, V., Blanchet, P. (2009) Nanocomposites coatings for wood industries, 2009 TAPPI International Conference on Nanotechnology for the Forest Products Industry, Edmonton, Alberta, Tappi Press, Atlanta.

[79] Jämsä, S., Kataja, K., Takala, S., Putkisto, K., Vastamäki, P., and Dyhr, H. (2011) Method for manufacturing paint or varnish, WIPO Patent Application WO/2011/124759, Assigned to VTT, Finland, October 13, 2011.

[80] Henriksson, M., Berglund, L.A., Isaksson, P., Lindström, T., and Nishino, T. (2008) Cellulose nanopaper structures of high toughness. *Biopolymers*, *9*, 1579.

[81] Zou, X., Tan, X., and Berry, R. (2010) Flexible, iridescent NCC film, and method for preparation, Assigned to FPInnovations, US Patent, PCT/CA2010/000638, November 4, 2010.

[82] http://www.vtt.fi/news/2012/20120223_birch_fibril_pulp.jsp (accessed April 2, 2013).

[83] Ankerfors, M., Hoc, M., Lindström, T., and Song, H. (2010) Reduction of the linting and dusting tendency by using microfibrillated cellulose coatings, 2010 TAPPI International Conference on Nanotechnology for the Forest Products Industry, Espoo, Finland, Tappi Press, Atlanta.

[84] Ankerfors, M., Lindstrom, T., Song, H., and Hoc, M. (2011) Composition for coating of printing paper, US Patent Application 2011/0081554, Assigned to Innventia AB, April 7, 2011.

[85] Svagan, A.J., Samir, M.A.S., and Berglund, L.A. (2008) Biomimetic foams of high mechanical performance based on nanostructured cell walls reinforced by native cellulose microfibrils. *Adv. Mater.*, *20*, 1263.

[86] Svagan, A.J., Jensen, P., Berglund, L.A., Furó, I., and Dvinskikh, S.V. (2010) Towards tailored hierarchical structures in starch-based cellulose nanocomposties prepared by freeze drying. *J. Mater. Chem.*, *20*, 6646.

[87] Aulin, C. (2009) Novel oil resistant cellulosic materials, PhD Thesis, Royal Institute of Technology, Stockholm.

[88] Svagan, A. (2008) Bio-inspired cellulose nanocomposites and foams based on starch matrix, PhD Thesis, Royal Institute of Technology, Stockholm.

[89] Heath, L., Thielemans, W. (2010) Cellulose nanowhisker aerogels. *Green Chem.*, *12*, 1448.

[90] Kelley, J., Taylor, M., Mekap, D., Simonsen, J., and Arey, B. (2009) Cellulose nanocrystal aerogels, The 2009 TAPPI Nanotechnology Conference, Edmonton, Alberta, June 24–26, 2009.

[91] Lapidot, S. (2011) Nano crystalline cellulose composite foam from renewable resources, 2011 Tappi International Conference on Nanotechnology for Renewable Resources, Washington, DC, Tappi Press, Atlanta.

[92] http://www.melodea.eu (accessed December 31, 2011).

[93] Eriksen, Ø., Gregersen, Ø.W., and Syverud, K. (2007) Characterization of nanofibrillar cellulose (NFC), COST Action E54. Characterisation of the fine structure and properties of papermaking fibres using new technologies. Programme and extended abstracts of the opening seminar, Riga, Latvia, 25–27 Apr. 2007, 15–18.

[94] Micro- and nanocellulose fiber filled engineering thermoplastic composites, US Department of Agriculture, http://www.reeis.usda.gov/web/crisprojectpages/0211928-micro-and-nanocellulose-fiber-filled-engineering-thermoplastic-composites.html (accessed April 2, 2013).

[95] Siddiqui, N. (2008) Characterization of mechanically and enzymatically produced cellulose nanofibers from wood pulp, M.Sc. Thesis, University of Maine, Orono.

[96] Hamada, H., Bousfield, D.W., (2010) Nano-fibrillated cellulose as a coating agent to improve print quality on synthetic fiber sheet, 11th TAPPI Advanced Coating Fundamentals Symposium, Munich, October 2010.

[97] Hamada, H., Beckvermit, J., and Bousfield, D.W. (2010) Nanofibrillated cellulose with fine clay as a coating agent to improve print quality, Paper presented at Papercon 2010, Tappi Press Atlanta.

[98] Heiskanen, I., Backfolk, K. (2011) A composition comprising microfibrillated cellulose and a process for the production of a composition, WIPO Patent Application WO/2011/141876, November 17, 2011, Assigned to StoraEnso Oyj.

[99] Aulin, C., Lindström, T. (2011) Biopolymer coatings for paper and paperboard. In: *Biopolymers—New Materials for Sustainable Films and Coatings*, ed. D. Plackett. Chichester, UK: J. Wiley & Sons.

[100] Jenkins, S. (2009) The future of functional and barrier coatings for paper and board, Report published by PIRA International Ltd., . 23.

[101] Aulin, C., Mikael Gällstedt, M., and Lindström, T. (2010) Oxygen and oil barrier properties of microfibrillated cellulose films and coatings. *Cellulose*, *17*, 559.

[102] Sanchez-Garcia, M.D., Lagaron, J.M. (2010) On the use of plant cellulose nanowhiskers to enhance the barrier propertie of poly lactic acid. *Cellulose*, *17*, 987.

[103] Sanchez-Garcia, M.D., Gimenez, E., and Lagaron, J.M. (2007) Novel PET nanocomposites of interest in food packaging applications and comparative barrier performance with biopolyester nanocomposites. *J. Plastic Film Sheet.*, *23*, 133.

[104] Mir, N., Beaudry, R.M. (2004) Modified atmosphere packaging. In: *The Commercial Storage of Fruits, Vegetables, and Florist and Nursery Stocks*, Agricultural Handbook 66. Washington, DC US Department of Agriculture.

[105] Syverud, K., Stenius, P. (2009) Strength and barrier properties of MFC films. *Cellulose*, *16*, 75.

[106] Viet, D., Beck-Candanedo, S., and Gray, D.G. (2007) Dispersion of cellulose nanocrystals in polar organic solvents. *Cellulose*, *17*, 109.

[107] Klungness, J.H., Caulfield, D.F. (1982) Mechanisms affecting fiber bonding during drying and aging of pulps. *TAPPI*, *65*(12), 94.

[108] Eyholzer, C., Bordeanu, N., Lopez-Suevos, F., Rentsch, D., Zimmermann, T., and Oksman, K. (2010) Preparation and characterization of water-dispersible nanofibrillated cellulose in powder form. *Cellulose*, *17*, 19.

[109] Boluk, Y., Zhao, L. (2010) Aircraft anti-icing fluids formulated with nanocrystalline cellulose, US Patent Application 2010/0327215, Assigned to Alberta Research Council, December 30, 2010.

[110] Avérous, L., Fringant, C., and Moro, L. (2001) Starch-based biodegradable materials suitable for thermoforming packaging. *Starch/Stärke*, *53*, 368.

[111] Nakagaito, A.N., Yano, H. (2005) Novel high-strength biocomposites based on microfibrillated cellulose having nano-order-unit web-like network structure. *Appl. Phys. A*, *80*, 155.

[112] Hajji, P., Cavaillé, J.Y., Favier, V., Gauthier, C., and Vigier, G. (1996) Tensile behaviour of nanocomposites from latex and cellulose whiskers. *Polym Compos.*, *17*, 612.

[113] Kim, J., Montero, G., Habibi, Y., Hinestroza, J., Genzer, J., Argyropoulos, D., and Rojas, O. (2009) Dispersion of cellulose crystalline by non-ionic surfactants in a hydrophobic polymer matrix. *Polym. Eng. Sci.*, *49*, 2054.

[114] Jovanovic, G., Atre, S.V., Paul, B.K., Simonsen, J., Remcho, V.T., and Chang, C.-H. (2011) Microfluidic devices, particularly filtration devices comprising polymeric membranes, and method for their manufacture and use, US Patent Application 20110253629, Assigned to the State University of Oregon, October 20, 2011.

[115] Li, S., Gao, Y., Bai, H., Zhang, L., Qu, P., and Bai, L. (2011) Preparation and characteristics of polysulfone dialysis membranes modified with nanocrystalline cellulose. *BioResources*, *6*, 1670.

[116] Azeredo, H.M.C., Mattoso, L.H.C., Wood, D., Williams, T.G., Avena-Bustillos, R.J., and Mchugh, T. (2009) Nanocomposite edible films from mango puree reinforced with cellulose nanofiber. *J. Food Sci.*, *74*, N31.

[117] Health impact of nanotechnologies in food production; Commission Directive 96/77/EC. of the European Parliament and of the council of 2 December 1996 laying down specific purity criteria on food additives other than colours and sweeteners. Official Journal of the European Union L 339, 30.12.1996, . 1.

[118] Grobe, A., Renn, O., and Jaeger, A., (2012) Risk governance of nanotechnology applications in food and cosmetics, International Risk Governance Council, http://www.nanowerk.com/nanotechnology/reports/reportpdf/report125.pdf (accessed April 2, 2013).

[119] Bouwmeester, H., Dekkers, S., Noordam, M., Hagens, H., Bulder, A., de Heer, C., ten Voorde, S., Wijnhoven, S., and Sips, A. (2007) Health impact of nanotechnologies in food production, RIKILT—Institute of Food Safety, Wageningen Univer-

sity and Research Centre; RIVM—National Institute for Public Health & the Environment, the Netherlands, http://edepot.wur.nl/120669 (accessed September 10, 2010).

[120] Bendahou, A., Kaddami, H., and Dufresne, A. (2010) Investigation on the effect of cellulosic nanoparticles' morphology on the properties of natural rubber-based nanocomposites. *Eur. Polym. J.*, *46*, 209.

[121] Bai, W. (2009) New application of crystalline cellulose in rubber composites, PhD Thesis, Oregon State University, November 2009.

[122] Wu, Q., Henriksson, M., Liu, X., and Berglund, L.A. (2007) A high strength nanocomposite based on microcrystalline cellulose and polyurethane. *Biomacromolecules*, *8*, 3687.

[123] Richardson, T.B. (2009) Nanoreinforced shape memory polyurethane, PhD Thesis, Auburn University, Auburn AL, December 2009.

[124] Iwatake, A., Nogi, M., and Yano, H. (2008) Cellulose nanofibre-reinforced polylactic acid. *Compos. Sci. Technol.*, *68*, 2103.

[125] Suryanegara, L., Nakagaito, A.N., and Yano, H. (2009) The effect of crystallization of PLA on the thermal and mechanical properties of microfibrillated cellulose-reinforced PLA composites. *Compos. Sci. Technol.*, *69*, 1187.

[126] Nakagaito, A.N., Fujimura, A., Sakai, T., Hama, Y., and Yano, H. (2009) Production of microfibrillated cellulose (MFC) reinforced polylactic acid (PLA) nanocomposites from sheets obtained by a papermaking-like process. *Compos. Sci. Technol.*, *69*, 1293.

[127] Hamad, W. (2006) On the development and applications of cellulosic nanofibrillar and nanocrystalline materials. *Can. J. Chem. Eng.*, *84*, 513.

[128] Landry, V., Blanchet, P., and Riedl, B. (2010) Mechanical and optical properties of clay-based nanocomposite coatings for wood flooring. *Prog. Org. Coat.*, *67*, 381.

[129] Sow, S., Riedl, B., and Blanchet, P. (2010) UV-waterborne polyurethane-acrylate nanocomposite coatings containing alumina and silica nanoparticles for wood: Mechanical, optical, and thermal properties assessment. *J. Coat. Technol. Res.*, *4*.

[130] Yano, S., Maeda, H., Nakajima, M., Hagiwara, T., and Sawaguchi, T. (2008) Preparation and mechanical properties of bacterial cellulose nanocomposites loaded with silica nanoparticles. *Cellulose*, *15*, 111.

[131] Shopsowitz, K., Qi, H., Hamad, W., and MacLachlan, M. (2010) Free-standing mesoporous silica films with tunable chiral nematic structures. *Nature*, *468*, 422.

[132] Qi, H., Shopsowitz, K.E., Hamad, W.Y., and MacLachlan, M.J. (2011) Chiral nematic assemblies of silver nanoparticles in mesoporous silica thin films. *J. Am. Chem. Soc.*, *133*, 3728.

[133] Jemaa, N., Paleologou, M., and Zhang, X. (2011) Fractionation of a waste liquor stream from nanocrystalline cellulose production, Assigned to FPInnovations, World Intellectual Property Organization Application WO 2011/017797, Published February 17, 2011.

[134] Hashhaikey, R., Hu, T.H., and Berry, R. (2010) Crystalline sulphated cellulose II and its production from sulphuric acid hydrolysis of cellulose, US Patent Application 2010/0286387, November 11, 2010, Assigned to FPInnovations.

[135] Aspler, J.S. (2004) Paper: The other flexible substrate, IDTechEx Printed Electronics Conference, New Orleans, December 2004.

[136] Nogi, M., Iwamoto, S., Nakagaito, A., and Yano, H. (2009) Optically transparent nanofiber paper. *Adv. Mater.*, *21*, 1595.

[137] Nogi, M., Yano, H. (1849) Transparent nanocomposites based on cellulose produced by bacteria offer innovation in the electronics device industry. *Adv. Mater.*, *20*, 2008.

[138] Sievanen, J., Torvinen, K., Hjelt, T., and Hellen, E. (2011) A new type of filler-nanocellulose composite substrate for printed electronics applications, The 2011 Tappi International Conference on Nanotechnology for Renewable Resources, Washington, DC, June 2011, Tappi Press, Atlanta.

[139] Lee, J.-H., Chun, S.-J., Lee, S.-Y., Doh, G.-H., and Jung, C. (2010) Cellulose nanofiber sheet as separator for Li ion battery, The 15th International Meeting on Lithium Batteries—IMLB 2010, Montreal.

[140] Padalkar, S., Capadona, J.R., Rowan, S.J., Weder, C., Moon, R.J., and Stanciu, L.A. (2011) Self-assembly and alignment of semiconductor nanoparticles on cellulose nanocrystals. *J. Mater. Sci.*, *46*, 5672.

CHAPTER 21

Spectroscopic Characterization of Renewable Nanoparticles and Their Composites

MIRTA I. ARANGUREN, MIRNA A. MOSIEWICKI, and NORMA E. MARCOVICH

The three most abundant renewable polysaccharides in nature share rather similar building blocks, but they exhibit quite different properties because the existing differences in molecular structure lead to different arrangements and supramolecular structures. Their morphologies in nature arc that of fibers in the case of cellulose and chitin and of granules in the case of starch, all of them showing crystalline and amorphous phases [1]. Furthermore, different treatments have been developed taking advantage of the large concentration of hydroxyl groups present in polysaccharides, an approach that can be applied for bulk and surface modifications [1]. In particular, surface modification is performed to improve interfacial compatibility and dispersion if the intended use of the nanoparticles is polymer reinforcement, or to achieve specific interactions with other material components if functional properties are required.

Spectroscopic techniques, such as Fourier transform infrared (FT-IR) and Raman spectroscopies, nuclear magnetic resonance (NMR), X-ray photoelectron spectroscopy (XPS), and X-ray diffraction allow investigation of the chemical composition of the original materials as well as the modifications introduced to tailor specific properties. Additionally, they can also supply information with respect to the degree and type of crystallinity in the particles, X-ray spectroscopy being the most frequently used technique for this purpose. Usually, characterization by FT-IR spectroscopy is most frequently reported. Although Raman spectroscopy is less used, it does not suffer interference from water as does infrared spectroscopy, which is particularly useful for the analysis of hygroscopic natural fibers. The combination of both techniques offers

Biopolymer Nanocomposites: Processing, Properties, and Applications, First Edition. Edited by Alain Dufresne, Sabu Thomas, and Laly A. Pothan.

complementary information on the chemical nature of the particles and their composites [2].

21.1 CELLULOSE NANOPARTICLES

Cellulose, the main building component of plant biomass, is the most abundant biopolymer on the planet, its production estimated at about 180 billion tons per year [3, 4]. Cellulose is a linear polysaccharide of (1→4)-linked β-D-glucopyranose units [3, 4], insoluble in water, and most organic solvents, and as it appears in nature, it has low chemical reactivity.

Cellulose nanofibers have excellent mechanical properties (high specific strength and modulus), large specific surface area, low coefficient of thermal expansion, high aspect ratio, environmental benefits, and low cost [5–10]. Several methods for extracting cellulose fibers from vegetable sources have been reported including mechanical, biological, and chemical approaches. Among the mechanical methods, cryo-crushing [11, 12], grinding [13, 14], and high pressure homogenizing [15, 16] can be mentioned. As an example of biological methods, enzyme-assisted hydrolysis has been reported by several authors [17–19]. However, the most widely used method to obtain nanofibers from different cellulose sources is the chemical treatment involving acid hydrolysis [20–22].

Besides plant cellulose, there are other sources of this material, such as bacterial cellulose, which is a representative example of a microbial extracellular polymer. One of its most important characteristics is its high chemical purity, contrary to plant cellulose that appears usually combined with wax, pectin, hemicelluloses, and lignin [23, 24]. Ribbon-like fibers from bacterial cellulose are considerably finer than plant cellulose fibers and form a dense reticulated structure stabilized by extensive hydrogen bonding. It also presents a high degree of polymerization and crystallinity, which is the reason for its very high Young's modulus [25, 26].

Independently of the source, cellulose has been treated to individualize crystals using several different methods. Probably, the most frequently reported method is the acidic treatment of cellulose sources [27–30], but other treatments or combination of them have been proposed: alkali treatment followed by sulfuric acid hydrolysis of branch-barks of mulberry [31], a combination of mechanical and high-ultrasonic treatments for poplar wood [10], hydrochloric acid hydrolysis of cotton fibers, followed by microbial attack with cellulolytic fungus *Trichoderma reesei* [32].

21.2 PLANT CELLULOSE

21.2.1 FT-IR Characterization

Figure 21.1 shows the FT-IR spectra of original mulberry barks, pretreated mulberry barks, cellulose fibers obtained by alkali treatment, and cellulose

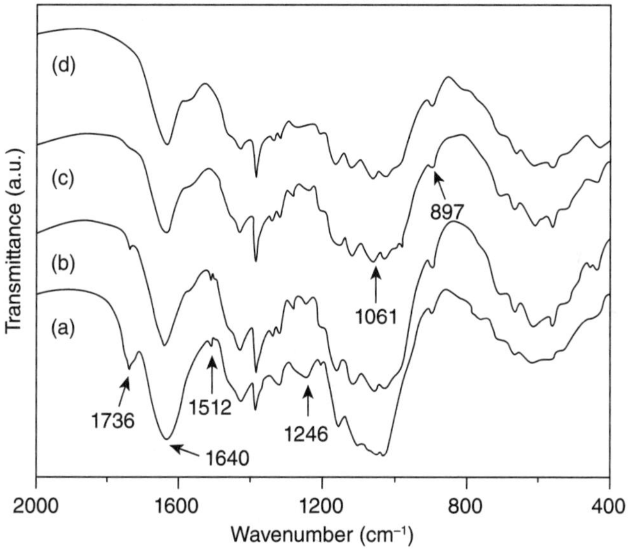

Figure 21.1 FT-IR spectra of (a) original mulberry barks, (b) pretreated barks with the NaOH solution, (c) cellulose fibers obtained at 130°C by alkali treatment of pretreated mulberry barks, and (d) cellulose whiskers. From Reference [31].

whiskers obtained from sulfuric acid hydrolysis. The peak centered at 1736 cm^{-1} in the spectrum of original mulberry barks (Fig. 21.1a) is attributed to the acetyl and uronic ester groups of hemicellulose or the ester linkage of the carboxylic group of ferulic, and p-coumaric acids of lignin and/or hemicellulose [33, 34]. It decreased in the spectrum of pretreated barks (Fig. 21.1b) and finally disappeared in the spectra of cellulose fibers obtained at 130°C by alkali treatment (Fig. 21.1c) and cellulose whiskers obtained after sulfuric acid hydrolysis (Fig. 21.1d), due to the removal of hemicellulose and lignin during the chemical extraction. The spectrum of the whiskers presents typical bands of cellulose including the band assigned to the vibrations of the H–O–H of the absorbed water at 1640 cm^{-1} associated with the bending vibrations of the hydrogen bonding of cellulose [29, 31, 35]. The spectra of the nanofibers also present a band in the range of 3000–4000 cm^{-1} (not shown) due to the stretching vibrations of –OH groups [29, 35]. The peaks at 1061 and 897 cm^{-1} are associated with the C–O stretching and β linkage in cellulose [30–35]. No difference was found between the spectra of cellulose whiskers and that of cellulose fibers, indicating that the molecular structure was preserved during hydrolysis.

21.2.2 Wide-Angle X-ray Diffraction Characterization

Many studies have reported on the WAXD of different cellulose sources and cellulose nanofibers extracted by different methods [10, 30, 31, 36]. As an

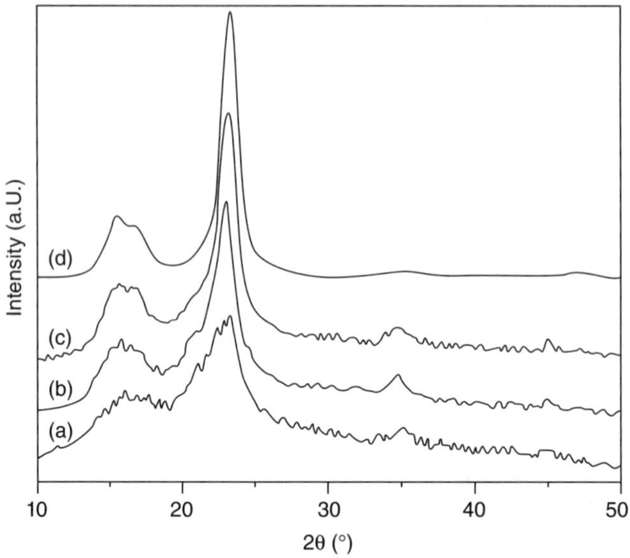

Figure 21.2 X-ray diffraction patterns of (a) original mulberry barks, (b) pre-treated barks with the NaOH solution, (c) cellulose fibers obtained at 130°C by alkali treatment of pretreated mulberry barks, and (d) cellulose whiskers. From Reference [31].

example, Figure 21.2 shows the X-ray patterns obtained using monochromatic CuKa radiation (λ = 1.54 Å) of mulberry barks, pretreated barks, cellulose fibers obtained by alkali treatment, and whiskers obtained by sulfuric acid hydrolysis. All the samples showed a typical of cellulose I pattern with well-defined peaks at 2θ = 23° and 16° [10, 36, 37]. The peaks became sharper with the treatment steps, from the original bark, to cellulose fibers and finally, to the whiskers, indicating an increase of crystallinity related to the removal of hemicellulose and lignin. Several authors have shown that although the mentioned peaks change in intensity, they remain during treatments, denoting that the crystal structure of cellulose is not changed during chemical and ultrasonic treatments [10, 29].

The percentage of crystallinity of the samples during the process to obtain cellulose nanofibers depends of several variables, such as raw cellulose source, type, and conditions of the treatment. Li et al. [31] observed a gradual increase of crystallinity from 46.9% for the original mulberry bark, to 58.8%, 63.7%, and 73.4% for the pretreated bark, cellulose fibers (obtained by alkali treatment), and cellulose whiskers, respectively. As another example, Marcovich et al. [30] found that hydrolysis of microcrystalline cellulose increased the percentage of crystallinity from 56% to 67%. In addition, Oksman et al. [36] reported that the crystallinity of a bioresidue from bioethanol production was 14.5%, and it increased to 73%, 77%, and 75% after sonication, homogenization, and hydrolysis, respectively. Crystallinity values in the range of 70–80% have also been reported for nanocrystals isolated by acid hydrolysis of different cellulose sources [38].

21.3 BACTERIAL CELLULOSE

Bacterial cellulose has already been identified as an useful material for biomedical applications because of the thin fibers obtained, its high purity, biocompatibility, high water holding capacity, great elasticity and high wet strength, and conformability [23, 26].

The effects of acid hydrolysis time, further neutralization, and dialysis on the properties of bacterial cellulose nanowhiskers have been reported by Martínez-Sanz and coworkers [39]. On the other hand, to improve rehydration of bacterial cellulose after drying (a problem caused by its high crystallinity), Huang and coworkers [26] added hydroxypropylmethyl cellulose into a fermentation culture medium, interfering with the formation of the bacterial cellulose and achieving *in situ* modification.

21.3.1 FT-IR Characterization

The spectra of the native bacterial cellulose and the nanowhiskers obtained after 2 hours and 48 hours of acid hydrolysis are shown in Figure 21.3. No significant differences were found between the chemical structures of plant and bacterial cellulose [26, 39, 40]. However, a shoulder appeared at 1720 cm^{-1}

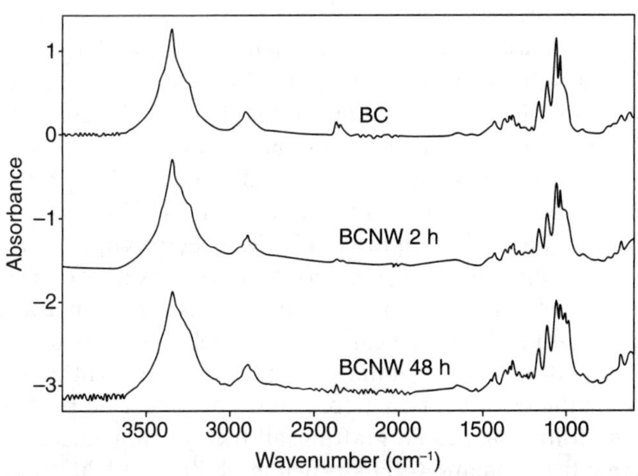

Figure 21.3 FT-IR spectra of bacterial cellulose (BC) and the obtained nanowhiskers (BCNW) after 2 hours and 48 hours of sulfuric acid hydrolysis. From Reference [39].

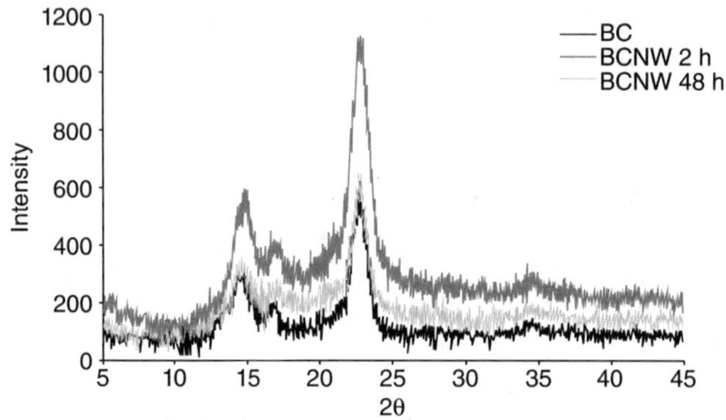

Figure 21.4 X-ray diffraction patterns of native bacterial cellulose (BC) and the obtained nanowhiskers (BCNW) after 2 hours and 48 hours of sulfuric acid hydrolysis. From Reference [39].

in the 48 hours spectrum, ascribed to carbonyl groups, and related to the incipient thermo-oxidative degradation of cellulose [39, 41].

21.3.2 XRD Characterization

The X-ray patterns obtained using monochromatic CuKα radiation, corresponding to untreated bacterial cellulose and the obtained nanowhiskers (Fig. 21.4), show two high intensity diffraction peaks at 14.5° and 22.5° and a smaller one at 16° [26, 39], corresponding to cellulose I crystals [29, 42, 43]. After 2 hours of acid hydrolysis (a typical time for treating plant cellulose) the intensity of the peaks is not significantly modified, indicating that that acid hydrolysis of bacterial cellulose may be slower than that for cotton or wood because of its higher crystallinity [39, 44]. On the other hand, the crystallinity appreciably increases to 90.31% after 48 hours and an additional 9% in the following 5 days, indicating that long hydrolysis times are required to produce a significant increase in the crystallinity of bacterial cellulose [39].

Other authors have also reported on the spectroscopic characterization of surface modified cellulose nanocrystals. For example, Yi et al. [45] grafted poly(styrene) (PS) onto the nanocrystals by atom transfer radical polymerization (ATRP). Esterification of the cellulose was carried out with 2-bromoisobutyryl-bromide to initiate the polymerization of PS in the following steps. After the esterification step, a new characteristic peak appeared at 1732 cm^{-1}, and following the PS grafting, all the typical peaks of the polymer were observed: the aromatic C–H stretching at 3059 and 3025 cm^{-1}, the C=C stretching bands for the aromatic ring at 1601, 1493, and 1452 cm^{-1}, the out-of-plane C–H bending of the aromatic ring at 756 cm^{-1} and the strong deformation vibration of the –CH– group of the monosubstituted benzene ring in PS at 696 cm^{-1}, all of them confirming the success of grafting.

21.4 SPECTROSCOPIC CHARACTERIZATION OF CELLULOSE NANOCOMPOSITES

The interest in micro- and nanocellulose fibers as reinforcing materials for polymer composites has been growing together with the trend of producing eco-friendly materials [3, 6, 46]. Cellulose nanocrystals from several sources such as tunicin, cotton, tunicate, algae, bacteria, ramie, and wood have been extensively investigated for the preparation of high performance composite materials made from natural and synthetic polymers [47–49]. Pereda et al. [50] used FT-IR spectroscopy in the characterization of films based on glycerol plasticized sodium caseinate with cellulose nanofibers obtained by acid hydrolysis of microcrystalline cellulose. The incorporation 3 wt% nanocellulose produced only a small increase of the peaks at 1040, 1110, and 1640 cm^{-1} in comparison to the spectrum of the sodium caseinate film, the three of them related to the absorptions of cellulose [35]. In the composite films, the band at 1640 cm^{-1} is due to the amide I vibrations of proteins [51], with the added contribution of adsorbed water in amorphous cellulose [35, 52].

Goetz et al. [53] studied the cross-linking of cellulose whiskers (obtained by acid hydrolysis of microcrystalline cellulose) with poly(methyl vinyl ether-co-maleic acid) and poly(ethylene glycol), PEG. FT-IR spectroscopy allowed confirming that cross-linking between the hydroxyl groups on the cellulose and the terminal hydroxyl groups of PEG took place via co-esterification with the acid.

ATR–FT-IR spectra of a polycaprolactone-based waterborne polyurethane (WPU) and its composite made with flax cellulose nanocrystals (obtained by acid hydrolysis) are shown in Figure 21.5 [47]. The spectrum of the unreinforced film presents a strong absorption peak centered at around 3363 cm^{-1}, assigned to the NH stretching vibration affected by the hydrogen bonding between NH and carbonyl groups, which increased the cohesion between the hard and soft segments; otherwise the free NH stretching vibration should appear at around 3420 cm^{-1}. Simultaneously, only a single prominent peak band centered at 1722 cm^{-1}, assigned to the stretching vibration of both urethane and ester carbonyl, was observed, confirming the existence of the integrated phase between hard and soft segments in the matrix. This peak shifts to 1731 cm^{-1} with the addition of cellulose nanocrystals, suggesting that its incorporation disturbs the hydrogen bonding between NH and CO, and favors microphase separation due to the strong interaction between cellulose nanocrystals and matrix molecules. A relative increase of the band located at 3200–3500 cm^{-1} with increasing cellulose nanocrystals content is also observed, while a shoulder appears at around 3300 cm^{-1}, which was attributed to the OH stretching in the cellulose nanocrystals.

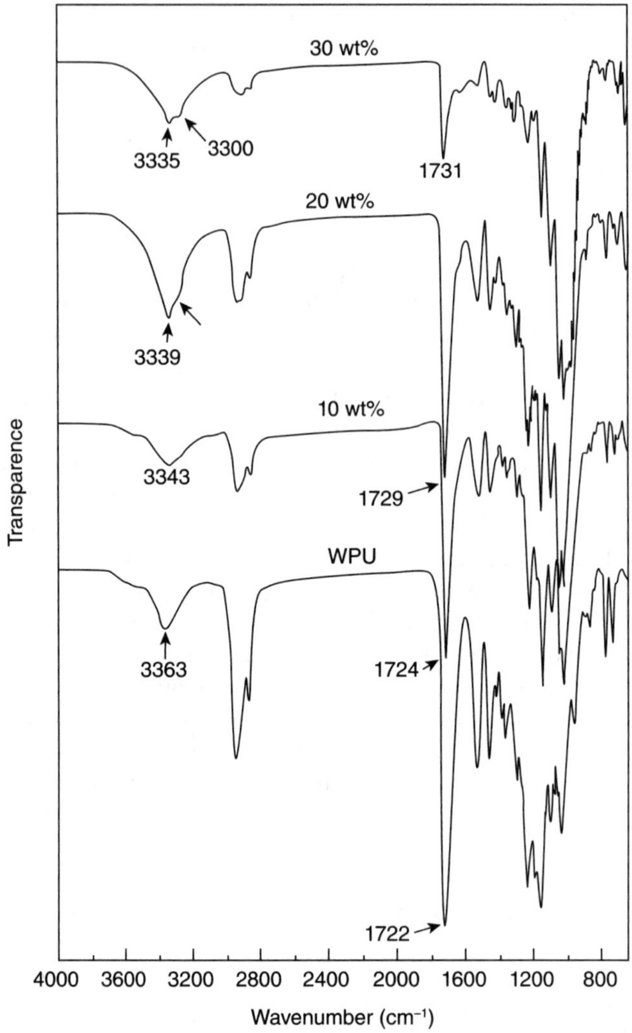

Figure 21.5 ATR–FT-IR spectra of waterborne polyurethane and its composites with flax cellulose nanocrystals. From Reference [47].

The same authors studied the X-ray patterns (CuKα source) of nanocomposites as a function of the flax cellulose nanocrystrals content (Fig. 21.6). Although poly(caprolactone) (PCL) is a semicrystalline thermoplastic polymer [47, 54], only a broad diffraction hump at $2\theta = 19°$ is observed, indicating that the film is amorphous. This was explained because the low molecular weight of the PCL does not allow chain folding [55, 56]. Additionally, diffraction peaks appear that increase with the cellulose nanocrystals content. At 20 wt%, three

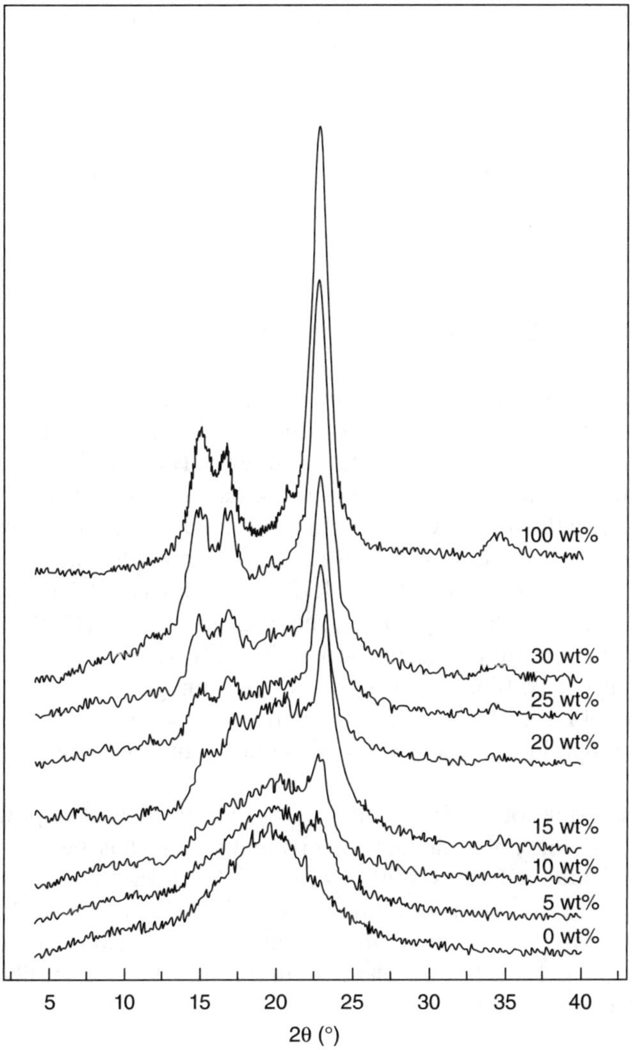

Figure 21.6 WAXD patterns of the waterborne polyurethane and its composites with flax cellulose nanocrystals. From Reference [47].

well-defined peaks at $2\theta = 14.7°$, $16.5°$, and $22.7°$ related to the crystalline nanoparticles are observed. The absence of any additional peak or peak shift indicates that the amorphous nature of matrix and the crystal structure of nanofibers are unaffected.

Auad et al. [57] studied the effect of adding cellulose nanocrystals to a semicrystalline shape memory polyurethane, finding a shift to higher angles in the X-ray diffraction peak of the soft segment crystalline phase. This effect was associated with improvement in soft and hard phase separation and the consequent formation of more perfect crystals promoted by the cellulose addition.

On the other hand, Stevanic et al. [58] found that the XRD patterns of arabinoxylan films with 5 wt% and 15 wt% of bacterial cellulose were just a combination of the results found for the individual patterns, with no apparent new diffractions or shifts that could indicate interactions between the components.

21.5 CHITIN AND CHITOSAN NANOPARTICLES

Chitin is a naturally abundant mucopolysaccharide, and the major structural component in the exoskeleton of arthropods and cell walls of fungi and yeast [59]. It is a white, hard, inelastic, nitrogenous polysaccharide and the major source of surface pollution in coastal areas. Its molecular structure is similar to cellulose with the hydroxyl at position C-2 replaced by an acetamido group. Because chitin is not readily dissolved in common solvents, it is often converted to its partially deacetylated derivative, chitosan [59–63]. Chitosan is often identified by its incomplete degree of deacetylation (DD), a percentual measure of the free amine groups along the chitosan backbone [59, 64]. Due to their nontoxicity, biocompatibility, low immunogenicity, biodegradability, and antimicrobial activity [60], both chitin and chitosan have present and potential interest in food, cosmetics, biomedical, and pharmaceutical applications [62].

The transformation of chitin and chitosan from flakes or powder to the nano scale, adding potential for novel applications, has been the focus of several publications. For example, Nair and Dufresne [65] obtained chitin whiskers by acid hydrolysis of chitin from crab shells, and used them as reinforcing material for natural rubber. Chemical modification of chitosan to obtain nanoparticles [66] and electrospinning to obtain nanofibers [67, 68] have also been reported.

21.6 SPECTROSCOPIC CHARACTERIZATION OF CHITIN NANOPARTICLES

Depending on its source, chitin can appear as α and β crystalline allomorphs [62], which can be differentiated by infrared and solid-state NMR spectroscopies together with X-ray diffraction. α-Chitin is by far the most abundant allomorph; it occurs in fungal and yeast cell walls, crab tendons and shells, and shrimp, krill, and lobster shells, as well as insect cuticle. Thus, the following discussion will be focused on this crystalline form.

The most common method to obtain chitin whiskers is the acid hydrolysis of a protein-free chitin source, followed by dialysis of the residues in distilled water until neutralization, and lyophilization [65, 69–76]. This procedure dissolves regions of low lateral order, and the insoluble, highly crystalline residue is then converted to a stable suspension by mechanical shearing [65, 76]. Most of the reported works used concentrated HCl solutions and relatively high temperature (~80 to 105°C), but sometimes more than one hydrolysis step was used in the preparation [70, 77], or different drying procedures were used [78]. More recently, other preparation techniques have been reported: Fan et al. [79] prepared chitin nanocrystals dispersed in water by 2,2,6,6-tetramethylpiperidine-1-oxyl radical (TEMPO)-mediated oxidation of α-chitin, followed by ultrasonic treatment, while Chang et al. [77] alternated hydrolysis procedures with mechanical sonication/disruption steps. However, in the last case, the obtained chitin nanoparticles exhibited lower crystallinity than that of conventional chitin whiskers or native chitin.

21.6.1 FT-IR Characterization

The FT-IR spectrum of chitin in Figure 21.7a shows all of the typical bands: 3430 and 3258 cm^{-1} (NH and OH stretching, respectively) [77, 80], 1659, 1625 cm^{-1} (amide I, CO stretching), 1563 cm^{-1} (amide II, NH deformation in the CONH plane), 1418 cm^{-1} (CH deformation), 1375 cm^{-1} (C–CH_3 amide stretching), 1314 cm^{-1} (amide III and >CH2 wagging), 1155 cm^{-1} (COC bridge stretching), 1073 cm^{-1} (COC stretching in ring), 1029 cm^{-1} (CO stretching), 896 cm^{-1} (b linkage), and 692 and 598 cm^{-1} [65, 71, 78]. Compared with the FT-IR spectrum of raw chitin (Fig. 21.7a), the one corresponding to chitin whiskers (Fig. 21.7b) shows sharper peaks in the –OH and amide regions. The

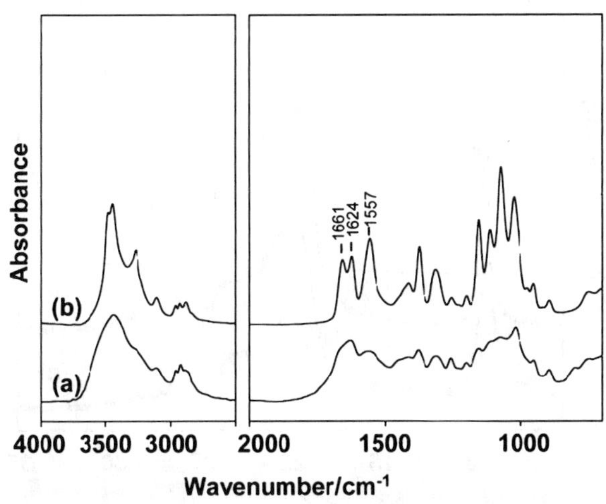

Figure 21.7 FT-IR spectra of (a) chitin from shrimp shells (in flake form) and (b) chitin whiskers. From Reference [70].

sharpening of the bands implies that chitin whiskers are more crystalline than original chitin [70, 71, 78], but also that the hydrolysis reaction may have changed the structure of chitin [77, 80] and the hydrogen-bonding interaction in chitin whiskers.

21.6.2 WAXD Characterization

The crystalline structure of α-chitin has been reported by many researchers [62, 81]. Figure 21.8 presents the XRD patterns of the original chitin (a) and chitin whiskers (b). The chitin structure contains extended polymer chains with anti-parallel arrangement and strong intermolecular hydrogen bonding [81, 82]. The XRD patterns of the stronger crystalline peaks have been indexed as 9.0 to 9.6° ([020] plane) and 19.1 to 19.7° ([110] plane), while the weaker ones have been reported at about 20–21°, 23°, and 26° [69, 77, 79, 81, 83–85]. The broad signal centered at $2\theta \sim 19°$ shown in Figure 21.8a is attributed to the N-glucosamine sequences (deacetylated unit blocks). Similarly, the intensity of the broad signal centered at $2\theta \sim 9°$ is due to N-acetyl-D-glucosamine sequences (acetylated unit blocks) [81, 86]. Figure 21.8b shows the WAXD pattern of chitin whiskers obtained by acid hydrolysis [70, 71], but the observed behavior can be extended to chitin nanofibers obtained by other procedures (i.e., TEMPO oxidation performed by Fan et al. [79]): the α-chitin structure is maintained, although the peak intensity due to the [020] plane clearly decreases as a result of the modification. Moreover, the intensity of the weaker peaks

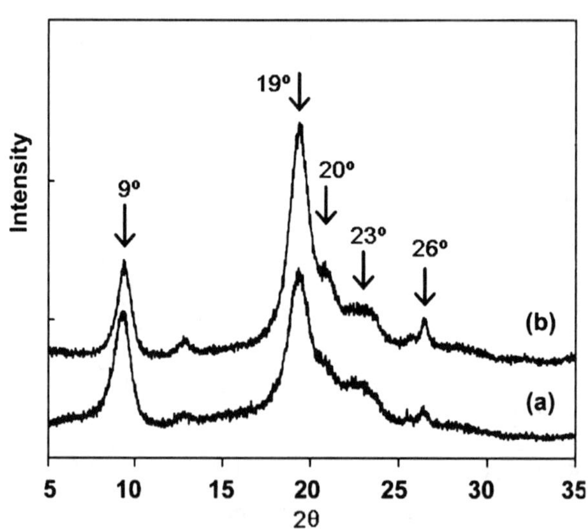

Figure 21.8 WAXD patterns of (a) chitin from shrimp shells (in flake form) and (b) chitin whiskers. From Reference [70].

increases slightly. Chang et al. [77], who used consecutive cycles of acidic hydrolysis and mechanical ultrasonication/ disruption, indicated that chitin nanoparticles of about 50–100 nm, rather than chitin whiskers, were obtained because amorphous and crystalline regions were attacked with this technique, and thus the crystallinity of these nanoparticles was lower than that of native chitin.

21.7 CHITOSAN NANOPARTICLES

Due to the inherent solubility of the chitosan, insoluble nanoparticles are usually synthesized by ionic gelation of chitosan with sodium tri-polyphosphate (TPP) [87–90], although other techniques have been reported. For example, chitosan nanoparticles with diameters less than 100 nm and cross-linked with glutaraldehyde have been prepared in a surfactant-mediated (sodium bis(ethylhexyl) sulfosuccinate)/n-hexane reverse micellar system by Banerjee et al. [91]. Conversely, Watthanaphanit et al. [76] prepared chitosan whiskers by deacetylation of chitin ones, and Osorio-Madrazo et al. [92] obtained chitosan crystalline particles by solid-state acid hydrolysis from a highly deacetylated semicrystalline chitosan.

21.7.1 FT-IR Characterization

FT-IR spectra of chitosan and chitosan nanoparticles are shown in Figure 21.9. The strong band in the range of 3400–3200 cm^{-1} corresponds to the combined peaks of the NH_2 and OH group stretching vibration in chitosan [87, 90, 91], with contribution from asymmetric and symmetric N–H stretching in primary amines at 3500 and 3400 cm^{-1} [91]. In TPP cross-linked nanoparticles, a broadening and simultaneous shift of the band from 3424–3434 cm^{-1} to 3400 cm^{-1} is noticed [87, 93], indicating increased hydrogen bonding. However, other authors [90] found the opposite behavior, that is, narrowing of the hydrogen bonding-related peak in the spectra of the nanoparticle complexes, and attributed it to the more open structure resulting from cross-linking with TPP, as compared to the compact bulk structure of chitosan. Moreover, for chitosan nanoparticles the peak at 1640–1650 cm^{-1} disappeared and a new sharp peak at 1630–1635 cm^{-1}, which corresponds to stretching vibrations of C=N bond, was detected [87, 90, 91]. When the cross-linkage was carried out using glutaraldehyde [91], this peak appeared as a strong band, indicating the formation of a Schiff's base as a result of the reaction between the carbonyl group of the aldehyde and the amine group of chitosan chains.

However, assignment is not easy because in the same region the peak corresponding to the NH_2 scissoring vibration (1640–1650 cm^{-1}) of the primary amines in chitosan chains appears. Additionally, the amine-bending peak in raw chitosan at 1602 cm^{-1} shifted to 1534 cm^{-1} in the nanoparticles [87], due to the linkage between the TPP and the ammonium group of chitosan

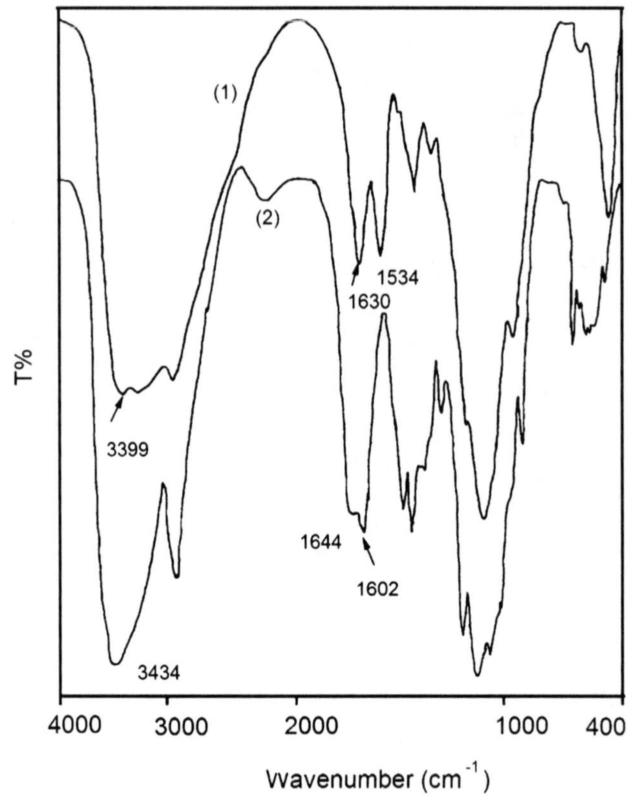

Figure 21.9 FT-IR of chitosan nanoparticles (1) and chitosan (2). From Reference [87].

nanoparticles [90]. A new peak at 1218 cm^{-1} due to P=O stretching was also identified in the FT-IR spectrum of the cross-linked chitosan nanoparticles [90].

21.7.2 XRD Characterization

The XRD patterns of chitosan (DD = of 82%) and chitosan nanoparticles cross-linked with TPP as determined by Wazed Ali et al. [90] are presented in Figure 21.10. There is a strong peak in the diffractogram of chitosan at $2\theta = 21.8°$, indicating its high crystallinity [90]. On the other hand, the diffractogram of the nanoparticles is characteristic of an amorphous structure. They are formed by a dense network structure of polymer chains cross-linked through TPP counterions [94] leading to a disarray in the chain alignment, which is revealed in the XRD pattern. Additionally, Osorio-Madrazo et al. [92] observed two main peaks in the WAXD patterns of fully deacetylated chitosan

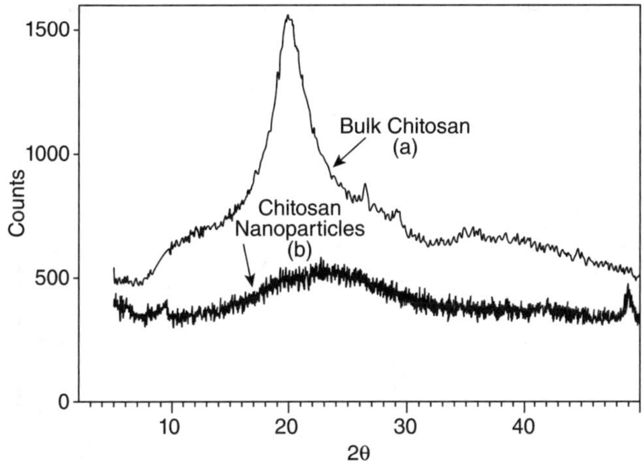

Figure 21.10 XRD patterns of (a) chitosan and (b) chitosan nanoparticles. Cu Kα radiation. From Reference [90].

samples, located at $2\theta = 10.80°$ for the [020] planes of the hydrated crystalline structure and around 20° for the contribution of two peaks at 20.16 and 21.80°, related to the reflections of the [200] and [220] planes of the hydrated polymorph, in agreement with the spectrum of the hydrated chitosan.

21.8 NANOCOMPOSITES

As in the case of cellulose, the use of chitin and chitosan whiskers as reinforcing nanofillers in a polymer matrix is appealing because of their high aspect ratios and highly crystalline nature [95, 96]. Chitin whiskers have been used as reinforcement of both synthetic [95, 96] and natural polymers [69, 72, 74]. As an example, Sriupayo et al. [69] prepared and characterized α-chitin whisker-reinforced poly(vinyl alcohol) (PVA) nanocomposite films by solution-casting technique. The α-chitin whiskers were prepared by acid hydrolysis of α-chitin from shrimp shells. Figure 21.11 illustrates the FT-IR spectra for pure PVA, as-prepared α-chitin whiskers, and resulting reinforced PVA films having whisker contents from 0.74 wt% to 29.6 wt.%. The absorption peaks at 1440 cm^{-1} (CH_2 bending) and 858 cm^{-1} (CH_2 rocking) characteristic to PVA were obvious in the spectra of pure and reinforced PVA films, while the characteristic amide I peak at 1655 cm^{-1} (–CONH–) was evidenced in the spectra of as-prepared α-chitin whiskers and reinforced PVA films [97]. It is obvious that the relative intensity of these characteristic peaks increased with increasing content of the corresponding component and that no additional peaks appeared in the spectra.

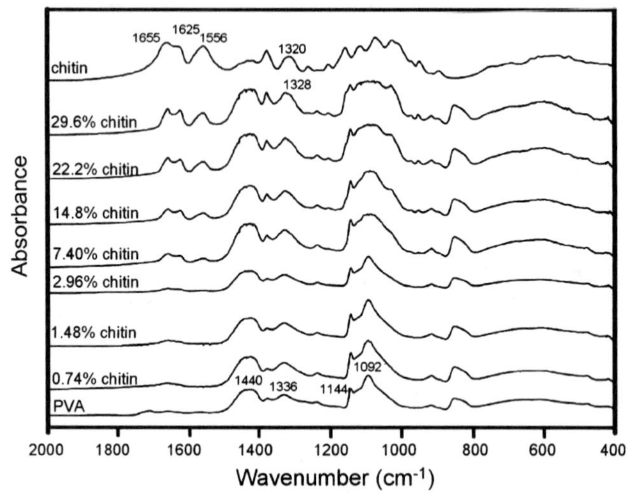

Figure 21.11 FT-IR spectra of pure PVA, as-prepared α-chitin whiskers, and α-chitin whisker-reinforced PVA films having different whisker contents ranging from 0.74 wt% to 29.6 wt%. From Reference [69].

In the same way, the WAXD patterns of the α-chitin whisker-reinforced PVA films resembled those of the pure components (see patterns (b)–(f) in Fig. 21.12), with the strong scattering peaks of α-chitin whiskers (i.e., at about 9 and 19°) increasing with the whisker content. Notice that the WAXD pattern for pure PVA film (i.e., pattern (a)) shows a single broad scattering peak centered at about 19°, suggesting that the film was essentially amorphous, in agreement with an earlier result for a film cast from a solution of PVA in formic acid [83].

21.9 STARCH

Starch is a renewable and largely available biodegradable polymer that is produced by plants as a means for storing energy. It is composed by two types of molecules, amylose (linear) and amylopectin (branched). In its native form, it has a semicrystalline complex hierarchichal structure from which different types of nanoparticles have been obtained.

Several routes have been used during the last years for that purpose, leading to two main groups of particles: nanocrystals obtained by acidic or enzymatic hydrolysis of starch and crystalline or amorphous/crystalline nanoparticles obtained from regeneration/ precipitation of starch [98]. Moreover, starch nanocrystals (StN) exhibit a platelet-like structure, similar to that of exfoliated layered silicates [99], usually forming clusters, whose separation and dispersion

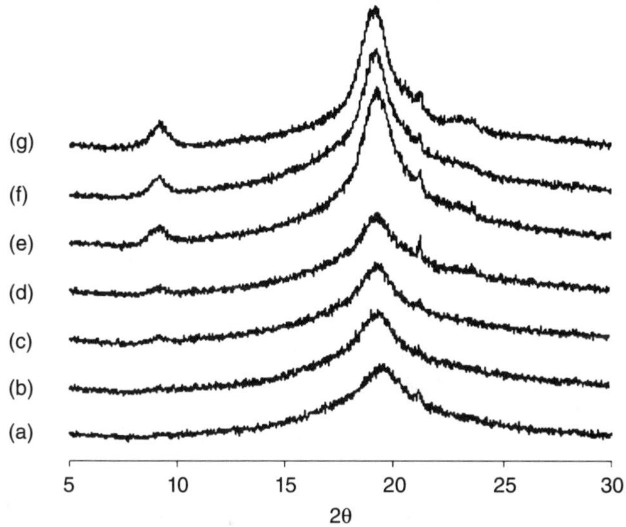

Figure 21.12 WAXD patterns of (a) pure PVA and α-chitin whisker-reinforced PVA films having different whisker contents of (b) 2.96 wt%, (c) 7.40 wt%, (d) 14.8 wt%, (e) 22.2 wt%, and (f) 29.6 wt%, respectively, and (g) as-prepared chitin whiskers. From Reference [69].

can be improved by chemical surface modification. On the other hand, precipitated particles can have different shapes, depending on the preparation method. In general, the reported efforts in producing nanocrystals were aimed at their use as fillers in polymeric matrices to improve mechanical and/or barrier properties in the final composite materials, while nanoparticles were mostly aimed at the transport and delivery of drugs and chemicals.

Usually, the characterization studies of the particles included infrared spectroscopy (FT-IR) as an initial approach to this goal. In native starch (Fig. 21.13a) the peaks at 1157, 1080, ca. 1000, and 927 cm^{-1} are due to CO bond stretching. In particular, the bands at 1080 and ca. 1000 cm^{-1} are assigned to the anhydroglucose ring O–C stretching. The peak at 1645 cm^{-1} is due to water bound to the starch and the strong and broad band centered at 3390 cm^{-1} is due to the absorption of the OH stretching.

Characterization by XRD, also frequently used, shows that StN or native starch patterns correspond to the A-type crystalline allomorph. Thus, if a CuKα source is used, the spectrum of the A-allomorph shows weak peaks at $2\theta = 20.05°$ and 11.5° (*d*-spacing = 8.7 Å and 7.6 Å, respectively), a strong signal at 15.3° (5.7 Å), a double strong peak at 17.2° (5.2 Å) and 18.0° (4.9 Å), and a strong peak at 23.5° (3.9 Å). In processed starch, B-type crystals that correspond to a more open arrangement of starch molecular helices appear (as

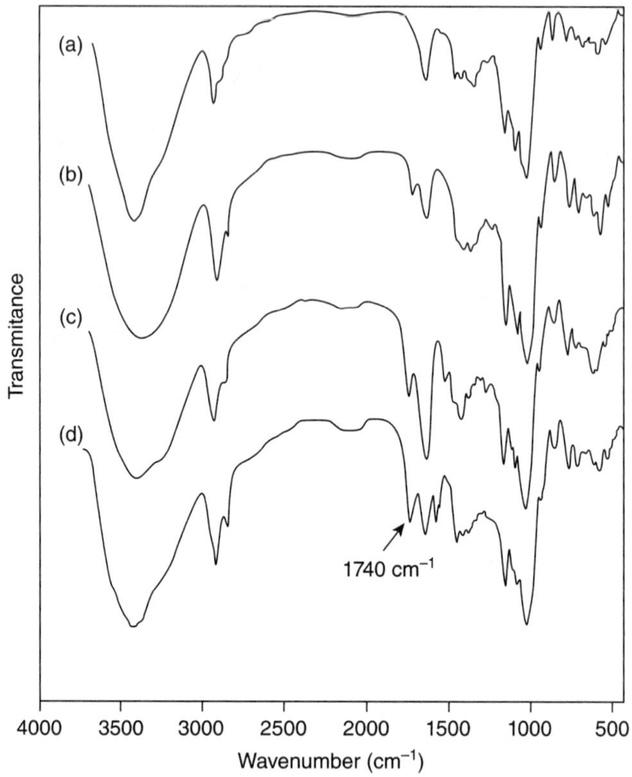

Figure 21.13 FT-IR spectrum of (a) unmodified starch nanocrystals, (b) decanoyl starch nanoparticles, (c) nonanoyl starch nanoparticles, and (d) octanoyl starch nanoparticles. From Reference [100].

compared with the closed packing in A-crystals), with intense peaks at 5.6 and 17.1° (*d*-spacing = 15.7 and 5.2 Å, respectively) [101, 102].

21.10 CHARACTERIZATION OF MODIFIED STARCH NANOPARTICLES

21.10.1 Spectroscopic Characterization of Esterified Starch Nanoparticles

Two of the most frequent modifications using the OH starch functional groups are esterification [11, 100, 103] and isocyanate-mediated reactions [104–106]. Scheme 21.1 illustrates these chemical routes.

A good example of characterization of esterification treatments is the work of Namazi and Dadkhah [100], who modified StN with different alkyd acid

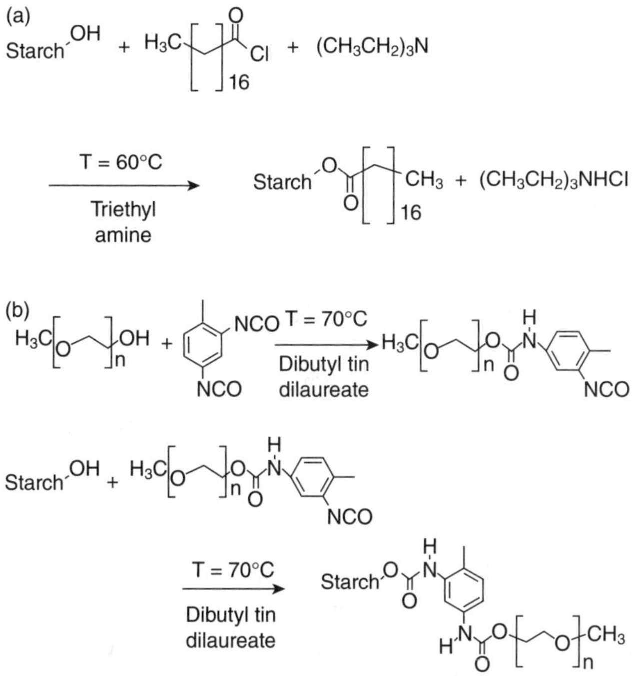

Scheme 21.1 Reaction scheme for the grafting of (a) stearates and (b) PEGME to the starch nanocrystal surface. From Reference [106].

chlorides. FT-IR transmission spectra of the different nanoparticles are shown in Figure 21.13. Corn starch and nanocrystals have similar spectra, while the spectra of the esterified crystals showed the characteristic peak of the ester groups at 1738 cm^{-1}. Another feature is the red shift of the 1017 cm^{-1} peak to 1026 cm^{-1}, which is the result of the ester bonds sterically hindering/weakening the hydrogen bond interaction of the oxygen in C–O–C groups of the glucose ring with the hydrogen of the starch hydroxyl groups. Similar observations were reported by Thielemans et al. [106], who modified StN with stearic acid, by Ma et al. [107], who studied the esterification of regenerated/precipitated corn starch nanoparticles with citric acid, and by Simi and Abraham [108], who prepared oleic and stearic esterified precipitated nanoparticles from cassava starch. In this last case, further cross-linking was performed to improve the stability of the nanoparticles under drug delivery conditions.

XRD spectra (Fig. 21.14) were obtained for the same materials using CuKα radiation [100]. The waxy corn starch shows A-type crystals, while the X-ray patterns of the modified nanocrystals show only two amorphous broad peaks in the ranges of 9°–15° and 16°–24°. However, calorimetry studies showed that the original crystallinity of starch was not affected, although the attached fatty

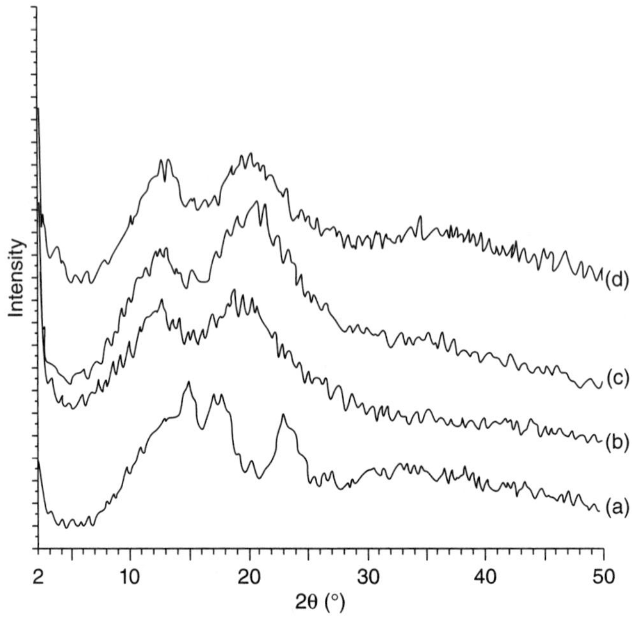

Figure 21.14 X-ray diffraction patterns of (a) unmodified starch nanocrystals and (b) octanoyl starch nanoparticles, (c) nonanoyl starch nanoparticles, and (d) decanoyl starch nanoparticles. From Reference [100].

acids did not crystallize. Similar findings on the inability of X-ray to detect starch crystal phases in treated nanocrystals have also been reported by other authors [105, 106], and it was attributed to the coating/shielding of the nanocrystals by the grafted layer of the fatty acid.

On the other hand, citric acid-modified regenerated nanoparticles [107] showed V-type crystallinity, corresponding to an inclusion complex formed by an amylose helix that includes other substances in its interior, such as the plasticizer ($2\theta = 13$ and 21.5°, using CuKα). As in the other cases, XRD did not show crystal patterns for the modified nanoparticles.

PCL has been frequently used in esterification reactions because it is known to be biocompatible, and thus the biodegradabilty and biocompatibility of starch is not affected. This modification introduces long tails on the surface of the crystals helping to individualize the crystals, improving the dispersion in nonpolar matrices and leading to a stronger interfacial adhesion by producing entanglements with bulk polymer chains.

Yu et al. [103] grafted poly(ε-caprolactone) onto the surface of StN obtained by acid hydrolysis using ring-opening polymerization assisted by microwave irradiation. Transmission FT-IR of the powders revealed the expected appearance of a high and sharp peak at 1728 cm^{-1} in the spectrum of the modified crystals (ester carbonyl group) and a distinct peak at 2866 cm^{-1} (methylene groups of the grafted PCL). On the other hand, the X-ray crystalline pattern of the grafted nanocrystals was essentially due to the crystalline ordering of the PCL attached chains ($2\theta = 21.2°$, 21.8° and 23.5°), completely obscuring the expected diffraction signals for the StN ($2\theta = 15.3°$, 17.2°, and 23.3°).

Stearate modified StN prepared by Thielemans et al. [106] have also led to high coverages and crystallinity of the attached chains. In that case, the X-ray showed signals corresponding to the crystals or stearate at 5.76° ($d = 1.53$ nm), 9.6° (0.92 nm), 1.44° (0.66 nm), 21.56° (0.41 nm), and 23.64° (0.38 nm). The first three spacings are similar to those for stearate crystals, while the latter two correspond to the [110] and [020] reflections in the orthorhombic subcell structure of paraffin crystals. Considering the high level of grafting in the esterified crystals, the authors proposed a brush-like structure extending from the nanoparticle surface outward, as it was also proposed by Habibi and Dufresne [104] for PCL-modified StN.

Chakraborty et al. [11] modified regenerated/precipitated starch nanoparticles by enzymatic esterifications after dispersing them in a reverse micelle media. The acyl donors were vinyl esters of fatty acids of different lengths, maleic anhydride, and ε-caprolactone. In this study, the characterization of the nanoparticles included the analysis of the ^{1}H NMR and ^{13}C-DEPT-135 NMR spectra of the original and modified starch (Fig. 21.15). The results indicated

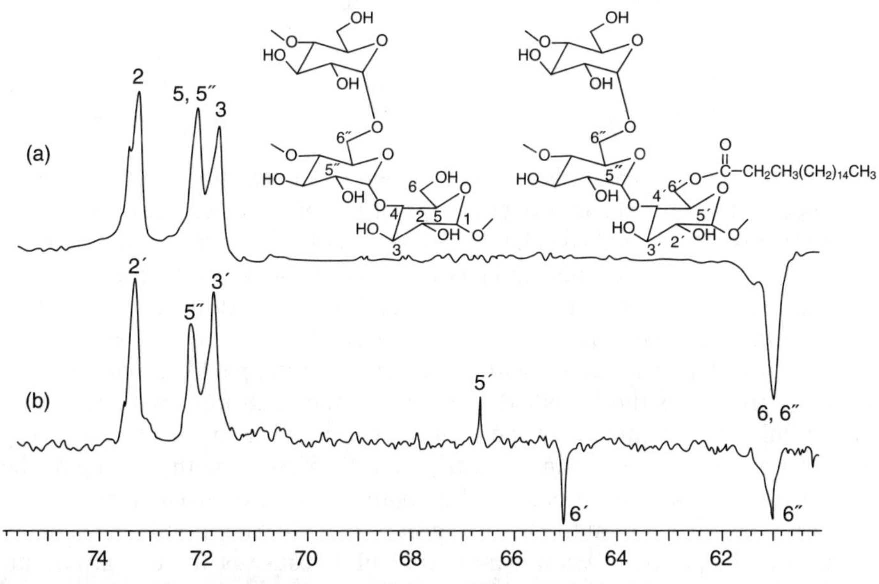

Figure 21.15 Expanded region of DEPT-135 (75 MHz, DMSO-d_6) spectra that shows the carbon signals for the sugar units of (a) native starch nanoparticles and (b) vinyl stearate-modified starch nanoparticles with DS = 0.8. From Reference [11].

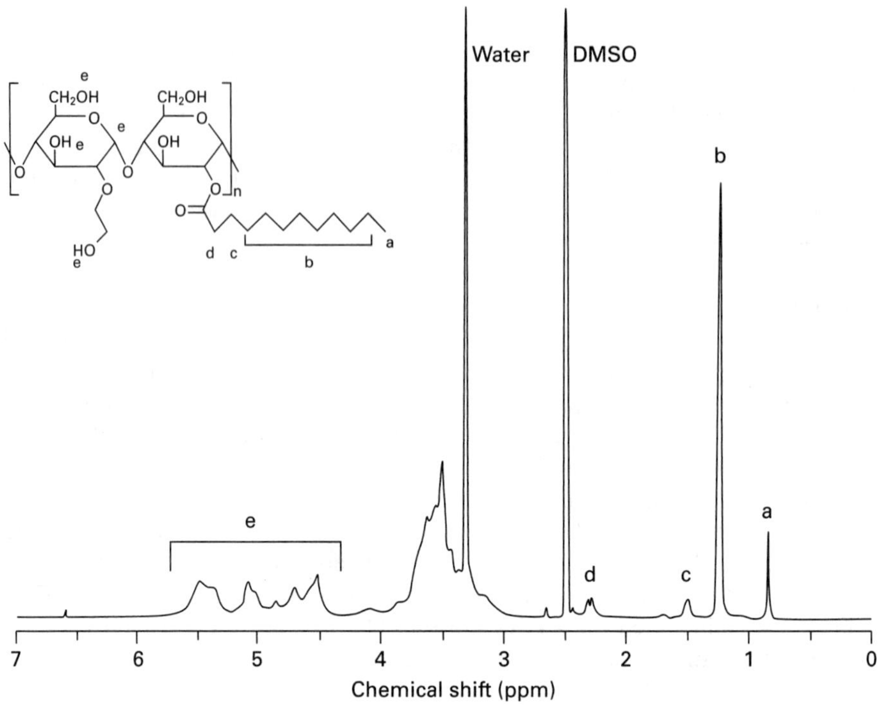

Figure 21.16 ^{1}H NMR spectrum of HES 200-L8.7. Inset shows the structure of HES laurate and peak assignment. From Reference [110].

that the enzymatic catalysis allowed selective substitution in C6 (carbon attached to the primary hydroxyl groups of the glucose residues), while the ^{1}H NMR results allowed calculating the degree of substitution from the ratio of the integrated areas corresponding to the protons of carbon C1 (at 5.27 ppm) and that of the CH_2 of stearate (at 1.53 ppm). Besheer et al. [110] presented an interesting spectroscopic study of nanovesicles from a commercial hydroxyethyl starch (HES) modified with fatty acids (lauric, palmitic, and stearic). Figure 21.16 shows the ^{1}H NMR spectrum of the HES modified with lauric acid. While the assignment of the peaks corresponding to the attached fatty acid chains was relatively simple: a triplet at 0.85 ppm (methyl group at the end of the chain), a sharp peak at 1.23 ppm (methylene groups), and broad peaks at 2.29 and 1.5 ppm (methylene groups close to ester bonds), the assignment of the peaks for HES was more difficult because of the overlap of different peaks. The peaks between 4.4 and 5.7 ppm were assigned to the protons bound to C1 and those in the hydroxyl groups of the glucose ring and the modifier moiety on HES.

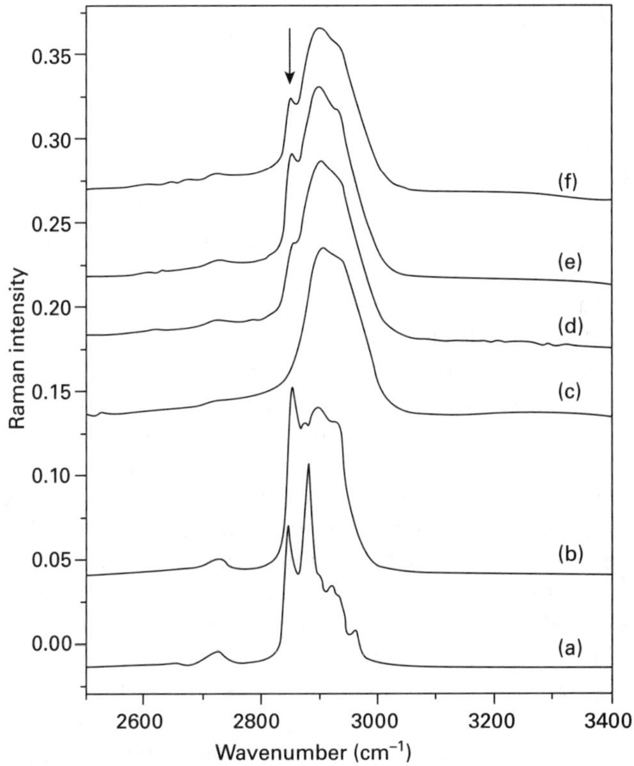

Figure 21.17 Raman spectra in the range from 2500 to 3400 cm^{-1} for (a) lauric acid measured at 23°C, (b) lauric acid at 55°C, (c) HES 450, (d) HES 450-L9.4, (e) HES 450-L23.3, and (f) HES 450-P12.6. The arrow at 2852 cm^{-1} is due to the symmetric CH_2 stretching of the alkyl chains of the fatty acid. From Reference [110].

The grafting of the HES was also confirmed by Raman spectroscopy (Fig. 21.17). The peak at 2852 cm^{-1} (sym. stretch of CH_2 groups of the attached alkyl chains) appears in the spectrum of the lauric acid and as a shoulder in that of the fatty acid modified HES. Additionally, Raman spectroscopy can also give information on the conformational states of the compounds; for example, the signals corresponding to the CH stretching in the 2800–3000 cm^{-1} differ depending on the crystalline or noncrystalline state of the fatty acids. The spectra were obtained from solubilized material (before solvent precipitation of the nanoparticles). The comparison between experimental and calculated simulated spectra (from weighed addition of the HES and lauric acid spectra) allowed concluding that the fatty acid does not crystallize when grafted to the starch, as it had also been reported for alkydic acid chloride modifications [100].

21.10.2 Spectroscopic Characterization of Starch Nanoparticles Modified through Urethane Linkages

Other frequent approach is to graft polymer chains through the formation of surface urethane bonds. Habibi and Dufresne [104] modified StN obtained by acid hydrolysis, by isocyanate-mediated reaction to graft PLC of different molecular weights. Wide-angle X-ray spectra of powdery dry samples were obtained using a Cu Kα source. The unmodifed crystals show the typical peaks of the A-type allomorph, as already discussed, but in this case the same peaks also appear in the patterns of the PLC-modified nanocrystals, indicating that its crystalline form is not changed by the modification. Additionally, a new peak at 21.3° is assigned to the crystalline structure of PLC induced by the nanocrystals, as it has already been found in the case of PCL grafted to nanocrystals by esterification [103]. Labet et al. [105] treated StN with poly(tetrahydrofuran), PTHF, PCL, and poly(propylene glycol) monobutylether (PPGBE), using (TDI) isocyanate-mediated grafting. The grafting efficiency decreased with the length of the attached chain. The FT-IR transmission spectra of the modified nanoparticles showed the band around 1700–1750 cm^{-1} corresponding to the urethane absorption in the grafted StN, and in the same region the ester absorption corresponding to PCL. Only surface OH groups participate in the grafting reaction, but bulk, as well as surface, OH contributes to the 3400 cm^{-1} absorption and so there was essentially no change in the band intensity due to the reaction. The X-ray pattern of the modified nanocrystals showed A-type crystallinity but less defined peaks because of the polymer layer covering the surface, similarly to other results in long chain grafted nanoparticles [100, 107].

Thielemans et al. [106] grafted poly(ethylene glycol) methyl ether (PEGME) through isocyanate co-reaction onto StN. Considering the high absorbance of urethane bonds and the small peak appearing in the transmission FT-IR spectrum at 1721 cm^{-1}, the coverage was considered limited to a small fraction of the available hydroxyl groups. The low coverage was also confirmed by X-ray spectroscopy since there was little crystalline contribution of the attached chains. In this particular example, the nanocrystal characterization was completed using XPS with a Mg Kα X-ray source, and locating the C1s signal at 285.0 kV. Figure 21.18 shows the spectra of the unmodified and modified StN, which were used to calculate the surface composition of the samples. The ratio of O/C determined for the unmodified particles was much lower than expected from the starch chemical structure (theoretical O/C = 0.83; experimental O/C = 0.56), in agreement with other works [109, 111], and was attributed to hydrocarbon impurities. In spite of the low coverage, the N1 signal and the analysis of the decomposition of the C1 signal confirmed the PEGME-TDI reaction (N/C = 0 and 0.06 for the unmodified and modified nanocrystals, respectively). Signals due to residual tin catalyst could also be observed in the XPS of the modified StN (spectrum c).

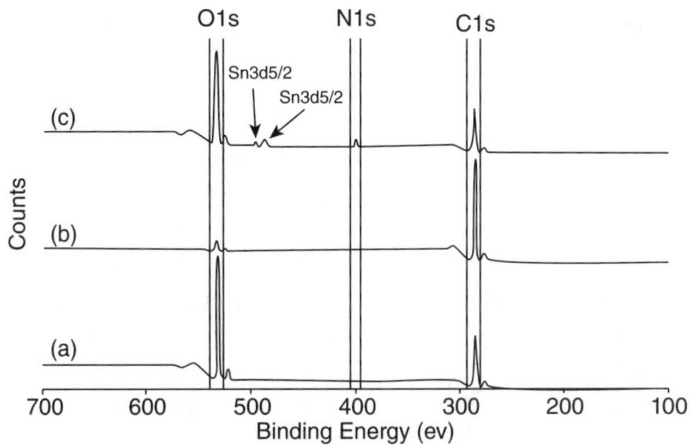

Figure 21.18 General XPS spectra of (a) unmodified and (b) stearate- and (c) PEGME-modified starch nanocrystal surfaces with signal assignments. From Reference [106].

21.11 NANOCOMPOSITIES FROM STARCH NANOPARTICLES

Starch nanoparticles have been incorporated into polymer matrices, generally, biopolymers, with the aim of improving mechanical properties, vapor barrier properties, or both. The effect of this addition is to reinforce the polymer matrix much in the same way as it occurs with inorganic nanofillers, with the additional advantages already discussed. For example, García et al. used ATR–FT-IR and X-ray spectroscopies to characterize starch–starch nanocomposite films [112, 113], obtained by adding 2.5 wt% nanocrystals in two different glycerol plasticized matrices, cassava and waxy starches. The spectra did not differ much, except for the relative intensity of the absorption peaks at 995 and 1022 cm^{-1}. These bands are sensitive to hydrogen bonding and, specifically, the intensity of the 1022 cm^{-1} band increases with decreasing crystallinity [114, 115].

The addition of nanocrystals interferes with the matrix crystallinity, reducing the local order since in both cases there was a small increase of the 1022 cm^{-1} band. The lower crystallinity of the composite films was confirmed by X-ray spectroscopy (Cu Kα radiation). In both films the peaks corresponding to the B-type of starch (matrix) appeared, although with reduced intensity in the composite films. Since the concentration of nanocrystals was low (2.5 wt.%), very small peaks corresponding to the A-crystals from waxy maize nanoparticles appear in the composite X-ray pattern, in agreement with the angles reported by Angellier et. al [116] for similar waxy maize StN.

Chang et al. [99] obtained nanocrystals by acid hydrolysis of pea starch and grafted PCL on their surface using ring-opening polymerization under microwave irradiation, achieving high levels of modification on platelet-like starch nanocrystals that did not exceed 100 nm [103]. These crystals were incorporated in WPU prepared from PCL and TDI by mixing the polymers with an acetone suspension of the nanoparticles, to obtain composite films (5–30 wt%) by casting. X-ray spectroscopy showed that the intensity of the crystalline peaks of the grafted PCL was even higher than in the neat PCL pattern displayed for comparison. The unreinforced films did not show crystalline peaks and the incorporation of 5 wt% of grafted nanocrystals did not change the X-ray pattern. However, as the concentration was further increased, the peaks of PCL crystals appeared and their intensity increased with the concentration, probably due to the aggregation of the nanoparticles that form crystalline domains.

The 1600–1800 cm^{-1} FT-IR region was carefully analyzed and modeled, and the conclusion was that the hydrogen bonding in the WPU matrix was partly destroyed by the addition of the StN. As more crystals were incorporated, the C=O contribution corresponding to the crystalline domain increased and the absorption of the amorphous region decreased. This is in agreement with X-ray spectroscopic results that showed high intensity PCL peaks in the composite films containing large concentrations of the nanocrystals.

Yu et al. [103] incorporated PCL-grafted StN to poly (lactic acid) (PLA). Figure 21.19 shows the ATR/FT-IR spectra obtained for the composite films with different nanocrystals concentration, as well as the spectra of the unfilled PLA films and the modified crystals. As the nanocrystals concentration increased

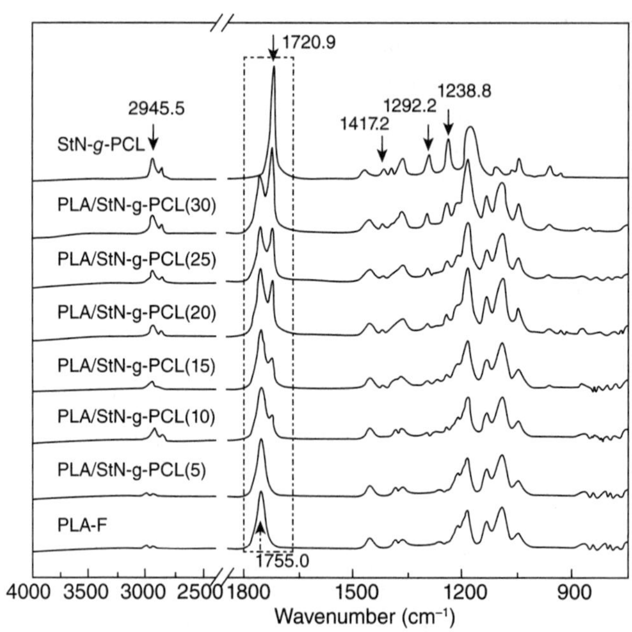

Figure 21.19 ATR–FT-IR spectra of the PLA/StN-g-PCL nanocomposite and PLA-F films, as well as the StN-g-PCL powder. From Reference [103].

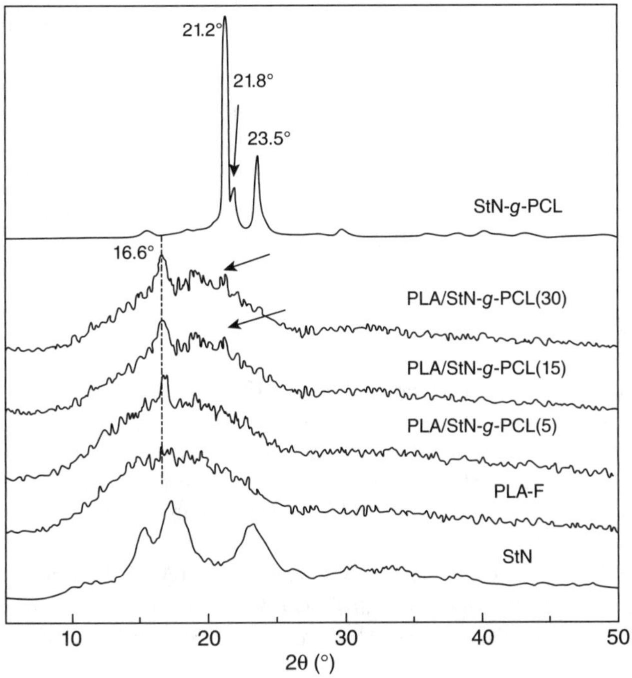

Figure 21.20 XRD patterns of the PLA/StN-*g*-PCL nanocomposite and PLA-F films, as well as the StN-*g*-PCL and StN powders. From Reference [103].

in the composite films, there was a large increase in the relative intensity of the 1720 cm^{-1} (C=O absorption in the crystalline phase of grafted PCL) and reduction of the of the 1750 cm^{-1} absorption of the PLA matrix, suggesting aggregation of the crystals at high concentrations (further confirmed by SEM) that form large crystalline regions for the PCL. XRD was also used to verify the above statements (Fig. 21.20); the unfilled film is mostly amorphous, while the StN showed strong peaks due to the crystalline arrangement of the grafted PCL. The incorporation of the StN to the PLA produces the appearance of different diffraction peaks, which were suggested to be due to the new crystalline forms of interfacial layers of PCL and PLA. However, as the concentration is further increased, a small diffraction peak appears at 21.2°, in the same location as the diffraction of PCL in the grafted nanocrystals. The concentration range is coincident with the appearance of aggregates observed by scanning electron microscopy (SEM), also in agreement with the FT-IR observations.

21.12 CONCLUDING REMARKS

The various examples presented along this Chapter on the use of spectroscopies applied to renewable nanoparticles and derived nanocomposites are not meant to be an exhaustive review of the subject, but just to give a glimpse into the different approaches to characterize materials and identify bulk and surface modifications. FT-IR and X-ray spectroscopies are by far the more used of the different techniques, because of the easiness of use, the amount of information returned and also the long tradition in analytical chemistry. The fact that water does not interfere with Raman absorption is an advantageous characteristic to be explored when working with natural nanoparticles, which are well known for being very hygroscopic in nature.

REFERENCES

[1] Gandini, A. (2008) *Macromolecules*, *41*, 9491.

[2] Mosiewicki, M.A., Marcovich, N.E., and Aranguren, M.I. (2011) Characterization of fiber surface treatments in natural fiber composites by infrared and Raman spectroscopy. In: *Interface Engineering of Natural Fibre Composites for Maximum Performance*, ed. N.E. Zafeiropoulos. Sawston, Cambridge, UK: Woodhead Publishing Limited, pp. 117–145.

[3] Hubbe, M., Rojas, O.J., Lucia, L.A., and Sain, M. (2008) *BioResources*, *3*, 929.

[4] Roman, M. (2009) Model cellulosic surfaces: History and recent advances. In: *Model Cellulosic Surfaces*, ed. M. Roman, ACS Symposium Series, vol. 1019, ed. M. Roman. Washington, DC.

[5] Azizi Samir, M.A.S., Alloin, F., and Dufresne, A. (2005) *Biomacromolecules*, *6*, 612.

[6] Eichhorn, S.J., Dufresne, A., Aranguren, M.I., Marcovich, N.E., Capadona, J.R., Rowan, S.J., Weder, C., Thielemans, W., Roman, M., Renneckar, S., Gindl, W., Veigel, S., Keckes, J., Yano, H., Abe, K., Nogi, M., Nakagaito, A.N., Mangalam, A., Simonsen, J., Benight, A.S., Bismarck, A., Berglund, L.A., and Peijs, T. (2010) *Journal of Materials Science*, *45*, 1.

[7] Siró, I., Plackett, D. (2010) *Cellulose*, *17*, 459.

[8] Nishino, T., Matsuda, I., and Hirao, K. (2004) *Macromolecules*, *37*, 7683.

[9] Orts, W.J., Shey, J., Imam, S.H., Glenn, G.M., Guttman, M.E., and Revol, J.-F. (2005) *Journal of Polymers and the Environment*, *13*, 301.

[10] Chen, W., Yua, H., Liua, Y., Chen, P., Zhang, M., and Hai, Y. (2011) *Carbohydrate Polymers*, *83*, 1804.

[11] Chakraborty, S., Sahoo, B., Teraoka, I., Miller, L.M., and Gross, R.A. (2005) *Macromolecules*, *38*, 61.

[12] Chakraborty, A., Sain, M., and Kortschot, M. (2006) *Holzforschung*, *60*, 53.

[13] Abe, K., Iwamoto, S., and Yano, H. (2007) *Biomacromolecules*, *8*, 3276.

[14] Abe, K., Yano, H. (2010) *Cellulose*, *17*, 271.

[15] Nakagaito, A.N., Yano, H. (2004) *Applied Physics A: Materials Science and Processing*, *78*, 547.

[16] Nakagaito, A.N., Yano, H. (2005) *Applied Physics A: Materials Science and Processing*, *80*, 155.

[17] Hayashi, N., Kondo, T., and Ishihara, M. (2005) *Carbohydrate Polymers*, *61*(2), 191.

[18] Henriksson, M., Henriksson, G., Berglund, L.A., and Lindström, T. (2007) *European Polymer Journal*, *43*(8), 3434.

[19] Pääkkö, M., Ankerfors, M., Kosonen, H., Nykänen, A., Ahola, S., Österberg, M., Ruokolainen, J., Laine, J., Larsson, P.T., Ikkala, O., and Lindström, T. (2007) *Biomacromolecules*, *8*(6), 1934.

[20] Elazzouzi-Hafraoui, S., Nishiyama, Y., Putaux, J.-L., Heux, L., Dubreuil, F., and Rochas, C. (2008) *Biomacromolecules*, *9*, 57.

[21] Liu, H., Liu, D., Yao, F., and Wu, Q. (2010) *Bioresource Technology*, *101*(14), 5685.

[22] Dong, X.M., Recol, J.F., and Gray, D.G. (1998) *Cellulose*, *5*, 19.

[23] Czaja, W., Krystynowicza, A., Bieleckia, S., and Brown, R.M. Jr. (2006) *Biomaterials*, *27*, 145.

[24] Lee, K.Y., Blaker, J.J., and Bismarck, A. (2009) *Composites Science and Technology*, *69*, 2724.

[25] Watanabe, K., Tabuchi, M., Morinaga, Y., and Yoshinaga, F. (1998) *Cellulose*, *5*, 187.

[26] Huang, H.-C., Chen, L.-C., Lin, S.-B., and Chen, H.-H. (2011) *Carbohydrate Polymers*, *83*, 979.

[27] Lima, M.M.D., Borsali, R. (2004) *Macromolecular Rapid Communications*, *25*, 771.

[28] Karaaslan, M.A., Tshabalala, M.A., Yelle, D.J., and Buschle-Diller, G. (2011) *Carbohydrate Polymers*, *86*, 192.

[29] Chen, Y., Liu, C., Chang, P.R., Cao, X., and Anderson, D.P. (2009) *Carbohydrate Polymers*, *76*, 607.

[30] Marcovich, N.E., Auad, M.L., Bellesi, N.E., Nutt, S.R., and Aranguren, M.I. (2006) *Journal of Materials Research*, *21*, 870.

[31] Li, R., Fei, J., Cai, Y., Li, Y., Feng, J., and Yao, J. (2009) *Carbohydrate Polymers*, *76*, 94.

[32] Satyamurthy, P., Jain, P., Balasubramanya, R.H., and Vigneshwaran, N. (2011) *Carbohydrate Polymers*, *83*, 122.

[33] Sain, M., Panthapulakkal, S. (2006) *Industrial Crops and Products*, *23*, 1.

[34] Sun, X.F., Xu, F., Sun, R.C., Fowler, P., and Baird, M.S. (2005) *Carbohydrate Research*, *340*, 97.

[35] Marcovich, N.E., Reboredo, M.M., and Aranguren, M.I. (1996) *Composite Interfaces*, *4*, 119.

[36] Oksman, K., Etang, J.A., Mathew, A.P., and Jonoobi, M. (2011) *Biomass and Bioenergy*, *35*, 146.

[37] Klemm, D., Heublein, B., Fink, H.P., and Bohn, A. (2005) *Angewandte Chemie International Edition*, *44*, 3358.

[38] Filson, P.B., Dawson-Andoh, B.E. (2009) *Bioresource Technology*, *100*, 2259.

[39] Martínez-Sanz, M., Lopez-Rubio, A., and Lagaron, J.M. (2011) *Carbohydrate Polymers*, *85*, 228.

[40] Maréchal, Y., Chanzy, H. (2000) *Journal of Molecular Structure*, *523*, 183.

[41] Kim, U.J., Kuga, S., Wada, M., Okano, T., and Kondo, T. (2000) *Biomacromolecules*, *1*, 488.

[42] Oh, S.Y., Yoo, D.I., Shin, Y., and Seo, G. (2005) *Carbohydrate Research*, *340*, 417.

[43] Uhlin, K.I., Atalla, R.H., and Thompson, N.S. (1995) *Cellulose*, *2*, 129.

[44] Yun, Y.S., Cho, S.Y., and Jin, H.J. (2010) *Molecular Crystals and Liquid Crystals*, *519*, 141.

[45] Yi, J., Xu, Q., Zhang, X., and Zhang, H. (2008) *Polymer*, *49*, 4406.

[46] Qiao, R., Brinson, L.C. (2009) *Composites Science and Technology*, *69*, 491.

[47] Cao, X., Dong, H., and Li, C.M. (2007) *Biomacromolecules*, *8*, 899.

[48] Mathew, A.P., Dufresne, A. (2002) *Biomacromolecules*, *3*, 609.

[49] Terech, P., Chazeau, L., and Cavaille, J.Y. (1999) *Macromolecules*, *32*, 1872.

[50] Pereda, M., Amica, G., Rácz, I., and Marcovich, N.E. (2011) *Journal of Food Engineering*, *103*, 76.

[51] Pereda, M., Aranguren, M.I., and Marcovich, N.E. (2008) *Journal of Applied Polymer Science*, *107*(2), 1080.

[52] Jayaramudu, J., Jeevan, D., Reddy, P., Guduril, B.R., and Varada, A. (2009) *Fibers and Polymers*, *10*, 338.

[53] Goetz, L., Mathew, A., Oksman, K., Gatenholm, P., and Ragauskas, A.J. (2009) *Carbohydrate Polymers*, *75*, 85.

[54] Sarasam, A.R., Krishnaswamy, R.K., and Madihally, V. (2006) *Biomacromolecules*, *7*, 1131.

[55] Ping, P., Wang, W., Chen, X., and Jing, X. (2005) *Biomacromolecules*, *6*, 587.

[56] Heijkants, R.G.J.C., van Calck, R.V., van Tienen, T.G., de Groot, J.H., Buma, P., Pennings, A.J., Veth, R.P.H., and Schouten, A.J. (2005) *Biomaterials*, *26*, 4219.

[57] Auad, M.L., Contos, V.S., Nutt, S., Aranguren, M.I., and Marcovich, N.E. (2008) *Polymer International*, *57*, 651–659.

[58] Stevanic, J.S., Joly, C., Mikkonen, K.S., Pirkkalainen, K., Serimaa, R., Rémond, C., Toriz, G., Gatenholm, P., Tenkanen, M., and Salmén, L. (2011) *Journal of Applied Polymer Science*, *122*, 1030.

[59] Jayakumar, R., Prabaharan, M., Nair, S.V., and Tamura, H. (2010) *Biotechnology Advances*, *28*, 142.

[60] Jayakumar, R., Menon, D., Manzoor, K., Nair, S.V., and Tamura, H. (2010) *Carbohydrate Polymers*, *82*, 227.

[61] Kurita, K. (2001) *Progress in Polymer Science*, *269*, 1921.

[62] Rinaudo, M. (2006) *Progress in Polymer Science*, *31*, 603.

[63] Pillai, C.K.S., Paul, W., and Sharma, C.P. (2009) *Progress in Polymer Science*, *34*, 641.

[64] Majeti, N.V., Kumar, R. (2000) *Reactive and Functional Polymers*, *46*, 1.

[65] Nair, K.G., Dufresne, A. (2003) *Biomacromolecules*, *4*, 657.

[66] Balthasar, S., Michaelis, K., Dinauer, N., Briesen, H.V., Kreuter, J., and Langer, K. (2005) *Biomaterials*, *26*, 2723.

[67] Min, B.-M., Leeb, S.W., Limb, J.N., Youb, Y., Leeb, T.S., Kangc, P.H., and Park, W.H. (2004) *Polymers*, *45*, 7137.

[68] Chen, Z., Mo, X., He, C., and Wang, H. (2008) *Carbohydrate Polymers*, *72*, 410.

[69] Sriupayo, J., Supaphol, P., Blackwell, J., and Rujiravanit, R. (2005) *Polymer*, *46*, 5637.

[70] Lertwattanaseri, T., Ichikawa, N., Mizoguchi, T., Tanaka, Y., and Chirachanchai, S. (2009) *Carbohydrate Research*, *344*, 331.

[71] Phongying, S., Aiba, S., and Chirachanchai, S. (2007) *Polymer*, *48*, 393.

[72] Lu, Y., Weng, L., and Zhang, L. (2004) *Biomacromolecules*, *5*, 1046.

[73] Tzoumaki, M.V., Moschakis, T., and Biliaderis, C.G. (2010) *Biomacromolecules*, *11*, 175.

[74] Mathew, A.P., Laborie, M.P.G., and Oksman, K. (2009) *Biomacromolecules*, *10*, 1627.

[75] Angulakshmi, N., Prem Kumar, T., Thomas, S., and Manuel Stephan, A. (2010) *Electrochimica Acta*, *55*, 1401.

[76] Watthanaphanit, A., Supaphol, P., Tamura, H., Tokura, S., and Rujiravanit, R. (2010) *Carbohydrate Polymers*, *79*, 738.

[77] Chang, P.R., Jian, R., Yu, J., and Ma, X. (2010) *Carbohydrate Polymers*, *80*, 420.

[78] Muzzarelli, R.A.A., Morganti, P., Morganti, G., Palombo, P., Palambo, M., Biagini, G., Belmonte, M.M., Giantomassi, F., Orlandi, F., and Muzzarelli, C. (2007) *Carbohydrate Polymers*, *70*, 274.

[79] Fan, Y., Saito, T., and Isogai, A. (2008) *Biomacromolecules*, *9*, 192.

[80] Hu, X.W., Du, Y.M., Tang, Y.F., Wang, Q., Feng, T., Yang, J., and Kennedy, J.F. (2007) *Carbohydrate Polymers*, *70*, 451.

[81] Zia, K.M., Zuber, M., Barikani, M., Jabbar, A., and Khosa, M.K. (2010) *Carbohydrate Polymers*, *80*, 539.

[82] Wada, M., Saito, Y. (2001) *Journal of Polymer Science Part B: Polymer Physics*, *39*, 168.

[83] Cho, Y.W., Jang, J., Park, C.R., and Ko, S.W. (2000) *Biomacromolecules*, *1*, 609.

[84] Al-Sagheer, F.A., Al-Sughayer, M.A., Muslim, S., and Elsabee, M.Z. (2009) *Carbohydrate Polymers*, *77*, 410.

[85] Li, X., Li, X., Ke, B., Shi, X., and Du, Y. (2011) *Carbohydrate Polymers*, *85*, 747.

[86] Jayakumar, R., Tamura, T. (2008) *International Journal of Biological Macromolecules*, *43*, 32.

[87] Xu, Y., Du, Y. (2003) *International Journal of Pharmaceutics*, *250*, 215.

[88] Kim, B.G., Kang, I.J. (2008) *Ultramicroscopy*, *108*, 1168.

[89] Chang, P.R., Jian, R., Yu, J., and Ma, X. (2010) *Food Chemistry*, *120*, 736.

[90] Wazed Ali, S., Rajendran, S., and Joshi, M. (2011) *Carbohydrate Polymers*, *83*, 438.

[91] Banerjee, T., Mitra, S., Singh, A.K., Sharma, R.K., and Maitra, A. (2002) *International Journal of Pharmaceutics*, *243*, 93.

[92] Osorio-Madrazo, A., David, L., Trombotto, S., Lucas, J.M., Peniche-Covas, C., and Domard, A. (2010) *Biomacromolecules*, *11*, 1376.

[93] Wu, Y., Yang, W., Wang, C., Hu, J., and Fu, S. (2005) *International Journal of Pharmaceutics*, *295*, 235.

[94] Tang, E.S.K., Huang, M., and Lim, L.Y. (2003) *International Journal of Pharmaceutics*, *265*, 103.

[95] Morin, A., Dufresne, A. (2002) *Macromolecules*, *35*, 2190.

[96] Paillet, M., Dufresne, A. (2001) *Macromolecules*, *34*, 6527.

[97] Janvikul, W., Thawornyutikarn, B. (2003) *Journal of Applied Polymer Science*, *90*, 4016.

[98] Le Corre, D., Bras, J., and Dufresne, A. (2010) *Biomacromolecules*, *11*, 1139.

[99] Chang, P.R., Ai, F., Chen, Y., Dufresne, A., and Huang, J. (2009) *Journal of Applied Polymer Science*, *111*, 619.

[100] Namazi, H., Dadkhah, A. (2010) *Carbohydrate Polymers*, *79*, 731.

[101] Imberty, A., Perez, S. (1988) *Biopolymers*, *27*, 1205.

[102] Imberty, A., Chanzy, H., Perez, S., Buléon, A., and Tran, V. (1987) *Macromolecules*, *20*, 2636.

[103] Yu, J., Ai, F., Dufresne, A., Gao, S., Huang, J., and Chang, P.R. (2008) *Macromolecular Materials and Engineering*, *293*, 763.

[104] Habibi, Y., Dufresne, A. (2008) *Biomacromolecules*, *9*, 1974.

[105] Labet, M., Thielemans, W., and Dufresne, A. (2007) *Biomacromolecules*, *8*, 2916.

[106] Thielemans, W., Belgacem, M.N., and Dufresne, A. (2006) *Langmuir*, *22*, 4804.

[107] Ma, X., Jian, R., Chang, P.R., and Yu, J. (2008) *Biomacromolecules*, *9*, 3314.

[108] Simi, C.K., Abraham, E. (2007) *Bioprocess and Biosystems Engineering*, *30*, 173.

[109] Varma, A.J. (1984) *Carbohydrate Polymers*, *4*, 473–480.

[110] Besheer, A., Hause, G., Kressler, J., and Mäder, K. (2007) *Biomacromolecules*, *8*, 359.

[111] Angellier, H., Molina-Boisseau, S., Belgacem, M.N., and Dufresne, A. (2005) *Langmuir: The ACS Journal of Surfaces and Colloids*, *21*, 2425–2433.

[112] García, N.L., Ribba, L., Dufresne, A., Aranguren, M.I., and Goyanes, S. (2009) *Macromolecular Materials and Engineering*, *294*, 169.

[113] García, N.L., Ribba, L., Dufresne, A., Aranguren, M.I., and Goyanes, S. (2011) *Carbohydrate Polymers*, *84*, 203.

[114] Vicentini, N.M., Dupuy, N., Leitzelman, M., Cereda, M.P., and Sobral, P.J.A. (2005) *Spectroscopy Letters*, *38*, 749.

[115] van Soest, J.J.G., Tournois, H., de Wit, D., and Vliegenthart, J.F.G. (1995) *Carbohydrate Research*, *279*, 201.

[116] Angellier, H., Molina-Boisseau, H.S., Dole, P., and Dufresne, A. (2006) *Biomacromolecules*, *7*, 531.

CHAPTER 22

Barrier Properties of Renewable Nanomaterials

VIKAS MITTAL

22.1 INTRODUCTION TO RENEWABLE MATERIALS AND BIONANOCOMPOSITES

Biopolymers (biobased and biodegradable) have become polymer matrices of interest in the recent years due to the concern of decreasing fossil resources, which are by far the biggest source of the monomers from which the majority of the polymers are generated. Also, increasing consciousness toward the use of more environmentally friendly materials and allied legislation have shifted the focus from the conventional nonbiodegradable polymers to "green" or biopolymers. The conventionally used polymers such as polypropylene, polystyrene, polyethylene, poly(methyl methacrylates), and so on are nonbiodegradable and their recycling or reuse is very challenging, thus contributing to piles of nonbiodegradable wastes all over the world. Similar is the fate of the composites generated from these polymers. As most of the biodegradable polymers find their origin in other alternate renewable sources, their use in the polymer composites technology is attractive; however, their use as an alternative to commercial nonbiodegradable polymers has emerged both as a goal and as a challenge. As already mentioned, the incorporation of layered silicate fillers caused significant enhancements in the properties of commercial nonbiodegradable polymers; however, it is important to develop this technology also for biodegradable polymers in order to generate high value materials from them. Thus, the potential to replace conventional polymers, especially thermoplastics, can be improved. Although it is important to study ways in

Biopolymer Nanocomposites: Processing, Properties, and Applications, First Edition. Edited by Alain Dufresne, Sabu Thomas, and Laly A. Pothan.

which the incorporation of the clay fillers in the biopolymer matrices can bring about the required improvements in performance and properties, the effect of clay on the biodegradability of the polymer in the composite also need to be studied.

A large number of biodegradable polymers exist and can be classified on the basis of their properties, origin, or synthesis methodology. Averous et al. [1] suggested the classification of biodegradable polymers depending on the synthesis as shown in Figure 22.1. According to this classification, the biodegradable polymers are classified into four categories:

- polymers from agro-resources
- polymers from microorganisms
- polymers chemically synthesized using biobased monomers
- polymers with monomers obtained from fossil resources.

The first three categories are biobased polymers, whereas the polymers in the last category are biodegradable, but have not biobased sources.

A number of research studies using the polymer systems mentioned in Figure 22.1 for the synthesis of clay-based nanocomposites have been reported in recent years.

The nanocomposites, by the virtue of the interactions between the organic and inorganic components, can have different morphologies as depicted in Figure 22.2 [2]. In the intercalated morphology, the clay layers are still ordered. However, the exfoliated or delaminated morphology provides the maximum benefit as the filler is completely delaminated and the platelet order is lost. This morphology has the maximum interfacial contacts with the polymer and thus the polymer properties like mechanical performance, barrier properties, biodegradability, and so on are significantly affected.

The nanocomposites have also been prepared by different synthetic routes [3,4]. *In situ* polymerization is based on the swelling of the clay in the monomer followed by the polymerization of monomer in the presence of clay, thus trapping the delaminated clay in the polymer structure. Melt intercalations is the method where the preformed high molecular weight can be directly used for the synthesis of the polymer nanocomposite. In this case, the polymer is melted at high temperature, and the filler is slowly added to this melt under shearing in the compounder. The filler platelets get sheared during compounding and kneading in the compounder and thus get well mixed with the polymer phase. Another method for the synthesis of the polymer nanocomposites is solvent intercalation. In this case, the polymer is dissolved in the same solvent in which the clay is also dispersed. Evaporation of the solvent leads to entrapment of the polymer chains in between the filler platelets. One can also synthesize thermoset polymer nanocomposites by this route. The prepolymer is dissolved in the solvent in which the filler is dispersed. The cross-linking is then initiated by the addition of cross-linker. The simultaneous evaporation of solvent and

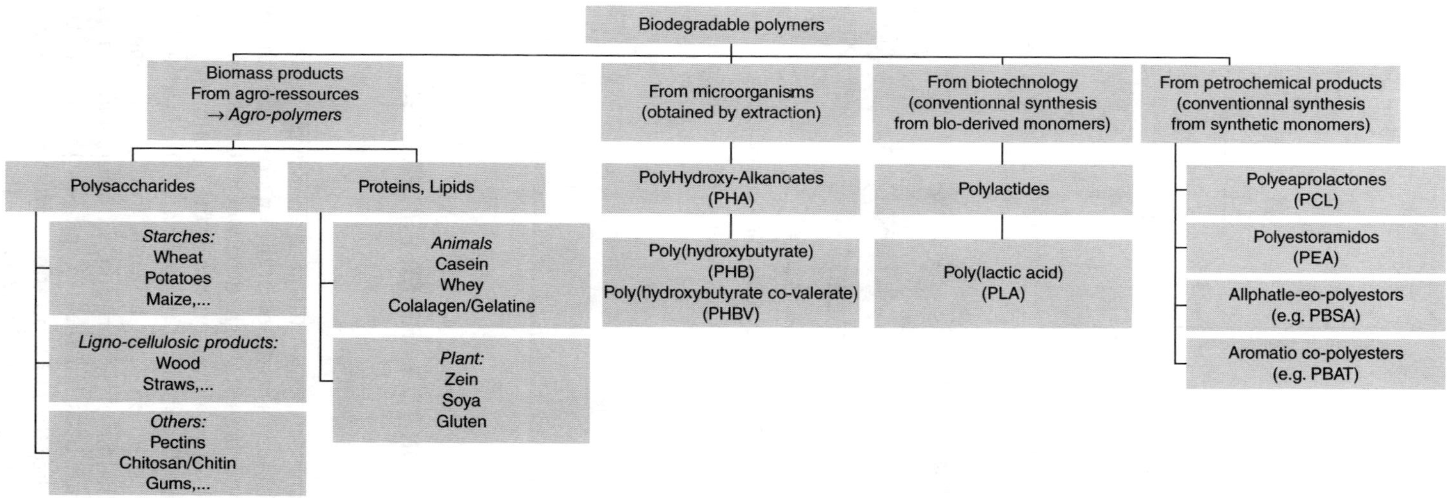

Figure 22.1 Classification of biodegradable polymers. Reproduced from Reference [1] with permission from Elsevier.

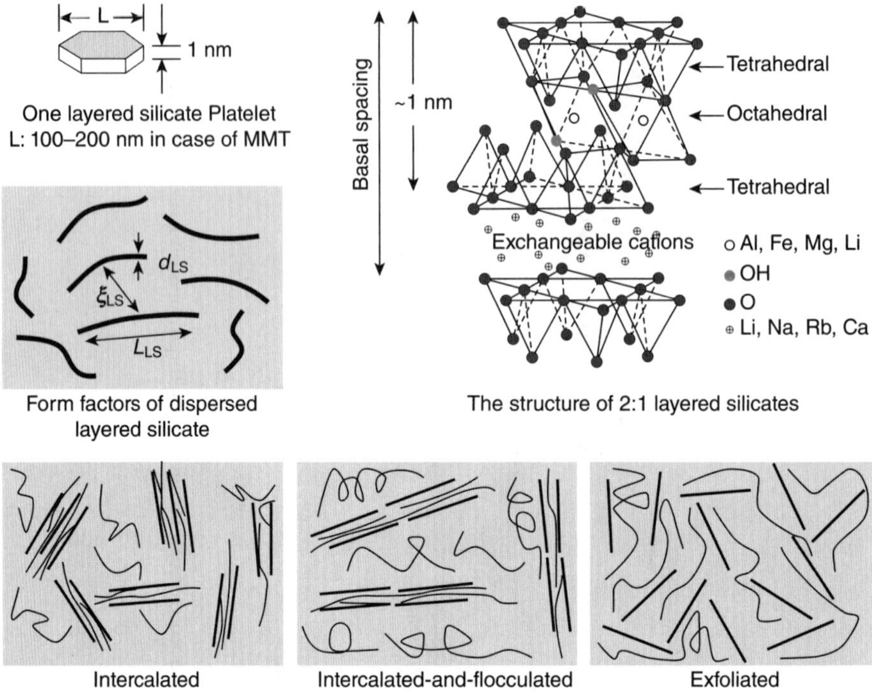

Figure 22.2 Representation of various morphologies of the polymer clay nanocomposites. Reproduced from Reference [2] with permission from Elsevier.

cross-linking of the polymer matrix leads to delaminated filler in the polymer matrix.

22.2 INTRODUCTION TO BARRIER PROPERTIES

22.2.1 Theory of Permeation

When a gas or vapor permeates through a polymer membrane, several processes are involved: the gas is sorbed at the entering face, dissolving there, with equilibrium rapidly being established between the two phases. The dissolved penetrant molecules then diffuse through the membrane, via a random walk hopping mechanism, desorbing at the exit face. The mechanism of permeation then involves both solution and diffusion [5]. The driving force behind the transport process that involves sorption, diffusion, and permeation is the concentration difference between the two phases [6, 7]. The transport process slowly tries to equalize the concentration difference or the chemical potential of the penetrant in the phases separated by the membrane. This process can be described in terms of Fick's first law of diffusion, according to which the flux J, in the direction of flow is proportional to the concentration gradient ($\delta c/\delta x$) and can be given as

$$J = \frac{D(c_1 - c_2)}{d} \tag{22.1}$$

where d is the membrane thickness and D is the diffusion coefficient. The penetrant distribution between the penetrant and the polymer phase is described by the Nernst distribution law.

$$c = KC \tag{22.2}$$

where c is the sorbed concentration; C, the ambient penetrant concentration in contact with the polymer surface; and K depends on temperature and c. In the case of transport of gases and vapors, pressure p is used instead of surface concentration. According to Henry's law

$$c = Sp \tag{22.3}$$

where S is the solubility coefficient. The combination of equations gives the well-known permeation equation

$$J = \frac{DS(p_1 - p_2)}{d} \tag{22.4}$$

where p_1 and p_2 are the ambient pressures on two sides of a film of thickness d. The product DS is called the permeability coefficient P, so that

$$P = DS. \tag{22.5}$$

In terms of permeability, the flux Equation (22.4) can be written as

$$J = \frac{P(p_1 - p_2)}{d}. \tag{22.6}$$

In dynamic measurements, first the sample is conditioned ($c_1 = c_2$). Then the concentration is step changed ($c_1 \neq c_2$), keeping c_1 and c_2 at a constant value. If we time-integrate the amount of gas passing the membrane, we observe that after a certain time the integrated curve, which describes the amount of gas per unit time, becomes linear with time. Extending the linear part of the curve (constant flow) and intercepting the time axis, one gets the time-lag value (t_L) [8]. Mathematically, t_L is given as:

$$t_L = \frac{d^2}{6 \cdot D}. \tag{22.7}$$

Hencc, the diffusion coefficient can be determined from the experimentally determined time-lag and a known thickness of the sample foil. For mineral polymer composites diffusion takes place only in the polymer phase (the diffusion through the mineral is orders of magnitude smaller). Measuring D and S may be error-prone, resulting in comparatively high values. Normally, the permeability coefficient is measured directly as a material property. In general the transmission rate r_T is determined, which is a measure for the volume of gas passing through a membrane of known area per unit time. With the thickness d of the sample and the pressure p, one obtains the permeability coefficient P, which is independent from the sample geometry, pressure, and time:

$$P = \frac{r_T \cdot d}{p} = \frac{(\text{quantity of permeant}) \cdot (\text{film thickness})}{(\text{area}) \cdot (\text{time}) \cdot (\text{pressure})}. \tag{22.8}$$

The permeability coefficient P is reported with a variety of different units in the literature. The most commonly used unit to present P is [$cm^3 \cdot \mu m/m^2 \cdot day \cdot mmHg$]. It can be converted to the SI unit [$cm^3 \cdot cm/cm^2 \cdot s \cdot Pa$] by multiplying with 8.68×10^{-16}.

22.2.2 Permeation through Heterogeneous Media

22.2.2.1 Laminates To calculate the diffusion coefficients through the laminates, it was reported that the neighboring parts of a composite medium may be considered to be in series or parallel [9]. A laminate composed of n sheets of thicknesses $l_1, l_2 \ldots l_n$ and diffusion coefficients $D_1, D_2 \ldots D_n$ placed in series has an effective diffusion coefficient D, given by

$$\frac{l_1}{D_1} + \frac{l_2}{D_2} + \ldots + \frac{l_n}{D_n} = \frac{l}{D} \tag{22.9}$$

where l is the total thickness of the laminate. The effective diffusion coefficient for a composite consisting of n sheets in parallel is given by

$$l_1 D_1 + l_2 D_2 + \ldots + l_n D_n = lD. \tag{22.10}$$

Both expressions are based on the assumption that the flow is unidirectional. If it is not, the relationships are approximations.

22.2.2.2 Materials with Plate-Like Inclusions The permeation of gases through mineral-filled polymers was first dealt with in the model by Nielsen [10]. The reduction in permeability in the composites by using the fillers of different geometrical dimensions is predicted in the model. The theory is based on the increase of the pathway of a gas molecule through a composite, caused by the filler. However, the model does not take into account Brownian motion and assumes impermeability of the filler for the gas as well as perfect alignment of the mineral inclusions. The ratio of the permeability coefficient of the composite P_c to one of the polymer matrix P_m, is given by:

$$\frac{P_c}{P_m} = \frac{\phi_m}{\tau} \tag{22.11}$$

where φ_m is the polymer volume fraction and τ is the increase of the pathway length of the diffusing particle, which is given by

$$\tau = \frac{d'}{d} \tag{22.12}$$

where d' describes the path length of the diffusing particle and d is the thickness of the membrane. With perfect alignment and a rectangular shape of the filler, d' can be described with

$$d' = d + d \cdot \frac{L}{2 \cdot W} \cdot \phi_f \tag{22.13}$$

where L is the length of the inclusion and W, its width. φ_f describes the filler volume fraction. Combining Equation (22.12) and Equation (22.13), τ can be written as

$$\tau = 1 + \frac{L}{2 \cdot W} \cdot \phi_f \tag{22.14}$$

where L/W equals the aspect ratio α of the filler. Using this and Equation (22.14) in Equation (22.11) we obtain an expression for the decrease of the permeability, depending on the filler volume fraction and the aspect ratio

$$\frac{P_c}{P_m} = \frac{\phi_m}{1 + \frac{\alpha \cdot \phi_f}{2}}. \tag{22.15}$$

With $\varphi_m = 1 - \varphi_f$ and $x = \alpha \cdot \varphi_f$, Equation (22.15) can be transformed into the following equation:

$$\frac{P_c}{P_m} = \frac{1 - \phi_f}{1 + \frac{x}{2}} \approx \frac{1}{1 + \frac{x}{2}} \text{(for small } \phi_f\text{)} \tag{22.16}$$

22.3 BARRIER PROPERTIES OF BIONANOCOMPOSITES

Thellen et al. [11] reported the synthesis and gas barrier properties of polylactic acid (PLA) nanocomposites as shown in Figure 22.3. The pure polymer as well as composites was processed using different screw speeds and feed rates.

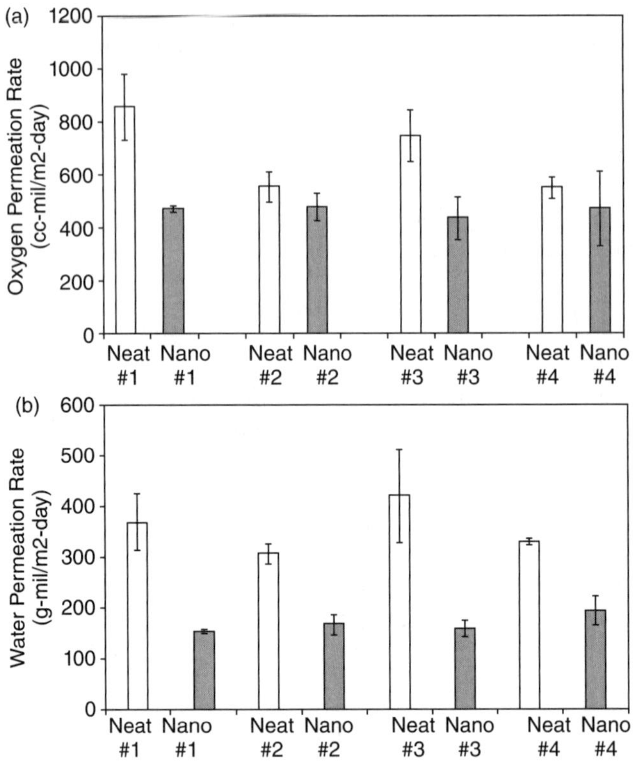

Figure 22.3 (a) Oxygen and (b) water vapor transmission through the pure PLA and PLA nanocomposite films generated by different processing conditions involving screw speed and feed rate. Reproduced from Reference [11] with permission from Elsevier.

Both oxygen and water vapor transmission through the nanocomposites films was observed to decrease as compared to pure polymer. In both cases, a maximum reduction of 50% was achieved. Another important observation was the impact of processing conditions on the barrier performance of the materials. The oxygen and water vapor permeation through the nanocomposite films was observed to be independent of the processing conditions, whereas the pure PLA polymer was observed to be sensitive to such changes.

Koh et al. [12] reported PLA/layered silicate nanocomposite membranes prepared by solution casting method for the application of dehumidification and gas separation. X-ray diffraction patterns of the nanocomposites showed the shifting of the diffraction peaks associated with the filler platelets to lower angles, thus confirming the polymer intercalation in filler interlayers and formation of intercalated nanostructures. As shown in Figure 22.4, the gas permeation through the nanocomposites decreased as compared to the pure polymer as a function of filler content. However, it was observed that a plateau

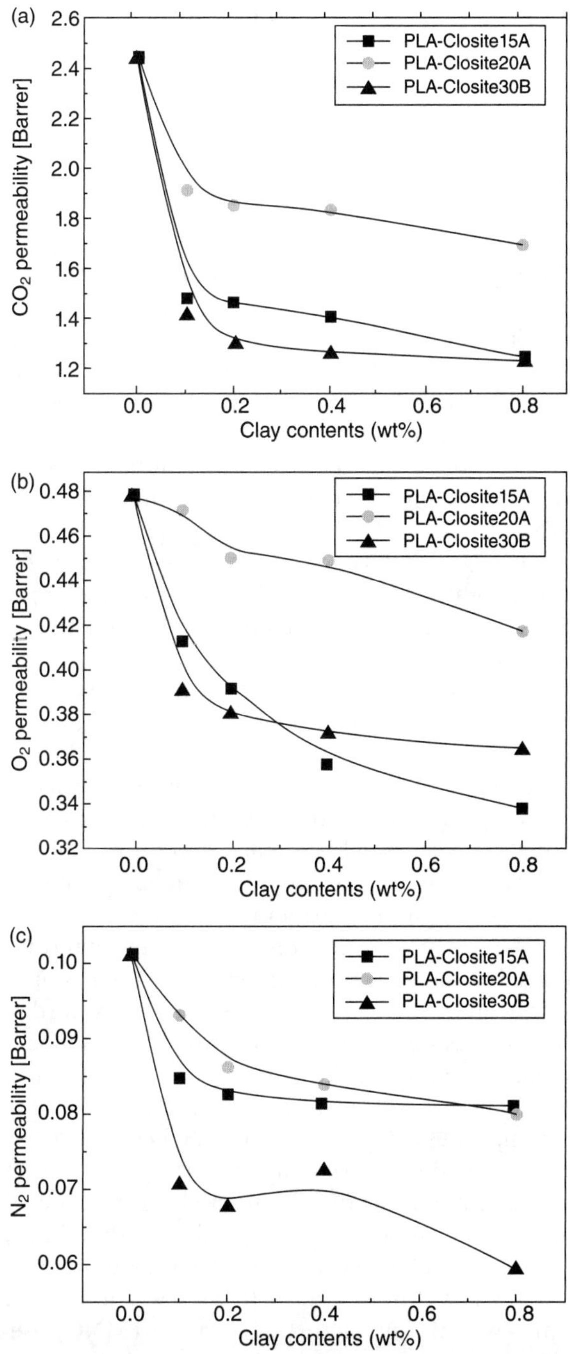

Figure 22.4 Gas permeation properties of the neat PLA as well as PLA nanocomposites as a function of different fillers and filler content: (a) CO_2, (b) O_2, and (c) N_2. Reproduced from Reference [12] with permission from Elsevier.

was reached at a certain weight percent of the filler, beyond which the permeation properties were not affected. Three different types of filler modifications were used to synthesize the nanocomposites. Cloisite 15A and 20A consisted of long alkyl chains as surface modifying molecules, whereas in Cloisite 30B, polar OH groups were also present in the chemical structure of the modifying molecules. When comparing the gas permeabilities of nanocomposites with different kinds of organoclay, the barrier performance of nanocomposites into which Closite30B fillers were incorporated was outstanding for all of the permeant gases. The authors also concluded from the microstructure analysis that the PLA/clay composite matrix was intercalated in the case of Closite15A, intercalated or exfoliated in the case of Closite20A, and exfoliated in the case of Closite30B, which resulted in the improved barrier resistance of the Cloisite30B nanocomposites. Figure 22.5 [12] also shows the gas permeability of the nanocomposites as a function of the kinetic diameters of the gas molecules. The authors reported that the gas permeabilities of the same composite material increased according to the order of the kinetic diameter of permeant gas molecules, (N_2 [3.64 A°] > O_2 [3.46 A°] > CO_2 [3.3 A°]). The least decrease of permeability for CO_2 was thus due to the smallest kinetic diameter and the largest molecular weight. The mechanical as well as thermal properties of the PLA nanocomposites were also reported to be enhanced on the incorporation of the organically modified clays (OMCs).

Ray et al. [13] reported the oxygen permeation properties of PLA nanocomposites. With 4 wt%, 5 wt%, and 7 wt% of the filler, the oxygen permeation through the polymer matrix was observed to decrease to 88%, 85%, and 81%, respectively. Chang et al. [14] also reported the oxygen permeation properties of the PLA nanocomposites. The authors synthesized the nanocomposites with three different types of organically modified montmorillonites (MMTs). Melt compounding approach was used for the nanocomposite generation. The oxygen permeation through the nanocomposite films decreased as a function of filler content. The oxygen permeation through the composites was reduced to half as compared to the pure polymer, when the filler content was 10 wt%; however, the permeation values were not normalized according to the thickness of the films. The permeation values through the nanocomposites were observed to be unaffected by the different types of organically modified fillers used in the study. Values of 327, 330, and 340 $cm^3/m^2 \cdot day$ were observed for 10 wt% filler nanocomposites using fillers C16MMT, DTA-MMT, and C25AMMT (C16: hexadecyl, DTA: dodecyltrimethyl, whereas C25A is the commercial product) as compared to a value of 777 $cm^3/m^2 \cdot day$ for the pure PLA polymer. The decrease in the oxygen permeation was also observed to reach a plateau value around 6 wt% as the extent of further decrease in oxygen permeation when the filler content was increased to 10 wt% was not significant.

Blends of chitosan and clay micro/nanoparticles (MNC) were prepared by dispersion of the clay particles in the polymer matrix and the films obtained were characterized in terms of water solubility, water vapor, oxygen, and carbon dioxide permeability by Casariego et al. [15]. The water solubility was

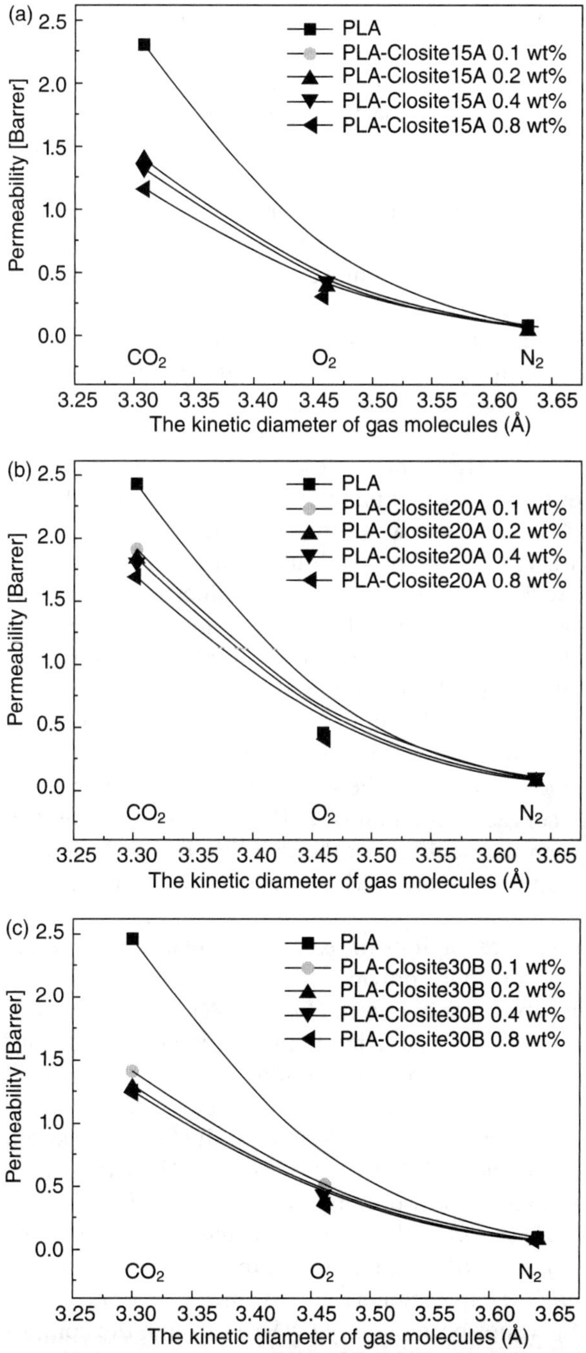

Figure 22.5 Gas permeability as a function of the kinetic diameters of gas molecules: (a) PLA/Closite15A, (b) PLA/Closite20A, and (c) PLA/Closite130B. Reproduced from Reference [12] with permission from Elsevier.

observed to decrease as the clay concentration increased keeping chitosan concentration constant. Figure 22.6 shows the barrier performance of the nanocomposites as a function of chitosan and MNC particles concentration. The water vapor barrier property of the films was observed to improve by incorporation of MNC in the film matrix. The highest values of water vapor permeation were obtained for those films with the lowest concentration of chitosan. Thus, both chitosan and clay concentrations affected the water vapor permeation through the nanocomposites. Casariego et al. report that the water vapor permeation decreased significantly by 9–32%, depending on the chitosan and clay concentrations.

The oxygen permeation through the nanocomposites decreased as a function of increased chitosan concentration. Chitosan and MNC particles did not significantly influence the CO_2 permeation.

Sanchez-Garcia et al. [16] presented the properties of nano-biocomposites of solvent cast polyhydroxybutyrate-co-valerate (PHBV) and polycaprolactone (PCL) containing carbon nanofiber or carbon nanotubes as a function of filler content. A reduction in water uptake of 45% was reported by the authors in the PCL with 1 wt% carbon nanotube content as compared with the pure polymer. At 3 wt% nanotube content, the reduction in water uptake reached up to 50%; however, it decreased on further increasing the content of nanotubes. For the PCL polymer with carbon nanofibers, the same behavior as with carbon nanotubes was observed. For the PHBV system, a reduction in the water uptake of 23% for the film of PHBV with 1 wt% CNT compared with the unfilled material; however, the water uptake increased on further increasing the nanotube content. For the PHBV with carbon nanofibers, a reduction of 49% was observed in the film of PHBV at 1 wt% content. Table 22.1 also shows the oxygen permeability of the PCL, PHBV, and their nanocomposites with carbon nanotubes and carbon nanofibers measured at 80% relative humidity (RH).

For PCL system, reductions of oxygen permeability of 52%, 38%, 8%, and 10%, as compared to pure polymer for the films with 1%, 3%, 5%, and 10% of carbon nanotubes, was observed (Fig. 22.7A). Increasing carbon nanotubes contents deteriorated the reduction in the oxygen barrier owing to agglomeration. In the case of carbon nanofibers, films of PCL with 1%, 3%, 5%, and 10 wt% of filler showed reduction in the oxygen permeability of 26%, 29%, 25%, and 25%, respectively. For PHBV system, reductions in oxygen permeability of 14%, 5%, 21%, and 58% for the films of PHBV with 1%, 3%, 5%, and 10 wt% of carbon nanofibers were observed (Fig. 22.7B).

Alves et al. [17] reported the the enhancement of the barrier to water vapor and gases (CO_2 and O_2) of a polymeric matrix composed by κ-carrageenan and pectin (66.7% κ-carrageenan) with the inclusion of mica flakes. Figure 22.8 shows the water vapor permeability (WVP) through the composites as a function of mica content as well as RH differences to which the two sides of the films were subjected. The water vapor permeation was observed to decrease as a function of mica content. Interesting differences were also observed at

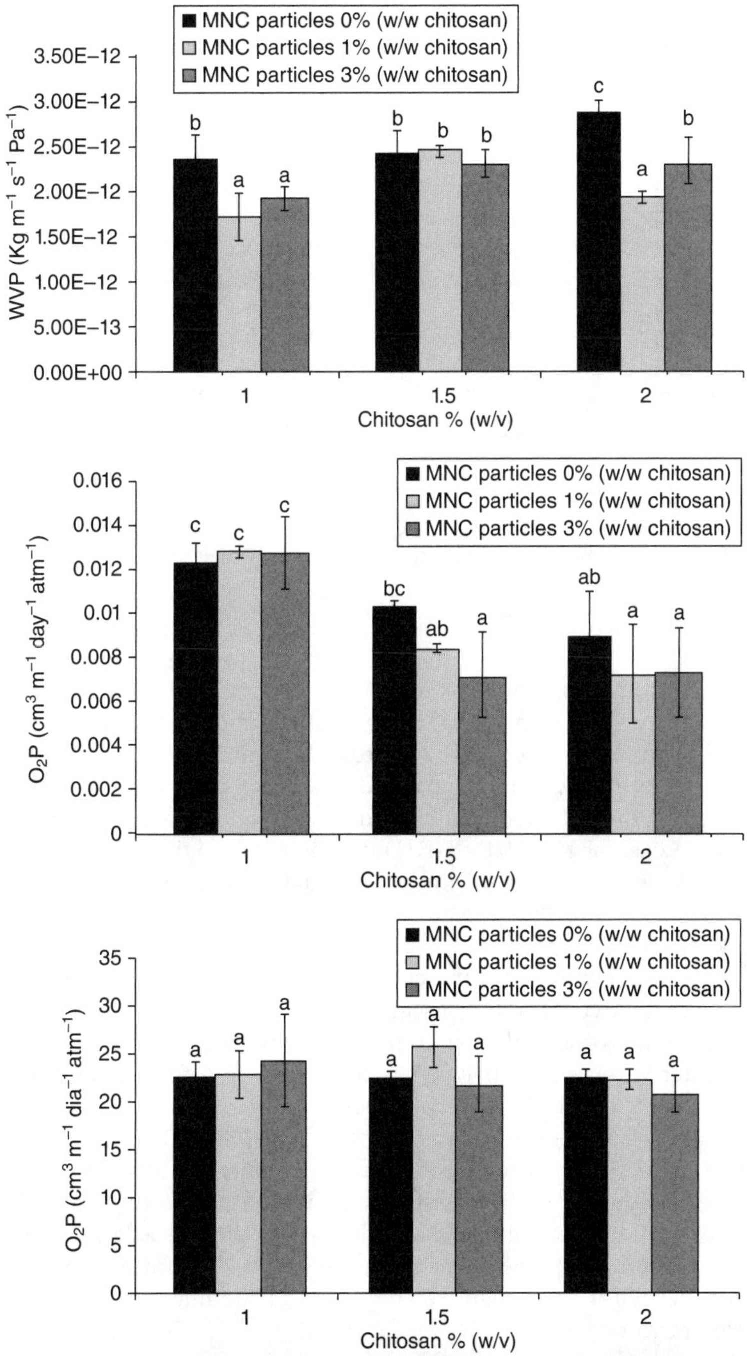

Figure 22.6 Barrier properties (water vapor, O_2, and CO_2) as a function of chitosan and MNC particles' concentration on chitosan/MNC particles' films produced with 1.0% (v/v) lactic acid. Reproduced from reference [15] with permission from Elsevier.

TABLE 22.1 Oxygen Permeability for PCL, PHBV and Their Nanocomposites. Reproduced from Reference [16] with Permission from Elsevier

	PO_2 (m^3 m/s m^2 Pa) at 80%RH		PO_2 (m^3 m/s m^2 Pa) at 80%RH
PCL	$7.06e^{-18}$	PHBV	$1.78e^{-18}$
PCL + 1%CNT	$3.39e^{-18}$	PHBV + 1%CNT	$0.67e^{-18}$
PCL + 3%CNT	$4.41e^{-18}$		
PCL + 5%CNT	$6.51e^{-18}$	PHBV + 5%CNT	$1.61e^{-18}$
PCL + 10%CNT	$6.36e^{-18}$	PHBV + 10%CNT	$1.18e^{-18}$
PCL + 1%CNT	$5.23e^{-18}$	PHBV + 1%CNF	$1.53e^{-18}$
PCL + 3%CNT	$5.03e^{-18}$	PHBV + 3%CNF	$1.70e^{-18}$
PCL + 5%CNT	$5.27e^{-18}$	PHBV + 5%CNF	$1.40e^{-18}$
PCL + 10%CNT	$5.33e^{-18}$	PHBV + 10%CNF	$0.75e^{-18}$
Literature value [24]PCL	$1.9e^{-18}$[a]	Literature value [25]PHB	$5.10e^{-18}$[b]

[a]At 0%RH (solvent casting).
[b]At 75%RH (commercial biobased materials).
CNF, carbon nanofiber.

different RH environments. The WVP was observed to be higher when the driving force of 75.4–60.5 (%RH) was applied. The authors opined that the observed effect was related to the hydrophilic character of the films material. The water entering the films acted as a diffusion species as well as a plasticizer, thus loosening the polymeric matrix and enhancing transmission. On the other hand, in the case of 100–10.8 (%RH) driving force, high water concentration gradient through the film thickness was suggested to limit the plasticizing effect, thus allowing the polymeric matrix to maintain its tightly packed network structure.

Figure 22.9 [17] illustrates the carbon dioxide and oxygen permeability as a function of mica content. The permeation values obtained for carbon dioxide and oxygen were observed to be four and six orders of magnitude lower than that of water vapor, respectively. A reduction of 73% for the nanocomposite film as compared with the pure polymer in the carbon dioxide permeability was observed for a mica content of 10%. However, the permeability was observed to increase on increasing the mica content further to 20%. Even though the polymeric matrix itself was a good barrier to oxygen, the barrier properties were further enhanced with the inclusion of mica flakes. The authors observed a decrease of the permeability of about 27% for a mica content of 10%, after which it rose again. Thus, 10 wt% of the filler was observed to be an optimum value.

Park et al. [18] reported the WVP through the thermoplastic starch nanocomposites generated by incorporating different types of clays. The permeation was observed to be reduced by half as compared to pure polymer on

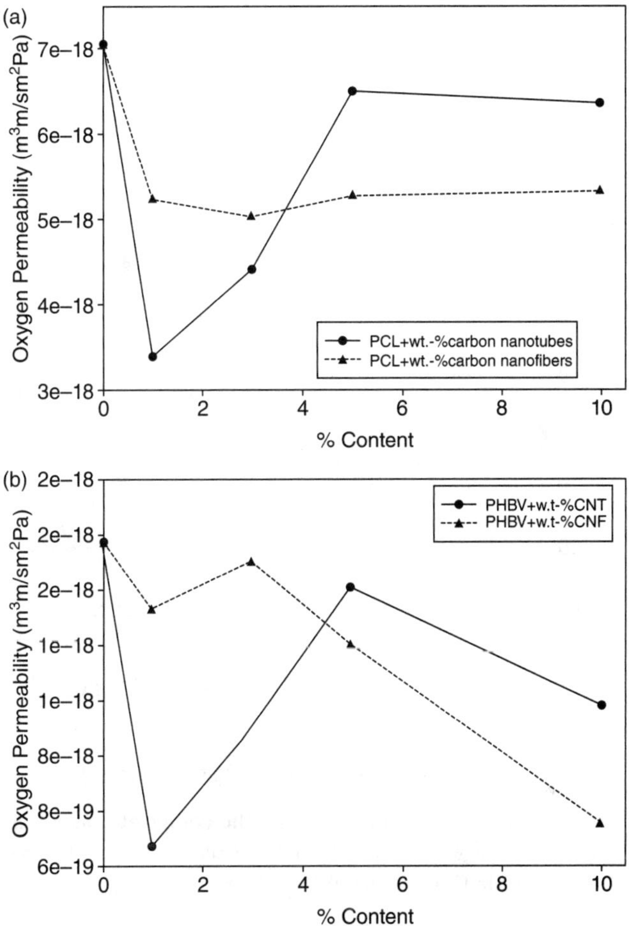

Figure 22.7 (a) Oxygen permeability through neat PCL polymer and nano-biocomposites with 1 wt%, 3 wt%, 5 wt%, and 10 wt% of the carbon nanotubes and carbon nanofibers measured at 80%RH; and (b) oxygen permeability through neat PHBV polymer and nano-biocomposites with 1 wt%, 3 wt%, 5 wt%, and 10 wt% of the carbon nanotubes and carbon nanofibers. Reproduced from reference [16] with permission from Elsevier.

addition of 5 wt% of the filler, especially Cloisite Na. Figure 22.10 also shows the permeability of the nanocomposites as a function of time. The nanocomposite with Cloisite sodium had the least decrease in the WVP,which was attributed to the better dispersion of this filler in the polymer matrix. Other fillers, which were organically modified by polar and nonpolar surface modifications, were worse in performance than Cloisite sodium; however, still reduced the increase in permeation over time as compared to pure polymer.

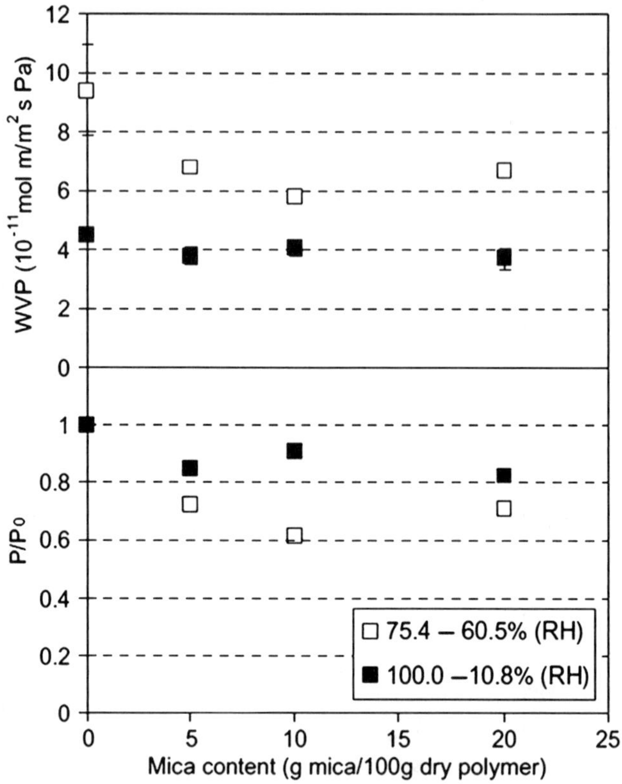

Figure 22.8 Water vapor permeability through the composites as a function of mica content and relative humidity. P/P_0 represents the normalized WVP values. Reproduced from reference [17] with permission from Elsevier.

Park et al. [19] also reported the synthesis of cellulose acetate nanocomposites using triethyl citrate as plasticizer and Cloisite 30B as filler. Different amounts of plasticizer as well as organoclays were used to synthesize the nanocomposites. Figure 22.11 shows the WVP of the different nanocomposites in comparison with the pure polymer. The permeability was observed to significantly decrease as a function of volume fraction of the filler. The filler platelets were observed to have an aspect ratio near to 150 at lower filler volume fractions, whereas the aspect ratio was observed to decrease to nearly 100 at higher filler volume contents owing to incomplete exfoliation. The polymer as well as nanocomposites containing higher amounts of plasticizer swas also observed to have a higher WVP than the systems with a lower plasticizer content. Reduction of the rigidity of the polymer matrix on incorporation of large amounts of plasticizers was opined to be the reason for enhanced WVP through the materials.

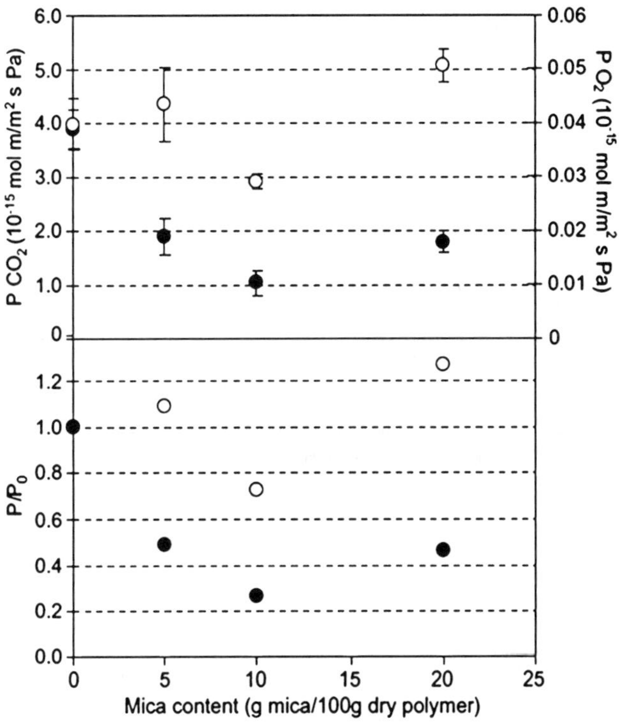

Figure 22.9 Carbon dioxide permeability (filled symbols) and oxygen permeability (open symbols), as a function of mica content. P/P_0 represents the normalized values. Reproduced from Reference [17] with permission from Elsevier.

Gorrasi et al. [20] reported studies on PCL/MMT composites, where the correlation between morphology and vapor barrier properties were presented. Different PCL/clay nanocomposites were synthesized by melt blending or catalyzed ring-opening polymerization of caprolactone. Microcomposites were obtained by direct melt blending of PCL and pristine MMT. Exfoliated nanocomposites were obtained by *in situ* ring-opening polymerization of CL with an OMC by using dibutyltin dimethoxide as an initiator/catalyst. Intercalated nanocomposites were formed either by melt blending with organically modified layered silicate (OMLS) or *in situ* polymerization within pristine MMT. Figure 22.12 shows water vapor diffusivity through these PCL nanocomposites as a function of clay content. The exfoliated nanocomposites were observed to have the least diffusivity, which was further reduced by increasing the amount of organic filler.

Kumar et al. [21] reported bionanocomposite films based on soy protein isolate (SPI) and MMT prepared using melt extrusion. Table 22.2 demonstrates the WVP of SPI–MMT films with 0%, 5%, 10%, and 15% MMT contents

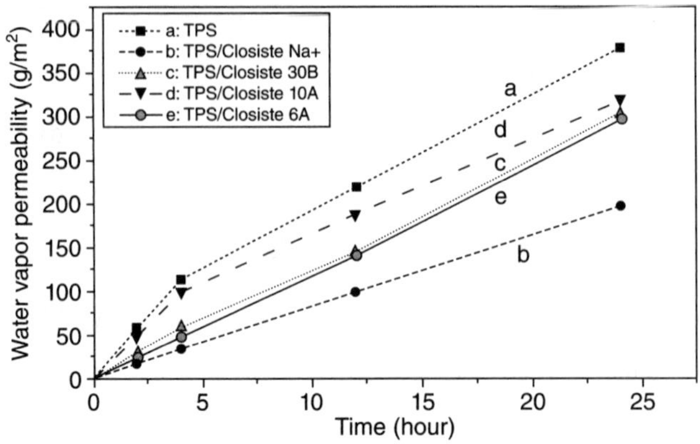

Figure 22.10 Water vapor permeability of thermoplastic starch/clay nanocomposites using different types of clays. Reprinted from Reference [18] with permission from Wiley.

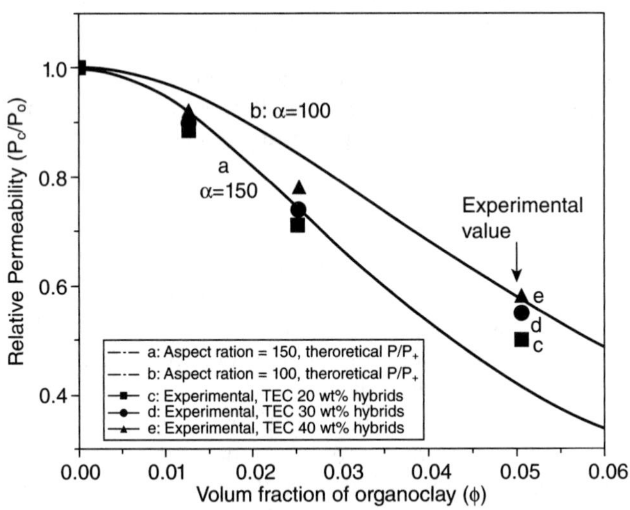

Figure 22.11 Water vapor relative permeability curve of the cellulose acetate (CA)/triethyl citrate (TEC) hybrid composites with different TEC plasticizer and organoclay contents: (c) CA/TEC (80/20 wt%), (d) CA/TEC (70/30 wt%), (e) CA/TEC (70/40 wt%) experimental relative permeability value of nanocomposites with clay contents 0–10 wt%. Curves (a) and (b) represent the theoretical predictions at different aspect ratios of the filler platelets. Reprinted from Reference [19] with permission from ACS.

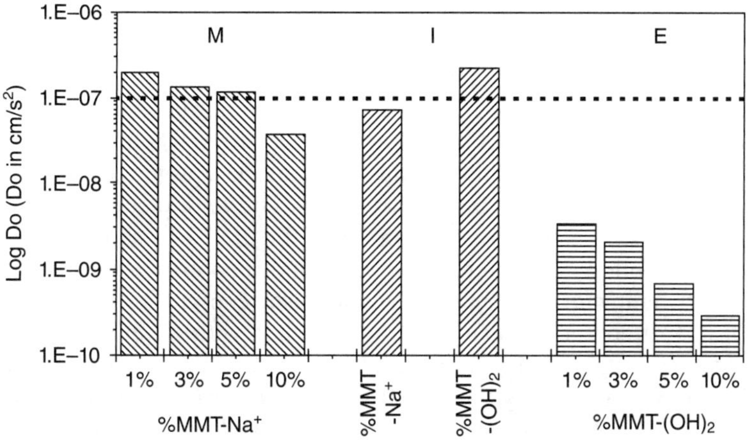

Figure 22.12 Water vapor diffusivity through the PCL nanocomposites as a function of clay content. M represents PCL microcomposite, E signifies PCL exfoliated nanocomposites, and the I denotes 3 wt% PCL intercalated nanocomposites. Reproduced from reference [20] with permission from Elsevier.

TABLE 22.2 Comparison of Water Vapor Permeability (WVP) of Soy Protein Isolate (SPI) Montmorillonite Films with Other Biopolymer, Bionanocomposite, and Plastic Films. Reproduced from Reference [21] with Permission from Elsevier

Films	WVP (g-mm/(m^2-h-kPa))
SPI	3.80 ± 0.11
SPI-5% MMT	2.96 ± 0.10
SPI-10% MMT	2.49 ± 0.08
SPI-15% MMT	2.17 ± 0.06
Soy protein	1.62–6.42
Whey protein	1.58–12.12
Chitosan	4.72
Chitosan-5% MMT	3.52
Starch	1.61
Starch-6% MMT	1.06
Cellophane	0.05–0.25
Polylactic acid (PLA)	0.06
Polyvinyl alcohol (PVOH)	3.15
Low density polyethylene (LOPE)	0.001
Polypropylene (PP)	0.02–0.04
Polyvinylidene chloride (PVDC)	0.01

compared with other biopolymers, bionanocomposites, and plastic films. Films containing MMT were observed to have significantly lower WVP as compared to that of pure SPI film. Five percent MMT inclusion was observed to reduce the permeability by 22.1%. Also, Kumar et al. reported that the WVP was reduced by 42.9% as the MMT content increased from 0% to 15%.

Kristo et al. [22] reported nanocomposite materials prepared using sorbitol-plasticized pullulan as the amorphous matrix and an aqueous suspension of starch nanocrystals as the reinforcing phase. Figure 22.13 shows the kinetics of water absorption and water uptake at equilibrium as well as the water diffusion coefficient values versus starch nanocrystal content, during conditioning at 94% RH of composites. The matrix was filled with different starch nanocrystal concentrations (ranging from 0% to 40%, w/w). The water uptake of samples showed a two-stage water absorption pattern. At lower times ($t < 50$ hours), the absorption kinetics was fast, followed by a slower absorption process, and finally achieved a plateau. The water uptake for a given time was lower as the starch nanoparticle content was higher. The unfilled sample showed the highest degree of hydration. The water uptake values at equilibrium were observed to range between 69.7% and 50.6% (w/w) for unfilled samples and nanocomposites filled with 40% (w/w) starch nanocrystals, respectively.

22.4 FACTORS AFFECTING BARRIER PERFORMANCE

As observed in the examples showed in the previous section, the barrier performance of the biopolymers is affected by a number of factors. First of all, it cannot be always assumed that the enhancement in the mechanical performance would lead to automatic improvement in barrier performance. The interfacial interaction between the filler and the polymer is the most important factor affecting the barrier performance. In the absence of positive interactions, the resulting phase mismatch at the interface can actually lead to enhancement of permeation rather than decreasing it [23]. The interactions can be improved by the addition of certain compatibilizers, but their impact on polymer symmetry, crystallinity, rigidity, and so on needs to be taken into account. Different surface modifications on the filler surface also perform differently owing to similar reasons of better surface polarity matching of the polymer with surface modification molecules in some cases than the others. The amount of the filler also has a significant impact on the barrier performance. In the case of exfoliated nanocomposites with good interfacial matching between the organic and inorganic phases, remarkable performance can be achieved at very low filler contents. The barrier properties also tend to reach a plateau value after a certain level of filler content is reached. The aspect ratio of the filler particles is also observed to generally decrease on enhancing the filler content owing to the increased interfacial contacts between the filler platelets. The barrier performance of the bionanocomposites is also affected

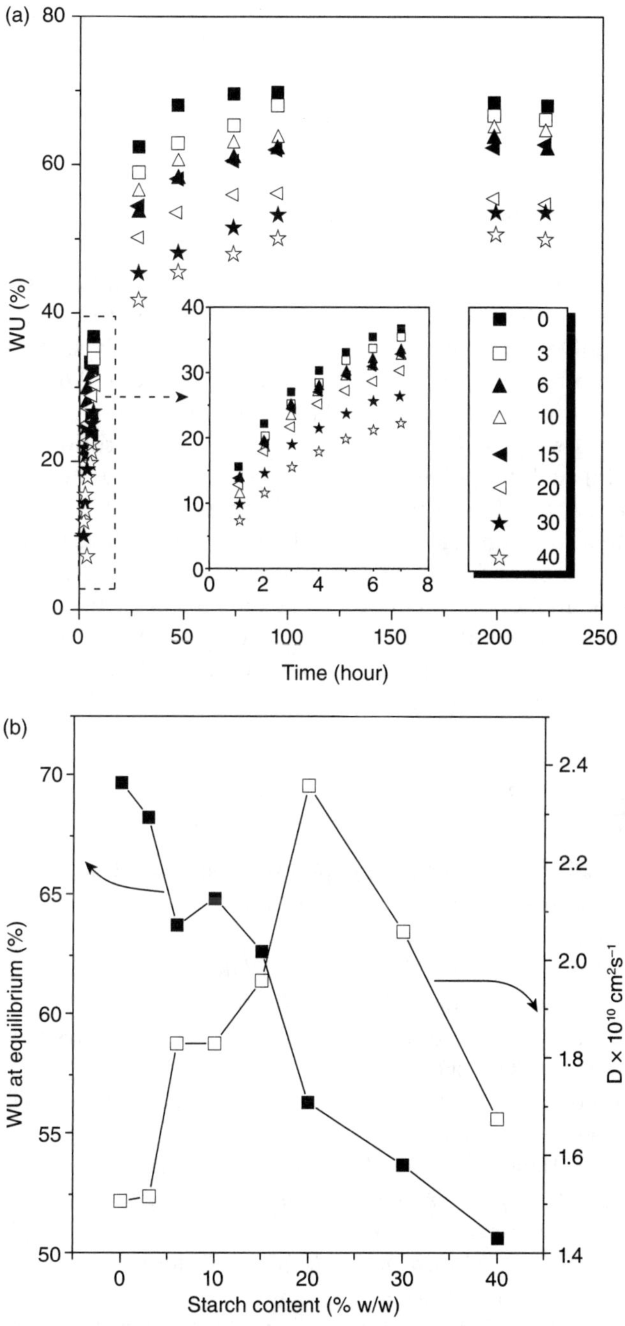

Figure 22.13 (a) Kinetics of water absorption and (b) water uptake at equilibrium as well as the water diffusion coefficient values versus starch nanocrystal content, during conditioning at 94% RH of composites filled with different starch nanocrystal concentrations (ranging from 0% to 40%, w/w). Reproduced from Reference [22] with permission from Elsevier.

by the processing conditions as well as humidity. The polymers also behave differently from different permeant gases owing to their specific interactions with these gases. For example, gases like oxygen do not interact with polymers, whereas water vapors are known to form hydrogen bonds with polymer chains.

22.5 CONCLUSIONS AND PERSPECTIVES

Increasing interest in the use of biopolymers for a number of applications brings additional challenges. Bionanocomposites lead to improved performance of biopolymers at very low filler contents. Generally, such materials are developed for enhancement of mechanical performance, and the improvement in mechanical properties is considered to automatically enhance other properties. However, owing to the complex nature of the sorption and diffusion phenomena through biopolymers, it is important to study the barrier properties separately. A number of biopolymers such as PLA, PCL, PHBV, and cellulose acetate have been reported in the literature to achieve reductions in the oxygen, water vapor, and carbon dioxide permeation by incorporation of fillers such as clay, nanofibers, and nanotubes. A number of factors including interfacial interactions, amount of filler, modification present on the surface of filler particles, humidity, processing conditions, and so on affect the barrier performance of the bionanocomposites.

REFERENCES

[1] Averous, L., Boquillon, N. (2004) *Carbohydr. Polym.*, *56*, 111.

[2] Ray, S.S., Yamada, K., Okamoto, M., Fujimoto, Y., Ogami, A., and Ueda, K. (2003) *Polymer*, *44*, 6633.

[3] Alexandre, M., Dubois, P. (2000) *Mater. Sci. Eng. R: Rep.*, *28*, 1.

[4] Pavlidoua, S., Papaspyrides, C.D. (2008) *Prog. Polym. Sci.*, *33*, 1119.

[5] Vieth, W.R. (1991) *Diffusion in and through Polymers: Principles and Applications*. Munich: Hanser Publishers.

[6] Cheremisinoff, N.P. (1989) *Handbook of Polymer Science and Technology*. New York: Dekker.

[7] George, S.C., Thomas, S. (2001) *Prog. Polym. Sci.*, *26*, 985.

[8] Elias, H.G. (1992) *Technologie Rohstoffe, Industrielle Synthesen, Polymere, Anwendungen*. Basel: Hthig & Wepf.

[9] Crank, J. (1998) *The Mathematics of Diffusion*. Oxford: Clarendon Press.

[10] Nielsen, L.E. (1967) *J. Macrromol. Sci.*, *A1*, 929.

[11] Thellen, C., Orroth, C., Froio, D., Ziegler, D., Lucciarini, J., Farrell, R., et al. (2005) *Polymer*, *46*, 11716.

[12] Koh, H.C., Park, J.S., Jeong, M.A., Hwang, H.Y., Hong, Y.T., Ha, S.Y., and Nam, S.Y. (2008) *Desalination*, *233*, 201.

[13] Ray, S.S., Yamada, K., Okamoto, M., and Ueda, K. (2003) *Polymer*, *44*, 857.

[14] Chang, J.-H., Uk–An, Y., and Sur, G.S. (2003) *J. Polym. Sci., Polym. Phys.*, *41*, 94.

[15] Casariego, A., Souza, B.W.S., Cerqueira, M.A., Teixeira, J.A., Cruz, L., Diaz, R., and Vicente, A.A. (2009) *Food Hydrocoll.*, *23*, 1895.

[16] Sanchez-Garcia, M.D., Lagaron, J.M., and Hoa, S.V. (2010) *Compos. Sci. Tech.*, *70*, 1095.

[17] Alves, V.D., Costa, N., and Coelhoso, I.M. (2010) *Carbohydrate Polym.*, *79*, 269.

[18] Park, H.M., Lee, X., Jin, C.Z., Park, C.Y., Cho, W.J., and Ha, C.S. (2002) *Macromol. Mater. Eng.*, *287*, 553.

[19] Park, H.M., Mohanty, A., Misra, M., and Drzal, L.T. (2004) *Biomacromolecules*, *5*, 2281.

[20] Gorrasi, G., Tortora, M., Vittoria, V., Pollet, E., Lepoittenvin, B., and Alexandre, M. (2003) *Polymer*, *44*, 2271.

[21] Kumar, P., Sandeep, K.P., Alavi, S., Truong, V.D., and Gorga, R.E. (2010) *J. Food Eng.*, *100*, 480.

[22] Kristo, E., Biliaderis, C.G. (2007) *Carbohydrate Polym.*, *68*, 146.

[23] Osman, M.A., Mittal, V., Morbidelli, M., and Suter, U.W. (2004) *Macromolecules*, *37*, 7250.

CHAPTER 23

Biocomposites and Nanocomposites Containing Lignin

CORNELIA VASILE and GEORGETA CAZACU

23.1 INTRODUCTION

In the scientific world, "bionanocomposite" has become a common term to designate those polymeric materials involving a biopolymer, produced from plants or microorganisms or by other biological processes, reinforced with inorganic/organic fillers showing at least one dimension on the nanometer scale [1]. Nanometer-sized fillers derived from plants show unique advantages over traditional inorganic nanoparticles by virtue of their biodegradability and biocompatibility. Currently, biomass-based nanofillers include cellulose, chitin, and starch nanowhiskers, and others derived from biomass, such as poly(lactic acid). Because of the particular structure of lignin and its main role in various systems as filler, these systems are considered as composites, although lignin is an organic component. In most cases, lignin contains an important amount of particles at the nanoscale level [2, 3].

Using strategies inspired by nature, different research teams have developed biopolymer-based nanocomposites that display both the well-known properties of nanocomposites derived from synthetic polymers (improved mechanical and gas barrier properties and higher thermal stability) and the remarkable features of either the biological or inorganic moieties, such as biocompatibility, biodegradability, and, in some cases, functional properties. These new materials have possible applications in biomedicine, agriculture, and the food and building industries, helping reduce the amount of waste products and lessen environmental pollution.

Biopolymer Nanocomposites: Processing, Properties, and Applications, First Edition. Edited by Alain Dufresne, Sabu Thomas, and Laly A. Pothan.

23.2 LIGNINS AND THEIR USE IN COMPOSITES

Lignocellulosic biomass offers a variety of potentially valuable by-products and their use in new eco-friendly composites is a quite challenging task. Lignin is one of the most abundant organic substances on earth and thus is the most promising feedstock for generation of the high value products with potential applications as components of polymer composites [4].

Lignin is primarily a structural material adding strength and rigidity to cell walls and constitutes between 15% and 40% of the dry matter of woody plants. Lignin in the scientific literature is known as "insoluble plant polymer" with an aromatic and highly cross-linked structure. From the chemical point of view, lignin can be defined as a very reactive macromolecular compound, because of its functional groups such as aromatic rings, phenolic and aliphatic alcohol groups, and methoxy groups that are possible sites for chemical modification and/or reaction. Lignin composition depends on its origin, the developmental stage of the plant, and environmental factors. Some properties are summarized in Table 23.1 for two main types of lignin [5].

Lignin is a fascinating molecule due to its diversity in both chemical composition and structure at a low or high level of order and degree of

TABLE 23.1 Main Characteristics of the Lignin from Wood [5]

Softwood lignin	Hardwood lignin
Lignin content is ~28%	Lignin content is ~20%
Lignin dissociates faster in solution	Lignin dissociates slower in solution
Lignin self-associates greater in solution	Lignin self-associates less in solution
Harder to break down lignocellulosic biomass	Easier to breakdown lignocellulosic biomass
Coniferyl alcohol primarily (~80%)	Coniferyl (~56%) and sinapyl (~40%) alcohols
Guaiacyl (coniferyl alcohol derived) G-lignin	Guaiacyl-syringyl (G-S) lignin; syringyl is sinapyl alcohol derived lignin
Gymnosperms	Angiosperms, dicotyledons
Molecular mass is larger than hardwood lignin	Molecular mass is lower than softwood lignin
Branching is higher	Branching is lower; lignin is more linear
Cross-links are greater	Cross-links are fewer
C–C bonds are greater	C–C bonds are fewer
5′ linkages more common	5′ linkages less common
-OCH_3 content is ~20%	-OCH_3 content is ~14%
β-O-4 ether bonds are lower	β-O-4 ether bonds are higher
β-β and β-5 bonds are higher	β-β and β-5 bonds are fewer
Deconstruction is harder	Deconstruction is easier
Lignin is condensed	

polymerization; thus, there may be no two identical lignin macromolecules with the same primary sequence of phenyl units.

The complexity of lignin macromolecules and their association with polysaccharides have made it impossible to isolate lignin and purify it from plant cell walls in a pure form since the various pulping techniques used cause substantial alterations to lignin structure. Only the ball milled wood lignin (MWL), isolated from finely powdered wood by the action of neutral solvents, is considered to be the closest to *in vivo* lignin. Lignins isolated from wood by classical delignification procedures are called "technical lignins" to distinguish them from *in vivo* native lignin. Technical lignins have undergone extensive depolymerization and chemical changes. Depending on the type of pulping, the lignins vary in molecular weight, functional groups, degree of condensation, types of intermonomer linkages, and the type and ratio of monomer units. Pronounced noncovalent attractive interactions also exist between individual molecular lignin species, resulting in enormous associated complexes [6, 7]. These supramacromolecular complexes are responsible for the cohesive nature of lignin-based materials [8]. Organosolv lignins differ significantly from other technical lignins. Structurally, organosolv lignins (OSL) have a higher relative amount of phenolic hydroxyl groups, a more oxidized structure (Hibbert's ketones), possess low glass transition temperatures T_g's, and are easier to thermally process than kraft lignins (KRL). As a result, OSL have advantages over other industrial lignins for composite-material applications.

Commercially, lignin plays a vital role in pulp and paper manufacturing as part of chemical recovery process. Lignin is primarily burnt as a fuel for energy; however, as a part of the biorefinery concept, value-added utilization of lignin is critical [9]. Among them, isolated lignin is industrially manufactured as an organic waste stream in the pulp and paper industry and in biomass conversion processes for biofuel production. Properties of lignin depend on their origin and extraction procedures [10–14].

23.3 OBTAINING LIGNIN NANOPARTICLES

Due to their accessibility and environmental compatibility, obtaining natural polymer nanoparticles constitutes a major concern for many research teams.

Self-assembly, self-organization, and aggregation give rise to a complex, three-dimensional network of lignin that displays randomly branched topology and fractal properties [5]. Also, lignins self-associate as well as interact with carbohydrates and proteins to form supramolecular structures, contributing to the difficulties in isolating "pure lignin." The supermolecular self-assembly of lignin at the nanoscale was studied *in vitro* dehydrogenative polymers (DHPs) synthesis of lignin model compounds by electronic microscopy.

Information on the synthesis of lignin-based nanoparticles is relatively limited and covered by patents [15]. Rojas [16] obtained either carbon nanofibers or lignin nanofibers by the electrospinning method (Fig. 23.1a) or by acetylation (Fig. 23.1b).

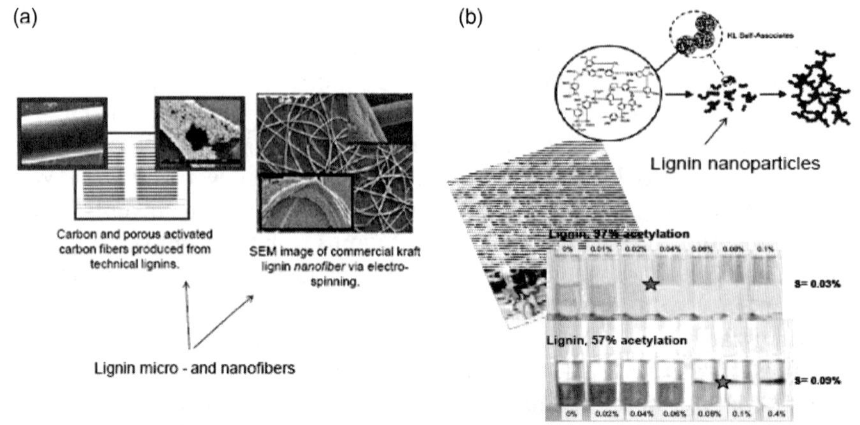

Figure 23.1 Possibilities of obtaining lignin-based nanofibers (adapted from Reference [16]). See color insert.

Popa et al. [17] obtained lignin nanoparticles from lignins in annual plants (wheat straw and Sarkanda grass) and commercial products (Protobind 1000, 2000, 3000), offered by Granit Recherche Developement SA (Lausanne, Switzerland), by hydroxymetylation (in the presence of formaldehyde and ammonium hydroxide), or by epoxydation (by reaction with epichlorohydrine) reactions. Under suitable synthesis conditions, nanoparticles have been obtained. The size distribution curves of grass lignin show a unimodal curve, while nanoparticles obtained from Protobind lignin are characterized by bimodal distribution and larger nanoparticles. The nanoparticle distribution domain for wheat straw lignin is between 50 and 500 nm, while Protobind samples, between 100 and 800 nm. It has been demonstrated that lignin complexes with copper could be used as biocides. Lignin nanoparticles have been also evidenced in alkaline solutions [18].

23.4 PREPARATION OF COMPOSITES AND BLENDS

Usually, the lignins used in preparation of blends or composites were produced industrially and obtained as a purified commercial polymer. Blends of various polymer/lignin ratios were then prepared by mechanical mixing followed by thermal extrusion using an mixing extruder with a single- or twin-screw configuration, rotation speed of 200–250 rpm, and a take-up speed of 100 m/min. Extrusion temperatures varied depending on the composite/blend composition, varying from 145 or 210 to 247°C.

To increase the interfacial bonding, sometimes modification of either thermoplastics by reactive extrusion [19, 20] or lignin (epoxidation, alkylation, etc.) is applied [21, 22]. Most compounding processes require the addition of compatibilizers, which circumvent incompatibility problems relating to poor interfacial adhesion between the filler and the polymer. The addition of maleic anhydride-modified polymers (e.g., maleic anhydride-grafted polymers) represents a prominent method for compatibilizing fillers and polymers. During compounding, the maleic anhydride-functionalized polymer reacts with the hydroxyl groups of the filler to form graft copolymers.

Lignin can be used as the main matrix in a new class of engineering materials based only on renewable resources. For example, the Arboform® (developed at the Fraunhofer Institute for Chemical Technology, Pfinztal, Germany) composites consist of isolated lignin, natural fibers, and natural additives that are elaborated through conventional thermoplastic processes such as injection molding [23]. The obtained composites have wood-like mechanical behavior and some properties in the range of those of polyamides. For certain practical applications (i.e., for the building industry), the thermal stability and fire resistance of these Arboform composites are key properties that need to be improved. The preparation of polymer–silicate clay nanocomposites has addressed these fireproofing issues. The nanoscale dispersion of clay layers forms a passive barrier, which hinders the out-diffusion of volatile decomposition products from thermal cracking [24]. Depending on the compatibility level between clay and organic polymer, the system exhibits an intercalated or exfoliated structure. The intercalated structure corresponds to insertion of the polymeric matrix between the layers of the clay, but these latter still exhibit spatial relationships with each other. In the exfoliated (or delaminated) structure, the layers are independently separated; they have no relationship with each other and are homogeneously distributed at nanoscale in the matrix. Both sepiolite and organically modified montmorillonite (Org-MMT) have been processed with lignin and natural fibers by shear mixing in a twin-screw extruder.

23.5 BIOBASED NANOCOMPOSITES CONTAINING LIGNIN

Biobased nanocomposites containing lignin were prepared from OSL/organoclay mixtures by mechanical mixing and subsequent melt intercalation [25]. Two Org-MMT organoclays with different ammonium cations (Cloisite 30B and Cloisite 20A, Southern Clay Co., Austin TX) were used. The Alcell™ lignin (organosolv lignin obtained by alcohol pulping, implemented by Repap Enterprises, Canada) and varying amounts of organoclay (1–10 wt%) mechanically mixed were then thermally extruded at 130°C and 150°C depending on the hybrid composition. Due to the chemical differences, the two organoclays will be dispersed differently within the polymer matrix, leading to different composite properties.

The effect of organoclays on the mechanical and thermal properties of the nanocomposites was studied. Thermal analysis revealed an increase in T_g for the nanocomposites as compared with the original OSL. The highest T_g for the Alcell/organoclay mixtures occurs under the most severe conditions tested (ball-milling for 30 minutes at 600 rpm). In general, thermal extrusion

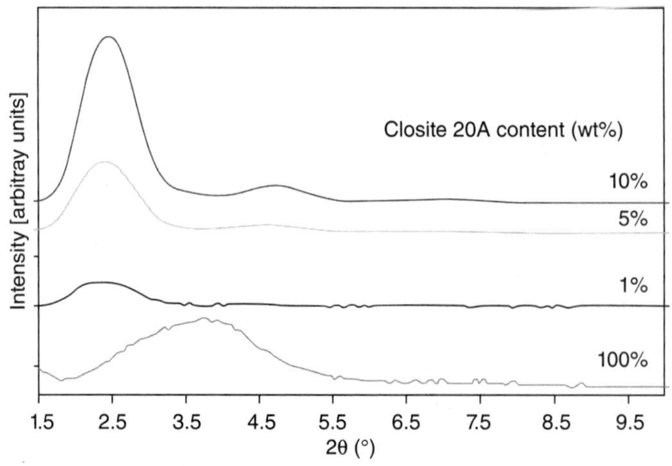

Figure 23.2 XRD patterns of Cloisite 20A and Alcell/Cloisite 20A hybrids with clay contents of 1 wt%, 5 wt%, and 10 wt% (adapted from Reference [25]).

of the lignin/organoclay mixtures led to a slight increase in T_g, the result of lignin condensation reactions rather than increased interaction with the organoclay.

The addition of organoclay to Alcell lignin improved its thermal processability; the increased organoclay content leads to decrease of the spinnability temperatures of the lignin/organoclay composite fibers. The Alcell/Cloisite 20A melt spinning temperatures were slightly higher than those of the corresponding Alcell/Cloisite 30B mixtures. For both organoclays, lignin intercalation into the silicate layers was observed using X-ray diffraction (XRD). The XRD pattern of lignin does not show any peaks due to its amorphous state. Cloisite 20A and Closite 30B, however, exhibit a single peak at $2\theta = 3.9°$ and $2\theta = 5.0°$, respectively. For all of the Alcell/Cloisite 20A mixtures (Fig. 23.2), a new peak appears at $2\theta = 2.4°$, corresponding to an increase in the *d*-spacing from ~2.4 to ~3.7 nm.

Similarly, Alcell/Cloisite 30B mixtures present a new peak at the lower 2θ corresponding to an increase in the *d*-spacing of the clay platelets from ~1.9 to ~3.2 nm. These observations indicate significant intercalation of Alcell into the galleries of the silicate layers of the organoclays.

The intercalated hybrids exhibited a substantial increase in tensile strength of lignin/organoclay composite fibers in comparison with the original Alcell lignin. Thus, incorporating just 5 wt% of Cloisite 20A led to a 54% increase in tensile strength, while the same amount of Cloisite 30B enhanced the tensile strength by more than 70%.

A new type of nanocomposite based on silicate clays, lignin, and natural fibers has been obtained by shear mixing in a twin-screw extruder. Lignin isolated from straw and grass crops by a soda pulping process and two types of nanoclays (nonmodified Pangel S9 sepiolite (TOLSA T Group) and Org-MMT) have been chosen for processing with lignin and natural fibers. The counterions of MMT have been exchanged with octadecyl ammonium cations. The morphological aspect of the obtained nanocomposites has been controlled by transmission electronic microscopy (TEM). For sepiolite-based nanocomposites, individual needles have been observed, while Org-MMT-based nanocomposites have exhibited partial delamination of silicate layers into intercalated and exfoliated nanostructures. Intercalated structures with higher interfoliar distance (~20–50 nm) than at 2% loading are observed together with individual exfoliated layers. Despite the higher Org-MMT concentration, the number of submicro aggregates has not increased. These observations indicate that presence of organophilic ammonium leads to a good compatibility of MMT layers with lignin and natural fibers. It is also noteworthy that the amphiphilic character of lignin has contributed to enhancing this compatibility. Moreover, the applied shear rate during the process was sufficient to obtain a relatively good delamination of MMT. The mechanical energy provided by the shear mixing has allowed the van der Waals interactions to be overcome between the MMT layers, leading to intercalated or exfoliated nanostructures. The properties of the specimens have been studied by thermogravimetry (TGA), dynamic mechanical analysis (DMA), and flexural tests. Incorporation of 2% or 5% w/w of sepiolite does not influence the mechanical and thermal behavior compared with the reference lignin/natural fiber composite, while Org-MMT-based nanocomposites show improved properties [26].

As shown in Table 23.2, the incorporation of nanoclays does not dramatically modify the mechanical properties of the sample [26].

It is generally accepted that incorporation of nanoclays in nonreinforced matrices increases stiffness and enhances ultimate stress. The improved mechanical properties are explained by the interfacial interactions between the matrix and nanofillers, which strengthen the nanohybrids [27].

TABLE 23.2 Results from Three-Points Bending Test and Dynamic Mechanical Analysis Conducted on Lignin/Natural Fibers Composite and Silicate Clay Nanocomposites at Various Loadings (2% and 5% w/w) [26]

Nanoclay (w/w loading)	Flexural Modulus (GPa)	Ultimate Stress (MPa) (std dev.)	Ultimate Strain (%) (std dev.)	T_g(max. of tan°δ) (°C)
None (ref.)	5.3	35	0.71	114
Sepiolite (2%)	5.3	36	0.71	114
Sepiolite (5%)	5.2	36	0.72	120
Org-MMT (2%)	5.7	37	0.71	115
Org-MMT (5%)	5.8	37	0.68	124

In the case of lignin/natural fiber composites, the influence of nanofillers is not so marked probably because the reference composite, which is reinforced with natural fibers, is already very stiff. Nevertheless, one can notice a slight increase in the flexural modulus after incorporation of Org-MMT. At 2% loading, the increase of modulus is about 7.5%, while for 5% Org-MMT-based nanocomposites, the flexural modulus increases about 9.4%. However, sepiolite-based nanocomposites do not show any significant variation of modulus and ultimate properties. These results may show evidence of anchoring between lignin chains and the organic surfactant (octadecyl ammonium) present in interlayer spaces of MMT. According to this anchoring, a better stress transfer from the matrix to inorganic MMT layer is expected compared with unmodified sepiolite needles. Enhanced stress transfer is accompanied by an increase in stiffness (i.e., flexural modulus) for Org-MMT nanocomposites. Thus, the MMT layers have developed good interactions with the lignin matrix and form passive protective barriers. Accordingly, dispersion of Org-MMT could be an interesting and eco-friendly approach to improve the fire resistance of these three-dimensional designed materials elaborated from renewable resources. The T_g increases ~6°C in the presence of 5% sepiolite and ~10°C for the 5% Org-MMT-based nanocomposite. Thermogravimetric data suggest that exfoliated or intercalated layered aluminosilicates act as a protective barrier against degradation of organic components.

23.6 LIGNIN IN THERMOPLASTIC COMPOSITES

Traditionally, a small amount of lignin, less than 2% of that produced, has been used as low cost filler or dispersants, adhesives, and surfactants. Recently, thermoplastics with high lignin content have been produced by alkylated KRL or blending KRL with synthetic polymers [8, 28, 29]. These new lignin-based materials possess mechanical properties comparable to polystyrene. Although 100% KRL and alkylated KRL films and fibers are quite brittle, they can be effectively plasticized by blending with suitable polyesters and polyethers [8, 28].

Because of its functionality, lignin is able to interact with many polymers, inducing relevant changes in their wettability, fire resistance, and mechanical properties. However, the complexity of the possible chemical interactions between lignin and polymers (in particular, natural polymers) remains an important challenge. In particular, several authors [4, 30, 31] investigated the effect of lignin–natural polymer interactions on the functional and mechanical properties of natural polymers. Lignin is extremely promising in its ability to facilitate the biodegradation of polyolefins [32–41]. In these studies, the variation/improvement of the characteristic is strongly dependent on the chemical functional groups of both lignin (alkaline lignin and lignosulfonate, epoxy modified lignin) and natural polymers (soy protein, starch, and others).

Commercially, lignin is obtained as a by-product of paper manufacturing. Some industrial processes produce standard well-defined lignins. As a result, lignin is utilized as a stabilizer (antioxidant) for plastics and rubber as well as in the formulation of dispersants, adhesives, and surfactants [42]. However, only a small amount of this amorphous polyphenolic macromolecule is used in tailored material applications. Many studies of lignin-filled thermoplastics including polyethylene [43], polypropylene (PP) [43–45], polystyrene [43, 46], polyethylene oxide (PO) [43], polyvinyl chloride [47, 48], polyethyleneterephthalate [49], polylactic acid [50, 51], polycaprolactone (PCL) [52], soy protein and other biodegradable polymers [3], and many others [53, 54] have already been reported.

Due to its phenolpropanoidic structure similar to that of hindered phenols, lignin can act as stabilizer [55] or initiator of PO degradation [56, 57]. Incorporation of lignin in the PO matrix influences the polymer's useful properties and biodegradation characteristics. Usually, the addition of lignin deteriorates the mechanical properties of plastics due to the poor miscibility.

Studies regarding the benefits of lignin incorporation into polyolefin blends have led to contradictory results. Gonzales-Sánchez and Espósito Alvarez [58] found that the incorporation of KRL (10–55 wt%) into polypropylene was accompanied by an increase in tensile modulus with lignin content and a decrease in maximum tensile strength, indicating poor adhesion between the nonpolar structure of PP and more polar structure of lignin and a poor dispersion into the matrix. However, Tudorachi et al. [32, 59] observed a reduction in physicomechanical properties with an increase in lignin content in the polymer matrix. Kharade and Kale [43] have incorporated up to 30 wt% dry lignin powder in low-density polyethylene (LDPE), high-density polyethylene (HDPE), and PP, and they observed that the tensile strength and elongation at break values decreased with lignin content for all three polymer blends, indicating poor compatibility or interaction between lignin and the matrix polymer. However, incorporation of lignin into a binary polyolefin mixture (70 polypropylene (PP)/30 polyethylene (PE)) in amounts of 2.5–40% (w/w) gave polymer blends with good thermal, physico-mechanical, and surface properties [35]. Levon et al. [60] have observed the thermal stabilizing effect of lignosulfonate when the blending with low-density polyethylene (LDPE) was carried out at temperatures over 190°C. Blending kraft lignin (10–60 wt%) with isotactic PP yielded a material with reduced tensile and flexural strength and improved moduli [61]. Canetti et al. [62] investigated the influence of lignin on the thermal degradation of isotactic polypropylene. An increase in the thermal degradation temperature of the blends was observed as a function of lignin content, in both oxidative and nonoxidative conditions. Also, the increase was pronounced for the experiments carried out in air atmospheres, where the interactions between the PP and lignin led to the formation of a protective surface able to reduce the oxygen diffusion toward the polymer bulk. Gregorová et al. [63] determined the Vicat softening temperature (VST) for the evaluation of the polypropylene and recycled polypropylene (rccPP) oxidative stability. It was shown that the incorporation of lignin in polyolefin matrices improves the heat resistance of PP and rccPP plaques.

Compatibility between PO and lignin can be improved by introduction of reactive functional groups in each or in only one blending compound.

Thus, Casenave et al. [64] developed a new technique to improve the interfacial adhesion by establishing the chemical bonds between the matrix and the additive and to ensure a better dispersion and adhesion to the synthetic matrix. In their method, ethylene or propylene monomers are first grafted by a catalytic reaction on the lignin surface, and then mixed with the polyethylene matrix. Incorporation of grafted lignin (L-g-PE) or ungrafted lignin (L) produced a decrease in mechanical properties with increasing lignin content. Young's modulus for PE/L-g-PE remains constant and close to that of pure PE. Scanning electron microscopy (SEM) revealed the different morphologies of the broken sections of PE/L-g-PE blends, showing either the particles that were extracted from the other half of the samples, but with enhanced roughness of the surface of spheres, or broken particles, indicating a better lignin-grafted/matrix adhesion.

Another way to improve the compatibility between lignin and polyolefins is to use a compatibilizing agent such as ethylene-vinyl acetate copolymer (EVA), ethylene-proylene rubber grafted with maleic anhydride (EP-MA) [65], or glycidylmethacrylate-grafted-polypropylene (PP-g-GMA) [61, 66]. Alkylating lignin with bromododecane improves compatibility with polypropylene and enables up to 70 wt% lignin to be incorporated into a thermally stable composite [67].

Pascu et al. [68, 69] showed that by a combination of the chemical procedures of component modification (e.g., epoxidation of lignin) with physical treatment of surfaces (e.g., photooxidation, plasma, electron beam treatment, UV-irradiation), bio/environmental disintegrable polyolefin materials can be obtained.

Other effects of lignin incorporation in PO are improvements in conductivity, hydrophilicity, biodegradability, and printability [70]. In some cases, lignin behaves as an antifungal [47] or nucleating agent [49] or improves the antioxidant [45] properties of the end-products.

The lignin in the concentration 1–2 wt% possessed the ability to act as a processing stabilizer and as an antioxidant during thermo-oxidative aging of polypropylene films [71]. The effects of polymer–polymer interactions, specifically hydrogen bonding, on lignin-based thermoplastics were studied in a series of lignin/synthetic polymer blends prepared by melt extrusion [28]. Thermal analysis revealed miscible blend behavior in the lignin blends containing poly(ethylene oxide) (PEO) and poly(ethylene terephthalate) (PET), whereas the polyvinyl alcohol (PVA) and PP lignin blends appeared immiscible. The glass transition T_g of the lignin/PEO and lignin/PET blends showed a negative deviation from a linear mixing rule, indicative of specific intermolecular interactions. Diffuse reflectance Fourier transform–infrared spectroscopy (DRFT-IR) analysis revealed the formation of a strong intermolecular hydrogen bond between lignin and PEO, but not with PET.

23.6.1 Poly(Ethylene Oxide)/Organosolv Lignin

Blends of PEO with OSL (Alcell) were prepared by thermal blending. Excellent fiber spinning with good thermal properties were achieved [72]. Alkoxyl

TABLE 23.3 DSC Results of the Alcell™ Lignin/PEO Blend Fibers [72]

Weight Fraction (w/w)						Weight Fraction in Amorphous Phase (w/w)	
Lignin	PEO	T_g (°C)	ΔC_p (J/g/C)	T_m (°C)	ΔH (J/g)	Lignin	PEO
1.00	0.00	80	0.69	X	0	1.00	0.00
0.95	0.05	76	0.74	X	0	0.95	0.05
0.87	0.13	49	0.72	X	0	0.87	0.13
0.75	0.25	25	0.74	X	0	0.75	0.25
0.62	0.38	−3	0.73	X	0	0.62	0.38
0.50	0.50	−22	0.82	53[a]	6[a]	x	x
0.40	0.60	−37	0.24	53	68[a]	x	x
0.35	0.65	−2	0.36	56	75	0.60	0.40
0.25	0.75	−4	0.30	60	111	0.4	0.36
0.10	0.90	−31	0.28	63	150	0.49	0.51
0.00	1.00	−50	0.12	67	168	0.00	1.00

[a]There was some difficulty in the determination of these values by DSC because of the recrystallization peak.

chains at the C_α and C_γ positions of the Alcell lignin side chain structure act as internal plasticizers and enhance the thermal mobility of the lignin. The addition of a small amount of Alcell lignin to PEO resulted in an increase in the PEO crystalline domain size (Table 23.3).

However, both PEO crystallinity and crystalline domain size decreased with lignin incorporation beyond 25 wt%. A negative polymer–polymer interaction energy density was observed, indicating weak specific intermolecular interactions.

Calculation of the fitting parameters *k* and *q* from the Gordon–Taylor and Kwei equations revealed the presence of the strong intermolecular interactions between blend components. FT-IR analysis confirmed the formation of strong hydrogen bonding system between Alcell lignin and PEO, being an important factor in the formation of miscible lignin-based polymer blends. Phenolic hydroxyl groups form stronger hydrogen bonds with the ether oxygen in poly(ethylene oxide) than with aliphatic hydroxyl groups. Moreover, the hydrogen bonds formed between lignin and poly(ethylene oxide) are significantly stronger than those within the lignin macromolecule itself. Kadla and Kubo [73] have shown that by incorporating 5% poly(ethylene oxide) into hardwood KRL it can be spun in continuous fibers, which exhibited a nearly 10% increase in tensile strength and Young's modulus.

23.6.2 Polycaprolactone/Lignin

Melt grafting of maleic anhydride (MA) onto PCL was the key to PCL-*g*-MA compatibilizers, which enhanced the mechanical properties of reinforced PCL compounds based on wood flour (WF) and lignin [74]. The grafting of MA onto PCL was achieved in high yields with a twin-screw extrusion process. Low concentrations of MA and appropriate MA/initiator ratios afforded PCL-*g*-MA. A low MA content of PCL-*g*-MA (1.44 wt%) and a small PCL-*g*-MA addition (2.5 wt%) were sufficient to achieve attractive property combinations. More than 70 wt% lignin was added without mechanical properties being impaired. Compositions containing 40 wt% lignin showed elongation at break exceeding 500%. The compatibilized compounds, up to a lignin content of 40 wt%, gave a slightly improved Izod impact strength. This is exceptional for such a high filler content. According to TEM analysis, very effective lignin dispersion was achieved within the PCL matrix. No biodegradation took place after 2 weeks of natural soil burial. This was likely to be associated with the phenolic nature of lignin, which prevents attacks by microorganisms and acid formation. Biodegradation studies revealed that lignin addition enhanced the biostability of PCL compounds. In conclusion, lignin can be used as a weak and nontoxic biostabilizer to enhance the product lifetime of PCL compounds in outdoor applications.

23.6.3 Furfuryl Alcohol/Lignin

Guigo et al. [75] presented a new polyfurfuryl alcohol (PFA)/lignin blend with thermomechanical performance. Furfuryl alcohol (FA) and lignin are, respectively, monomeric and polymeric precursors issued from biomass feedstock. The morphologies of the combined PFA/lignin systems present a monophasic structure. The DMA results are in good agreement with SEM observations, showing presence of a unique relaxation peak. TGA reveals that the PFA/PL thermo-oxidative degradation occurs at higher temperatures compared with the natural (PL) component system, together with a lower rate of decomposition. This underlines a good interpenetration of lignin within the furanic matrix.

23.6.4 Protein/Lignin Bionanocomposites

It is well known that polysaccharides and proteins are biodegradable and/or compostable under specific environmental conditions. Compared with oil-based polymers, they are inferior in terms of processing and functional and structural performance. The processing problems are mainly due to reduced plastic flow properties and poor reproducibility, and control over the molecular architecture and spatial conformation of natural macromolecules is difficult. Production of plastics from plant materials at an industrial scale has the effect of increasing food prices. One alternative is the use of biological feedstocks that are the by-product of the food and agricultural industries, such as corn zein, wheat gluten, feather keratin, egg albumin, and lactalbumin [76]. Zein is found in corn endosperm, which has been commercially available since 1938 for its film-forming ability and its unique hydrophobicity, which is due to its high content of nonpolar amino acids. Zein shows high density and brittle

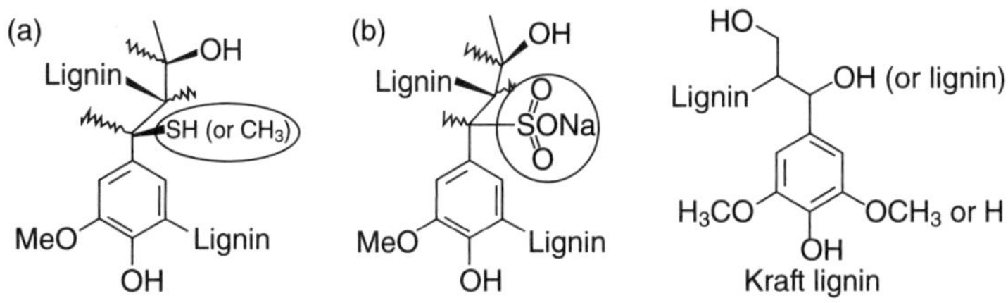

Figure 23.3 (a) Alkali lignin (AL) and (b) sodium lignosulfonate (LSS) and Kraft lignin.

behavior, with a high modulus and stress to break and a low strain to break. It can be plasticized by using small polar molecules such as water, glycerol, and ethylene glycol, which reduce the protein–protein hydrogen bonding interactions and increase the free volume and molecular mobility.

Soy protein isolate (SPI)/hydroxypropyl alkaline lignin (HPL), content lower than 6 wt% composites have been prepared by mixing in aqueous solution containing a small amount of glutaraldehyde as compatibilizer, and then compression molding to obtain plastic sheets. The HPL domain in SPI/HPL composites is about 50 nm, indicating a high interfacial activity. The glass transition temperature of the SPI/HPL sheets increased from 62.5°C to 70.4°C with an increase of HPL content from 0 wt% to 6 wt%. The tensile strength of the SPI/HPL nanocomposite sheets with 6 wt% HPL and 3.3 wt% glutaraldehyde was enhanced from 8.4 to 23.1 MPa compared with that of the SPI sheets, because of the SPI reinforcement effect [3].

The effect of alkaline lignin (AL) and sodium lignosulfonate (LSS) (Fig. 23.3) on the structure of thermoplastic zein (TPZ) was investigated [77] by means of XRD and Fourier transform infrared (FT-IR) spectroscopy and correlated with physical properties. At low AL concentration, the strong hydrogen bondings between the functional groups of AL and the amino acids in zein induced a destructuring of inter- and intramolecular interactions in α-helix, β-sheet, and β-turn secondary structures. This destructuring allowed for an extensive protein conformational modification that, in turn, resulted in a strong improvement of the physical properties of the bionanocomposite. By programming weak forces such as hydrogen bondings, hydrophobic interactions, and ionic interactions, biopolymers may achieve mechanical properties similar to those of synthetic polymers. Bionanocomposites maize zein (TPZ)/various kinds of lignin have been prepared by melt mixing in presence of ethylene glycol by Oliviero et al. [77].

The lignin content influenced, in different ways, the arrangements of secondary structure of zein in TPZ. At lower concentrations (especially 1 wt%) of lignins, specific functional groups are able to create strong hydrogen bonding with the polar group of amino acids in the zein protein, thereby inducing new arrangements of the whole secondary structure (Fig. 23.4), which affect mechanical properties (Table 23.4) mainly in the glass transition region. In

TABLE 23.4 X-Ray *d*-Spacings and Glass Transition Temperatures for Neat TPZ and TPZ-Based Nanocomposites [77]

TPZ-Base Composites	D_o (+)	d_i (+)	T_g (°C)
TPZ	4.6	8.3	46.1
TPZ—1AL	3.9	11.3	42.2
TPZ—3AL	4.6	10.0	37.8
TPZ—10AL	4.5	8.6	45.6
TPZ—1LSS	4.6	9.2	40.1
TPZ—3LSS	4.6	9.3	39.5
TPZ—10LSS	4.6	8.2	44.0

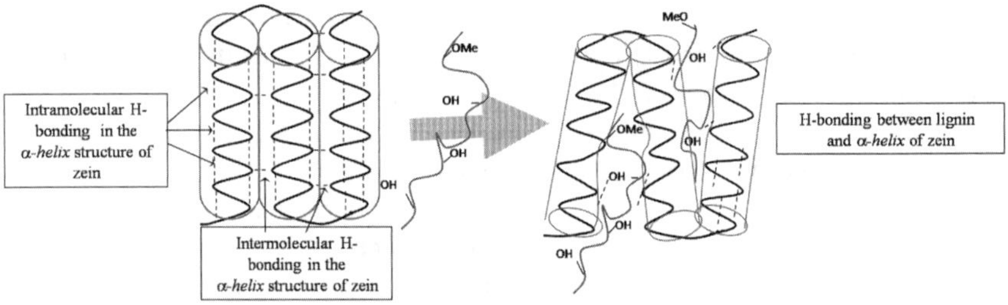

Figure 23.4 Proposed mechanism of interaction between lignin and α-helix structure of zein. (adapted from Reference [77]).

particular, at low lignin (AL and LSS) content, samples presented a remarkable decrease of the T_g toward lower temperatures because of the reduction of hydrogen bonding in the secondary structure of the zein through the insertion of the lignin (see Fig. 23.4).

The presence of LSS did not reduce the water sensitivity of TPZ because the sulfonic acid groups of LSS are hydrophilic. Conversely, the introduction of hydrophobic AL efficiently restricted the water uptake, especially at an AL content of 1 wt%. At this AL concentration between the hydroxyl groups of zein and OH and/or SH groups of AL, a significant decrease in the concentration of hydrophilic amino acids residues in zein resulted in reduced sorption sites for water molecules [78, 81]. The value of tensile strength increased with AL content, and a significant improvement was observed, in particular, at low AL content (1 wt%) from 1.79 MPa for neat TPZ to 4.53 MPa for the TPZ/1AL bionanocomposite. However, a further increase of AL content resulted in a slight decrease of the tensile strength. A significant reduction in the percentage elongation at break values was observed only when the content of AL was 1 wt%. This variation can be attributed to the optimal interaction between lignin and protein molecules, in particular, to the formation of the strong hydrogen bonding between the amino acids of the zein (-C=O, -OH, -NH) and the functional groups (-OH and, mainly, -SH) of AL. At 10 wt%, the AL can be simply considered as a reinforcing filler for TPZ. As the AL content increased, phase separation occurred in the bionanocomposite because protein–protein interactions were again favored with respect to lignin–protein interactions. An improvement of the mechanical properties of TPZ/LSS nanocomposites was observed only for LSS content from 5 to 10 wt%. Compared with LSS, AL exhibited a greater reinforcement effect on TPZ, mainly due to the presence of SH groups, capable of forming stronger hydrogen bonding rather than simple protein–protein interactions. The PEG in which the lignin has been dispersed reduced the degree of self-association between lignin molecules and favored a good mixing and interaction of AL with zein molecules [79].

Comparison of the mechanical properties of TPZ/AL nanocomposites and those of other zein/ MMT nanocomposites, prepared by solvent casting by Luecha et al. [80], film blowing of a resin obtained by the precipitation of zein from an aqueous ethanol solution, and melt mixing methods, clearly shows the high efficiency of AL with respect to MMT, even at a very low lignin content (1 wt%).

Huang et al. [30, 81] investigated the effect of AL and lignosulfonate on properties of soy protein plastics and observed only a 50% increase in tensile strength with the addition of 30 wt% lignin. They concluded that the observed behavior might have been due to inadequate mixing of lignin with protein and suggested improvement of the affinity of the two phases by using methylene diphenyl diisocyanate as a compatibilizer or by using a chemically modified lignin [82].

By modifying the hierarchical structure of zein with the use of a highly interactive additive such as lignin, it is possible to achieve a supramolecular design of the protein and obtain bionanonocomposites characterized by enhanced mechanical performance and specific functional properties (from hydrophilic to hydrophobic behavior) suitable for specific applications (from biomedical uses to food packaging).

23.6.5 Lignin/Starch Composites

The use of lignin as filler in starch films has several goals: the utilization of industrial lignins, the improvement of the starch matrix hydrophobicity, and to learn more aabout starch–lignin interactions. Baumberger et al. [31, 83, 84] found that the composition and molecular weight of various fractions of lignin influence the compatibility and properties of starch–lignin films. Films prepared by thermal molding or by cast from wheat starch and lignosulfonate in the presence of glycerol show the modified mechanical properties with respect to standard plasticized starch films. Thus, ultimate stress decreases and elongation at break increases independently of the lignosulfonate content, while the dynamic mechanical thermal behavior is as a function of lignin content [84].

Pine KRL heterogeneity was shown to play an important role in the behavior of starch–lignin composites. By sequential extraction with organic solvents, three fractions with different carbohydrate contents and molecular size distributions have been obtained. A minor portion of the low molecular weight fractions of KRL seems to be primarily involved in plasticization or compatibilization phenomena [83].

Vengal and Srikumar [85] have explored the possibility of preparation of novel lignin-based biomaterials. The polymeric films obtained by blending lignin with starch or gelatin at various proportions and with starch–gelatin mixture and a plasticizer (glycerol) presented mechanical properties comparable to normal films. The water absorption tests indicated that the starch–lignin and gelatin–lignin films are highly hydrophilic. The swelling degree of films increases with the increase in lignin content and pH.

Other starch-based blends that contain lignin or lignin derivatives were reported by Ghosh et al. [86]. They blended in various ratios lignin or its ester (lignin butyrate, LB) with a starch–caprolactone copolymer (SCL) by solvent casting or melt-processing procedure.

Presence of L or LB had a significant effect on crystallinity and melting temperatures of the PCL component, revealing polymer–polymer interactions between SCL and lignin. Unmodified lignin causes an increase in the crystallinity degree of PCL during the melt process, while LB has a more pronounced effect during solvent casting processing. These results suggest that high T_g of nonmodified lignin plays a role as a nucleating agent for PCL, enhancing crystallization and melting, and low T_g of lignin esters (LB) shows a significant compatibility with PCL, reducing crystallization and fusion. Addition of lignin or LB up to a concentration of 10–20% by weight improves the mechanical properties of SCL.

Starch (S) and lignin (LI) have been used as biodegradation inductors, increasing the biodegradation rate of PVA/protein hydrolysate in activated sludge while preserving or improving the technical and usage properties of blends [87]. To prepare the blends, a plasticizer (glycerol) has been used, allowing good processing. Glycerol produced a PVA breakdown greater with the increase in lag phase. Combining LI and S distinctly accelerated the degradation of a blend, although the final biodegradation percentage is lower. The PVA/G PH S blend presents a breakdown time lower than that of pure PVA, and the films produced have the convenient mechanical properties.

23.6.6 Lignin–Rubber Composites

Unmodified and modified lignins (15, 30, 45, and 60 phr) were used in rubber formulations with the specific aim of reducing rolling resistance. It was found that lignin shows significant reinforcing behavior with increased values of mechanical properties (modulus, elongation at break, and tensile strength). In addition, it caused a reduction in the cure time of rubber compounds that was attributed to the participation of lignin in the cross-linking reaction via the reactive functional groups and sulfur present in its structures [88]. The use of lignin derivatives influenced the physical and mechanical properties of lignin-filled styrene butadiene rubber (SBR) (i.e., lignosulfonate decreased the cure time and acted as secondary accelerator, while methylated lignin and thiolignin increased the crosslink density) [89]. The addition of lignin in carbon black filled natural rubber exerts a stabilizing effect comparable to that of commercial antioxidants used in the rubber industry [90]. Besides their use as filler, lignins in the form of low to higher molecular weight, meltable, or soluble products can be specially tailored for plastics and rubber compounding [91]. Natural rubber (NR) was mixed with curatives and other compounding ingredients, on an open two-roll mill of 170 mm diameter and 300 mm working distance. The incorporation of either soda lignin or thiolignin into NR mixes results in increasing the maximum torque of the composites obtained. The great increase in maximum torque in the case of thiolignin can be attributed to the sulfur present in thiolignin. Also, the incorporation of either soda lignin or thiolignin into NR mixes results in shorter cure times with higher rate indices. NR/soda lignin and NR–thiolignin composites possessed lower tensile strengths and slightly fewer elongations at break than the blank (NR) vulcanizate. The NR–thiolignin composite possessed the best thermal stability, and the optimum thiolignin loading was found to be 20 phr [92]. Incorporation of either soda lignin or thiolignin into NR improves the dielectric properties.

23.7 LIGNIN IN THERMOSETS

Besides polyurethanes and polyesters, phenolic and epoxy resins are the most prominent applications of technical lignins in thermosetting materials [93–95].

Vegetable oils from soybean, flax, and palm are some of the most versatile renewable substrates and can be converted into fuels, polymers, and specialty products [96, 97]. Epoxidation of the unsaturated bonds along the fatty acid chains of the seed oils produces intermediary derivatives used then for polymers, coatings, adhesives, and similar industrial products [98].

The use of the vegetable oil esters for new products was recently reported [99–101]. Holser [102] investigated the preparation of epoxidized methyl esters by transesterification of epoxidized soybean oils during mild reaction conditions with preservation of the epoxide function. Epoxidized methyl esters can be readily converted into surfactants, fuel additives, and other industrial products. Although the vegetable oil-based epoxy resins have a positive effect on the environmental, it is difficult them to applied in industrial fields where epoxy resins are generally used in combination with curing agents, because they have no aromatic rings in their backbone structures and have no performance mechanical properties. Wool [103, 104] investigated possibility of incorporation of unmodified/modified KRL (ethoxylated and maleated KRL) in different vegetable oil-based resins (epoxidized soybean oil, acrylated epoxidized soybean oil, hydroxylated soybean oil, soybean oil monoglyceride) and obtained polymer networks whose properties are influenced by the lignin content.

Addition of butyrated KRL to an unsaturated thermosetting resin, consisting of a mixture of acrylated epoxidized soybean oil and styrene, improved the adhesion of the resin to the reinforcing flax fibers [105]. The flexural strength increased by 40% for a 5 wt% butyrated lignin addition.

Many researchers have made an effort to use lignin in the production of epoxy resin, due to its aromatic structure, expecting the performance properties of the lignin-based epoxy resin to be equivalent to those of conventional epoxy resins [106–108].

Thus, Nonaka [109] prepared lignin–epoxy resin by mixing the alkaline solution of KRL and water-soluble epoxy compounds (polyethylene glycol diglycidyl ether) and curing agents. The cured resins had a compatibility to form a kind of interpenetrating polymer network (IPN). Simionescu et al. [110] obtained a liquid lignin–epoxy resin by treatment of calcium lignosulfonate with epichlorohydrin in alkaline aqueous medium.

By simply blending the KRL powder (up to 20% lignin) with the liquid epoxy prepolymer and hardener, an adhesive system has been obtained. For 5–20% lignin content, epoxy resin–lignin blends showed a single T_g, but two T_g's for blends with a higher content of lignin [111]. The significant improvement in the adhesion of the lignin–epoxy resin blends (L-ER) to an aluminum substrate could be explained by an interaction on the one hand between lignin–polyamine hardener and on the other hand between lignin–epoxy prepolymer [112, 113].

The studies on the epoxy resin–lignin blends derived from lignin types showed that those blends containing hardwood lignin (Tomlinite™, Tembec, Canada; Eucalin™, Borreggard-Lignotech Group; or steam explosion lignin) impart a better adhesion than those having softwood lignin (Indulin AT, Westvaco) [114]; this can be explained by the differences in chemical structural and molecular weight of lignins.

Functionalization of lignin by hydroxyalkylation or phenolation, followed by reaction with epichlorohydrin, allowed epoxy-functional lignins to be obtained, which may be cured with diamines or anhydrides [95, 115, 116].

Tomita et al. [117] described a new L-ER adhesive in which lignin was modified by ozonization. Hofman and Glasser [118, 119] prepared lignin–epoxy resins from hydroxyalkyl lignin derivatives with varying degrees of alkoxylation. It was shown that these lignin-based epoxy resins can be crosslinked with aromatic diamine (*m*-phenylene diamine [*m*-PDA]) to form strong thermosets whose properties depend on lignin content and polyether chain length and which have T_g's ranging from 0°C to 100°C. Ito and Shiraishi [120] treated KRL with bisphenol A and then with epichlorohydrine in the presence of a catalyst, which yields a waterproof adhesive with improved glueability. The increase in degree of lignin phenolation led to an increase in T_g and the storage modulus in the rubbery plateau region of the cured films, developing a better three-dimensional structure.

Simionescu et al. [21, 121–123] reported synthesis of lignin epoxy resins by coupling reactions with simple (phenol, β-naphthol, bisphenol) and macromolecular (novolak) phenols of lignosulfonate (ammonium, calcium, iron, and chromium lignosulfonate) followed by reaction with epichlorohydrine in alkaline medium. Incorporation of phenolated lignin–epoxy compounds in epoxy resin matrix led to uniform and homogeneous polymer network having a glass transition temperature of 52°C. Simionescu et al. [22] showed that significant lignin loads (<50% w/w) could be incorporated into the epoxy resin while maintaining good mechanical and electrical properties (volume and surface resistivity, dielectric constant, loss in dielectric tangent angle, and dielectric rigidity) and high impact toughness (Charpy impact strength) compared with the lignin-free phenol epoxy resins. Also, including some fillers (lead soap, alum earth, aluminum trioxide, talc, chalk, sand, silica, glass fibers), plasticizer (polyester C_6, dibutylphthalate), and pigments (iron oxide, titanium oxide) in lignin–epoxy blends in a proportion of 10–40 wt% permits epoxy-modified lignin–epoxy composite materials to be obtained without affecting cross-linking reactions and properties [22, 124]. A significant improvement in the adhesive strength and weatherability of epoxy was observed by the introduction of organosilanes (2 wt% epoxyalkylsilane) into lignin–epoxy blends (10% lignin) [60].

In 2001, Kosbar et al. [125] published the results of research carried out at IBM on incorporation of lignin–epoxy resin for fabrication of printed circuit boards. The method of fabrication of a cross-linked polymer composition formed from lignin (~50 wt%), an epoxy resin cross-linking agent (diglycidyl ether benzylamine, epoxy novolac or epoxy cresol novolac), and an initiator have been described by Afzali-Ardakani et al. [126]. The materials obtained had physical and electrical properties similar to those of common laminate resins.

A process for making lignin–epoxide resins from lignin separated from the waste cooking liquor from a paper pulping process comprising the steps of lignin isolation with a chemical reaction to produce unsaturated end groups on the lignin molecule that are epoxidized to produce the resin has been patented by Holsopple et al. [127]. Lenox Resources Limited developed technology that makes an epoxy resin from lignin through the selective addition of epoxide groups to the lignin molecule without the use of epichlorohydrin. The production of Lenox® (Lenox Resources, Port Huron, MI) lignin–epoxy resins for molded composites to industrial scale achieved ~350 kL/annum (prior to liquidation of the company in 2000).

Methylation and phenolation are the two most studied modification approaches aimed at introducing reactive functional groups to lignin molecules. Modified lignin from these two methods could partially replace phenol in PF resin synthesis. Other methods including reduction, oxidation, and hydrolysis improve the reactivity of lignin as well as producing phenolic compounds from lignin [128].

The phenolation of lignin may take place in the presence of both catalysts (HCl, H_2SO_4, or BF_3) [114] and alkaline hydroxide at elevated temperatures, obtaining the soluble and fusible phenol-modified lignin resin, which, by treatment with formaldehyde, is transformed into thermoreactive resins. However, chemical heterogeneity and polydispersity limited the application of lignin in phenol formaldehyde resin production. This can be countered in several ways: biochemical modification of the lignin [129, 130] for increasing its reactivity, the addition of filler agents (starch) to reduce the level of condensation [131], and the use of OSL, which can be effective phenol diluents in phenol-based resin systems [132–134].

KRL, acid hydrolysis, organosolv, steam-explosion lignins, and lignosulfonates as lignin components, and phenol, cresol, resorcinol, or bisphenol as phenol have been used for manufacturing lignin–phenolic resins and obtaining adhesives with satisfactory strength properties. Lignin–polyphenolic resins are predominantly used for production of weather-resistant wood products, such as plywood, particleboard, fiberboard, flake board, and strand board. Generally, these board products are manufactured using lignocellulosic materials (wood veneer, random geometric configuration wood chips, or wood fibers) as an adhesive.

The pyrolytic lignin was smoothly incorporated into phenol–formaldehyde resins at 30%, 40%, and 50% phenol replacement levels, and the resultant resins were evaluated as oriented stand board core-level binder. The lignin was effective for up to about 40% replacement of phenol in synthesizing wood adhesive-type resins [135].

Thielemans et al. [136] reported an exploratory work regarding the use of lignin as fillers and comonomers in thermosetting unsaturated polyesters and vinyl esters. The different lignins (pine kraft, hardwood, ethoxylated, and maleinated) were incorporated in different resin systems (acrylated epoxidized soybean oil, hydroxylated soybean oil, soy oil monoglyceride, and a commercial vinyl ester). An increase in the glass transition temperature and a decrease in modulus at 20°C was observed. This was explained by the plasticizing effect of lignin. Lignin modified by adding double bond functionality was introduced in resins by reaction with maleic anhydride and epoxidized soybean oil. This modification improved the solubility of lignin in styrene-containing resins, as well as the chemical incorporation of lignin in the resin. Moreover, lignin was used to treat the surfaces of natural hemp fibers to cover surface defects in the natural fibers and to increase the bonding strength between the resin and fiber.

A degradative potential lignin–polyurethane blend was synthesized from poly(ester-urethane) elastomers containing lactate segments and lignin [137]. After treatment with fungal peroxidase (E.C. 1.11.1.7.) and laccase (benzenediol:oxygen oxidoreductase, EC 1.10.3.2.) isolated from *Aspergillus* sp., it was observed that polyurethane degradation was strongly influenced by lignin addition and enzyme type, being accompanied by significant changes in the thermomechanical behavior.

23.8 LIGNIN AND OTHER POLYMER BLENDS

Various lignin-containing composites are summarized in Table 23.5.

Ghosh et al. [86] blended in melt and solution OSL and its ester (lignin butyrate (LB)) with polyhydroxy-alkanoate, such as poly-3-hydroxybutyrate (PHB)

TABLE 23.5 Various Kinds of Lignin-Containing Biocomposites and Bionanocomposites

System	Preparation	Properties	Reference
PE/lignin bio-based hybrid materials	Ethylene vinyl alcohol copolymer—compatibilizer	TGA, DSC, SEM, mechanical properties—compatible blends	[149]
PET/lignin	Single-screw extruder	Lignin particles acting as nucleating agent in the composite increased the crystallization rate. Lignin presence produced a noticeable enhancement of PET crystallinity and crystal dimensions.	[49]
Polysiloxane–lignin composites	Cross-linking	Polydimethylsiloxane-α,ω-diol (PDMS) as polymeric matrix, silica aerogel as reinforcing filler, and the lignin powder, as bulking filler	[150]
PHB/Lignin	Poly (3-hydroxybutyrate)/soda lignins (having distinct chemical group functionalities) were produced through the delignification of sugarcane fiber (i.e., bagasse) and subsequent sequential solvent fractionation.	Good compatibility for a content up to 30% lignin	[151]
Polypropylene/ lignin composites	Pre-graft (alkylation) of lignin	Alkylate lignin has better compatibility with PP (10 wt% is the optimum content of alkylated lignin). Lignin played a role of heterogeneous nucleation when the PP crystallization. After 50 days' biological degradation, the maximum weight loss rate of composites was up to 9.61% (70 wt% content of alkylate lignin)	[71, 152]
HDPE/lignin from bioethanol production (straw and stalks)	By extruded pellet fabrication and thermoform process.	With increase of the lignin content from 55% to 75%, the static flexural strength of the resulting blends was hardly influenced, but the tensile strength decreased 42%.	[153]

(Continued)

TABLE 23.5 (*Continued*)

System	Preparation	Properties	Reference
Cellulose (85–34 wt%/Starch (10–62 wt%)/lignin (1–5 wt%)	From an ionic liquid (IL), 1-allyl-3-methylimidazolium chloride (AmimCl) by coagulating in a nonsolvent condition	Excellent mechanical properties in dry and wet states owing to the mutual property supplement of different components. Good thermal stability and high gas barrier capacity and a CO2:O2 permeability ratio close to 1.	[139]
Chitosan/lignin	Solution casting technique	Lignin well dispersed in chitosan when the content of lignin is below 20 wt% due to the strong interfacial interaction by hydrogen bonding. Storage modulus, thermal degradation temperature, and glass transition temperature of chitosan have been largely improved by adding lignin. Fully degradable composites	[154]
PVC/lignin	Lignin was treated with a polyacrylate, poly(ethyl acrylate-co-acrylic acid)	Polyacrylate improved the adhesion between lignin particles and PVC matrix. The tensile and impact strengths of the composites are improved.	[155]
PVC/wood-flour/lignin	Lignin was aminated by Mannich reaction	The lignin amine treatment improved the mechanical performance.	[156]
	The composites were prepared by melt compounding	Interfacial interactions of wood with PVC matrix are strengthened.	
Different types of lignins/poly(vinyl chloride) composites	Melt compounding	Lignin influences the processability of the blends and mechanical properties and weathering.	[157]
		Interactions between lignin and PVC depend of types of lignin.	
PVA/kraft lignin or modified kraft lignin	Solution casting from dimethyl sulfoxide (DMSO)	15–25% lignin content; PVA/lignin films present high resistance to photochemical attack and thermal stability in comparison with PVA films	[158, 159]
PVA/lignin	Thermal extrusion	PVA/lignin blend is immiscible in the bulk. The results indicate the formation of specific intermolecular interaction between components	[29]
Lignin/wool	50% shoddy wool; 25% viscose; 25% lignin	Oil sorbents; optimal composition 90 wt% wool/10 wt% lignin	[160]

and poly-3-hydroxybutyrate-co-3-hydroxyvalerate (PHBV). The presence of lignin, in proportion of 10–20%, inhibits or retards crystallization PHB at any cooling rate. Glass transition temperature T_g values of the PHB phase shifted toward the glass transition of L or LB, indicating a polymer–polymer interaction between blend components. Incorporation of LB in the PHBV led to some interaction and an important reduction of crystallinity in the samples. Generally, the presence of lignin improved tensile strength and elongation at break.

Blends containing poly(hydroxybutyrate) (PHB) and L-ER have been prepared on a Brabender plastograph [138]. The L-ER content in the binary blends was 10 wt%, 15 wt%, 20 wt%, and 35 wt%. The epoxy groups of L-ER interact either with the end groups of the PHB or with diaminodiphenylmethane (a cross-linking agent), achieving a satisfactory homogeneity of the blends. Some blends with good properties are totally degradable after 45 days.

New biocomposites containing starch, cellulose, and lignin were prepared from ionic liquid (IL) by coagulating in nonsolvent conditions [139]. Results showed that the content of cellulose, lignin, and starch had a significant influence on the mechanical properties of composite films. The results obtained showed that the composite films presented good thermal stability, high gas barrier capacity, and a CO_2:O_2 permeability ratio close to 1.

Another family of thermoplastic polysaccharides such as starch/cellulose-2,5-acetate were produced by reactive processing technology that grafted cyclic lactones simultaneously onto polysaccharide, hydroxyfunctional plasticizer, and hydroxyfunctional fillers [140]. Organosolv lignin, cellulose, starch, and chitin were added to effect reinforcement of the polymer matrix. Compatibility between oligolactone-modified cellulose acetate and fillers was improved when fillers were added during the reactive extrusion process.

A series of blends composed of cellulose or hydroxypropyl cellulose (HPC) and an unmodified OSL or ethylated lignin (E_tL) have been prepared by solvent casting or by injection molding procedure [141, 142]. Ghosh et al. [143] reported a certain degree of compatibility between modified cellulose and modified lignin due to the presence of secondary interactions between the two components. When mixing of hydroxypropyl cellulose and OSL was carried out in the melt followed by extrusion, partially miscible blends displaying a mesomorphic type behavior were obtained [142]. The lignin component reinforced the amorphous cellulose derivative matrix, forming an oriented crystal mesophase structure, which behaves as a nanocomposite structure.

In the case of a blend formed from ethyl cellulose (EC) and lignin, addition of L (up to 40%) led to the appearance of two separate phases with creation of a liquid crystal mesophase [143].

To increase the phase compatibility between lignin and cellulose (derivatives), block copolymers were synthesized containing covalent bonds between lignin and cellulose ester segments [144, 145]. Lignin–thermoplastic cellulose derivative copolymer exhibited phase distinctions with the individual block having molecular weights as low as 10^3 Da [146]. Differential scanning calorimetry (DSC) data revealed that lignin–cellulose propionate (CP) block copolymer was unable to improve lignin–cellulose (derivatives) blend compatibility. Results show that lignin favors the formation of an amorphous or liquid crystalline mesophase structure through establishing a strong interactive association between lignin and the polysaccharide component.

The addition of lignin reduced the dynamic elastic modulus of cellulose ester solutions, but the modulus of the cellulose solutions (in DMAc/LiCl) is increased at all shear frequencies. This is explained by the formation of secondary interactions of lignin with cellulose in the DMAc/LiCl solvent system [146]. However, when cellulose ester–cellulose acetate butyrate (CAB)/L mixtures were spun into continous fibers from DMAc solution, both fiber tensile strength and modulus increased significantly. This increase was explained by the positive effect of lignin (even at small amounts of lignin, i.e., 4%) on the molecular order of the cellulose derivatives in solution and in solid state.

A lignin–polymer complex of lignosulfonate from softwood delignification and sawdust was tested as soil conditioner with good results [147]. Such a composite also exhibits the ability to repel moisture and water [148].

23.9 CONCLUSIONS AND OUTLOOK

The past 30-year history of lignin-containing polymeric materials has been characterized by changing perceptions of macromolecular lignin structure. The first formulations originated in the view of lignins as a three-dimensionally branched network of polymers. Whether incorporated covalently or noncovalently into polymeric materials, lignin preparations were, with few exceptions, limited to maximum contents of 30–45% (w/w). However, once the significance of noncovalent interactions between the constituent molecular components had been recognized, it was possible to generate promising polymeric materials composed exclusively of ethylated and methylated KRL, and other equally simple lignin derivatives. Plasticization of these materials can be achieved in a predictable way by blending with miscible low-T_g polymers. The field of lignin bionanocomposites is still developing. The future of novel bionanocomposites with improved properties and multifunctionality can be envisaged as an emerging, open field of research, with plenty of possibilities because of the great abundance and diversity of biopolymers in nature, as well as the advantage of their synergistic combination with inorganic nanosized solids. On the other hand, biocompatibility is a critical property for the application of these biohybrids in food packaging or tissue engineering in regenerative medicine.

REFERENCES

[1] Darder, M., Aranda, P., and Ruiz-Hitzky, E. (2007) Bionanocomposites: A new concept of ecological, bioinspired, and functional hybrid materials. *Adv. Mater.*, *19*, 1309–1319.

[2] Shulga, G., Skudra, S., Shakels, V., Belkova, L., Cazacu, G., Vasile, C., and Nita, L. (2011) Lignin separated from the hydrolyzate of hydrothermal treatment of birch wood and its surface properties. *Cell. Chem. Technol.*, *46*(5–6), 307–318.

[3] Chen, P., Zhang, L.N., Peng, S.P., and Liao, B. (2006) Effects of nanoscale hydroxypropyl lignin on properties of soy protein plastics. *J. Appl. Polym. Sci.*, *101*, 334–341.

[4] Doherty, W.O.S., Mousavioun, P.C., and Fellows, M. (2010) Value adding to cellulosic ethanol: Lignin polymers. *Ind. Crops Prod.*, *33*, 259–276.

[5] Achyuthan, K.E., Achyuthan, A.M., Adams, P.D., Dirk, S.M., Harper, J.C., Simmons, B.A., and Singh, A.K. (2010) Supramolecular self-assembled chaos: Polyphenolic lignin's barrier to cost-effective lignocellulosic biofuels. *Molecules*, *15*, 8641–8688.

[6] Sarkanen, S., Teller, D.C., Stevens, C.R., and McCarthy, J.L. (1984) Associative interactions between kraft lignin components. *Macromolecules*, *17*, 2588–2597.

[7] Sarkanen, S., Teller, D.C., Hall, J., and McCarthy, J.L. (1981) Associative effects among organosolv lignin components. *Macromolecules*, *14*, 426–434.

[8] Li, Y., Mlynar, J., and Sarkanen, S. (1997) The first 85% kraft lignin-based thermoplastics. *J. Polym. Sci. B*, *35*, 1899–1910.

[9] Pye, E.K. (2006) Industrial lignin production and applications. In: *Biorefineries-Industrial Processes and Product*, Vol. 2, eds. B. Kamm, P.R. Gruber, and M. Kamm. Weinhcim, Germany: Wiley-VCH Verlag GmbH & Co. KGaA, pp. 165–200.

[10] Vasile, C., Popescu, M.C., Stoleriu, A., and Gosselink, R. (2006) Thermal characterization of lignins. In: *New Trends in Natural and Synthetic Polymers*, eds. C. Vasile, G.E. Zaikov. New York: Nova Science, pp. 135–165.

[11] Brebu, M., Vasile, C. (2010) Thermal degradation of lignin—A review. *Cell. Chem. Technol.*, *44*(9), 353–363.

[12] Popescu, C.M., Vasile, C., Popescu, M.C., Singurel, G.H., Popa, V.I., and Munteanu, B.S. (2006) Analytical methods for lignin characterization. II. Spectroscopical studies. *Cell. Chem. Technol.*, *40*(8), 597–622.

[13] Vasile, C., Gosselink, R., Quintus, P., Koukios, E.G., Koullas, D.P., Avgerinos, E., and Abacherli, A. (2006) Analytical methods for lignin characterization. I. Thermogravimetry. *Cell. Chem. Technol.*, *40*(6), 421–429.

[14] Koullas, D.P., Koukios, E.G., Avgerinos, E., Abaecherli, A., Gosselink, R., Vasile, C., Lehnen, R., Saake, B., and Suren, J. (2006) Analytical methods for lignin characterization, Differential scanning calorimetry. *Cell. Chem. Technol.*, *40*(9–10), 719–725.

[15] Schilling, P. (1993) US Patent no. 5,192,361, March 9.

[16] Rojas, O. (2008) Nanotechnology in lignocellulosic applications. Americas Innovation Forum, March 30–April 1, 2008, Punta del Este, FIA/AIF 2008, Uruguay Available at: http://www.anii.org.uy/FIA_AIF2008/ponencias/ponencias/ROJASFIA2008_Uruguay.pdf.

[17] Popa, V.I., Căpraru, A.M., Grama, S., and Măluţan, T. (2011) Nanoparticles based on modified lignins with biocide properties. *Cell. Chem. Technol.*, *45*(3–4), 221–226.

[18] Cazacu, G., Vasile, C., Popescu, M.C., and Darie, R.N. (2010) Lignin characterization for its use in complex polymeric systems. FP7 NAPOLYNET "Setting up Research-intensive Clusters Across the EU on Characterization of Polymer

Nanostructures" and COST ACTION FA0904 "Eco-sustainable Food Packaging based on Polymer Nanomaterials, Polymer Nanomaterials for Food Packaging, Characterization Needs, Safety and Environmental Issues," September 1–2, 2010, London, United Kingdom.

[19] Misra, M., Vivekanandhan, S., Mohanty, A., and Sahoo, S. (2010). Lignin. Recent development and new opportunities. *CIPET-TIMES*, A CIPET Inhouse Publication, January–April 2010, 8–14; http://www.cipet.gov.in/pdfs/CIPET_TIMES_jan-april-2010.pdf (accessed 2011).

[20] Darie, R., Vasile, C., and Kozlowski, M. (2011) *Compatibilization of Complex Polymeric Systems, Reactive Compatibilization-cum-Impact Modification Applied to Virgin Polymer and Waste*. Saarbrucken, Germany: LAP Lambert Academic Publishing. ISBN 978-3-8443-1225-6.

[21] Simionescu, C.I., Turt, C.I., Bucova, S.A., Macoveanu, M.M., Cazacu, G., and Vasile, C. (1994) Sinteza, caracterizarea fizico-chimic(si spectrala a unor rasini lignin epoxidice. *Buletinul Acad. St. a Republicii Moldova. Stiinte Biologice si Chimice*, *3*(270), 62–73.

[22] Simionescu, C.I., Macoveanu, M.M., Cazacu, G., Vasile, C., and Stoleriu, A. (1993) Lignin-epoxy composites. *Compos. Sci. Technol.*, *48*, 317–323.

[23] Nägele, H., Pfitzer, J., Näegele, E., Inone, E.R., Eisenreich, N., Eckl, W., and Eyerer, P. (2002) Arboform: a thermoplastic, processable material from lignin and natural fibers. In: *Chemical Modification, Properties, and Usage of Lignin*, ed. T.Q. Hu. New York: Kluwer Academic/Plenum, pp. 101–120.

[24] Leszczynska, A., Njuguna, J., Pielichowski, K., and Banerjee, J.R. (2007) Polymer/montmorillonite nanocomposites with improved thermal properties: Part I. Factors influencing thermal stability and mechanisms of thermal stability improvement. *Thermochim. Acta*, *453*(2), 75–96.

[25] Sevastyanova, O., Qin, W., and Kadla, J.F. (2010) Effect of nanofillers as reinforcement agents for lignin composite fibers. *J. Appl. Polym. Sci.*, *117*(5), 2877–2881.

[26] Guigo, N., Vincent, L., Mija, A., Nägele, H., and Sbirrazzuoli, N. (2009) Innovative green nanocomposites based on silicate clays/lignin/natural fibres. *Compos. Sci. Technol.*, *69*(11–12), 1979–1984.

[27] Okada, A., Usuki, A. (2006) Twenty years of polymer–clay nanocomposites. *Macromol. Mater. Eng.*, *291*(12), 1449–1476.

[28] Kadla, J.F., Kubo, S. (2004) Lignin-based polymer blends: Analysis of intermolecular interactions in lignin-synthetic polymer blends. *Compos. Part A Appl. Sci. Manuf.*, *35*, 395–400.

[29] Kubo, S., Kadla, J.F. (2003) The formation of strong intermolecular interactions in immiscible blends of poly(vinyl alcohol) (PVA) and lignin. *Biomacromolecules*, *4*, 561–567.

[30] Huang, J., Zhang, L., and Chen, P. (2003) Effects of lignin as a filler on properties of soy protein plastics. II. Alkaline lignin. *J. Appl. Polym. Sci.*, *88*, 3291–3297.

[31] Baumberger, S., Lapierre, C., Monties, B., and Valle, G.D. (1998) Use of kraft lignin as filler for starch films. *Polym. Degrad. Stab.*, *59*, 273–277.

[32] Tudorachi, N., Cascaval, C.N., and Rusu, M. (2000) Biodegradable polymer blends based on polyethylene and natural polymers. Degradation in soil. *J. Polym. Eng.*, *20*(4), 287–304.

[33] Vasile, C., Iwanczuk, A., Frackowiak, S., Cazacu, G., Constantinescu, G., and Kozlowski, M. (2006) Modified lignin/polyethylene blends. *Cell. Chem. Technol.*, *40*(5), 345–351.

[34] Cazacu, G., Mihaies, M., Pascu, M.C., Profire, L., Kowarski, A.I., and Vasile, C. (2004) Polyolefins /lignosulfonate blends. IX. Functionalized polyolefins/lignin blends. *Macromol. Mater. Eng. Sci.*, *289*, 880–889.

[35] Cazacu, G., Pascu, M.C., Profire, L., Kowarski, A.I., and Vasile, C. (2004) Lignin role in a complex polyolefin blend. *Ind. Crops Prod.*, *20*(204), 205–221.

[36] Vasile, C. (2002) Role of polymer degradation processes in environmental pollution and protection. *Polymery (Poland)*, *XLVII*(7/8), 517–522.

[37] Cazacu, G., Pascu, M.C., Profire, L., and Vasile, C. (2002) Environmental friendly polymer materials I. Polyolefins-lignin based materials. *Environ. Protect. Ecol.*, *3*(1), 242–248.

[38] Macoveanu, M.M., Constantin, L., Manoliu, A.L., Pascu, M.C., Profire, L., Cazacu, G., and Vasile, C. (2001) Polyolefins/lignosulfonates blends. IV. Bio-and environmental degradation testing of polyolefins/lignosulfonates blends. *Cell. Chem. Technol.*, *35*(1–2), 197–203.

[39] Macoveanu, M.M., Georgescu-Buruntea, N., Pascu, M.C., Casariu, M., Ioanid, A., Vidrascu, P., and Vasile, C. (2001) Polyolefins/lignosulfonates blends. V. Bio and environmental degradation. *Cell. Chem. Technol.*, *35*, 3–4, 263–273.

[40] Pascu, M.C., Macoveanu, M.M., Ioanid, A., and Vasile, C. (2001) Polyolefins/lignosulfonates blends VI. Bio and environmental behaviour of photo-oxidized blends. *Cell. Chem. Technol.*, *35*(5–6), 513–522.

[41] Vasile, C., Downey, M., Wong, B., Macoveanu, M.M., Pascu, M.C., Choi, J.H., Sung, C., and Baker, W. (1998) Polyolefins/lignosulfonates blends. II. Isotactic polypropylene/epoxi-modified lignin. *Cell. Chem. Technol.*, *32*(1–2), 61–88.

[42] Bozell, J.J., Holladay, J.E., Johnson, D., and White, J.F. (2007). Top value added chemicals from biomass. Vol. II: Results of screening for potential candidates from biorefinery. Available at http://wwwl.pnl.gov/main/publications/external/technical_reports/PNNL-16983.pdf (accessed 2011).

[43] Kharade, A.Y., Kale, D.D. (1999) Lignin-filled polyolefins. *J. Appl. Polym. Sci.*, *72*(10), 1321–1326.

[44] Pouteau, C., Baumberger, S., Cathala, B., and Dole, P. (2004) Lignin-polymer blends: Evaluation of compatibility by image analysis. *C. R. Biol.*, *327*(9–10), 935–943.

[45] Pouteau, C., Dole, P., Cathala, B., Averous, L., and Boquillon, N. (2003) Antioxidant properties of lignin in polypropylene. *Polym. Degrad. Stabil.*, *81*(1), 9–18.

[46] de Oliveira, W., Glasser, W.G. (1994) Multiphase materials with lignin. XIV. Starlike copolymers with styrene. *J. Wood Chem. Technol.*, *14*(1), 119–126.

[47] El-Aghoury, A., Vasudeva, R., Banu, D., Elektorowicz, M., and Feldman, D. (2006) Contribution to the study of fungal attack on some plasticized vinyl formulations. *J. Polym. Environ.*, *14*(2), 135–147.

[48] Feldman, D., Banu, D., Lora, J., and El-Raghi, S. (1996) Rigid poly(vinyl chloride)–organosolv lignin blends for applications in building. *J. Appl. Polym. Sci.*, *61*(12), 2119–2128.

[49] Canetti, M., Bertini, F. (2007) Supermolecular structure and thermal properties of poly(ethylene terephthalate)/lignin composites. *Compos. Sci. Technol.*, *67*(15–16), 3151–3157.

[50] Graupner, N. (2008) Application of lignin as natural adhesion promoter in cotton fibre-reinforced poly(lactic acid) (PLA) composites. *J. Mater. Sci.*, *43*(15), 5222–5229.

[51] Li, J.C., He, Y., and Yoshio, I. (2003) Thermal and mechanical properties of biodegradable blends of poly(L-lactic acid) and lignin. *Polym. Int.*, *52*(6), 949–955.

[52] Hansjorg, N., Hinnerk, S., Rudiger, L., and Rolf, M. (2001) Reactive extrusion of polycaprolactone compounds containing wood flour and lignin. *J. Appl. Polym. Sci.*, *81*(8), 1972–1984.

[53] Kelley, S.S., Glasser, W.G., and Ward, T.C. (1989) Multiphase materials with lignin: 9. Effect of lignin content on interpenetrating polymer network properties. *Polymer*, *30*(12), 2265–2268.

[54] Chen, Y.R., Sarkanen, S. (2006) From the macromolecular behavior of lignin components to the mechanical properties of lignin-based plastics. *Cell. Chem. Technol.*, *40*(3–4), 149–163.

[55] Vasile, C., Macoveanu, M.M., Cazacu, G., and Popa, V.I. (1998) 5th Mediterranean School on Science and Technology of Advanced Polymer-Based Materials, eds. E. Martuscelli, P. Musto, and G. Ragosta. October 4–9, 1997, Capri, Italy, UNESCO OFFICE, 185–209.

[56] Košiková, B., Damianova, V., and Kacurocova, M.J. (1993) Sulfur-free as composites of polypropylene films. *J. Appl. Polym. Sci.*, *47*, 1065–1073.

[57] Košiková, B., Miklesova, K., and Damianova, V. (1993) Characteristics of free radicals in composite lignin/polypropylene films studied by the EPR method. *Eur. Polym. J.*, *29*, 1495–1497.

[58] Gonzales-Sánchez, C., Expósito Alvarez, L.A. (1999) Micromechanics of lignin/polypropylene composites suitable for industrial applications. *Angew. Makromol. Chem.*, *272*, 65–70.

[59] Tudorachi, N., Rusu, M., Cascaval, N.C., Constantin, L., and Rugina, V. (2000) Biodegradable polymeric materials. I. Polyethylene–natural polymer blends. *Cell. Chem. Technol.*, *34*, 101–111.

[60] Levon, K., Huhtala, J., Malm B., and Lindberg, J.J. (1987). Improvement of thermal stabilization of polyethylene with lignosulfonate. *Polymer*, *28*(5), 745–750.

[61] Toriz, G., Denes, F., and Young, R.A. (2002). Lignin-polypropylene composites: Part 1: Composites from unmodified lignin and polypropylene. *Polym Composites*, *23*(5), 806–813.

[62] Canetti, M., Bertini, F., De Chirico, A., and Audiso, G. (2006) Thermal degradation behaviour of isotactic polypropylene blended with lignin. *Polym. Degrad. Stab.*, *91*, 494–498.

[63] Gregorová, A., Cibulková, Z., Košiková, B., and Šimon, P. (2005) Stabilization effect of lignin in polypropylene and recycled polypropylene. *Polym. Degrad. Stab.*, *89*, 553–558.

[64] Casenave, S., Aït-Kadfi, A., and Riedl, B. (1996) Mechanical behaviour of highly filled lignin/polyethylene composites made by catalytic grafting. *Can. J. Chem. Eng.*, *74*(2), 308–315.

[65] Darie, R., Cazacu, G., Vasile, C., and Kozlowski, M. (2007) Blends with polypropylene matrix and lignin additive, Italic 4—Science & Technology of Biomass: Advances and Challenges, May 8–10, 2007, Rome, Italy.

[66] Simionescu, C.I., Macoveanu, M.M., Vasile, C., Ciobanu, F.L., and Cazacu, G. (1995) Lignin containing blends and composites. In: *High Tchnology Composites in Modern Applications*, eds. S.A. Paipetas, A.G. Youtsos. Greece: University of Patras, pp. 222–232.

[67] Zhong, M., Dai, H., Yao, H., Dai, D., Zhou, Y., Yang, J., and Chen, F. (2011) Strong, flexible high-lignin polypropylene blends. Society of Plastics Engineers (SPE). *Plast. Res.*, 24.06.2011, DOi: 10.2417/spepro.003642.

[68] Pascu, M.C., Macoveanu, M.M., Vasile, C., Ioanid, A., and Oghina, R.C. (2000) Polyolefins/lignosulfonates blends. III. Photo-oxidation of isotactic polypropylene/epoxidized lignosulfonate blends. *Cell. Chem. Technol.*, *34*, 399–420.

[69] Pascu, M., Vasile, C., Popa, G., Mihaila, I., and Pohoata, V. (2002) Modification of polymer blends properties by plasma/electron beam treatment.I. Plasma diagnosis and bulk properties of plasma treated blends. *Intern. J. Polym. Mater.*, *51*(½), 181–192.

[70] Košiková, B., Demianova, V., and Kacurakova, M. (1993) Lignin biopolymers in the composite polypropylene films. In: *Cellulosics Chemical, Biochemical Material Aspects*, eds. J. Kennedy, G.O. Phillips, and D.A. Williams. New York: Ellis Horwood, pp. 537–542.

[71] Košiková, B., Lábaj, J. (2009) Lignin-stimulated protection of polypropylene films and DNA in cells of mice against oxidation damage. *BioResources*, *4*(2), 805–815.

[72] Kubo, S., Kadla, J.F. (2004) Poly(ethylene oxide)/organosolv lignin blends: Relationship between thermal properties, chemical structure and blend behavior. *Macromolecules*, *37*, 6904–6911.

[73] Kadla, J.F., Kubo, S. (2003) Miscibility and hydrogen bonding in blends of poly(ethylene oxide) and kraft lignin. *Macromolecules*, *36*, 7803–7811.

[74] Nitz, H., Semke, H., Landers, R., and Mülhaupt, R. (2001) Reactive extrusion of polycaprolactone compounds containing wood flour and lignin. *J. Appl. Polym. Sci.*, *81*(8), 1972–1984.

[75] Guigo, N., Mija, A., Vincent, L., and Sbirrazzuoli, N. (2010) Eco-friendly composite resins based on renewable biomass resources: Polyfurfuryl alcohol/lignin thermosets. *Eur. Polym. J.*, *46*(5), 1016–1023.

[76] Athamneh, A.I., Griffin, M., Whaley, M., and Barone, J.R. (2008) Conformational changes and molecular mobility in plasticized proteins. *Biomacromolecules*, *9*, 3181–3187.

[77] Oliviero, M., Verdolotti, L., Di Maio, E., Aurilia, M., and Iannace, S. (2011) Effect of supramolecular structures on thermoplastic zein-lignin bionanocomposites. *J. Agric. Food Chem.*, *59*(18), 10062–10070.

[78] Huang, J., Zhang, L., and Chen, P. (2003) Effects of lignin as a filler on properties of soy protein plastics. II. Alkaline lignin. *J. Appl. Polym. Sci.*, *88*, 3291–3297.

[79] Oliviero, M., Verdolotti, L., Nedi, I., Docimo, F., Di Maio, E., and Iannance, S. (2012) Effect of two kinds of lignins, alkaline lignin and sodium lignossulfonate, on the foamibility of thermoplastic zein-based biocomposites. *J. Cellular Plastics*, *48*(6), 516–525.

[80] Luecha, J., Sozer, N., and Kokini, J.L. (2010) Synthesis and properties of corn zein/montmorillonite nanocomposite films. *J. Mater. Sci.*, *45*, 3529–3537.

[81] Huang, J., Zhang, L., and Chen, F. (2003) Effects of lignin as a filler on properties of soy protein plastics. I. Lignosulfonate. *J. Appl. Polym. Sci.*, *88*, 3284–3290.

[82] Wei, M., Fan, L., Huang, J., and Chen, Y. (2006) Role of star-like hydroxylpropyl lignin in soy-protein plastics. *Macromol. Mater. Eng.*, *291*, 524–530.

[83] Baumberger, S., Lapierre, C., and Monties, B. (1998) Utilization of pine kraft lignin in starch composites: Impact of structural heterogeneity. *J. Agric. Food Chem.*, *46*, 2234–2240.

[84] Ming, S., Baumberger, S., Lapierre, C., Monties, B., Lourdin, D., and Colonna, P. (1997) Preparation and properties of thermally moulded and cast lignosulfonates-starch blends. *Ind. Crops Prod.*, *6*, 253–258.

[85] Vengal, J.C., Srikumar, M. (2005) Processing and study of novel lignin-starch and lignin-gelatin biodegradable polymeric films. *Trends Biomater. Artif. Organs*, *18*(2), 237–241.

[86] Ghosh, I., Jain, R.K., and Glasser, W.G. (2000) Blends of biodegradable thermoplastics with lignin esters. In: *Lignin: Historical, Biological, and Materials Perspectives*, ACS Symp. Ser. 742, eds. W.G. Glasser, R.A. Northey, T.P. Schultz, Washington, DC: American Chemical Society, pp. 331–350.

[87] Julinova, M., Kupec, J., Alexy, P., Hoffmann, J., Sedlarık, V., Vojtek, T., Chromcakova, J., and Bugaj, P. (2010) Lignin and starch as potential inductors for biodegradation of films based on poly(vinyl alcohol) and protein hydrolysate. *Polym. Degrad. Stab.*, *95*, 225–233.

[88] Bahl, K. (2011) *Lignin based rubber composites towards production of energy efficient compounds for tire industry*. American Chem. Soc. Rubber Division, October 12, 2011. http://rubber.confex.com/rubber/180/webprogram/Paper11314.html.

[89] Botros, S.H., Eid, M.A.M., and Nageeb, Z.A. (2005) Thermal stability and dielectric relaxation of NR/soda lignin and NR/thiolignin composites. *Egypt. J. Solids*, *28*(1), 67–83.

[90] Gregorova, A., Kosikova, B., and Moravcik, R. (2006). Stabilization effect of lignin in natural rubber. *Polym. Degrad. Stab.*, *91*(2), 229–233.

[91] Nitz, H., Semke, H., Mülhaupt, R., and Abächerli, A. (2001) Lignin-tailor made use as non-toxic biostabiliser. *Kunstst-Plast. Eur.*, *11*, 40–42.

[92] Botros, S.H., Eid, M.A.M., and Nageeb, Z.A. (2006) Thermal stability and dielectric relaxation of natural rubber/soda lignin and natural rubber/thiolignin composites. *J. Appl. Polym. Sci.*, *99*(5), 2504–2511.

[93] El Mansouri, N.E., Yuan, Q., and Huang, F. (2011) Characterization of alkaline lignins for use in penol-formaldehyde and epoxy resins. *BioResources*, *6*(3), 2647–2662.

[94] Simionescu, C. I, Rusan, V., Macoveanu, M.M., Cazacu, G., Lipsa, R., Vasile, C., and Stoleriu, A. (1993) Lignin-epoxy composites. In: *Microphenomena in Advanced Composites*, eds. H.S. Wagner, G. Marom, and B. Haris. Elsevier Appl. Sci., pp. 317–323.

[95] Simionescu, C.I., Vasile, C., Macoveanu, M.M., Cazacu, G., and Ioanid, A. (1994) Behaviour by burial in soil of epoxy resins, lignin-epoxy resins and their mixtures. *Cell. Chem. Technol.*, *28*, 517–539.

[96] Gervasio, G.C. (1996) Fatty acids and derivatives from coconut oil. In *Bailey's Industrial Oil & Fat Products*, Vol. 3, ed. Y.H. Hui. New York: John Wiley & Sons, pp. 33–83.

[97] Khot, S.N., Lascala, J.J., Can, E., Morye, S.S., Williams, G.I., Palmese, G.R., Kusefoglu, S.H., and Wool, R.P. (2001) Development and application of triglyceride-based polymers and composites. *J. Appl. Polym. Sci.*, *82*, 703–723.

[98] Harry-O'kuru, R.E., Holser, R.A., Abbott, T.P., and Weisleder, D. (2002) Synthesis and characteristics of polyhydroxy triglycerides from milkweed oil. *Ind. Crops Prod.*, *15*, 51–58.

[99] Doll, K.M., Erhan, S.Z. (2005) Synthesis of carbonated fatty methyl ester using supercritical carbon dioxide. *J. Agric. Food Chem.*, *53*, 9608–9614.

[100] Moser, B.R., Erhan, S.Z. (2006) Synthesis and evaluation of a series of hydroxy ethers derived from isopropyl oleate. *J. Am. Oil Chem. Soc.*, *83*, 959–963.

[101] Kurth, T.L., Sharma, B.K., Doll, K.M., and Erhan, S.Z. (2007) Adsorption behavior of epoxidized fatty esters via boundary lubrication coefficient of friction measurements. *Chem. Eng. Comm.*, *194*, 1065–1077.

[102] Holser, R.A. (2008) Transesterificarea of epoxidized soybean oil to prepared epoxy methyl esters. *Ind. Crops Prod.*, *27*, 130–132.

[103] Wool, R.P. (2005) Lignin polymers and composites. In *Bio-based Polymers and Composites*, eds. R.P. Wool, X.S. Sun. London: Elsevier Academic Press, pp. 551–558.

[104] Thielemans, W., Wool, R.P. (2005) Kraft lignin as fiber treatment for natural fiber-reinforced composites. *Polym. Compos.*, *26*(5), 695–705.

[105] Thielemans, W., Wool, R.P. (2004) Butyrated kraft lignin as compatibilizing agent for natural fiber reinforced thermoset composites. *Compos. Part A*, *35*, 327–338.

[106] Lee, H.J., Tomita, B., and Hosoya, S. (1991) Development of ozonized kraft lignin/epoxy resin adhesives. *Wood Ind.*, *46*, 412–417.

[107] Hirose, S., Hatakeyama, T., and Hatakeyama, H. (2005) Glass transition and thermal decomposition of epoxy resins from the carboxylic acid system consisting of ester-carboxylic acid derivatives of alcoholysis lignin and ethylene glycol with various decarboxylic acids. *Thermochim. Acta.*, *431*, 76–80.

[108] Ismail, T.N.M.T., Hassan, H.A., Hirose, S., Taguchi, Y., Hatakeyama, T., and Hatakeyama, H. (2010) Synthesis and thermal properties of ester-type crosslinked epoxy resins derived from lignosulfonate and glycerol. *Polym. Int.*, *59*, 181–186.

[109] Nonaka, Y., Tomita, B., and Hatano, Y. (1997) Synthesis of lignin/epoxy resins in aqueous systems and their properties. *Holzforschung*, *51*, 183–187.

[110] Simionescu, C.I., Cazacu, G., and Macoveanu, M.M. (1987) Lignin-epoxy resins. II. Physical and chemical characterization. *Cell. Chem. Technol.*, *21*(5), 525.

[111] Feldman, D., Khoury, M.J. (1988) Epoxy-lignin polyblends. Part II. Adhesive behavior and weathering. *J. Adhes. Sci. Technol.*, *2*(2), 107–116.

[112] Feldman, D., Banu, D. (1988) Kinetic data on the curing of an epoxy polymer in the presence of lignin. *J. Polym Sci. Part A. Polym Chem.*, *26*(4), 973–983.

[113] McCarthy, J.L., Islam, A. (2000) Lignin chemistry, technology, and utilization. A brief history. In: *Lignin: Historical, Biological and Materials Perspectives*, ACS Symposium Series 742, eds. W.G. Glasser, R.A. Northey, and T.P. Schultz. Washington, DC: American Chemical Society, pp. 2–99.

[114] Shiraishi, N. (1989) Recent progress in wood dissolution and adhesives from kraft lignin. In: *Lignin, Properties and Materials*, ACS Symposium Series 397, eds. W.G. Glasser, S. Sarkanen. Washington, DC: American Chemical Society, pp. 488–495.

[115] Gargulak, J.D., Lebo, S.E. (2000) Commercial use of lignin-based materials. In: *Lignin: Historical, Biological and Materials Perspectives*, *ACS* Symposium Series

742, eds. W.G. Glasser, R.A. Northey, and T.P. Schultz. Washington, DC: American Chemical Society, pp. 304–312.

[116] Feldman, D., Banu, D., Luchian, C., and Wang, J. (1991) Epoxy-lignin blends: Correlation between polymer interaction and curing temperature. *J. Appl. Polym. Sci.*, *42*(5), 1307–1318.

[117] Tomita, B., Kurozumi, K., Takemura, A., and Hosoya, S. (1989) Ozonized lignin-epoxy resins. In: *Lignin, Properties and Materials*, ACS Symp. Ser. 397, eds. W.G. Glasser, S. Sarkanen. Washington, DC: American Chemical Society, pp. 496–505.

[118] Hofmann, K., Glasser, W.G. (1993) Engineering plastics from lignin. 21. Synthesis and properties of epoxidized lignin-poly(propylene oxide) copolymers. *J. Wood Chem. Technol.*, *13*(1), 73–95.

[119] Hofmann, K., Glasser, W.G. (1993) Engineering plastics from lignin. 22. Cure of lignin based epoxy resins. *J. Adhesion*, *40*(2), 229–241.

[120] Ito, H., Shiraishi, N. (1987) Epoxy resin adhesives from thiolignin. *Mokuzai Gakkaishi*, *33*(5), 393–399.

[121] Simionescu, C.I., Rusan, V., Macoveanu, M.M., Cazacu, G., Lipsa, R., Stoleriu, A., and Ioanid, A. (1991) Crosslinked lignin-epoxy resins. IV. *Cell. Chem. Technol.*, *25*, 5–6, 355–368.

[122] Simionescu, C.I., Rusan, V., Turta, C.I., Bobcova, S.A., Macoveanu, M.M., Cazacu, G., and Stoleriu, A. (1993) Synthesis and characterization of some iron-lignosulfonate based lignin-epoxy resins. *Cell. Chem. Technol.*, *27*(6), 627–644.

[123] Simionescu, C.I., Rusan, V., Cazacu, G., Macoveanu, M.M., Lipsa, R., and Stoica, G. (1990) Lignin-epoxy resins. III. Physical and chemical characterization of solid products. *Cell. Chem. Technol.*, *24*(3), 397–405.

[124] Vasile, C., Macoveanu, M.M., Popa, V.I., Pascu, M.C., and Cazacu, G. (1998) Lignin and environment. *Roum. Che. Quart. Rev.*, *6*(2), 85–111.

[125] Kosbar, L.L., Gelorme, J., Japp, R.M., and Fotorny, W.T. (2001) Introducing bio-based materials into the electronics industry. *J. Ind. Ecol.*, *4*, 93–98.

[126] Afzali-Ardakan, A., Gelorme, J.D., and Kosbar, L.C. (2002) Methods of fabricating cross-linked biobased materials and structures fabricated therewith, US Patent 6,339,116.

[127] Holsopple, D.B., Kurlpe, W.W., Kurlpe, W.M., and Kurlpe, K.R. (1981) Method of making epoxide-lignin resins, US Patent 4,265,809.

[128] Hu, L., Pan, H., Zhou, Y., and Zhang, M. (2011) Methods to improve lignin's reactivity as a phenol substitute and as replacement for other phenolic compounds: A brief review. *BioResources*, *6*(3), 3515–3525.

[129] Popp, J.L., Kirk, T.K., and Dordick, J.S. (1991) Enzymic modification of lignin for use as a phenolic resin. *Enzyme Microb. Technol.*, *13*, 964–968.

[130] Doering, G.A. (1993) Lignin modified phenol-formaldehyde resins, US Patent 5,202,403.

[131] Peng, W., Riedl, B. (1994) The chemorheology of phenol–formaldehyde thermoset resin and mixtures of the resin with lignin fillers. *Polymer*, *35*, 1280–1286.

[132] Sellers, T., Lora, J.H., and Okuma, M. (1994) Organosolv-lignin-modified phenolics as strandboard binder I. Organosolv lignin and modified phenolic resin. *Mokuzai Gakkaishi*, *40*, 1079–1084.

[133] Sellers, T., Kim, M.G., Miller, G.D., Haupt, R.A., and Strichlande, R.C. (1994) Comparison of strandboards made with phenol–formaldehyde resin and modified with TVA acid hydrolysis lignin. *Forest Prod. J.*, *44*, 63–68.

[134] Cetin, N.S., Ozman, N. (2002) Use of organosolv lignin in phenol–formaldehyde resins for particleboard production I. Organosolv lignin modified resins. *Int. J.*, *22*, 477–480.

[135] Sukhbaatar, B., Steele, P.H., and Kim, M.G. (2009) Use of lignin separated from bio-oil in oriented strand board binder phenol-formaldehyde resins. *BioResources*, *4*(2), 789–804.

[136] Thielemans, W., Can, E., Morye, S.S., and Wool, R.P. (2002) Novel applications of lignin in composite materials. *J. Appl. Polym. Sci.*, *83*(2), 323–331.

[137] Ignat, L., Ignat, M., Stoica, E., Ciobanu, C., and Popa, V.I. (2011) Lignin blends with polyurethane-containing lactate segments. Properties and enzymatic degradation effects. *Cell. Chem. Technol.*, *45*(3–4), 233–243.

[138] Pascu, M.C., Vasile, C., Agafitei, G.E., Gazacu, G., and Stoleriu, A. (2010) Compatibility and degradability of the polyalkanoates/epoxy modified lignin blends. In: *Monomers, Oligomers, Polymers, Composites and Nanocomposites Research: Synthesis, Properties and Applications*, eds. R.A. Pethrick, P. Petkov, A. Zlatarov, G.E. Zaikov, S.K. Rakovski. Nova Science, pp. 265–282.

[139] Wu, R.L., Wang, X.L., Li, H.Z., and Wang, Y.Z. (2009) Green composite films prepared from cellulose, starch and lignin in room-temperature ionic liquid. *Biores. Technol.*, *100*(9), 2569–2574.

[140] Warth, H., Mülhaupt, R., and Schätzle, J. (1997) Thermoplastic cellulose acetate and cellulose acetate compounds prepared by reactive processing. *J. Appl. Polym. Sci.*, *64*(2), 231–242.

[141] Glasser, W.G., Rials, T.G., Kelly, S.S., and Dave, V. (1998) Studies of the molecular interaction between cellulose and lignin as a model for the hierarchical structure of wood. In: *Cellulose Derivatives. Modification, Characterisation and Nanostructure*, ACS Symposium, Series 688, eds. T.J. Heinze, W.G. Glasser. Washington, DC.: American Chemical Society, pp. 265–282.

[142] Rials, T.G., Glasser, W.G. (1989) Multiphase materials with lignin. IV. Blends of hydroxylpropyl cellulose with lignin. *J. Appl. Polym. Sci.*, *37*, 2399–2415.

[143] Ghosh, I., Jain, R.K., and Glasser, W.G. (1999) Multiphase materials with lignin. XV. Blends of cellulose acetate butyrate with lignin esters. *J. Appl. Polym. Sci.*, *74*, 448–457.

[144] Oliveira, W.D., Glasser, W.G. (1994) Multiphase materials with lignin. 13. Block copolymers with cellulose propionate. *Polymer*, *35*(9), 1977–1985.

[145] Oliveira, W.D., Glasser, W.G. (1994) Multiphase materials with lignin. 11. Starlike copolymers with caprolactone. *Macromolecules*, *27*(1), 5–11.

[146] Dave, V., Glasser, W.G. (1997) Cellulose-based fibres from liquid crystalline solutions. 5. Processing and morphology of CAB blends with lignin. *Polymer*, *38*, 2121–2126.

[147] Shulga, G., Betkers, T., Shakels, V., Neiberte, B., Verovkins, A., Brovkina, J., Belous, O., Ambrazaintene, D., and Žukauskaite, A. (2007) Effect of the modification of lignocellulosic materials with a lignin-polymer complex on their mulching properties. *BioResources*, *2*(4), 572–582.

[148] Li, J., Li, S., Wang, H., Yang, Y., and Guo, G. (2011) Preparation of a lignin-based composite and its properties. *BioResources*, *6*(2), 1532–1542.

[149] Samal, S.K., Fernandes, E.G., Corti, A., and Chiellini, E. (2009) Hybrid polymeric composites based on polyethylene and lignin. *Intern. J. Mater. Prod. Technol.*, *36*(1–4), 62–72.

[150] Stiubianu, G., Cazacu, M., Cristea, M., and Vlad, A. (2009) Polysiloxane-lignin composites. *J. Appl. Polym. Sci.*, *113*(4), 2313–2321.

[151] Mousavioun, P., Doherty, W.O.S., George, G., and Halley, P. (2009) Thermal behaviour of PHB/lignin composites, 11th Pacific Polymer Conference, Dec. 6–11, 2009, Cairns, Australia. Available at https://leishman.conferenceservices.net/resources/266/1604/pdf/PPC2009_0794.pdf.

[152] Li, D.X. (2010) *Preparation and Property of Biodegradable Polypropylene/Lignin Composites*. Posted on March 30, 2011, by China Papers, http://mt.china-papers.com/?p=205818.

[153] Liu, T., Wang, Q.W., and Zhao, H. (2011) Physical properties of waste lignin/HDPE composites. *Adv. Mater. Res.*, *221*, 60–65.

[154] Chen, L., Tang, C., Ning, N., Wang, C., Fu, Q., and Zhang, Q. (2009) Preparation and properties of chitosan/lignin composite films. *Chin. J. Polym. Sci.*, *27*(5), 739–774.

[155] Liu, F., Xu, K., Chen, M., and Cao, D. (2011) The roles of polyacrylate in poly(vinyl chloride)/lignin composites. *Polym. Compos.*, *32*(9), 1399–1407.

[156] Yue, X., Chen, F., and Zhou, X. (2011) Improved interfacial bonding of PVC/wood-flour composites by lignin amine modification. *BioResources*, *6*(2), 2022–2034.

[157] Feldman, D., Banu, D.J. (1997) Contribution to the study of rigid PVC polyblends with different lignins. *J. Appl. Polym. Sci.*, *66*, 1731–1744.

[158] Stival Bittencourt, P.R., dos Santos, G.L., Gómez Pineda, E.A., and Winkler Hechenleitner, A.A. (2005) Studies on the thermal stability and film irradiation effect of poly(vinyl alcohol)/kraft lignin blends. *J. Thermal Anal. Calorim.*, *79*, 371–374.

[159] Fernandes, D.M., Winkler Hechenleitner, A.A., Job, A.E., Radovanocic, E., and Gómez Pineda, E.A. (2006) Thermal and photochemical stability of poly(vinyl alcohol)/modified lignin blends. *Polym. Degrad. Stab.*, *91*, 1192–1201.

[160] Nenkova, S. (2007) Study of sorption properties of lignin-derived fibrous composites for the remediation of oil polluted receiving waters. *BioResources*, *2*(3), 408–418.

CHAPTER 24

Preparation, Processing and Applications of Protein Nanofibers

MEGAN GARVEY, MADHUSUDAN VASUDEVAMURTHY, SHIVA P. RAO, HEATH ECROYD, JULIET A. GERRARD, and JOHN A. CARVER

24.1 INTRODUCTION

Protein nanofibers are formed as a result of ordered protein misfolding wherein the native structure of a protein rearranges itself, by adopting an intermediately folded conformer, into highly ordered β-sheet containing nanofibrillar structures [1]. The nanofibrillar structure is proposed to be accessible to all proteins, under appropriate conditions, and is formed by many proteins *in vitro* [2]. *In vivo*, these nanofibers are frequently associated with cellular damage in the tissues within which they are formed. Their cytotoxicity arises from the formation of toxic precursors and the deposition of large conglomerates composed primarily of nanofibers (known as amyloid fibrils) arranged into amyloid plaques [3]. In addition, biologically active nanofiber species (known as "functional amyloid" or amyloidin) exist in nature, with varied and specific biological functions. For example, curlin protein amyloid in *Escherichia coli* assists in surface binding [4]; Pmel 17 protein amyloid in melanosomes in humans (and other species) provides surfaces that enable small molecule conversion into melanin [5]; and fungal hydrophobins form into β-scaffolds, which are critical to the formation of arial and other fungal structures [6].

The formation, characteristics, and effects of protein nanofibers are of intense research interest, due to their roles in a wide range of disease states associated with protein misfolding and aggregation (e.g., Alzheimer's, Parkinson's and Huntington's diseases) and their possible nanotechnological uses

Biopolymer Nanocomposites: Processing, Properties, and Applications, First Edition. Edited by Alain Dufresne, Sabu Thomas, and Laly A. Pothan.

[7–10]. The latter topic is the focus of this chapter, which explores the production, characterization, and applications of protein nanofiber structures formed from a variety of peptides and proteins.

24.2 PROTEIN NANOFIBERS: STRUCTURE, FORMATION, AND APPLICATIONS

24.2.1 Protein Nanofiber Characteristics

Protein nanofibers all contain similar structural characteristics, regardless of which native peptide or protein they are derived from Reference [11]. They are typically long, filament-like structures which are unbranched, 6 to 12 nm in diameter (larger for more complex multifibril structures) and may extend up to micrometers in length (Fig. 24.1). They are resistant to proteolysis, have high kinetic and thermodynamic stability, and feature a characteristic "cross-β-sheet" fibril core (more below) [12–14]. Protein nanofiber structure has been investigated using a wide variety of techniques including dye binding (e.g., thioflavin T [ThT]), microscopy (e.g., transmission electron, cryo electron, or atomic force), X-ray fiber diffraction and various spectroscopic techniques [15]. A battery of techniques is used because each provides complementary structural information. Detailed, atomic resolution information about nanofiber structure has been obtained only on small nanofiber-forming peptides using solid state nuclear magnetic resonance and X-ray crystallographic techniques, with the findings being extrapolated to the structures of protein-containing amyloid nanofibers [2].

The intrinsic cross-β-sheet structure of all protein nanofibers has two major features, that is, β-strands and β-sheet filaments, which are independent of the protein incorporated into the nanofiber (Fig. 24.1) [16]. The β-strands, which lie perpendicular to the fibril axis, are formed by the polypeptide backbone folding upon itself at a distance of 4.7 to 4.8 Å. The β-sheet filaments run parallel to the fibril axis and are composed of interwoven β-strands 8 to 10 Å

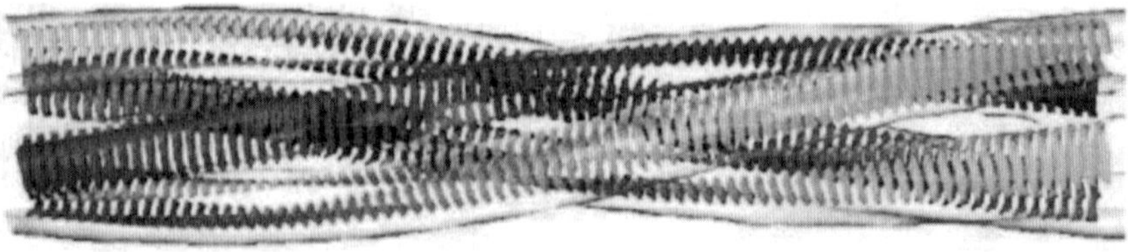

Figure 24.1 Cross-β sheet structural core of an insulin protein nanofiber, modeled from cryo-electron microscopic analysis [16]. The four protofilaments, individually containing aligned β-strands that form β-sheets, interact together to form a complete nanofiber.

apart (a distance that is dependent upon the number of protofilaments) [2, 14, 17]. Each nanofiber generally contains between four and six protofilaments, which are wound together to form the complete fibril assembly [14, 16, 18]. The intertwining of these filaments causes a twisted appearance of the fibril when seen by transmission electron microscopy (TEM). Some regions of the polypeptide chain, particularly in larger proteins, are not incorporated into the β-strands. They may arise from flexible loops and turns (e.g., from unstructured regions of the protein), which remain tightly associated with the fibrillar assemblies and are resistant to proteolysis. Such regions are deemed part of the fibrillar core [19]. Other regions of the native protein may remain unassociated with the fibril structure, being external to the β-strands and β-sheets and thus can be proteolytically degraded [20]. One example of this is the short, flexible C-terminal extension of αB-crystallin, a small heat-shock protein, which retains its mobility even when the protein is incorporated into a fibrillar structure and is therefore external to the fibril core [21].

The length of the native protein chain, or of those sections of native protein not incorporated into the fibrillar core, plays a role in the stiffness of the nanofiber produced [8, 12]. Knowles et al. [12] established that while the majority of protein nanofibers have a single relatively rigid structure, some exhibit greater flexibility. The latter flexible (curly) protein nanofiber structures appear to be stabilized by different types of intermolecular interactions compared with the more commonly observed rigid (straight) nanofibers. These differences suggest that the curly structures are a late-stage protofibrillar structure, rather than complete nanofibers. Further support for this hypothesis is provided by Fourier transform infrared spectroscopy in which structural differences are observed between straight and curly fibrils of β2-microglobulin, prepared under different nanofiber-forming conditions [22]. The different morphological properties of protein nanofibers, while of uncertain disease relevance, may have a significant role to play in the nanobiotechnological potential of these species [8]. In this context, morphology may be controlled by chemically derivitizing the starting protein [23] or by carefully adjusting the manner in which the fibrils are prepared from crude sources [24].

24.2.2 Protein Nanofiber Formation

Formation of protein nanofibers often involves a significant rearrangement of native protein structure. For this to occur in a well-structured globular protein, the polypeptide chain must undergo structural destabilization, which may be triggered *in vivo* or *in vitro* by physiological stress factors such as altered pH, increased temperature, breakage of disulfide linkages, or by alterations such as truncation, destabilizing mutations, or post-translational modifications [3]. Irrespective of the cause, the destabilization leads to the partial unfolding of the protein or, in the case of unstructured or natively unfolded proteins, partial folding of the protein, followed by the production of stable intermediately folded species. These species interact in an ordered manner, in contrast to amorphous aggregation, leading to the stacking of the polypeptide backbone of the proteins into β-sheet structures [1]. Even in cases where the protein is unstructured, the formation of a stable intermediate structure is still thought to be a prerequisite for the formation of nanofibers [25]. This stable intermediate forms the basis of the nanofiber nucleus, or seed fibril. Nanofiber formation *in vitro* can be instigated or increased (both in rate and in total yield) by the addition of protein nanofiber seeds from the same, or a highly homologous protein [26–29].

24.2.3 Nanobiotechnological Potential of Protein Nanofibers

Potentially, protein nanofibers could have great use in nanobiotechnology because of the following desirable characteristics: (1) their ability to self assemble, (2) their similarity in size to currently used nanofibers (such as carbon nanotubes), (3) their ability to adopt a variety of morphological forms, and (4) their ability to host attached functional molecules [8, 30, 31]. Due to the difficulties in manipulating nanometer-sized products, self assembly is a highly desirable trait in nanotechnology. There are frequently physical constraints on our ability to efficiently produce nanostructures from larger starting materials [32]. An alternative assembly system is the self-association of small molecules, such as proteins. Identifying mechanisms to induce proteins to self-assemble into ordered structures will potentially allow for large-scale production of desirable base-nanomaterials (e.g., nanotubes or nanofibers). Some examples of where these could be used include in drug delivery systems, as protein scaffolds and in electronic devices [33].

The self-assembling characteristic of amyloid and amyloid-like nanofibers has an additional benefit from a nanotechnological point of view: it affords the potential to form different morphological forms [34]. Amyloid-like nanofibers have been observed to occur in a range of alternative structures, including the classic straight nanofibers, twisted nanofibers, linear assemblies (such as ribbon structures), curled nanofibers, ring structures, and spherulites [8] (Fig. 24.2). In some cases, multiple structures may be produced from the same protein, simply by changing the solution conditions, such as by altering the pH [22] or the addition of modifying substances, for example, lipids [35]. In other cases, it has been observed that a range of fibrillar structures may occur in the same sample [16]. The proposal that a nanofiber structure is a conformation accessible to many (arguably all) proteins and peptides [19, 36] has led to the development of a range of peptides designed specifically for their ability to form nanofibers, and thus be of potential use as bionanostructures [18, 32, 37–39]. The production of reliable, robust, and reproducible structures remains one of the challenges in the field of bionanomaterials, and is an area in which designed or specifically selected peptides holds much promise [34]. Progress has also been recently made in controlling the length of protein nanofibers, for example, for application in nanoelectronics and biosensing [40].

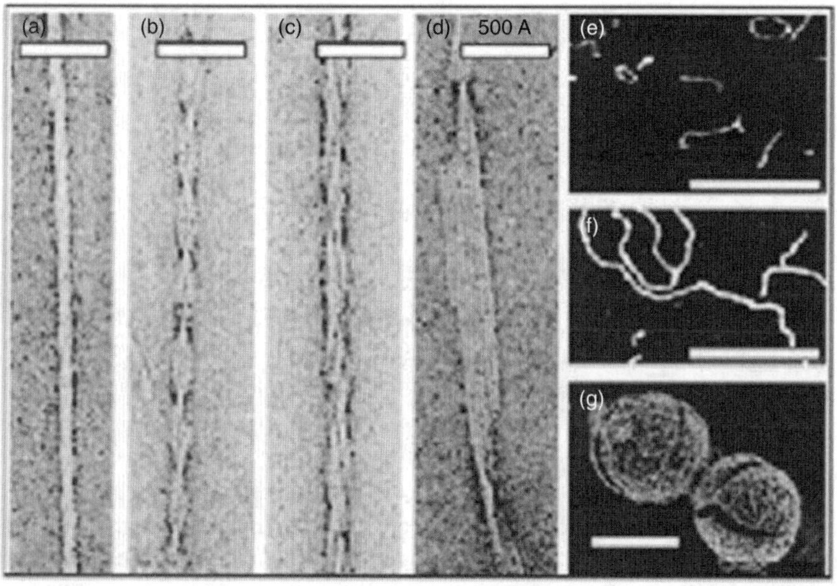

Figure 24.2 Morphological variation of protein nanofibers, including straight (a); twisted or intertwined (b, c); linear assemblies (d); curly or loop (e, f); and spherulite assemblies (g). Structures are composed of insulin (a, b, c, d) [16]—scale bars: 50 nm; αB-crystallin (e, f) [21]—scale bars: 500 nm; and insulin (g) [47]—scale bar is 50 μm.

Research into the use of functionalized protein nanofibers is based, in part, on the widespread presence in nature of amyloidin, amyloid-like nanofibers with native functionality [5]. It has also been established that it is possible to incorporate folded and functional domains within designed protein nanofiber structures [8, 41]. Designed peptides have been used to create nanofibers which incorporate functional fluorophores, enzymes or ligand binders [7, 30, 31, 42], showing that specific functionalization of the end-product is possible when using amyloid-like nanofibers. Further, conversion into the nanofiber structure increases the stability and longevity of some functional elements [42, 43]. Together with being able to self assemble into a variety of morphological species, the ability to functionalize amyloid-like nanofibers greatly enhances their potential use as protein nanomaterials.

24.3 PREPARATION AND PROCESSING OF PROTEIN NANOFIBERS

24.3.1 Protein Nanofibers Formed from Lens Crystallin Proteins

As discussed previously, protein nanofiber formation from full length, well-ordered proteins requires the destabilization of the native structure. The lens

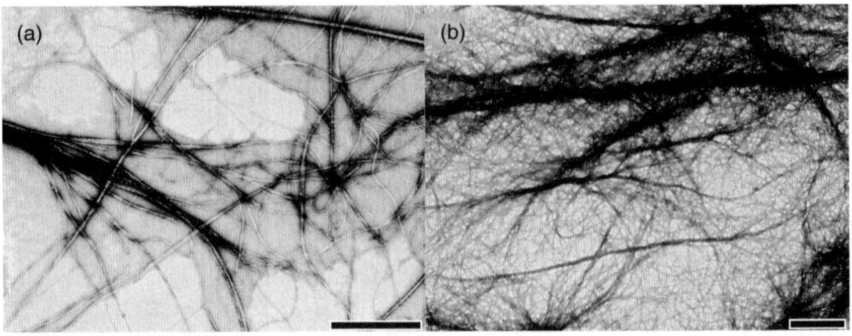

Figure 24.3 Fish crystallin nanofibers formed at 60°C (a) and 90°C (b), as assessed by transmission electron microscopy [45]. Scale bars are 1 μm.

contains a very high concentration of crystallin proteins arranged in a well-ordered manner to ensure lens transparency and the correct refraction of light. Crystallin proteins contain a high proportion of native β-sheet structure [44]; however, an intermediately folded structure is required for ordered aggregation to produce crystallin protein nanofibers. As described in Healy et al. [45], a mixture of fish crystallin proteins (i.e., containing α-, β- and γ-crystallins), upon destabilization with trifluoroethanol (TFE) and incubation at elevated temperatures, forms protein nanofibers. Altering the incubation temperature produces variations in the supramolecular structure of the nanofibers (Fig. 24.3).

Fish crystallin proteins can be readily extracted from the lenses of fish eyes. Once the eye has been extracted from the fish head, a scalpel is used to carefully slit the membrane of the eye and the lens is extracted by placing a small amount of force on the opposite side of the eye. Homogenize each lens in approximately 3 mL of homogenizing buffer, containing 20 mM Tris, 1 mM Ethylene-diamine-tetraacetic acid (EDTA) at pH 2. The lenses are homogenized at very low speed for 5 minutes, using a homogenizer or blender with an external speed controller (which is able to run at 2000 rpm or less). The mixture is filtered through a sieve to remove the hard (central) core of the fish lenses. Homogenization is continued, still at very low speed (~1000 rpm), for a further 5 minutes. The homogenate is centrifuged at 13,000 *g* for 30 minutes at room temperature. The supernatant contains the soluble crystallin proteins, while the pellet is discarded. The supernatant (fish lens homogenate) can be used immediately for nanofiber formation or frozen with liquid nitrogen and stored at –20°C.

Nanofiber formation from fish crystallins is achieved by the addition of 10% v/v TFE to the fish lens homogenate. It is most effective when the homogenate is diluted to attain a final crystallin protein concentration of 10 mg/mL. For the production of nanofibers with different morphologies, the sample can be

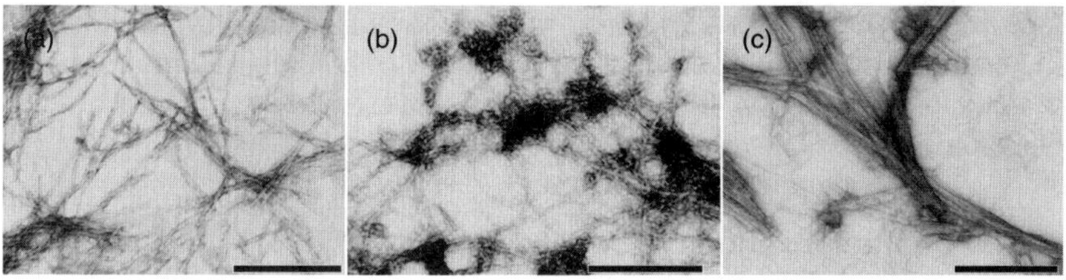

Figure 24.4 Insulin nanofibers formed in the absence (a) and presence (b, c) of chymotrypsin, as assessed by transmission electron microscopy [23]. Scale bars are 250 nm.

divided, providing one sample (frozen with liquid nitrogen) as a control and two which can be incubated at 60°C and 90°C for 24 hours to attain protein nanofibers. These samples can then be frozen to halt further nanofiber formation and assessed for their characteristics (Fig. 24.3). The sample incubated at 60°C should contain high numbers of individual nanofiber structures, approximately 10 to 20 nm in width and 0.5 to 1.0 μm in length. In the sample of nanofibers formed at 90°C, these smaller nanofibers will intertwine to form large nanofiber-bundles, up to 500 nm wide and 10 μm in length.

24.3.2 Protein Nanofibers Formed from Insulin

The nanofibril formation of insulin is well characterized; it forms nanofibers of various morphologies under a range of solution conditions [46]. Additionally, higher order structures, now known as spherulites, are formed under certain solution conditions [47, 48]. Both chains of the insulin monomer can participate in nanofiber formation [46, 49] and a model of the insulin fibril structure (Fig. 24.1), derived from cryo electron microscopy, implies that both chains are incorporated into the nanofiber [16]. Separation of the insulin chains or alteration of their structures by other means, such as chemical modifications or enzymatic digestion, lead to significant differences in the morphology of the nanofibers formed [23, 49].

Bovine pancreatic insulin in the presence or absence of the protease α-chymotrypsin at neutral pH can rapidly form protein nanofibers with multiple morphological characteristics (Fig. 24.4) [23]. A solution of 5.7 mg/mL insulin in 25 mM HCl, 100 mM NaCl, pH 8 is divided into three: one is frozen to act as a non-aggregated control. To one of the remaining insulin solutions, α-chymotrypsin is added at a 1:50 enzyme/insulin w/w ratio. The pure insulin and the insulin–chymotrypsin mixtures are both incubated at 60°C for 3 hours. They are then removed from the heat and frozen to halt further nanofiber formation. The samples are assessed for nanofiber characteristics and their morphological differences are compared. Insulin alone at pH 8 produces long straight fibrils with clearly visible twists (Fig. 24.4a) caused by the intertwining of protofilaments (Fig. 24.2b). After incubation with chymotrypsin, there is a change in morphology of the nanostructures, with a combination of long thin curvy nanofibers (Fig. 24.4b) and bundles of short straight nanofibers observed (Fig. 24.4c).

24.3.3 Characterization of Protein Nanofibers

Thioflavin T (ThT) is a fluorescent molecule that is frequently used to assess or monitor fibril formation. ThT can bind to both native and nanofibrillar β-sheet structures; however, even in cases of native proteins that contain a large amount of β-sheet (such as crystallin proteins), significant differences in ThT fluorescence are observed after nanofiber formation. To examine whether nanofibers have been formed, ThT fluorescence spectra of the initial and final samples are acquired [50]. The samples arising from nanofiber formation should be completely thawed and then prepared (in duplicate) at 0.5 mg/mL protein, 20 μM ThT, and 50 mM glycine–sodium hydroxide, pH 9. The samples are incubated for exactly 10 minutes at room temperature and their fluorescence spectrum is acquired using a fluorescence spectrophotometer, exciting at a wavelength of 440 nm and acquiring the emission spectrum at ~482 nm. In comparing the fluorescence spectrum before and after nanofiber formation, a significant increase in fluorescence will be observed when nanofibers are present (Fig. 24.5).

Transmission electron microscopy (TEM) is a technique used to image at the nanoscale (nanometers to microns). It is based on the scattering of electrons and utilizes a high voltage electron source to image-coated surfaces.

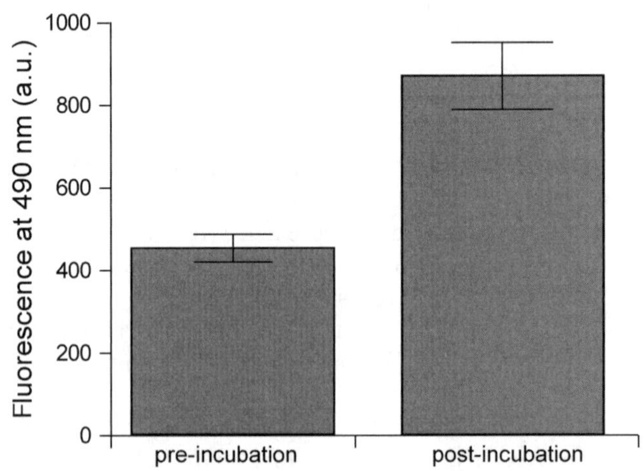

Figure 24.5 ThT fluorescence intensity at 490 nm for bovine crystallin proteins after 18 hours of incubation at pH 2 and 60°C. Error bars show standard deviation of three replicates [24].

Contrast and hence enhanced resolution can be gained by the addition of a negative stain, such as uranyl acetate. TEM is commonly used for investigating the length and width features of protein nanofibers, in some cases even revealing aspects of their protofibrillar structure [16]. Following the method of Thorn et al. [50], TEM grids can be prepared by the addition of dilute nanofiber solutions (~0.5 mg/mL) onto formvar and carbon-coated copper or nickel electron microscopy grids. After rinsing with water, the nanofibers are negatively stained using uranyl acetate* (2% w/v) and any excess stain is removed with filter paper. Samples can be magnified 20,000 to 125,000 times, using an excitation voltage of 80 to 120 kV, depending on the recommendations for the transmission electron microscope used. Variations will be apparent in the nanofiber morphologies and in the formation of macrostructures (Fig. 24.3 and Fig. 24.4).

24.4 APPLICATIONS OF PROTEIN NANOFIBERS

24.4.1 Protein Nanofibers as Nanoscaffolds for Enzyme Immobilization

The ability to functionalize protein nanofibers, by enzyme immobilization, makes them one of the most attractive nanomaterials for various applications that utilize enzymatic activity. Glucose oxidase, an enzyme well known for potential biosensor applications, has been successfully immobilized on insulin fibrils and retains activity upon attachment [30]. The resulting functionalized enzyme scaffold has been incorporated into a model poly(vinyl alcohol) (PVOH) film to create a new bionanomaterial. The functionalized film inhibits the growth of *E. coli*, proving its antibacterial effectiveness [30]. The potential application of protein nanofibers in bioremediation and chemical detoxification has also been shown recently by successfully immobilizing the enzyme organophosphate hydrolase on insulin protein nanofibers [31].

24.4.2 Protein Nanofibers to Form Nanowires and Nanogaps

Protein nanofibers, because of their almost unequalled aspect ratio (i.e., their ratio of diameter to length), have great potential in the field of templated one-dimensional objects such as conducting nanowires [40]. Insulin protein nanofibers can be coated with the highly conducting polymer alkoxysulfonate Poly (3,4-ethylenedioxythiophene), or PEDOT, through molecular self-assembly. The resulting networks of conducting nanowires are electrically and electrochemically active [51]. A recent study also explored whether fish lens crystallin protein nanofibers (see Section 24.3) could be used as a template for the growth

* Uranyl acetate is toxic. The material safety data sheet (MSDS) protocols should be read carefully for appropriate usage and disposal guidelines.

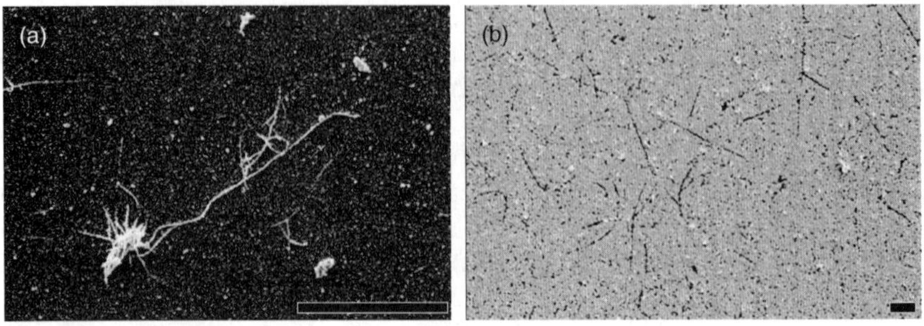

Figure 24.6 Scanning electron micrographs of nanowires and nanogaps. (a) A gold nanowire prepared by electroless plating of fish lens crystallin nanofibers, on a carbon surface. Scale bar is 10 μm. (b) Thin gold layer with nanogaps corresponding to crystallin nanofibers, on a silicon surface [45]. Scale bar is 1 μm.

of gold nanowires using electroless plating [45]. When the underlying carbon surface was exchanged for silicon, radically different observations were made wherein gold seeds appeared to have a greater affinity for silicon than for carbon. On silicon surfaces, nanogaps (small spaces of nanoscaled size) were observed repeatedly across the surface (Fig. 24.6). Such dramatically different nanostructures show promise in the fabrication of variety of nanodevices, either utilizing the nanowire potential of these nanofibers, or using them to produce nanogaps for applications in biosensing and similar fields.

24.4.3 Composite Materials Containing Protein Nanofibers

The use of protein nanofibers as novel bionanomaterials is one of the most promising of all potential applications. Welland and coworkers have recently reported a new approach to making multifunctional biomaterials, by exploiting the self-assembling properties of insulin nanofibers by combining insulin nanofibers with a silicone elastomer [52]. The resultant protein nanofiber composite was twice as stiff as a carbon nanotube elastomeric composite of the same filling ratio, with the elastic modulus of the protein nanofiber composite being equivalent to that of the carbon nanotube composite. Protein nanofibers can also impart useful changes to mechanical and morphological properties of polymeric materials. For example, when insulin nanofibers were combined with PVOH at a filling level of 0.6 wt%, the nanofiber-reinforced PVOH was 15% stiffer than a PVOH control containing unfibrillized insulin [53].

24.5 CONCLUSIONS AND FUTURE PROSPECTS

The practical use of protein nanofibers in bionanotechnology is gaining in popularity, primarily due to their qualities such as self-assembly and their ability to be readily functionalized. Currently, these aspects are significant stumbling blocks to other nanotechnological approaches. It is now apparent that inexpensive and bulk-scale methods of protein nanofiber production are feasible. Thus, it is foreseeable that a greater role for proteins will develop as an emergent source for future materials, standing at the interface of biology and nanotechnology.

ACKNOWLEDGMENTS

The authors thank Dr. Clement L. Roux, University of Canterbury, for the scanning electron micrographs of nanowires and nanogaps.

REFERENCES

[1] Uversky, V.N., Fink, A.L. (2004) Conformational constraints for amyloid fibrillation: The importance of being unfolded. *Biochimica et Biophysica Acta*, *1698*, 131–153.

[2] Chiti, F., Dobson, C.M. (2006) Protein misfolding, functional amyloid and human disease. *Annual Reviews of Biochemistry*, *75*, 333–366.

[3] Dobson, C.M. (2006) Protein aggregation and its consequences for human disease. *Protein & Peptide Letters*, *13*, 219–227.

[4] Chapman, M.R., Robinson, L.S., Pinkner, J.S., Roth, R., Heuser, J., Hammar, M., Normark, S., and Hultgren, S.J. (2002) Role of *Escherichia coli* curlin operons in directing amyloid fiber formation. *Science*, *295*, 851–855.

[5] Fowler, D.M., Koulov, A.V., Alory-Jost, C., Marks, M.S., Balch, W.E., and Kelly, J.W. (2006) Functional amyloid formation within mammalian tissue. *PLoS Biology*, *4*, e6.

[6] Kwan, A.H., Winefield, R.D., Sunde, M., Matthews, J.M., Haverkamp, R.G., Templeton, M.D., and Mackay, J.P. (2006) Structural basis for rodlet assembly in fungal hydrophobins. *Proceedings of the National Academy of Sciences of the United States of America*, *103*, 3621–3626.

[7] Cherny, I., Gazit, E. (2008) Amyloids: Not only pathological agents but also ordered nanomaterials. *Angewandte Chemie International Edition*, *47*, 4062–4069.

[8] Gras, S.L. (2007) Amyloid fibrils: From disease to design. New biomaterial applications for self-assembling cross-β fibrils. *Australian Journal of Chemistry*, *60*, 333–342.

[9] Hamley, I.W. (2007) Peptide fibrillization. *Angewandte Chemie International Edition*, *46*, 8128–8147.

[10] Waterhouse, S.H., Gerrard, J.A. (2004) Amyloid fibrils in bionanotechnology. *Australian Journal of Chemistry*, *57*, 519–523.

[11] Hoang, T.X., Marsella, L., Trovato, A., Seno, F., Banavar, J.R., and Maritan, A. (2006) Common attributes of native-state structures of proteins, disordered

proteins and amyloid. *Proceedings of the National Academy of Sciences of the United States of America*, *103*, 6883–6888.

[12] Knowles, T.P., Fitzpatrick, A.W., Meehan, S., Mott, H.R., Vendruscolo, M., Dobson, C.M., and Welland, M.E. (2007) Role of intermolecular forces in defining material properties of protein nanofibrils. *Science*, *318*, 1900–1903.

[13] Carulla, N., Caddy, G.L., Hall, D.R., Zurdo, J., Gairi, M., Feliz, M., Giralt, E., Robinson, C.V., and Dobson, C.M. (2005) Molecular recycling within amyloid fibrils. *Nature*, *436*, 554–558.

[14] Serpell, L.C., Sunde, M., Benson, M.D., Tennent, G.A., Pepys, M.B., and Fraser, P.E. (2000) The protofilament substructure of amyloid fibrils. *Journal of Molecular Biology*, *300*, 1033–1039.

[15] Nielsen, L., Khurana, R., Coats, A., Frokjaer, S., Brange, J., Vyas, S., Uversky, V.N., and Fink, A.L. (2001) Effect of environmental factors on the kinetics of insulin fibril formation: Elucidation of the molecular mechanism. *Biochemistry*, *40*, 6036–6046.

[16] Jimenez, J.L., Nettleton, E.J., Bouchard, M., Robinson, C.V., Dobson, C.M., and Saibil, H.R. (2002) The protofilament structure of insulin amyloid fibrils. *Proceedings of the National Academy of Sciences of the United States of America*, *99*, 9196–9201.

[17] Sunde, M., Serpell, L.C., Bartlam, M., Fraser, P.E., Pepys, M.B., and Blake, C.C. (1997) Common core structure of amyloid fibrils by synchrotron X-ray diffraction. *Journal of Molecular Biology*, *273*, 729–739.

[18] Lashuel, H.A., Labrenz, S.R., Woo, L., Serpell, L.C., and Kelly, J.W. (2000) Protofilaments, filaments, ribbons and fibrils from peptidomimetic self-assembly: Implications for amyloid fibril formation and materials science. *Journal of the American Chemical Society*, *122*, 5262–5277.

[19] Fändrich, M., Dobson, C.M. (2002) The behaviour of polyamino acids reveals an inverse side chain effect in amyloid structure formation. *EMBO Journal*, *21*, 5682–5690.

[20] Zurdo, J., Guijarro, J.I., and Dobson, C.M. (2001) Preparation and characterization of purified amyloid fibrils. *Journal of the American Chemical Society*, *123*, 8141–8142.

[21] Meehan, S., Knowles, T.P., Baldwin, A.J., Smith, J.F., Squires, A.M., Clements, P., Treweek, T.M., Ecroyd, H., Tartaglia, G.G., Vendruscolo, M., MacPhee, C.E., Dobson, C.M., and Carver, J.A. (2007) Characterisation of amyloid fibril formation by small heat-shock chaperone proteins human αA-, αB- and R120G αB-crystallins. *Journal of Molecular Biology*, *372*, 470–484.

[22] Jahn, T.R., Tennent, G.A., and Radford, S.E. (2008) A common β-sheet architecture underlies *in vitro* and *in vivo* β2-microglobulin amyloid fibrils. *Journal of Biological Chemistry*, *283*, 17279–17286.

[23] Rao, S.P., Meade, S.J., Joyce, N.I., Healy, J.P., Sutton, K.H., Larsen, N.G., and Gerrard, J.A. (2011) Amyloid fibril formation from crude protein mixtures. *Biotechnology Progress*, *27*, 1768–1776.

[24] Garvey, M., Gras, S.L., Meade, S.J., Carver, J.A., and Gerrard, J.A. (2009) Protein nanofibres of defined morphology prepared from mixtures of crude crystallins. *International Journal of Nanotechnology*, *6*, 258–273.

[25] Ecroyd, H., Carver, J.A. (2008) Unraveling the mysteries of protein folding and misfolding. *IUBMB Life*, *60*, 769–774.

[26] Gsponer, J., Vendruscolo, M. (2006) Theoretical approaches to protein aggregation. *Protein & Peptide Letters*, *13*, 287–293.

[27] Lundmark, K., Westermark, G.T., Olsén, A., and Westermark, P. (2005) Protein fibrils in nature can enhance amyloid protein A amyloidosis in mice: Cross-seeding as a disease mechanism. *Proceedings of the National Academy of Sciences of the United States of America*, *102*, 6098–6102.

[28] Ohhashi, Y., Kihara, M., Naiki, H., and Goto, Y. (2005) Ultrasonication-induced amyloid fibril formation of β2-microglobulin. *Journal of Biological Chemistry*, *280*, 32843–32848.

[29] O'nuallain, B., Williams, A.D., Westermark, P., and Wetzel, R. (2004) Seeding specificity in amyloid growth induced by heterologous fibrils. *Journal of Biological Chemistry*, *279*, 17490–17499.

[30] Pilkington, S.M., Roberts, S.J., Meade, S.J., and Gerrard, J.A. (2010) Amyloid fibrils as a nanoscafold for enzyme immobilization. *Biotechnology Progress*, *26*, 93–100.

[31] Raynes, J.K., Pearce, G.F., Meade, S.J., and Gerrard, J.A. (2011) Immobilization of organophosphate hydrolase on an amyloid fibril nanoscaffold: Towards bioremediation and chemical detoxification. *Biotechnology Progress*, *27*, 360–367.

[32] Hamada, D., Yanagihara, I., and Tsumoto, K. (2004) Engineering amyloidogenicity towards the development of nanofibrillar materials. *Trends in Biotechnology*, *22*, 93–97.

[33] Rajagopal, K., Schneider, J.P. (2004) Self-assembling peptides and proteins for nanotechnological applications. *Current Opinion in Structural Biology*, *14*, 480–486.

[34] Zhang, S. (2003) Fabrication of novel biomaterials through molecular self-assembly. *Nature Biotechnology*, *21*, 1171–1178.

[35] Griffin, M.D., Mok, M.L., Wilson, L.M., Pham, C.L., Waddington, L.J., Perugini, M.A., and Howlett, G.J. (2008) Phospholipid interaction induces molecular-level polymorphism in apolipoprotein C-II amyloid fibrils via alternative assembly pathways. *Journal of Molecular Biology*, *375*, 240–256.

[36] Fändrich, M., Fletcher, M.A., and Dobson, C.M. (2001) Amyloid fibrils from muscle myoglobin. *Nature*, *410*, 165–166.

[37] Baldwin, A.J., Bader, R., Christodoulou, J., MacPhee, C.E., Dobson, C.M., and Barker, P.D. (2006) Cytochrome display on amyloid fibrils. *Journal of the American Chemical Society*, *128*, 2162–2163.

[38] Deechongkit, S., Powers, E.T., You, S.L., and Kelly, J.W. (2005) Controlling the morphology of cross β-sheet assemblies by rational design. *Journal of the American Chemical Society*, *127*, 8562–8570.

[39] Fezoui, Y., Hartley, D.M., Walsh, D.M., Selkoe, D.J., Osterhout, J.J., and Teplow, D.B. (2000) A *de novo* designed helix-turn-helix peptide forms nontoxic amyloid fibrils. *Nature Structural Biology*, *7*, 1095–1099.

[40] Domigan, L.J., Healy, J.P., Meade, S.J., Blaikie, R.B., and Gerrard, J.A. (2012) Controlling the dimensions of amyloid fibrils: Towards homogenous components for bionanotechnology. *Biopolymers*, *97*, 123–133.

[41] Sackewitz, M., Von Einem, S., Hause, G., Wunderlich, M., Schmid, F.X., and Schwarz, E. (2008) A folded and functional protein domain in an amyloid-like fibril. *Protein Science*, *17*, 1044–1054.

[42] Gras, S.L., Tickler, A.K., Squires, A.M., Devlin, G.L., Horton, M.A., Dobson, C.M., and MacPhee, C.E. (2008) Functionalised amyloid fibrils for roles in cell adhesion. *Biomaterials*, *60*, 333–342.

[43] Yu, L., Banerjee, I.A., Gao, X., Nuraje, N., and Matsui, H. (2005) Fabrication and application of enzyme-incorporated peptide nanotubes. *Bioconjugate Chemistry*, *16*, 1484–1487.

[44] Bloemendal, H., De Jong, W., Jaenicke, R., Lubsen, N.H., Slingsby, C., and Tardieu, A. (2004) Ageing and vision: Structure, stability and function of lens crystallins. *Progress in Biophysical and Molecular Biology*, *86*, 407–485.

[45] Healy, J., Wong, K., Sawyer, E.B., Roux, C., Domigan, L., Gras, S.L., Sunde, M., Larsen, N.G., Gerrard, J.A., and Vasudevamurthy, M. (2012) Polymorphism and higher order structures of protein nanofibres from crude mixtures of fish lens crystallins: Towards useful materials. *Biopolymers*, *97*, 595–606.

[46] Hong, D.-P., Ahmad, A., and Fink, A.L. (2006) Fibrillation of human insulin A and B chains. *Biochemistry*, *45*, 9342–9353.

[47] Krebs, M.R., MacPhee, C.E., Miller, A.F., Dunlop, I.E., Dobson, C.M., and Donald, A.M. (2004) The formation of spherulites by amyloid fibrils of bovine insulin. *Proceedings of the National Academy of Sciences of the United States of America*, *101*, 14420–14424.

[48] Waugh, D.F. (1946) A fibrils modification of insulin. I. The heat precipitate of insulin. *Journal of the American Chemical Society*, *68*, 247–250.

[49] Devlin, G.L., Knowles, T.P.J., Squires, A., McCammon, M.G., Gras, S.L., Nilsson, M.R., Robinson, C.V., Dobson, C.M., and MacPhee, C.E. (2006) The component polypeptide chains of bovine insulin nucleate or inhibit aggregation of the parent protein in a conformation-dependent manner. *Journal of Molecular Biology*, *360*, 407–509.

[50] Thorn, D.C., Meehan, S., Sunde, M., Rekas, A., Gras, S.L., MacPhee, C.E., Dobson, C.M., Wilson, M.R., and Carver, J.A. (2005) Amyloid fibril formation by bovine milk κ-casein and its inhibition by the molecular chaperones αs- and β-casein. *Biochemistry*, *44*, 17027–17036.

[51] Hamedi, M., Herland, A., Karlsson, R.H., and Inganas, O. (2008) Electrochemical devices made from conducting nanowire networks self-assembled from amyloid fibrils and alkoxysulfonate PEDOT. *Nano Letters*, *8*, 1736–1740.

[52] Oppenheim, T., Knowles, T., Lacour, S.P., and Welland, M.E. (2010) Fabrication and characterization of protein fibril-elastomer composites. *Acta Biomaterialia*, *6*, 1337–1341.

[53] Rao, S.P., Meade, S.J., Healy, J.P., Sutton, K.H., Larsen, N.G., Staiger, M.P., and Gerrard, J.A. (2012) Amyloid fibrils as functionalizable components of nanocomposite materials. *Biotechnology Progress*, *28*(1), 248–256.

CHAPTER 25

Protein-Based Nanocomposites for Food Packaging

HÉLÈNE ANGELLIER-COUSSY, PASCALE CHALIER, EMMANUELLE GASTALDI, VALÉRIE GUILLARD, CAROLE GUILLAUME, NATHALIE GONTARD, and STÉPHANE PEYRON

25.1 INTRODUCTION

Over the last decade, the development of new sustainable, recyclable, renewable, and/or biodegradable packaging materials constitutes one of the major driving forces to increase the overall presence of eco-friendly packaging solutions across the food chain. In parallel, as a response to consumer trends toward lightly preserved, fresh, tasty, and convenient packaged food products, growing attention has also been given to new food packaging technologies, such as active* packaging [1], which have been recently introduced in Food Contact Materials regulation (1935/2004 framework regulation and related directive). Thus, modern concepts of food packaging are expected to combine renewable materials with the latest innovative technologies, such as nanotechnologies [2, 3].

Among renewable sources, proteins are agricultural products that have long and empirically been used as raw materials in a wide range of nonfood applications [4]. In contrary to most other biosourced polymers, proteins are heteropolymers constituted of different amino acids, which offer a large spectrum of chemical functionalities and, thus, various polymer network structures.

* Active packaging deliberately incorporates active components intended to release or to absorb substances into, onto, or from the packaged food or the environment surrounding the food with their conditions of use. They act on different reactions of degradation of food or as vector of compounds of interest.

Biopolymer Nanocomposites: Processing, Properties, and Applications, First Edition. Edited by Alain Dufresne, Sabu Thomas, and Laly A. Pothan.

Many proteins stemming from either animal (casein, whey proteins, gelatin, collagen, egg white, keratin, and fish myofibrillar proteins) or vegetable (soy proteins, wheat gluten, corn zein, and pea protein isolates) sources display good film-forming properties and can be used for the production of amorphous films or coatings. In addition to their large availability, full biodegradability, and non-ecotoxicity, protein-based materials have interesting barrier and release properties for food packaging applications.

A downside to their sustainable and extensive use is their intrinsic reactivity and thus lower inertia when compared with most conventional petrochemical-based plastics. They are known to be sensitive to microbial spoilage and also to water due to the hydrophilic nature of many amino acids constituting their primary structure and to the substantial amount of hydrophilic plasticizer required to impart thermo-processability and film flexibility. As a consequence, their mechanical properties and water vapor barrier properties in high moisture conditions are poor compared with synthetic films such as low-density polyethylene.

Efforts to develop eco-friendly packaging for food, such as protein-based materials, tend to be oriented toward the replacement of petrochemical plastics with materials displaying similar high inertial and barrier characteristics. However, it is important to understand that packaging properties, and especially barrier properties, need to be tailored to food requirements and shelf life [5]. As largely described in the literature, one route to significantly modulating the functional properties of packaging materials consists in developing nanocomposite structures by incorporating fillers at a nanometric scale within the polymer matrix. Fillers are nanoparticles, that is, particles with at least one dimension between 1 and 100 nm, and those properties depend on their nature (inorganic or organic), reactivity, and morphology (dimensions, aspect ratio). This strategy has begun to be explored in the field of protein-based materials for gelatin less than 10 years ago (first publications in 2002) with the introduction of layered silicates in the formulation. The development of protein-based nanocomposites has been devoted to almost all proteins previously mentioned with the intention of mainly modulating mechanical and barrier properties. The use of protein-based nanocomposites has also been investigated for the controlled release of active agents, for example, for antimicrobial applications [2].

This chapter focuses on the ability of protein-based materials for preserving food quality and safety. The first part of the chapter is devoted to presenting the structure of proteins and the possible processing routes to preparing protein-based nanocomposites. Then, the two next sections of the chapter focus on how the incorporation of nanofillers can modulate mechanical and barrier properties of protein-based materials. After that, some examples of the use of protein-based nanocomposites as controlled delivery systems for active packaging are given. Finally, the chapter ends with food safety aspects by presenting the possibility of considering protein-based nanocomposites as food contact materials.

25.2 PREPARATION OF PROTEIN-BASED NANOCOMPOSITE MATERIALS

25.2.1 Structure of Proteins

Proteins are heteropolymers constituted of 20 different amino acids and characterized by different levels of structure that permit various interactions and bindings, differing in position, type, and energy. The functionalities of protein-based materials depend mainly on the composition of the amino acids constituting their primary sequence and more precisely on the chemical reactivity of the amino acids lateral groups, that is, polar or hydrophobic, ionized or not, acid or basic, aromatic or aliphatic (Table 25.1). According to the chemical specificity of the amino acids constituting proteins, different types of interactions could be established leading to intra- and inter-polypeptidic chain linkages. Among them, the most representative are noncovalent bonds since they are considered as critical in stabilizing the protein three-dimensional structure that can be described at four levels, that is, from primary structure to secondary, tertiary, and quaternary structure (Fig. 25.1).The primary structure corresponds to the chemical composition, that is, the sequence of a chain of amino acids. At this level, the amino acids that make up the protein are linked together by covalent bonds. The secondary structure of proteins corresponds to the spatial arrangement of the polypeptide chains, which results from the establishment of periodic hydrogen bonding involving carbonyl and amine groups (CO–NH) and leads to either α-helices or β-sheets structures. The tertiary structure of proteins results from many interactions (hydrogen, hydrophobic, electrostatic, covalent, and so on) between amino acids belonging to the same polypeptide chain but being not neighbors. Finally, the quaternary structure results from generally noncovalent interactions between polypeptide chains. This latter structure is on the base of the functional properties of the final protein-based materials.

Amino acids are generally classified in groups depending on their ability to establish interactions via either hydrogen bonds (involving nonionized polar amino acids, 8–40 kJ/mol), ionic interactions (implying ionized polar amino acids, 42–84 kJ/mol), hydrophobic interactions (between apolar amino acids, 4–12 kJ/mol), or covalent linkages (amino acids involved in disulfide bonds, 330–380 kJ/mol) (Table 25.1). Although the noncovalent bonds are weak chemical forces ranging from 4 to 30 kJ/mol [6], they impart stability to the structures generated by their collective action. Van der Waals forces (involving both polar and apolar nonionised amino acids, 0.4–4 kJ/mol) are also implicated, but their effect on protein structure is minor. In addition to noncovalent interactions, covalent disulfide bonds formed by oxidation of two thiol groups also play an important role in the stability of the proteins containing cysteine residues.

The functional properties of the final materials will depend on the respective contribution of these different kinds of interactions that are closely related

TABLE 25.1 Biochemical Characteristics of the Main Proteins Used to Prepare Protein-Based Nanocomposites

	Amino Acid[a] Ratios				Subunits		
Proteins	Ionized Polar	Nonionized Polar	Apolar	Cysteine Involved in Disulfide Bonds	Name	Rate (%)	Molecular Weight (kDa)
Gelatin [102, 103]	29.1	27.6	43.3	0			3–200
Whey proteins [104, 105]	28.5	25	42.5	4	α-Lactalbumin	60	14
					β-Lactoglobulin	20	18
Soy proteins [106]	47	16.7	35	1.3	Glycinin	40	350
					β-Conglycinine	35	185
Wheat gluten [107]	18	41	38.3	2.7	Gliadin	45	30–80
					Glutenin	45	200–2000
Corn zein [107]	15	38	45	2	Zein	80	23–27

[a]Ionized polar :Glu, Asp, Lys, Arg, His; Nonionized polar: Gly, Asn, Gln, Ser, Thr, Tyr; Apolar : Ala, Val, Leu, Ileu, Pro, Phe, Met, Trp.

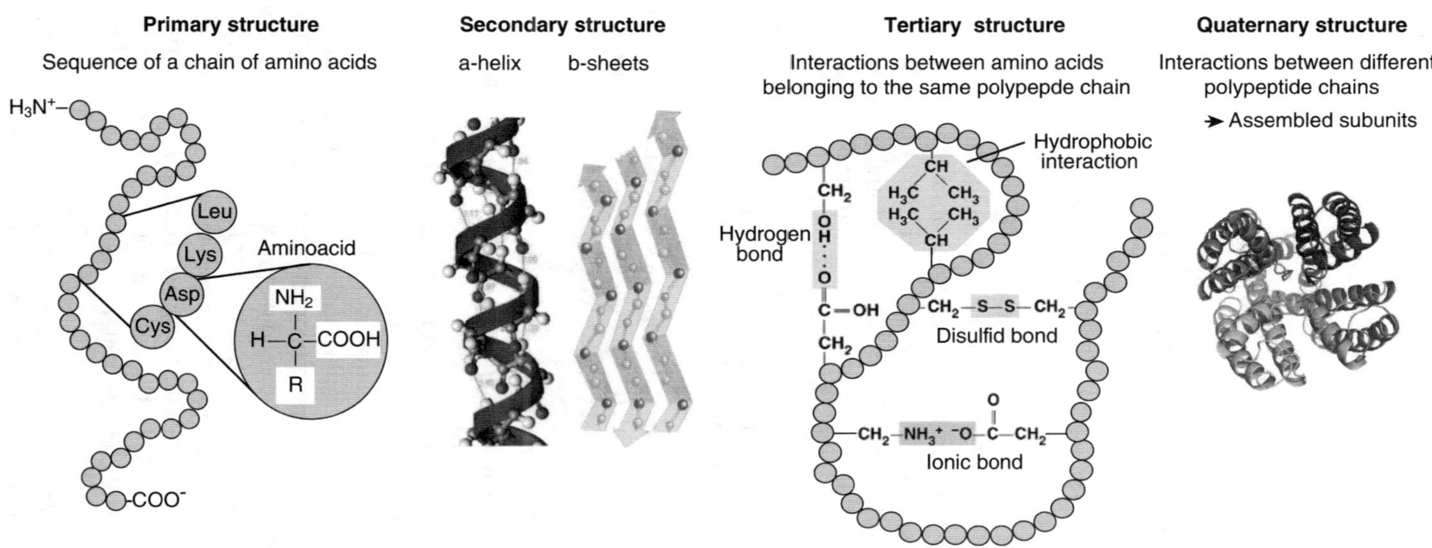

Figure 25.1 Different levels of protein structure.

to the characteristics of the protein raw material but that can also be modulated by the conditions used to process and store the protein-based materials. Furthermore, this heterogeneous structure offers many opportunities for cross-linking or chemical grafting.

25.2.2 Main Proteins Used as Matrices

Most studies devoted to the development of protein-based nanocomposites involve the use of either gelatin, whey proteins, soy protein isolates (SPI), corn zein, or wheat gluten as protein matrices.

25.2.2.1 Gelatin Gelatin is a protein resulting from partial thermal or chemical hydrolysis of collagen polypeptide chains stemming from animal skin, bones, and cartilages. For commercial production of gelatin, two processes are mainly used differing by the pH conditions; type A gelatin involves an acid extraction whereas type B gelatin is obtained by an alkaline lime followed by a solubilization at near neutral pH at 60–90°C [7]. Gelatin was one of the first food proteins to be used as raw material not only for food applications but also in the pharmaceutical and photographic film industries due to large availability, low production cost (between 1 and 5 €/kg depending on the animal origin and the targeted application, food or nonfood uses), and excellent film forming ability. Gelatin has been successfully used to form films by casting that are transparent, flexible, water-resistant, and impermeable to oxygen [8]. Although gelatin can be used as a valuable biopolymer in tissue engineering, its poor mechanical properties, especially in the wet state, and its high water sensitivity, revealed by a high water swelling, restrict its application as a structural biomaterial such as osteosynthetic devices.

25.2.2.2 Whey Proteins Together with caseins, whey proteins are the second category of proteins contained in milk. Milk has a protein content of about 33 g/L,whey proteins representing 20% of the total. Whey proteins are by-products of the cheese or casein manufacturing that remain in milk serum after pH/rennet-induced casein coagulation. Increased production of whey because of expanding cheese manufacture has led to identification of new uses for whey proteins. This also implies that whey proteins are available at lower price than caseins. Film-forming properties of milk proteins have been indirectly demonstrated by films formed on processing equipment surfaces and at air–water interfaces during heating of nonfat milk [8]. Whey proteins have been used to produce transparent, flexible, colorless, and odorless edible films and coatings displaying excellent oxygen aroma and oil barrier properties at low relative humidity (RH), but poor moisture barrier properties as related to their high hydrophilicity and water solubility.

25.2.2.3 Soy Protein Isolates Soybean meal, also known as soybean oil cake, is a flour obtained by grinding the solid residue of soybean oil production. Protein meal can further be concentrated for the production of commercially available soy protein concentrates and isolates. Because SPI are a highly refined class of soybean proteins that contains more than 90% protein and 18 different amino acids, they have become the favorite choice of researchers. The film-forming properties of soy proteins have been traditionally used in Asia to obtain edible films by collecting the lipo-proteic skin formed after boiling soya milk [8]. SPI-based films alone possess poor flexibility. Therefore, plasticizers are usually used to overcome their brittleness, leading unavoidably to a significant decrease in their tensile strength and barrier properties [9].

25.2.2.4 Corn Zein Zein is the major corn protein fraction (prolamin fraction) commercially produced from corn gluten meal, a by-product of corn wet-milling processing. Corn gluten meal contains about 55–60% proteins and is available at relatively low price; 600–700 dollar/metric ton depending on the grade (food or not). Zein-based materials can be formed by different processes including casting and thermoplastic processing. Due to its low content in polar amino acids and high level in apolar amino acids, zein-based films are not water soluble and exhibit remarkable grease-resistance properties [8]. However, corn zein-based materials, although less water sensitive than other biopolymers still show high water vapor permeability and low tensile strength when compared with commodity polymers.

25.2.2.5 Wheat Gluten Wheat gluten has been widely investigated as a protein source because it is annually renewable and readily available at a reasonable cost (1.3 €/kg). Wheat gluten proteins consist of monomeric gliadins with a molecular weight ranging from 15,000 to 85,000 and a mixture of glutenin polymers with a molecular weight between about 80,000 to several million (Table 25.1). Wheat gluten proteins can undergo disulfide interchange upon heating, which leads to the formation of a covalent three-dimensional macromolecular network. With respect to its unique viscoelastic and film-forming properties, wheat gluten is an interesting raw material that can be used as a food packaging material. Wheat gluten based-films present an attractive combination of strength and flexibility [10, 11], a high gas (oxygen and carbon dioxide) permeability in dry condition and a significant gas perm selectivity at high RH [12–14], and good grease [14] and aroma barrier properties [15], which are key functional properties for food quality preservation. They also are translucent [10] and can be heat sealed [16]. A downside to the use of wheat gluten-based materials is their per se reactivity and thus lower inertia than most conventional petrochemical-based plastics. As mentioned above for other protein-based films, wheat gluten-based films are sensitive toward water due to the hydrophilic nature of many amino acids constituting the protein chains and to the need for adding hydrophilic plasticizer.

25.2.3 Main Nanoparticles Used as Fillers in Protein-Based Matrices

The most common fillers used as reinforcing agents in protein-based matrices are by far layered silicates, and, in a less extend carbon nanotubes, polysaccharide (cellulose, chitin, starch) nanocrystals, silica, and TiO_2 nanoparticles. These different fillers can be classified into different categories depending on their nature, inorganic or organic, but also on their particle size and shape. Indeed, we can typically distinguish three major categories based on the dimensional characteristics of the fillers.

25.2.3.1 Isodimensional Particles They have the same size in all directions and their aspect ratio is usually close to unity. Examples include spherical silica, TiO_2 nanoparticles, carbon black, and fullerenes. Up to now, only silica and TiO_2 nanoparticles have been used as fillers in protein-based matrices.

Silica In the field of nanocomposite materials, silica (SiO_2) is mainly used as colloidal suspensions of amorphous spherical nanoparticles whose properties and structure depend directly on their synthesis process. Three different structures can be distinguished: (1) elementary nanoparticles with a diameter ranging from 10 to 40 nm, (2) indivisible aggregates constituted of elementary nanoparticles with a size ranging from 50 to 500 nm, and (3) agglomerates of important size (0.2–30 μm) corresponding to the association of aggregates that can be fractionated. The aggregate size is tailored by the pH or the ionic force of the suspension. The surface of silica nanoparticles, which are naturally hydrophilic, can easily be chemically modified to bring new functionalities or improve their compatibility with a hydrophobic polymer matrix. For the moment, colloidal suspensions of silica nanoparticles have only been tested as fillers in a soy protein-based matrix [17].

TiO_2 Nanoparticles Titanium dioxide, TiO_2, occurs in nature in the form of well-known minerals that are rutile, anatase, and brookite. It is the most widely used white pigment because of its unique physical and chemical properties such as high refractive index, brightness, opacity, and excellent optical transmittance in the visible. In the food packaging area, the main purpose of TiO_2 incorporation is the reduction of ultraviolet (UV) damage of foodstuff in transparent packaging. Its surface irradiation is used as scratchproof coating and its antimicrobial properties are also extensively investigated. Regarding their photocatalytic effects, the incorporation of TiO_2 nanoparticles was recently prospected as catalyst for the biodegradation of organic packaging materials. TiO_2 nanoparticles have already been used as fillers in whey protein-based matrices [18, 19].

25.2.3.2 Elongated Particles They consist of fibrils with diameter ranging between 1 and 100 nm and length of several hundred of nanometers. They include carbon nanotubes, which are inorganic fillers, as well as cellulose and chitin nanofibrils, which are organic fillers.

Carbon Nanotubes Carbon nanotubes are molecular-scale tubes formed by rolled-up graphene sheets. We can distinguish two kinds of nanotubes depending on their structure: (1) single-walled carbon nanotubes (SWCNT), which consist of a single sheet of graphene rolled seamlessly to form a cylinder with diameter of the order of 1 nm and length of up to several centimeters, and (2) multiwalled carbon nanotubes (MWCNT), which consist of an array of such cylinders formed concentrically and separated by 0.36 nm, similar to the basal plane separation in graphite. MWCNTs can have outer diameters of about 30 nm, and lengths of 1–100 μm [20]. Up to now, SWCNTs have not been used as reinforcing fillers in protein-based matrices owing to the difficulty to be dispersed within a polymer matrix. Indeed, such carbon nanotubes tend to aggregate into large bundles due to their high flexibility and surface energy [20]. Even if MWCNTs are more rigid, their effective utilization as fillers strongly depends on their ability to be homogeneously distributed and dispersed within the polymer matrix without destroying their integrity. To improve their dispersibility, their surface can be chemically functionalized to optimize the interfacial interaction between carbon nanotubes and the polymer matrix [20]. Up to now, the reinforcing effect of MWCNTs has only been evaluated in gelatin [21] and SPI [22] matrices.

Cellulose and Chitin Nanocrystals Polysaccharides are among the most promising organic and renewable sources for the production of nanoparticles with various different morphologies and geometrical characteristics providing a wide range of nanoparticle properties. Aqueous suspensions of cellulose or chitin nanocrystals can be prepared by acid hydrolysis of native substrates. The object of this treatment is to dissolve away regions of low lateral order so that the water-insoluble, highly crystalline residue may be converted into a stable suspension. The resulting nanocrystals occur as rod-like particles or whiskers, whose dimensions depend on the nature of the substrate. They are typically a few hundred nanometers long and in the order of a few nanometers in diameter [23]. The aspect ratio of these whiskers is defined as the ratio of the length to the diameter. Their surface can be modified using well-established carbohydrate chemistry to offer new functionalities.

25.2.3.3 Layered Particles This third category comprises fillers characterized by one dimension, often the thickness, ranging from several angstroms to several nanometers. It includes layered silicates and starch nanocrystals.

Layered Silicates Smectites, and in particular montmorillonites (MMT), are the layered silicates most largely used to modulate the properties of protein-based materials. They belong to the phyllosilicate family and display a perfect crystalline structure consisting of two-dimensional layers where a central octahedral sheet of alumina or magnesia is linked with two external silica tetrahedrons [24]. The layer thickness is around 1 nm and the lateral dimensions of these layers vary from 30 nm to several micrometers [24]. The limiting aspect of the use of such nanofillers is the difficulty to correctly disperse them within a polymeric matrix due to their tendency to form agglomerated tactoïds.

Indeed, the achievement of a well-exfoliated nanocomposite structure, rather than a conventional microcomposite structure or even an intercalated nanocomposite structure, is required to significantly modulate mechanical and barrier properties of protein-based materials. One important feature of clays is that the space between layers contains hydrated cations such as Na^+ or K^+ that can undergo exchange reaction with organic cations such as alkylammoniums. Thus, organically modified layered silicates can be purposely prepared to impart functionality and compatibility with polymers.

Starch Nanocrystals Similarly to cellulose and chitin whiskers, starch nanocrystals are obtained from mild acid hydrolysis of native starch granules and correspond to the highly crystalline water-insoluble lamellae of the native starch granules. Up to now, only pea starch nanocrystals have been tested as reinforcing agents in a SPI-based matrix [25]. Starch nanocrystals are characterized by a length ranging from 60 to 150 nm, a width ranging from 15 to 30 nm, and an estimated thickness of about 5–7 nm [26]. Starch nanocrystals, as cellulose and chitin whiskers, possess a highly reactive surface covered with hydroxyl groups, which provides the possibility of chemical modification in order to create new specific functions and thus expand their applications.

25.2.4 Processing Protein-Based Nanocomposite Materials

Proteins are required to unfold and realign before a new three-dimensional network can be formed, and then stabilized by new inter- and intramolecular interactions. The formation of a macromolecular network from proteins requires three steps: first the rupture of low-energy intermolecular bonds that stabilize polymer chains in the native state, secondly, the arrangement and the orientation of polymer chains (shaping), and finally the formation of a three-dimensional network stabilized by new interactions and bonds after the agent that breaks intermolecular bonds is removed [8]. Due to the high glass transition temperature of proteins (above their temperature of degradation), their formulation often requires the addition of plasticizers to allow their thermoprocessability and impart material flexibility. Besides water molecules, which could be considered as natural plasticizers of many proteins, the most usual plasticizers are polyols (glycerol, sorbitol), mono, di-, and oligosaccharides, and fatty acids. Whatever the type of nanoparticle used to prepare protein-based nanocomposites, two processing methods can be considered (Fig. 25.2).

25.2.4.1 Wet Process, That is, Dispersion of Nanoparticles from Solution The wet process, commonly called casting, is based on the dispersion or solubilization of proteins in a solvent (water, ethanol, and occasionally acetone). This process is suitable for all proteins. Depending on the amino acid sequence, dispersing proteins in solution may also involve the addition of disruptive agents, pH adjustment by the addition of acids or bases, and/or ionic strength control by electrolyte addition [8]. In aqueous medium, proteins

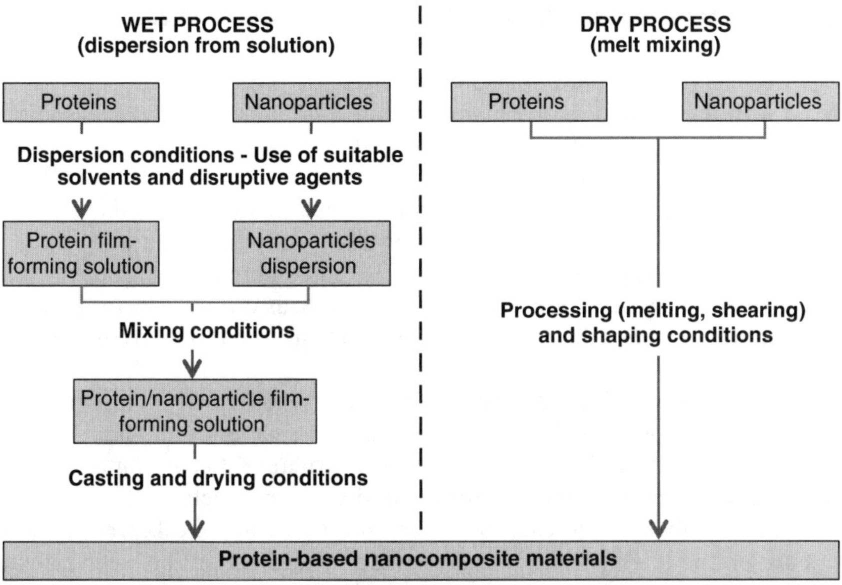

Figure 25.2 Schematic representation of the two main and derivated technological processes used to prepare protein-based nanocomposite materials.

generally adopt a conformation in which the polar amino acids are located at the interface with the solvent, whereas the hydrophobic parts of the polypeptide chains tend to avoid solvent contact by moving within the structure.

In case of layered silicates, the challenge is to intercalate protein chains within the galleries [24]. For this purpose, clay nanoparticles are first exfoliated into single layers using a solvent in which the proteins are easy to disperse. Proteins then adsorb onto the delaminated surfaces of clay and when the solvent is evaporated, the sheets reassemble, sandwiching proteins. To facilitate introduction of proteins into clay galleries, it is possible to modify the pH value of the film-forming solution to positively increase the protein net charge in favor of electrostatic interactions established with the negatively charged clay surfaces.

So far, the dispersion of nanoparticles in solution was largely preferred for making protein-based nanocomposites for several reasons, including the non-necessity to use plasticizers, the easiness of processing, and the possibility to achieve well-dispersed nanocomposite structures. In some cases, solution intercalation followed by solvent evaporation (often by freeze-drying) is only considered as a preliminary step leading to good nanoparticle dispersion, and the resulting dried powders being then shaped using a dry process (Fig. 25.2). It is generally the case for polysaccharide (cellulose, chitin, starch) nanofillers, which are available in the form of aqueous colloidal suspensions [25, 27, 28].

25.2.4.2 Dry Process, That is, Dispersion at the Melt State Dry processes, including extrusion and thermomolding, are thermomechanical processes based on the thermoplastic properties of proteins. Nanoparticles are mixed with the protein matrix in the molten state under heating and shearing (Fig. 25.2). Thus, such processes require the formation of a homogeneous protein melt, implying processing temperatures above the protein softening point. For this purpose, chemical additives, such as sodium sulfite, sodium dodecyl sulfate, and urea, which are known to disrupt covalent and noncovalent bonds, are often used as processing aids in combination with plasticizers. Although most proteins are suitable for the extrusion process, to date, only wheat gluten and SPI have led to nanocomposite materials by using a dry process.

In case of layered silicates, when their surface is sufficiently compatible with the protein, polypeptide chains can crawl into the galleries and lead to either an intercalated or exfoliated nanocomposite structure. According to Vaia and Giannelis [29], an interplay of entropic and energetic factors determines the outcome of polymer intercalation. Exfoliation or complete layer separation depends on the establishment of very favorable interactions between polymer and clay surfaces to overcome the entropy loss associated with polymer confinement and penetration between the clay layers. The kinetics of intercalation of the polymer chains between the clay layers also depend on the polymer mobility that is related with the temperature difference between the processing temperature and the polymer glass temperature [30].

Although still of limited use to prepare protein-based nanocomposites, the polymer melt intercalation is considered as a more efficient and environmentally benign alternative than the polymer intercalation from solution since it requires no solvent. Moreover, dry processes are all available at the industrial scale and commonly used forindustrial plastic processing.

25.3 MECHANICAL PROPERTIES OF PROTEIN-BASED NANOCOMPOSITE MATERIALS

As mentioned previoulsy, all protein-based films are sensitive to water due to the hydrophilic nature of many amino acids constituting the protein chains and to the substantial amount of hydrophilic plasticizers required to impart thermo-processability and film flexibility. As a consequence, in high moisture conditions, their mechanical properties are poor as compared to synthetic films such as low-density polyethylene. Maintaining physical film integrity is essential to guaranteeing the barrier properties of food packaging and the handling of the packages. Mechanical properties are thus essential requirements for food protection and commercialization.

Layered silicates are the most largely used reinforcing nanoparticles in protein-based matrices, with first papers published in 2002 for gelatin. Most studies are devoted to the introduction of unmodified clays, especially montmorillonite (MMT) [10, 31–40]. Thanks to the hydrophilic nature of both proteins and unmodified clays, this type of nanofiller was expected to give an enhanced nanodipersion state, that is, intercalated or partially exfoliated nanocomposite structures, resulting in the improvement of mechanical properties.

Zheng et al. [31] were the first to report the reinforcing effect of unmodified MMT in a gelatin-based matrix, which was revealed by an improvement in Young's modulus and the tensile strength (Table 25.2). The significant enhancement of both Young's modulus and the tensile strength of gelatin-based materials by a factor of about 2 without sacrificing toughness was further confirmed by many authors for low clay particles contents (about 5 wt%) and was ascribed to the nanodispersion of clays within the matrix [32, 33, 41] (Table 25.2). The reinforcing effect of layered silicates was also reported for corn zein [34], and overall for soy proteins [42–44] and wheat gluten [10, 35] -based matrices (Table 25.2). All studies demonstrated that the introduction of unmodified MMT nanoparticles in such matrices resulted in significant improvements in the tensile strength and Young's modulus with a simultaneous decrease in the elongation at break due to increased interfacial interactions (Table 25.2). To our knowledge, only one article has been recently published on the incorporation of layered silicates to improve the mechanical properties of corn zein-based films [34]. Authors reported that the improvement of the mechanical properties of zein-based films (Table 25.2), which was ascribed to the achievement of partially exfoliated structures, did not evolve linearly with the MMT content, indicating that there was a critical MMT percentage (i.e., 5 wt%) where the dispersion during processing was better. An optimal filler content of around 5 wt% was also required to achieve the greatest improvement of mechanical properties of wheat gluten-based nanocomposites, probably due to the possibility of layered silicates to form a three-dimensional network [35]. In the case of soy proteins-based nanocomposites, whatever the MMT type and loading (from 5 to 15 wt%), highly exfoliated and intercalated structures were successfully prepared either via a solution intercalation process in a neutral aqueous media without any special aid [42–44] or via melt extrusion [39, 40]. According to Chen and Zhang [43], the high affinity between soy proteins and the MMT layers was due to two kinds of interactions: hydrogen bonding between the –NH and Si–O groups and electrostatic interactions between the positive charge-rich domains of soy protein and the negatively charged MMT layers as evidenced by zeta-potential measurements and Fourier transform infrared (FT-IR) analysis. However, structures and properties were shown to strongly depend on the filler content used. An increase in the filler content up to 12 wt% led to a significant increase in both Young's modulus and the tensile strength (Table 25.2). It is accompanied by a drop in the deformation at break due to the restricted segmental motion of soy proteins and the establishment of favorable interactions between the clay sheets and the protein chains. For filler contents as high as 12–16 wt%, all the mechanical properties deteriorated, due to some agglomeration of

TABLE 25.2 Mechanical Properties of Protein/Clay-Based Nanocomposites: Young's Modulus (E), Tensile Strength (TS), and Deformation at Break (ε)

Matrix	Sample	E (MPa)	TS (MPa)	ε (%)	Ref.
Gelatin	Bovine skin gelatin (type B) (25 wt% glycerol)	900	49.3	–	[29]
	–5 wt% Na+ MMT	1600	78.9	–	
	–17 wt% Na+ MMT	2000	89.1	–	
	Gelatin (type 4) (no plasticizer)	3300	88.9	9.6	[32][a]
	–5 wt% Cloisite Na+	5900	97.4	3	
	–10 wt% Cloisite Na+	8300	110.8	0.9	
	Bloom fish gelatin (20 wt% glycerol)	–	30.3 ± 2.4	48 ± 4	[33][a]
	–5 wt% Cloisite Na+	–	40.9 ± 5.3	38 ± 3	
	Porcine skin gelatin (type A)	2100 ± 200	–	3.3	[41]
	–5 wt% sepiolite	3300 ± 100	–	2.5	
Whey proteins	Whey (33 wt% glycerol)	172 ± 14	3.4 ± 0.6	51 ± 13	[37][a]
	–5 wt% Cloisite Na+	109 ± 18	3.0 ± 0.3	42 ± 8	
	–5 wt% Cloisite C30B	162 ± 38	3.3 ± 0.1	52 ± 5	
	–5 wt% Cloisite 20A	115 ± 13	1.5 ± 0.3	29 ± 9	
Soy protein isolates (SPI)	SPI (glycerol)	–	7.4	68	[42]
	–10 wt% Na^+ MMT	–	5.7	16	
	–10 wt% OMMT	–	9.4	54	
	–10 wt% bentonite	–	9.7	58	
	SPI (30 wt% glycerol)	180	8.7	88	[43][b]
	–4 wt% Na^+ MMT	260	10.3	63	
	–12 wt% Na^+ MMT	420	14	15	
	–20 wt% Na^+ MMT	587	14.5	6	
	SPI (30 wt% glycerol)	180	6.8	108	[44][b]
	–4 wt% rectorite	360	9	57	
	–12 wt% rectorite	560	12.8	12	
	–20 wt% rectorite	580	9.4	7	
	SPI (15 wt% glycerol)	–	2.3 ± 0.5	11.8 ± 0.4	[49][c]

	–5 wt% Na+ MMT	–	6.3 ± 0.9	65 ± 5	
	–10 wt% Na+ MMT	–	12.6 ± 0.6	24 ± 5	
	–15 wt% Na+ MMT	–	15.6 ± 1.7	18 ± 3	
	SPI (15 wt% glycerol)	–	2.3 ± 0.5	11.8 ± 0.4	[40][c]
	–5 wt% Cloisite 30B	–	15.1 ± 0.9	82 ± 3	
	–10 wt% Cloisite 30B	–	16.2 ± 0.8	104 ± 5	
	–15 wt% Cloisite 30B	–	18.6 ± 0.3	55 ± 3	
Corn zein	Zein	×1	×1	×1	[34][d,g]
	–3 wt% OMMT	×1.2	×1.5	×0.4	
	–5 wt% OMMT	×1.6	×2.4	×0.3	
	–10 wt% OMMT	×2.9	×1.5	×0.1	
Wheat gluten	WG (10 wt% glycerol)	3.7	1.9	58	[35][e]
	–2.5 wt% Na^+ MMT	5.6	2.4	55	
	–5 wt% Na^+ MMT	10.6	4.7	16	
	–7.5 wt% Na^+ MMT	11.4	3.6	15	
	WG (T = 80°C; 37.2 wt% glycerol)	1.1	0.31	33	[10][f]
	–2 wt% Na+ MMT	1.7	0.32	30	
	–5 wt% Na+ MMT	6.4	0.66	23	
	WG (12 wt% water + 15 wt% glycerol)	65.4	5.8	106	[47][a]
	–3 wt% C30B	91.8	9	78	

Mechanical properties evaluated at
[a]50%RH;
[b]0%RH;
[c]65%RH;
[d]3%RH;
[e]70%RH;
[f]75%RH. Such details were not specified for other data.
[g]In the study of Luecha et al. [34], only relative mechanical properties are available.

MMT that occurred in the soy protein matrix as a phase separation phenomenon [43, 44].

Since mechanical properties of nanocomposite materials are directly related to their structure and, in this respect, to the state of dispersion (intercalation/exfoliation) of layered silicates [45], the reinforcing effect of layered silicates is sometimes poor, and even absent. The level of exfoliation is known to highly depend on the nature of the MMT [24, 45]. Some authors attempted to use organically modified nanoclays, but it often resulted in a negative effect on mechanical properties [34, 36–38, 46, 47]. In the case of wheat gluten reinforced with organically modified MMT, it was shown that the quaternary alky ammonium could act as a plasticizer, which might play a negative role in achieving higher strength enhancement in the nanocomposites [47].

Even if well-dispersed nanocomposite structures were obtained for gelatin, corn zein, soy proteins, and wheat gluten-based materials, it is worth noting that it was not the case for whey protein-based nanocomposites. Data from Sothornvit et al. [37, 38] and Hedenqvist et al. [36] all reported a bad dispersion of clays within the whey-based matrices, leading to the absence of a reinforcing effect of layered silicates. In Sothornvit et al. studies, different types of MMT were tested: an unmodified clay (Cloisite Na^+) and two organically modified MMT (Cloisite 30B and Cloisite 20A) with a clay loading of 5 wt% [37] or up to 20 wt% [38]. The fact that mechanical properties (tensile strength and elongation at break) of nanocomposites were either unchanged (Cloisite Na^+ and Cloisite 30B) or decreased (Cloisite 20A) as compared to the neat matrix (Table 25.2) suggested that a microcomposite rather than a nanocomposite structure was achieved [37]. In the work reported by Hedenqvist et al. [36] the MMT used were previously intercalated with poly(vinylidene pyrrolidone), which is water soluble and acts as surfactant by easily adsorbing onto the unmodified MMT surface. Authors also noted the absence of a significant increase in relaxed rubber modulus in the nanocomposites due to an incomplete exfoliation of nanoparticles, as confirmed by transmission electron microscopy (TEM) analysis.

Other fillers such as chitin or cellulose whiskers and starch nanocrystals, silica, and carbon nanotubes were tentatively used to improve mechanical properties of SPI-based materials (Table 25.3). Polysaccharide-based fillers display a similar reinforcing effect than unmodified MMT. Indeed, the introduction of increasing contents (up to 20 wt%) of either chitin [27] or cellulose whiskers [28] led to a strengthening of SPI-based materials, characterized by a significant increase in both Young's modulus and the tensile strength, together with a decrease in the elongation at break (Table 25.3). For example, for SPI/chitin whiskers nanocomposites, increasing chitin content from 0 wt% to 20 wt% led to an increase in the tensile strength and Young's modulus from 3.3 to 8.4 MPa and from 26 to 158 MPa, respectively, whereas the elongation at break of the filled composites decreased from 205% to 29%. This reinforcing effect was ascribed to the establishment of hydrogen bonding interactions between the whiskers and the SPI matrix, which restricted the motion of soy proteins.

TABLE 25.3 Mechanical Properties of Protein/(TiO_2, Silica, Carbon Nanotubes, or Polysaccharide Nanofillers)-Based Nanocomposites: Young's Modulus (E), Tensile Strength (TS) and Deformation at Break (ε)

System	Sample	E (MPa)	TS (MPa)	ε (%)	Ref.
Protein/TiO_2	Whey (33 wt% glycerol)	–	6	95	[21][a]
	–0.1 wt% TiO_2	–	10	95	
	–0.25 wt% TiO_2	–	10.2	98	
	–0.5 wt% TiO_2	–	9.2	96	
	–1 wt% TiO_2	–	8.4	75	
	–2 wt% TiO_2	–	6.9	27	
	Whey (50 wt% glycerol)	31 ± 3	1.7 ± 0.1	55 ± 1	[18][b]
	–0.5 wt% TiO_2	66 ± 3	2.4 ± 0.1	54 ± 1	
	–1 wt% TiO_2	63 ± 2	2.2 ± 3.1	40 ± 1	
	–2 wt% TiO_2	44 ± 3	1.9 ± 0.2	15 ± 1	
	–4 wt% TiO_2	39 ± 4	1.8 ± 0.1	12 ± 1	
Protein/silica	SPI (30 wt% glycerol)	174	6.8	108	[17][c]
	–4 wt% SiO_2	310	11.3	133	
	–8 wt% SiO_2	195	7.9	175	
	–12 wt% SiO_2	295	9.9	59	
	–16 wt% SiO_2	275	9.6	37	
	–20 wt% SiO_2	240	9.3	27	
	–24 wt% SiO_2	115	9.2	11	
Protein/carbon nanotubes	SPI (30 wt% glycerol)	88	6.9	171	[22][d]
	–0.25 wt% MWCNT	261	11.6	193	
	–0.5 wt% MWCNT	252	10.7	153	
	–0.75 wt% MWCNT	248	9.1	148	
	–1 wt% MWCNT	101	6.2	158	
	–2 wt% MWCNT	118	7.8	160	
	–3 wt% MWCNT	177	9.8	165	

(Continued)

TABLE 25.3 (*Continued*)

System	Sample	E (MPa)	TS (MPa)	ε (%)	Ref.
Protein/chitin or cellulose whiskers	SPI (30 wt% glycerol)	26	3.3	205	[27][d]
	–5 wt% chitin whiskers	32	3.9	134	
	–10 wt% chitin whiskers	34	5	86	
	–15 wt% chitin whiskers	45	5.7	82	
	–20 wt% chitin whiskers	158	8.4	29	
	–30 wt% chitin whiskers	106	6.3	25	
	SPI (30 wt% glycerol)	44	5.8	188	[28][d]
	–5 wt% cellulose whiskers	51	6.3	133	
	–10 wt% cellulose whiskers	62	6.9	115	
	–15 wt% cellulose whiskers	74	7.6	95	
	–20 wt% cellulose whiskers	90	8.1	61	
	–30 wt% cellulose whiskers	133	8.1	36	
Protein/starch nanocrystals	SPI (30 wt% glycerol)	108	6.8	171	[25][c]
	–1 wt% starch nanocrystals	259	9.5	60	
	–2 wt% starch nanocrystals	313	10.2	51	
	–4 wt% starch nanocrystals	270	8.8	34	
	–8 wt% starch nanocrystals	248	8.5	68	
	–16 wt% starch nanocrystals	205	6.1	12	

Mechanical properties evaluated at
[a]56%RH;
[b]50%RH;
[c]35%RH and
[d]43%RH.

Composites containing more than 20 wt% chitin or cellulose whiskers exhibited a decrease in all mechanical properties, due to a phase separation phenomenon, as supported by scanning electron microscopy (SEM) pictures [27, 28]. Starch nanocrystals displayed a similar reinforcing effect, but with an optimal filler content around 2 wt% [25]. Indeed, nanocomposites containing 2 wt% of starch nanocrystals showed the highest strength and rigidity, which was enhanced by approximately 50% and 200%, respectively, in contrast to the neat matrix (Table 25.3), due to hydrogen interactions between both components. With an increase in filler content, the increasing number and size of starch nanocrystals domains lowered the effective active filler surface for interacting with protein chains and destroyed the ordered structure in the SPI matrix [25]. This caused a gradual decrease of the tensile strength and Young's modulus.

The reinforcing effect of silica was characterized by an increase in all mechanical properties with increasing filler content, with maximum values of rigidity and strength for 4 wt% of silica and a maximum value of elongation at break for 8 wt% of silica [17] (Table 25.3). The reinforcing effect of silica was ascribed to the competition between three kinds of interactions: original interactions in the SPI matrix, interfacial interactions between silica and the SPI matrix, and the self-aggregated affinity among silica nanoparticles [17]. The improvement of mechanical properties was mainly attributed to homogeneous dispersion of nanoclusters as well as to strong interfacial interactions. In the case of high filler contents, silica nanoparticles self-aggregated as great domains, leading to a microphase separation.

The particularity of the reinforcement with carbon nanotubes (MWCNT) is the use of very low filler contents as compared to other fillers (10 times lower). The incorporation of very low MWCNT contents (0.25 wt%) in a SPI matrix led to a simultaneous reinforcing and toughening with maximum strength, Young's modulus, and elongation at break values (Table 25.3) [22]. Thereafter, the tensile strength, strain at break, and Young's modulus decreased with increasing MWCNT content up to 1 wt%. Then, with a continuous increase in MWCNT content, all mechanical properties demonstrated increasing tendency once more. In case of the SPI/MWCNT nanocomposites, the SPI chains can wrap carbon nanotubes or penetrate into the internal channels of MWCNT if the internal diameter is large enough [22]. Although a high MWCNT loading did not favor enhancement of mechanical properties, a number of aggregated MWCNT bundles can also reinforce and toughen the material to some extent. On addition, the restriction of SPI motion caused by relatively rigid MWCNTs and their aggregates can be associated with three kinds of structures: the wrapping of SPI chains on MWCNTs, the spatial hindrance of MWCNT aggregates, and the penetration of SPI chains into MWCNT channels.

Finally, works of Zhou et al. [28] and Li et al. [29] showed that the incorporation of low amounts (<0.5 wt%) of TiO_2 nanoparticles in whey-based materials led to a significant increase (factor of about 2) of Young's modulus and the tensile strength, without affecting too much the strain at break, as already reported for silica reinforcement (Table 25.3). The addition of more than 1 wt% of TiO_2 nanoparticles induced a drastic drop in all mechanical properties, resulting from an inhomogeneous distribution of TiO_2 in the whey-based matrix, the increase in size, and crystallization degree of TiO_2 agglomerates and the discontinuity of the film network microstructure, mainly due to the low affinity between the hydrophobic TiO_2 particles and the hydrophilic matrix [18, 19].

To conclude, the affinity between proteins and fillers appears to be a key parameter governing the state of dispersion of the fillers. For example, the use of unmodified MMT, which are naturally hydrophilic and, thus, compatible with proteins, seems to be more appropriate than organically modified MMT to improve mechanical properties of protein-based materials. Furthermore, for many systems, reinforcement seems to be more effective at low filler loadings, which suggests that specific filler–protein interactions and filler dispersion play a significant role in obtaining structurally highly efficient nanocomposites. In the case of layered silicates, the optimal amount seems to be around 5 wt% [34, 35, 41], whereas it seems to be about 2 wt% for starch nanocrystals [25], between 4 wt% and 8 wt% for silica [17], and 0.5 wt% for TiO_2 nanoparticles [18, 19].

25.4 BARRIER PROPERTIES OF PROTEIN-BASED NANOCOMPOSITE MATERIALS

25.4.1 Defining Barrier Properties of Protein-Based Materials from Food Requirements

In addition to cost, environmental, and marketing aspects, as well as protection against mechanical damage, the main role of packaging is to preserve food quality and safety. Shelf life of food is related to degradation reactions occurring at different levels: physical–chemical and microbial reactions for nonliving products but also physiological reactions for living products. The quality is first of all dependent of the initial quality of raw materials. Then degradation reactions will occur during the processing of the food and of course during its storage and distribution. During these last steps, cold storage (chill chain) is known to slow down food degradation reactions, but the control of mass transfers between the food, the packaging, and the atmosphere is also crucial [48].

Food packaging has to control the transfer of the main gases (oxygen, carbon dioxide, ethylene, etc.) and vapors (moisture, aroma compounds, etc.) involved in food degradation. Moisture barrier property is an essential basic requirement when packing many foods, whether dry or moist, to preserve the texture (crispiness, softness, firmness, etc.) and control microbial development of spoilage and pathogen species. Packaging plays also a crucial role for protecting food from oxygen, which is the key element of many degradation reactions occurring in food, that is, oxidations. Oxygen acts as the main factor of organoleptic and nutritional quality degradation of food through the oxidation of vitamins, aroma compounds, pigments, or lipidic compounds. Oxygen is also involved in microbial development and respiration rate of fresh living products such as fresh fruits and vegetables or fermented cheeses. Carbon dioxide needs to be preserved for many carbonated drinks. This gas is involved in the inhibition of respiration rate of living produces (but may be detrimental to some CO_2-sensitive commodities) and is also used as bacteriostatic agent. Aroma compounds barrier properties are also key elements for preserving flavor and taste of many foods such as coffee. Other mass transfer properties such as ethylene (for ripening of fresh fruits and vegetables) or ammoniac (for maturation of cheeses) transfer properties also need to be taken into account.

As regard food requirements, we can distinguish nonrespiring and respiring products. They require completely different O_2/CO_2 and water vapor barrier properties (Fig. 25.3). For example, a whole mushroom, which is a respiring product with a very short shelf life, needs O_2 and CO_2 permeabilities ranging from respectively $(1000–5000) \times 10^{-18}$ mol/(m.s.Pa) to $(5000–10{,}000) \times 10^{-18}$ mol/(m·s·Pa).* On the contrary, pasteurized orange juice, which is an O_2-sensitive nonrespiring product with a long shelf life, requires an O_2 permeability lower than 1×10^{-17} mol/(m·s·Pa) for ascorbic acid being preserved, at least, at half its initial concentration after 6 months of storage [49, 50]. Thus, it is obvious that protein-based films, because of their relatively high O_2 and CO_2 permeabilities, are definitively dedicated to the packing of respiring products such as fresh fruits and vegetables. Moreover, their high water sensitivity and biodegradability make them unsuitable for the packing of intermediate and long shelf-life products and for water-sensitive products. For instance, crispy dry biscuits require a water vapor permeability lower than that of low-density polyethylene (0.05×10^{-12} mol/(m·s·Pa) [51]) for preserving their texture during at least 4–6 months of storage, whereas the water vapor permeability of wheat gluten-based films is 100-fold greater, around 5×10^{-12} mol/(m·s·Pa) [52].

The particular interest of protein-based films compared with usual plastic films is thus their gas permeability and selectivity (CO_2 versus O_2), which are sensitive to temperature but more particularly to RH [13]. For instance, wheat gluten-based films exhibit a large range of selectivity values (3 to 28) as a function of moisture content, contrary to the selectivity of most synthetic films, which usually varies between 4 and 6 [13]. The high selectivity value of wheat gluten-based films (28°C at 24°C and 100% RH) and, moreover, a high ethylene permeability [53] could be very interesting for the preservation of fresh or minimally processed fruits and vegetables under modified atmosphere (Fig. 25.4).

* O_2 and CO_2 permeabilities were predicted by the web application http://www.tailorpack.org for a mass of 250 g of mushrooms packed in a pouch of 50 microns of thickness with a total volume of 1.0 L and 676 cm^2 of total surface area. Respiratory parameters and sugar catabolism of mushroom (*Agaricus bisporus* Lange) are from Varoquaux et al. [90].

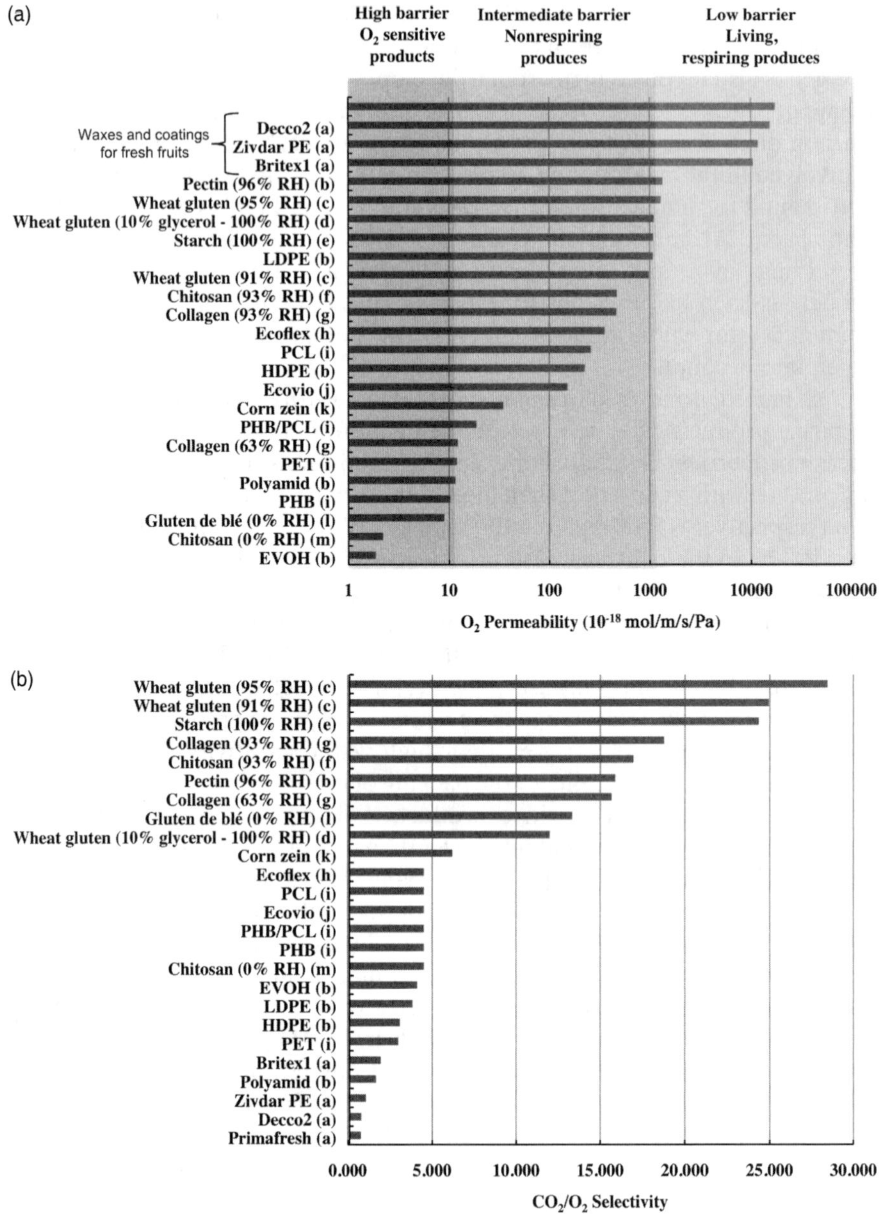

Figure 25.3 O_2 permeability (amol = 10^{-18} mol) (A) and CO_2/O_2 selectivity (B) for a selection of commercial bioplastics, common synthetic materials, and polysacharide- and protein-based materials: suitability for food packaging applications (adapted from (a) Reference [91]; (b) Reference [92]; (c) Reference [12]; (d) Reference [35]; (e) Reference [93]; (f) Reference [94]; (g) Reference [95]; (h) Reference [96]; (i) Reference [97]; (j) Reference [98]; (k) Reference [99]; (l) Reference [100]; (m) Reference [101]).

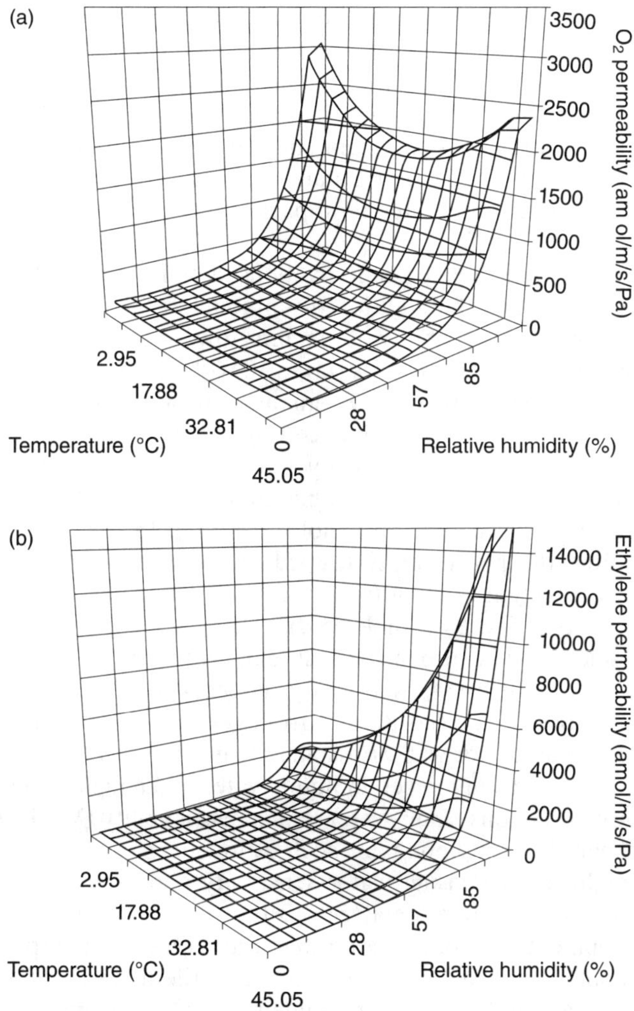

Figure 25.4 Evolution of O_2 permeability (a) and ethylene (b) permeability of a wheat gluten film as a function of temperature and relative humidity (adapted from References [13] and [53]).

25.4.2 Modulating Barrier Properties of Protein-Based Materials by Nanofillers Incorporation

As mentioned just above, the particular interest of protein-based films as food packaging materials is their high gas permeability and selectivity (CO_2 versus O_2) especially at high humidity and temperature. These characteristics are very interesting for the packing of respiring products such as fresh fruits and vegetables. However, their high water sensitivity requires to associate the protein layer to a mechanical and water barrier support or to decrease the water sensitivity of the protein matrix itself. This could be achieved by using thermal treatment, cross-linking treatment or addition in the neat matrix of impermeable nanoparticles inducing a more tortuous pathway for the permeant. The addition of nanoparticles is the main strategy used to modulate gas barrier properties of protein-based materials, and especially to improve them in order to extend the number of targeted food packaging applications to O_2-sensitive products.

Layered silicates have been tentatively used as filler to modulate O_2 and CO_2 barrier properties. The addition of unmodified MMT into wheat gluten had no effect on the permeability of the films toward gases (O_2 and CO_2) over the range of studied MMT contents (up to 10 wt%) (Table 25.4) [35]. The fact the permeability toward O_2 and CO_2 remained constant meant that no change in diffusivity or in solubility was induced by the presence of nanoclays [35]. Authors concluded that the tortuosity induced by the dispersion of MMT layers within the wheat gluten matrix had no effect on gas barrier properties, probably due to the small size of molecules [35]. In case of gelatin-based nanocomposites, the permeability toward O_2 was significantly reduced, but very slightly (reduction ranging from only 20% [19] up to a factor of 2 [33] for filler contents of 5 wt%) (Table 25.4).

Layered silicates were also incorporated in protein-based matrices with the objective to reduce their water sensitivity. Indeed, because of their hydrophilic nature, protein-based materials (either from gelatin, whey, wheat gluten, corn zein or SPI) display a poor water resistance, which is revealed by an important swelling when they are immersed in liquid water and a high water sorption in high moisture conditions. Their water vapor permeability (WVP) is also very high as compared to synthetic films such as low-density polyethylene (about 100 times higher). Decreasing water sensitivity and especially water vapor permeability of protein-based materials is thus one of the main critical issues in the development of protein-based materials for extending potential food packaging applications. Whatever the nature of the matrix, water vapor permeability and water uptakes (either liquid or vapor) were significantly reduced with the introduction of about 5 wt% of unmodified MMT, but at best by a factor of 2–2.5 (Table 25.4). The improvement of water vapor barrier properties is generally ascribed to the impermeable nature of MMT layers distributed in the polymer matrix, thus increasing the effective diffusion pathway [34, 39, 54]. However, since O_2 and CO_2 permeabilities of wheat gluten-based materials were not affected by the introduction of unmodified MMT, the reduction in WVP was more likely ascribed to a change in solubility rather than to a change in diffusivity [35]. The decrease in WVP with increasing MMT loading may be related to the establishment of hydrophilic interactions between proteins and nanoclays, which resulted in a lower availability of the hydrophilic sites for water vapor [35]. This assumption was further confirmed by Chevillard et al. [55]. It appears that an optimal low amount of MMT is often necessary to achieve the greatest improvement of WVP: 5 wt% in case

TABLE 25.4 Water Vapor Permeability (WVP), Liquid Water Uptake (WU), Water Vapor Uptake Measured at 98%RH (WVU), and Permeability Toward Oxygen (PO_2) of Protein-Based Nanocomposites

System	Sample	WVP (mol/(m·s·Pa)	WU (%)	WVU (%)	P_{O2} (mol/(m·s·Pa))	Ref.
Protein/clay	Bovine skin gelatin (type B) (25 wt% glycerol)	–	778	–	–	[21]
	–5 wt% Na+ MMT	–	599	–	–	
	–17 wt% Na+ MMT	–	376	–	–	
	Bloom fish gelatin (20 wt% glycerol)	17.3×10^{-12}	–	–	2.08×10^{-18}	[33][a]
	–5 wt% Cloisite Na+	7.6×10^{-12}	–	–	9.8×10^{-19}	
	Bovine hide gelatin (type B)	13.7×10^{-12}	–	–	6.87×10^{-17}	[54][b]
	–5 wt% Na+ MMT	7.7×10^{-12}	–	–	5.58×10^{-17}	
	SPI (glycerol)	17×10^{-11}	–	–	–	[42]
	–10 wt% Na^+ MMT	20×10^{-11}	–	–	–	
	–10 wt% OMMT	16×10^{-11}	–	–	–	
	–10 wt% bentonite	8.4×10^{-11}	–	–	–	
	SPI (15 wt% glycerol)	$5.85 \pm 0.17 \times 10^{-11}$	–	–	–	[39][c]
	–5 wt% Na+ MMT	$4.56 \pm 0.15 \times 10^{-11}$	–	–	–	
	–10 wt% Na+ MMT	$3.85 \pm 0.12 \times 10^{-11}$	–	–	–	
	–15 wt% Na+ MMT	$3.34 \pm 0.09 \times 10^{-11}$	–	–	–	
	WG (10 wt% glycerol)	$14.6 \pm 0.63 \times 10^{-12}$	Disruption of the film	(–)[f]	$1.10 \pm 0.09 \times 10^{-15}$	[35][d]
	–2.5 wt% Na^+ MMT	$14.3 \pm 0.40 \times 10^{-12}$	151 ± 5	109 ± 8	$1.12 \pm 0.07 \times 10^{-15}$	
	–5 wt% Na^+ MMT	$6.59 \pm 0.14 \times 10^{-12}$	113 ± 16	105 ± 6	$1.20 \pm 0.02 \times 10^{-15}$	
	–7.5 wt% Na^+ MMT	$7.22 \pm 0.28 \times 10^{-12}$	92 ± 4	94 ± 9	$1.22 \pm 0.04 \times 10^{-15}$	
	–10 wt% Na^+ MMT	$6.32 \pm 0.04 \times 10^{-12}$	82 ± 3	91 ± 2	$1.28 \pm 0.06 \times 10^{-15}$	
	Corn zein	$7.44 \pm 1.02 \times 10^{-12}$	–	–	–	[34][e]
	–1 wt% OMMT	$4.61 \pm 0.52 \times 10^{-12}$	–	–	–	
	–3 wt% OMMT	$2.93 \pm 1.02 \times 10^{-12}$	–	–	–	
	–5 wt% OMMT	$6.55 \pm 1.03 \times 10^{-12}$	–	–	–	
	–10 wt% OMMT	$9.51 \pm 0.55 \times 10^{-12}$	–	–	–	

(Continued)

TABLE 25.4 (*Continued*)

System	Sample	WVP (mol/(m·s·Pa)	WU (%)	WVU (%)	P_{O2} (mol/(m·s·Pa))	Ref.
Protein/ polysaccharide nanofillers	SPI (30 wt% glycerol)	–	–	39.5	–	[27]
	–5 wt% chitin whiskers	–	–	35	–	
	–10 wt% chitin whiskers	–	–	33	–	
	–15 wt% chitin whiskers	–	–	31.5	–	
	–30 wt% chitin whiskers	–	–	23	–	
	SPI (30 wt% glycerol)	–	–	36	–	[28]
	–5 wt% cellulose whiskers	–	–	32	–	
	–10 wt% cellulose whiskers	–	–	31.5	–	
	–15 wt% cellulose whiskers	–	–	30	–	
	–30 wt% cellulose whiskers	–	–	26	–	
	SPI (30 wt% glycerol)	–	–	29	–	[25]
	–3 wt% starch nanocrystals	–	–	29	–	
	–8 wt% starch nanocrystals	–	–	31	–	
	–16 wt% starch nanocrystals	–	–	32	–	
Protein/carbon nanotubes	SPI (30 wt% glycerol)	–	–	35	–	[22]
	–0.75 wt% MWCNT	–	–	30	–	
	–3 wt% MWCNT	–	–	30	–	

[a] P_{O2} measured at 50%RH.
[b] WVP measured between 0% and 65%RH and P_{O2} at 65%RH.
[c] WVP between 65% and 100%RH.
[d] WVP measured between 0% and 100%RH.
[e] WVP measured between 0% and 100%RH.
[f] No correct value was obtained for WG (10 wt% glycerol) in Reference [35] because films were very sticky.

of wheat gluten-based nanocomposites [35] and 3 wt% in case of corn zein-based nanocomposites [34]. This could be explained by the formation of a percolating nanoclays network within the matrix [35], and/or also, in case of higher amounts of nanoclays, to a competition between the positive effect of impermeable nanoclays and the negative effect of an only partially exfoliated nanocomposite structure obtained in that case [34].

In case of soy protein-based materials, other fillers such as chitin [27] or cellulose whiskers [28], starch nanocrystals [25], and carbon nanotubes [22] have been tentatively used to reduce the water sensitivity of the matrix (Table 25.4). Globally, the effect of polysaccharide (chitin, cellulose or starch) nanofillers on the water sensitivity of soy protein-based materials was very low as compared to the effect of layered silicates [25, 27, 28]. Indeed, considering a conventional filler content of 5 wt%, a reduction of only 10% was noted when incorporating chitin [27] or cellulose [28] whiskers (Table 25.4). In case of starch nanocrystals, their effect was even negative since all nanocomposite materials displayed slightly higher values of water vapor uptake than the unfilled matrix (Table 25.4) [25]. This poor effect was ascribed to the hydrophilic character of such nanofillers, which was sometimes more pronounced than those of the protein matrix, as it was the case for the soy protein/starch nanocrystals system [25]. However, it is worth noting that a significant decrease in the water vapor uptake can be observed for very high amounts of chitin [27] or cellulose [28] whiskers (reduction by factors of up to 1.7 and only 1.4 for a chitin or cellulose whisker content of 30 wt%, respectively) (Table 25.4). This reduction was ascribed to the formation of a rigid cellulose or chitin network resulting from the three-dimensional hydrogen bonding between chitin or cellulose fillers during the film formation [25, 28]. In case of systems reinforced with carbon nanotubes, all nanocomposites showed a lower water vapor uptake than the unfilled matrix (Table 25.4). This effect, even very low, was attributed to the restriction of the protein chain motion due to the incorporation of relatively rigid fillers [22].

Even if layered silicates are the best candidates for improving barrier properties of protein-based films, it is worth noting that an overall gain in barrier properties of 50% maximum can be reached, which is not sufficient to extend the use of protein-based materials to O_2- or water-sensitive products. This poor effect could be explained by the low size aspect ratio (length to width) of the nanofiller after material processing. Indeed, even if the initial nanoclay exhibits a high size aspect ratio (e.g., >100), this ratio may be completely different when characterized *in situ* in the nanocomposite material because of the presence of stacks of several layers, even for a well-exfoliated structure. For instance, a mean size aspect ratio of 20–25 is usually obtained in nanocomposite materials [56]. By using this value in the mathematical model of Bharadwaj [57], which models the effect of ribbon-like nanoparticles addition on the diffusivity of a permeant (Eq. 25.1), prediction indicates that the permeant diffusivity could be decreased at the most by a factor of 2 (in case of a maximum 10.8 vol% of MMT particles) (Fig. 25.5).

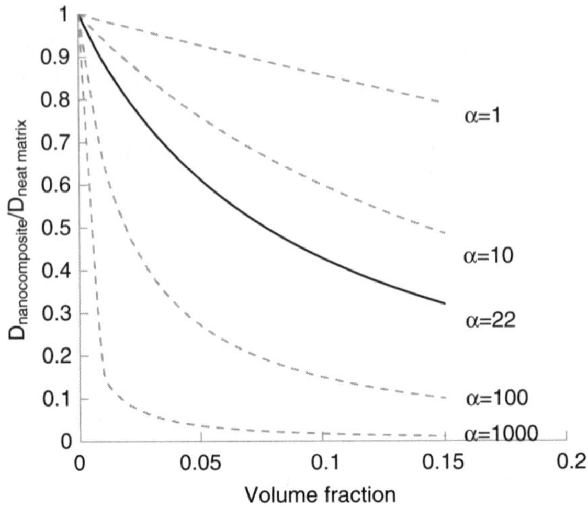

Figure 25.5 Prediction of the diffusivity in nanocomposite materials calculated from the diffusivity in the neat matrix and from the volume fraction, aspect ratio, and orientation of the nanoparticles added in the matrix. Simulations were carried out by using the Bharadwaj equation [57].

$$\frac{D_{composite}}{D_{matrix}} = \frac{1}{1+\frac{\bar{\alpha}}{2}\phi\left(\frac{2}{3}\right)\left(S+\frac{1}{2}\right)}. \tag{25.1)*$$

With the aim of substituting petrochemical plastics with protein-based materials, similar barrier properties need to be achieved. Taking an example: the water vapor permeability of wheat gluten-based materials for a 0–100% RH difference ranges from 5×10^{-12} mol/m/s/Pa to 6.2×10^{-11} mol/m/s/Pa depending on the film preparation process [52, 58], whereas it equals to 0.05×10^{-12} mol/m/s/Pa in the case of low-density polyethylene [59]. This means that, water permeability of wheat gluten-based films should be reduced by a factor of 10 to 100 if large applications in the food packaging sector are targeted. Based on (Eq. 25.1), it can already be concluded that such a reduction factor is not feasible with the type of nanofillers usually employed

* where $\bar{\alpha}$ is the mean aspect ratio of each particle or tactoïd), ϕ the volume fraction of particles and S, an order parameter defined as: $S = \frac{1}{2}\langle 3\cos^2\theta - 1\rangle$
Where θ represents the angle between the direction of preferred orientation and the surface of the film. The orientation parameter S ranges from a value of –0.5 for a system where the long axis of the filler is oriented parallel to the flux direction (no tortuosity), to a value of 1, where it is oriented perpendicular to the flux direction (maximum tortuosity), a value of 0 representing random orientation.

(e.g., nanoclays such as MMT) and within the processable filler content range. This leads to the conclusion that future research must focus on the use of nanoparticles displaying higher aspect ratio values (greater than 100 when dispersed within the polymer matrix) in order to sufficiently improve barrier properties of protein-based materials.

25.5 CONTROLLED DELIVERY SYSTEMS FOR ACTIVE PACKAGING

As mentioned above, active packaging deliberately incorporates active components intended to release or to absorb substances into, onto or from the packaged food or the environment surrounding the food in their conditions of use. The main objectives are to prevent or, at least, delay microbial growth (antimicrobial packaging) and to reduce or delay oxidation of nutritional compounds (antioxidant packaging), both by reducing the dose of additives used in food items and in order to extend food shelf life. Antimicrobial or antioxidant effect of packaging materials could be achieved either by [6]:

- direct contact with food surfaces, using nonmigratory molecules or controlled diffusion systems for solute compounds,
- or by indirect contact, using controlled release systems for volatile compounds that create active modified atmosphere in the headspace of packaging.

Many preservatives, such as bacteriocins, organic acids, various plant extracts, volatiles from essential oils and silver ions, have been incorporated in packaging materials to bring antimicrobial property [58–64] or, to a minor extent, antioxidant capacity [65–69]. Due to their film forming ability and unique properties as vehicles of active agents in controlled delivery systems of solute or volatile compounds, biomolecules such as polysaccharides and proteins from animal (whey proteins) or vegetal origin (soy proteins, corn zein, and wheat gluten) might be used as polymer matrices to produce packaging materials with antimicrobial or antioxidant properties [70–72]. If proteins are already used in pharmaceutical applications for drug delivery, mainly in the form of capsules [73, 74],they are less studied for packaging purposes.

The efficiency of a controlled delivery system depends on the concentration of the active agent on the food surface (solute compounds) or in the surrounding atmosphere (volatile compounds). Therefore, the antimicrobial or antioxidant capacity of the active agent must be proved and the efficient dose determined, it must be retained at a sufficient amount when nonused and delivered in conditions of use (packed food), that is, in response to a stimulus as change in relative humidity, pH, temperature, and so on. The ability of protein-based materials to modify their transfer properties as a function of moisture and temperature seemed worthy of investigation for developing controlled delivery materials triggered by moisture and temperature variation. This is mainly due to a modification in the network structure and polymeric chain mobility when increasing relative humidity and/or temperature, corresponding to a change from a glassy to a viscoelastic state. As a consequence, diffusivity of the active agent is greatly affected. As an example, effective diffusivity of carvacrol included in soy protein-coated paper varied from 1.71×10^{-16} to 138×10^{-16} m^2/s at 30°C when increasing RH from 60% to 100%RH and from 7.50×10^{-16} to 138×10^{-16} m^2/s at 100% RH when passing from 5°C to 30°C [75]. The development of protein-based nanocomposites could be of great interest for modulating the active agent diffusion/release due to structural modifications in the polymeric network when properly exfoliated. However, nanoparticles may interact with the active agent, leading to a change in viscosity of the blend that could make the processability of the packaging material difficult, as well as modify the morphology of the final network acting on the retention capacity and diffusivity of the active agent. When increasing MMT content in wheat gluten [35] or fish gelatin [33] solutions, the viscosity of the blend dramatically increased at low shear rates for nanoparticle loadings higher than 5 wt%. It was consistent with previous results obtained at the molten state, where a higher amount of plasticizer was needed to reduce the increase in both the torque and the temperature during mixing of wheat gluten and MMT [10]. However, the viscosity of such a preparation was restored when introducing a volatile active agent as carvacrol; this was a consequence of both a plasticizing effect of the aroma compound but also to the formation of clay aggregates within the polymeric network when nanoparticles content was higher than 5% [76]. Then it was possible to obtain active wheat gluten-based materials as casted and self-supported films [76] or coated onto paper support [77]. Up to now these authors are the only ones who proposed and developed an antimicrobial delivery system based on a protein matrix containing an active agent and reinforced by nanoparticles. In self-supported active films, the aggregated structures were able to entrap carvacrol molecules, leading to a better retention of the active agent: 76.4% of retention without MMT and 90.3% of retention when MMT content exceeded 5 wt% [76]. Similar results were obtained when this active solution was coated onto paper support: 77.5 % of retention without MMT and 90.7% of retention when MMT content exceeded 5 wt% [77]. During storage, the carvacrol release from wheat gluten coated paper was controlled by relative humidity (from 40% to 100%RH) and MMT content (from 0 to 7 wt%). The presence of high amount of MMT (≥5 wt%) accentuated the RH-induced effect as demonstrated by the high diffusivity values found in the presence of 5 wt% MMT at 100%RH (Fig. 25.6). However, the presence of well-dispersed nanofillers such as MMT is known to extend the diffusion pathway in the material and, consequently, to decrease the apparent diffusivity by a "tortuosity" effect [78]. The unexpected effect of MMT in this study was explained by the formation of an aggregated structure in the presence of carvacrol and 5 wt% MMT specifically. This type of structure could create both a better carvacrol retention during the process-

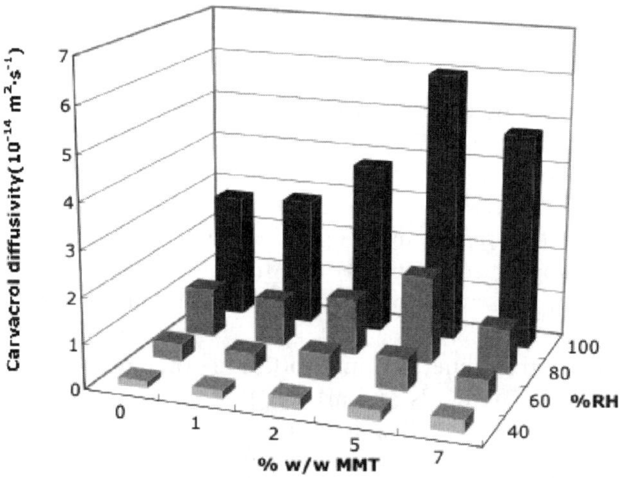

Figure 25.6 Apparent diffusivity values of carvacrol in the wheat gluten coated paper as a function of the % of MMT (adapted from Chalier et al. [75]).

ing step of the coated paper and a preferential pathway for carvacrol diffusion. Such behavior appeared particularly suitable for an antimicrobial food packaging system and the antimicrobial efficiency of this system was assessed toward *Escherichia coli* and *Botrytis cinerea.*

Nanoparticles have also been used in protein-based nanocomposites for their intrinsic antimicrobial properties: TiO_2 added in whey proteins isolate films [18] and organic modified clays introduced in whey proteins [37, 38]. The antimicrobial properties of TiO_2 is due to its photocacatalytic activity, and it acts by direct contact, whereas for modified clays, the bacterial effect is induced by both electrostatic adsorption onto the bacterial cell (direct contact) and the release of the quaternary ammonium group introduced in the silicate layer, which disrupts the bacterial cell [79, 80]. While the impact of TiO_2 nanoparticle incorporation on water barrier and tensile properties has been demonstrated, the antimicrobial effect has not yet been studied. In contrast, for whey protein isolate/Cloisite 30B films, a bacteriostatic effect against *Listeria monocytogenes* was clearly evidenced. Compared with TiO_2, clay minerals seem more adaptable to being introduced in protein-based materials due to their nontoxic and environmentally friendly characteristics. Moreover, indirect contact appeared more attractive since potential migration of nanoparticles could be limited.

25.6 SAFETY ASPECTS OF PROTEIN-BASED NANOCOMPOSITES USED AS FOOD CONTACT MATERIALS

Like any other food packaging material, protein-based materials lie within the scope European food contact material (FCM) legislation as well as conventional material. This requires that they must prove compliance with the general requirements laid down by the regulation EC 1935/2004 [81] and meet the *inertia* criteria specified therein. Regulation 1935/2004 includes several provisions requested by every FCM: (1) FCM shall not transfer their components into the food in quantities that could endanger human health; (2) FCM shall not change the composition of the food in an unacceptable way; and (3) FCM shall not cause deterioration in the taste, odor, or texture of the food. The first requirement only deals with migration (contamination of food by packaging components) and food safety, and the other two are related not only to migration but also to food quality (e.g., sorption of aroma compounds).

In addition to this general framework, the community regulation establishes 17 groups of materials and articles that are covered by specific measures, but no specific directive is still available for biopolymer-based or bionanocomposite materials. By default and according to the expected type of usage of finished packaging, such material could be considered as plastic material and be submitted to the same compliance tests. However, the procedure of migration test recommended for plastics appears specially designed for water-resistant materials since it consists of putting a sample of packaging in contact with the food product or with food simulating liquids (FSL) at defined conditions of time and temperature [82]. The requirement of inertia for the material becomes an overall migration limit (OML) defined as the total mass released by the packaging material during a migration test (60 mg/kg of food for plastic material according to the Regulation EU No. 10/2011, Article 12 [82]). Furthermore, the directive establishes a specific migration limit (SML) for all approved substances (monomers and additives) referenced in a *positive list* in respect of toxicological data. It must be pointed out that such an approach does not take into account the loss of barrier and mechanical properties nor even the possible degradation of material under humid conditions. In the case of water-sensitive materials such as protein-based materials, migration values tend to be particularly high, preventing their use in contact with high water activity foodstuffs. Because protein-based materials are exclusively composed of substances authorized as food additives, they are not submitted to any SML. However, when used as a packaging film demonstrating a high migration level of protein material or edible coating, protein-based material must be additionally considered with respect to their allergenic potential and should be legitimately covered by the amending European Directive 2000/13/EC [83], which relates to the labeling of the ingredients present in foodstuffs. The directive recognize cereals containing gluten, soybeans, milk, and products thereof as belonging to common food allergens and are referred to as ingredients subject to labeling rules, giving appropriate information to consumers suffering from food allergy. This issue is especially important for gluten-based materials considering the risk-related to gluten intolerance (the so-called celiac disease).

The incorporation of nanoparticles to protein-based materials only added further complexity in terms of safety concerns. To date, there is no mention of engineered nanomaterial (ENM) in European legislation related to FCM, except in the Regulation EC 450/2009 [84] (articles 5.2 10), which relates to active and intelligent material, stating that ENM must be separated from the food by a functional barrier, which should completely prevent potential migration into the food product. Little is known about the risk that could be linked to the use of ENM in FCM, and a recent scientific opinion of the European Food Safety Authority (EFSA) highlights that a suitable risk assessment approach should take into account the "specific properties of the ENM in addition to those common to the equivalent non-nanoforms (dissolved or macroscale) substances" [85]. It is important to note that the toxicity of ENM depends on a large number of factors including their structure, surface area, particle number, charge, chemistry, size and size distribution, state of aggregation, shape, and elemental composition [86]. Therefore, a sound approach to determine ENM migration levels and assess their toxicity should provide information on those parameters (at least size distribution, state of aggregation, and elemental composition). Such a performance is still hardly achieved with the analytical methods currently available to characterize ENM in the environment and food [87]. Until now very few studies have been conducted about the release of nanoparticle and their subsequent effect on foodstuff. From a purely theoretical approach, the migration of ENM from common packaging materials is assumed to be negligible given the slow rate of transfer due to the large size of ENM [88]. Such a statement cannot be extended to protein-based materials. A recent study was conducted about the release of nanoclays from wheat-gluten-based material where aluminum and silicon were distinctly used as tracers of MMT particles [89]. The detected migration level of these two elements was below the limit recommended by EFSA but the respective amounts of released aluminum and silicon did not match their initial proportion in the MMT. The fact that uneven quantities of aluminum and silicon were found suggests that MMT would be released not intact but after chemical modificationsThis question is of high importance since toxicity of nanoparticles is allegedly related to their high specific surface more than to their constituting elements. Beyond the safety issues related to the exposure of consumer via migration of the nanoparticles themselves, the nanoparticles are known to influence to some extent the stability of protein material in an aqueous medium what consequently could modulate the OML values. In this way, Tunc et al. [35] showed that an increasing amount of MMT decreases the water uptake of wheat gluten films. The improved water sensitivity would explain the fact that MMT addition decreases the amount of protein migration in water and in the 15% ethanol FSL [89]. On the basis of this unique investigation, it is difficult to establish a general rule applicable to all other protein-based materials and as for any new technology that is coming into the stage of application, a sound evaluation and a risk–benefit analysis needs to be carried out.

25.7 CONCLUSIONS AND PERSPECTIVES

General trends in the food packaging field leads to the development of more and more diverse and complex packaging materials and systems. Polymer nanotechnology provides huge opportunities impacting across all sectors of the global polymer markets. Nanocomposites offer a wide range of opportunity for active and biodegradable food packaging development. However, tackling food packaging issues, and especially through the development of nanotechnology- and agropolymer-based materials, demands a multidisciplinary and integrated approach involving a large panel of scientific competences. Research in the field of food packaging is still too disperse because it is driven by disconnected scientific objectives and developments in the fields of materials, chemistry, or food science. Beyond the pursuance and deepening of knowledge about the objectives and activities that are mentioned in this paper, it is essential to set up new approaches and related tools, allowing a rational integrated approach to relationships between material structure and properties, especially mass transfer. The other essential step is to combine this knowledge with the knowledge of targeted food degradation reactions with the help of modeling/prediction tools. The main expected result is the development of strategies (reverse engineering) for designing valuable food packaging materials applicable in a sustainable, safe, and competitive context.

REFERENCES

[1] Danielli, D., Gontard, N., Spyropoulos, D., Zondervan Van den beuken, E., and Tobback, P. (2008) Active and intelligent food packaging: Legal aspects and safety concerns. *Trends in Food Science & Technology*, *19*, 99–108.

[2] Guillaume, C., Chalier, P., and Gontard, N. (2008) Modified atmosphere packaging using environmentally compatible and active food packaging materials. In: *Environmentally Compatible Food Packaging*, ed. E. Chiellini. Boca Raton, FL: CRC press.

[3] Gontard, N., Angellier-Coussy, H., Chalier, P., Gastaldi, E., Guillard, V., Guillaume, C., and Peyron, S. (2010) Food packaging applications of biopolymer-based films. In: *Films and Coatings from Renewable Resources—An Applications Perspective*, ed. D. Placket. New York: John Wiley and Sons.

[4] Angellier-Coussy, H., Guillard, V., Guillaume, C., and Gontard, N. (2010) Wheat gluten-based materials for food packaging. In: *Multifunctional and Nanoreinforced Polymers for Food Packaging*, ed. J.M. Lagaron. Cambridge, UK: Woodhead Publishing Limited.

[5] Gontard, N., Guillaume, C. (2009) Packaging and the shelf life of fruits and vegetables. In: *Food Packaging and Shelf Life: A Practical Guide*, ed. G.L. Robertson. Boca Raton, FL: CRC press.

[6] Garrett, R.H., Grisham, C.M. (1996) Part I: Molecular components of cells. In: *Biochemistry*, 3rd ed, ed. T. Brooks/Cole. Belmont CA: Harcourt Brace, Custom Publishers, pp. 1–375.

[7] Arvanitoyannis, I. (2002) Formation and properties of collagen and gelatin films and coatings. In: *Protein Based Films and Coatings*, ed. A. Gennadios. Boca Raton, FL: CRC Press, pp. 275–304.

[8] Cuq, B., Gontard, N., and Guilbert, S. (1998) Proteins as agricultural polymers for packaging production. *Cereal Chemistry*, *75*(1), 1–9.

[9] Wang, S., Sue, H.J., and Jane, J. (1996) Effects of polyhydric alcohols on the mechanical properties of soy protein plastics. *Journal of Macromolecular Science-Part A*, *33*(5), 557–569.

[10] Angellier-Coussy, H., Torres-Giner, S., Morel, M.H., Gontard, N., and Gastaldi, E. (2008) Functional properties of thermoformed wheat gluten/montmorillonite materials with respect to formulation and processing conditions. *Journal of Applied Polymer Science*, *107*(1), 487–496.

[11] Cuq, B., Boutrot, F., Redl, A., and Lullien-Pellerin, V. (2000) Study of the temperature effect on the formation of wheat gluten network: Influence on mechanical properties and protein solubility. *Journal of Agricultural and Food Chemistry*, *48*, 2954–2959.

[12] Gontard, N., Thibault, R., Cuq, B., and Guilbert, S. (1996) Influence of relative humidity and film composition on oxygen and carbon dioxide permeabilities of edible films. *Journal of Agricultural and Food Chemistry*, *44*, 1064–1069.

[13] Mujica Paz, H., Gontard, N. (1997) Oxygen and carbon dioxide permeability of wheat gluten film: Effect of relative humidity and temperature. *Journal of Agricultural and Food Chemistry*, *45*, 4101–4105.

[14] Guillaume, C., Pinte, J., Gontard, N., and Gastaldi, E. (2010) Wheat gluten-coated papers for bio-based food packaging: Structure, surface and transfer properties. *Food Research International*, *43*(5), 1395–1401.

[15] Chalier, P., Peyches-Bach, A., Gastaldi, E., and Gontard, N. (2007) Effect of concentration and relative humidity on the transfer of alkan-2-ones through paper coated with wheat gluten. *Journal of Agricultural and Food Chemistry*, *55*, 867–875.

[16] Cho, S.W., Ullsten, H., Gallstedt, M., and Hedenqvist, M.S. (2007) Heat-sealing properties of compression-molded wheat gluten films. *Journal of Biobased Materials and Bioenergy*, *1*, 56–63.

[17] Ai, F., Zheng, H., Wei, M., and Huang, J. (2007) Soy protein plastics reinforced and toughened by SiO_2 nanoparticles. *Journal of Applied Polymer Science*, *105*, 1597–1604.

[18] Zhou, J.J., Wang, S.Y., and Gunasekaran, S. (2009) Preparation and characterization of wheyprotein film incorporated with TiO_2 nanoparticles. *Journal of Food Science*, *74*(7), 50–56.

[19] Li, Y., Jiang, Y., Liu, F., Ren, F., Zhao, G., and Leng, X. (2011) Fabrication and characterization of TiO_2/whey protein isolate nanocomposite film. *Food Hydrocolloids*, *25*, 1098–1104.

[20] Coleman, J.N., Khan, U., Blau, W.J., and Gun'ko, Y.K. (2007) Small but strong: A review of the mechanical properties of carbon nanotube-polymer composites. *Carbon*, *44*, 1624–1652.

[21] Li, P., Zheng, J.P., Ma, Y.L., and Yao, K.D. (2003) Gelatin/montmorillonites hybrid nanocomposite. II. Swelling behaviour. *Journal of Applied Polymer Science*, *88*, 322–326.

[22] Zheng, H., Ai, F., Wei, M.W., Huang, J., and Chang, P.R. (2007) Thermoplastic soy protein nanocomposites reinforced by carbon nanotubes. *Macromolecular Materials and Engineering*, *292*, 780–788.

[23] Dufresne, A. (2008) Polysaccharide nanocrystal reinforced nanocomposites. *Canadian Journal of Chemistry*, *86*(6), 484–494.

[24] Alexandre, M., Dubois, P. (2000) Polymer-layered silicate nanocomposites: Preparation, properties and uses of a new class of materials. *Material Science and Engineering: R: Reports*, *28*, 1–63.

[25] Zheng, H., Ai, F., Chang, P.R., Huang, J., and Dufresne, A. (2009) Structure and properties of starch nanocrystal-reinforced soy protein plastics. *Polymer Composites*, *30*(4), 474–480.

[26] Lin, N., Huang, J., Chang, P.R., Anderson, D.P., and Yu, J. (2011) Preparation, modification and application of starch nanocrystals in nanomaterials: A review. *Journal of Nanomaterials*. doi: 10.1155/2011/573687.

[27] Lu, Y., Weng, L., and Zhang, L. (2004) Morphology and properties of soy protein isolate thermoplastics reinforced with chitin whiskers. *Biomacromolecules*, *5*, 1046–1051.

[28] Wang, Y., Cao, X., and Zhang, L. (2006) Effects of cellulose whiskers on properties of soy protein thermoplastics. *Macromolecular Bioscience*, *6*, 524–531.

[29] Vaia, R.A., Giannelis, E.P. (1997) Lattice model of polymer melt intercalation in organically-modified layered silicates. *Macromolecules*, *30*, 7990–7999.

[30] Vaia, R.A., Jandt, K.D., Kramer, E.J., and Giannelis, E.P. (1995) Kinetics of polymer melt intercalation. *Macromolecules*, *28*, 8080–8085.

[31] Zheng, J.P., Li, P., Ma, Y.L., and Yao, K.D. (2002) Gelatin/montmorillonites hybrid nanocomposite. I. Preparation and properties. *Journal of Applied Polymer Science*, *86*, 1189–1194.

[32] Rao, Y.Q. (2007) Gelatin-clay nanocomposites of improved properties. *Polymer*, *48*, 5369–5375.

[33] Bae, H.J., Park, H.J., Hong, S.I., Byun, Y.J., Darby, D.O., Kimmel, R.M., and Whiteside, W.S. (2009) Effect of clay content, homogenization RPM, pH, and ultrasonication on mechanical and barrier properties of fish gelatin/montmorillonitenanocomposite films. *LWT-Food Science and Technology*, *42*(6), 1179–1186.

[34] Luecha, J., Sozer, N., and Kokini, J.L. (2010) Synthesis and properties of corn zein/montmorillonites nanocomposite films. *Journal of Materials Science*, *45*(13), 3529–3537.

[35] Tunc, S., Angellier, H., Cahyana, Y., Chalier, P., Gontard, N., and Gastaldi, E. (2007) Functional properties of wheat gluten/montmorillonitenanocomposite films processed by casting. *Journal of Membrane Science*, *289*(1–2), 159–168.

[36] Hedenqvist, M.S., Backman, A., Gällstedt, M., Boyd, R.H., and Gedde, U.W. (2006) Morphology and diffusion properties of whey/montmorillonite nanocomposites. *Composites Science and Technology*, *66*(13), 2350–2359.

[37] Sothornvit, R., Rhim, J.W., and Hong, S.I. (2009) Effect of nano-clay type on the physical and antimicrobial properties of whey protein isolate/clay composite films. *Journal of Food Engineering*, *91*(3), 468–473.

[38] Sothornvit, R., Hong, S.I., An, D.J., and Rhim, J.W. (2010) Effect of clay content on the physical and antimicrobial properties of whey protein isolate/organo-clay composite films. *LWT-Food Science and Technology*, *43*(2), 279–284.

[39] Kumar, P., Sandeep, K.P., Alavi, S., Truong, V.D., and Gorga, R.E. (2010) Preparation and characterization of bio-nanocomposite films based on soy protein isolate and montmorillonites using melt extrusion. *Journal of Food Engineering*, *100*, 480–489.

[40] Kumar, P., Sandeep, K.P., Alavi, S., Truong, V.D., and Gorga, R.E. (2010) Effect of type and content of modified montmorillonite on the structure and properties of bio-nanocomposite films based on soy protein isolate and montmorillonite. *Journal of Food Science*, *75*(5), 46–56.

[41] Fernandes, F.M., Ruiz, A.I., Darder, M., Aranda, P., and Ruiz-Hitzky, E. (2009) Gelatin-clay bio-nanocomposites: Structural and functional properties as advanced materials. *Journal of Nanoscience and Nanotechnology*, *9*, 221–229.

[42] Rhim, J.W., Lee, J.H., and Kwak, H.S. (2005) Mechanical and water barrier properties of soy protein and clay mineral composite films. *Food Science and Biotechnology*, *14*(1), 112–116.

[43] Chen, P., Zhang, L. (2006) Interaction and properties of highly exfoliated soy protein/montmorillonite nanocomposites. *Biomacromolecules*, *7*, 1700–1706.

[44] Yu, J., Cui, G., Wei, M., and Huang, J. (2007) Facile exfoliation of rectoritenanoplatelets in soy protein matrix and reinforced bionanocomposites thereof. *Journal of Applied Polymer Science*, *104*, 3367–3377.

[45] Ray, S.S., Bousmina, M. (2005) Biodegradable polymers and their layered silicate nanocomposites: In greening the 21st century materials world. *Progress in Materials Science*, *50*, 962–1079.

[46] Olabarrieta, I., Gällstedt, M., Ispizua, I., Sarasua, J.R., Hedenqvist, M.S. (2006) Properties of aged montmorillonite-wheat gluten composite films. *Journal of Agricultural and Food Chemistry*, *54*, 1283–1288.

[47] Zhang, X., Do, M.D., Dean, K., Hoobin, P., and Burgar, I.M. (2007) Wheat gluten-based natural polymer nanoparticle composites. *Biomacromolecules*, *8*, 345–353.

[48] Arvanitoyannis, I.S. (1999) Totally and partially biodegradable polymer blends based on natural synthetic macromolecules: Preparation, physical properties, and potential as food packaging materials. *Journal of Macromolecular Science, Reviews in Macromolecular Chemistry and Physics*, *C39*(2), 205–271.

[49] Berlinet, C., Brat, P., Brillouet, J.M., and Ducruet, V. (2006) Ascorbic acid, aroma compounds and browning of orange juices related to PET packaging materials and pH. *Journal of the Science of Food and Agriculture*, *86*(13), 2206–2212.

[50] Berlinet, C., Brat, P., and Ducruet, V. (2008) Quality of orange juice in barrier packaging material. *Packaging Technology and Science*, *21*(5), 279–286.

[51] Myers, A.W., Meyer, J.A., Rogers, C.E., Stannet, V., and Szwarc, M. (1961) Studies in the gas and vapor permeability of plastic films and coated papers. *Tappi Journal*, *45*, 58–67.

[52] Gontard, N., Guilbert, S., and Cuq, J.L. (1993) Water and glycerol as plasticizers affect mechanical and water vapor barrier properties of an edbile wheat gluten film. *Journal of Food Science*, *58*, 206–211.

[53] Mujica Paz, H., Guillard, V., Reynes, M., and Gontard, N. (2005) Ethylene permeability of wheat gluten film as a function of temperature and relative humidity. *Journal of Membrane Science*, *256*(1–2), 108–115.

[54] Martucci, J.F., Ruseckaite, R.A. (2010) Biodegradable three-layer film derived from bovine gelatin. *Journal of Food Engineering*, *99*, 377–383.

[55] Chevillard, A., Angellier-Coussy, H., Cuq, B., Cesar, G., Guillard, V., Gontard, N., and Gastaldi, E. (2011) How the biodegradability of wheat gluten based agromaterials can be modulated by adding nanoclays. *Polymer Degradation and Stability*, *96*(12), 2008–2097.

[56] Picard, E., Gerard, J.F., and Espuche, E. (2008) Water transport properties of polyamide 6 based nanocomposites prepared by melt blending: On the importance of the clay dispersion state on the water transport properties at high water activity. *Journal of Membrane Science*, *313*(1–2), 284–295.

[57] Bharadwaj, R.K. (2001) Modeling the barrier properties of polymer-layered silicate nanocomposites. *Macromolecules*, *34*(26), 9189–9192.

[58] Pommet, M., Redl, A., Morel, M.H., and Guilbert, S. (2003) Study of wheat gluten plasticization with fatty acids. *Polymer*, *44*, 115–122.

[59] Plastics Design Library Staff (1995) *Permeability and Other Film Properties of Plastics and Elastomers*, PDL Handbook Series. Norwich, NY: Plastics Design Library (PLD).

[60] Coma, V., Sebti, I., Pardon, P., Deschamps, A., and Pichavant, F.H. (2001) Antimicrobial edible packaging based on cellulosic ethers, fatty acids, and nisin incorporation to inhibit *Listeria innocua* and *Staphylococcus aureus*. *Journal of Food Protection*, *64*(4), 470–475.

[61] Appendini, P., Hotchkiss, J.H. (2002) Review of antimicrobials food packaging. *Innovative Food Science and Emerging Technologies*, *3*, 113–126.

[62] Quintavalla, S., Vicini, L. (2002) Antimicrobial food packaging in meat industry. *Meat Science*, *62*(3), 373–380.

[63] Lee, C.H., Park, H.J., and Lee, D.S. (2004) Influence of antimicrobial packaging on kinetics of spoilage microbial growth in milk and orange juice. *Journal of Food Engineering*, *65*(4), 527–531.

[64] Pranoto, Y., Rakshit, S.K., and Salokhe, V.M. (2005) Enhancing antimicrobial activity of chitosan films by incorporating garlic oil, potassium sorbate and nisin. *Lwt-Food Science and Technology*, *38*(8), 859–865.

[65] Bentayeb, K., Rubio, C., Batlle, R., and Nerin, C. (2007) Direct determination of carnosic acid in a new active packaging based on natural extract of rosemary. *Analytical and Bioanalytical Chemistry*, *389*(6), 1989–1996.

[66] Soto-Cantu, C.D., Graciano-Verdugo, A.Z., Peralta, E., Islas-Rubio, A.R., Gonzalez-Cordova, A., Gonzalez-Leon, A., and Soto-Valdez, H. (2008) Release of butylatedhydroxytoluene from an active film packaging to Asadero cheese and its effect on oxidation and odor stability. *Journal of Dairy Science*, *91*(1), 11–19.

[67] Gemili, S., Yemenicioglu, A., and Altinkaya, S.A. (2010) Development of antioxidant food packaging materials with controlled release properties. *Journal of Food Engineering*, *96*(3), 325–332.

[68] Lopez de Dicastillo, C., Nerin, C., Alfaro, P., Catala, R., Gavara, R., and Hernandez-Munoz, P. (2011) Development of new antioxidant active packaging films based on ethylene vinyl alcohol copolymer (EVOH) and green tea extract. *Journal of Agricultural and Food Chemistry*, *59*(14), 7832–7840.

[69] Manzanarez-Lopez, F., Soto-Valdez, H., Auras, R., and Peralta, E. (2011) Release of alpha-Tocopherol from poly(lactic acid) films, and its effect on the oxidative stability of soybean oil. *Journal of Food Engineering*, *104*(4), 508–517.

[70] Cha, D.S., Chinnan, M.S. (2004) Biopolymer-based antimicrobial packaging: A review. *Critical Reviews in Food Science and Nutrition*, *44*(4), 223–237.

[71] Spizzirri, U.G., Parisi, O.I., Iemma, F., Cirillo, G., Puoci, F., Curcio, M., and Picci, N. (2010) Antioxidant-polysaccharide conjugates for food application by eco-friendly grafting procedure. *Carbohydrate Polymers*, *79*(2), 333–340.

[72] Kuorwel, K., Cran, M.J., Sonneveld, K., Miltz, J., and Bigger, S.W. (2011) Antimicrobial activity of biodegradable polysaccharide and protein-based films containing active agents. *Journal of Food Science*, *76*(3), 90–102.

[73] Yeo, Y., Park, K.N. (2004) Control of encapsulation efficiency and initial burst in polymeric microparticle systems. *Archives of Pharmacal Research*, *27*(1), 1–12.

[74] Livney, Y.D. (2010) Milk proteins as vehicles for bioactives. *Current Opinion in Colloid and Interface Science*, *15*(1–2), 73–83.

[75] Chalier, P., Ben Arfa, A., Guillard, V., and Gontard, N. (2009) Moisture and temperature triggered release of a volatile active agent from soy protein coated paper: Effect of glass transition phenomena on carvacrol diffusion coefficient. *Journal of Agricultural and Food Chemistry*, *57*(2), 658–665.

[76] Mascheroni, E., Chalier, P., Gontard, N., and Gastaldi, E. (2010) Designing of a wheat gluten/montmorillonite based system as carvacrol carrier: Rheological and structural properties. *Food Hydrocolloids*, *24*(4), 406–413.

[77] Mascheroni, E., Guillard, V., Gastaldi, E., Gontard, N., and Chalier, P. (2011) Anti-microbial effectiveness of relative humidity-controlled carvacrol release from wheat gluten/montmorillonite coated papers. *Food Control*, *22*(10), 1582–1591.

[78] Choudalakis, G., Gotsis, A.D. (2009) Permeability of polymer/clay nanocomposites: A review. *European Polymer Journal*, *45*(4), 967–984.

[79] Hong, S.I., Rhim, J.W. (2008) Antimicrobial activity of organically modified nanoclays. *Journal of Nanoscience and Nanotechnology*, *8*(11), 5818–5824.

[80] Wu, T., Xie, A.G., Tan, S.Z., and Cai, X. (2011) Antimicrobial effects of quaternary phosphonium salt intercalated clay minerals on *Escherichia coli* and *Staphylococci aureus*. *Colloids and Surfaces B-Biointerfaces*, *86*(1), 232–236.

[81] European Commission (2004) EC Framework Regulation No 1935/2004 of the European Parliament and of the Council of 27 October 2004 on materials and articles intended to come into contact with food.

[82] European Commission (2011) EC Commission Regulation No. 10/2011 of 14 January 2011, on plastic materials and articles intended to come into contact with food.

[83] European Commission (2000) EC Directive 2000/13/EC of 20 March 2000 on the labelling of foodstuffs includes general provisions on the labelling of foodstuffs to be delivered to the consumer.

[84] European Commission (2009) EC Commission Regulation No 450/2009 of 29 May 2009 on active and intelligent materials and articles intended to come into contact with food.

[85] European Food Safety Authority (EFSA) (2011) Scientific opinion on guidance on the risk assessment of the application of nanoscience and nanotechnologies in the food and feed. *The EFSA Journal*, *9*(5), 1–36.

[86] Oberdörster, G., Maynard, A., Donaldson, K., Castranova, V., Fitzpatrick, J., Ausman, K., Carter, J., Karn, B., Kreyling, W., Lai, D., Olin, S., Monteiro-Riviere, N., Warheit, D., and Yang, H. (2005) Principles for characterizing the potential human health effects from exposure to nanomaterials: Elements of a screening strategy. *Particle and Fibre Toxicology*, *2*(8), 1–35.

[87] Tiede, K., Boxall, A.B.A., Tear, S.P., Lewis, J., David, H., and Hassello, M. (2008) Detection and characterization of engineered nanoparticles in food and the environment. *Food Additives and Contaminants*, *25*, 795–821.

[88] Simon, P., Chaudry, Q., and Bakos, D. (2008) Migration of engineered nanoparticles from polymer packaging to food—A physicochemical view. *Journal of Food and Nutrition Research*, *47*, 105–113.

[89] Mauricio-Iglesias, M., Peyron, S., Guillard, V., and Gontard, N. (2010) Wheat gluten nanocomposite films as food-contact materials: Migration tests and impact of a novel food stabilization technology (high pressure). *Journal of Applied Polymer Science*, *116*(5), 2526–2535.

[90] Varoquaux, P., Gouble, B., Barron, C., and Yildiz, F. (1999) Respiratory parameters and sugar catabolism of mushroom (*Agaricus bisporus* Lange). *Postharvest Biology and Technology*, *16*(1), 51–61.

[91] Mannheim, C.H., Soffer, T. (1996) Permeability of different wax coatings and their effect on citrus fruit quality. *Journal of Agricultural and Food Chemistry*, *44*(3), 919–923.

[92] Guilbert, S., Cuq, B. (1998) Les films etenrobages comestibles. In: *L'emballage des denréesalimentaires de grandeconsommation*, eds. G. Bureau, J.L. et Multon. Paris: Lavoisier, Tec et Doc, pp. 471–530.

[93] Gennadios, A., Weller, C.L., and Testin, R.F. (1993) Temperature effect on oxygen permeability of edible protein-based films. *Journal of Food Science*, *58*(1), 212–214.

[94] Butler, B.L., Vergano, P.J., Testin, R.F., Bunn, J.M., and Wiles, J.L. (1996) Mechanical and barrier properties of edible chitosan films as affected by composition and storage. *Journal of Food Science*, *61*(5), 953–961.

[95] Lieberman, E.R., Gilbert, S.G. (1973) Gas permeation of collagen films as affected by cross-linkage, moisture, and plasticizer content. *Journal of Polymer Science, Symposium*, *41*, 33–43.

[96] Product information, Ecoflex® F Blend C1200 Biodegradable polyester for compostable film (2011) http://www.bioplastics.basf.com/pdf/Product_information_Ecoflex_F_Blend_A1200V1-3.pdf (accessed March 2013).

[97] Sanchez-Garcia, M.D., Gimenez, E., and Lagaron, J.M. (2008) Morphology and barrier properties of solvent cast composites of thermoplastic biopolymers and purified cellulose fibers. *Carbohydrate Polymers*, *71*(2), 235–244.

[98] Product information, Ecovio® L BX 8180, Biodegradable polyester for compostable film with 80% of renewable resources (2009) http://www.plasticsportal.

net/wa/plasticsEU~en_GB/function/conversions:/publish/common/upload/biodegradable_plastics/Ecovio_L_BX_8180.pdf (accessed March 2013).

[99] Aydt, T.P., Waller, C.L., and Testin, R.F. (1991) Mechanical and barrier properties of edible corn and wheat protein films. *Transactions of the ASAE*, *34*(1), 207–211.

[100] Park, H.J., Testin, R.F., Chinnan, M.S., and Park, J.W., eds. (2001) Active biopolymer films and coatings for food and biotechnological uses. Pre-Congress Short Course of IUFoST. Korea University, Seoul, Korea. p. 250.

[101] Kato, Y., Kaminaga, J., Matsuo, R., and Isogai, A. (2005) Oxygen permeability and biodegradability of polyuronic acids prepared from polysaccharides by TEMPO-mediated oxidation. *Journal of Polymers and the Environment*, *13*(3), 261–265.

[102] Rose, P.I. (1987) Gelatin. In: *Encyclopedia of Polymer Science and Engineering*, Vol. 7, eds. H.F. Mark, N.M. Bilakes, C.G. Overberger, and G. Menges. New York: John Wiley & Sons, pp. 488–513.

[103] FAO (Food and Agriculture Organization of the United Nations) (1981) Amino acid content of foods and biological data on proteins. http://www.fao.org/DOCREP/005/.../AC854T00.htm (accessed March 2013).

[104] Braunitzer, G., Chen, R., Schrank, B., and Stangl, A. (1972) Automatische sequenzanalyse eines protein ((-lactoglobulin AB). *Hoppe-Seyler's Z Physiology and Chemistry*, *353*, 832–834.

[105] Vanaman, T.C., Brew, K., and Hill, R.L. (1970) The disulfide bonds of bovine (-lactalbumin. *Journal of Biological Chemistry*, *245*, 4583–4590.

[106] Boldwell, C.E., Hopkins, D.T. (1985) Soy protein products: characteristics, nutritional aspects, and utililization. In: *New Protein Foods*, Vol. 2, eds. A.M. Altschul, H.L. Wilcke. Orlando, FL: Academic Press, p. 152.

[107] Solsuski, F.W., Imafidon, G.I. (1990) Amino acid composition and nitrogen-to-protein conversion factors for animal and plant foods. *Journal of Agricultural and Food Chemistry*, *38*, 1351–1356.

INDEX

Biopolymer Nanocomposites: Processing, Properties, and Applications, First Edition. Edited by Alain Dufresne, Sabu Thomas, and Laly A. Pothan.